Heat Transfer

SECOND EDITION

P.S. GHOSHDASTIDAR

Professor
Department of Mechanical Engineering
Indian Institute of Technology Kanpur

OXFORD
UNIVERSITY PRESS

Oxford University Press is a department of the University of Oxford.
It furthers the University's objective of excellence in research, scholarship
and education by publishing worldwide. Oxford is a trade mark of
Oxford University Press in the UK and in certain other countries.

Published in India
by Oxford University Press
22 Workspace, 2nd Floor, 1/22 Asaf Ali Road, New Delhi 110002, India

First Edition published in 2004
Second Edition published in 2012
Digitally Printed in 2024

ISBN-13: 978-0-19-807997-2
ISBN-10: 0-19-807997-4

Typeset in Times New Roman
by Anvi Composers, New Delhi 110063
Printed in India by Manipal Technologies limited, Manipal

To
my parents
Prof. Mihir Kumar Ghoshdastidar
and
Rina Ghoshdastidar

Preface to the Second Edition

The first edition of Heat Transfer received warm response from reputed universities and institutes in India and abroad. It gives me great pleasure to present this second edition to the engineering academic community. I sincerely hope that this revised and enlarged edition of the book will aid students in understanding the basic concepts and principles of the subject and awake in them the interest to know more of the vast field of heat and mass transfer.

Heat transfer refers to energy transport due to a temperature difference in a medium or between media. Heat is not a storable quantity and is defined as energy in transit due to a temperature difference. The primary aim of studying the science of heat transfer is to understand the mechanism of heat transfer processes and to predict the rate at which heat transfer takes place.

Whenever we refer to heat transfer we actually imply the heat transfer rate. It is this 'rate' that differentiates the field of heat transfer from thermodynamics. Climatic changes, formation of rain and snow, heating and cooling of the earth's surface, the origin of dew drops and fog, spreading of forest fires are some of the natural phenomena wherein heat transfer plays a dominant role. The applications of heat transfer are diverse, both in nature and in industry.

About the Book

The main objective of this text is to lay emphasis on the fundamental principles of heat and mass transfer and to equip the students and other users to solve engineering heat transfer problems efficiently. It will serve as a useful text for undergraduate as well as postgraduate students. Researchers and practicing professionals will also benefit from this book.

Instructors can use this book for teaching a one-semester undergraduate course on heat and mass transfer. Some advanced topics have also been included in every chapter for students who intend to study the subject further.

Written in a lucid style, the text contains a large number of solved and unsolved problems to make a student-friendly book. A large number of illustrations aid the explanations provided in the text.

Key Features

- Emphasizes the fundamental concepts with easy-to-understand mathematics
- Provides detailed solution methodologies
- Includes a detailed coverage of the computer methods in heat transfer
- Reinforces theoretical concepts through numerous solved examples

New to the Second Edition

The book has been revised extensively on the basis of the feedback received from faculty members and students.

1. **New Chapter**   A new chapter on solidification and melting has been added.
2. **New Topics**   Many new sections such as heat generation in a solid sphere, heat conduction in a plane wall with temperature-dependent thermal conductivity, graphical method and conduction shape factor, flow regimes in free convection over a vertical plate, review of phase change processes of pure substances, formation of vapour bubbles, detailed analysis of radiation exchange in two- and three-surface enclosures, and one-dimensional steady diffusion through a stationary medium have been introduced.
3. **Solved Examples**   A considerable number of solved examples have been added in different chapters.
4. **Titles to Solved Examples**   Each example problem has been given a short title so that the students can immediately see what kind of problem is being solved.
5. **Review Questions**   New review questions have been added in all chapters.
6. **Recapitulation**   At the end of every chapter, important concepts and formulae have been recapitulated in order to help the students get a quick overview of the chapter.
7. **Answers and Hints**   At the end of the book, answers and hints have been given for exercise problems of every chapter.
8. **New Appendices**   The thermophysical properties of water at atmospheric pressure and the solutions of finite-difference problems in heat conduction using C are listed in the new appendices.

Extended Chapter Material

Chapter 2: Steady-state Conduction: One-dimensional Problems   Plane wall with variable conductivity and heat generation in a solid sphere have been added.

Chapter 3: Steady-state Conduction: Two- and Three-dimensional Problems   It includes graphical methods and conduction shape factor. Detailed concepts of isotherms and heat flux lines aided by illustrations have also been added.

Chapter 4: Unsteady-state Conduction   Concluding remarks on Heisler charts have been added.

Chapter 5: Forced Convection Heat Transfer   This chapter includes several new sections, for example, energy integral solution for uniform heat flux (q_s'' = constant) at the wall, the physical aspects of turbulent boundary layer, basic approach in solving turbulent heat transfer on a flat plate, effect of axial conduction in the fluid in laminar tube flow, salient features of liquid metal heat transfer in turbulent tube flow, insertion of new material in appropriate places for better understanding of physics, solution procedure for constant heat flux flat plate thermal boundary layer, proof of $\dfrac{\partial T}{\partial z} = \dfrac{dT_s}{dz} = \dfrac{dT_m}{dz}$ = constant for heat transfer in tube flow subject to constant heat flux, derivation of the expression for T_m vs. z for heat transfer in tube flow subject to constant wall temperature.

Chapter 6: Natural Convection Heat Transfer Physical mechanism of natural convection, flow regimes in free convection over a vertical plate, genesis of the physical meaning of Gr, Re, and Gr/Re^2 from dimensional analysis, correlations for free convection over a vertical plate subjected to uniform heat flux, physical aspects of free convection over inclined walls, free convection flow patterns when a horizontal hot plate faces upward or downward, or when a horizontal cold plate faces downward or upward have been explained in this chapter.

Chapter 7: Boiling and Condensation This chapter includes review of phase change processes of pure substances, the formation of vapour bubbles, bubble departure diameter and frequency of bubble release, heat transfer mechanism in nucleate boiling: Rohsenow's model and its basis, modification of Nusselt's correlation by Chen (1961) for laminar film condensation on a vertical tier of n horizontal tubes.

Chapter 8: Thermal Radiation New addition in this chapter is three- and two-surface enclosure A general revision by inserting new materials in the existing text and addition of new figures for enhancing physical understanding of the various topics have also been done.

Chapter 9: Heat Exchangers A general revision has been done by adding new figures and solved examples.

Chapter 11: Mass Transfer This chapter includes one-dimensional steady diffusion through a stationary medium: plane wall, cylindrical and spherical shell, and the proof of $D_{AB} = D_{BA}$.

Chapter 12 This new chapter provides definitions of solidification and melting, and their applications in industry and nature. It provides exact solutions of solidification—Stefan's problem for steady 1D analysis and Neumann's problem for unsteady 1D analysis. It also offers an exact solution of melting unsteady 1D analysis.

Appendix A New tables on thermophysical properties of water at atmospheric pressure and the solutions of finite-difference problems in heat conduction using C have been added.

Online Resources

Three new C and four new C++ programs demonstrating applications of finite-difference methods in solving 1D unsteady heat conduction in Cartesian and cylindrical coordinates have been provided in the Online Resource Centre of the book.

Content and Structure

The contents of this textbook have been developed assuming that the readers have an adequate background in calculus, thermodynamics, fluid mechanics, and computer programming in C, C++, and FORTRAN 77. There are 12 chapters and 8 appendices.

Chapter 1 is an introduction to the subject. Beginning with the aims of studying heat transfer, it discusses the applications and basic modes of heat transfer. It also defines thermal conductivity and provides value of this parameter for various materials.

Chapter 2 discusses one-dimensional steady-state conduction covering Fourier's law, initial and boundary conditions, overall heat transfer coefficient, critical thickness of insulation, and extended surfaces. 2D and 3D problems in steady-state conduction, presenting the method of separation of variables, methods of superposition and imaging are discussed in Chapter 3. It also discusses isotherms and heat flux lines.

Chapter 4 covers unsteady-state conduction. It explains lumped and distributed systems, and introduces the concept of Biot and Fourier numbers. It also discusses the application of Heisler's charts. Chapter 5 presents forced convection heat transfer over a flat plate and inside a tube. It introduces the concepts of hydrodynamic and thermal boundary layers, laminar and turbulent flow, and external flows over cylinders, spheres, and banks of tubes. It also defines the Nusselt, Prandtl, and Peclet numbers and explains their physical significance. Natural convection heat transfer including free convection from a vertical plate and other geometries as well as mixed convection are introduced in Chapter 6.

Chapter 7 covers boiling and condensation, including nucleate pool boiling, boiling modes, flow boiling, critical boiling states, dropwise and film condensation of pure vapour, Nusselt's theory of film condensation, and condensation of flowing vapours in tubes and heat pipes. Chapter 8 explains radiation heat transfer, covering the laws of black body radiation such as Planck's law, Wien's displacement law, Stefan–Boltzmann law, and Kirchhoff's law, and radiation characteristics of non-black surfaces. Also included are discussions on radiation exchange in enclosures, radiation shields, solar radiation, and the greenhouse effect. Heat exchangers providing their classification have been discussed in Chapter 9. It also covers the fouling factor, log-mean temperature difference, the correction factor approach, the effectiveness–NTU method, and design considerations for heat exchangers.

Chapter 10 introduces finite-difference methods in heat conduction. It presents a very interesting application of computational heat transfer in cryosurgery. Basic concepts of mass transfer, Fick's law of diffusion, heat and mass transfer analogy, boundary conditions in mass transfer, and evaporative cooling have been explained in Chapter 11.

Chapter 12 discusses solidification and melting.

The appendices contain useful and relevant information required to work out the solutions to heat transfer problems.

The Online Resource Centre of the book contains computer programs for solution of some problems in Chapter 10 will help in appreciating the applications of finite-difference methods in heat conduction. There are ten programs—three C, four C++, and three FORTRAN 77 programs. Sufficient comment lines have been given in the programs to familiarize the readers with the programs. The programs can be run on a PC or a workstation (such as SUN or SILICON GRAPHICS) having a FORTRAN, C, or C++ compiler. The readers should see the comment lines of each program for specific instructions.

Acknowledgements

I am grateful to the anonymous reviewers whose valuable comments and suggestions for improvement have gone a long way in shaping the second edition of the book. I also thank the Curriculum Development Cell under the Quality Improvement Programme at IIT Kanpur for giving me adequate financial support for revision of this book.

My special thanks are due to postgraduate student Radhe Shyam of IIT Kanpur, and undergraduate students Anuj Kumar Garg of IIT Kanpur and Nilanjan Sen of NIT Rourkela, who have assisted me in the preparation of the new C and C++ programs for the Online Resource Centre.

Sumita, my wife, has given her unstinted support and encouragement without which the present edition would not have been completed. She often tolerated my long hours of absence from home in the evenings and weekends during the period of this book revision exercise with patience and a smiling face.

I wish to acknowledge the support provided by the editorial and production team at the Oxford University Press India for bringing out the revised edition professionally and in a very elegant format.

Comments and suggestions for the improvement of the book are welcome. Please send them to me at psg@iitk.ac.in.

P. S. GHOSHDASTIDAR

Preface to the First Edition

It was nearly six years ago when I embarked upon the task of writing the textbook on heat transfer. While teaching this subject at the Indian Institute of Technology Kanpur, I had felt that there was a tremendous need for a sufficiently up-to-date textbook that was concise and yet exhaustive. When I took up this challenge, I found that the most difficult part in drafting a text on such a vast and diverse topic was to decide what is to be included and, more importantly, what is to be excluded. If one tries to put everything on the subject in a book, it no longer remains a textbook but becomes a kind of a reference handbook. On the other hand, if one makes it too concise, readers may not find the book useful. Therefore, at every stage I had to be very careful about the choice of topics included in this book. My guiding principle was to always emphasize the fundamental concepts supported by plenty of solved examples, so that students get a feel of the subject and are able to tackle complex real problems later in their career as engineers. This book is primarily intended for undergraduate students of engineering. However, postgraduate students and practising engineers will also find it useful.

Contents and Structure

Heat transfer is a compulsory subject taught in the third year of an undergraduate curriculum primarily in the field of mechanical engineering as also in the fields of chemical, metallurgical, and aerospace engineering.

The contents of this textbook have been developed assuming that the readers have an adequate background in calculus, thermodynamics, fluid mechanics, and computer programming in FORTRAN. There are 11 chapters, 6 appendices, and a CD containing computer programs written in FORTRAN 77. A prospective instructor can use this book for teaching a one-semester undergraduate course on heat and mass transfer. Some advanced topics have also been included in every chapter for students who intend to study the subject further.

Unlike in the currently available textbooks on heat transfer, this volume exhaustively covers two- and three-dimensional heat conduction, forced and free convection, boiling heat transfer, finite-difference methods in heat conduction, and heat exchangers. A very interesting application of computational heat transfer in cryosurgery is presented in Chapter 10. Also included are discussions on solar radiation and the greenhouse effect in Chapter 8. There are a large number of solved examples, presented through simple mathematical expressions and illustrations, which students will find extremely useful to understand the basic concepts. There are a large number of exercise problems. Answers to selected exercise problems in each chapter have also been provided. To impart a sense of the history of heat transfer as a subject, short biographies of famous scientists in the field of heat and mass transfer have been included.

A CD containing computer programs for solution of some problems in Chapter 10 will help in appreciating the application of finite-difference methods in heat conduction. There are eight programs, one subroutine (TDMA), and eight sample output files, one for each program, included in the CD. The program files including the subroutine can be viewed with Notepad, while the output files can be opened with WordPad. Sufficient comment lines have been given in the programs to familiarize the readers with the programs. Written in FORTRAN 77, the programs can be run on a PC or a workstation (such as SUN or SILICON GRAPHICS) having a FORTRAN 77 compiler. In programs related to the additional solved examples of Chapter 10, the input data are to be entered in an interactive manner. The readers should see the comment lines of each program for specific instructions.

Acknowledgements

Throughout the text, acknowledgements have been made to the authors of books and research papers that form the sources from which some data and figures have been taken. I would like to acknowledge the interaction with my students, both in the classroom and outside, which has greatly contributed to the development of this book. My special thanks are due to postgraduate students Nirmal Kumar Pathak and Surendra Kumar Singh who have assisted me in the preparation of solutions for the additional solved examples in Chapter 10.

I wish to acknowledge the support and encouragement provided by the editorial and production team at Oxford University Press India. I am also grateful to the anonymous reviewers whose valuable comments and suggestions for improvement have gone a long way in shaping the final version of this book.

The typing was carried out with great care and patience by Mr Yash Pal. Figures were competently drawn by Mr B.N. Srivastava, Mr Ashwani Kumar, and Mr J.C. Verma. The publication of this book would not have been possible without the generous financial support of the Curriculum Development Cell under the Quality Improvement Programme at IIT Kanpur.

I am fortunate to be working as a faculty member in IIT Kanpur, which has provided me with a lively and supportive environment and, above all, academic freedom.

Sumita, my wife, as always is my main inspiration and counsellor. I owe every bit of my achievement to her. My lovely daughter Shreya ungrudgingly accepted my long hours of absence from home in the evenings and weekends during the period of writing this text. This book would not have been possible without their love, encouragement, and sacrifice. Last but not least, I would like to express my love, regard, and gratitude to my parents, who have always showered their love and blessings on me and who have taught me how to face life.

P. S. GHOSHDASTIDAR

Brief Contents

Detailed Contents

Nomenclature

The symbols listed here are not exhaustive but the main ones used in the text. Other symbols have been defined at appropriate places in the book.

Symbols

A area, m^2

Bi   Biot number

B_{ij} absorption factor for radiation from the ith surface to the jth surface

c specific heat, J/kg °C; speed of light in a given medium

c_i molar concentration of species i, mol/m^3

c_p specific heat at constant pressure, J/kg °C

C_f skin friction coefficient

D diameter, m

D_{AB} binary diffusivity of species A into species B, m^2/s

E emissive power, W/m^2

E_b black body emissive power

Ec   Eckert number

f non-dimensional stream function in connection with similarity analysis

F_{ij} view factor for radiation from the ith surface to the jth surface

Fo   Fourier number

g acceleration due to gravity ($= 9.8 \ m/s^2$)

G irradiation, W/m^2

Gr   Grashof number

h heat transfer coefficient, W/m^2 °C or W/m^2 K; Planck's constant

h_{fg} latent heat of vapourization, J/kg

h_m local mass transfer coefficient, m/s

I intensity of radiation, W/m^2 sr

J radiosity, W/m^2

k thermal conductivity, W/m °C or W/m K; Boltzmann's constant

K absolute temperature scale, Kelvin ($= °C + 273$)

L length, m

Le   Lewis number

m fin parameter (m^{-1})

n unit vector normal to the boundary; index of refraction

Nu   Nusselt number

p pressure, N/m^2; perimeter, m

Pr   Prandtl number

Pr_t turbulent Prandtl number

q heat transfer, W

q'' heat flux, W/m^2

q''' heat generation rate per unit volume, W/m^3

r radial coordinate in cylindrical and spherical geometry

Ra   Rayleigh number

Re   Reynolds number

Sc   Schmidt number

Sh   Sherwood number

St   Stanton number

t time, s

T temperature, °C or K;

T_∞ ambient temperature; free stream temperature

T_i initial temperature

T_m mean temperature

T_s surface temperature

T_{sat} saturation temperature

T_w wall temperature

ΔT_w wall superheat or excess temperature ($= T_w - T_{sat}$)

ΔT_m log-mean-temperature difference

u velocity component in the x-direction, m/s

u_∞ free stream velocity

v velocity component in the y-direction

v_m mean velocity

v_r velocity component in the radial-direction

v_q velocity component in the circumferential direction

v_z velocity component in the axial direction

w velocity component in the z-direction

x, y, z coordinate distances

x_i mole fraction of species i

Y_i mass fraction of species i

Greek letters

α thermal diffusivity, m²/s; relaxation factor; absorptivity

β volumetric thermal expansion coefficient, $-(1/\rho)(\partial r/\partial T)_p$, K⁻¹

δ velocity or momentum boundary layer thickness, m

δ_c concentration boundary layer thickness, m

δ_t thermal boundary layer thickness, m

η similarity variable

η_f fin efficiency

λ wavelength, m

μ coefficient of viscosity, kg/s m

ν kinematic viscosity, m²/s; frequency of radiation

ε emissivity; tolerance limit; effectiveness of heat exchanger

ϕ circumferential angle in cylindrical and spherical coordinate systems; fin effectiveness

ρ density, kg/m³; reflectivity

ρ_i mass concentration of species i, kg/m³

ψ azimuthal angle in the spherical coordinate; stream function

σ Stefan–Boltzmann constant; surface tension, N/m

τ dimensionless time; transmissivity; shear stress, N/m²

θ dimensionless temperature; temperature difference; circumferential angle in the cylindrical coordinate system

Subscripts

∞ ambient

b blackbody

c concentration

f fin; friction

i initial; species

i,j,k grid point numbers

cr   critical

m mean

max   maximum

min   minimum

0   base of a fin; wall

p constant pressure

s surface

sat   saturation

t thermal; turbulent

w wall

Superscripts

p at present time; previous iteration number

$p+1$ at some future time; current iteration number

$p+1/2$ halfway between p and $p+1$

*   intermediate step in the ADI method

/   differentiation with respect to the similarity variable

Special symbols

Δ increment

Σ summation

∇^2 Laplacian operator

∇ gradient operator

Abbreviations

1D   one-dimensional

2D   two-dimensional

3D   three-dimensional

BC   boundary condition

CHF   critical heat flux

GDE   governing differential equation

GE   Gaussian elimination

GS   Gauss–Seidel iterative method

IC   initial condition

LMTD   log-mean-temperature difference

NTU   number of transfer units

TDM   tri-diagonal matrix

TDMA   tri-diagonal matrix algorithm

Chapter 1

Introduction

1.1 Aims of Studying Heat Transfer

Heat transfer is the energy interaction due to a temperature difference in a medium or between media. Heat is not a storable quantity and is defined as energy in transit due to a temperature difference. The primary aims of studying the science of heat transfer are: (i) to understand the mechanism of heat transfer processes and (ii) to predict the rate at which heat transfer takes place. Whenever we refer to heat transfer we actually imply the heat transfer rate. It is this 'rate' that differentiates the field of heat transfer from thermodynamics. Thermodynamics essentially deals with systems in equilibrium. It may be used to predict the amount of energy required to change a system from one equilibrium state to another. But, thermodynamics may not be used to calculate the rate at which this change takes place, since the system is not in equilibrium during the process. Heat transfer utilizes the first and second laws of thermodynamics in addition to the rate laws such as Fourier's law of heat conduction, Newton's law of cooling, and the Stefan–Boltzmann law of radiation.

To show the difference between heat transfer and thermodynamics, let us take a practical application such as quenching of a hot steel bar in an oil bath. Thermodynamics will help us calculate the final equilibrium temperature of the steel bar–oil combination. However, it will not be helpful in predicting the time that will be taken by the steel bar–oil combination to reach the steady state or the rate at which the temperatures of the steel bar and oil will change with time. On the other hand, heat transfer will help us chart out the time–temperature history of both the bar and oil.

Before the basic modes of heat transfer (conduction, convection, and radiation) are discussed in this chapter, an overview of the applications of heat transfer follows.

1.2 Applications of Heat Transfer

The applications of heat transfer are diverse, both in nature and in industry. Climatic changes, formation of rain and snow, heating and cooling of the earth's surface, the origin of dew drops and fog, spreading of forest fires are some of the natural phenomena wherein heat transfer plays a dominant role. The existence of living beings is possible due to the supreme heat source, the Sun.

The importance of heat transfer in industry, including medical applications, can be seen by focusing on the following classes of problems.

Thermal insulations In this class, the maximum and minimum temperatures (T_{max} and T_{min}) experienced by a heat transfer medium are usually fixed. The main objective is to reduce the 'heat loss' or 'heat leak'. The thermal design involves judicious changes in the constitution of insulation (i.e., its size, material, shape, structure, flow pattern) so that the heat transfer indeed decreases while T_{min} and T_{max} remain fixed. Typical examples where thermal insulations are used are Thermos flask, ice box, hot box, building walls, steam pipes, and cryogenics.

Heat transfer enhancement (augmentation) The main application is in the design of heat exchangers where the total heat transfer rate (q) between the hot and cold fluid streams separated by solid surfaces is usually a prescribed quantity. The objective is to transfer q across a minimum temperature difference. This can be done by changing the flow patterns of the two streams and by using finned or extended solid surfaces over which the fluid streams flow.

Temperature control In many areas, overheating of a heat-generating body is not permissible. Examples of temperature control applications include cooling of electronic equipments such as personal computers and supercomputers, cooling of nuclear reactor cores, and the cooling of the outer surface of space vehicles during re-entry. Cooling of high heat flux surfaces such as electronic chips in a tightly packaged set of electronic circuits is quite challenging because of the size limitations. The temperature of the electronic chips cannot rise much above the ambient temperature, because high temperatures drastically reduce their performance. Another important application of temperature control is the film cooling of gas turbine blades by routing air through channels within the blades.

Bioheat transfer Heat transfer plays a very important role in living systems as it affects the temperature and its spatial distribution in tissues. The primary role of temperature is the regulation of a plethora of rate processes that govern all aspects of the life process. These thermally driven rate processes define the differences between sickness and health, injury and successful therapy, comfort and pain, and accurate and limited physiological diagnosis. Typical applications of bioheat transfer include human thermoregulation, thermal surgical procedures such as microwave, ultrasound, radio frequency and laser, cryo-preservation of living cells, and thermal burn injury (Diller and Ryan, 1998).

Materials processing Recent years have seen surging interest among researchers to understand heat transfer aspects in various material processing systems such as solidification and melting, metal cutting, welding, rolling, extrusion, plastic and food processing, and laser cutting of materials. This has led to improved designs of material processing systems.

Other areas of heat transfer applications are in power production, chemical and metallurgical industries, heating and air conditioning of buildings, design of internal combustion engines, design of electrical machinery, weather prediction and environmental pollution, oil exploration, drying, and processing of solid and liquid wastes. The list is endless.

It is no wonder that J.B. Joseph Fourier, the father of the theory of heat diffusion, made this remark in 1824: 'Heat, like gravity, penetrates every substance of the universe; its rays occupy all parts of space. The theory of heat will hereafter form one of the most important branches of general physics.'

1.3 Basic Modes of Heat Transfer

We know that there are basically three fundamental modes of heat transfer, namely, conduction, convection, and radiation.

Conduction Heat conduction is essentially the transmission of energy by molecular motion. When one part of a body is at a higher temperature than the other, energy transfer takes place from the high-temperature region to the low-temperature region. In this case the energy is said to be transferred by conduction. Higher temperatures are associated with higher molecular energies, and when neighbouring molecules collide, a transfer of energy from the more energetic to the less energetic molecules must occur. In the presence of a temperature gradient, energy transfer by conduction must occur in the direction of decreasing temperature.

Generally speaking, a liquid is a better conductor than a gas and a solid is a better conductor than a liquid. This is because molecules in a gas are spaced relatively wide apart and their motion is random. This means that the energy transfer by molecular impact is much slower than in the case of a liquid, in which motion is still random but the molecules are more closely packed. The same is true concerning heat conduction in solids and liquids; however, other factors such as whether the solid is crystalline or amorphous become important when a solid state is formed.

The rate equation for conduction is given by Fourier's law. For heat conduction in the x-direction, normal to area A (Fig. 1.1), assuming the material is isotropic and homogeneous, the rate of heat flow (q) is described by

$$q = -kA\frac{\partial T}{\partial x} \tag{1.1}$$

where q is the rate of heat flow in the x-direction by conduction (in W), k is the thermal conductivity (in W/m°C or W/mK), A is the area normal to the x-direction through which heat flows (in m^2), T is the temperature (in °C), and x is the length variable (in m). Equation (1.1) may be considered as the defining equation for thermal conductivity, which is the physical property denoting the ease with which heat is transmitted through a

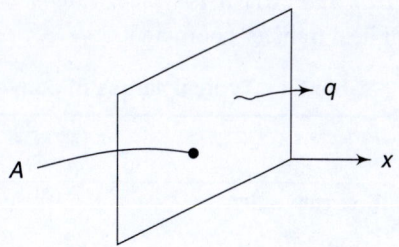

Fig. 1.1 Conduction in the x-direction, normal to area A

particular substance. The minus sign is inserted so that q is positive (since $\partial T/\partial x$ is negative in the increasing x-direction). In the case when T increases in the increasing x-direction, the minus sign is not necessary because $\partial T/\partial x$ will be positive.

Convection Convection is a process by which thermal energy is transferred between a solid and a fluid flowing past it. Strictly speaking, convection is not a separate mode of heat transfer. It denotes a fluid system in motion and heat transfer occurs by the mechanism of conduction alone. Obviously, we must allow for the motion of the fluid system in writing the energy balance, but there is no new basic mechanism of heat transfer involved.

If the fluid motion involved in the process is induced by some external means (pump, blower, wind, vehicle motion, etc.), the process is generally called *forced convection*. If the fluid motion arises from external force fields, such as gravity, acting on density gradients induced by the transport process itself, we usually call the process *free convection*. When both free and forced convection effects are significant and neither of the two can be neglected, the process is called *mixed convection*.

The rate equation used to describe the mechanism of convection is given in Eq. (1.2) and is sometimes called Newton's law of cooling when the solid surface is cooled by a fluid (Fig. 1.2),

$$q_c = hA(T_s - T_\infty) \tag{1.2}$$

where q_c is the rate of heat flow by convection (in W), h is the heat transfer coefficient (in W/m^2°C or W/m^{2}K), $T_s - T_\infty$ is the temperature potential difference for heat flow away from the surface (in °C), and A is the surface area through which heat flows (in m^2). It should be noted that Eq.

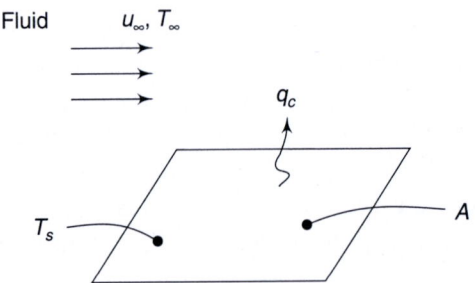

(1.2) is really a definition of h. The heat transfer coefficient depends on the space, time, geometry, orientation of the solid surface, flow conditions, and fluid properties. h is either constant or a function of the temperature difference. It is important to observe that h being a function of temperature difference does not destroy the validity of Eq. (1.2). Table 1.1 lists typical values of heat transfer coefficients.

Fig. 1.2 Convection from surface area A at T_s to cool flowing fluid at T_∞

Table 1.1   Typical values of convection heat transfer coefficient

Process	*h (W/m^{2}K or W/m^2°C)*	*Process*	*h (W/m^{2}K or W/m^2°C)*
Free convection		Liquids	50–20,000
Gases	2–25	**Convection with phase change**	
Liquids	50–1000	Boiling	2500 –
Forced convection			100,000
Gases	25–250	Condensation	4000–25,000

Source: Incropera and Dewitt (1998)

Radiation   Radiation is a mode of heat transfer which is distinctly different from conduction and convection. Whereas a material medium is a must for conduction and convection, heat may also be transferred through perfect vacuum. The mechanism in this case is electromagnetic radiation travelling at the speed of light. The electromagnetic radiation which is propagated as a result of a temperature difference between the heat-exchanging bodies is called thermal radiation, which we loosely refer to as radiation.

Thermodynamic considerations show that an ideal thermal radiator, or black body, will emit energy at a rate proportional to the fourth power of the absolute temperature of the body, to its surface area (A) and to the square of the refractive index (n) of the bounding medium. Thus

$$q_{emitted} = \sigma A n^2 T^4 \tag{1.3}$$

where $n = 1$ for vacuum and $n \approx 1$ for gases. Therefore, when the bounding medium is either vacuum or gaseous, Eq. (1.3) takes the form

$$q_{emitted} = \sigma A T^4 \tag{1.4}$$

Equation (1.4) is commonly called the Stefan–Boltzmann law of thermal radiation, although Eq. (1.3) is its original form. σ is called the Stefan–Boltzmann constant, having the value of 5.669×10^{-8} W/m^{2}K^4. Equation (1.3) or (1.4) governs only the radiation emitted by a black body. A black body is a body that radiates energy according to the T^4 law. We call such a body black because black surfaces, such as a piece of metal coated with lamp black, approximate this type of behaviour. Other types of surfaces, such as a glossy painted surface or a polished metal plate, do not radiate as much energy as the black body. However, the total radiation emitted by these bodies still follows the T^4 proportionality. To take into account the non-black nature of such surfaces, the definition of 'emissivity', ε, which relates the radiation from a non-black surface to that of an ideal black surface, is introduced.

The radiation heat loss from a hot surface to the cool air (Fig. 1.3) is given by Eq. (1.5):

$$q_r = \sigma \varepsilon A(T_s^4 - T_\infty^4) \tag{1.5}$$

where q_r is the rate of heat flow by radiation (in W), ε is the emissivity of the surface ($=1$ for black body, <1 for non-black body), σ is the Stefan–Boltzmann constant (5.669×10^{-8} W/m^{2}K^4), A is the surface area through which heat flows (in m^2), T_s is the absolute surface temperature [in K ($= °C + 273$)], and T_∞ is the absolute ambient temperature [in K ($= °C + 273$)].

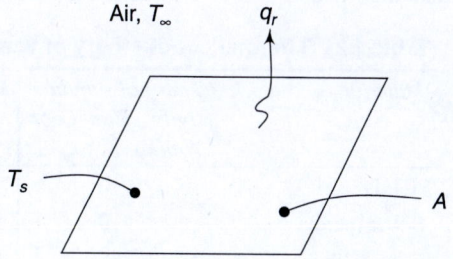

Fig. 1.3 Radiation heat loss from a hot surface at T_s to cool air at T_∞

From the above discussion, it is evident that the importance of radiation becomes intensified at high absolute temperature levels because of the T^4 term. Consequently, radiation contributes substantially to the heat transfer in furnaces and combustion chambers and in the energy emission from a nuclear explosion. Furthermore, radiation heat transfer does not require the presence of an intervening medium between heat-exchanging bodies. Some common instances are the heat leakage through the evacuated walls of a Dewar flask or Thermos bottle, or the heat dissipation from the filament of a vacuum tube. Radiation is the only mode through which heat is lost by a power plant operating in space.

1.4 Thermal Conductivity

From Fourier's law, as defined in Eq. (1.1), thermal conductivity can be expressed as

$$k = -\frac{q/A}{\partial T/\partial x} \tag{1.6}$$

It is evident from Eq. (1.6) that, for a specified temperature gradient, the conduction heat flux increases with increase in thermal conductivity. Generally, the thermal conductivity of a solid is larger than that of a liquid, which is larger than that of a gas. Table 1.2 lists typical values of thermal conductivities for various materials. At one end of the spectrum, there is silver having $k = 410$ W/m°C and at the other end, there is carbon dioxide, CO_2, with $k = 0.0146$ W/m°C. Usually, metals are better conductors than non-metals—a notable exception being diamond, which has extremely high thermal conductivity (2300 W/m°C).

In Eq. (1.6), the material is supposed to be homogeneous and isotropic, that is, when a point within it is heated, heat spreads out equally well in all directions. Such substances (which are called isotropic) are opposed to crystalline and anisotropic substances, in which certain directions are more favourable for the conduction of heat than others. There are also heterogeneous solids, in which the conditions of conduction vary from point to point as well as in direction at each point. Gases, most liquids, and amorphous solids are isotropic. Wood and other ordered fibrous materials have higher thermal conductivities parallel to the grain than perpendicular to it. Graphite is another example of an anisotropic solid. Quartz is a crystalline substance which shows a directional variation of thermal conductivity.

Table 1.2   Thermal conductivity of various materials at 0°C

Material	Thermal conductivity (W/m°C or W/m K)	Material	Thermal conductivity (W/m°C or W/m K)
Metals		**Non-metallic solids**	
Silver (pure)	410	Diamond	2300
Copper (Pure)	385	Quartz, parallel to axis	41.6
Aluminium (pure)	202	Magnesite	4.15
Nickel (pure)	93	Sand stone	1.83
Iron (pure)	73	Glass, window	0.78
Carbon steel, 1% C	43	Saw dust	0.059
Lead (pure)	35	Glass wool	0.038
Chrome–nickel steel (18% Cr, 8% Ni)	16.3	Ice	2.22
		Snow, firm	0.46
Zinc	122	Sugar (fine)	0.58

(Contd)

(Contd)

Salt (rock salt)	7	**Gases**	
Maple or oak	0.17	Hydrogen	0.175
Liquids		Helium	0.141
Mercury	8.21	Air	0.024
Water	0.556	Water vapour	0.0206
Ammonia	0.54	(saturated)	
Lubricating oil, SAE 50	0.147	Carbon dioxide	0.0146
Freon 12, CCl_2F_2	0.073		

Source: Holman (1981); Bejan (1993)

Thermal conductivity is measured in units of W/m°C or W/mK temperature difference. It may be noted that the temperature difference may be expressed in °C or K, both being the same when the difference is considered. Basically, the numerical value of thermal conductivity indicates how fast heat will flow in a given material. In the molecular model of heat conduction discussed in Section 1.3, it is clear that the faster the molecules move, the faster they will transfer energy. Therefore, the thermal conductivity of a gas should be dependent on temperature and, indeed, this is verified through a simplified analytical treatment which shows the thermal conductivity of a gas to vary with the square root of the absolute temperature.

The physical mechanism of heat conduction in liquids is qualitatively the same as in gases. Thermal conduction in solids is due to two effects: the transport by free electrons and lattice vibration. It may be noted that a solid may be composed of free electron and of atoms bound in a periodic arrangement called the lattice. The conductivity (k) of a solid is the sum of the electronic component (k_e) and the lattice component (k_l):

$$k = k_e + k_l$$

To a first approximation, k_e is inversely proportional to the electrical resistivity ρ_e. For pure metals, which are of low ρ_e, k_e is much greater than k_l. In contrast, for alloys, which are of substantially larger ρ_e, the contribution of k_l to k is no longer small. For non-metallic solids, k is determined primarily by k_l.

The regularity of lattice arrangement has an important effect on k_l, with crystalline (well-ordered) materials such as quartz having a higher thermal conductivity than amorphous materials such as glass. The reason why diamond, which is a crystalline substance, has a high conductivity as compared to that of good conductors such as silver or aluminium is now evident.

Strictly speaking, thermal conductivity is not constant for the same substance, but depends on temperature. But, when the range of temperature is limited, this change in k may be neglected. Thus,

$$k = k_0 (1 + \beta T) \tag{1.7}$$

where β is small and negative for most solids and liquids and positive for gases. Conductivity may also be a function of time, either through a dependence on the local temperature or through a change in the local material state or condition with time.

Thermal conductivities of a wide variety of solids, liquids, and gases are listed in Appendix A1. Several insulation materials are also listed in Appendix A1. Some typical values are 0.23 W/mK for Bakelite, 0.037 W/mK for glass wool, 0.032–0.04 W/mK for felt, hair. It is important to recognize that at high temperatures, the energy transfer through insulating materials may involve conduction through a fibrous or porous solid material, through the air trapped in the void spaces, and at sufficiently high temperatures, through radiation. The effective thermal conductivity accounts for all of these processes.

There are several types of insulations, namely, fibre, powder, or flake-type insulations in which the solid material is finally dispersed throughout an air space; cellular insulation (such as foamed systems, particularly those made from plastic and glass materials) in which small voids or hollow spaces are formed by bonding or fusing portions of the solid material; reflective insulations which comprise multilayered, parallel, thin sheets or foils of high reflectivity, which are spaced to reflect radiant heat back to the source. Multilayered insulations are most effective in insulating storage tanks of cryogenic liquids such as liquid hydrogen over extended periods of time. These insulations (also known as super insulations) can be used at very low temperatures (down to about −250 °C). The space between the sheets is evacuated to minimize air conduction, and thermal conductivities as low as 0.3 W/m K are possible.

Important Concepts and Formulae

Basic Difference Between Heat Transfer and Thermodynamics

Heat transfer is the energy interaction due to a temperature difference in a medium or between media. Heat is not a storable quantity and is defined as energy in transit due to a temperature difference. Thermodynamics essentially deals with systems in equilibrium. It may be used to predict the amount of energy required to change a system from one equilibrium state to another. But, thermodynamics may not be used to calculate the rate at which this change takes place, since the system is not in equilibrium during the process. Heat transfer utilizes the first and second laws of thermodynamics in addition to the rate laws such as Fourier's law of heat conduction, Newton's law of cooling, and the Stefan-Boltzmann law of radiation.

Basic Modes of Heat Transfer

Conduction

Heat conduction is essentially transmission of energy by molecular motion. When one part of a body is at a higher temperature than the other, energy transfer takes place from the high-temperature region to the low-temperature region. In this case the energy is said to be transferred by conduction.

The rate equation for conduction is given by Fourier's law. For heat conduction in the x-direction, normal to area A, assuming that the material is isotropic and homogeneous, the rate of heat flow (q) is described by

$$q = -kA\frac{\partial T}{\partial x}$$

where k is the thermal conductivity (W/mK or W/m°C).

The negative sign is given to make q positive, since $\partial T/\partial x$ is negative in the direction of increasing x.

Convection

Convection is a process by which thermal energy is transferred between a solid and a fluid flowing past it. Strictly speaking, convection is not a separate mode of heat transfer. It denotes a fluid system in motion and heat transfer occurs by the mechanism of conduction alone. Obviously, we must allow for the motion of fluid system in writing the energy balance, but there is no new basic mechanism of heat transfer involved.

If the fluid motion involved in the process is induced by some external means (pump, blower, wind, vehicular motion, etc.), the process is generally called forced convection. If the fluid motion arises from external force fields, such as gravity, acting on density gradients induced by the transport process itself, we usually call the process free convection. When both free and forced convection effects are significant and neither of the two can be neglected, the process is called mixed convection.

The rate equation used to describe the mechanism of convection is called Newton's law of cooling when the solid surface is cooled by a fluid.

$$q_c = hA\,(T_s - T_\infty)$$

where h is the heat transfer coefficient (W/m^{2}K or W/m^2°C).

The heat transfer coefficient depends on the space, time, geometry, orientation of the solid surface, flow conditions, and fluid properties.

Radiation

Radiation is a mode of heat transfer which is distinctly different from conduction and convection. The electromagnetic radiation which is propagated as a result of temperature difference between the heat-exchanging bodies is called thermal radiation, which we loosely refer to as radiation.

The emission of radiation from a black surface of area A is given by Stefan-Boltzmann law:

$$q_{emitted} = \sigma A n^2 T^4$$

where n is the refractive index of the bounding medium. $n = 1$ for vacuum and $n \approx 1$ for gases. T is in Kelvin (K). σ is called the Stefan-Boltzmann constant having the value of 5.669×10^{-8} W/m^2 K^4.

For non-black surfaces,

$$q_{emitted} = \varepsilon\sigma A n^2 T^4$$

where ε is the emissivity of the surface. $\varepsilon = 1$ for a black surface and $\varepsilon < 1$ for a non-black surface. A black body is a perfect emitter.

The radiation heat loss from a hot surface to cool air ($n = 1$) is given by

$$q_r = \sigma\varepsilon A(T_s^4 - T_\infty^4)$$

Thermal Conductivity

Basically, the numerical value of thermal conductivity indicates how fast heat will flow in a given material. Faster the molecules move, the faster they will transfer energy. Therefore, the thermal conductivity of a gas is dependent on temperature.

The physical mechanism of heat conduction in liquids is qualitatively the same as in gases. Thermal conduction in solids is due to two effects: the transport by free electrons and lattice vibration.

For metals, free electron movement plays a major role. For non-metals, conductivity is primarily determined by the effect of lattice vibration.

The regularity of lattice arrangement has an important effect on the lattice component of conductivity. This is the reason why crystalline (well-ordered) materials such as quartz have a higher thermal conductivity than amorphous materials such as glass.

Applications of Heat Transfer

The applications of heat transfer are diverse, both in nature and industry. Some important applications in industry are: design of thermal insulations, heat transfer enhancement, temperature control, bioheat transfer, and materials processing. Other areas are in power production, chemical and metallurgical industries, heating and air conditioning of buildings, design of internal combustion engines, design of electrical machinery, weather prediction and environmental pollution, oil exploration, drying, and processing of solid and liquid wastes. The list is endless.

Review Questions

1.1 Define heat transfer.
1.2 What is the difference between heat transfer and thermodynamics?
1.3 Name five important applications of heat transfer.
1.4 How is convection different from conduction?
1.5 Write Fourier's law of heat conduction, Newton's law of cooling and Stefan-Boltzmann law of thermal radiation.
1.6 Define thermal conductivity and convective heat transfer coefficient.
1.7 Explain the physical mechanism of heat conduction in solids.
1.8 Diamond has a very high thermal conductivity. Explain why.
1.9 Define homogeneous, heterogeneous, isotropic, and anisotropic solids in terms of their thermal conductivities.
1.10 What kind of materials is used for insulating storage tanks of cryogenic liquids?

Chapter 2

Steady-state Conduction: One-dimensional Problems

2.1 Introduction

Steady-state heat conduction is defined as the condition prevailing in a heat-conducting body when temperatures at all points inside the body do not change with time. The term 'one-dimensional' as applied to a heat conduction problem means that only one space coordinate is required to describe the temperature distribution in a body. Although, in practice, such a situation rarely exists, considerable simplification in the analysis of such problems can be achieved by assuming one-dimensional conduction.

The flow of heat through a plane wall having a finite thickness, but large dimensions in the other two directions, is essentially dependent on the coordinate measured normal to the plane of the wall. A typical example is the wall of a house. Here, the assumption of one-dimensionality is valid at regions far away from the edges. Thus, the edge effects are neglected. Similarly, heat transfer in a long, hollow cylinder which is maintained at uniform but different temperatures on its inner and outer surfaces may be assumed to be taking place in the radial direction only. A very thin rod, or wire, having fixed different temperatures on its ends may be considered to conduct heat along the axial direction only, neglecting the temperature variation in the radial direction as the cross-section is very small. Thus the temperature of the rod may be taken as uniform over any cross-section. It is also assumed that the ambient temperature is uniform on the periphery of the rod or wire. Another example of one-dimensional heat conduction is that in a hollow spherical shell, the inner and outer surfaces of which are maintained at uniform but different temperatures. In this case, heat transfer occurs in the radial direction only.

Before we proceed to present the solutions of one-dimensional steady-state conduction problems, the general theory of heat conduction will be discussed.

2.2 Fourier's[1] Law of Heat Conduction

The theory of the conduction of heat is said to be founded upon a hypothesis suggested by the following experiment.

---

[1]Jean Baptiste Joseph Fourier (1768–1830) was a French mathematician and, undoubtedly, the father of the science of heat transfer. He developed the general methodology for solving problems of heat conduction. The famous 'Fourier series' is named after him. Fourier also greatly influenced other areas of applied mathematics.

Consider a flat plate of thickness L as shown in Fig. 2.1. Part of this plate is assumed to be bounded by an imaginary cylinder of small cross-section A whose axis is normal to the surfaces of the plate. This cylinder is supposed to be so far from the ends of the plate that no heat crosses its peripheral surface; in other words, heat transfer is essentially one-dimensional along the axis of the cylinder. The two surfaces are kept

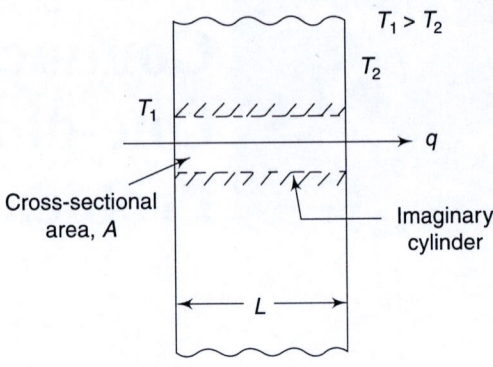

Fig. 2.1 Heat conduction under steady state in a plane wall partly bounded by an imaginary cylinder

at different temperatures, T_1 and T_2 ($T_1 > T_2$), the difference being not so great as to cause any considerable change in the properties of the solid. For example, the left surface may be kept at a fixed temperature by having a stream of warm water continuously flow over it while the right surface is maintained at the temperature of melting ice by a supply of pounded ice packed upon it. When these conditions have persisted for a long time the temperature of the different points of the solid settles down towards its steady value.

According to the first law of thermodynamics, under steady conditions heat flows at a constant rate (q) through any cross-section of the cylinder parallel to the surfaces of the plate. From the second law of thermodynamics, we know that the direction of this heat flow is from a higher temperature to a lower temperature, that is, from the left to the right, in this case.

The results of experiments upon different solids suggest that, when the steady state of temperature has been reached, the quantity Q of heat which flows through the plate in t seconds over the cross-sectional area A is equal to

$$Q = \frac{k\,(T_1 - T_2)At}{L} \tag{2.1}$$

where k is a constant, the so-called thermal conductivity of the material of the plate.

The rate of heat transfer per unit time is denoted by q W, where 1 W = 1 J/s. Therefore, Eq. (2.1) may be written as

$$q = \frac{k\,(T_1 - T_2)A}{L} \tag{2.2}$$

In other words, the flow of heat between these two surfaces is proportional to the difference of temperatures of the surfaces.

Equation (2.2) may be expressed in the following form:

$$q_n'' = k\left(\frac{T_1 - T_2}{L}\right) \tag{2.3}$$

where q_n'' is the heat flux (i.e., heat transfer per unit area) due to conduction in a direction normal to the surfaces of the plate. The above equation is known as *Fourier's law for homogeneous isotropic continua*. This brings us to the definitions of four important terms: homogeneous, heterogeneous, isotropic, and anisotropic. A continuum is said to be *homogeneous* if its conductivity does not vary from point to point. A continuum is called *heterogeneous* if there is such a variation. A continuum is termed as *isotropic* if its conductivity is same in all directions. A continuum is termed as *anisotropic* if there exists directional variation of conductivity.

Let us suppose now that a plate is isotropic but heterogeneous. Let the temperatures of two isothermal surfaces corresponding to the positions x and $x + \Delta x$ be T and $T + \Delta T$, respectively (Fig. 2.2). Since, this plate may be assumed to be locally homogeneous in the small layer of thickness Δx, Fourier's equation, Eq. (2.3), can be used for this region as $\Delta x \to 0$. Thus it becomes possible to state the differential form of Fourier's law of conduction, giving the heat flux at x in the direction of increasing x, as follows:

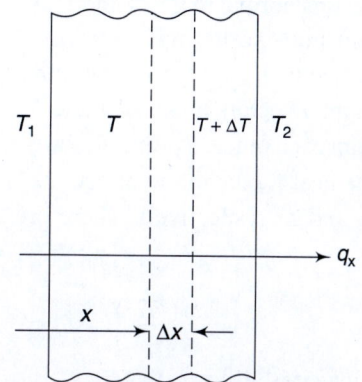

$$q''_x = -k \lim_{\Delta x \to 0} \left(\frac{\Delta T}{\Delta x} \right)$$

$$\Rightarrow \qquad q''_x = -k \left(\frac{\partial T}{\partial x} \right) \qquad (2.4)$$

Fig. 2.2 Heat conduction in a small layer of thickness Δx in a heterogeneous, isotropic continua

Equation (2.4) is called *Fourier's law for heterogeneous isotropic continua*. By introducing a minus sign, q''_x is made positive in the direction of increasing x [Fig. 2.3(a)]. It may be noted, however, that when heat conduction occurs in the direction opposite to the direction of increasing x as shown in Fig. 2.3(b), the minus sign is not necessary, as $\partial T/\partial x$ is positive. The convention is that q_x should be positive in the direction of increasing x. Equation (2.4) may be readily extended to any isothermal surface if it is stated that the heat flux across an isothermal surface is

$$q''_n = -k \left(\frac{\partial T}{\partial n} \right) \qquad (2.5)$$

where $\partial/\partial n$ represents differentiation along the normal to the surface.

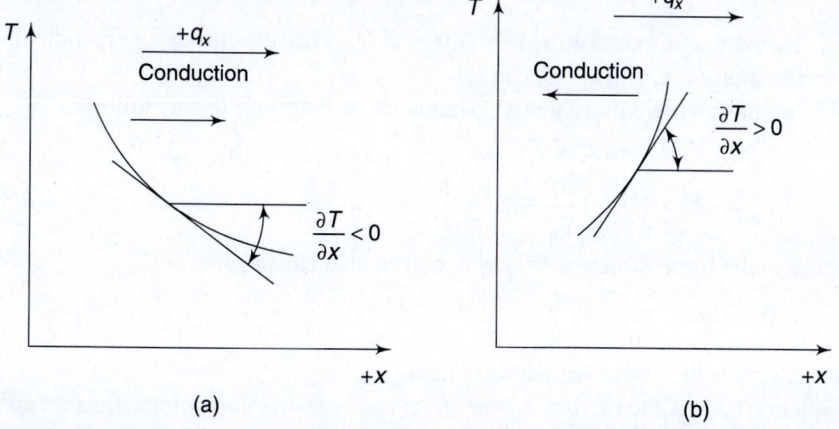

Fig. 2.3 (a) T vs x graph showing $\partial T/\partial x < 0$, (b) T vs x graph showing $\partial T/\partial x > 0$

Can Fourier's law be extended to non-isothermal surfaces? The answer is 'yes'. Let us examine Fig. 2.4, which shows a non-isothermal surface intersecting an isothermal surface at P. $\hat{n}$ and $\hat{s}$ are the unit direction vectors normal to the isothermal and non-isothermal surfaces, respectively, at P. The two unit normal vectors make an angle θ with each other. q_n'' and q_s'' are the heat fluxes in the directions $\hat{n}$ and $\hat{s}$, respectively. Therefore,

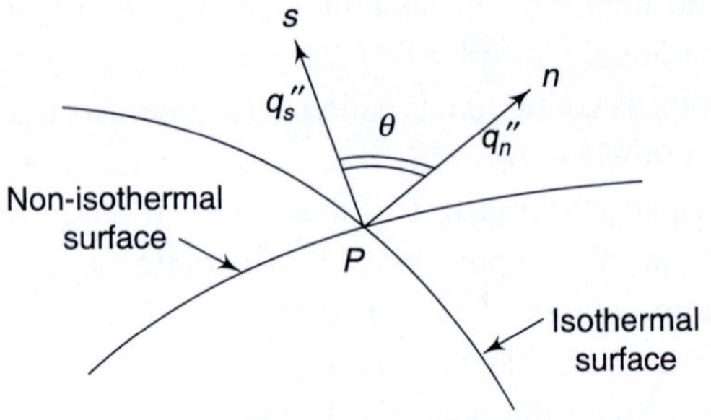

Fig. 2.4 The directions of heat fluxes and normal vectors at the intersection of isothermal and non-isothermal surfaces

$$q_s'' = q'' \cdot \hat{s} = q_n'' \hat{n} \cdot \hat{s} = q_n'' \cos\theta$$

$$= -k\left(\frac{\partial T}{\partial n}\right)\cos\theta$$

Since it is also true that

$$\left(\frac{\partial T}{\partial n}\right)\cos\theta = \frac{\partial T}{\partial s}$$

therefore,

$$q_s'' = -k\left(\frac{\partial T}{\partial s}\right) \tag{2.6}$$

where $\partial/\partial s$ represents differentiation in the direction of the normal s.

In the Cartesian coordinate system, the three components of the heat flux vector q'' are given by

$$q_x'' = -k\left(\frac{\partial T}{\partial x}\right) \tag{2.7a}$$

$$q_y'' = -k\left(\frac{\partial T}{\partial y}\right) \tag{2.7b}$$

$$q_z'' = -k\left(\frac{\partial T}{\partial z}\right) \tag{2.7c}$$

which are the magnitudes of the heat fluxes at P across the surfaces perpendicular to the directions x, y, and z, respectively.

We can now write a more general statement of Fourier's law as follows:

$$q'' = -k\nabla T = -k\,\mathrm{grad}\,T$$

$$= -k\left(\hat{i}\frac{\partial T}{\partial x} + \hat{j}\frac{\partial T}{\partial y} + \hat{k}\frac{\partial T}{\partial z}\right) \tag{2.8}$$

where ∇ is the three-dimensional *del* operator, also called *gradient*,

$$\nabla = \hat{i}\frac{\partial}{\partial x} + \hat{j}\frac{\partial}{\partial y} + \hat{k}\frac{\partial}{\partial z}$$

and $T(x, y, z)$ is the scalar temperature field.

Equation (2.8) is the *vectorial form of Fourier's law for heterogeneous isotropic continua*.

2.3 Fourier's Law in Cylindrical and Spherical Coordinates

In cylindrical coordinates (Fig. 2.5) Fourier's law takes the following form:

$$q_r'' = -k\left(\frac{\partial T}{\partial r}\right) \qquad (2.9a)$$

$$q_\phi'' = -k\frac{1}{r}\left(\frac{\partial T}{\partial \phi}\right) \qquad (2.9b)$$

$$q_z'' = -k\left(\frac{\partial T}{\partial z}\right) \qquad (2.9c)$$

ϕ is called the polar angle ($0° \le \phi \le 360°$).

In spherical coordinates (Fig. 2.6), Fourier's law can be expressed as

$$q_r'' = -k\left(\frac{\partial T}{\partial r}\right) \qquad (2.10a)$$

$$q_\psi'' = -k\frac{1}{r}\left(\frac{\partial T}{\partial \psi}\right) \qquad (2.10b)$$

$$q_\phi'' = -k\frac{1}{r\,\sin\psi}\left(\frac{\partial T}{\partial \phi}\right) \qquad (2.10c)$$

ψ is called the zenith angle ($0° \le \psi \le 180°$) and ϕ is called the azimuthal angle ($0° \le \phi \le 360°$).

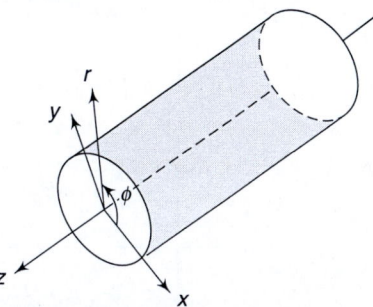

Fig. 2.5 Cylindrical coordinate system

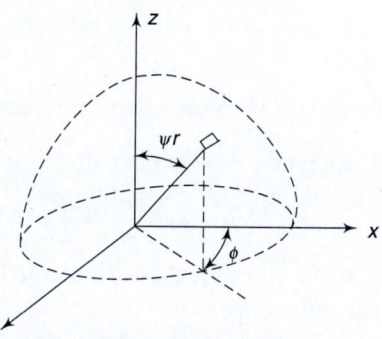

Fig. 2.6 Spherical coordinate system

2.4 Heat Conduction Equation for Isotropic Materials

In this section the heat conduction equation (also called the energy equation) in the differential form in Cartesian coordinates will be derived. The differential form of the heat conduction equation is most useful. It will be assumed that the material is isotropic.

Consider an infinitesimal volume element of dimensions Δx, Δy, and Δz in the Cartesian coordinate system as shown in Fig. 2.7. Here, the unsteady condition of temperature variation with time t is also taken into account. According to Fourier's law of heat conduction, the heat flowing into the left face of the element in the x-direction can be expressed as

$$q_x = -k\,\Delta y\,\Delta z\frac{\partial T}{\partial x}$$

The magnitude of heat flow out of the right face of the element can be obtained by expanding q_x in a Taylor series and retaining only the first two terms as a reasonable approximation:

$$q_{x+\Delta x} = q_x + \frac{\partial}{\partial x}(q_x)\Delta x + \cdots$$

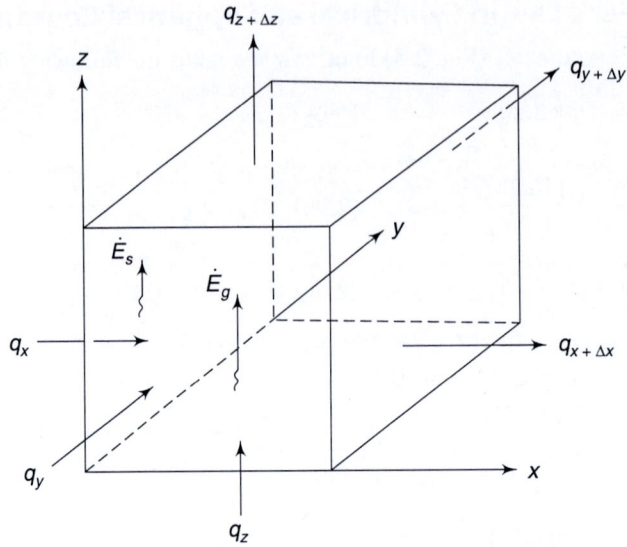

Fig. 2.7   The volume element for derivation of the heat conduction equation

The net heat flow in the x-direction is therefore

$$q_x - q_{x+\Delta x} = \frac{\partial}{\partial x}\left(k\frac{\partial T}{\partial x}\right)\Delta x\,\Delta y\,\Delta z \tag{2.11}$$

The two other equations similar to Eq. (2.11) for the y and z directions can be written in the same way:

$$q_y - q_{y+\Delta y} = \frac{\partial}{\partial y}\left(k\frac{\partial T}{\partial y}\right)\Delta x\,\Delta y\,\Delta z \tag{2.12}$$

$$q_z - q_{z+\Delta z} = \frac{\partial}{\partial z}\left(k\frac{\partial T}{\partial z}\right)\Delta x\,\Delta y\,\Delta z \tag{2.13}$$

The sum of the net quantities of heat is the net difference between the incoming and the outgoing heat transfer by conduction in the volume element:

$$(q_x - q_{x+\Delta x}) + (q_y - q_{y+\Delta y}) + (q_z - q_{z+\Delta z})$$

$$= \left[\frac{\partial}{\partial x}\left(k\frac{\partial T}{\partial x}\right) + \frac{\partial}{\partial y}\left(k\frac{\partial T}{\partial y}\right) + \frac{\partial}{\partial z}\left(k\frac{\partial T}{\partial z}\right)\right]\Delta x\,\Delta y\,\Delta z \tag{2.14}$$

If $q'''(x, y, z, t)$ is the rate of heat generation per unit volume (in W/m^3) then the generation of heat in the element is

$$\dot{E}_g = q'''\Delta x\,\Delta y\,\Delta z \tag{2.15}$$

The heat remaining in the volume element owing to conduction [Eq. (2.14)] and the heat generated within the volume element [Eq. (2.15)] together serve to increase the internal energy of the volume element. The time rate of change of internal energy within the volume element can be written as

$$\dot{E}_s = \rho c\,\Delta x\,\Delta y\,\Delta z\left(\frac{\partial T}{\partial t}\right) \tag{2.16}$$

where c is the specific heat (in J/kgK), ρ is the density (in kg/m^3), and t is the time (in s).

An energy balance can be made on the volume element to equate the time rate of change of internal energy to the net heat flowing into the element due to conduction and to the heat generated within the element to yield the expression

$$\dot{E}_s = (q_x - q_{x+\Delta x}) + (q_y - q_{y+\Delta y}) + (q_z - q_{z+\Delta z}) + \dot{E}_g$$

$$\Rightarrow \quad \rho c \frac{\partial T}{\partial t} = \left[\frac{\partial}{\partial x}\left(k \frac{\partial T}{\partial x}\right) + \frac{\partial}{\partial y}\left(k \frac{\partial T}{\partial y}\right) + \frac{\partial}{\partial z}\left(k \frac{\partial T}{\partial z}\right) + q''' \right] \qquad (2.17)$$

It should be noted here that

$$k = k(x, y, z, t), \quad c = c(x, y, z, t), \quad \rho = \rho(x, y, z, t)$$

so that Eq. (2.17) is valid for isotropic, heterogeneous media. Equation (2.17) can be further simplified for an isotropic, homogeneous material, that is, for the case of constant thermal conductivity k. Therefore, for this case, the heat conduction equation [Eq. (2.17)] takes the form

$$\frac{k}{\rho c}\left[\frac{\partial^2 T}{\partial x^2} + \frac{\partial^2 T}{\partial y^2} + \frac{\partial^2 T}{\partial z^2} \right] + \frac{q'''}{\rho c} = \frac{\partial T}{\partial t}$$

$$\Rightarrow \quad \frac{\partial^2 T}{\partial x^2} + \frac{\partial^2 T}{\partial y^2} + \frac{\partial^2 T}{\partial z^2} + \frac{q'''}{k} = \frac{1}{\alpha}\frac{\partial T}{\partial t} \qquad (2.18)$$

Equation (2.18) can be written in a compact form as

$$\nabla^2 T + \frac{q'''}{k} = \frac{1}{\alpha}\frac{\partial T}{\partial t} \qquad (2.19)$$

where ∇^2 is called the Laplacian operator:

$$\nabla^2 = \frac{\partial^2}{\partial x^2} + \frac{\partial^2}{\partial y^2} + \frac{\partial^2}{\partial z^2}$$

In cylindrical coordinates,

$$\nabla^2 = \frac{1}{r}\frac{\partial}{\partial r}\left(r\frac{\partial}{\partial r}\right) + \frac{1}{r^2}\frac{\partial^2}{\partial \phi^2} + \frac{\partial^2}{\partial z^2}$$

In spherical coordinates,

$$\nabla^2 = \frac{1}{r^2}\frac{\partial}{\partial r}\left(r^2\frac{\partial}{\partial r}\right) + \frac{1}{r^2 \sin\psi}\frac{\partial}{\partial \psi}\left(\sin\psi\frac{\partial}{\partial \psi}\right) + \frac{1}{r^2 \sin^2\psi}\frac{\partial^2}{\partial \phi^2}$$

The term $k/\rho c$ has the dimension of length squared per time (m^2/s) and is referred to as the thermal diffusivity α, which is a property of the conducting material and basically signifies the rate at which heat diffuses into the medium during changes in temperature with time. A material that has a high thermal conductivity or a low heat capacity (ρc) will have a large thermal diffusivity. The larger the thermal diffusivity, the faster is the propagation of heat into the medium. A low value of thermal diffusivity means that heat is mostly absorbed by the material and a small amount of heat will be conducted further.

The heat conduction equation, Eq. (2.19), has been developed with reference to a solid material in the shape of a rectangular parallelepiped. The same equation also applies to the case where the parallelopiped is actually a column of *incompressible and motionless liquid or gas.*

2.4.1 Heat Conduction Equation in a Cylindrical Coordinate System

Using a coordinate transformation, Eq. (2.18) can be expressed in a form suitable for a cylindrical system. Thus, from Fig. 2.5, for $x = r\cos\phi$, $y = r\sin\phi$, and $z = z$,

$$\frac{\partial^2 T}{\partial r^2} + \frac{1}{r}\frac{\partial T}{\partial r} + \frac{1}{r^2}\frac{\partial^2 T}{\partial \phi^2} + \frac{\partial^2 T}{\partial z^2} + \frac{q'''}{k} = \frac{1}{\alpha}\frac{\partial T}{\partial t} \tag{2.20}$$

Equation (2.20) can also be obtained by considering a small volume element in cylindrical coordinates and performing an energy balance on it. The energy-in and energy-out terms on a volume element cut out from a cylinder for one-dimensional heat conduction in the radial direction are shown in Fig. 2.8.

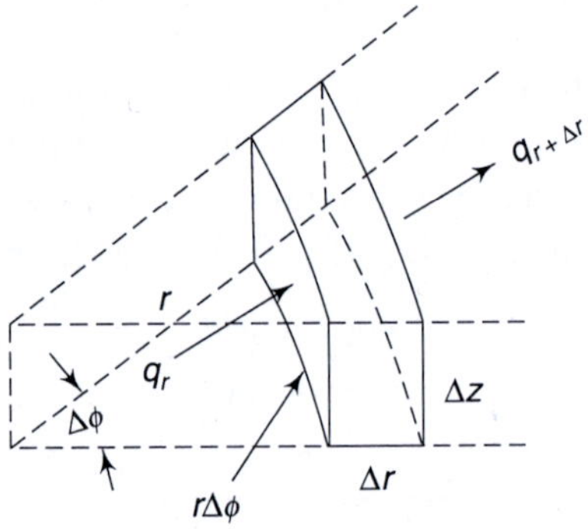

2.4.1.1 Derivation of Eq. (2.20) using coordinate transformation

Fig. 2.8 The energy balance on a volume element cut out from a cylinder

The starting equation is the heat conduction equation in Cartesian coordinates, Eq. (2.18), which is

$$\frac{\partial^2 T}{\partial x^2} + \frac{\partial^2 T}{\partial y^2} + \frac{\partial^2 T}{\partial z^2} + \frac{q'''}{k} = \frac{1}{\alpha}\frac{\partial T}{\partial t} \tag{2.21}$$

Using coordinate transformation (see Fig. 2.5),

$$x = r\cos\phi$$

$$y = r\sin\phi$$

$$z = z$$

Since $T = T(x, y, z, t) = T(x(r, \phi), y(r, \phi), z, t)$, applying the chain rule of differentiation,

$$\frac{\partial T}{\partial x} = \frac{\partial T}{\partial r}\frac{\partial r}{\partial x} + \frac{\partial T}{\partial \phi}\frac{\partial \phi}{\partial x} \tag{2.22}$$

where $r = (x^2 + y^2)^{1/2}$ and $\tan\phi = y/x$. Now,

$$\frac{\partial r}{\partial x} = \frac{1}{2}(x^2 + y^2)^{-1/2}\, 2x$$

$$= \frac{x}{\sqrt{x^2 + y^2}} = \frac{x}{r} = \cos\phi$$

$$\Rightarrow \qquad \frac{\partial r}{\partial x} = \cos\phi \tag{2.23a}$$

Now, $\tan\phi = y/x$ (2.23b)

Therefore, differentiating both sides of Eq. (2.23b) with respect to x,

$$\sec^2\phi\,\frac{\partial\phi}{\partial x} = -\frac{y}{x^2}$$

$$\Rightarrow \quad \frac{\partial\phi}{\partial x} = \frac{-\dfrac{y}{x^2}}{\sec^2\phi} = \frac{-\dfrac{y}{x^2}}{1+\left(\dfrac{y^2}{x^2}\right)}$$

$$= -\frac{y}{r^2} = -\frac{r\sin\phi}{r^2} = -\frac{\sin\phi}{r}$$

$$\Rightarrow \quad \frac{\partial\phi}{\partial x} = -\frac{\sin\phi}{r} \tag{2.24}$$

Substituting Eqs (2.23) and (2.24) into Eq. (2.22), we get

$$\frac{\partial T}{\partial x} = \frac{\partial T}{\partial r}\cos\phi + \frac{\partial T}{\partial\phi}\left(-\frac{\sin\phi}{r}\right)$$

Therefore,

$$\frac{\partial^2 T}{\partial x^2} = \frac{\partial}{\partial x}\left(\frac{\partial T}{\partial x}\right)$$

$$= \frac{\partial}{\partial r}\left(\frac{\partial T}{\partial x}\right)\frac{\partial r}{\partial x} + \frac{\partial}{\partial\phi}\left(\frac{\partial T}{\partial x}\right)\frac{\partial\phi}{\partial x}$$

$$= \frac{\partial}{\partial r}\left[\frac{\partial T}{\partial r}\cos\phi - \frac{\partial T}{\partial\phi}\frac{\sin\phi}{r}\right]\cos\phi + \frac{\partial}{\partial\phi}\left[\frac{\partial T}{\partial r}\cos\phi - \frac{\partial T}{\partial\phi}\frac{\sin\phi}{r}\right]\left(-\frac{\sin\phi}{r}\right)$$

$$= \cos^2\phi\,\frac{\partial^2 T}{\partial r^2} - 2\frac{1}{r}\frac{\partial^2 T}{\partial r\partial\phi}\sin\phi\cos\phi + \frac{1}{r}\frac{\partial T}{\partial r}\sin^2\phi$$

$$\qquad + \frac{1}{r^2}\frac{\partial^2 T}{\partial\phi^2}\sin^2\phi + \frac{2}{r^2}\frac{\partial T}{\partial\phi}\sin\phi\cos\phi \tag{2.25}$$

Similarly,

$$\frac{\partial T}{\partial y} = \left(\frac{\partial T}{\partial y}\right)\frac{\partial r}{\partial y} + \left(\frac{\partial T}{\partial\phi}\right)\frac{\partial\phi}{\partial y}$$

$$= \left(\frac{\partial T}{\partial r}\right)\sin\phi + \left(\frac{\partial T}{\partial\phi}\right)\frac{\cos\phi}{r}$$

Therefore,

$$\frac{\partial^2 T}{\partial y^2} = \frac{\partial}{\partial y}\left(\frac{\partial T}{\partial y}\right) = \sin^2\phi\,\frac{\partial^2 T}{\partial r^2} + \frac{2}{r}\sin\phi\cos\phi\,\frac{\partial^2 T}{\partial r\partial\phi} - \frac{1}{r^2}\frac{\partial T}{\partial\phi}\cos\phi\sin\phi$$

$$\qquad + \frac{1}{r^2}\frac{\partial^2 T}{\partial\phi^2}\cos^2\phi + \frac{1}{r}\frac{\partial T}{\partial r}\cos^2\phi - \frac{1}{r^2}\frac{\partial T}{\partial\phi}\sin\phi\cos\phi \tag{2.26}$$

Adding Eqs (2.25) and (2.26) and using $\cos^2\phi + \sin^2\phi = 1$, we obtain

$$\frac{\partial^2 T}{\partial x^2} + \frac{\partial^2 T}{\partial y^2} = \frac{\partial^2 T}{\partial r^2} + \frac{1}{r}\frac{\partial T}{\partial r} + \frac{1}{r^2}\frac{\partial^2 T}{\partial \phi^2} \tag{2.27}$$

Therefore, substituting Eq. (2.27) into Eq. (2.21), finally the heat conduction equation in cylindrical coordinates is obtained:

$$\frac{\partial^2 T}{\partial r^2} + \frac{1}{r}\frac{\partial T}{\partial r} + \frac{1}{r^2}\frac{\partial^2 T}{\partial \phi^2} + \frac{\partial^2 T}{\partial z^2} + \frac{q'''}{k} = \frac{1}{\alpha}\frac{\partial T}{\partial t} \tag{2.28}$$

2.4.2 Heat Conduction Equation in a Spherical Coordinate System

Using a coordinate transformation, Eq. (2.18) can be expressed in a form suitable for spherical systems. Thus, from Fig. 2.6, for $x = r\sin\psi \cos\phi$, $y = r\sin\psi \sin\phi$, and $z = r\cos\psi$,

$$\frac{1}{r^2}\frac{\partial}{\partial r}\left(r^2\frac{\partial T}{\partial r}\right) + \frac{1}{r^2\sin\psi}\frac{\partial}{\partial \psi}\left[\sin\psi \frac{\partial T}{\partial \psi}\right] + \frac{1}{r^2\sin^2\psi}\frac{\partial^2 T}{\partial \phi^2} + \frac{q'''}{k}$$

$$= \frac{1}{\alpha}\frac{\partial T}{\partial t} \tag{2.29a}$$

Alternatively, Eq. (2.29a) can be written as

$$\frac{1}{r}\left(\frac{\partial^2 (rT)}{\partial r^2}\right) + \frac{1}{r^2\sin\psi}\frac{\partial}{\partial \psi}\left[\sin\psi \frac{\partial T}{\partial \psi}\right] + \frac{1}{r^2\sin^2\psi}\frac{\partial^2 T}{\partial \phi^2} + \frac{q'''}{k}$$

$$= \frac{1}{\alpha}\frac{\partial T}{\partial t} \tag{2.29b}$$

Equation (2.29a) or (2.29b) can also be obtained by considering a small volume element in spherical coordinates and performing an energy balance on it.

2.5 Heat Conduction Equation for Anisotropic Materials

In the preceding sections the heat conduction equation for isotropic media was derived. Certain industrially important materials and laminates have direction-dependent thermal conductivity. Such materials are called anisotropic materials. Included in this category of materials are crystalline substances, woods, laminated plastics, and laminated metals such as are used in transformer cores and plywood.

In the case of anisotropic heat conduction, the component of heat flux in any direction, for example, q_x'' in the x-direction, depends on the temperature gradients in each of the three coordinate directions. That is,

$$q_x'' = -\left[k_{11}\frac{\partial T}{\partial x} + k_{12}\frac{\partial T}{\partial y} + k_{13}\frac{\partial T}{\partial z}\right] \tag{2.30}$$

Three conductivity coefficients may also arise in the y and z directions. Then the thermal conductivity becomes the following second-order tensor quantity:

$$k_{ij} = \begin{bmatrix} k_{11} & k_{12} & k_{13} \\ k_{21} & k_{22} & k_{23} \\ k_{31} & k_{32} & k_{33} \end{bmatrix} \tag{2.31}$$

The heat flux component in the x_i direction, q''_{x_i}, is then

$$q''_{x_i} = -k_{ij} \frac{\partial T}{\partial x_j} \tag{2.32}$$

Here, x_1 corresponds to x, x_2 to y, and x_3 to z. Recall that one sums on any repeated subscript, for example, j in Eq. (2.32). Thus, putting $i = 1, 2, 3$, all three heat flux components are generated. The general anisotropic formulation includes the isotropic case if $k_{ij} = 0$ for $i \neq j$ and if the remaining $k_{ii} = k$, that is, if all non-diagonal terms in the term formulation given in Eq. (2.31) are zero and all the remaining diagonal terms are equal. In other words, for isotropic materials,

$$k_{ij} = \begin{bmatrix} k & 0 & 0 \\ 0 & k & 0 \\ 0 & 0 & k \end{bmatrix} \tag{2.33}$$

For orthotropic materials (e.g., wood, fibrous materials, and numerous crystalline substances), $k_{ii} = k_i$ and $k_{ij} = 0$ for $i \neq j$. In that case,

$$k_{ij} = \begin{bmatrix} k_1 & 0 & 0 \\ 0 & k_2 & 0 \\ 0 & 0 & k_3 \end{bmatrix} \tag{2.34}$$

The general heat conduction equation which takes care of anisotropic as well as isotropic materials then becomes

$$\frac{\partial}{\partial x_i} \left(k_{ij} \frac{\partial T}{\partial x_j} \right) + q''' = \rho c \frac{\partial T}{\partial t} \tag{2.35}$$

Detailed discussions regarding conduction in anisotropic materials are somewhat beyond the scope of the present text. Interested readers are, however, encouraged to refer to Carslaw and Jaeger (1959), Eckert and Drake (1959), and Gebhart (1993).

2.6 Initial and Boundary Conditions

Before we discuss the mathematical formulation of conduction problems, it is necessary to know how to express mathematically the initial and boundary conditions that the temperature satisfies.

It is assumed that in the interior of a solid, temperature T is a continuous function of x, y, z, and t, and that this holds also for the first derivative of T with respect to t and for the first and second derivatives of T with respect to x, y, and z. At the boundary of the solid, and at the instant at which heat flow is supposed to start, these assumptions do not hold.

2.6.1 Initial Condition

For an unsteady problem, the temperature throughout the body must be known at some instant of time. In many cases, this instant is most conveniently taken to be the beginning of the heat flow. If this arbitrary temperature is a continuous function, so that

$$T = T(x, y, z)$$

our solution of Eq. (2.19)

$$\nabla^2 T + \frac{q'''}{k} = \frac{1}{\alpha} \frac{\partial T}{\partial t}$$

must be such that

$$\lim_{t \to 0} T(x, y, z, t) = T(x, y, z)$$

at all points of the solid.

If the initial temperature distribution is discontinuous at points or surfaces, these discontinuities must disappear after so short a time, and in this case our solution must converge to the value given by the initial temperature at all points where this distribution is continuous.

2.6.2 Boundary Conditions

The surface or boundary conditions usually encountered in the theory of heat conduction are as follows.

A. Prescribed surface temperature   The surface temperature may be constant, or a function of time or space, or both. This is also known as the Dirichlet condition and is the easiest boundary condition to work with. But it must be remembered that in practice it is often difficult to prescribe the surface temperature, and actual conditions may be better represented by a boundary condition of type D (i.e., convective condition) discussed later in this section.

B. Prescribed heat flux   The heat flux across the boundaries is specified to be a constant or a function of time or space, or of both. The mathematical description of this condition may be given in the light of Kirchhoff's current law: *the algebraic sum of heat fluxes at a boundary must be equal to zero.* The following sign conventions are used.

- Heat flux to the boundary is positive.
- Heat flux from the boundary is negative.

The next four cases illustrate the application of Kirchhoff's current law. In Figs 2.9(a)–(d), q_n'' is the conductive heat flux and q'' is the prescribed heat flux acting on the surface. $\hat{n}$ is the normal to the boundary. The subscript σ indicates the boundary or surface.

Case I
Referring to Fig. 2.9(a),

$$q_n'' + (-q'') = 0$$

or

$$q_n'' - q'' = 0$$

or

$$-k\left(\frac{\partial T}{\partial n}\right)_\sigma - q'' = 0$$

(a)

Case II

Referring to Fig. 2.9(b),

$$q_n'' + q'' = 0$$

or

$$-k\left(\frac{\partial T}{\partial n}\right)_\sigma + q'' = 0$$

Case III

Referring to Fig. 2.9(c),

$$-q_n'' - q'' = 0$$

or

$$-\left[-k\left(\frac{\partial T}{\partial n}\right)_\sigma\right] - q'' = 0$$

Case IV

Referring to Fig. 2.9(d),

$$-q_n'' + q'' = 0$$

or

$$-\left[-k\left(\frac{\partial T}{\partial n}\right)_\sigma\right] + q'' = 0$$

C. No heat flux across the surface (insulation) $(\partial T/\partial n)_\sigma = 0$ at all points of the surface. Here $\partial T/\partial n$ denotes differentiation in the direction of the outward normal to the surface. The conditions B and C are also called Neumann conditions.

D. Heat transfer to the surroundings by convection Referring to Fig. 2.10,

$$q_c'' = h\,(T_\sigma - T_\infty)$$

Therefore,

$$q_n'' - q_c'' = 0$$

or

$$-k\left(\frac{\partial T}{\partial n}\right)_\sigma - q_c'' = 0$$

or

$$-k\left(\frac{\partial T}{\partial n}\right)_\sigma = h\,(T_\sigma - T_\infty)$$

h is called the convective heat transfer coefficient and depends on the space, time, geometry, flow conditions, physical properties of the fluid, and the orientation of the surface. As $h \to 0$, the boundary condition tends to condition C and as $h \to \infty$, it tends to condition A. Condition D is also known as the Robbins condition.

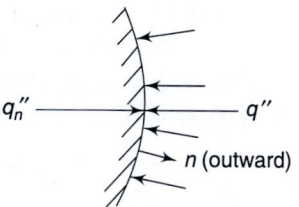

(b)

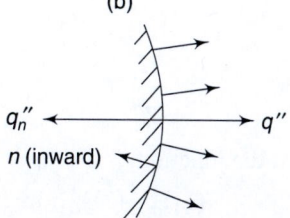

(c)

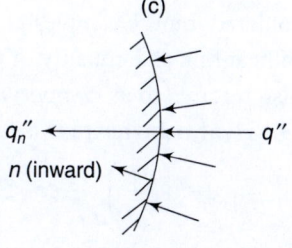

(d)

Fig. 2.9 (a)–(d) Prescribed heat flux conditions for cases I–IV

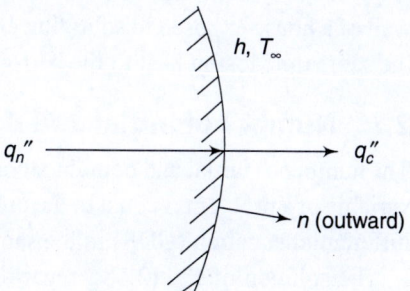

Fig. 2.10 Convective boundary condition

E. **Interface of two media of different conductivities k_1 and k_2** Let T_1 and T_2 denote the temperatures in two media. When the two media have a common boundary, the heat flux across this boundary evaluated from both media, regardless of the direction of the normal, gives

$$q_1'' - q_2'' = 0$$

or $$-k_1 \left(\frac{\partial T_1}{\partial n} \right)_\sigma - \left[-k_2 \left(\frac{\partial T_2}{\partial n} \right)_\sigma \right] = 0$$

or $$k_1 \left(\frac{\partial T_1}{\partial n} \right)_\sigma = k_2 \left(\frac{\partial T_2}{\partial n} \right)_\sigma$$

At the junction,

$$(T_1)_\sigma = (T_2)_\sigma$$

This assumption is valid if the media are solid and in intimate contact, such as a soldered joint. Examples are composite walls and insulated tubes. The continuity of heat flux and equality of temperature at the interface of two different media are also referred to as *compatibility conditions*.

F. **Heat transfer to the surroundings by radiation**

$$-k \left(\frac{\partial T}{\partial n} \right)_\sigma = \varepsilon \sigma (T_\sigma^4 - T_\infty^4)$$

This is a non-linear boundary condition because of the T^4 term.

G. **Prescribed heat flux acting at a distance**

Referring to Fig. 2.11,

$$q_n'' + q'' - q_c'' = 0$$

or $$-k \left(\frac{\partial T}{\partial n} \right)_\sigma + q'' - h(T_\sigma - T_\infty) = 0$$

or $$-k \left(\frac{\partial T}{\partial n} \right)_\sigma = h\,(T_\sigma - T_\infty) - q''$$

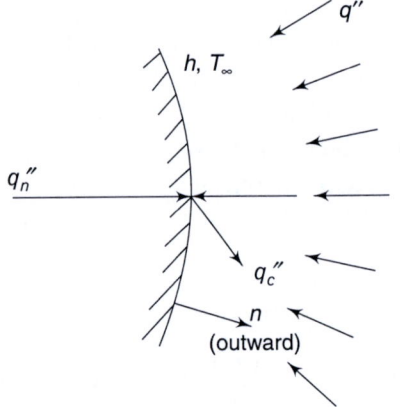

Fig. 2.11 Prescribed heat flux acting at a distance

A typical example is the outer surface of the wall of a house exposed to solar flux and at the same time losing heat to the surroundings by convection.

2.7 Number of Initial and Boundary Conditions

The number of initial and boundary conditions in the direction of each independent variable of a problem is equal to the order of the highest derivative of the governing differential equation (GDE) in the same direction.

The solution of Eq. (2.18), regardless of the mathematical method employed, requires a single integration in time and a double integration in each of the three space variables involved. Thus, the number of initial conditions:

$$\partial T/\partial t \rightarrow \text{order } 1 \rightarrow 1$$

The number of boundary conditions in the *x*-direction is equal to the order of the highest derivative:

$$\partial^2 T/\partial x^2 \rightarrow \text{order } 2 \rightarrow 2$$

The number of boundary conditions in the *y*-direction:

$$\partial^2 T/\partial y^2 \rightarrow \text{order } 2 \rightarrow 2$$

The number of boundary conditions in the *z*-direction:

$$\partial^2 T/\partial z^2 \rightarrow \text{order } 2 \rightarrow 2$$

The total number of boundary conditions is 6. The total number of initial conditions is 1. Therefore, to solve a three-dimensional unsteady-state conduction problem, one initial condition and six boundary conditions are required.

2.8 Simple One-dimensional Steady Conduction Problems

Whenever the geometry of the heat-conducting body is simple and heat transfer is in one direction only and is steady (i.e., invariant with time), the heat conduction equation can be greatly simplified. Some very practical heat conduction problems fall into this category, for example, the plane wall, the hollow cylindrical tube, and the hollow sphere.

2.8.1 Plane Wall

The term 'plane wall' is applied to a system which is finite in one direction but extends to infinity in the other two directions. For engineering purposes, a large wall with finite thickness generally meets the criterion of one-dimensional (1D) heat transfer because the edge effects can be neglected.

An example might be to calculate the steady-state heat loss through a brick wall of thickness *L* and constant conductivity in a house as a function of the outside temperature (T_2) when the inside temperature is maintained at T_1 (Fig. 2.12). Thus, from the

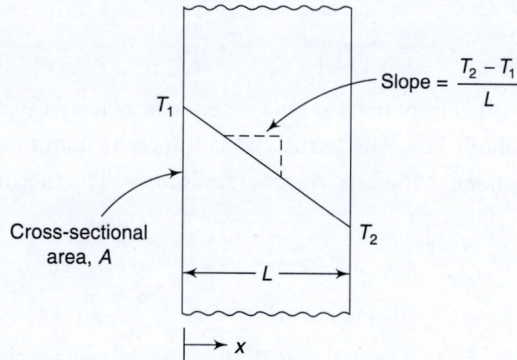

Fig. 2.12 Steady heat conduction in a plane wall

knowledge of the steady-state heat loss, the heat transfer analyst can determine the rate of heat input to be given to the inner wall to maintain it at a fixed temperature. The energy equation, Eq. (2.17), becomes

$$d^2 T/dx^2 = 0 \qquad (2.36)$$

Integrating Eq. (2.36) twice yields

$$T = C_1 x + C_2$$

The constants C_1 and C_2 can be evaluated from the boundary conditions which prescribe the temperature of the surface $x = 0$ and $x = L$. The boundary conditions (BCs) are mathematically written as

$$\text{BC-1: at } x = 0, \ T = T_1 \tag{2.37a}$$

$$\text{BC-2: at } x = L, \ T = T_2 \tag{2.37b}$$

The application of BC-1 gives

$$C_2 = T_1$$

The application of BC-2 gives

$$C_1 L + C_2 = T_2$$

$$C_1 = \frac{T_2 - C_2}{L} = \frac{T_2 - T_1}{L}$$

Therefore, the temperature distribution in the plane wall is

$$T = \frac{T_2 - T_1}{L} x + T_1 \tag{2.38}$$

Hence, T versus x is a straight line with a negative slope.

The heat flow through a wall of cross-sectional area A can be obtained from Fourier's law of heat conduction:

$$q = -kA \frac{dT}{dx}$$

$$= -kA \frac{T_2 - T_1}{L}$$

$$= \frac{T_1 - T_2}{L/kA} \tag{2.39}$$

It is important to note here the similarity of Eq. (2.39) to the usual statement of Ohm's law. The term L/kA is the equivalent of electrical resistance and is appropriately called the *thermal resistance*. The thermal circuit is shown in Fig. 2.13.

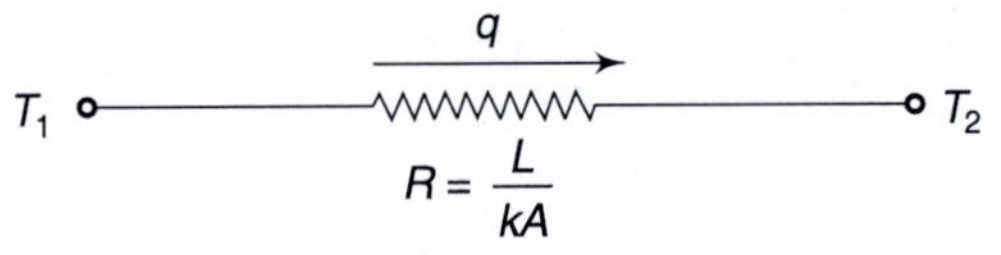

Fig. 2.13 Thermal circuit for 1D heat flow in a plane wall

If the outer wall of the house is losing heat to the surroundings by convection, then

$$q_c = hA(T_2 - T_\infty)$$

$$= \frac{T_2 - T_\infty}{1/hA} \tag{2.40}$$

$1/hA$ is termed as convective resistance for the heat transfer to the fluid adjacent to the outer wall. In case the inner fluid temperature is at $T_{\infty 1}$ ($> T_{\infty 2}$, the temperature of the fluid adjacent to the outer wall), the heat flow through the wall can be written as

$$q = \frac{T_{\infty 1} - T_{\infty 2}}{\dfrac{1}{h_1 A} + \dfrac{L}{kA} + \dfrac{1}{h_2 A}}$$

$$(2.41)$$

where h_1 and h_2 are the inner and outer heat transfer coefficients, respectively (Fig. 2.14).

The equivalent electric circuit for this problem is shown in Fig. 2.15. In this circuit there are three resistances in series. The aforesaid problem refers to the case where the inner and outer wall temperatures are not specified, which is more realistic.

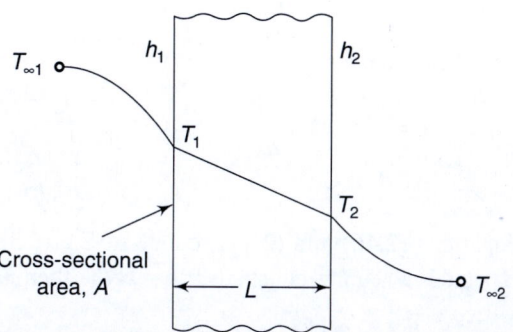

Fig. 2.14 Steady heat conduction in a plane wall bounded by fluids having different temperatures

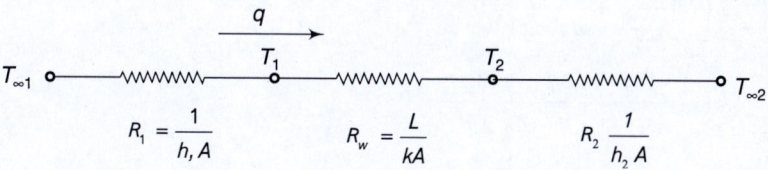

Fig. 2.15 Equivalent series circuit for steady heat conduction in a plane wall bounded by fluids having different temperatures

2.8.1.1 Plane wall with variable thermal conductivity

If $k = k(x)$, then from Eq. (2.17) for steady state, the one-dimensional heat conduction equation is given by

$$\frac{d}{dx}\left[k(x)\frac{dT}{dx}\right] = 0$$

$$(2.42)$$

Integrating the equation twice, we get

$$T = C_1 \int_0^x \frac{dx}{k(x)} + C_2$$

$$(2.43)$$

BC-1: $T(0) = T_1$ $$(2.44)$$
BC-2: $T(L) = T_2$ $$(2.45)$$

Using BC-1 and BC-2 in Eq. (2.43), we obtain

$$C_1 = -\frac{T_1 - T_2}{\displaystyle\int_0^L \frac{dx}{k(x)}}$$

$$C_2 = T_1$$

Substituting C_1 and C_2 into Eq. (2.43) the final expression for T is arrived at as

$$T(x) = T_1 - \frac{T_1 - T_2}{L} \int_0^x \frac{dx}{k(x)} \tag{2.46}$$

Now, heat transfer q is

$$q = -k(x) A \frac{dT}{dx} = -AC_1 = \frac{A(T_1 - T_2)}{\int_0^L \frac{dx}{k(x)}} \tag{2.47}$$

Equations (2.46) and (2.47) reduce to Eq. (2.38) and Eq. (2.39), respectively as expected, when k = constant. If $k = k(T)$, then Eq. (2.17) becomes

$$\frac{d}{dx}\left[k(T)\frac{dT}{dx}\right] = 0 \tag{2.48}$$

Expanding Eq. (2.48), we get

$$k(T)\frac{d^2 T}{dx^2} + \frac{dk}{dx}\frac{dT}{dx} = 0$$

or $$k(T)\frac{d^2 T}{dx^2} + \left(\frac{dk}{dT}\frac{dT}{dx}\right)\frac{dT}{dx} = 0$$

Finally,

$$\Rightarrow \quad k(T)\frac{d^2 T}{dx^2} + \frac{dk}{dT}\left(\frac{dT}{dx}\right)^2 = 0 \tag{2.49}$$

Hence, Eq. (2.48) is a non-linear differential equation since in its expanded form the coefficient of $\frac{d^2 T}{dx^2}$ in Eq. (2.49) is a function of temperature. It is thus non-linearity which makes analytical solution difficult to obtain.

Now, integrating Eq. (2.48) once results in

$$k(T)\frac{dT}{dx} = C_1 = -\frac{q}{A}. \tag{2.50}$$

Integrating again from $x = 0$ to $x = L$ [see Eqs (2.44) and (2.45)], we finally get

$$q = \frac{A}{L}\int_{T_2}^{T_1} k(T)\,dT \tag{2.51}$$

In terms of a mean thermal conductivity, k_m, q can also be expressed as

$$q = k_m A \frac{T_1 - T_2}{L} \tag{2.52}$$

where $$k_m = \frac{1}{T_1 - T_2}\int_{T_2}^{T_1} k(T)\,dT \tag{2.53}$$

Equation (2.53) represents nothing but the integrated average of thermal conductivity over the temperature range $(T_1 - T_2)$. The expression for temperature distribution can be obtained by integrating Eq. (2.50). The result is as follows:

$$\int_{T_1}^{T(x)} k(T)\,dT = -\frac{q}{A}\int_0^x dx = -\frac{q}{A}x \qquad (2.54)$$

The above expression, however, cannot be written explicitly for $T(x)$ unless the exact relation $k = k(T)$ is given.

If k is a linear function of temperature, then the following relation can be used:

$$k(T) = k_o\,[1 + \beta\,(T - T_0)] \qquad (2.55)$$

If $\beta > 0$, k increases with T.

If $\beta < 0$, k decreases with T.

Putting this in Eq. (2.53), the mean thermal conductivity becomes

$$k_m = \frac{1}{T_1 - T_2}\int_{T_2}^{T_1} k_o\left[1 + \beta(T - T_o)\right]dT = k_o\left[1 + \beta\left(\frac{T_1 + T_2}{2} - T_o\right)\right] \qquad (2.56)$$

From Eqs (2.52) and (2.54) the temperature distribution is obtained as

$$T^2(x) + \frac{2}{\beta}\left[1 - \beta T_o\right]T(x) + \frac{2}{\beta}\left[T_1\left(\beta T_o - \frac{\beta}{2}T_1 - 1\right)\right] + \frac{k_m}{k_o}(T_1 - T_2)\frac{x}{L} = 0 \qquad (2.57)$$

It is evident from Eq. (2.57) that the temperature distribution is not linear. For each value of x, a quadratic equation in T is to be solved and two roots will be obtained. However, at a particular x, the solution is unique as temperature cannot have two values at one point. One of the roots will turn out to be physically unrealistic and will be discarded.

The temperature distribution for positive and negative β are shown in Fig. 2.16 by dashed lines. The solid line indicates the case of $\beta = 0$, that is, for constant conductivity.

For $\beta > 0$, it is clear from Eq.(2.49) that $\dfrac{d^2 T}{dx^2} < 0$ and for $\beta < 0$ $\dfrac{d^2 T}{dx^2} > 0$.

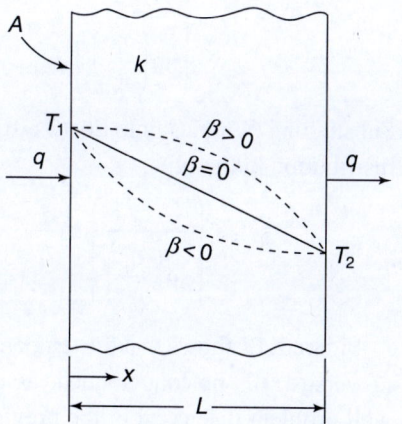

Fig. 2.16 Temperature distribution in a plane wall with $k = k_0\,[1 + \beta\,(T - T_0)]$ for $\beta > 0$, $\beta < 0$, $\beta = 0$

2.8.2 Hollow Cylinder

Consider a very long thick-walled hollow cylinder or tube (Fig. 2.17) having the inside surface temperature maintained at T_i (at r_i) and the outer surface temperature maintained at T_o (at r_o). Note that $T_i > T_o$. The heat flow occurs only in the radial direction because the tube is very long (typically, for $L/D > 3$, the cylinder may be treated as long) and hence axial conduction effects may be neglected. Furthermore, the inside and outside temperatures are uniform in the circumferential direction and, therefore, there cannot be any circumferential variation of temperature in the cylinder wall. This kind of geometry is encountered in any pipe flow situation.

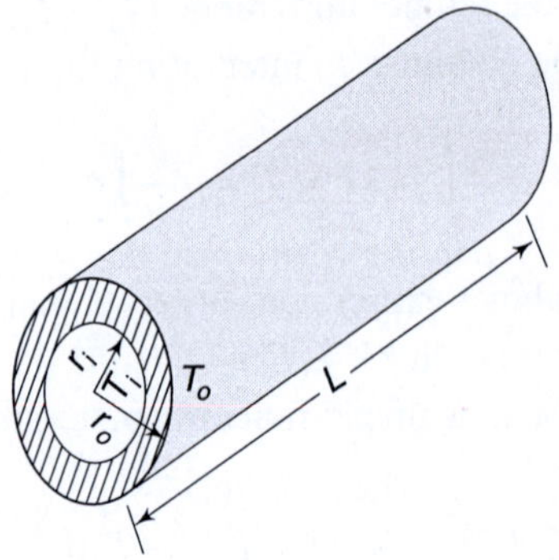

For the steady-state case, Eq. (2.20) reduces to the ordinary differential equation

$$\frac{d^2 T}{dr^2} + \frac{1}{r}\frac{dT}{dr} = 0 \qquad (2.58)$$

The boundary conditions are

$$\text{BC-1: } r = r_i, \quad T = T_i \qquad (2.59a)$$
$$\text{BC-2: } r = r_o, \quad T = T_o \qquad (2.59b)$$

The solution of Eq. (2.58) is

$$T = C_1 \ln r + C_2 \qquad (2.60)$$

Applying BC-1 and BC-2 to Eq. (2.60), we get

$$C_1 = \frac{T_i - T_o}{\ln\left(\dfrac{r_i}{r_o}\right)} \qquad (2.61a)$$

$$C_2 = T_i - \left(\frac{T_i - T_o}{\ln\left(\dfrac{r_i}{r_o}\right)}\right)\ln r_i \qquad (2.61b)$$

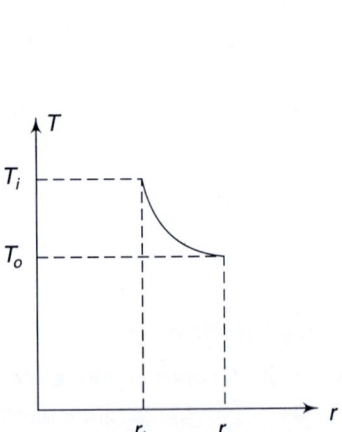

Fig. 2.17 Steady heat conduction in a long, thick-walled tube

Substituting C_1 and C_2 into Eq. (2.60) yields the expression for the radial temperature distribution in the tube:

$$T = T_i + \left(\frac{T_i - T_o}{\ln\left(\dfrac{r_i}{r_o}\right)}\right)\ln\frac{r}{r_i} \qquad (2.62)$$

Figure 2.18 shows that the temperature profile (T versus r) is no longer linear as in the plane wall problem discussed in the previous section. This is because the area normal to the heat flow vector now increases with increase in the radius. Since, for the steady state, the heat transfer rate is constant, this implies that dT/dr must decrease as r increases.

The heat transfer rate through the cylinder wall can be found by evaluating q at any r from Fourier's law. Thus

$$q = -\left[kA\frac{dT}{dr}\right]$$

Fig. 2.18 Radial temperature distribution in the tube wall

$$= -k \left[2\pi rL \left(\frac{T_i - T_o}{r \ln \left(\dfrac{r_i}{r_o} \right)} \right) \right]$$

$$= \frac{2\pi kL}{\ln \left(\dfrac{r_o}{r_i} \right)} (T_i - T_o) \tag{2.63}$$

Equation (2.63) has a form similar to Eq. (2.39) for the plane wall, except that here the thermal resistance is

$$\frac{1}{2\pi kL} \ln \left(\frac{r_o}{r_i} \right)$$

2.8.3 Composite Tube

Physically this may be a pipe with layers of various types of insulation (Fig. 2.19). The inner and outer surfaces are maintained at T_1 and T_4 ($T_1 > T_4$). The heat loss is then computed as

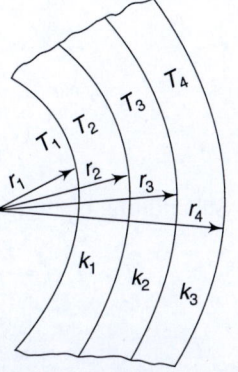

$$q = \frac{T_1 - T_4}{R_1 + R_2 + R_3}$$

$$= \frac{T_1 - T_4}{\dfrac{1}{2\pi k_1 L} \ln \dfrac{r_2}{r_1} + \dfrac{1}{2\pi k_2 L} \ln \dfrac{r_3}{r_2} + \dfrac{1}{2\pi k_3 L} \ln \dfrac{r_4}{r_3}} \tag{2.64}$$

Fig. 2.19 Steady heat conduction in a composite tube

Also, the intermediate temperatures T_2 and T_3 can be easily determined. The equivalent series circuit is shown in Fig. 2.20. The analysis for a composite plane wall is similar (see Example 2.5).

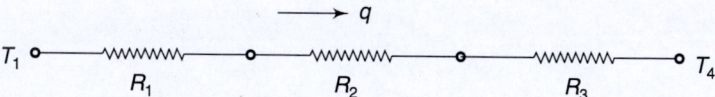

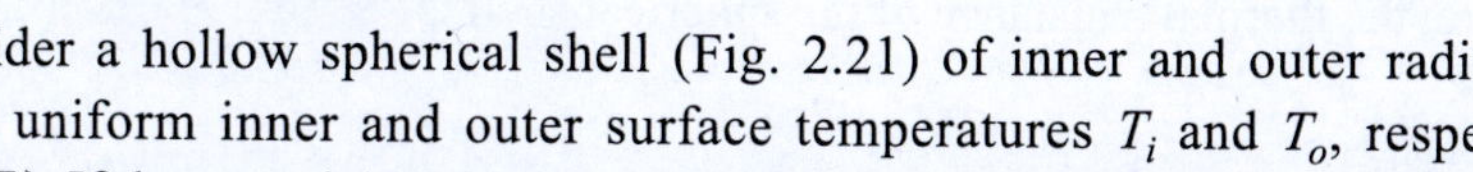

Fig. 2.20 Equivalent series circuit representation of heat flow in a composite tube

2.8.4 Hollow Sphere

Consider a hollow spherical shell (Fig. 2.21) of inner and outer radii r_i and r_o, at uniform inner and outer surface temperatures T_i and T_o, respectively ($T_i > T_o$). If the material of the sphere is homogeneous and the heat transfer steady, then the temperature distribution in the shell will be a function of r only, that is, $T = T(r)$, and Eq. (2.29a) takes the form

$$\frac{1}{r} \left(\frac{d^2(rT)}{dr^2} \right) = 0 \tag{2.65}$$

Equation (2.65) can also be expressed as

$$\frac{d}{dr}\left(r^2\frac{dT}{dr}\right) = 0 \qquad (2.66)$$

The boundary conditions are

BC-1: at $r = r_i$, $T = T_i$ (2.67a)

BC-2: at $r = r_o$, $T = T_o$ (2.67b)

Integrating Eq. (2.66) twice yields

$$T = -\frac{C_1}{r} + C_2 \qquad (2.68)$$

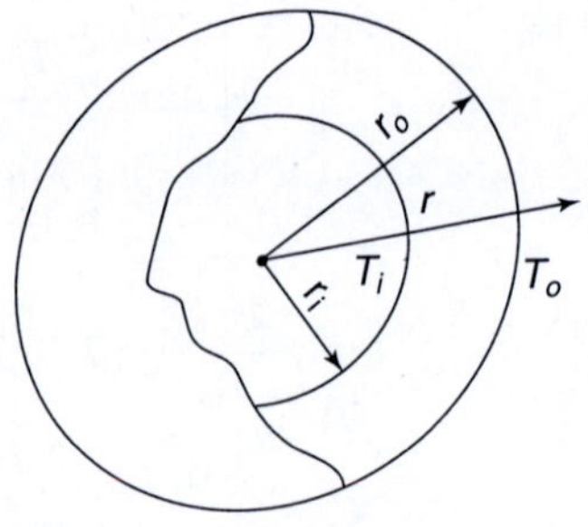

Imposing the boundary conditions in Eq. (2.68), we obtain

$$C_1 = \frac{T_i - T_o}{\dfrac{1}{r_o} - \dfrac{1}{r_i}}$$

$$C_2 = T_i + \frac{T_o - T_i}{r_i\left(\dfrac{1}{r_o} - \dfrac{1}{r_i}\right)}$$

Substituting C_1 and C_2 into Eq. (2.68),

$$T = T_i + \frac{T_i - T_o}{\left(\dfrac{1}{r_o} - \dfrac{1}{r_i}\right)}\left(\dfrac{1}{r_i} - \dfrac{1}{r}\right) \qquad (2.69)$$

Fig. 2.21 Steady heat conduction in a hollow sphere

The heat transfer rate through the spherical shell is

$$q = -k(4\pi r^2)\frac{dT}{dr}$$

$$= -k(4\pi r^2)\frac{T_i - T_o}{r^2\left(\dfrac{1}{r_o} - \dfrac{1}{r_i}\right)}$$

$$= \frac{T_i - T_o}{\dfrac{1}{4\pi k}\left(\dfrac{1}{r_i} - \dfrac{1}{r_o}\right)} \qquad (2.70)$$

Therefore, the thermal resistance of the spherical shell is

$$\frac{1}{4\pi k}\left(\frac{1}{r_i} - \frac{1}{r_o}\right)$$

The *composite sphere* case can be readily formulated from Eq. (2.70) in the manner used to obtain Eq. (2.64).

2.9   Overall Heat Transfer Coefficient

In the analysis of heat transfer through a plane wall bounded by hot and cold fluids (Fig. 2.14) as described in Section 2.8.1, q is written as

$$q = \frac{T_{\infty_1} - T_{\infty_2}}{\dfrac{1}{h_1 A} + \dfrac{L}{kA} + \dfrac{1}{h_2 A}} \tag{2.71}$$

Equation (2.71) can also be written in terms of the so-called overall heat transfer coefficient U as follows:

$$q = AU\,(T_{\infty 1} - T_{\infty 2}) \tag{2.72}$$

A comparison of Eqs (2.71) and (2.72) reveals that

$$\frac{1}{U} = \frac{1}{h_1} + \frac{L}{k} + \frac{1}{h_2} \tag{2.73}$$

Equation (2.72) defines U in terms of a heat transfer area A. In the cylindrical wall case, A is not a constant, but varies from $2\pi r_i L$ to $2\pi r_o L$ (Fig. 2.22). Therefore, the definition of U in this case depends on the area selected. q in this case can be expressed as

$$q = \frac{T_{\infty_i} - T_{\infty_o}}{\dfrac{1}{h_i A_i} + \dfrac{\ln \dfrac{r_o}{r_i}}{2\pi kL} + \dfrac{1}{h_o A_o}} \tag{2.74}$$

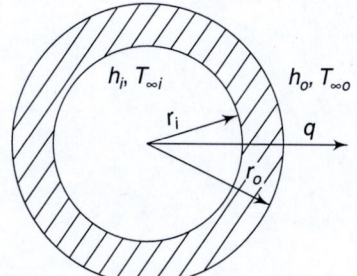

Also, $\quad q = U_i A_i \left(T_{\infty i} - T_{\infty o} \right) \tag{2.75}$

Since $A_i = 2\pi r_i L$,

$$\frac{1}{U_i} = \frac{1}{h_i} + \frac{r_i}{k} \ln \frac{r_o}{r_i} + \frac{r_i}{h_o r_o} \tag{2.76}$$

U could also have been based on A_o. However, in any case,

Fig. 2.22 Steady heat conduction in a cylindrical wall bounded by fluids having different temperatures

$$A_i U_i = A_o U_o \tag{2.77}$$

The concept of the overall heat transfer coefficient is widely applied in heat exchangers where a metallic wall separates the hot and cold streams of fluids.

Example 2.1 Overall Heat Transfer Coefficient, Plane Wall Approximation

The large bucket shown in Fig. E2.1 is designed to keep water in it at a high temperature, $T_h = 60\,°C$. The outside temperature is $T_c = 12\,°C$. To keep the temperature of the water constant, the heat leak through the insulation, q, is made up by an electrical resistance heater placed in the centre of the bucket. Calculate the electrical power dissipated in the heater. The dimensions of the bucket are $D = 1$ m and $L = 4$ m. The insulating wall consists of a 10-cm-thick layer of polyvinyl chloride (PVC). The heat transfer coefficients on the internal and external surfaces of the wall are $h_i = 15$ W/m^{2}K and $h_o = 10$ W/m^{2}K, respectively. The thermal conductivity of PVC is 0.15 W/mK.

Solution

To keep the water temperature inside the bucket at the designated temperature, heat leak through the insulation must be replenished by the heat generation of the electrical heater. Therefore, $\quad q_{generation} = q_{leak} = q$

Now, let us look at the geometry of the bucket. The bucket is a long cylinder since its $L/D > 3$. Therefore, heat conduction can be assumed to be one-dimensional, that is, $T = T(r)$ only. But its thickness is much smaller than its radius. The question is: Can we neglect the effect of the curvature? In such a situation, the wall of the bucket can be approximated as a plane wall.

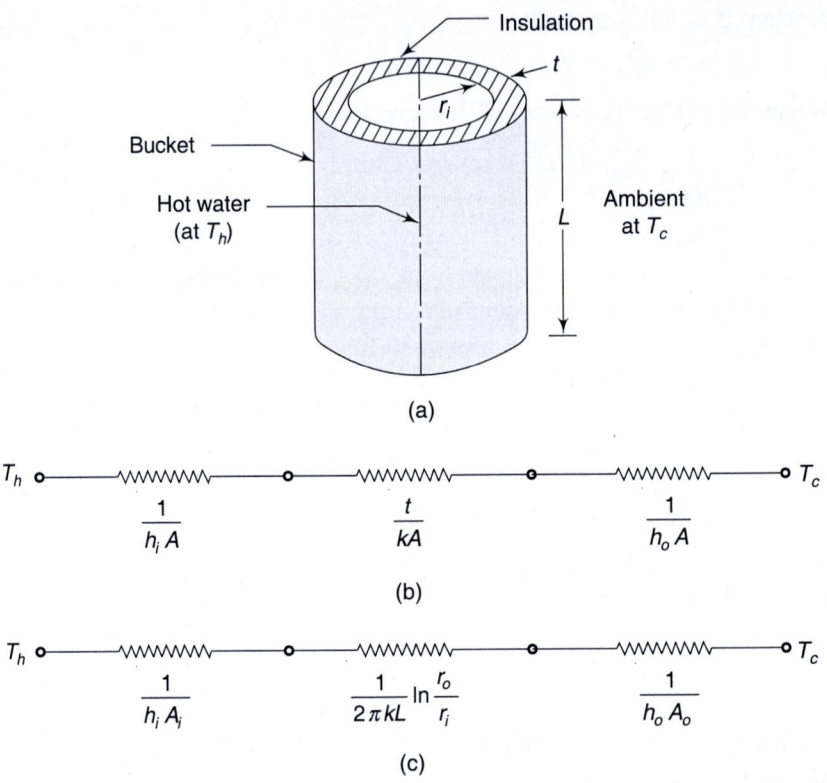

(a)

(b)

(c)

Fig. E2.1 (a) The insulated cylindrical bucket containing hot water. (b) Equivalent electrical network showing individual resistances for plane wall approximation. (c) Equivalent electrical network showing individual resistances for cylindrical geometry (no approximation).

Now, the heat flow from the inside of the bucket to the outside faces three resistances, namely, (i) convection from the hot water to the inner wall, (ii) conduction within the wall, and (iii) convection from the outer surface to the surroundings. The equivalent electrical network is shown in Fig. E2.1(a). The heat flux through the wall, q'' is

$$q'' = U\,(T_h - T_c)$$

where
$$\frac{1}{U} = \frac{1}{h_i} + \frac{t}{k} + \frac{1}{h_o}$$

$$= \frac{1}{15} + \frac{0.1}{0.15} + \frac{1}{10}$$

$$= 0.067 + 0.67 + 0.1 \quad = 0.837 \text{ m}^2\text{K/W}$$

Therefore, $q = UA\,(T_h - T_c)$

$$= (1.194)\,(2\pi r_i\, L)\, 60 - 12)$$

$$= (1.194)(2\pi \times 0.5 \times 4)(48)$$

$$= 720.18 \text{ W}$$

If we do not make the plane wall approximation and solve the problem using the original cylindrical geometry, let us see what value of 'q' we will obtain.

For the cylindrical geometry, let us calculate the overall heat transfer coefficient based on the inner area A_i, where $A_i = 2\pi r_i L$. The individual resistances are shown in Fig. E2.1(b). Therefore, $q = U_i A_i (T_h - T_c)$, where

$$\frac{1}{U_i} = \frac{1}{h_i} + \frac{r_i}{k} \ln \frac{r_o}{r_i} + \frac{r_i}{h_o r_o}$$

$$= \frac{1}{15} + \frac{0.5}{0.15} \ln \frac{0.6}{0.5} + \frac{0.5}{(10)(0.6)}$$

$$= 0.067 + 0.607 + 0.083$$

$$= 0.757 \text{ m}^2\text{K/W}$$

Therefore, $U_i = 1.32 \text{ W/m}^2\text{K}$

Hence, $q = (1.32)(2\pi \times 0.5 \times 4)(60 - 12)$

$$= 796.18 \text{ W}$$

Thus, we see that in this case the plane wall approximation of the bucket roughly underpredicts the heat dissipated in the electric heater by 10%.

So we conclude that to neglect the effect of curvature, the thickness to radius ratio should be smaller than taken here (which is 20%). Typically, for a thickness to radius ratio of 5% or less, the plane wall approximation is reasonable.

2.10 Critical Thickness of Insulation

A special application of the thermal resistance formulae developed so far is in determining the thickness of the annular insulation that should be applied to the outer surface of a small-diameter circular tube wall of a known wall temperature. A practical application is the problem of insulating electrical wires where the objective would be the provision of adequate electrical insulation, at the same time providing for maximum wire cooling. Another example is the design of the annular layer of foam insulation wrapped around a pipe carrying steam or hot water.

The reason why this problem has technical importance is as follows. In the case of small-diameter circular tubes the application of insulating material to the outer surface may in special instances increase the heat loss from the surface. As insulation is added to the pipe (Fig. 2.23), the temperature of the outer surface will decrease; but at the same time the surface area for convective heat transfer

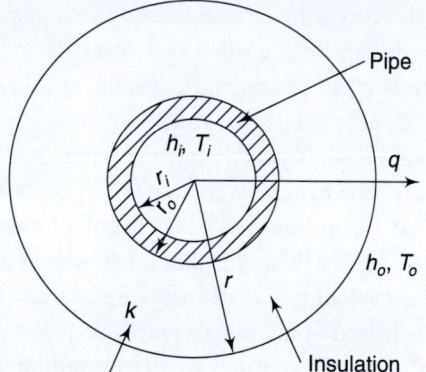

Fig. 2.23 Critical thickness of pipe insulation

will increase. It is, therefore, possible that some optimum thickness of insulation exists due to these opposing effects.

Let us consider the case of a steam-carrying pipe of a fixed outer radius, r_o. r denotes the radius of insulation. So the thickness of insulation is $(r - r_o)$. $k_{pipe} \gg k_{ins}$ and $h_i \gg h_o$. The implication of the foregoing first assumption is that the relative thermal resistance of the pipe is so small that there will be virtually no temperature drop in the wall of the pipe. The assumption of a very high inner heat transfer coefficient, h_i, implies negligible inner convective resistance. Therefore, out of the four resistances to the heat flow path from T_i to T_o, the first two can be neglected. Hence, the heat flow per unit length of the pipe is

$$\frac{q}{L} = \frac{T_i - T_o}{\dfrac{1}{2\pi k}\ln\left(\dfrac{r}{r_o}\right) + \dfrac{1}{h_o(2\pi r)}}$$

$$= \frac{2\pi(T_i - T_o)}{\dfrac{[\ln(r/r_o)]}{k} + \dfrac{1}{h_o r}} \tag{2.78}$$

The rate of heat flow will be a maximum when the denominator becomes a minimum. The existence of a minimum for the denominator can be readily seen from Eq. (2.78), which indicates that when the thickness of the insulation is varied, the first and the second terms of the denominator vary inversely.

The minimum value from the denominator can be calculated by taking the derivative of the denominator with respect to r while r_o is held as a constant parameter and setting the result equal to zero. This gives

$$\frac{1}{kr} - \frac{1}{h_o r^2} = 0 \tag{2.79}$$

from which

$$r_c = \frac{k}{h_o} \tag{2.80}$$

where r_c is the critical radius of insulation.

It can be seen that this result is independent of r_o. The heat transfer coefficient h_o is considered constant in this calculation. Although h_o is a function of the outer radius r as well as the outer surface temperature, the foregoing approximation is reasonable for many practical cases when the variation in r is small.

If one evaluates $[d^2/dr^2(q/L)]_{r=r_c}$ one will see that it is negative, which shows that the optimum radius r_c is one of maximum heat loss and not minimum.

The conclusion is that tubes whose outer radii (in this case r_o) are smaller than the critical radius of insulation, r_c, as calculated here, can have their heat losses increased by adding insulation up to the value of the critical thickness [Fig. 2.24(a)]. This usually requires small tube radii, relatively large thermal conductivities of the insulation, and small heat transfer coefficients. Further increase in the thickness of insulation will cause the heat loss to decrease from this peak value, but until a certain amount of insulation, denoted by r^* at b, is added, the heat loss is still greater than that for the bare pipe. Thus, an insulation thickness in excess of $(r^* - r_o)$ must be added to reduce the heat loss below the uninsulated rate.

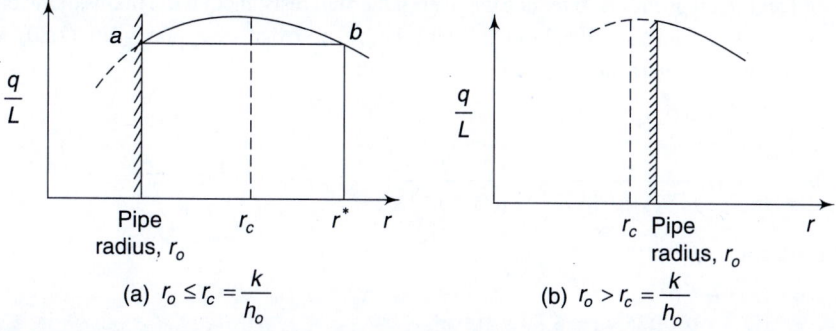

$$\text{(a) } r_o \le r_c = \frac{k}{h_o}$$

$$\text{(b) } r_o > r_c = \frac{k}{h_o}$$

Fig. 2.24 (a) Small pipe case, (b) large pipe case

Figure 2.24(b) illustrates the case of large pipes, in which the outside pipe radius r_o is larger than the critical radius r_c, and any insulation will decrease the heat loss.

Example 2.2 Cooling of Electrical Wire, Critical Radius of Insulation

An uninsulated wire suspended in air produces electrical heating at the rate of $q' = 2$ W/m. The wire is a bare cylinder of radius $r_i = 0.5$ mm, and the temperature difference between it and the atmosphere is 25 °C. It is recommended that this wire be covered with a plastic (PVC) sleeve of electrical insulation, the outer radius of which is $r_o = 1$ mm. The thermal conductivity of the plastic material is $k = 0.15$ W/mK.

(a) Will the plastic sleeve produce a heat transfer augmentation effect or will it provide a thermal insulation effect?

(b) To verify your answer, calculate the new wire–surroundings temperature difference when the wire is encased in plastic.

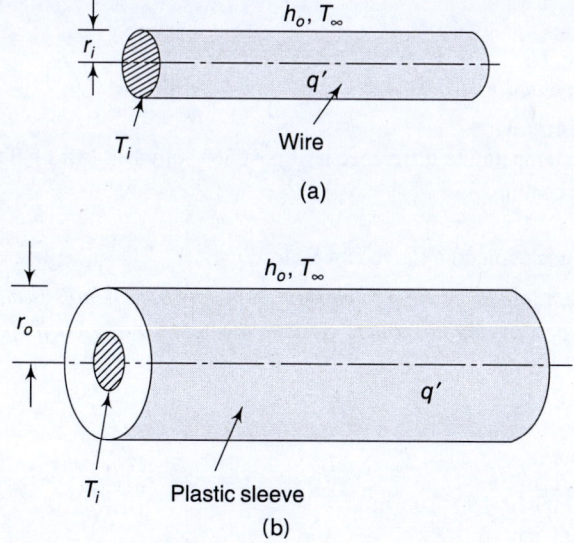

Fig. E2.2 (a) Bare wire, (b) insulated wire

Solution

(a) The plastic sleeve of electrical insulation promises to have a heat transfer enhancement effect (i.e., it promises to reduce the overall thermal resistance) if the radius of the bare wire is less than the critical radius of insulation. To calculate r_c, using Eq. (2.80), we must first calculate the heat transfer coefficient:

$$q' = 2\pi r_i h_o (T_i - T_\infty)$$

$\Rightarrow \quad h_o = \dfrac{q'}{2\pi r_i (T_i - T_\infty)} = \dfrac{2}{2\pi (0.5 \times 10^{-3})(25)} = 25.46 \text{ W/m}^2 \text{ K}$

The critical radius

$r_c = k/h_o$
$\quad = 0.15/25.46 = 5.89 \times 10^{-3} \text{ m}$
$\quad = 5.89 \text{ mm}$

Since r_c is much greater than r_i (the radius of the bare wire) and greater than the outer radius of the plastic sleeve, we can expect a heat transfer augmentation effect (a decrease in $T_i - T_\infty$) from the presence of the plastic sleeve.

(b) The new wire–ambient temperature difference $T_i - T_\infty$ follows from the definition of R_{th}:

$$T_i - T_\infty = R_{th}q = R_{th}Lq'$$

Now, $R_{th}L = \dfrac{\ln(r_o/r_i)}{2\pi k} + \dfrac{1}{2\pi r_o h}$

$\quad = \dfrac{\ln(1/0.5)}{2\pi \times 0.15} + \dfrac{1}{2\pi \times 1 \times 10^{-3} \times 25.46}$

$\quad = 0.735 + 6.25$
$\quad = 6.985 \text{ mK/W}$

Therefore,

$T_i - T_\infty = R_{th}Lq'$
$\quad = 6.985 \times 2$
$\quad = 13.97 \,°\text{C}$

In conclusion, the temperature difference is almost 56% of what it was before the installation of the plastic coating.

Example 2.3 Insulation on a Large Hot Water Pipe

Will the rate of heat loss decrease if foam insulation, k = 0.09 W/mK, is added to a 5-cm-outer-diameter pipe carrying hot water? Assume the heat transfer coefficient on the outer surface is $h_o = 10 \text{ W/m}^2 K$.

Solution

$r_c = k/h_o$
$\quad = 0.09/10 = 9 \times 10^{-3} \text{ m}$
$\quad = 0.9 \text{ cm}$

Since, the outer radius of the pipe is $r_o = 5$ cm, $r_o > r_c$, heat loss from the pipe will decrease.

2.11 Heat Generation in a Body: Plane Wall

There are many practical situations in which heat is generated in a heat-conducting body. An important application is the case of electric current flowing through an electrical conductor in which the dissipated electrical energy is transformed into heat. The calculation of the temperatures which originate through this heat generation is of specific interest in the design of electrical machinery such as electric motors, transformers, etc. Other areas of application lie in the chemical and nuclear fields. A typical example is an exothermic chemical reaction distributed throughout a body. In a nuclear reactor, heat is generated due to nuclear reactions in a fissionable material. The biological problem of fermentation is another area involving generation of heat. Special refrigeration systems must be designed in order to prevent production of inadmissibly high temperatures by the generation of heat during the setting of concrete. The flow losses in fluids are also transformed into heat. A high-speed aircraft or a re-entry spacecraft encounters tremendous frictional heating due to dissipation of mechanical energy into heat. Considerable temperature increases occur from frictional heating in the oil films used for lubrication of fast-running bearings. Viscous dissipation of heat also occurs in polymer and food processing in screw extruders.

In this section, the analysis will be restricted to solid bodies, and for the sake of simplicity the initial consideration will involve a plane wall or slab (Fig. 2.25). Uniformly distributed heat sources are present in a wall of thickness $2L$ and, therefore, a quantity of heat at the rate of q''' W/m^3 is generated. Hence q''' is independent of the space coordinate x. On each of the exposed surfaces the slab is bounded by a circulating fluid of temperature T_∞. The convective heat transfer coefficient for both the surfaces is h. For constant conductivity of the wall and steady-state heat conduction, Eq. (2.18) reduces to

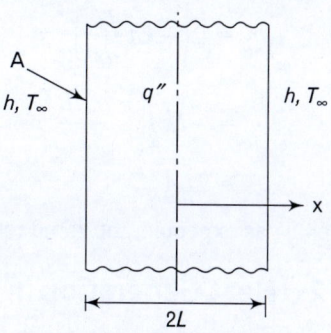

Fig. 2.25 Heat generation in a plane wall

$$\frac{d^2T}{dx^2} + \frac{q'''}{k} = 0 \qquad (2.81)$$

BC-1: at $x = 0$, $dT/dx = 0$

(due to thermal and geometric symmetry) $\qquad (2.82a)$

BC-2: at $x = L$, $-k(dT/dx) = h(T - T_\infty)$ $\qquad (2.82b)$

The solution of Eq. (2.81) is

$$T - T_\infty = \frac{q'''L^2}{2k}\left[1 - \left(\frac{x}{L}\right)^2\right] + \frac{q'''L}{h} \qquad (2.83)$$

The maximum temperature occurs at the mid-plane ($x = 0$):

$$T_c - T_\infty = \frac{q'''L^2}{2k} + \frac{q'''L}{h} \qquad (2.84)$$

The surface temperature is evaluated from

$$T_s - T_\infty = \frac{q'''L}{h} \tag{2.85}$$

The temperature drop from the mid-plane to the surface is computed from

$$T_c - T_s = \frac{q'''L^2}{2k} \tag{2.86}$$

Equation (2.86) is obtained by subtracting Eq. (2.85) from Eq. (2.84). The parabolic temperature profile in the slab is depicted in Fig. 2.26.

The total rate of heat generated in the wall should be equal to the rate of heat loss to the surrounding fluid. Therefore, the heat lost to the surrounding fluid per unit time is given by

$$q_L = (2LA)q''' \tag{2.87}$$

The heat lost to the surrounding fluid per unit time can also be calculated as follows:

$$q_L = 2\left(-kA\frac{dT}{dx}\right)_{x=L}$$

$$= -2kA\left(-\frac{q'''L}{k}\right)$$

$$= 2LAq''' \tag{2.88}$$

which is, as expected, the same result as in Eq. (2.87).

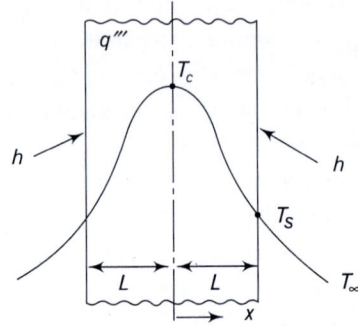

Fig. 2.26 Temperature profile in the plane wall with heat sources

2.12 Heat Generation in a Solid Cylinder

The physical problem is similar to heat generation in a plane wall. The difference is the geometry. Figure 2.27 shows the physical domain and the temperature profile. The governing differential equation for this problem is

$$\frac{d^2T}{dr^2} + \frac{1}{r}\frac{dT}{dr} + \frac{q'''}{k} = 0 \tag{2.89}$$

BC-1: at $r = 0$, $dT/dr = 0$ (axial symmetry) or $T =$ finite $\hspace{1cm}$ (2.90a)

BC-2: at $r = r_o$, $-k(dT/dr) = h(T - T_\infty)$ $\hspace{1cm}$ (2.90b)

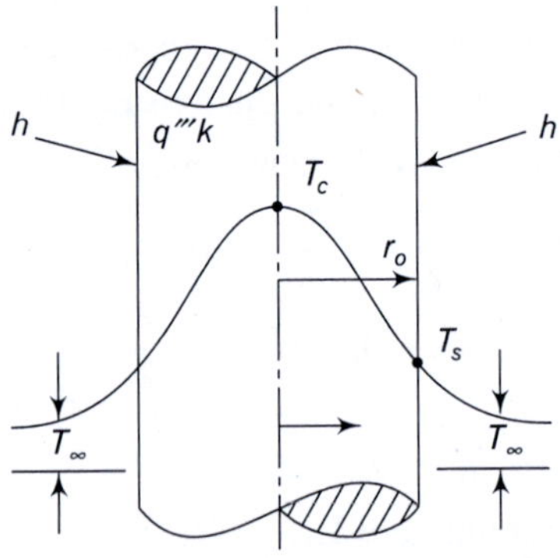

Fig. 2.27 Temperature profile in a solid cylinder with heat sources

The general solution of Eq. (2.89) is

$$T = -\frac{q'''r^2}{4k} + C_1 \ln r + C_2 \tag{2.91}$$

The application of BC-1 and BC-2 to Eq. (2.91) yields

$$C_1 = 0$$

$$C_2 = T_\infty + \frac{q'''r_o^2}{4k}\left(1 + \frac{2k}{hr_o}\right)$$

Therefore, the final solution is

$$T = T_\infty + \frac{q'''r_o^2}{4k}\left[1 - \left(\frac{r}{r_o}\right)^2 + \frac{2k}{hr_o}\right] \tag{2.92}$$

The temperature drop from the centre line to the surface of the cylinder is then given by

$$(\Delta T)_{max} = T_c - T_s = \frac{q'''r_o^2}{4k} \tag{2.93}$$

Example 2.4   Heat Generation in a Transformer Coil

A cylindrical transformer coil made of insulated copper wire has an inside diameter of 16 cm and an outside diameter of 24 cm. A fraction $\phi = 0.6$ of the total cross-section of the coil is copper, and the rest insulation (mica, glue). The density of the current in the conductors is $j = 208$ A/cm^2; the specific resistance of copper is $\rho = 194.8 \times 10^{-6}$ $\Omega\,cm^2/m$. The heat transfer coefficient on both surfaces in the coil, which are cooled by air at 25°C, is $h = 22$ W/m^{2}K. The thermal conductivity of the coil is $k = 0.346$ W/mK. Calculate the temperature at the centre of the spool.

Solution

Heat generation per unit volume in the coil $(q''') = \phi j^2 \rho$

$$= 0.6 \times (208)^2 \text{ A}^2/\text{cm}^4 \times 194.8 \times 10^{-6} \text{ } \Omega\,\text{cm}^2/\text{m}$$
$$= 5.056 \text{ A}^2 \text{ } \Omega/\text{cm}^2\text{m}$$
$$= 5.056 \text{ W}/(10^{-2})^2 \text{ m}^3$$
$$= 5.056 \times 10^4 \text{ W/m}^3$$

If we consider the coil in a first approximation as a plane wall with its thickness $2L = (24 - 16)/2 = 4$ cm $= 4 \times 10^{-2}$ m, then we obtain the temperature at the centre of the spool (the maximum temperature) from Eq. (2.84):

$$T_c = T_\infty + \frac{q'''L^2}{2k} + \frac{q'''L}{h}$$

$$= 25 + \frac{5.056 \times 10^4 \times (2 \times 10^{-2})^2}{2 \times 0.346} + \frac{5.056 \times 10^4 \times 2 \times 10^{-2}}{22}$$

$$= 25 + 29.22 + 45.96$$
$$= 100.18°C$$

A more accurate calculation of the thermal field can be obtained if the coil is considered as a hollow cylinder. Readers are encouraged to do this exercise to compare the two results and find out whether the plane wall approximation is adequate.

2.13   Heat Generation in a Solid Sphere

The problem concerns steady state heat conduction in a solid sphere of radius r_0, having constant thermal conductivity k and uniform volumetric heat generation at the rate of W/m^3. The outer surface of the sphere is losing heat to the surrounding by convection. The ambient temperature is T_∞ and heat transfer coefficient is h.

The governing differential equation is the reduced form of Eq. (2.19) for one-dimensional steady state heat conduction with heat generation in spherical coordinates, which is

$$\text{GDE:} \quad \frac{1}{r^2}\frac{d}{dr}\left[r^2 \frac{dT}{dr}\right] + \frac{q'''}{k} = 0 \tag{2.94}$$

The boundary conditions are

$$\text{BC-1: at } r=0, \quad \frac{dT}{dr} = 0 \tag{2.95a}$$

$$\text{BC-2: at } r = r_0, \quad -k\frac{dT}{dr} = h(T - T_\infty) \tag{2.95b}$$

Integrating Eq. (1) once, we get

$$r^2 \frac{dT}{dr} = -\frac{q'''}{3k}r^3 + C_1 \tag{2.96}$$

Application of BC-1 in Eq. (2.96) yields $C_1 = 0$
Integration of Eq. (2.96) results in

$$T = -\frac{q'''}{6k}r^2 + C_2 \tag{2.97}$$

Now, the application of BC-2 in Eq. (2.97) gives

$$-\frac{q'''r_0}{3} + h\left(-\frac{q'''r_0^2}{6k} + C_2\right) = hT_\infty$$

$$\Rightarrow \quad C_2 = \frac{q'''r_0}{3h} + \frac{q'''r_0^2}{6k} + T_\infty$$

Therefore, the temperature distribution in the sphere is

$$T = \frac{q'''r_0^2}{6k}\left[1 - \left(\frac{r}{r_0}\right)^2\right] + \frac{q'''r_0}{3h} + T_\infty \tag{2.98}$$

Special Cases:
I. For $h \to \infty$, the solution reduces to

$$T = \frac{q'''r_0^2}{6k}\left[1 - \left(\frac{r}{r_0}\right)^2\right] + T_\infty \tag{2.99}$$

which is basically the same problem when the outer surface of the sphere is at T_∞.
II. For $h \to 0$, the outer surface of the sphere becomes insulated, and the problem has no steady state solution because heat generated has no way to escape from the solid.

2.14 Thin Rod

Another simple but important solution of Eq. (2.18) is that for a thin rod connected to the base of a heated wall and which is transferring heat from its tip and periphery to a surrounding fluid. The system is shown in Fig. 2.28. The base temperature is T_0, the cross-sectional area of the rod is A, its perimeter is p, and its length is L. The convection on the surface is such that it results in a constant heat transfer coefficient over the entire surface. The area A and the perimeter p are constant along the length of the rod.

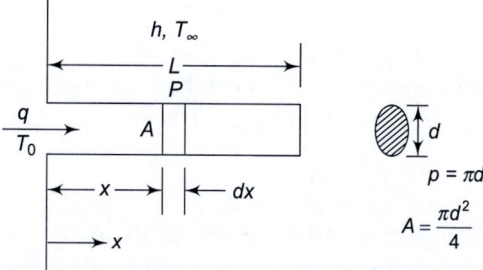

Fig. 2.28 Steady heat conuction in a thin rod

If the diameter of the rod is small as compared with its length and if the convection essentially controls the heat flow, there will be no radial temperature distribution in the rod; but there will be a large axial temperature distribution. The fact that the heat being conducted along the rod from the base is being lost to the surrounding fluid by convection suggests that the problem can be solved by reducing Eq. (2.18) to the terms describing axial conduction and a distributed heat sink which is equal to the convection loss. Therefore, Eq. (2.18) becomes

$$\frac{d^2 T}{dx^2} + \frac{q'''}{k} = 0 \tag{2.100}$$

Note that q''' is a heat sink (or negative source) per unit volume, which must be evaluated in terms of the convection loss. Therefore,

$$q''' = -\frac{hp\,dx\,(T - T_\infty)}{A\,dx}$$

$$= -\frac{hp\,(T - T_\infty)}{A} \tag{2.101}$$

Substituting q''' from Eq. (2.101) into Eq. (2.100), we obtain

$$\frac{d^2 T}{dx^2} - \frac{hp}{kA}(T - T_\infty) = 0 \tag{2.102}$$

which is the governing differential equation for the problem. Equation (2.102) would have been also obtained by equating the net heat conduction in a volume to the convection loss by the same volume. Let $\theta = T - T_\infty$. Then Eq. (2.102) becomes

$$\frac{d^2 \theta}{dx^2} - \frac{hp}{kA}\theta = 0 \tag{2.103}$$

The general solution of Eq. (2.103) is

$$\theta = C_1 e^{mx} + C_2 e^{-mx} \tag{2.104}$$

where $m = (hp/kA)^{1/2}$. The boundary conditions are

BC-1: at $x = 0$, $T = T_0$ (2.105a)

BC-2: at $x = L$, $-k(dT/dx) = h(T - T_\infty)$ (2.105b)

Now, writing the boundary conditions in terms of θ,

At $x = 0$, $\theta = \theta_0 = T_0 - T_\infty$ (2.106a)

At $x = L$, $d\theta/dx = -(h/k)\theta$ (2.106b)

If the rod is very long so that the tip of the fin almost assumes the temperature of the surroundings, then the heat loss from the tip can be neglected. Therefore, the boundary condition at $x = L$ becomes

$$\frac{d\theta}{dx} = 0$$ (2.107)

Using the boundary conditions in Eq. (2.104), C_1 and C_2 can be found out. Finally, the solution is

$$\frac{\theta}{\theta_0} = \frac{T - T_\infty}{T_0 - T_\infty} = \frac{e^{m(L-x)} + e^{-m(L-x)}}{e^{mL} + e^{-mL}}$$ (2.108)

It is rather cumbersome to make calculations with the result in the above form since numerous exponentials must be computed and then added and divided to obtain the temperature. The computations can be simplified by using hyperbolic functions. Recall that

$$\cosh mx = \frac{1}{2}(e^{mx} + e^{-mx})$$

$$\sinh mx = \frac{1}{2}(e^{mx} - e^{-mx})$$

$$\frac{\theta}{\theta_0} = \frac{T - T_\infty}{T_0 - T_\infty} = \frac{\cosh m(L - x)}{\cosh mL}$$ (2.109)

The heat flow through the base of the rod ($x = 0$) is

$$q = -kA\left(\frac{d\theta}{dx}\right)_{x=0}$$

$$= mkA\theta_0\left(\frac{\sinh m(L - x)}{\cosh mL}\right)_{x=0}$$

$$= \sqrt{hpkA}\,\theta_0 \tanh mL$$ (2.110)

which is the same as that convected by the entire rod. The excess temperature at the end of the rod ($x = L$) is

$$\theta_L = \frac{\theta_0}{\cosh mL}$$ (2.111)

The two functions cosh mL and tanh mL are listed in Table 2.1. It is readily seen that as the length L increases, the heat flow increases rapidly at first; but the incremental increase becomes smaller and smaller, and finally the heat flow approaches an asymptotic value. The excess temperature at the end of a very long rod is zero.

Table 2.1 Hyperbolic functions for heat conduction in a rod

mL	0	0.5	1	1.5	2	3	4	5	6
cosh mL	1	1.1276	1.543	2.352	3.762	10.07	27.31	74.21	201.7
tanh mL	0	0.4621	0.7616	0.9052	0.9640	0.9951	0.9993	0.9999	1

The solution of Eq. (2.103) with the boundary condition expressed by Eq. (2.106b) (i.e., the case of heat loss from the end of the rod) is as follows:

$$\frac{\theta}{\theta_0} = \frac{T - T_\infty}{T_0 - T_\infty} = \frac{\cosh m(L - x) + (h_e/mk)\sinh m(L - x)}{\cosh mL + (h_e/mk)\sinh mL} \qquad (2.112)$$

The heat flow through the base of the rod ($x = 0$) becomes

$$q = mkA\theta_0 \frac{(h_e/mk) + \tanh mL}{1 + (h_e/mk)\tanh mL} \qquad (2.113)$$

The temperature excess at the end of the rod ($x = L$) becomes

$$\theta_L = \frac{\theta_0}{\cosh mL + (h_e/mk)\sinh mL} \qquad (2.114)$$

In Eqs (2.112)–(2.114) the value h_e is the heat transfer coefficient at the end of the rod. h_e is generally different from the heat transfer coefficient h along the rod surface. This is because the heat transfer coefficient depends on the orientation of the surface. Equation (2.113) reduces to Eq. (2.110) for $h_e = 0$.

The following section illustrates an important technical application of heat conduction in a thin rod, that is, calculation of error in the measurement of the temperature of a fluid flowing in a tube by a thermometer or thermocouple put into a well which is welded into the tube wall as shown in Fig. 2.29.

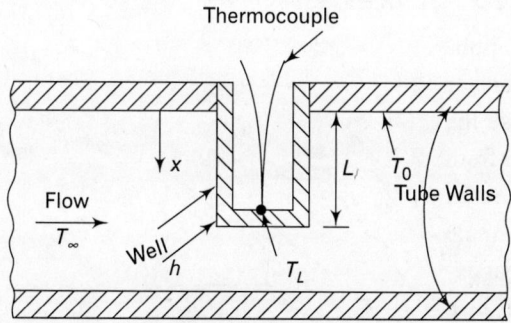

Fig. 2.29 Thermometer well conduction error

2.15 Thermometer Well Errors Due to Conduction

With respect to Fig. 2.29, if the fluid temperature T_∞ differs greatly from the outside temperature, then the tube wall has a lower temperature than the gas and heat flows by conduction from the well to the tube wall. The end of the well, where the thermometer bulb or thermocouple junction is placed, may become colder than the fluid, and the indicated temperature will not be the true fluid temperature.

If the thermometer well is assumed to have good thermal contact with the tube wall, at temperature T_0, then it may be treated as a thin rod of uniform cross-section and finite length L. The thermometer or thermocouple placed in the well will be assumed to have perfect contact with the bottom of the well, so the indicated temperature T_L may be assumed to be that of the end of the rod. The thin rod was treated in the previous section and when losses at the end of rod are neglected, Eq. (2.111) shows that the temperature at the end is given by

$$T_L - T_\infty = \frac{T_0 - T_\infty}{\cosh mL} \tag{2.115}$$

where $m = \left(\dfrac{hp}{kA}\right)^{\frac{1}{2}}$. In Eq. (2.115), T_∞ is the fluid temperature, which is to be calculated. Therefore, $(T_L - T_\infty)$ is the error in the reading given by the thermocouple. Also, in Eq. (2.115) A is the cross-sectional area for heat flow in the thermocouple well wall and k is the thermal conductivity of the well material. p denotes the perimeter of the outer well surface, and h is the convective heat transfer coefficient there.

The calculation of the error in the thermocouple reading requires the knowledge of the tube surface temperature T_0. The error is directly proportional to $(T_0 - T_\infty)$ and hence the error may be reduced by insulating the outside pipe surface in the vicinity of the well in order to reduce this difference. Also, any means of making mL as large as possible (since $1/\cosh mL$ decreases with increasing mL) will reduce the error. One way of doing this is to increase L. If the length of the well is greater than the tube diameter, it is necessary to locate the well obliquely in the tube.

In the present analysis, heat radiation between the end of the well and the tube wall, which may cause an additional error in temperature measurement, has been neglected.

2.16 Extended Surfaces: Fins

One of the main applications of heat transfer study is to increase the rate of heat transfer from a heated surface to a cool fluid. A typical example is conventional heat exchangers in which heat is transferred from one fluid to another through a metal wall. The rate of heat transfer is directly proportional to the surface area of the wall and the temperature difference between hot and cold fluids. In most cases, however, the temperature difference cannot be changed. Therefore, the only way to increase the rate of

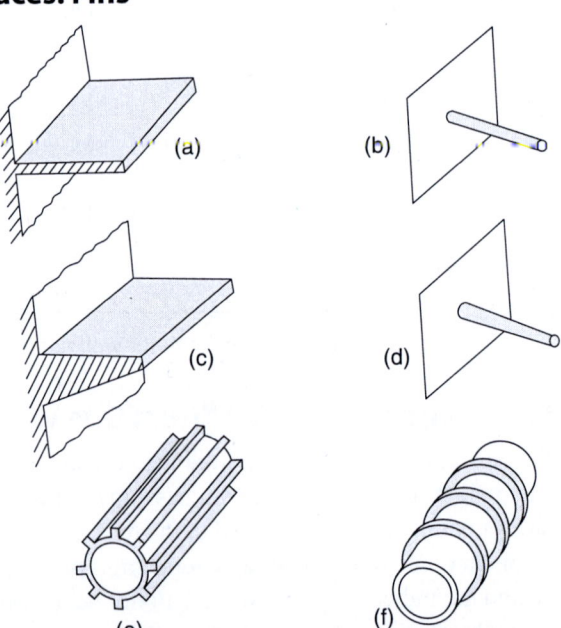

Fig. 2.30 Several types of extended surfaces: (a) longitudinal fin of rectangular profile, (b) cylindrical spine, (c) longitudinal fin of trapezoidal profile, (d) truncated conical spine, (e) cylindrical tube equipped with straight fins of rectangular profile, (f) cylindrical tube equipped with annular fins of rectangular profile

heat transfer is to increase the effective heat transfer area. The effective heat transfer area on a solid surface can be enhanced by attaching thin metal strips, called fins or spines (thin cylindrical or tapered rods), to the surface. Although attaching such extended surfaces effectively increases the heat transfer area, there is also a price to be paid. That is, these extended surfaces also act as additional resistances to heat transfer and as a result, there is a temperature drop in the fins or spines. This means that the average surface temperature of the fins or spines will not be the same as the original surface temperature of the wall, but will be closer to the fluid temperature. This causes the rate of heat transfer to be less than proportional to the extent of the total heat transfer area. Some commonly used fins and spines are shown in Fig. 2.30. The extended surfaces can be attached to the base material by pressing, soldering, or welding. In some cases, they may be integral parts of the base material obtained by a casting or extruding process.

Finned surfaces are widely used on car radiators and heating units, heat exchangers, air-cooled engines, electrical transformers, motors, electronic transistors, etc. In this section, the analysis is limited to one-dimensional extended surfaces with the following assumptions.

(a) Heat flow in the extended surface is steady.

(b) Thermal conductivity of the fin material is constant.

(c) The thickness of the extended surface is so small compared to its length that the temperature gradients normal to the surface may be neglected. Also, the side areas of the fin are very small and heat losses from the sides are closer to zero. So, effectively, the sides are treated as insulated. All this makes the heat flow in the fin one-dimensional.

(d) The convective heat transfer coefficient between the fin and the surroundings is constant. Although the value of the heat transfer coefficient on the surface of the fin or spine varies from point to point, the use of a circumferentially and axially averaged value in analytical studies gives heat transfer results that are, in most cases of practical interest, in good agreement with experimental measurements.

(e) The base temperature is constant. This is, however, a questionable assumption but, nevertheless, used for the sake of simplicity.

(f) The temperature of the surrounding fluid is uniform and constant.

Let us now consider the conduction of heat in the extended surface as shown in Fig. 2.31(a). Note that the fin is of variable cross-sectional area. An energy balance (i.e., the first law of thermodynamics) when applied to the system as shown in Fig. 2.31(b) gives

$$q_x'' A = q_c'' p \Delta x + q_x'' A + \frac{d}{dx}(q_x'' A)\Delta x \qquad (2.116)$$

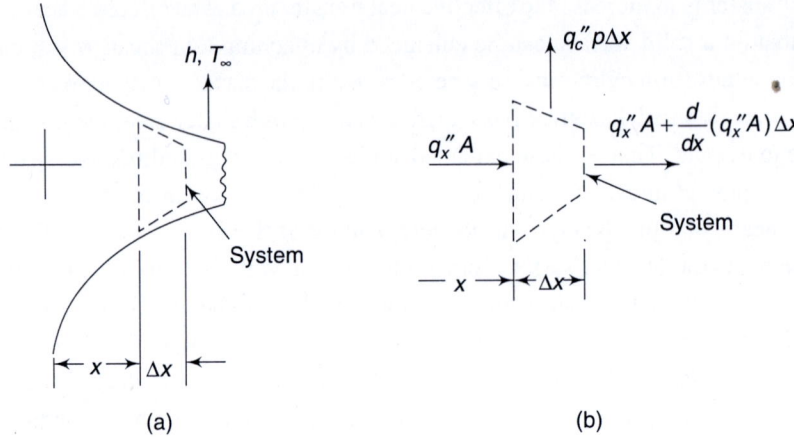

Fig. 2.31 (a) Extended surface of variable cross-section, (b) energy balance on a system in a 1D fin

But from Fourier's law,

$$q_x'' = -k\frac{dT}{dx} \tag{2.117}$$

Substituting Eq. (2.117) into Eq. (2.116), we obtain

$$\frac{d}{dx}\left(-kA\frac{dT}{dx}\right)\Delta x = -q_c''p\Delta x$$

or

$$\frac{d}{dx}\left(kA\frac{dT}{dx}\right) = q_c''p \tag{2.118}$$

But, from Newton's law of cooling,

$$q_c'' = h(T - T_\infty) \tag{2.119}$$

where q_c'' is the heat transfer per unit circumferential area by convection. Substituting Eq. (2.119) into Eq. (2.118), we get

$$\frac{d}{dx}\left(kA\frac{dT}{dx}\right) = hp(T - T_\infty) \tag{2.120}$$

Since, k is constant, Eq. (2.120) may be written as

$$\frac{d}{dx}\left(A\frac{dT}{dx}\right) - \frac{hp}{k}(T - T_\infty) = 0 \tag{2.121}$$

Let $\theta = T - T_\infty$. Then Eq. (2.121) transforms to

$$\frac{d}{dx}\left(A\frac{d\theta}{dx}\right) - \frac{hp}{k}\theta = 0 \tag{2.122}$$

This is the governing differential equation for 1D heat transfer from a fin. Since Eq. (2.122) is of second order, two boundary conditions are needed in the x-direction; one at the base and the other at the tip of the fin.

2.16.1 Extended Surfaces with Constant Cross-sections

For an extended surface with constant cross-section, Eq. (2.122) reduces to

$$\left(\frac{d^2\theta}{dx^2}\right) - m^2\theta = 0 \qquad (2.123)$$

where $m^2 = hp/kA$. The general solution of Eq. (2.123) can be written as

$$\theta(x) = C_1 e^{mx} + C_2 e^{-mx} \qquad (2.124a)$$

or $\theta(x) = C_3 \sinh mx + C_4 \cosh mx$ (2.124b)

where C_1 and C_2 or C_3 and C_4 are constants of integration to be determined from the boundary conditions. Since the base temperature T_0 is constant, the boundary condition at $z = 0$ is

     BC-1: at $x = 0$, $T = T_0$

or $\theta = T_0 - T_\infty = \theta_0$ (2.125)

The second boundary condition (BC-2) depends on the nature of the problem as discussed next.

Case A: Infinitely long fin

The extended surface shown in Fig. 2.32 is very long. In this case, the temperature at the tip is essentially equal to the temperature of the surrounding fluid. The second boundary condition can therefore be written as

     BC-2: at $x \to \infty$, $T \to T_\infty$

Therefore,

     At $x \to \infty$, $\theta = T - T_\infty \to 0$

Therefore,

$$\lim_{x \to \infty} \theta(x) \to 0 \quad (2.126)$$

The application of BCs [Eqs (2.125) and (2.126)] to Eq. (2.124a) gives $C_1 = 0$ and $C_2 = \theta_0$. Hence, the temperature distribution is found to be

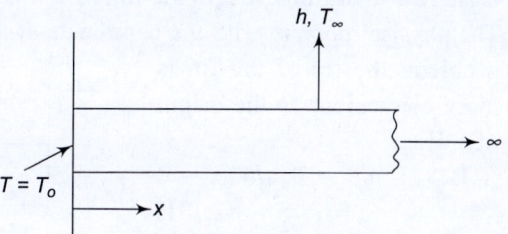

Fig. 2.32 Infinite fin

$$\theta(x) = \theta_0 e^{-mx} \qquad (2.127)$$

or $\dfrac{T - T_\infty}{T_0 - T_\infty} = e^{-mx}$ (2.128)

The temperature profile T versus x is shown in Fig. 2.33.

The heat transfer from the fin can now be calculated by integrating the local convective heat transfer over the whole periphery (note that no heat loss occurs at the tip of the fin):

$$q = \int_0^\infty hp\,dx\,(T - T_\infty) = hp \int_0^\infty \theta(x)\,dx$$

$$= hp\theta_0 \int_0^\infty e^{-mx}\,dx = \frac{hp\theta_0}{m}$$

$$= \sqrt{hpkA}\ \theta_0 \qquad\qquad\qquad (2.129)$$

If one considers the entire fin as the system, then it is easy to see that the heat transferred from the fin by convection to the surrounding fluid must be equal to the heat conducted to the fin at the base. Hence, we may also evaluate the heat transfer from the fin by applying Fourier's law at the base:

$$q = -kA\left(\frac{dT}{dx}\right)_{x=0}$$

$$= -kA\left(\frac{d\theta}{dx}\right)_{x=0}$$

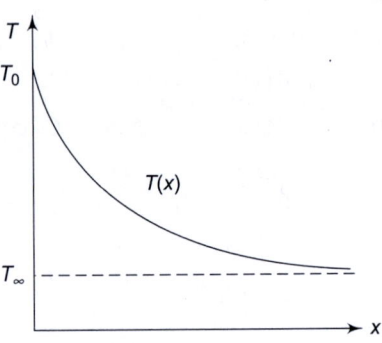

Fig. 2.33 Temperature distribution in the infinite fin

$$= -kA\theta_0\left[\frac{d}{dx}(e^{-mx})\right]_{x=0}$$

$$= kA\theta_0 m$$

$$= \sqrt{hpkA}\ \theta_0 \qquad\qquad\qquad (2.130)$$

Since this involves differentiation, this method is more convenient to use than the first method which requires integration.

Case B: Fin of finite length having insulated tip
The physical domain with the coordinate system is shown in Fig. 2.34. For this problem, the tip of the fin is more convenient as the origin of x. Here,

BC-2: at $x = 0$, $d\theta/dx = 0$ $\qquad (2.131)$

Applying, BCs [Eqs (2.125) and (2.131)] to Eq. (2.124b), we get

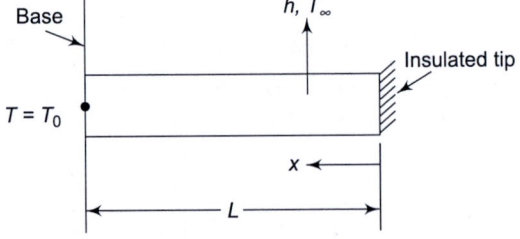

Fig. 2.34 The physical domain and coordinate system for a fin of finite length with insulated tip

$$C_4 = \theta_0/\cosh mL$$
$$C_3 = 0$$

Note that Eq. (2.124b) is more convenient to use in the case of a finite fin. Therefore, the temperature distribution in the fin is

$$\theta = \frac{\theta_0}{\cosh mL}\cosh mx \qquad\qquad (2.132)$$

Heat transfer from the fin is

$$q = -\left[-kA\left(\frac{d\theta}{dx}\right)\right]_{x=L}$$

$$= \frac{kA\theta_0}{\cosh mL}(m\sinh mx)_{x=L}$$

$$= kA\theta_0(\tanh mL)\,m$$

$$= \sqrt{hpkA}\,\theta_0\,\tanh mL \tag{2.133}$$

Since $\tanh ml \to 1$ as $mL \to \infty$, $q_{\text{finite-fin}}$ approaches that for an infinite fin. This statement is independent of the boundary condition employed at the tip of the fin, since the effect of the tip diminishes as $L \to \infty$. The condition $mL \to \infty$ may also be interpreted as $m \to \infty$ for a given L. That means $(hp/kA)^{1/2} \to \infty$, or $h \to \infty$, or $k \to 0$.

Case C: Fin of finite length with convecting tip
The temperature distribution and heat transfer can be calculated from Eqs (2.112) and (2.113), respectively, as the physical problem is identical with the thin rod heat transfer discussed in Section 2.14.

2.17 Evaluation of Fin Performance

Two yardsticks are used to compare and evaluate extended surfaces in augmenting heat transfer from the base area. They are (a) fin efficiency and (b) fin effectiveness.

2.17.1 Fin Efficiency

Fin efficiency (η_f) is defined as the ratio of the actual heat transfer to the heat that would be transferred if the entire fin were at the base temperature.
For Case A:

$$\eta_f = \lim_{L\to\infty} \frac{\theta_0\,(hpkA)^{1/2}}{\theta_0\,hpL}$$

$$= \lim_{L\to\infty} \left(\frac{kA}{hp}\right)^{1/2}\frac{1}{L}$$

$$= \lim_{L\to\infty}\left(\frac{1}{mL}\right) \to 0 \tag{2.134}$$

Thus, for an infinite fin, the efficiency tends to zero.
For Case B:

$$\eta_f = \frac{\theta_0(hpkA)^{1/2}\tanh mL}{\theta_0\,hpL}$$

$$= \left(\frac{kA}{hp}\right)^{1/2}\frac{1}{L}\tanh mL$$

$$= \left(\frac{1}{mL}\right)\tanh mL \tag{2.135}$$

$$L = 0 \quad\Rightarrow\quad mL = 0$$

Therefore,

$$\eta_f = \frac{\tanh(0)}{0} = \frac{0}{0}$$

Let $mL = x$. Applying L'Hôpital's rule,

$$\eta_f = \lim_{x \to 0} \frac{\frac{d}{dx}(\tanh x)}{\frac{d}{dx}(x)} = \lim_{x \to 0} \frac{\operatorname{sech}^2 x}{1}$$

$$= 1/1 = 1$$

Therefore $\eta_f = 1$ at $mL = 0$ or $L = 0$. As mL increases η_f decreases.

It is interesting to note that the fin efficiency reaches its maximum value for the trivial case of $L = 0$ or no fin at all. Therefore, we should not expect to be able to maximize fin efficiency with respect to the fin length. It is, however, possible to maximize the efficiency with respect to the quantity of fin material (mass, volume, or cost).

For a fin of given material and dimensions, the efficiency decreases as h increases. For example, a fin that is highly efficient when used with a gas coolant will usually be found inefficient when used with water where the value of h is usually much higher.

2.17.2 Total Efficiency of a Finned Surface

The fin efficiency η_f is concerned with expressing the performance of the fin itself. However, most applications employing extended surfaces involve the use of an array of fins attached to the primary surface or base surface. Figure 2.35 depicts such an array for straight fins. In such applications, it proves useful to define a total efficiency which gives a measure of the performance of an entire array. Let A_f be the surface area of the fins only, A be the total exposed surface area, including the fins and the unfinned primary surface, η_f be the efficiency of a single fin, and ξ be the total efficiency of the finned array.

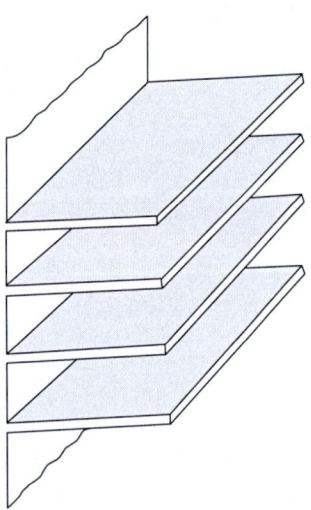

Fig. 2.35 A finned array

$$\xi = \left(\frac{\text{Heat rejected by an array}}{\begin{array}{c}\text{Heat the array would reject if the entire surface were}\\\text{maintained at the base temperature}\end{array}} \right)$$

$$= \frac{\eta_f A_f h\theta_0 + (A - A_f) h\theta_0}{A h\theta_0}$$

Note that the first term in the numerator indicates the heat given up by fins, while the second term signifies the heat given up by the exposed portion of the primary surface or base surface. Continuing further, we get

$$\xi = \frac{\eta_f A_f}{A} + \frac{A - A_f}{A}$$

$$= 1 - \frac{A_f}{A}(1 - \eta_f) \tag{2.136}$$

Since $A_f/A < 1$ and $\eta_f \leq 1$, one deduces that $\xi \leq 1$.

2.17.3 Fin Effectiveness

Fin effectiveness ϕ is defined as the ratio of the actual heat transfer to the heat that would be transferred from the same base area A_0 without the fin, with the base temperature T_0 remaining constant. Therefore (e.g., for a fin with an insulated tip),

$$\phi = \frac{q_{\text{fin}}}{q_{\text{base}}} = \frac{\int_{A_f} h\theta(x)dA}{h\theta_0 A_0}$$

$$= \frac{\int_{A_f} \theta(x)dA}{\theta_0 A_0} \tag{2.137}$$

Note that

$$\eta_f = \frac{\int_{A_f} h\theta(x)dA}{hA_f \theta_0}$$

$$= \frac{\int_{A_f} \theta(x)dA}{A_f \theta_0} \tag{2.138}$$

where A_f is the total surface area over which the fin transfers heat to the surrounding fluid. A comparison of the expressions for ϕ and η_f reveals that

$$\phi = \frac{A_f}{A_0}\eta_f \tag{2.139}$$

2.17.4 Conditions Under Which the Addition of a Fin to a Solid Surface Decreases the Heat Transfer Rate

When h is high as compared to k/δ for a straight rectangular fin where 2δ is the fin thickness, the width of the fin is unity and the length is L, the addition of such a fin to a solid surface may decrease the heat transfer rate. This means that the fin effectiveness ϕ would be less than unity.

2.18 Straight Fin of Triangular Profile

For the design of cooling devices on vehicles, especially aircraft, the problem of exchanging the greatest amount of heat with the least amount of weight in the heat exchanger is of utmost importance. The minimum weight–maximum heat transfer fins are called optimum fins. In determining the optimum fin, a question arises: can a significant weight advantage be gained by using a profile other than a rectangular one for the fin cross-section? In the following discussion, a straight fin of triangular cross-section will be considered. Such a fin is shown in Fig. 2.36. The mathematical treatment in this case is similar to the case of the fins of a rectangular profile except that the area normal to the heat flow is a function of the distance along the fin, decreasing as the fin length increases. It is assumed that $b/L \ll 1$ and $l/L \gg 1$ (or ends in the direction normal to the page are insulated).

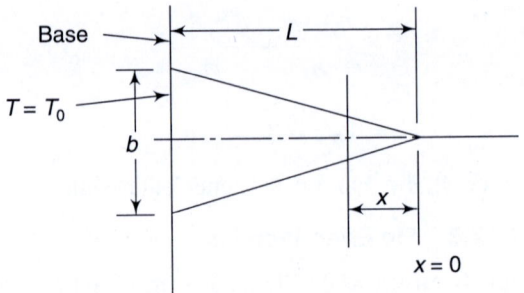

Fig. 2.36   A fin of triangular profile

We see from Fig. 2.36 that

$$A(x) = \frac{bx}{L}l$$

$$p(x) = 2\left(\frac{bx}{L} + l\right)$$

where l is the width of the fin. If we assume that $b \ll l$, then $p(x) \approx 2l$. Inserting it into the general heat conduction equation for fins of variable cross-section, that is, Eq. (2.122), and using $\theta = T - T_\infty$,

$$\frac{d}{dx}\left[A(x)\frac{d\theta}{dx}\right] - \frac{hp(x)}{k}\theta = 0$$

$$\Rightarrow \qquad \frac{d}{dx}\left[\frac{bx}{L}l\frac{d\theta}{dx}\right] - \frac{h(2l)}{k}\theta = 0$$

$$\Rightarrow \qquad \frac{d}{dx}\left(x\frac{d\theta}{dx}\right) - \frac{2hL}{bk}\theta = 0$$

$$\Rightarrow \qquad \frac{d}{dx}\left(x\frac{d\theta}{dx}\right) - m^2\theta = 0$$

$$\Rightarrow \qquad x\frac{d^2\theta}{dx^2} + \frac{d\theta}{dx} - m^2\theta = 0 \qquad\qquad (2.140)$$

where $m^2 = 2hL/bk$. Multiplying both sides of Eq. (2.140) by x, we get

$$x^2\frac{d^2\theta}{dx^2} + x\frac{d\theta}{dx} - m^2 x\theta = 0 \qquad\qquad (2.141)$$

Let us recall the following differential equation:

$$x^2\frac{d^2 y}{dx^2} + x\frac{dy}{dx} - (m^2 x^2 + v^2)y = 0 \qquad\qquad (2.142)$$

which is called the modified Bessel's differential equation of order v. Equation (2.142) is a linear second-order ordinary differential equation with variable coefficients. The general solution of Eq. (2.142) is

$$y(x) = C_1 I_v(mx) + C_2 K_v(mx) \qquad\qquad (2.143)$$

where the functions $I_v(mx)$ and $K_v(mx)$ are known as the modified Bessel functions of the first kind and the second kind of order v, respectively. To make Eq. (2.141) similar to Eq. (2.142), we define a new independent variable

$$\eta = \sqrt{x} \tag{2.144}$$

Using Eq. (2.144), Eq. (2.141) is transformed into

$$\eta^2 \frac{d^2\theta}{d\eta^2} + \eta \frac{d\theta}{d\eta} - 4m^2\eta^2\theta = 0 \tag{2.145}$$

Now, Eq. (2.145) is analogous to Eq. (2.142). Therefore, by analogy, from Eq. (2.143), we can write the solution for θ, i.e.,

$$\theta(\eta) = C_1 I_0(2m\eta) + C_2 K_0(2m\eta)$$

or $\qquad \theta(x) = C_1 I_0(2m\sqrt{x}) + C_2 K_0(2m\sqrt{x}) \tag{2.146}$

The boundary conditions are

$\qquad$ BC-1: at $x = 0$, $T =$ finite

or $\qquad \theta =$ finite $= \theta_m \tag{2.147a}$

$\qquad$ BC-2: at $x = L$, $T = T_0$

or $\qquad \theta = T_0 - T_\infty = \theta_0 \tag{2.147b}$

Note that $x = 0$ is taken at the tip of the fin for the sake of mathematical convenience. Applying BC-1,

$$\theta = \theta_m = C_1 I_0(0) + C_2 K_0(0)$$

Since $K_0(0) \to \infty$ and $\theta(0)$ is finite, therefore $C_2 = 0$. Applying BC-2,

$$\theta_0 = C_1 I_0(2mL^{1/2})$$

Therefore,

$$C_1 = \frac{\theta_0}{I_0(2mL^{1/2})}$$

Therefore, the solution is

$$\theta(x) = \frac{\theta_0}{I_0(2mL^{1/2})} I_0(2mx^{1/2}) \tag{2.148}$$

The heat transfer from the fin is

$$q = -\left[-kA\frac{d\theta}{dx} \right]_{x=L}$$

$$= kA\frac{d\theta}{dx}\bigg|_{x=L}$$

where A is the area of the base of the fin. Therefore,

$$q = \frac{kA\theta_0}{I_0(2mL^{1/2})} \frac{m}{\sqrt{L}} I_1(2ml^{1/2})$$

$$= \frac{kA\theta_0 \, mL^{-1/2} I_1(2mL^{1/2})}{I_0(2mL^{1/2})} \tag{2.149}$$

and
$$\frac{q}{kA\theta_0/L} = \frac{(mL^{1/2})I_1(2mL^{1/2})}{I_0(2mL^{1/2})}$$
(2.150)

For equal heat flow, triangular fins require less thickness than rectangular fins, indicating the weight advantage of the former. However, triangular fins are difficult to manufacture since the tip has zero surface area and has a tendency to break. That is the reason why trapezoidal and parabolic fins are so widely used.

2.19 Thermal Contact Resistance

In the analysis of composite walls it was assumed that a perfect interface exists between two adjoining walls. In practice, there is a finite contact resistance due primarily to surface roughness effects. Therefore, the temperature drop across the interface between the materials may be appreciable.

There are two principal contributions to the heat transfer at the interface. They are (a) the solid-to-solid conduction at the spots of contact and (b) the conduction and/or radiation through entrapped gases in the void spaces created by the contact. The second factor is believed to comprise a large part of the total resistance to heat flow, because the thermal conductivity of a gas is quite small in comparison with that of a solid.

If h_c is the contact coefficient (W/m^{2}K), then the quantity $1/h_cA$ is called the thermal contact resistance. A is the total contact area. The contact resistance should increase with a decrease in the ambient gas pressure below a threshold value where the mean free path of the molecules is large compared with a characteristic dimension of the void space. The contact resistance should decrease for an increase in the joint pressure since this results in deformation of high spots of the contact surfaces, thereby creating a greater contact area between the solids. Thermal contact resistance can be reduced to a great extent by the use of a 'thermal grease' such as Dow-340. The consideration of thermal contact resistance becomes of importance in many industrial heat transfer applications where mechanical joining of two materials is involved, such as welding, soldering, etc.

Additional Examples

Example 2.5 Heat Transfer in a Composite Wall

The outer wall of a house is composed of two parallel slabs (Fig. E2.5). The thermal conductivities and the thicknesses of the slabs are k_1, k_2 and L_1, L_2, respectively. The inside and outside ambient temperatures and heat transfer coefficients are T_∞ and h, respectively. The net radiation between the sun and the outer surface of the wall is q''. Calculate

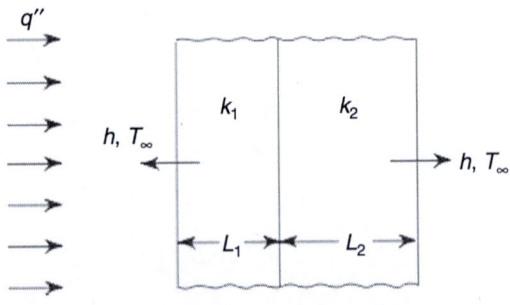

Fig. E2.5 Composite wall of a house

the heat transfer to the house (a) by an equivalent electric circuit method and (b) by solving energy equations for the slabs.

Solution

This is a composite plane wall problem.

   (a)   *Equivalent electric circuit method*

$$q''A = q_1 + q_2 = \frac{T_1 - T_\infty}{\dfrac{1}{hA}} + \frac{T_1 - T_\infty}{\dfrac{L_1}{k_1 A} + \dfrac{L_2}{k_2 A} + \dfrac{1}{hA}}$$

Let $\quad C = \dfrac{1}{h}$

and $\quad D = \dfrac{L_1}{k_1} + \dfrac{L_2}{k_2} + \dfrac{1}{h}$

Therefore, $T_1 - T_\infty = \dfrac{q''}{\dfrac{1}{C} + \dfrac{1}{D}}$

Now, q_2 is the heat transfer to the house and is given by

$$q_2 = \frac{T_1 - T_\infty}{\dfrac{L_1}{k_1 A} + \dfrac{L_2}{k_2 A} + \dfrac{1}{hA}}$$

or $\quad q_2'' = \dfrac{T_1 - T_\infty}{D} = \dfrac{q''}{\left(\dfrac{1}{C} + \dfrac{1}{D}\right)D} = \dfrac{q''}{\dfrac{D}{C} + 1} = \dfrac{q''}{2 + \dfrac{L_1 h}{k_1} + \dfrac{L_2 h}{k_2}}$

Therefore, $q_2 = \dfrac{q''A}{2 + \dfrac{L_1 h}{k_1} + \dfrac{L_2 h}{k_2}}$

   (b)   *Energy equation method*

Taking the origin $x = 0$ at the left end of the slab, we can write the following:

GDE for slab 1: $\dfrac{d^2 T_1}{dx^2} = 0$

GDE for slab 2: $\dfrac{d^2 T_2}{dx^2} = 0$

BC-1: at $x = 0$, $-k_1 \dfrac{dT_1}{dx} = q'' - h(T_1 - T_\infty)$

BC-2: at $x = L_1$, $-k_1 \dfrac{dT_1}{dx} = -k_2 \dfrac{dT_2}{dx}$

BC-3: at $x = L_1$, $T_1 = T_2$

BC-4: at $x = L_1 + L_2$, $-k_2 \dfrac{dT_2}{dx} = h(T_2 - T_\infty)$

Integrating twice GDEs for slab 1 and slab 2, we obtain

$$T_1 = c_1 x + c_2$$

$$T_2 = c_3 x + c_4$$

Applying BCs 1, 2, 3, and 4 gives the four integration constants and, hence, the temperature distributions in slab 1 and slab 2. The net heat transfer to the house is

$$q''A - hA(T_{1x=0} - T_\infty)$$

Example 2.6 Distributed Heat Generation in a Wall

Consider a shielding wall for a nuclear reactor. The wall receives a γ-ray flux such that heat is generated within the wall according to the relation $q''' = q_0''' e^{-ax}$, where q_0''' is the heat generated per unit volume at the inner face of the wall exposed to γ-ray flux and a is a constant. Derive an expression for the temperature distribution in a wall of thickness L, where the inside and outside temperatures are maintained at T_i and T_o, respectively. Also obtain an expression for the maximum temperature in the wall.

Solution

Assume steady 1D heat conduction and constant properties.

GDE: $\dfrac{d^2 T}{dx^2} + \dfrac{q'''}{k} = 0$

BC-1: at $x = 0$, $T = T_i$
BC-2: at $x = L$, $T = T_o$

Integrating GDE twice, we obtain

$$T = -\dfrac{q_0'''}{a^2 k} e^{-ax} + c_1 x + c_2$$

Using BC-1 and BC-2 in the above equation gives the constants c_1 and c_2. Finally,

$$T = -\dfrac{q_0'''}{a^2 k} e^{-ax} + \left[\left(\dfrac{T_o - T_i}{L} \right) - \dfrac{1}{L} \dfrac{q_0'''}{a^2 k} (1 - e^{-aL}) \right] x + T_i + \dfrac{q_0'''}{a^2 k}$$

To obtain the maximum temperature in the wall, set $dT/dx = 0$ and find the corresponding x and substitute it into the expression for T.

Example 2.7 Critical Radius of Insulation for a Spherical Shell

Derive an expression for the critical radius of insulation for a sphere.

Solution

Assumptions: (1) $k_{shell} \gg k_{insulation}$

(2) $h_i \gg h$

Therefore, heat transfer from the inside fluid to the surroundings (neglecting inner convective resistance and conduction resistance in the shell by invoking the above assumptions) in a spherical shell (of outer radius r) covered by a layer of insulation of outer radius r_o may be written as

$$q = \frac{T_i - T_o}{\dfrac{r_o - r}{4\pi r_o\, r k_{\text{insulation}}} + \dfrac{1}{4\pi r_o^2\, h}}$$

For maximum heat loss,

$$\frac{dq}{dr_o} = 0$$

which gives

$$r_{oc} = \frac{2\,k_{\text{insulation}}}{h}.$$

Example 2.8   Heat Transfer from an Annular Fin

For the steady state, 1D heat transfer from an annular fin (Fig. E2.8) of rectangular profile, obtain an expression for the temperature distribution in the fin and the heat transfer. Take the fin tip to be insulated.

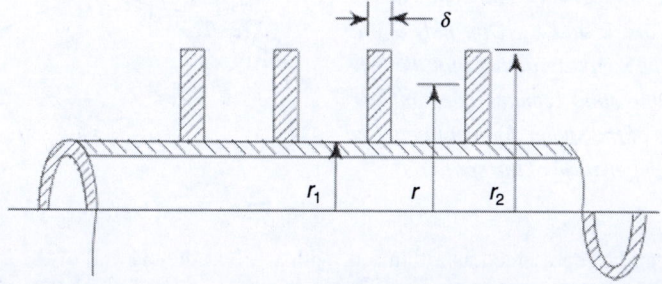

Fig. E2.8   Annular fin

Solution

The GDE for variable cross-sectional area fins will apply here. For one of the fins, we have

$$A\,(r) = 2\pi r \delta$$
$$p\,(r) = 4\pi r$$

$$\text{GDE: } \frac{d}{dr}\left[A\,(r)\frac{d\theta}{dr}\right] - \left[\frac{h p\,(r)}{k}\theta\right] = 0$$

Substituting $A(r)$ and $p(r)$ in the above GDE, we obtain

$$r^2 \frac{d^2\theta}{dr^2} + r\frac{d\theta}{dr} - m^2\, r^2\, \theta = 0 \tag{A}$$

where $m^2 = 2h/k\delta$ and $\theta = T - T_\infty$. If the heat loss from the tip of the fins is assumed to be negligible, then the boundary conditions can be stated as follows:

BC-1: at $r = r_1$, $\theta = \theta_0 = T_0 - T_\infty$

BC-2: at $r = r_2$, $\dfrac{d\theta}{dr} = 0$

The general solution of Eq. (A) is

$$\theta\,(r) = c_1 I_0\,(mr) + c_2 K_0\,(mr)$$

After obtaining the two integration constants, the solution is obtained as shown below:

$$\frac{\theta(r)}{\theta_0} = \frac{I_0(mr)K_1(mr_2) + K_0(mr)I_1(mr_2)}{I_0(mr_1)K_1(mr_2) + K_0(mr_1)I_1(mr_2)}$$

The heat transfer from the fin is

$$q = -k\, 2\pi r_1 \delta \left(\frac{d\theta}{dr}\right)_{r=r_1}$$

$$= 2\pi r_1 \sqrt{2hk\delta}\,\theta_0 \frac{I_1(mr_2)K_1(mr_1) - I_1(mr_1)K_1(mr_2)}{I_0(mr_1)K_1(mr_2) + I_1(mr_2)K_0(mr_1)}$$

Example 2.9 Temperature Distribution in a Spoon in a Soup Bowl

A spoon in a soup bowl may be approximated as a rod of constant cross-section (Fig. E2.9). The thermal conductivity, length, periphery, and cross-sectional area of the spoon are k, 2L, p, and A, respectively. The heat transfer coefficients are h and h_o. One-half of the spoon is in the soup. Assuming that the temperature of the soup remains constant and that the ends of the spoon are insulated, find the steady temperature of the spoon.

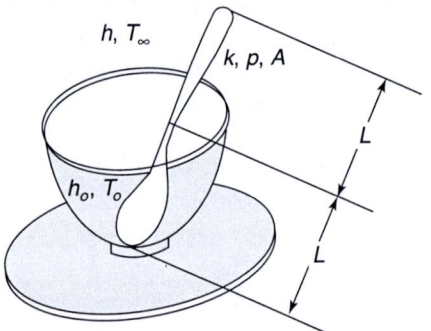

Fig. E2.9 Heat transfer in a spoon

Solution

The spoon can be represented as a thin rod, with $x = 0$ at the big end of the spoon. Note that in the first half $[T_1 = T_1(x)]$, heat is transferred from tea to the spoon $(T_0 > T_1)$ while in the second half $[T_2 = T_2(x)]$, transfer of heat takes place from the spoon to the atmosphere $(T_2 > T_\infty)$. Therefore, GDE for domain 1 is

$$\frac{d^2\theta_1}{dx^2} - m_1^2\,\theta_1 = 0$$

where $m_1^2 = h_0 p/kA$ and $\theta_1 = T_0 - T_1$. GDE for domain 2 is

$$\frac{d^2\theta_2}{dx^2} - m_2^2\,\theta_2 = 0$$

where $m_2^2 = hp/kA$ and $\theta_2 = T_2 - T_0$.

BC-1: at $x = 0$, $\dfrac{d\theta_1}{dx} = 0$

BC-2: at $x = L$, $k\dfrac{dT_1}{dx} = -k\dfrac{dT_2}{dx}$ or $\dfrac{d\theta_1}{dx} = -\dfrac{d\theta_2}{dx}$

(Note that in the first half the temperature increases with x, while in the second half the temperature decreases with x.)

BC-3: at $x = L$, $T_1 = T_2$ or $\theta_1 + \theta_2 = T_0 - T_\infty$

BC-4: at $x = 2L$, $\dfrac{d\theta_2}{dx} = 0$

Integrating twice the GDE for domain 1, we get

$$\theta_1 = c_1 e^{m_1 x} + c_2 e^{-m_1 x} \tag{A}$$

Integrating twice the GDE for domain 2, we get

$$\theta_2 = c_3 e^{m_2 x} + c_4 e^{-m_2 x} \tag{B}$$

Applying BCs 1, 2, 3, and 4 to Eqs (A) and (B) yields the four integration constants. Therefore, the temperature distribution in the spoon can be obtained.

$$c_1 = -\left(\frac{T_0 - T_\infty}{M}\right)\left(\frac{m_2}{m_1}\right)\left(\frac{e^{-3 m_2 L} - e^{-m_2 L}}{e^{m_1 L} - e^{-m_1 L}}\right)$$

$$c_1 = c_2$$

$$c_3 = \frac{e^{-4 m_2 L}(T_0 - T_\infty)}{M}$$

$$c_4 = \frac{T_0 - T_\infty}{M}$$

where
$$M = e^{-3 m_2 L} + e^{-m_2 L} - \left(\frac{m_2}{m_1}\right)\left(\frac{e^{-3 m_2 L} - e^{-m_2 L}}{e^{m_1 L} - e^{-m_1 L}}\right)$$

Example 2.10 Thermocouple Measurement Error

A thermocouple in a cylindrical well is inserted into a gas stream (see Fig. 2.29). Estimate the true temperature of the gas stream if T_L (the temperature indicated by the thermocouple) = 260°C, T_0 (wall temperature) = 177°C, h = 680 W/m²K, k = 103.8 W/mK, t = 2 mm, and L = 6 cm.

Solution

The thermocouple well wall of thickness t (see Fig. 2.29) is in contact with the gas stream on one side only, and the tube thickness is small compared with the diameter. Hence the temperature distribution along this wall will be nearly the same as that along a bar of thickness $2t$, in contact with the gas stream on both sides. According to Eq. (2.115), the temperature at the end of the well (that registered by the thermocouple) is

$$\frac{T_L - T_\infty}{T_0 - T_\infty} = \frac{1}{\cosh mL} \tag{A}$$

where
$$m = \left(\frac{hp}{kA}\right)^{1/2}$$

Now, $A = (2t)(1)$ and $p = 2 + 4t = 2 + 4(2 \times 10^{-3}) = 2.008$. The width of the bar is taken as 1 m for the calculation of A and p. Therefore,

$$m = \left(\frac{(680)(2.008)}{(103.8)(2)(2 \times 10^{-3})}\right)^{1/2} = 81.1$$

$$mL = (81.1)(6 \times 10^{-2}) = 4.866$$

$$T_L = 260°C$$

$$T_0 = 177°C$$

Hence, from Eq. (A)

$$\frac{260 - T_\infty}{177 - T_\infty} = \frac{1}{\cosh 4.866} = 0.0154$$

or, $T_\infty = 261.3\,°C$, which is the true temperature of the gas stream. Therefore, the reading of the thermocouple is $1.3\,°C$ too low.

Example 2.11 Plane Wall with Temperature Dependent Thermal Conductivity

A plane wall of 40 cm thickness is made of a material whose thermal conductivity varies linearly with temperature according to the relation, where T is in °C and in W/m K. The left side (of the wall is kept at 500°C and the other side at 0°C. Calculate the temperature at x = 20 cm.

Solution

We know that the temperature distribution at any location in a plane wall with $k = k(T)$ can be obtained by solving Eq. (2.57) given as

$$T^2 + \frac{2}{\beta}\left[1 - \beta T_o\right]T + \frac{2}{\beta}\left[T_1\left(\beta T_o - \frac{\beta}{2}T_1 - 1\right) + \frac{k_m}{k_o}(T_1 - T_2)\frac{x}{L}\right] = 0 \qquad (A)$$

From the input data of the problem

$$\beta = 0.0012 \text{ K}^{-1}$$
$$L = 40 \text{ cm}$$
$$T_1 = 500 \text{ °C}$$
$$T_2 = 0\text{°C}$$
$$x = 20 \text{ cm}$$
$$k_0 = 1 \text{ W/m K}$$

Also, from Eq. (2.53), we know

$$k_m = \frac{1}{T_1 - T_2}\int_{T_2}^{T_1} k(T)\,dT$$

$$k_m = \frac{1}{500}\int_0^{500}(1 + 0.0012\,T)\,dT = \frac{1}{500}\left[|T|_0^{500} + 0.0012\frac{T^2}{2}\Big|_0^{500}\right] = 1.3\,\text{W/m K}$$

Substituting k_m and the input data into Eq. (A) we get the following quadratic equation:

$$T^2 + 1666.67T - 541667.75 = 0 \qquad (B)$$

The roots of Eq. (B) are: $T = 278.8°C, -1945.5°C$

Obviously, $T = 278.8°C$ is the solution to our problem. The other solution ($-1945.5°C$) is discarded as the value is unrealistic since temperature at any location inside the wall must be lying between $0°C$ and $500°C$.

Example 2.12 Heat Generation in a Plastic Insulated Electrical Wire

A long electrical stainless steel wire of diameter 0.4 cm and thermal conductivity 15.1 W/mK carries electric current resulting in a uniform heat generation inside it at the rate of 1000 kW/m³. The wire is encased in a 0.5 cm thick layer of plastic (k = 0.15 W/m K). If the outer surface of the plastic sleeve is measured to be 50°C, determine the temperature at the centre-line of the wire, the interface of the wire and plastic at the steady state.

Solution

This is a two-domain problem with an interface which is assumed to be perfect (Fig. E2.12). At the interface the compatibility conditions (i.e., continuity of heat flux and equality of temperature) are to be satisfied.

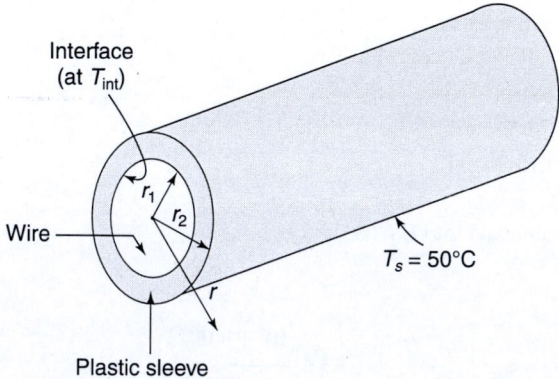

Fig. E2.12 Heat generation in a stainless steel wire encased in plastic

In domain-1 (wire) the governing differential equation is

$$\frac{1}{r}\frac{d}{dr}\left(r\frac{dT_{\text{wire}}}{dr}\right)+\frac{q'''}{k}=0$$

BC: $T_{\text{wire}}(r_1)=T_{\text{int}}$

$$\frac{dT_{\text{wire}}}{dr}(0)=0$$

The solution is: $T_{\text{wire}}(r)=T_{\text{int}}+\dfrac{q'''}{4k_{\text{wire}}}\left(r_1^2-r^2\right)$ (A)

In domain-2 (plastic) the governing differential equation is

$$\frac{1}{r}\frac{d}{dr}\left(r\frac{dT_{\text{plastic}}}{dr}\right)=0$$

BC: $T_{\text{plastic}}(r_1)=T_{\text{int}}$
$T_{\text{plastic}}(r_2)=T_s=50°C$

The solution is

$$T_{\text{plastic}}(r)=\frac{\ln\left(\dfrac{r}{r_1}\right)}{\ln\left(\dfrac{r_2}{r_1}\right)}(T_s-T_{\text{int}})+T_{\text{int}}$$ (B)

At the interface the condition of continuity of heat flux must be satisfied (the equality of temperature has been already satisfied as seen in the first boundary condition for domain-1 and domain-2, respectively). Therefore,

$$-k_{\text{wire}}\frac{dT_{\text{wire}}}{dr}(r_1)=-k_{\text{plastic}}\frac{dT_{\text{plastic}}}{dr}(r_1)$$ (C)

Using Eqs (A) and (B) in Eq. (C), we obtain

$$T_{int} = \frac{q''' r_1^2}{2 k_{plastic}} \ln\left(\frac{r_2}{r_1}\right) + T_s \qquad (D)$$

Now, in this problem the input data are

$r_1 = 0.2$ cm $= 0.002$ m

$r_2 = 0.2 + 0.5 = 0.7$ cm $= 0.007$ m

$q''' = 10^6$ W/m^3

$k_{wire} = 15.1$ W/m K

$k_{plastic} = 0.15$ W/m K

$T_s = 50°C$

Substituting the input data into Eq. (D), we get

$T_{int} = 66.69°C$

From Eq. (A)

$$T_{wire}(0) = T_{int} + \frac{q''' r_1^2}{4 k_{wire}} = 63.3 + \frac{(10^6)(0.002)^2}{4(15.1)} = 66.76°C$$

Thus the centre-line temperature of the wire is 63.4°C and the wire-plastic interface temperature is 63.3°C.

Example 2.13   Parallel and Series-Parallel Thermal Resistance Networks

Draw the thermal resistance network for heat conduction through the multi-layered walls as shown in Fig. E2.13(a) and Fig. E2.13(b). Obtain also an expression for equivalent thermal resistance in each case.   State clearly the assumptions.

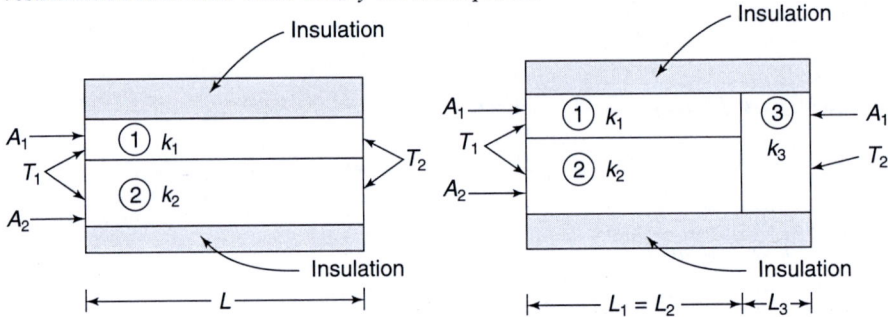

Fig. E2.13(a)   Two-layered wall   **Fig. E2.13(b)**   Three-layered wall

Solution

Assumptions

1. Steady state one-dimensional heat conduction since top and bottom surfaces are insulated and lateral surfaces are very long as compared to the overall thickness of the body so that the end effects can be neglected.

2. Surfaces are isothermal since ID heat conduction is assumed.

3. Heat transfer between layers 1 and 2 is neglected.

Case (a)

Figure E2.13(a) shows a composite wall of two parallel layers. The thermal resistance network comprises two parallel resistances as represented in Fig. E2.13(c). Since total heat transfer is the sum of the heat transfer through each layer, we can write

$$q = q_1 + q_2 = \frac{T_1 - T_2}{R_1} + \frac{T_1 - T_2}{R_2} = (T_1 - T_2)\left(\frac{1}{R_1} + \frac{1}{R_2}\right) \qquad \text{(A)}$$

Using electrical analogy, we get

$$q = \frac{T_1 - T_2}{R_{eq}} \qquad \text{(B)}$$

Comparing Eqs (A) and (B)

$$\frac{1}{R_{eq}} = \frac{1}{R_1} + \frac{1}{R_2}$$

Or, $\qquad R_{eq} = \dfrac{R_1 R_2}{R_1 + R_2}$ (C)

where $\quad R_1 = \dfrac{L}{k_1 A_1}$ and $R_2 = \dfrac{L}{k_2 A_2}$

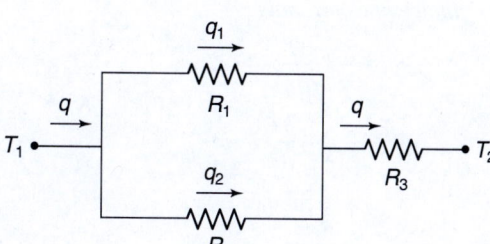

Fig. E2.13(c) Parallel resistance network

Case (b)

This is a series-parallel arrangement as shown in Fig. E2.13(b). The thermal resistance network for the same is depicted in Fig. E2.13(d). The total heat transfer through this composite wall can be expressed as

$$q = \frac{T_1 - T_2}{R_{eq}} \qquad \text{(D)}$$

Fig. E2.13(d) Series-parallel resistance network

where $R_{eq} = R_{12} + R_3 \dfrac{R_1 R_2}{R_1 + R_2} + R_3$

and $\qquad R_1 = \dfrac{L_1}{k_1 A_1} \qquad R_2 = \dfrac{L_1}{k_2 A_2} \qquad R_3 = \dfrac{L_3}{k_3 A_3}$

Important Concepts and Formulae

Steady State Heat Conduction

Steady-state heat conduction is defined as the condition prevailing in a heat conducting body when temperatures at all points inside the body do not change with time.

One-dimensional (1D) Heat Conduction

The term 'one-dimensional' as applied to a heat conduction problem means that only one space coordinate is required to describe the temperature distribution in a body. In 1D heat conduction heat flux lines (which are normal to isotherms in isotropic media) always point in one direction (longitudinally, or radially). In multi-dimensional heat conduction, heat flux lines flare out in two or more directions.

Isotropic Medium A continuum is said to be isotropic if its conductivity is same in all directions.

Anisotropic Medium A continuum is termed as anisotropic if there is a directional variation of conductivity exists in it.

Homogeneous Medium   A continuum is said to be homogeneous if its conductivity does not vary from point to point.

Heterogeneous Medium   A continuum is termed as heterogeneous if there is a variation of its conductivity from point to point.

Fourier's Law of Heat Conduction in Isotropic, Heterogeneous Media

Cartesian Coordinates

$$q''_x = -k\frac{\partial T}{\partial x}$$

$$q''_y = -k\frac{\partial T}{\partial y}$$

$$q''_z = -k\frac{\partial T}{\partial z}$$

Cylindrical Coordinates

$$q''_r = -k\frac{\partial T}{\partial r}$$

$$q''_\phi = -k\frac{1}{r}\frac{\partial T}{\partial \phi}$$

$$q''_z = -k\frac{\partial T}{\partial z}$$

Spherical Coordinates

$$q''_x = -k\frac{\partial T}{\partial r}$$

$$q''_\psi = k\frac{1}{r}\frac{\partial T}{\partial \psi}$$

$$q''_\phi = -k\frac{1}{r\sin\psi}\frac{\partial T}{\partial \phi}$$

Note: Although Fourier's law was founded upon a hypothesis suggested by a steady state experiment it is also used in unsteady problems as a valid particular law as it has never been refuted.

Heat Conduction Equation for Isotropic, Homogeneous Materials

$$\nabla^2 T + \frac{q'''}{k} = \frac{1}{\alpha}\frac{\partial T}{\partial t}$$

where
In Cartesian Coordinates

$$\nabla^2 = \frac{\partial^2}{\partial x^2} + \frac{\partial^2}{\partial y^2} + \frac{\partial^2}{\partial z^2}$$

In Cylindrical Coordinates

$$\nabla^2 = \frac{1}{r}\frac{\partial}{\partial r}\left(r\frac{\partial}{\partial r}\right) + \frac{1}{r^2}\frac{\partial^2}{\partial \phi^2} + \frac{\partial^2}{\partial z^2}$$

In Spherical Coordinates

$$\nabla^2 = \frac{1}{r^2}\frac{\partial}{\partial r}\left(r^2\frac{\partial}{\partial r}\right) + \frac{1}{r^2 \sin\psi}\frac{\partial}{\partial\psi}\left(\sin\psi\frac{\partial}{\partial\psi}\right) + \frac{1}{r^2 \sin^2\psi}\frac{\partial^2}{\partial\phi^2}$$

Energy Balance at a Boundary

Kirchhoff's Current Law

The algebraic sum of heat fluxes at a boundary must be equal to zero.
Sign Conventions
- Heat flux to the boundary is positive.
- Heat flux from the boundary is negative.

Thermal Resistance Formulae in Steady One-dimensional Conduction

Plane Wall: $R_{cond} = \dfrac{L}{kA}$

Hollow Cylinder: $R_{cond} = \dfrac{1}{2\pi kL}\ln\left(\dfrac{r_o}{r_i}\right)$

Spherical Shell: $R_{cond} = \dfrac{1}{4\pi k}\left(\dfrac{1}{r_i} - \dfrac{1}{r_o}\right)$

Convective Resistance:

$$R_{conv} = \frac{1}{hA}$$

Note: Thermal resistance concept is strictly applicable to steady one-dimensional problem without heat generation.

Critical Radius of Insulation for Cylindrical Pipe

A special application of the thermal resistance formulae is to determine the thickness of the annular insulation that should be applied to the outer surface of a small-diameter circular pipe of a known wall temperature. As insulation is added to the pipe the conduction resistance increases. But at the same time the surface area for convective heat transfer will also increase and hence the convective resistance will decrease. Since the total resistance in the heat flow path is the sum of these two resistances, it is possible that a minimum total resistance exists at a particular radius corresponding to which the heat transfer will be maximum. Thus, adding insulation to a circular pipe may increase heat loss up to a certain thickness of insulation. The radius at which the heat loss is maximum is called the critical radius of insulation. The expression for the critical radius of insulation is

$$r_c = \frac{k}{h_o}$$

where k is the conductivity of the insulation and h_0 is the outside heat transfer coefficient.

If the outer radius of a pipe is less than the critical radius of insulation, then heat losses can be increased by adding insulation up to the value of the critical thickness. Further increase in the thickness of insulation will cause the heat loss to decrease from this peak value, but until a certain amount of insulation is added, the heat loss is still greater than that for the bare pipe.

In the case of large pipes in which the outside pipe radius is larger than the critical radius, any amount of insulation will decrease the heat loss.

Heat Generation in a Body

Plane Wall (Convective environment on both sides)

$$T = T_\infty + \frac{q''' L^2}{2k}\left[1 - \left(\frac{x}{L}\right)^2\right] + \frac{q''' L}{h}$$

Solid Cylinder (surrounded by a convective environment)

$$T = T_\infty + \frac{q''' r_o^2}{4k}\left[1 - \left(\frac{r}{r_o}\right)^2 + \frac{2k}{hr_o}\right]$$

Solid Sphere (surrounded by a convective environment)

$$T = T_\infty + \frac{q''' r_o^2}{6k}\left[1 - \left(\frac{r}{r_o}\right)^2\right] + \frac{q''' r_o}{3h}$$

Extended Surfaces: Fins

Fins are thin metal strips attached to a primary or base surface. The objective is to enhance heat transfer by increasing the effective area of the base or primary surface with small temperature drop in the protrusions.

Governing Energy Equation

$$\frac{d^2 \theta}{dx^2} - m^2 \theta = 0$$

where $\theta = T - T_\infty$ and $m^2 = \dfrac{hp}{kA}$

Infinitely Long Fin

$$\frac{T - T_\infty}{T_o - T_\infty} = e^{-mx}$$

$$q = \sqrt{hpkA}\,\theta_o$$

Fin of Finite Length having Insulated Tip

$$\theta = \frac{\theta_o}{\cosh mL}\cosh mx$$

$$q = \sqrt{hpkA}\,\theta_o \tanh mL$$

Heat transfer from a fin of finite length with insulated tip approaches that from an infinitely long fin as $mL \to \infty$. In practice, for $mL \geq 2$, the aforesaid condition is almost reached.

Fin Efficiency

Infinitely Long Fin

$$\eta_f = \frac{1}{mL}$$

As $L \to \infty$, $\eta_f \to 0$.

Fin of Finite Length having Insulated Tip

$$\eta_f = \frac{\tanh mL}{mL}$$

Total Efficiency of a finned surface

$$\xi = 1 - \frac{A_f}{A}\left(1 - \eta_f\right)$$

Fin Effectiveness

$$\phi = \frac{A_f}{A_o}\eta_f$$

It is desirable to obtain a fin efficiency of 90% and fin effectiveness greater than 2. A fin having an efficiency of less than 60% is never used.

Review Questions

2.1 Write the vectorial form of Fourier's law of heat conduction in heterogeneous, isotropic media.

2.2 Why is there a negative sign on the RHS of the heat flux expression?

2.3 Write Fourier's law of heat conduction in Cartesian, cylindrical, and spherical coordinates for heterogeneous isotropic solids.

2.4 What are the basic laws used in deriving the heat conduction equation?

2.5 Under what conduction will the heat conduction equation be applicable also to liquids and gases?

2.6 Write the Fourier's law of heat conduction and heat conduction equation for anisotropic solids. Give an example of orthotropic materials.

2.7 For solving two-dimensional unsteady state heat conduction in a solid, how many initial and boundary conditions are required?

2.8 Write the conduction resistance expressions for a plane wall, hollow cylinder, and hollow sphere.

2.9 Write an expression for convective resistance.

2.10 Give an example each of series network and series-parallel thermal resistance network in 1D heat conduction.

2.11 Define overall heat transfer coefficient for a plane wall and a cylindrical tube.

2.12 Why is it important to know critical radius of insulation for designing insulation for steam pipe? Will there be a critical thickness of insulation for a channel?

2.13 Give three examples of heat generation in a solid.

2.14 What is a fin or an extended surface? Where are they used?

2.15 What is the advantage of using a triangular fin? Is there also a disadvantage?

2.16 What is the value of mL corresponding to which heat transfer from a rectangular fin of finite width with insulated tip will be equal to that from an infinitely long fin?

2.17 Define fin efficiency and fin effectiveness? Can fin effectiveness be less than 1?

2.18 Define total efficiency and total effectiveness of a finned surface.

2.19 Will water be a better coolant than air for a fin?

2.20 Give an example of one industrial application where thermal contact resistance will be of importance.

Problems

2.1 A plastic panel of area $A = 0.093$ m^2 and thickness $L = 0.64$ cm is found to conduct heat at the rate of 3 W at steady state with temperatures $T_1 = 26\,°C$ and $T_2 = 24\,°C$ on the left and right surfaces, respectively. What is the thermal conductivity of the plastic at $25\,°C$?

2.2 Derive the heat conduction equation in cylindrical coordinates using an elemental volume for a stationary, isotropic solid.

2.3 Obtain the energy equation for heat flow in a solid moving with velocity $V = \hat{i}u + \hat{j}v + \hat{k}w$.

2.4 A plane wall of 50 cm thickness is constructed from a material of thermal conductivity bearing a relation with temperature as $k = 1 + 0.0015T$, where T is in °C and k in W/mK. Calculate the rate of heat transfer through this wall per unit area if one side of the wall is maintained at 1000°C and the other at 0°C. Assume steady-state conditions.

2.5 Show that for one-dimensional, steady heat conduction in a hollow cylinder and a sphere, the heat transfer q can be written as $q = k\bar{A} \, (T_1 - T_2)/(s_2 - s_1)$ where s_1, s_2 are the coordinates in cylindrical and spherical coordinate systems, and $\bar{A}$ is the equivalent area which is $\bar{A} = (A_2 - A_1)/[\ln(A_2/A_1)]$ in cylindrical coordinates and $\bar{A} = (A_1 A_2)^{1/2}$ in spherical coordinates. Note that for a plane wall, $\bar{A} = A$ and s is replaced by x.

2.6 Draw the analogous electric circuit for heat conduction through the composite wall shown in Fig. Q2.6. State clearly all assumptions.

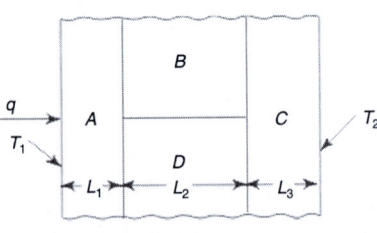

Fig. Q2.6

2.7 The fuel elements in a nuclear reactor are in the form of hollow tubes of inner and outer radii r_i and r_o, respectively. The neutron flux results in uniformly distributed heat sources of strength q''' W/m³ in the fuel elements. The cooling fluid temperatures on the inside and outside of the fuel elements are T_{∞_1} and T_{∞_2}, and the respective convective heat transfer coefficients are h_1 and h_2. Obtain expressions for the surface temperatures. What is the maximum temperature within the fuel elements?

2.8 A copper sphere ($k = 386$ W/m°C) having a diameter of 5 cm is exposed to a convection environment at 25°C, $h = 15$ W/m²°C. Heat is generated uniformly in the sphere at the rate of 1.0 MW/m³. Calculate the steady centre temperature of the sphere.

2.9 Consider a long nuclear fuel rod, which is surrounded by an annular layer of aluminium 'cladding'. Within the fuel rod heat is produced by fission; this heat source is dependent on position, with a source strength varying approximately as $q''' = q_0'''$ $[1 + b \, (r/r_1)^2]$ where q_0''' is q''' at $r = 0$ and r_1 is the radius of the rod. b is a constant. Calculate the maximum temperature in the fuel rod, if the outer surface of the cladding is in contact with a liquid coolant at T_∞, the heat transfer coefficient being h. The thermal conductivities of the fuel rod and the cladding are k_F and k_C, respectively.

2.10 A Cr–Ni steel wire, 2.5 mm in diameter and 30 cm in length, has a voltage of 10 V applied to it, while its surface is maintained at 90°C. Assuming that the resistivity of the wire is 70×10^{-6} ohm cm and the thermal conductivity is 17.3 W/m°C, calculate the centre-line temperature.

2.11 At what radius of asbestos insulation, $k = 0.151$ W/m°C, will the heat loss from a pipe of 0.5 inch outer radius be the same as the heat loss without insulation? Assume that the heat transfer coefficient on the outer surface is the same in both the cases and given by $h = 6$ W/m²°C.

2.12 A very long rod of 2.5 cm diameter is heated at one end. Under steady state conditions, the temperatures at two different locations along the rod, which are 7.5 cm apart, are measured to be 125°C and 90°C, while the surrounding air temperature is 25°C. Assuming that the heat transfer coefficient is 20 W/m²K, estimate the value of the thermal conductivity of the rod.

2.13 An iron well with inner diameter $d = 1.5$ cm for a thermometer is placed in a tube of 9 cm diameter in which flows superheated steam. The wall thickness of the well is 0.09 cm. The heat transfer coefficient between the steam and the tube wall is 105 W/m²°C. Calculate the length of the well, which gives an error of less than 0.5% of the difference between the gas temperature and the tube-wall temperature. How will the well be positioned in the tube—vertically or obliquely? Justify your answer.

2.14 Obtain an expression for the efficiency of a 1D straight fin with convective tip. Assume same heat transfer coefficients at the tip and over the periphery.

2.15 Show that the effectiveness of a finned wall can be expressed as $\phi_{\text{fin-array}} = (A_w + \eta_f A_f)/(A_b + A_w)$, where A_w is the total wall surface area between the fins, A_f is the total heat transfer surface area of the fins, and A_b is the total base area.

2.16 A thin rod (of constant conductivity k) containing uniform heat source per unit volume q''' is connected to two walls having temperatures T_1 and T_2. The length of the rod is L. The rod is exposed to an environment with convection heat transfer coefficient h and temperature T_∞. Obtain an expression for the steady-state temperature distribution in the rod.

2.17 The wall of a furnace has a thermal conductivity of 1.15 W/mK. If the inner surface is at 1100°C and the outer surface is at 350°C, what thickness of the wall should be used if the heat flux through the furnace wall is limited to 2500 W/m²?

2.18 A long, hollow cylinder of inner and outer radii r_1 and r_2, respectively, is heated such that its inner and outer surfaces are at uniform temperatures T_1 and T_2. The thermal conductivity of the cylinder material varies with temperature as $k = k_0(1 + bT)$. Find the rate of heat flow through the cylinder.

2.19 Find the relation for the rate of heat flow through a single-layered plane wall composed of a material whose thermal conductivity varies as $k = k_0(1 + bT + cT^2)$.

2.20 A large slab of concrete, 1 m thick, has both surfaces maintained at 25°C. During the curing process a uniform heat generation of 60 W/m³ occurs throughout the slab. If the thermal conductivity of concrete is 1.1 W/mK, find the steady temperature at the centre of the slab.

2.21 A copper rod ($k = 401$ W/mK) 0.5 cm in diameter and 35 cm long has its two ends maintained at 20°C. The lateral surface of the rod is perfectly insulated, so conduction may be taken as one-dimensional along the length of the rod. Find the maximum electrical current that the rod may carry if the temperature is not to exceed 100°C at any point and the electrical resistivity (resistance × cross-section/length) is 1.73×10^{-6} Ω cm.

2.22 A bare copper wire, 0.2 cm in diameter, has its outer surface maintained at 20°C while carrying an electrical current. The electrically generated heat is conducted one-dimensionally in the radial direction. If the center-line temperature of the wire is not to exceed 100°C, find the maximum current the wire can carry. Use the thermal and electrical properties for copper given in Question 2.21.

2.23 A spine protruding from a wall at temperature T_0 has the shape of a circular cone. The radius of the cone base is R. The spine comes to a point at its tip and its length is L.

(a) If T_∞ denotes the temperature of the ambient fluid, h the surface heat transfer coefficient, and k the thermal conductivity of the cone, show that the governing differential energy equation assuming steady state is as follows (x is the distance measured from the cone tip):

$$\frac{d^2\theta}{dx^2} + \frac{2}{x}\frac{d\theta}{dx} - l^2\frac{\theta}{x} = 0$$

where

$$l^2 = \frac{2hL}{kR}\sqrt{1 + \left(\frac{R}{L}\right)^2}$$

and $\theta = T - T_\infty$.

(b) Show that the general solution of this equation is

$$\theta = \frac{BI_1(2lx^{1/2}) + DK_1(2lx^{1/2})}{x^{1/2}}$$

2.24   For the conical spine described in Question 2.23 show that the temperature distribution in the spine is given by

$$\frac{\theta}{\theta_0} = \left(\frac{L}{x}\right)^{1/2}\frac{I_1(2lx^{1/2})}{I_1(2lL^{1/2})}$$

2.25   Show that the rate of heat flow from the spine in Questions 2.23 and 2.24 is expressed as

$$q = k\pi R^2\theta_0\left[\frac{l}{L^{1/2}}\frac{I_0(2lL^{1/2})}{I_1(2lL^{1/2})} - \frac{1}{L}\right]$$

2.26   Consider a straight fin of uniform thickness t and length L. For a fixed amount of fin material per unit width of the fin (that is, Lt = constant), show that if the heat loss from the fin tip is negligible, the fin dissipates the maximum amount of heat if its length L and thickness t are related by the following condition:

$$\tanh\xi = 3\xi\,\text{sech}^2\,\xi$$

where

$$\xi = L\sqrt{\frac{2h}{kt}}$$

Show also that the solution of the above transcendental equation is $\xi = 1.4192$.

2.27   Two circular rods, both of diameter D and length L, are joined at one end and both heated to the same temperature, T_0, at the free ends. The heat transfer coefficient is the same for all the surfaces. If T_∞ is the temperature of the surrounding fluid and if the thermal conductivities of the two rods are k_a and k_b, show that the temperature of the junction, T_j, is given by

$$\frac{T_j - T_\infty}{T_0 - T_\infty} = \frac{\sqrt{(k_a/k_b)}\,\sinh m_b L + \sinh m_a L}{\sqrt{(k_a/k_b)}\,(\cosh m_a L)(\sinh m_b L) + (\sinh m_a L)(\cosh m_b L)}$$

where

$$m_a = \sqrt{hp_a/k_a A} \quad\text{and}\quad m_b = \sqrt{hp_b/k_b A}$$

2.28   Write an expression for the overall heat transfer coefficient of a two-layer sphere with inside and outside convective heat transfer coefficients. Base the overall heat transfer coefficient on the outer surface area.

2.29   A plane surface is equipped with an array of straight fins of rectangular profile. The fins are 1.9 cm long, 0.15 cm thick, and the distance between the center lines of successive fins is 2 cm. The fins are made of aluminium and the surface heat transfer coefficient is 142 W/m^{2}K. What is the total surface effectiveness?

2.30 The main steam line of a proposed power plant will carry steam at 113 bar and 400°C. To insulate the pipe, 85% magnesia ($k = 0.078$ W/mK) will be used. Since magnesia is not an effective insulator at temperatures above 300°C, it is recommended that a layer of an expensive high-temperature insulation ($k = 0.2$ W/mK) be placed between the pipe and the magnesia layer. Enough insulation must be used so that the outside surface temperature of the magnesia layer is 48°C. The outer diameter of the pipe is 12.75 inch and its wall thickness is 1.312 inch. The pipe material has a conductivity of 40 W/mK. The heat transfer coefficients on the steam and air sides may be taken as 4500 W/m^{2}K and 12 W/m^{2}K, respectively. What thicknesses of the high-temperature insulation and magnesia layers would you recommend for an ambient air temperature of 30°C? **Note:** 1 inch = 2.54 cm.

2.31 A thin wire of cross-section A and perimeter P is extruded at a fixed velocity U through an extrusion die. The temperature of the wire at the die exit is T_0. The extruded wire passes horizontally through air at T_∞ for some distance L and then is rolled onto a large spool, where the temperature reduces to T_L. The heat transfer coefficient from the wire to the surroundings is h.

  (a) Using a control volume approach, derive a differential equation that governs the variation of the steady temperature of the wire as a function of distance x from the die exit.

  (b) Solve the problem formulated in (a) and obtain an expression for the variation of wire temperature as a function of x.

2.32 Indicate the shape of the temperature profile for steady-state heat conduction in a plane wall whose surfaces are maintained at T_1 and T_2 if the thermal conductivity k increases with T ($T_2 > T_1$).

2.33 A cylindrical battery of diameter $D = 3$ cm and height $L = 6$ cm is placed on a plate heater of temperature $T_0 = 50$°C. The ambient temperature is $T_\infty = 20$°C, and the heat transfer coefficient at the outer cylindrical surface is $h = 5$ W/m^{2}K. The outer skin of the battery is made of stainless steel ($k = 15$ W/mK) sheet with thickness $t = 0.5$ mm. The interior of the battery is such a poor conductor that the heat transfer between it and the stainless steel shell can be neglected. Take the heat transfer coefficient at the top edge of the shell to be the same as that at the outer cylindrical surface.

  (a) Determine whether the stainless steel shell can be treated as a 1D fin of infinite length. (*Hint:* Check whether $mL \gg 1$.)

  (b) Calculate the temperature at the top edge of the shell, that is, at $x = L$.

2.34 Electric current $I = 500$ A flows through a carbon steel conductor of length $L = 1$ m and diameter $D = 5$ mm having an electric resistance of $R = 5 \times 10^{-4}$ ohm/m. The ambient temperature is 0°C and the heat transfer coefficient between the wire and the ambient temperature is 40 W/m^{2}K. The thermal conductivity of the conductor is 60 W/mK. Calculate the centre and surface temperatures of the cable.

2.35 A plane wall of thickness L and constant thermal conductivity k has both its boundary surfaces kept at 0°C. Heat is generated in the plate at the rate of Ax^2 W/m^3, where A is a constant and x is measured from the left end of the wall.

  (a) Develop an expression for the temperature distribution $T(x)$ in the slab.

  (b) Calculate the maximum temperature in the slab for $k = 30$ W/mK,
  $q''' = 1000 \, x^2$ W/m^3, $L = 50$ cm.

2.36 Develop an expression for the steady temperature distribution $T(r)$ in a solid sphere of radius b, in which heat is generated at a rate of

$$q''' = q_0'''\left(1 - \frac{r}{b}\right) \text{W/m}^3$$

where q_0''' is a constant and the boundary surface at $r = b$ is maintained at T_w.

2.37 Plot the logarithm of thermal conductivity versus temperature (as possible) in the range of $0°C$ to $500°C$ for the following substances: air, liquid sodium, brass, low-pressure steam, concrete, plain carbon steel, carbon dioxide gas, building brick, tin, water, glass, copper, mercury, ice, silver. See Appendix A for conductivity values.

2.38 A nuclear fuel element of spherical form consists of a fissionable material with radius r_1 surrounded by a spherical shell of aluminium cladding with outer radius r_2. Due to nuclear fission reaction thermal energy is generated in the fuel element. This heat source is given by a simple parabolic function:

$$q''' = q_0'''\left[1 + b\left(\frac{r}{r_1}\right)^2\right]$$

Here q_0''' is the rate of heat generation per unit volume at the centre of the sphere, and b is a dimensionless constant between 0 and 1. The temperature at the outer surface of the cladding (which is cooled by a coolant) is T_0. Obtain expressions for the temperature profiles in the fuel and the cladding.

2.39 A copper wire has a radius of 2 mm and a length of 5 m. For what voltage drop would the temperature rise at the wire axis be $10°C$, if the surface temperature of the wire is $20°C$? For copper, $k/k_e T_0$ (which is also called the Lorenz number) is 2.23×10^{-8} $V^2 K^{-2}$, where k is the thermal conductivity of copper (W/mK), k_e is the electrical conductivity of copper (ohm^{-1}m^{-1}) and T_0 is the surface temperature of the wire (in kelvin). The current density (A/m^2) is related to the voltage drop V over a length L as $I = k_e(V/L)$.

2.40 A heated sphere of radius R is suspended in a large, motionless body of fluid. It is desired to study the heat conduction in the fluid surrounding the sphere. It is assumed in this problem that free convection effects can be neglected. The boundary conditions are as follows:

BC-1: at $r = R$, $T = T_R$
BC-2: at $r = \infty$, $T = T_\infty$

(a) Obtain the temperature profile in the fluid.
(b) Obtain an expression for the heat flux at the surface of the sphere.
(c) Show that the dimensionless heat transfer coefficient (known as the *Nusselt number*) is given by
Nu $= hD/k = 2$, where h is the heat transfer coefficient, D is the diameter of the sphere, and k is the conductivity of the fluid. The above result is the limiting value of Nu for heat transfer from spheres at low Reynolds or Grashof numbers (see Chapters 5 and 6), e.g., for small spheres.

2.41 A copper (impure) wire 1 mm in diameter is insulated uniformly with plastic to an outer diameter of 3 mm and is exposed to surroundings at $38°C$. The heat transfer coefficient from the outer surface of the plastic to the surroundings is 8.5 W/m^{2}K. What is the maximum steady current, in amperes, that this wire can carry without heating any part of the plastic above its operating limit of $93°C$? The thermal and electrical conductivities may be assumed constant at the values listed below:

	k (W/mK)	k_e (Ω^{-1} cm^{-1})
Copper:	380	5.1×10^5
Plastic:	0.346	0

2.42 Liquefied gases are sometimes stored in well-insulated spherical vessels vented to the atmosphere.

(a) Develop an expression for the steady-state heat transfer rate through walls of such a vessel. The radii of the outer and inner walls are r_0 and r_1, respectively. It is assumed that the temperatures T_0 and T_1 are known. The thermal conductivity of the insulation varies according to the relation

$$k = k_0 + (k_1 - k_0)\left(\frac{T - T_0}{T_1 - T_0}\right)$$

where k_0 and k_1 are constants with units of thermal conductivity.

(b) Estimate the rate of evaporation (in kg/h) of liquid oxygen from a spherical container of 1.8 m inside diameter covered with 0.3 m of asbestos insulation. The following data are available:

Temperature at the inner surface of insulation $= -183\,°C$
Temperature at the outer surface of insulation $= 0\,°C$
Boiling point of $O_2 = -183\,°C$
Latent heat of vaporization of $O_2 = 1.636$ kcal/gmole
Thermal conductivity of insulation at $0\,°C = 0.156$ W/mK
Thermal conductivity of insulation at $-183\,°C = 0.125$ W/mK

2.43 Calculate the heat loss from a rectangular fin for the following conditions:

Air temperature $= 177\,°C$
Wall temperature $= 260\,°C$
Thermal conductivity of the fin $= 103.8$ W/mK
Heat transfer coefficient $= 680$ W/m^{2}K
Length of the fin $= 6$ cm
Width of the fin $= 30$ cm
Thickness of the fin $= 4$ mm

Chapter 3

Steady-state Conduction: Two- and Three-dimensional Problems

3.1 Introduction

In the preceding chapter, one-dimensional, steady-state heat conduction problems were discussed. The readers must have noted that the governing differential equations for such cases are ordinary differential equations. On the other hand, two- or three-dimensional steady-state problems necessitate the solution of partial differential equations, thus making the solution procedure lot more involved. Typical examples of two-dimensional problems are a thin plate in which temperature gradients in the z-direction are negligible or a long bar in which the thermal picture is identical in all planes parallel to the plane under consideration. Thus, $T(x, y)$ is sufficient to describe the temperature field. In a typical three-dimensional heat conduction problem, the dimensions of the body are comparable and hence the temperature field is dependent on all three coordinates. Thus, the temperature of the body is described by $T(x, y, z)$. All physical problems are actually three-dimensional problems reducible to one- or two-dimensional problems in many instances.

Another aspect in which multi-dimensional problems differ from one-dimensional problems is in terms of the direction of heat flux. While in the case of 1D heat conduction, heat flux lines always point in one direction (longitudinally, or radially), heat flux lines flare out in two or more directions in the case of multi-dimensional conduction.

3.2 Steady Two-dimensional Problems in Cartesian Coordinates

Illustration 3.1 *Consider a solid bar of rectangular cross-section as shown in Fig. 3.1. The bar is free of internal heat sources and has a constant thermal conductivity. If there are no temperature gradients in the z-direction, because the end effects can be neglected as the bar is long or its surfaces perpendicular to the z-direction at the two ends are perfectly insulated, then under steady-state conditions the temperature distribution T(x, y) in the bar must*

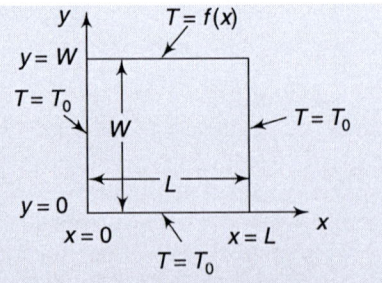

Fig. 3.1 2D heat conduction in a solid bar of rectangular cross-section

satisfy the Laplace equation in two dimensions:

$$\frac{\partial^2 T}{\partial x^2} + \frac{\partial^2 T}{\partial y^2} = 0 \tag{3.1}$$

Assume that the surfaces at x = 0, x = L, and y = 0 are maintained at T = T_0, while the temperature at the surface at y = W is given as a function of the x-coordinate, that is, T(x, W) = f(x). The boundary condition at y = W may result from non-uniform heating by a flame or exposure of a part of the surface to sunlight. The boundary conditions can then be written as follows:

$$\text{At } x = 0, \ T = T_0 \tag{3.2a}$$

$$\text{At } x = L, \ T = T_0 \tag{3.2b}$$

$$\text{At } y = 0, \ T = T_0 \tag{3.2c}$$

$$\text{At } y = W, \ T = f(x) \tag{3.2d}$$

The formulation of the problem with the second-order partial differential equation [Eq. (3.1)] and the boundary conditions [Eqs 3.2(a–d)] is now complete. The objective is to find the temperature of the bar, T(x, y), by solving the aforesaid Laplace equation, subject to the prescribed boundary conditions.

Solution

Before we proceed, let us make the boundary conditions at $x = 0$, $x = L$, and $y = 0$ homogeneous by using $\theta = T - T_0$, which transforms Eqs (3.1) and (3.2a–d) as follows:

$$\frac{\partial^2 \theta}{\partial x^2} + \frac{\partial^2 \theta}{\partial y^2} = 0 \tag{3.3}$$

$$\text{At } x = 0, \ \theta = 0 \tag{3.4a}$$

$$\text{At } x = L, \ \theta = 0 \tag{3.4b}$$

$$\text{At } y = 0, \ \theta = 0 \tag{3.4c}$$

$$\text{At } y = W, \ \theta = f(x) - T_0 \tag{3.4d}$$

We now seek a solution by the *method of separation of variables*, which requires the assumption of the existence of a product solution of the form

$$\theta(x, y) = X(x)Y(y) \tag{3.5}$$

where X is a function of x alone and Y is a function of y alone. Introducing Eq. (3.5) into Eq. (3.3), we get

$$-\frac{1}{X}\frac{d^2 X}{dx^2} = \frac{1}{Y}\frac{d^2 Y}{dy^2} \tag{3.6}$$

Since each side of Eq. (3.6) involves only one of the independent variables, they may be equal only if they are both the same constant. Calling this constant λ^2,

$$-\frac{1}{X}\frac{d^2 X}{dx^2} = \frac{1}{Y}\frac{d^2 Y}{dy^2} = \pm\lambda^2 \tag{3.7}$$

This brings us to the vital question: what sign of λ^2 is to be chosen? Before the question is answered, let us say a few words about the applicability of the method of separation of variables in the following.

The method of separation of variables is applicable to steady two-dimensional problems if and when (i) one of the directions of the problem is expressed by a homogeneous differential equation subject to homogeneous boundary conditions (the homogeneous direction) while the other direction is expressed by a homogeneous differential equation subject to one homogeneous and one non-homogeneous boundary condition (the non-homogeneous direction) and (ii) the sign of λ^2 is chosen such that the boundary-value problem of the homogeneous direction leads to a characteristic-value problem.

A boundary-value problem is a characteristic-value problem when it has particular solutions that are periodic in nature. A typical example of a characteristic equation is

$$\frac{d^2 y}{dx^2} + \lambda^2 y = 0$$

whose general solution is $y = C_1 \sin \lambda x + C_2 \cos \lambda x$.

Now, let us see what is meant by homogeneous differential equation. A linear differential equation is homogeneous when all its terms include either the unknown function or one of its derivatives. Similarly, a boundary condition is homogeneous when an unknown function or its derivatives or any linear combination of its function and its derivatives vanishes at the boundary. In the present problem, Eqs (3.4a – c) satisfy the homogeneity of boundary conditions, while Eq. (3.4d) does not. Thus, x is the homogeneous direction, while y is the non-homogeneous direction.

Now, applying the rule regarding the choice of the sign of λ^2, we readily decide on $+\lambda^2$ as the homogeneous x-direction leads to a characteristic-value problem. Thus, we have

$$-\frac{1}{X}\frac{d^2 X}{dx^2} = \frac{1}{Y}\frac{d^2 Y}{dy^2} = +\lambda^2 \tag{3.8}$$

$$\frac{d^2 X}{dx^2} + \lambda^2 X = 0 \tag{3.9}$$

$$\frac{d^2 Y}{dy^2} - \lambda^2 Y = 0 \tag{3.10}$$

The general solutions of Eqs (3.9) and (3.10) are, respectively,

$$X = B_3 \sin \lambda x + B_4 \cos \lambda x \tag{3.11}$$

$$Y = B_1 \sinh \lambda y + B_2 \cosh \lambda y \tag{3.12}$$

Substituting X and Y in the product solution [Eq. (3.5)], we obtain

$$\theta = (B_1 \sinh \lambda y + B_2 \cosh \lambda y)(B_3 \sin \lambda x + B_4 \cos \lambda x) \tag{3.13}$$

Now, imposing the boundary condition [Eq. (3.42)], that is, at $y = 0$, $\theta = 0$, we get

$$0 = B_2(B_3 \sin \lambda x + B_4 \cos \lambda x)$$

Therefore, $B_2 = 0$. Using Eq. (3.4a), that is, at $x = 0$, $\theta = 0$,

$$0 = (B_1 \sinh \lambda y) B_4$$

Therefore, $B_4 = 0$. Hence, $\theta = (\sinh \lambda y) B \sin \lambda x$, where $B = B_1 B_3$. At $x = L$, $\theta = 0$ [Eq. (3.4b)], therefore, $0 = (\sinh \lambda y) B \sin \lambda L$. The only way that this may be satisfied for all values of y is for

$$\sin \lambda L = 0$$

or $\quad \sin \lambda L = \sin n\pi$

where $n = 0, 1, 2, 3, \ldots,$

or $\quad \lambda = \dfrac{n\pi}{L}$ $\qquad\qquad\qquad\qquad\qquad\qquad\qquad$ (3.14)

Note that negative integers of n have been discarded because they do not give any new solution. Each of the λ's in Eq. (3.14) gives rise to a separate solution of $\theta = (\sinh \lambda y)\, B \sin \lambda x$ and since the general solution will be the sum of the individual solutions (since the problem is linear), one has

$$\theta = \sum_{n=0}^{\infty} B_n (\sinh \lambda_n y)(\sin \lambda_n x) \qquad\qquad\qquad (3.15)$$

The symbol B_n represents the constant B for each solution. Since $\lambda_n = 0$ for $n = 0$, no contribution is made by the first term. Therefore,

$$\theta = \sum_{n=1}^{\infty} B_n (\sinh \lambda_n y)(\sin \lambda_n x) \qquad\qquad\qquad (3.16)$$

Evaluation of B_n

Applying the last boundary condition [Eq. (3.4d)], that is, at $y = W$, $\theta = f(x) - T_\infty$, we have

$$f(x) - T_0 = \sum_{n=1}^{\infty} B_n \sinh \lambda_n W \sin \lambda_n x \qquad\qquad (3.17)$$

$$\lambda_n = \frac{n\pi}{L}, \quad n = 1, 2, 3, \ldots$$

$$0 \le x \le L$$

Before we detail the method of evaluation of B_n, it might be worthwhile to recall the definition of orthogonal functions. Given an infinite set of functions, that is, $g_1(x), g_2(x), g_3(x), \ldots, g_n(x), \ldots, g_m(x)$, the functions are termed orthogonal in the interval $a \le x \le b$, if

$$\int_a^b g_m(x)\, g_n(x)\, dx = 0, \quad \text{for } m \ne n \qquad\qquad (3.18)$$

The word *orthogonality* comes from vector analysis. Let $g_m(x_i)$ denote a vector in 3D space whose rectangular components are $g_m(x_1), g_m(x_2)$, and $g_m(x_3)$. Two vectors $g_m(x_i)$ and $g_n(x_i)$ are said to be orthogonal or perpendicular to each other, if

$$g_m(x_i) \cdot g_n(x_i) = \sum_{i=1}^{3} g_m(x_i) g_n(x_i) = 0 \qquad\qquad (3.19)$$

If $f(x)$ denotes an arbitrary function, consider the possibility of expressing it as a linear combination of the orthogonal functions:

$$f(x) = C_1 g_1(x) + C_2 g_2(x) + \cdots + C_n g_n(x) + \cdots + C_m g_m(x) + \cdots$$

$$= \sum_{n=1}^{\infty} C_n g_n(x) \qquad\qquad\qquad\qquad (3.20)$$

where C's are constants to be determined. Multiplying both sides of Eq. (3.20) by $g_n(x)$ and integrating between the limits $x = a$ and $x = b$,

$$\int_a^b f(x) g_n(x)\, dx = C_1 \int_a^b g_1(x) g_n(x)\, dx + C_2 \int_a^b g_2(x) g_n(x)\, dx + \cdots$$

$$+ C_n \int_a^b g_n^2(x)\, dx + \cdots + C_m \int_a^b g_m(x) g_n(x)\, dx$$

or $\quad C_n \int_a^b g_n^2(x)\, dx = \int_a^b f(x) g_n(x)\, dx$

Therefore, $\quad C_n = \dfrac{\displaystyle\int_a^b f(x)\, g_n(x)\, dx}{\displaystyle\int_a^b g_n^2(x)\, dx}$ $\qquad\qquad$ (3.21)

Now, consider the following set of functions in the interval $0 \le x \le L$:

$$\sin \frac{\pi x}{L}, \sin \frac{2\pi x}{L}, \sin \frac{3\pi x}{L}, \ldots, \sin \frac{n\pi x}{L}, \ldots$$

These may also be expressed as

$$\sin \lambda_1 x, \sin \lambda_2 x, \sin \lambda_3 x, \ldots, \sin \lambda_n x, \ldots$$

$$\lambda_n = n\pi/L, \quad n = 1, 2, 3, \ldots,$$

It can be easily shown that

$$\int_0^L \sin \lambda_n x \sin \lambda_m x\, dx = 0$$

Thus, $\sin \lambda_n x$ is an orthogonal set. Returning to Eq. (3.20), if we replace $g_n(x)$ by $\sin \lambda_n x$ [or $\sin (n\pi/L)x$], then

$$f(x) = \sum_{n=1}^{\infty} C_n \sin \left(\frac{n\pi}{L} \right) x$$

where $0 \le x \le L$. The above is the *Fourier sine series* of $f(x)$ over an interval $(0, L)$. Let us now go back to Eq. (3.17), which reads

$$f(x) - T_0 = \sum_{n=1}^{\infty} B_n \sinh \lambda_n W \sin \lambda_n x$$

where $\quad \lambda_n = \dfrac{n\pi}{L}, \quad n = 1, 2, 3, \ldots$

$$0 \le x \le L$$

Taking $B_n \sinh \lambda_n W = C_n$, we can write, using Eq. (3.21),

$$C_n = B_n \sinh \lambda_n W = \frac{\displaystyle\int_0^L [f(x) - T_0] \sin \lambda_n x\, dx}{\displaystyle\int_0^L \sin^2 \lambda_n x\, dx} \qquad\qquad (3.22)$$

Now, using integration by parts, it can be shown that

$$\int_0^L \sin^2 \lambda_n x\, dx = \frac{L}{2}$$

Therefore, $C_n = \dfrac{2}{L}\displaystyle\int_0^L [f(x) - T_0]\sin \lambda_n x\, dx$

$$= B_n \sinh \lambda_n W$$

Therefore, $B_n = \dfrac{2}{L}\dfrac{1}{\sinh \lambda_n W}\displaystyle\int_0^L [f(x) - T_0]\sin \lambda_n x\, dx$ $\qquad$ (3.23)

Substituting B_n from Eq. (3.23) into Eq. (3.16), we get

$$\theta = \frac{2}{L}\sum_{n=1}^{\infty}\frac{\left[\displaystyle\int_0^L [f(x) - T_0]\sin \lambda_n x\, dx\right]}{\sinh \lambda_n W}\sin \lambda_n x \sinh \lambda_n y \qquad (3.24)$$

where $\lambda_n = n\pi/L$, $n = 1, 2, 3, \ldots$

The edge at $y = W$ at a uniform temperature Let $f(x) = T_c$ (Fig. 3.2). From Eq. (3.24),

$$\theta = T - T_0 = \frac{2}{L}\sum_{n=1}^{\infty}\frac{\sinh\left(\dfrac{n\pi y}{L}\right)}{\sinh\left(\dfrac{n\pi W}{L}\right)}\sin\left(\frac{n\pi x}{L}\right)\int_0^L (T_c - T_0)\sin\frac{n\pi x}{L}\, dx$$

Therefore, $\dfrac{T - T_0}{T_c - T_0} = \dfrac{2}{L}\displaystyle\sum_{n=1}^{\infty}\dfrac{\sinh\left(\dfrac{n\pi y}{L}\right)}{\sinh\left(\dfrac{n\pi W}{L}\right)}\sin\left(\dfrac{n\pi x}{L}\right)\int_0^L \sin\dfrac{n\pi x}{L}\, dx$

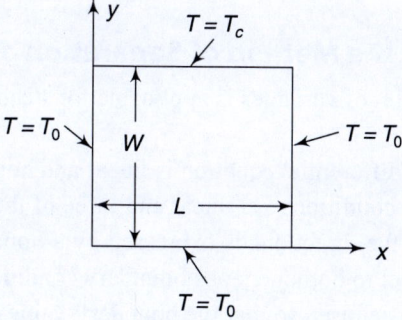

Fig. 3.2 The upper edge at $T = T_c$, all other edges being at $T = T_0$

Now, $\displaystyle\int_0^L \sin\frac{n\pi x}{L}\, dx = \left.-\frac{\cos\dfrac{n\pi x}{L}}{\dfrac{n\pi}{L}}\right|_0^L = -\frac{L}{n\pi}[\cos n\pi - \cos 0]$

$$= \frac{L}{n\pi}[1 - (-1)^n]$$

Therefore, $\dfrac{T - T_0}{T_c - T_0} = 2\displaystyle\sum_{n=1}^{\infty}\dfrac{1 - (-1)^n}{n\pi}\dfrac{\sinh\left(\dfrac{n\pi y}{L}\right)}{\sinh\left(\dfrac{n\pi W}{L}\right)}\sin\left(\dfrac{n\pi x}{L}\right)$ $\qquad$ (3.25)

It may be noted that except for very small values of y/L, the series in Eq. (3.25) converges rapidly, and only the first few terms would be sufficient to calculate the temperature at any point numerically. The following example illustrates this.

Example 3.1   Centre-line Temperature in a Rod of Rectangular Cross-section

Referring to Fig. 3.2, let us take $T_0 = 0\,°C$ and $T_c = 100\,°C$. Assuming $L = 2W$, calculate the centre temperature.

Solution

At the centre,

$$x = L/2, \quad y = W/2 = L/4$$

Using Eq. (3.25),

$$T\left(\frac{L}{2}, \frac{L}{4}\right) = \frac{2 \times 100}{\pi} \sum_{n=1}^{\infty} \frac{1-(-1)^n}{n} \frac{\sinh\left(\dfrac{n\pi}{4}\right)\sin\left(\dfrac{n\pi}{2}\right)}{\sinh\left(\dfrac{n\pi}{2}\right)}$$

$$= 48.061 - 3.987 + 0.502 - \ldots$$

$$= 44.576\,°C$$

It is evident from the above that computation of only three terms is necessary as the series converges rather rapidly at $x = L/2$ and $y = L/4$.

Test for convergence

The readers may recall that an infinite series $\sum\limits_{n=1}^{\infty} a_n$ converges if $a_n \to 0$ as n increases and also the ratio a_{n+1}/a_n is less than 1 in absolute value.

3.3   Summary of the Method of Separation of Variables

The method of separation of variables is applicable for steady 2D problems if and when

(a) the governing differential equation is linear and homogeneous.

(b) four boundary conditions are linear and three of them are homogeneous, so that one of the directions is expressed by a homogeneous differential equation subject to homogeneous boundary conditions.

(c) the sign of λ^2 is chosen so that the boundary-value problem in the homogeneous direction becomes a characteristic-value problem.

It may be noted that characteristic functions form an orthogonal set (the mathematical proof not given here). Since the homogeneous direction gives a characteristic equation, orthogonality always exists in the homogeneous direction. Furthermore, separation of variables requires finiteness in the homogeneous direction.

3.4   Isotherms and Heat Flux Lines

In a two-dimensional problem the temperature field and heat transfer can be best visualised by drawing isothermal lines (or isotherms) and heat flux lines in the computational plane. The method may be extended to three-dimensional problems by drawing isothermal surfaces.

Consider a solid with a distribution of temperature at a time t given by $T = T(x, y, z, t)$. A surface may exist in the solid such that at every point upon it the temperature at this instant is the same, say T_1. Such a surface is called *isothermal surface* for temperature T_1, and it may be thought of as separating the parts of a body which are hotter than T_1 from the parts which are colder than T_1. *No two isotherms can cut each other, since no part of a body can have two temperatures at the same time.* The solid is thus pictured as divided up into thin shells of its isotherms. In a 2D problem, since the temperature is invariant in the z-direction, drawing of isothermal lines in the x–y plane is sufficient.

In an isotropic solid heat spreads out equally in all directions as opposed to crystalline and anisotropic solids, in which certain directions are more favourable for the conduction of heat than others. Because of the symmetry in isotropic solids, the flux vector at a point must be along the normal to the isothermal surface through the point, and in the direction of falling temperature. In other words, *heat flux lines and isotherms intersect at right angles.*

Figure 3.3 shows a qualitative sketch of isotherms and heat flux lines for the problem discussed in Example 3.1. The directions of heat flux vectors are represented by heat flux lines. The heat flux lines (also called heat tubes) are denser near the upper-left and upper-right corners, because in those corner regions, two isothermal boundaries intersect (note the finite temperature difference between two intersecting boundaries).

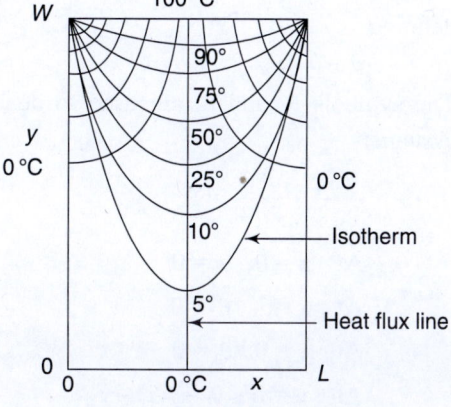

The total heat transfer rate through the $y = W$ side is obtained by integrating the heat flux from $x = 0$ to $x = L$:

$$q' = +k \int_0^L \left(\frac{\partial T}{\partial y} \right)_{y=W} dx \tag{3.26}$$

Fig. 3.3 Isotherms and heat flux lines in a 2D rectangular domain

This quantity is expressed in W/m. Note the positive sign as temperature increases in the direction of increasing y.

3.5 Method of Superposition

The method of superposition is used when the separation of variables method cannot be directly applied because (i) both the boundary conditions in one or more directions are non-homogeneous and neither of them is made homogeneous by any transformation, or (ii) the governing equation is linear but non-homogeneous. In such cases, the main problem is divided into several sub-problems so that the solution of each sub-problem is added to each other to obtain the desired solution. Generally, each sub-problem can be solved by the separation of variables approach.

3.5.1   Rectangular Plate with a Specified Temperature Distribution on More than One Edge

With respect to the boundary conditions shown in Fig. 3.4, the governing differential equation for the 2D heat conduction problem is

$$\frac{\partial^2\theta}{\partial x^2} + \frac{\partial^2\theta}{\partial y^2} = 0 \qquad (3.27)$$

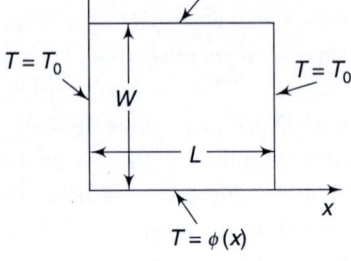

where $\theta = T - T_0$. The boundary conditions are as follows:

At $x = 0,\ \theta = 0$ (3.28a)

At $x = L,\ \theta = 0$ (3.28b)

At $y = 0,\ \theta = \phi(x) - T_0$ (3.28c)

At $y = W,\ \theta = f(x) - T_0$ (3.28d)

Fig. 3.4   2D rectangular plate with non-homogeneous boundary conditions in the y-direction

Since, Eq. (3.27) is linear, it can be reduced to two simpler sub-problems by defining

$$\theta = \theta_1 + \theta_2 \qquad (3.29)$$

The symbols θ_1 and θ_2 are used to denote the solutions to the following two systems:

$$\frac{\partial^2\theta_1}{\partial x^2} + \frac{\partial^2\theta_1}{\partial y^2} = 0 \qquad (3.30)$$

At $x = 0,\ \theta_1 = 0$ (3.31a)

At $x = L,\ \theta_1 = 0$ (3.31b)

At $y = 0,\ \theta_1 = 0$ (3.31c)

At $y = W,\ \theta_1 = f(x) - T_0$ (3.31d)

$$\frac{\partial^2\theta_2}{\partial x^2} + \frac{\partial^2\theta_2}{\partial y^2} = 0 \qquad (3.32)$$

At $x = 0,\ \theta_2 = 0$ (3.33a)

At $x = L,\ \theta_2 = 0$ (3.33b)

At $y = 0,\ \theta_2 = \phi(x) - T_0$ (3.33c)

At $y = W,\ \theta_2 = 0$ (3.33d)

The pictorial representation of the solution methodology is given in Fig. 3.5. Note that both sub-problems can now be solved by the method of separation of variables. The solution of Eq. (3.30) can be written straight from Eq. (3.24). Therefore,

$$\theta_1 = \frac{2}{L}\sum_{n=1}^{\infty} \frac{\sinh\left(\dfrac{n\pi y}{L}\right)}{\sinh\left(\dfrac{n\pi W}{L}\right)} \sin\left(\frac{n\pi x}{L}\right) \int_0^L [f(x) - T_0]\sin\frac{n\pi x}{L}\,dx \qquad (3.34)$$

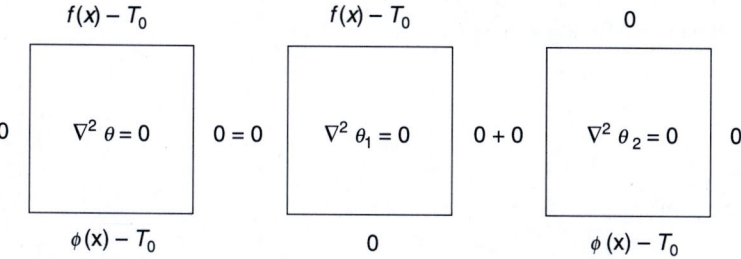

Fig. 3.5 Pictorial representation of the method of superposition

The solution of Eq. (3.32) can be arrived at by transforming the y-coordinate, since the non-homogeneous boundary condition appears on the lower side of the rectangle in this case. If $y' = W - y$, then we have the following:

$$\text{At} \quad y = W, y' = 0, \quad \theta_2 = 0$$
$$\text{At} \quad y = 0, y' = W, \quad \theta_2 = \phi(x) - T_0$$

Therefore,

$$\theta_2 = \frac{2}{L} \sum_{n=1}^{\infty} \frac{\sinh\left(\dfrac{n\pi(W-y)}{L}\right)}{\sinh\left(\dfrac{n\pi W}{L}\right)} \sin\left(\frac{n\pi x}{L}\right) \int_0^L [\phi(x) - T_0] \sin\frac{n\pi x}{L}\, dx \quad (3.35)$$

The final solution is

$$\theta = \theta_1 + \theta_2$$

where θ_1 and θ_2 are obtained from Eqs (3.34) and (3.35), respectively.

Heat transfer configurations with more complicated boundary conditions than in Fig. 3.4 may require using more than two sub-solutions ($\theta_1, \theta_2, \theta_3, ...$) to construct the desired temperature field (θ). The readers will gain confidence in this respect by trying to solve various exercise problems in this area at the end of this chapter.

3.5.2 2D Heat Conduction with Uniform Heat Generation

Illustration 3.2 *Consider uniform heat generation (q''') in a 2D rectangular bar of ($2L \times 2l$) cross-section. The sides of the bar are exposed to a fluid having temperature T_∞. The heat transfer coefficient is large. Find the steady temperature distribution $T(x, y)$ in the bar. See Fig. 3.6.*

Solution

Since the heat transfer coefficient is large, the boundaries of the bar will assume the temperature of the fluid to be T_∞. As the problem is thermally and geometrically symmetric, computation in only one-quarter of the physical domain is sufficient. In the present case, we take the upper-right quarter of the rectangle. The coordinate system is placed at the centre of the physical

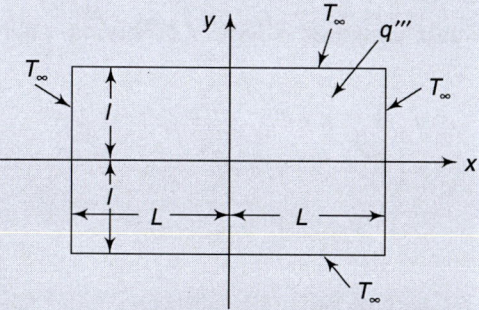

Fig. 3.6 The physical domain of the rectangular bar with heat generation

domain [i.e., the lower-left corner of the computational domain (Fig. 3.7)]. The governing differential equation for the problem is

$$\frac{\partial^2 \theta}{\partial x^2} + \frac{\partial^2 \theta}{\partial y^2} + \frac{q'''}{k} = 0 \qquad (3.36)$$

where $\theta = T - T_\infty$. The boundary conditions are as follows:

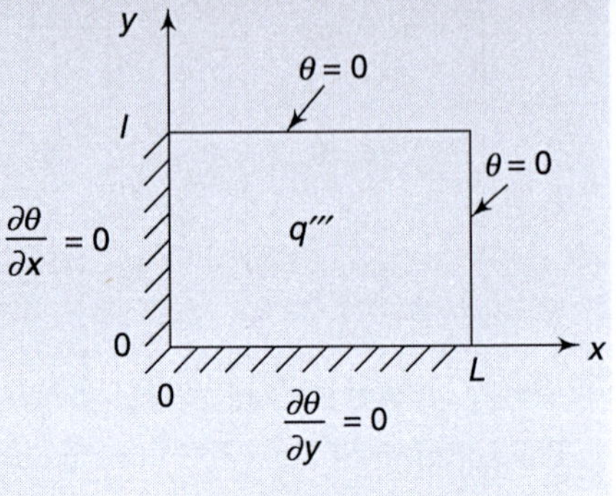

At $x = 0$, $\dfrac{\partial \theta}{\partial x} = 0$ at any y

$$(3.37a)$$

At $x = L$, $\theta = 0$ at any y

$$(3.37b)$$

At $y = 0$, $\dfrac{\partial \theta}{\partial y} = 0$ at any x $\qquad\qquad (3.37c)$

At $y = l$, $\theta = 0$ at any x $\qquad\qquad (3.37d)$

Fig. 3.7 The computational domain and the coordinate system

Since Eq. (3.36) is non-homogeneous, it is not separable. The non-homogeneity arises from the q'''/k term. The solution of the problem is now assumed to be

$$\theta(x, y) = \psi(x, y) + \phi(x) \qquad\qquad (3.38)$$

$$\theta(x, y) = \psi(x, y) + \phi(y) \qquad\qquad (3.39)$$

The use of either of these forms is arbitrary in this case.

Now, including the heat generation term q''' in the formulation of the 1D problem, $\phi(x)$ or $\phi(y)$, we will see that the governing equation for the 2D problem, $\psi(x, y)$, can be made homogeneous. Thus, $\psi(x, y)$ will be suitable for separation of variables. However, the complete formulation of ϕ, say, $\phi(x)$ and of $\psi(x, y)$ requires that the boundary conditions of these be specified. Hence,

$$\frac{d^2\phi}{dx^2} + \frac{q'''}{k} = 0 \qquad\qquad (3.40)$$

subject to

$$\frac{d\phi}{dx}(0) = 0 \qquad\qquad (3.41a)$$

$$\phi(L) = 0 \qquad\qquad (3.41b)$$

Substituting Eqs (3.38) and (3.40) into Eq. (3.36), we get

$$\frac{\partial^2 \psi}{\partial x^2} + \frac{\partial^2 \psi}{\partial y^2} + \left(\underbrace{\frac{d^2\phi}{dx^2} + \frac{q'''}{k}}_{0} \right) = 0$$

or $$\frac{\partial^2 \psi}{\partial x^2} + \frac{\partial^2 \psi}{\partial y^2} = 0 \qquad\qquad (3.42)$$

The corresponding boundary conditions for $\psi(x, y)$ problem are

$$\frac{\partial \psi}{\partial x}(0, y) = \frac{\partial \theta}{\partial x}(0, y) - \frac{d\phi}{dx}(0) = 0 - 0 = 0 \qquad\qquad (3.43a)$$

$$\psi(L, y) = \theta(L, y) - \phi(L) = 0 - 0 = 0 \tag{3.43b}$$

$$\frac{\partial \psi}{\partial y}(x, 0) = \frac{\partial \theta}{\partial y}(x, 0) = 0 \tag{3.43c}$$

$$\psi(x, l) = \theta(x, l) - \phi(x) = 0 - \phi(x) = -\phi(x) \tag{3.43d}$$

Thus, the solution of the non-separable problem of $\theta(x, y)$ is the superposition of the solutions of the separable problem $\psi(x, y)$ and the problem of plane wall with heat generation $\phi(x)$. Now, the solution of the 1D problem, that is, the solution of Eq. (3.40) subject to the boundary conditions Eqs (3.41a) and (3.41b), is

$$\phi(x) = \frac{1}{2} \frac{q'''}{k} L^2 \left[1 - \left(\frac{x}{L} \right)^2 \right] \tag{3.44}$$

Next, we have to find the solution of the $\psi(x, y)$ problem using the method of separation of variables. It is clear that x is the homogeneous direction. The details of the solution methodology will not be repeated here. Finally,

$$\psi(x, y) = -2 \frac{q''' L^2}{k} \sum_{n=0}^{\infty} \frac{(-1)^n}{(\lambda_n L)^3} \left(\frac{\cosh \lambda_n y}{\cosh \lambda_n l} \right) \cos \lambda_n x \tag{3.45}$$

where $\lambda_n L = (2n + 1)(\pi/2)$, $n = 0, 1, 2, \dots$. Therefore, the solution is

$$\frac{\theta(x, y)}{\dfrac{q''' L^2}{k}} = \frac{1}{2} \left[1 - \left(\frac{x}{L} \right)^2 \right] - 2 \sum_{n=0}^{\infty} \frac{(-1)^n}{(\lambda_n L)^3} \left(\frac{\cosh \lambda_n y}{\cosh \lambda_n l} \right) \cos \lambda_n x \tag{3.46}$$

where $\lambda_n L = (2n + 1)(\pi/2)$ and $n = 0, 1, 2, \dots$.

3.6 Method of Imaging

Another technique, called the method of imaging, is used to solve a class of 2D heat conduction problems. The concept is demonstrated with the following illustration.

Illustration 3.3 *Consider an infinitely long rod of triangular cross-section. One surface of the rod has a uniform temperature T_0, the second zero, while the hypotenuse is insulated (Fig. 3.8). It is desired to find the steady temperature at point P of the hypotenuse.*

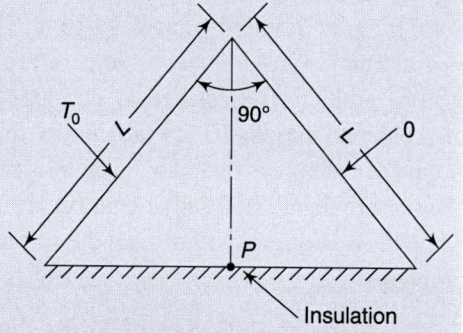

Fig. 3.8 Infinitely long rod of triangular cross-section and the boundary conditions

Solution

Since the triangle is right-angled isosceles and its hypotenuse is insulated, the problem can be modelled as if the heat conduction is occurring in an infinite rod of square cross-section, having thermal and geometric symmetry about one of its diagonals (Fig. 3.9).

It is evident that both x and y directions are non-homogeneous. Since the governing equation $\nabla^2 T = 0$ is linear, the method of superposition can be used to split the problem into two sub-problems, each of which can be individually solved by the separation of variables method. The solution procedure is visually depicted in Fig. 3.10. Therefore, $T = T_1 + T_2$. Hence, the steady temperature at the location P, that is, at $(L/2, L/2)$, can be easily found out.

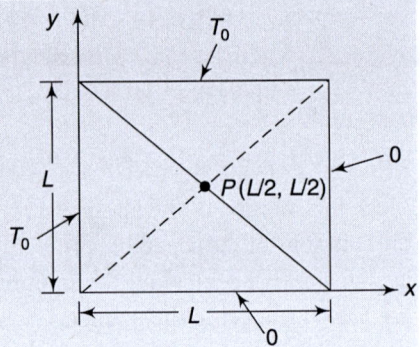

Fig. 3.9 Equivalent representation of Fig. 3.8 by the method of imaging

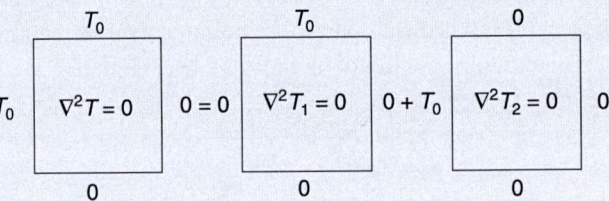

Fig. 3.10 Visual depiction of the solution procedure

3.7 Steady 2D Problems in Cylindrical Geometry

The heat conduction equation in the cylindrical coordinates with constant thermal conductivity under steady-state conditions and without heat sources is given by

$$\frac{\partial^2 T}{\partial r^2} + \frac{1}{r}\frac{\partial T}{\partial r} + \frac{1}{r^2}\frac{\partial^2 T}{\partial \phi^2} + \frac{\partial^2 T}{\partial z^2} = 0 \tag{3.47}$$

where $T = T(r, \phi, z)$. The above relates to a three-dimensional heat conduction problem. It is obvious that two-dimensional problems occur when $T = T(r, \phi)$ or $T = T(r, z)$ or $T = T(\phi, z)$. The $T(r, \phi)$ problem arises when there is a circumferential surface temperature or surface heat flux variation. This type of problem is classified as non-axisymmetric problem. The second group of problems, that is, $T(r, z)$ falls under the category of axisymmetric problems. A typical example is that of a 2D cylindrical fin. The third category, that is, $T(\phi, z)$ class of problems, has really no physical significance except in thin-walled tubes.

In this section, we will discuss the solution methodologies of $T(r, z)$ and $T(r, \phi)$ problems.

3.7.1 Circular Cylinder of Finite Length Having no Circumferential Variation of Temperature: $T(r, z)$ Problem

Illustration 3.4 *A circular cylinder of finite length refers to a short cylinder, that is, a cylinder whose L/D < 3. The physical problem and the boundary conditions are shown in Fig. 3.11. The objective is to find out T(r, z).*

Solution The governing differential equation for the problem is

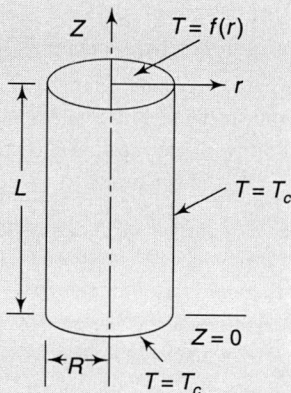

$$\frac{\partial^2 \theta}{\partial r^2} + \frac{1}{r}\frac{\partial \theta}{\partial r} + \frac{\partial^2 \theta}{\partial z^2} = 0 \qquad (3.48)$$

where $\theta = T - T_c$. The boundary conditions are as follows:

At $z = 0$, $T = T_c$ or $\theta = 0$ (3.49a)

At $z = L$, $T = f(r)$ or $\theta = f(r) - T_c$ (3.49b)

At $r = R$, $T = T_c$ or $\theta = 0$ (3.49c)

At $r = 0$, $T = $ finite or $\theta = $ finite

$$\frac{\partial T}{\partial r} = 0 \quad \text{or} \quad \frac{\partial \theta}{\partial r} = 0 \qquad (3.49d)$$

Fig. 3.11 The geometry and the boundary conditions for the $T(r, z)$ problem

From the above equations, it can be inferred that r is the homogeneous direction as at $r = R$, $\theta = 0$ and at $r = 0$, $\partial\theta/\partial r = 0$. The last boundary condition, that is, Eq. (3.49d) which says that at $r = 0$, $\partial\theta/\partial r = 0$ arises from the fact that the problem is axisymmetric as there is no circumferential variation of temperature.

Now, since the governing equation is linear and homogeneous and the boundary condition in one direction (the r-direction) is homogeneous while that in the other direction (the z-direction) is non-homogeneous, the problem can be solved by the method of separation of variables. The temperature θ can be written as a product solution of the form

$$\theta = \mathcal{R}(r)\, Z(z) \qquad (3.50)$$

Substitution of Eq. (3.50) into Eq. (3.48) results in

$$\frac{1}{\mathcal{R}}\frac{d^2\mathcal{R}}{dr^2} + \frac{1}{\mathcal{R}}\frac{1}{r}\frac{d\mathcal{R}}{dr} = -\frac{1}{Z}\frac{d^2 Z}{dz^2} = \pm\lambda^2 \qquad (3.51)$$

Since r is the homogeneous direction, orthogonality exists in the r-direction. Taking negative sign for λ^2 gives a characteristic equation (Bessel's equation of zero order) in the r-direction:

$$\frac{1}{\mathcal{R}}\frac{d^2\mathcal{R}}{dr^2} + \frac{1}{\mathcal{R}}\frac{1}{r}\frac{d\mathcal{R}}{dr} = -\frac{1}{Z}\frac{d^2 Z}{dz^2} = -\lambda^2 \qquad (3.52)$$

Therefore, in the r-direction, we get

$$\frac{d^2\mathcal{R}}{dr^2} + \frac{1}{r}\frac{d\mathcal{R}}{dr} + \lambda^2 \mathcal{R} = 0$$

or $$r\frac{d}{dr}\left(r\frac{d\mathcal{R}}{dr}\right) + \lambda^2 r^2 \mathcal{R} = 0 \qquad (3.53)$$

In the z-direction, we obtain

$$\frac{d^2 Z}{dz^2} - \lambda^2 Z = 0 \qquad (3.54)$$

The general solution of Eq. (3.53) is

$$\mathcal{R}(r) = B_1 J_0(\lambda r) + B_2 Y_0(\lambda r) \qquad (3.55)$$

The general solution of Eq. (3.54) is

$$Z(z) = B_3 \sinh \lambda z + B_4 \cosh \lambda z \qquad (3.56)$$

Therefore, from Eq. (3.50),

$$\theta = [B_1 J_0 (\lambda r) + B_2 Y_0 (\lambda r)][B_3 \sinh \lambda z + B_4 \cosh \lambda z] \qquad (3.57)$$

where $J_0(\lambda r)$ and $Y_0(\lambda r)$ are the Bessel functions of the first and second kind of order zero, respectively. See Appendix A2 for tabular listing of these functions. It might be worthwhile to recall at this stage the general form of a Bessel equation. Equation (3.58) is a Bessel equation:

$$x \frac{d}{dx}\left(x \frac{dy}{dx} \right) + (m^2 x^2 - k^2) y = 0 \qquad (3.58)$$

where m is a parameter and k may be zero, a fractical number, or an integer. Note that k is non-negative. The general solution of Eq. (3.58) is

$$y(x) = A J_k (mx) + B Y_k (mx) \qquad (3.59)$$

Comparing Eq. (3.53) with Eq. (3.58), we see that

$$y = \Re, \quad x = r, \quad m = \lambda, \quad k = 0$$

Therefore, the solution of Eq. (3.53) can be written as

$$\Re(r) = B_1 J_0 (\lambda r) + B_2 Y_0 (\lambda r)$$

Now, coming back to the task of finding the constants $B_1, B_2, B_3,$ and B_4, we apply the first boundary condition, that is, Eq. (3.49a):

$$0 = [B_1 J_0 (\lambda r) + B_2 Y_0 (\lambda r)] [B_4]$$

Therefore, $B_4 = 0$. Applying the last boundary condition, that is, Eq. (3.49d) which says, at $r = 0$, $\theta =$ finite, we get

$$\text{finite} = [B_1 J_0 (\lambda r) + B_2 Y_0 (\lambda r)] [B_3 \sinh \lambda z]$$

Since $Y_0(0) \to -\infty$, $B_2 = 0$. Therefore, $\theta = B_1 B_3 \sinh \lambda z \, J_0 (\lambda r)$

or $\qquad \theta = B \sinh \lambda z \, J_0 (\lambda r) \qquad (3.60)$

where $B = B_1 B_3$. Applying the third boundary condition, that is, Eq. (3.49c), we obtain

$$B \sinh \lambda z \, J_0 (\lambda R) = 0 \qquad (3.61)$$

The only way Eq. (3.61) can be satisfied for all values of z between 0 and L is for

$$J_0(\lambda R) = 0 \qquad (3.62)$$

Examination of tables of $J_0(\lambda R)$ shows that J_0 has a succession of zeros that differ by an interval approaching π as $\lambda R \to \infty$. Hence, there are an infinite number of λ's satisfying the defining relation. Thus,

$$J_0(\lambda_n R) = 0 \qquad (3.63)$$

The first five are $\lambda_1 R = 2.4048$, $\lambda_2 R = 5.5201$, $\lambda_3 R = 8.6537$, $\lambda_4 R = 11.7915$, and $\lambda_5 R = 14.9309$. The successive differences are 3.1153, 3.1336, 3.1378, 3.1394, which approach π as $\lambda R \to \infty$. Hence, the general solution is the sum of all the solutions corresponding to each of the λ_n's:

$$\theta = \sum_{n=1}^{\infty} (B_n \sinh \lambda_n z) \, J_0 (\lambda_n r) \qquad (3.64)$$

Finally, the application of the second boundary condition, that is, Eq. (3.49b), results in

$$f(r) - T_c = \sum_{n=1}^{\infty} (B_n \sinh \lambda_n L) \, J_0 (\lambda_n r) \qquad (3.65)$$

Note that, as always, the non-homogeneous boundary condition is applied last. Now, recall the Fourier–Bessel series which is of the form

$$f(x) = \sum_{n=1}^{\infty} C_n \, J_0 (\lambda_n x), \qquad 0 \le x \le R \qquad (3.66)$$

where $\quad C_n = \dfrac{\displaystyle\int_0^R x\, f(x)\, J_0\,(\lambda_n x)\, dx}{\displaystyle\int_0^R x\, J_0^2\,(\lambda_n x)\, dx}$ $\qquad\qquad$ (3.67)

Comparing Eq. (3.65) with Eq. (3.66),

$$f(x) = f(r) - T_c$$

$$x = r$$

Therefore, $\quad C_n = \dfrac{\displaystyle\int_0^R x\,[\,f(r) - T_c\,]\, J_0\,(\lambda_n r)\, dr}{\displaystyle\int_0^R r\, J_0^2\,(\lambda_n r)\, dr}$ $\qquad\qquad$ (3.68)

The denominator of Eq. (3.68) can be evaluated by integration by parts[1] and using differentiation and integration rules of Bessel functions. For ready reference, some rules are given below.

$$\frac{d}{dx}\,[\,J_0\,(\lambda_n x)\,] \;=\; -\lambda_n J_1\,(\lambda_n x)$$

$$\frac{d}{dx}\,[\,x J_1\,(\lambda_n x)\,] \;=\; (\lambda_n x)\, J_0\,(\lambda_n x)$$

$$\int J_1\,(\lambda_n x)\, dx \;=\; -\frac{1}{\lambda_n}\, J_0\,(\lambda_n x)$$

$$\int x\, J_0\,(\lambda_n x)\, dx \;=\; \frac{x}{\lambda_n}\, J_1\,(\lambda_n x)$$

Hence, $\quad \displaystyle\int_0^R r\, J_0^2\,(\lambda_n r)\, dr \;=\; \frac{R^2}{2}\, J_1^2\,(\lambda_n R)$ $\qquad\qquad$ (3.69)

Therefore, $\quad C_n = B_n \sinh \lambda_n L = \dfrac{\displaystyle\int_0^R r\,[\,f(r) - T_c\,]\, J_0\,(\lambda_n r)\, dr}{\dfrac{R^2}{2}\, J_1^2\,(\lambda_n R)}$

$$B_n = \dfrac{\displaystyle\int_0^R r\,[\,f(r) - T_c\,]\, J_0\,(\lambda_n r)\, dr}{(\sinh \lambda_n L)\,\dfrac{R^2}{2}\, J_1^2\,(\lambda_n R)} \qquad\qquad (3.70)$$

Hence, the final solution is found by substituting B_n in Eq. (3.64). Therefore,

$$\theta = \frac{2}{R^2} \sum_{n=1}^{\infty} \frac{\sinh \lambda_n z}{\sinh \lambda_n L}\, \frac{J_0\,(\lambda_n r)}{J_1^2\,(\lambda_n R)}\, \int_0^R r\,[\,f(r) - T_c\,]\, J_0\,(\lambda_n r)\, dr \qquad (3.71)$$

---

[1] $\displaystyle\int uv\, dx = u \int v\, dx - \int \left(\frac{du}{dx} \int v\, dx\right) dx$

One end at uniform temperature

Referring to Fig. 3.11, let $f(R) = T_0$. Then, the integral in Eq. (3.71) is evaluated as

$$\int_0^R r\,[T_0 - T_c]\,J_0\,(\lambda_n r)\,dr \;=\; [T_0 - T_c]\int_0^R r\,J_0\,(\lambda_n r)\,dr$$

$$= [T_0 - T_c]\left.\frac{r}{\lambda_n}\,J_1\,(\lambda_n r)\right|_0^R$$

$$= [T_0 - T_c]\frac{R}{\lambda_n}\,J_1\,(\lambda_n R)$$

Therefore, the temperature distribution can be expressed as

$$\frac{T - T_c}{T_0 - T_c} = 2\sum_{n=1}^{\infty}\frac{1}{\lambda_n R}\frac{\sinh \lambda_n z}{\sinh \lambda_n L}\frac{J_0\,(\lambda_n r)}{J_1\,(\lambda_n R)} \tag{3.72}$$

3.7.2 Long Circular Cylinder Having Circumferential Surface Temperature Variation: $T(r, \phi)$ Problem

Figure 3.12 shows a long solid cylinder of circular cross-section with an arbitrary surface temperature $f(\phi)$. The governing differential equation for the steady-state problem is

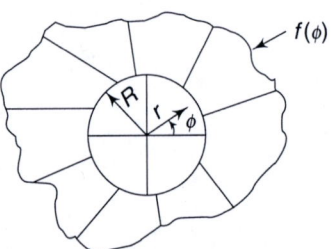

$$\frac{\partial^2 T}{\partial r^2} + \frac{1}{r}\frac{\partial T}{\partial r} + \frac{1}{r^2}\frac{\partial^2 T}{\partial \phi^2} = 0 \qquad (3.73)$$

The boundary conditions are

BC-1: $T(0,\phi) = $ finite $\qquad$ (3.74a)

BC-2: $T(R,\phi) = f(\phi)$ $\qquad$ (3.74b)

BC-3: $T(r,\phi) = T(r, \phi + 2\pi)$ $\qquad$ (3.74c)

BC-4: $\quad -\dfrac{k}{r}\dfrac{\partial T}{\partial \phi}(r, \phi) = -\dfrac{k}{r}\dfrac{\partial T}{\partial \phi}(r, \phi + 2\pi)$

or $\qquad \dfrac{\partial T}{\partial \phi}(r, \phi) = \dfrac{\partial T}{\partial \phi}(r, \phi + 2\pi) \qquad\qquad$ (3.74d)

Fig. 3.12 Cross-section of a long cylinder with arbitrary circumferential surface temperature variation

It is evident that the problem is non-axisymmetric. Special attention should be given to BC-3 and BC-4, that is, Eqs (3.74c) and (3.74d). They are called *periodic boundary conditions*. Note that the r-direction cannot be made homogeneous by any transformation. This leaves ϕ as the only possible homogeneous direction. Using the product solution

$$T(r, \phi) = \Re\,(r)\,\Phi(\phi) \tag{3.75}$$

Equation (3.73) takes the form

$$\frac{1}{\Re(r)}\left[r^2\frac{d^2\Re}{dr^2} + r\frac{d\Re}{dr}\right] = -\frac{1}{\Phi}\frac{d^2\Phi}{d\phi^2} = \pm\lambda^2 \tag{3.76}$$

$+\lambda^2$ is taken because that makes the equation in ϕ a characteristic equation. Now, the equation in ϕ becomes

$$\frac{d^2\Phi}{d\phi^2} + \lambda^2\Phi = 0 \tag{3.77}$$

The general solution of Eq. (3.77) is

$$\Phi = A\cos\lambda\phi + B\sin\lambda\phi \tag{3.78}$$

The corresponding boundary conditions become

$$\Phi(\phi) = \Phi(\phi + 2\pi) \tag{3.79a}$$

and $$\frac{d\Phi}{d\phi}(\phi) = \frac{d\Phi}{d\phi}(\phi + 2\pi) \tag{3.79b}$$

Equation (3.77) belongs to a class of problems called *Sturm–Liouville* problems, the general form of which is given as follows:

$$\frac{d}{dx}\left(p(x)\frac{dy}{dx}\right) + [q(x) + \lambda w(x)]\,y = 0 \tag{3.80}$$

If $p(a) = p(b)$, $y(a) = y(b)$, and $(dy/dx)(a) = (dy/dx(b))$ then orthogonality of y is ensured. λ is real and non-negative. Hence, λ can be replaced by λ^2 with no loss in generality of the problem.

Since, Eq. (3.77) and the corresponding boundary conditions satisfy the above requirements, the ϕ-direction is called the homogeneous direction. It may be noted that in the present case, $p(a) = p(b) = 1$, $w(x) = 1$, and $q(x) = 0$. The equation in r becomes

$$r^2\frac{d^2\Re}{dr^2} + r\frac{d\Re}{dr} - \lambda^2\Re = 0 \tag{3.81a}$$

$$\Re(0) = \text{finite} \tag{3.81b}$$

Equation (3.81a) is called *equidimensional equation* or *Cauchy–Euler equation*, the general solution of which is

$$\Re = Cr^\lambda + Dr^{-\lambda} \tag{3.82}$$

provided $\lambda \neq 0$. Therefore, from Eq. (3.75),

$$T(r, \phi) = (Cr^\lambda + Dr^{-\lambda})(A\cos\lambda\phi + B\sin\lambda\phi) \tag{3.83}$$

To evaluate λ, we will now have to impose the periodic boundary conditions, Eq. (3.74c) and Eq. (3.74d). Imposing $T(r, \phi) = T(r, \phi + 2\pi)$ on Eq. (3.83), we have

$$[A\cos\lambda\phi + B\sin\lambda\phi] = [A\cos\lambda(\phi + 2\pi) + B\sin\lambda(\phi + 2\pi)]$$

or $$[\sin\lambda\phi - \sin\lambda(\phi + 2\pi)]B + [\cos\lambda\phi - \cos\lambda(\phi + 2\pi)]A = 0 \tag{3.84}$$

Imposing $\dfrac{\partial T}{\partial\phi}(r, \phi) = \dfrac{\partial T}{\partial\phi}(r, \phi + 2\pi)$ on Eq. (3.83), we have

$$[-A\lambda\sin\lambda\phi + B\lambda\cos\lambda\phi] = [-A\lambda\sin\lambda(\phi + 2\pi) + B\lambda\cos\lambda(\phi + 2\pi)]$$

or $$[\cos\lambda\phi - \cos\lambda(\phi + 2\pi)]B = -[\sin\lambda\phi - \sin\lambda(\phi + 2\pi)]A = 0 \tag{3.85}$$

In order to have a non-trivial solution for B and A, the determinant of the co-efficient matrix must vanish. Therefore,

$$\begin{vmatrix} \sin \lambda\phi - \sin \lambda(\phi + 2\pi) & \cos \lambda\phi - \cos \lambda(\phi + 2\pi) \\ \cos \lambda\phi - \cos \lambda(\phi + 2\pi) & -\{\sin \lambda\phi - \sin \lambda(\phi + 2\pi)\} \end{vmatrix} = 0 \qquad (3.86)$$

or $\sin^2 \lambda\phi + \cos^2 \lambda\phi + \sin^2 \lambda(\phi + 2\pi) + \cos^2 \lambda(\phi + 2\pi)$

$$-2[\cos \lambda\phi \cos \lambda(\phi + 2\pi) + \sin \lambda\phi \sin \lambda(\phi + 2\pi)] = 0$$

or $1 + 1 - 2[\cos \{\lambda\phi + 2 \lambda\pi - \lambda\phi\}] = 0$

or $\cos 2\pi \lambda = 1$ (3.87)

Equation (3.87) is possible only when $\lambda = n$, where $n = 0, 1, 2, \ldots$. Therefore, $\lambda_n = n$, $n = 0, 1, 2, \ldots$. Hence,

$$\Phi = \sum_{n=1}^{\infty} (A_n \cos n\phi + B_n \sin n\phi)$$

$$\Phi = A_0 \quad \text{when } n = 0$$

$$\mathfrak{R} = \sum_{n=1}^{\infty} C_n r^n + D_n r^{-n}$$

$$n = 1, 2, 3, \ldots$$

For $\lambda \neq 0$, that is $n \neq 0$, the above solution for $\mathfrak{R}$ valid. For $\lambda = 0$ or $n = 0$, Eq. (3.81) reduces to

$$r^2 \frac{d^2\mathfrak{R}}{dr^2} + r \frac{d\mathfrak{R}}{dr} = 0$$

or $$\frac{1}{r} \frac{d}{dr}\left(r \frac{d\mathfrak{R}}{dr}\right) = 0$$

$\Rightarrow$ $$\mathfrak{R} = C_1 \ln r + C_0$$

Therefore,

$$T(r, \phi) = A_0 (C_1 \ln r + C_c) + \sum_{n=1}^{\infty} (C_n r^n + D_n r^{-n})(A_n \cos n\phi + B_n \sin n\phi) \quad (3.88)$$

Now, applying BC-1, that is, $T(0, \phi) = $ finite, we infer that

$$C_1 = 0$$
$$D_n = 0$$

Therefore, Eq. (3.88) transforms to

$$T(r, \phi) = A_0 C_0 + \sum_{n=1}^{\infty} (A_n C_n r^n \cos n\phi + B_n C_n r^n \sin n\phi) \quad (3.89)$$

Designating $a_0 = A_0 C_0$, $a_n = A_n C_n$, and $b_n = B_n C_n$, Eq. (3.89) becomes

$$T(r, \phi) = a_0 + \sum_{n=1}^{\infty} (a_n \cos n\phi + b_n \sin n\phi) r^n \quad (3.90)$$

Now applying BC-2, that is, $T(R, \phi) = f(\phi)$, we get

$$f(\phi) = a_0 + \sum_{n=1}^{\infty} (a_n \cos n\phi + b_n \sin n\phi) R^n$$

$$= a_0 + \sum_{n=1}^{\infty} a_n R^n \cos n\phi + \sum_{n=1}^{\infty} b_n R^n \sin n\phi \quad (3.91)$$

The above is a complete Fourier series in $f(\phi)$.

Recall that the general form of a complete Fourier series for $f(x)$ in the interval $-L < x < L$ having a period of $2L$ is

$$f(x) = a_0 + \sum_{n=1}^{\infty} \left[a_n \cos\left(\frac{n\pi}{L}\right) x + b_n \sin\left(\frac{n\pi}{L}\right) x \right] \tag{3.92}$$

where

$$a_0 = \frac{1}{2L} \int_{-L}^{L} f(x)\, dx \tag{3.93a}$$

$$a_n = \frac{1}{L} \int_{-L}^{L} f(x) \cos\left(\frac{n\pi}{L}\right) x\, dx \tag{3.93b}$$

$$b_n = \frac{1}{L} \int_{-L}^{L} f(x) \sin\left(\frac{n\pi}{L}\right) x\, dx \tag{3.93c}$$

Equation (3.91) is analogous to Eq. (3.92). Therefore, by a direct comparison of the terms of the two equations, we get the constants in Eq. (3.91) as

$$a_0 = \frac{1}{2\pi} \int_0^{2\pi} f(\phi)\, d\phi \tag{3.94a}$$

$$a_n R^n = \frac{1}{\pi} \int_0^{2\pi} f(\phi) \cos n\phi\, d\phi \tag{3.94b}$$

$$b_n R^n = \frac{1}{\pi} \int_0^{2\pi} f(\phi) \sin n\phi\, d\phi \tag{3.94c}$$

Let us solve the problem with $f(\phi)$ specified as shown in Fig. 3.13. The boundary condition at $r = R$ implies that the upper-half surface temperature is T_0 while the lower-half surface temperature is zero. Mathematically, this can be represented as

$$T(R, \phi) = T_0, \quad 0 < \phi < \pi \tag{3.95a}$$
$$T(R, \phi) = 0, \quad \pi < \phi < 2\pi \tag{3.95b}$$

Therefore,
$$a_0 = \frac{1}{2\pi} \int_0^{\pi} f(\phi)\, d\phi + \frac{1}{2\pi} \int_{\pi}^{2\pi} f(\phi)\, d\phi$$

$$= \frac{1}{2\pi} \int_0^{\pi} f(\phi)\, d\phi + \frac{1}{2\pi} \int_{\pi}^{2\pi} (0)\, d\phi$$

$$= \frac{1}{2\pi} \int_0^{\pi} T_0\, d\phi = \frac{1}{2\pi} T_0 [\pi] = \frac{T_0}{2}$$

$$a_n R^n = \frac{1}{\pi} \int_0^{\pi} T_0 \cos n\phi\, d\phi$$

$$= \frac{T_0}{\pi} \left. \frac{\sin n\phi}{n} \right|_0^{\pi}$$

$$= \frac{T_0}{\pi} [\sin n\pi - \sin 0] = 0$$

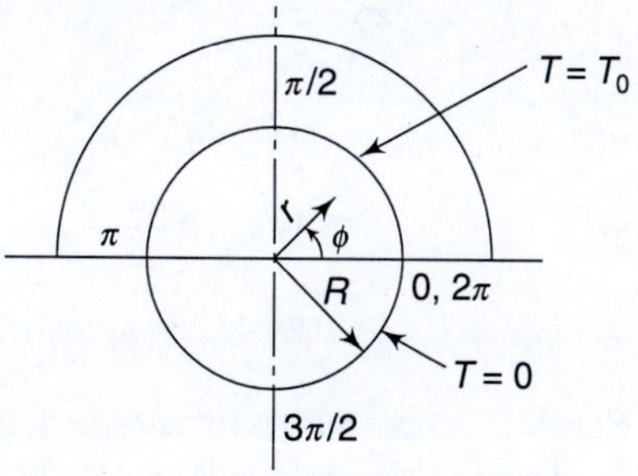

Fig. 3.13 Example where $f(\phi)$ at $r = R$ is specified

$$b_n R^n = \frac{1}{\pi} \int_0^\pi T_0 \sin n\phi \, d\phi$$

$$= -\frac{T_0}{\pi} \left. \frac{\cos n\phi}{n} \right|_0^\pi$$

$$= -\frac{T_0}{\pi} \left[\frac{\cos n\pi - \cos 0}{n} \right]$$

$$= -\frac{T_0}{\pi} [-1-1] = \frac{2T_0}{n\pi}$$

for $n = 1, 3, 5, \dots$. Hence,

$$T(r, \phi) = \frac{T_0}{2} + \sum_{n=1,3,5}^\infty \frac{2T_0 \, r^n}{n\pi \, (R)^n} \sin n\phi$$

or
$$\frac{T(r, \phi)}{T_0} = \frac{1}{2} + 2 \sum_{n=1,3,5}^\infty \frac{1}{n\pi} \left(\frac{r}{R} \right)^n \sin n\phi \qquad (3.96)$$

Example 3.2   Temperature Distribution in a Television Antenna Rod

One-half of a television antenna rod of radius R is exposed to the radiation heat flux from the sun while the other half radiation heat flux received is negligible (Fig. E3.2). The rod is also losing heat to the surrounding atmosphere (which is at T_∞) from its entire periphery by convection. Find the steady temperature of the rod.

Given: $q''(R, \phi) = q_0'' \sin \phi, \, 0 < \phi < \pi$
$$= 0, \; \pi < \phi < 2\pi$$

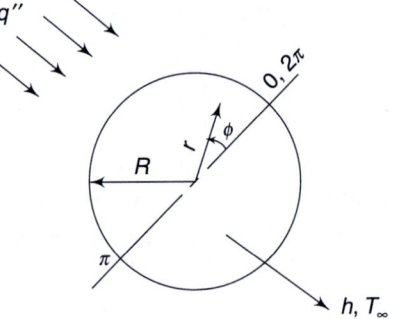

Fig. E3.2   The television antenna rod problem

Solution

The formulation of this problem is identical with that discussed in the previous section except for the surface boundary condition. Let $\theta = T - T_\infty$. To write the mathematical representation of the boundary condition at $r = R$, we have to apply Kirchhoff's current law (see Section 2.6.2) as follows:

$$q'' - q_n'' - q_c'' = 0$$

$$\Rightarrow \qquad q'' = q_n'' + q_c''$$

$$\Rightarrow \qquad q''(\phi) = k \frac{\partial \theta}{\partial r} + h\theta$$

$$\Rightarrow \qquad k \frac{\partial \theta}{\partial r} = q''(\phi) - h\theta \qquad (3.97)$$

Note that temperature is increasing in the direction of increasing r in the region of $0 < \phi < \pi$, whereas in the other half $(\pi < \phi < 2\pi)$ where $q''(\phi) = 0$, temperature is decreasing with r. Now, from Eq. (3.90), by analogy, we can write

$$\theta(r, \phi) = a_0 + \sum_{n=1}^\infty r^n (a_n \cos n\phi + b_n \sin n\phi) \qquad (3.98)$$

Also, $\quad k\dfrac{\partial \theta}{\partial r}(R,\phi)=q''(\phi)-h\theta(R,\phi)$ $\qquad\qquad$ (3.99)

Therefore, evaluating $\partial\theta/\partial r$ from Eq. (3.98) and substituting in Eq. (3.99), we get

$$q''(\phi)=h\left[a_0+\sum_{n=1}^{\infty}R^n\left(1+\frac{nk}{hR}\right)(a_n\cos n\phi+b_n\sin n\phi)\right] \qquad (3.100)$$

which is the complete Fourier series of $q''(\phi)$ in the period 0 to 2π. The rest of the solution procedure is left as an exercise to the reader. The final solution is

$$\frac{\theta(r,\phi)}{q_0''/h}=\frac{1}{\pi}-\frac{1}{\pi}\sum_{n=2,4,6,\dots}^{\infty}\frac{1+(-1)^n}{(n^2-1)\left(1+\dfrac{nk}{hR}\right)R^n}r^n\cos n\phi$$

$$+\frac{1}{2}\frac{1}{\left(1+\dfrac{k}{hR}\right)R}r\sin\phi \qquad (3.101)$$

Example 3.3 Temperature Distribution in a Semi-circular Rod

Consider a long solid cylinder of semicircular cross-section as shown in Fig. E3.3. The cylindrical surface at $r = R$ is held at an arbitrary temperature $f(\phi)$ and the planar surfaces at $\phi = 0$ and $\phi = \pi$ are both maintained at the same constant and uniform temperature T_0. Find the steady temperature of the rod.

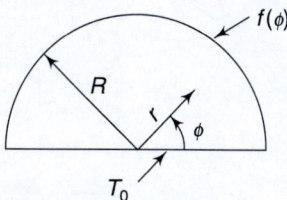

Fig. E3.3 Long solid cylinder of semicircular cross-section

Solution

The governing differential equation is

$$\frac{\partial^2 T}{\partial r^2}+\frac{1}{r}\frac{\partial T}{\partial r}+\frac{1}{r^2}\frac{\partial^2 T}{\partial \phi^2}=0 \qquad (3.102)$$

The boundary conditions are

BC-1: $T(0,\phi)=T_0$ $\qquad\qquad$ (3.103a)
BC-2: $T(R,\phi)=f(\phi)$ $\qquad\qquad$ (3.103b)
BC-3: $T(r,0)=T_0$ $\qquad\qquad$ (3.103c)
BC-4: $T(r,\pi)=T_0$ $\qquad\qquad$ (3.103d)

Defining a new temperature $\theta = T - T_0$, and assuming a product solution of the form $\theta(r,\phi)=\Re(r)\,\Phi(\phi)$, we obtain from Eq. (3.102)

$$r^2\frac{d^2\Re}{dr^2}+r\frac{d\Re}{dr}-\lambda^2\Re=0 \qquad (3.104)$$

and $\qquad\qquad\qquad \dfrac{d^2\Phi}{d\phi^2}+\lambda^2\Phi=0$ $\qquad\qquad\qquad$ (3.105)

where the sign of the separation constant is taken as $+\lambda^2$ as ϕ is the homogeneous direction for $\theta(r,\phi)$. Therefore,

$$\theta(r,\phi)=(C_1 r^\lambda+C_2 r^{-\lambda})(C_3\cos\lambda\phi+C_4\sin\lambda\phi) \qquad (3.106)$$

Since, at $r = 0$, $\theta = 0$, therefore, $C_2 = 0$. Hence,

$$\theta(r,\phi)=r^\lambda(A\cos\lambda\phi+B\sin\lambda\phi) \qquad (3.107)$$

Applying $\theta(r, 0) = 0$ to Eq. (3.107), we get

$$A = 0$$

Thus, $\theta(r, \phi) = r^\lambda (B \sin \lambda \phi)$ (3.108)

Applying $\theta(r, \pi) = 0$ to Eq. (3.108), we get

$$0 = r^\lambda (B \sin \lambda \pi)$$

The above equation will be true if

$$\sin \lambda \pi = \sin n\pi, \quad n = 0, 1, 2, 3,\ldots$$

or $\lambda = (n\,\pi/\pi) = n, \quad n = 0, 1, 2, 3, \ldots$ (3.109)

It may be noted that $\lambda = 0$ does not contribute to the solution since $A = 0$. Therefore, $\lambda = n, n = 1, 2, 3,\ldots$ is used. Now, at $r = R$, $q = f(f) - T_0$. Therefore,

$$f(\phi) - T_0 = \sum_{n=1}^{\infty} R^n B_n \sin n\phi$$ (3.110)

Equation (3.110) is the Fourier sine series of $f(\phi) - T_0$ over the interval $(0, \pi)$. Then,

$$B_n R_n = \frac{1}{\left(\dfrac{\pi}{2}\right)} \int_0^\pi [f(\phi) - T_0] \sin n\phi \, d\phi$$

$$= \frac{2}{\pi} \int_0^\pi [f(\phi) - T_0] \sin n\phi \, d\phi$$

or $B_n = \dfrac{2}{\pi R^n} \displaystyle\int_0^\pi [f(\phi) - T_0] \sin n\phi \, d\phi$ (3.111)

Thus, the solution for the temperature distribution, $\theta(r, \phi)$, can be written as

$$\theta(r, \phi) = \frac{2}{\pi} \sum_{n=1}^{\infty} \left(\frac{r}{R}\right)^n \sin n\phi \int_0^\pi [f(\phi) - T_0] \sin n\phi \, d\phi$$ (3.112)

or $T(r, \phi) = T_0 + \dfrac{2}{\pi} \displaystyle\sum_{n=1}^{\infty} \left(\frac{r}{R}\right)^n \sin n\phi \int_0^\pi [f(\phi) - T_0] \sin n\phi \, d\phi$ (3.113)

3.8   Steady Three-dimensional Conduction in Cartesian Coordinates

Illustration 3.5   *Consider a semi-infinite rod of rectangular cross-section $(2L \times 2l)$. The base temperature of the rod is T_0 and the ambient temperature is T_∞. The heat transfer coefficient is large. The objective is to find the steady temperature of the rod. The object may be thought of as a 3D infinite fin of which the height and the width are comparable (Fig. 3.14).*

Solution

Using a new temperature $\theta = T - T_\infty$, the governing differential equation for the problem is

$$\frac{\partial^2 \theta}{\partial x^2} + \frac{\partial^2 \theta}{\partial y^2} + \frac{\partial^2 \theta}{\partial z^2} = 0 \quad (3.114)$$

The boundary conditions are

$$\frac{\partial \theta}{\partial x}(0, y, z) = 0, \quad \theta(L, y, z) = 0$$

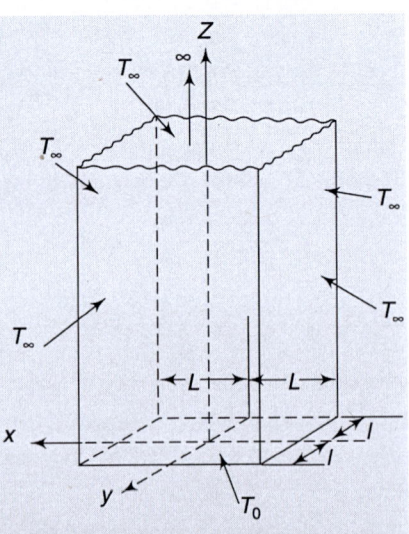

Fig. 3.14   The geometry and the boundary conditions for the $T(x, y, z)$ problem

$$\frac{\partial \theta}{\partial y}(x, 0, z) = 0, \quad \theta(x, l, z) = 0$$

$$\theta(x, y, 0) = \theta_0, \quad \theta(x, y, \infty) = 0$$

Note that the boundary conditions in x and y at $x = 0$ and $y = 0$ have arisen due to thermal and geometric symmetry in those directions. That is the reason why the coordinate system x-y-z is located at the centre of the base of the semi-infinite rod. The problem is homogeneous in both the x and y directions. Using the product solution $\theta(x, y, z) = X(x)Y(y)Z(z)$, we may separate Eq. (3.114) in the form

$$-\frac{1}{X}\frac{d^2 X}{dx^2} = \frac{1}{Y}\frac{d^2 Y}{dy^2} + \frac{1}{Z}\frac{d^2 Z}{dz^2} = \pm \lambda^2 \tag{3.115}$$

$+\lambda^2$ is taken to produce characteristic equations in x and y directions. Therefore, in the x-direction,

$$\frac{d^2 X}{dx^2} + \lambda^2 X = 0 \tag{3.116}$$

Subject to
$$\frac{dX(0)}{dx} = 0 \tag{3.117a}$$

$$X(L) = 0 \tag{3.117b}$$

In the y-direction, re-arranging the second equality,

$$-\frac{1}{Y}\frac{d^2 Y}{dy^2} = \frac{1}{Z}\frac{d^2 Z}{dz^2} - \lambda^2 = \mu^2$$

$$\Rightarrow \qquad \frac{d^2 Y}{dy^2} + \mu^2 Y = 0 \tag{3.118}$$

Subject to $\dfrac{dY(0)}{dy} = 0 \tag{3.119a}$

$$Y(l) = 0 \tag{3.119b}$$

Thus, the z-direction, from the second equality, is found to satisfy

$$\frac{d^2 Z}{dz^2} - (\lambda^2 + \mu^2)Z = 0 \tag{3.120}$$

$$Z(\infty) = 0 \text{ (finite)} \tag{3.121}$$

Note that the second and non-separable boundary condition, $\theta(x, y, 0) = \theta_0$, is left to the end of the solution as before. Now, the solution of Eq. (3.116) is

$$X_n(x) = A_n \phi_n(x), \quad \phi_n(x) = \cos \lambda_n x$$

$$\lambda_n L = (2n + 1)\frac{\pi}{2}, \quad n = 0, 1, 2, 3,...$$

Similarly, the solution of Eq. (3.118) is

$$Y_n(y) = B_n \psi_m(y), \quad \psi_m(y) = \cos \mu_m y$$

$$\mu_m l = (2m + 1)\frac{\pi}{2}, \quad m = 0, 1, 2, 3,...$$

Finally, the solution of Eq. (3.120) is

$$Z_{mn}(z) = C_{mn} e^{-(\lambda_n^2 + \mu_m^2)^{1/2} z}$$

Thus, the product solution leads to

$$\theta(x, y, z) = \sum_{n=0}^{\infty}\sum_{m=0}^{\infty} a_{mn} e^{-(\lambda_n^2 + \mu_m^2)^{1/2} z} \cos \lambda_n x \cos \mu_m y$$

where $a_{mn} = A_n B_m C_{mn}$. Applying the other boundary condition,

$$\theta_0 = \sum_{n=0}^{\infty} \sum_{m=0}^{\infty} a_{mn} \cos \lambda_n x \cos \mu_m y \qquad (3.122)$$

Equation (3.122) represents the double Fourier cosine series expansion of θ_0 over the cross-section of the rod:

$$a_{mn} = \theta_0 \frac{\int_0^L \int_0^l \cos \lambda_n x \cos \mu_m y \, dx \, dy}{\int_0^L \int_0^l \cos^2 \lambda_n x \cos^2 \mu_m y \, dx \, dy} \qquad (3.123)$$

which results in $\quad a_{mn} = \dfrac{4\theta_0 (-1)^{n+m}}{(\lambda_n L)(\mu_m l)}$

Finally,

$$\frac{\theta(x,y,z)}{\theta_0} = 4 \sum_{n=0}^{\infty} \sum_{m=0}^{\infty} \frac{(-1)^{n+m} \, e^{-(\lambda_n^2 + \mu_m^2)^{1/2} z}}{(\lambda_n L)(\mu_m l)} \cos \lambda_n x \cos \mu_m y \qquad (3.124)$$

3.9   Graphical Method and Conduction Shape Factor

The graphical method can be used for plotting isotherms and heat flux lines for steady two-dimensional problems. It involves the insulated and isothermal boundaries in an isotropic medium of a complicated geometry, and calculates the rate of heat flow through it. This technique has been surpassed today by computer methods such as finite difference, finite element and finite volume (see Chapter 10) but nevertheless may still be applied to get a first estimate of the temperature distribution and to develop a physical feel for the nature of the temperature field and heat transfer in a body of complex shape.

3.9.1   Basic Principles

The basic principle of the graphical method originates from the fact that isotherms must be normal to the direction of heat flow (i.e., heat flux lines) for isotropic media (Section 3.4). Using the graphical method a network of isotherms and heat flux lines is systematically constructed. This network which is traditionally referred to as a flux plot, is the basis for obtaining the rate of heat flow through the system. The following should be remembered while constructing a flux plot.

1. **Lines of symmetry** Identify the lines of symmetry arising out of thermal and geometrical conditions.
2. **No heat flow in a direction normal to the lines of symmetry** There can be no heat flow in a direction perpendicular to the lines of symmetry. Thus, lines of symmetry are adiabatic. Hence, they should be treated as heat flow lines.
3. **Sketch lines of constant temperatures** After isotherms related to the object boundaries have been identified, draw isotherms within the domain. Note that isotherms should be always perpendicular to insulation or adiabatic boundaries or lines.
4. **Draw curvilinear squares** The network should be made of curvilinear squares. Make sure that heat flow lines and isotherms intersect at right angles and all

sides of each square are of approximately same length. It is often difficult to meet the aforesaid requirement and several iterations need to be made. The main requirement is that each curvilinear loop must look like a 'square'. In certain places, such as corners, it may be virtually impossible to obtain curvilinear squares. However, such inaccuracies have little effect on the overall result obtained from the flux plot.

5. **Temperature distribution and heat transfer** Once the flux plot is completed, it may be used to infer the temperature distribution in the object. From a simple analysis (shown next in this section), the heat transfer can be computed.

3.9.2 Calculation of Heat Flow Rate

Figure 3.15 shows the rectangular cross-section of a plane wall which is very long in the direction normal to the plane of the figure and whose two lateral surfaces are perfectly insulated. The top surface is a temperature, T_1 while the bottom surface is at temperature, T_2 $(T_1 > T_2)$. Evidently, one-dimensional heat transfer is taking place from the top to bottom surface of the wall. The heat transfer is assumed to be steady. This figure also shows isotherms and heat flux lines.

The network of heat flux lines and isotherms of Fig. 3.15 is constructed by first dividing the total heat flow, 'q' into 'n' increments of heat flows of equal amount. Thus,

$$q_i = \frac{q}{n} \tag{3.125}$$

where $i = 1, 2, 3, \ldots \ldots n$. Each q_i flows through a heat tube, that is, the space between two adjacent flux lines. Each such space can be called a 'lane'. That each increment of heat flow (i.e., q_i) is equal indicated by the fact that the heat flux lines are equidistant, i.e., each lane has the same thickness.

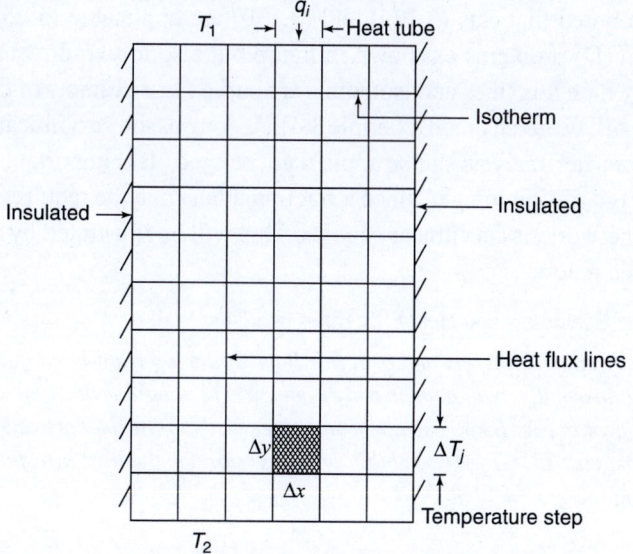

Fig. 3.15 Network of orthogonal heat flux lines and isotherms: square grids of equal size

The second step in the plotting of the network is the drawing of sufficient number of isotherms so that each loop is a square. Each square is subjected to the same temperature differential in the vertical direction, namely,

$$\Delta T_j = \frac{T_1 - T_2}{m} \text{ where } j = 1, 2, \ldots\ldots m \tag{3.126}$$

where m is the number of temperature increments between the top and bottom surfaces. According to the Fourier's law of heat conduction, q_i that passes through the (i, j) square is

$$q_i = k\Delta x W \frac{\Delta T_j}{\Delta y} = kW\Delta T_j \text{ (since } \Delta x = \Delta y) \tag{3.127}$$

where k = thermal conductivity of the material (W/m K)

Δx = space increment in x-direction (m)

Δy = space increment in y-direction (m)

W = width of the wall measured in the direction normal to the plane of Fig. 3.15.

From Eqs (3.125) – (3.127) it can be easily shown that

$$q = \frac{n}{m} Wk(T_1 - T_2) \tag{3.128}$$

Equation (3.128) can also be written as

$$q = sk(T_1 - T_2) \tag{3.129}$$

where $\quad s = \dfrac{n}{m} W$ (3.130)

and is called "conduction shape factor" having a unit of metre.

Thus it is seen that the total heat transfer can be calculated by simply counting the number of heat lanes or tubes (n) and the temperature steps (m).

It may be noted that Eqs (3.129) and (3.130) are applicable to not only one-dimensional (1D) problems as shown in figure but also to two-dimensional (2D) problems in which flux lines and isotherms are curved. An example of flux plot in a 2D problem will be given next (Example 3.4). As long as the curvilinear loops look "square" the earlier analysis can be applied unchanged. It is important to note that 'n' is not necessarily an integer, since a fractional lane may be required to obtain a satisfactory network of curvilinear squares. This will be illustrated by an example problem given below.

Example 3.4 Isotherms and Heat Flux Lines in a Duct Wall

Show the steady state isotherms and heat flux lines in the top right-hand quadrant of the wall of a long, thermally and geometrically symmetric rectangular duct, the cross-section of which is shown in Fig. E3.4. The inner and outer walls of the duct are at T_1 and T_2, respectively. Note that $T_1 > T_2$. Also, obtain an expression for the heat transfer through the duct wall using the graphical method.

Solution

Note that the lines of symmetry are the left and bottom surfaces of the computational domain (Fig. E3.4). Lines of symmetry are heat flux lines and the isotherms must hit them at right

angles. The inner walls are at T_1 and
hence the whole of them is an isotherm,
and the same is true for the outer walls
at T_2. Therefore, heat flux lines must
be normal to the inner and outer walls.
Keeping in mind the isotherms and
heat flux lines have been sketched. A
typical curvilinear square (indicated
by hatched lines) formed by heat flux
lines and isotherms is shown in the
computational domain. An expanded
view of the same is also visible outside
Fig. E3.4. Observe that it is impossible
to obtain a square loop near the upper
right-hand corner. The heat transfer
across the curvilinear section shown is
given by Fourier's law, assuming unit
width of the material.

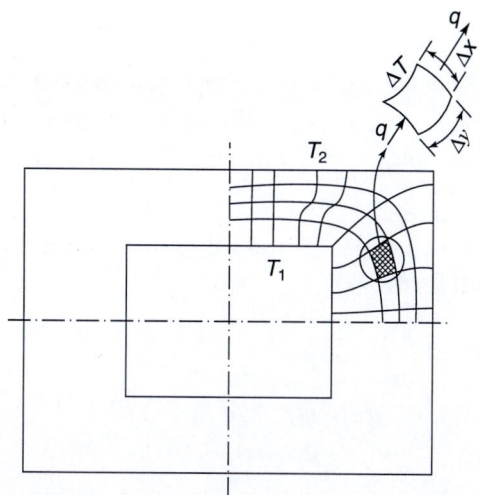

Fig. E3.4 Sketch showing curvilinear square formed by heat flux lines and isotherms

$$q = k\Delta x (1)\frac{\Delta T}{\Delta y} \tag{A}$$

The above heat flow rate will be same through each section within this heat tube, and the total heat transfer will be the sum of the heat transfer through all the heat tubes. Since $\Delta x \approx \Delta y$, the heat flow rate is proportional to the temperature differential, ΔT across the section. Now that this heat flow rate is constant, the ΔT across each loop must be same within the same heat tube. Thus,

$$\Delta T = \frac{\Delta T_{overall}}{m} \tag{B}$$

where m is the number of temperature increments between the inner and outer surfaces.

It is to be noted further that the heat transfer through each lane or tube is the same since it is independent of the values of Δx and Δy as they are constructed equal. Thus, the heat transfer in the quadrant, $q_{quadrant}$ is

$$q_{quadrant} = nk\Delta T = \frac{n}{m}k\Delta T_{overall} = \frac{n}{m}k(T_1 - T_2) \tag{C}$$

where n is the number of heat flow lanes or heat tubes.

In summary, it may be said that to calculate the heat transfer, the flux plot has to be constructed first. Then the number of heat flow lanes and temperature increments are to be counted. In the present example, the upper right-hand quadrant, $m = 4$ and $n = 8.2$. Note that n in this case is fractional. The total number of heat flow lanes in the entire object = 4 × 8.2 = 32.8 and therefore, the total heat transfer is $4q_{quadrant}$.

The accuracy of this method depends entirely on the skill of the draftsman drawing the square loops. Even a rough sketch can give a fairly good estimate of the heat transfer. Use of a durable paper, an H grade pencil and a good eraser is recommended. Now-a-days, one can draw the flux plots on computer using graphics software.

Additional Examples

Example 3.5   3D Heat Conduction with Heat Generation in a Rectangular Body

Find the temperature distribution at the steady state for the problem of three-dimensional heat conduction with uniform heat generation. See Fig. E3.5.

Solution

In this problem, thermal and geometric symmetries exist with respect to x and y.

Let $\theta = T - T_\infty$

GDE: $\dfrac{\partial^2 \theta}{\partial x^2} + \dfrac{\partial^2 \theta}{\partial y^2} + \dfrac{\partial^2 \theta}{\partial z^2} + \dfrac{q'''}{k} = 0$ \hfill (A)

BC-1: $\theta(L, y, z) = 0$

BC-2: $\theta(x, l, z) = 0$

BC-3: $\dfrac{\partial \theta}{\partial x}(0, y, z) = 0$

BC-4: $\theta(x, 0, z) = 0$

BC-5: $k\dfrac{\partial \theta}{\partial z}(x, y, 0) = h_2 \theta$

BC-6: $-k\dfrac{\partial \theta}{\partial z}(x, y, \delta) = h_1 \theta$

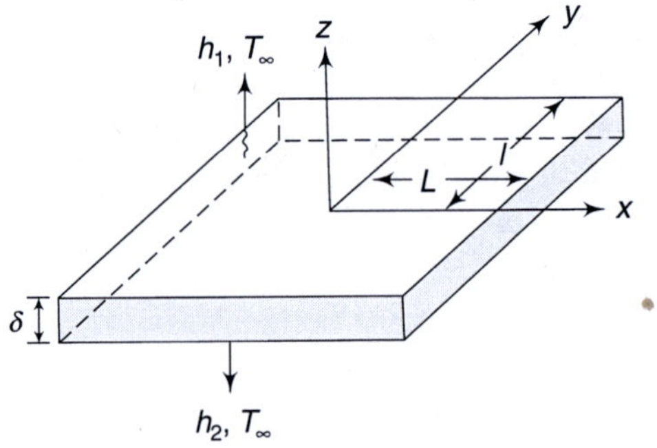

Since the governing equation [Eq. (A)] is non-homogeneous due to the generation term, we can solve the problem by the method of superposition as described next.

Fig. E3.5

$$\theta(x, y, z) = \psi(x, y, z) + \phi(x, y) \tag{B}$$

The first term in Eq. (B) represents a 3D problem without heat generation, whereas the second term represents a 2D problem with heat generation.

GDE for ϕ : $\dfrac{\partial^2 \phi}{\partial x^2} + \dfrac{\partial^2 \phi}{\partial y^2} + \dfrac{q'''}{k} = 0$ \hfill (C)

BCs for ϕ:

$\phi(L, y) = 0$

$\dfrac{\partial \phi}{\partial x}(0, y) = 0$

$\phi(x, l) = 0$

$\dfrac{\partial \phi}{\partial y}(x, 0) = 0$

The solution for $\phi(x, y)$ (see Section 3.5.2) is

$$\frac{\phi(x, y)}{(q''' L^2 / k)} = \frac{1}{2}\left[1 - \left(\frac{x}{L}\right)^2\right] - 2\sum_{n=0}^{\infty} \frac{(-1)^n}{(\lambda_n L)^3} \frac{\cosh \lambda_n y}{\cosh \lambda_n l} \cosh \lambda_n x$$

where $\lambda_n L = (2n + 1)\pi/2$, $n = 0, 1, 2, \ldots$.

GDE for ψ:

$$\frac{\partial^2 \psi}{\partial x^2} + \frac{\partial^2 \psi}{\partial y^2} + \frac{\partial^2 \psi}{\partial z^2} = 0 \tag{D}$$

BCs for ψ:

$$\psi(L, y, z) = \theta(L, y, z) - \phi(L, y) = 0 - 0 = 0$$

$$\frac{\partial \psi}{\partial x}(0, y, z) = \frac{\partial \theta}{\partial x}(0, y, z) - \frac{\partial \phi}{\partial x}(0, y) = 0 - 0 = 0$$

$$\psi(x, l, z) = \theta(x, l, z) - \phi(x, l) = 0 - 0 = 0$$

$$\frac{\partial \psi}{\partial y}(x, 0, z) = \frac{\partial \theta}{\partial y}(x, 0, z) - \frac{\partial \phi}{\partial y}(x, 0) = 0 - 0 = 0$$

$$k \frac{\partial \psi}{\partial z}(x, y, 0) = h_2 \psi + h_2 \phi(x, y) \quad \text{(obtained from BC-5)}$$

$$-k \frac{\partial \psi}{\partial z}(x, y, \delta) = h_1 \psi + h_1 \phi(x, y) \quad \text{(obtained from BC-6)}$$

We see, however, that along the z-direction both BCs are non-homogeneous. Therefore, we can solve the problem by the method of superposition. Therefore,

$$\psi = \psi_1 + \psi_2$$

GDE for ψ_1: $\nabla^2 \psi_1 = 0$

BCs for ψ_1:

$$\psi_1(L, y, z) = 0$$

$$\frac{\partial \psi_1}{\partial x}(0, y, z) = 0$$

$$\psi_1(x, l, z) = 0$$

$$\frac{\partial \psi_1}{\partial y}(x, 0, z) = 0$$

$$k \frac{\partial \psi_1}{\partial z}(x, y, 0) = h_2 \psi_1$$

$$-k \frac{\partial \psi_1}{\partial z}(x, y, \delta) = h_2 \psi_1 + h_1 \phi$$

GDE for ψ_2: $\nabla^2 \psi_2 = 0$

BCs for ψ_2:

$$\psi_2(L, y, z) = 0$$

$$\frac{\partial \psi_2}{\partial x}(0, y, z) = 0$$

$$\psi_2(x, l, z) = 0$$

$$\frac{\partial \psi_2}{\partial y}(x, 0, z) = 0$$

$$k \frac{\partial \psi_2}{\partial z}(x, y, 0) = h_2 \psi_2 + h_2 \phi$$

$$-k \frac{\partial \psi_2}{\partial z}(x, y, \delta) = h_1 \psi_2$$

Solutions for ψ_1 and ψ_2 can be obtained by the separation of variables method.

Example 3.6   Temperature Distribution in a Boiler Tube Wall

One-half of a thick-walled boiler tube receives uniform heat flux q'' while the other half is insulated. The inner and outer radii are R_i and R_o, respectively. The temperature of the inside fluid is T_∞ and the inside heat transfer coefficient is large (boiling). Find the steady temperature of the tube wall. See Fig. E3.6.

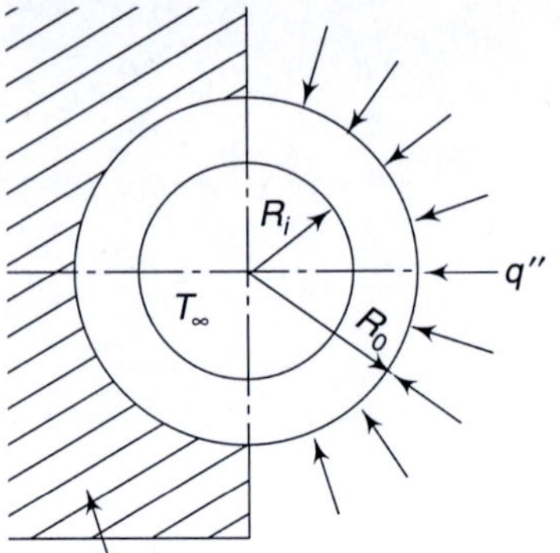

Solution

The formulation of this problem is identical with that discussed in Section 3.7.2 except for the inner and outer surface boundary conditions, which are given below. Let $\phi = T - T_\infty$.

Fig. E3.6

$$\text{At } r = R_o, \quad q''(\theta) = k\frac{\partial \phi}{\partial r}, \quad 0 < \theta < \pi$$

$$= 0, \qquad \pi < \theta < 2\pi$$

$$\text{At } r = R_i, \quad \phi = 0 \quad (\text{since } T = T_\infty \text{ as h is very large})$$

Therefore, $\phi(r, \theta) = A_0 C_1 \ln r + A_0 C_0 + \displaystyle\sum_{n=1}^{\infty} (C_n r^n + D_n r^{-n}) A_n \cos n\theta$

$$+ \sum_{n=1}^{\infty} (C_n r^n + D_n r^{-n}) B_n \sin n\theta$$

If $A_0 C_0 = a_0$ and $A_0 C_1 = a_1$, then

$$\phi(r, \theta) = a_1 \ln r + a_0 + \sum_{n=1}^{\infty} (A_n \cos n\theta + B_n \sin n\theta)(C_n r^n + D_n r^{-n})$$

Now, $\phi(R_i, \theta) = 0$.

Therefore, $0 = a_0 + a_1 \ln R_i + \displaystyle\sum_{n=1}^{\infty} (A_n \cos n\theta + B_n \sin n\theta)(C_n R_i^n + D_n R_i^{-n})$ (A)

Again, $k(\partial \phi / \partial r)_{r=R_o} = q''$.

Therefore, $\dfrac{R_o q''(\theta)}{k} = a_1 + \displaystyle\sum_{n=1}^{\infty} (A_n \cos n\theta + B_n \sin n\theta)(nC_n R_o^n - nD_n R_o^{-n})$ (B)

Equation (B) is the complete Fourier series of $R_o q''(\theta)/k$.

Hence, $a_1 = \dfrac{1}{2\pi} \displaystyle\int_0^{2\pi} \dfrac{R_o q''(\theta)}{k} d\theta = \dfrac{1}{2\pi} \dfrac{R_o}{k} \left[\int_0^{\pi} q'' d\theta + \int_0^{2\pi} q'' d\theta \right]$

$$= \dfrac{R_o q''}{2k}$$

Note that q'' is constant in $0 < \theta < \pi$ and is 0 in $\pi < \theta < 2\pi$.

$$nA_n C_n R_o^n - nA_n D_n R_o^{-n} = \dfrac{1}{\pi} \int_0^{2\pi} \dfrac{R_o}{k} q'' \cos n\theta \, d\theta = 0$$

or $\qquad nA_n \left(C_n R_o^n - D_n R_o^{-n} \right) = 0$ $\hfill$ (C)

$$B_n \left(nC_n R_o^n - nD_n R_o^{-n} \right) = \frac{1}{\pi} \int_0^{2\pi} \frac{R_o q''}{k} \sin n\theta \, d\theta$$

$$= \frac{R_o q''}{\pi k n} [1 - (-1)^n] \hfill (D)$$

From Eqs (C) and (D), it is evident that $A_n = 0$. Again, Eq. (A) is a complete Fourier series. Therefore,

$$a_0 + a_1 \ln R_i = 0$$

or $\qquad a_0 = -a_1 \ln R_i = -\dfrac{R_o q''}{2k} \ln R_i$

Also, $A_n C_n R_i^n + A_n D_n R_i^{-n} = 0$, but this is an identity since $A_n = 0$, and

$$B_n C_n R_i^n + B_n D_n R_i^{-n} = 0 \hfill (E)$$

From Eq. (D), we write

$$nB_n C_n R_o^n - nB_n D_n R_o^{-n} = \frac{R_o q''}{\pi k n} [1 - (-1)^n] \hfill (F)$$

Solving Eqs (E) and (F) simultaneously, we obtain

$$B_n D_n = \frac{-R_o q'' [1 - (-1)^n]}{\pi k n^2 [R_i^{-2n} R_o^n + R_o^{-n}]} \hfill (G)$$

and $\qquad B_n C_n = \dfrac{R_o q'' [1 - (-1)^n] R_i^{-2n}}{\pi k n^2 [R_i^{-2n} R_o^n + R_o^{-n}]}$ $\hfill$ (H)

Finally, substituting a_0, a_1, $B_n C_n$, $B_n D_n$ into the equation for $\phi(r, \theta)$ and noting that for $n = 2, 4, 6, \ldots,$

$$1 - (-1)^n = 0$$

and for $n = 1, 3, 5, \ldots,$

$$1 - (-1)^n = 2$$

we obtain

$$\phi(r, \theta) = \frac{R_o q''}{2k} \ln\left(\frac{r}{R_i}\right) + \frac{R_o q'' \displaystyle\sum_{n=1,3,5,\ldots}^{\infty} 2\sin n\theta \, (R_i^{-2n} r^n - r^{-n})}{\pi k n^2 (R_i^{-2n} R_o^n + R_o^{-n})}$$

Important Concepts and Formulae

Method of Separation of Variables

The method of separation of variables is applicable to steady two-dimensional conduction problems if and when

   (i)   the governing differential equation is linear and homogeneous;

   (ii)   one of the directions of the problem is expressed by a homogeneous differential equation subject to homogeneous boundary conditions (the homogeneous direction) while the other direction is expressed by a homogeneous differential equation subject to one homogeneous and one non-homogeneous boundary condition (the non-homogeneous direction);

(iii) the sign of λ^2 is chosen such that the boundary-value problem of the homogeneous direction leads to a characteristic-value problem.

It may be noted characteristic functions always form an orthogonal set. Since the homogeneous direction gives a characteristic equation, orthogonality always exists in the homogeneous direction. Furthermore, separation of variables requires finiteness in the homogeneous direction.

For three-dimensional problems, the method of separation of variables can be applied if two directions are homogeneous while the other direction is non-homogeeous.

Method of Superposition

The method of superposition is used when the separation of variables method cannot be directly applied because

(i)   both the boundary conditions in one or more directions are non-homogeneous and neither of them can be made homogeneous by any transformation; or

(ii) the governing equation is linear but non-homogeneous.

In such cases, the main problem is divided into several sub-problems so that the solution of each sub-problem is added to each other to obtain the desired solution. Generally, each sub-problem can be solved by the separation of variables method.

Isotherms and Heat Flux Lines

(i)   No two isotherms can cut each other, since no part of a body can have two temperatures at the same time.

(ii) For isotropic solids heat flux lines and isotherms intersect at right angles.

Steady 2D and 3D Problems

Detailed solution methodologies have been shown for steady 2D problems with and without heat sources in Cartesian coordinates. The techniques of superposition and imaging have been demonstrated. Next, axisymmetric $[T(r, z)]$ and non-axisymmetric $[T(r, \theta)]$ problems in cylindrical geometry have been solved. Finally, the chapter ends with the demonstration of the solution procedure for a 3D conduction problem in Cartesian coordinates, bringing in the concept of double Fourier series.

Review Questions

3.1 What are the pre-requisites for application of separation of variables to solve a steady 2D heat conduction problem?

3.2 What is an orthogonal function?

3.3 Explain the method of superposition.

3.4 How will you graphically represent 2D temperature distribution in a solid?

3.5 Define an isotherm.

3.6 Isotherms and heat flux lines intersect at right angles to each other in an isotropic solid. Why?

3.7 Can two isotherms cut each other?

3.8 What is a double Fourier series?

3.9 Define conduction shape factor.

3.10 What is the advantage of a graphical method? Can any heat conduction problem be solved by the graphical method?

Problems

3.1 A semi-infinite plate with edges at $x = 0$, $x = L$, $y = 0$, and $y = \infty$ is subjected to the following boundary conditions at steady state:

At $x = 0$, $T = 0$

At $x = L$, $T = 0$

At $y = \infty$, $T = 0$

At $y = 0$, $T = T_A \sin(\pi x/L)$ ($T_A =$ constant)

(a) Write an expression for $T(x, y)$.

(b) Evaluate the heat flux at $y = 0$ for any x.

3.2 Consider the steady heat conduction problem with the boundary conditions as shown in Fig. Q3.2. Indicate how you will solve this problem. Show the solution procedure neatly in a pictorial manner.

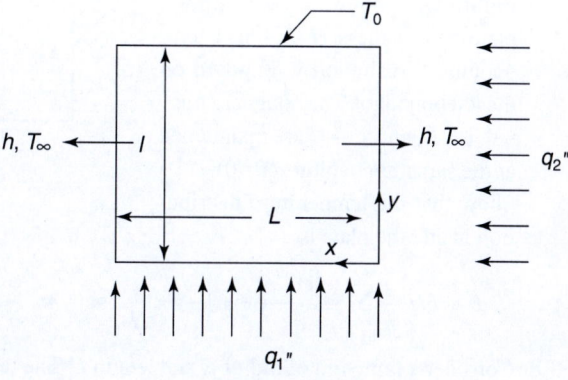

Fig. Q3.2

3.3 An infinitely long rod of square cross-section ($L \times L$) floats in a fluid (Fig. Q3.3). The heat transfer coefficient between the rod and the fluid is large as compared to that between the rod and the surroundings. The fluid and the ambient temperatures are T_0 and T_∞, respectively. Find the steady temperature of the rod.

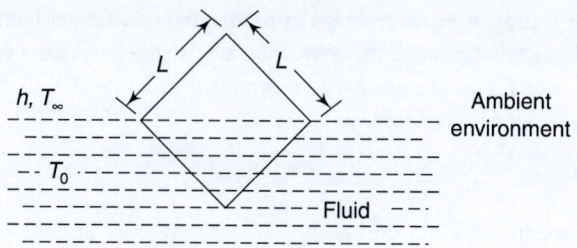

Fig. Q3.3

3.4 Obtain an expression for the steady-state temperature distribution for the problem shown in Fig. Q3.4.

3.5 A solid cylinder of radius R and length L has its two circular ends maintained at 0°C. On the periphery of the cylinder, the temperature distribution is a function of only the axial coordinate z. Show that the steady-state solution for the temperature distribution is given by

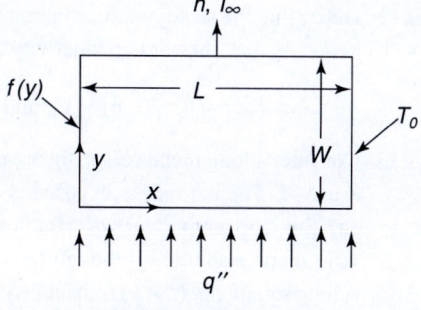

Fig. Q3.4

$$T = \frac{2}{L} \sum_{n=1}^{\infty} \frac{I_0\,(\lambda_n r)}{I_0\,(\lambda_n R)} \sin \lambda_n z \int_0^L f(z) \sin \lambda_n z\, dz$$

where $\lambda_n = n\pi/L$.

3.6 A 5 cm × 10 cm rectangular plate has its two 10 cm sides maintained at 100°C, one of its 5 cm sides maintained at 300°C, and its other 5 cm side maintained at 500°C. Find the temperature at the centre of the plate. Also, draw the isothermal lines within the plate.

3.7 Figure Q3.7 shows a semi-infinite plate of thickness H. A linear temperature distribution is imposed on the left boundary. The boundaries at $y = 0$, $y = H$, and $x \to \infty$ are maintained at the same temperature $(\theta = 0)$. Show that the temperature distribution inside the plate is

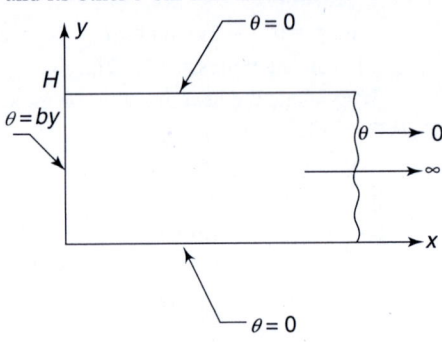

Fig. Q3.7

$$\theta = bH\,\frac{2}{\pi} \sum_{n=0}^{\infty} \frac{(-1)^{n+1}}{n} \exp\left(-n\pi\,\frac{x}{H}\right) \sin\left(n\pi\,\frac{y}{H}\right)$$

3.8 Consider a finite rod of radius R and length L. The temperature of the one-half of the peripheral surface is kept at the uniform temperature θ_0, while the other half and the ends are at zero. Find the steady temperature of the rod, $\theta(r, \phi, z)$.

3.9 Develop an expression for the steady-state temperature distribution $T(x, y)$ in a long bar of rectangular cross-section for the following boundary conditions:

$$T(x, b) = T_1,\ T(x, 0) = T_2,\ T(a, y) = T_3,\ T(0, y) = T_4$$

3.10 Consider a straight fin of rectangular profile and of constant thermal conductivity. The fin has a thickness a in the x-direction and is very long in the y-direction. Obtain an expression for the steady-state temperature distribution $T(x, y)$ in this fin under the following boundary conditions:

$$T(0, y) = 0,\ \frac{\partial T(a, y)}{\partial x} = 0,\text{ and } T(x, 0) = T_0$$

3.11 Obtain an expression for the steady-state temperature distribution $T(r, z)$ in a solid cylinder of length L and radius r_0 for the following boundary conditions:

$$T(r_0, z) = f_1(z),\ T(r, 0) = f_2(r),\ T(r, L) = f_3(r)$$

3.12 Determine the steady-state temperature distribution $T(r, z)$ in a solid rod of radius r_0, height H, and constant conductivity k under the following boundary conditions:

$$\frac{\partial T(r_0, z)}{\partial r} = 0,\ T(r, 0) = T_1,\text{ and } T(r, H) = T_2$$

3.13 Consider a long metal rod of square cross-section $(L \times L)$. The upper and lower faces are at T_0. The left face is exposed to a uniform heat flux q''. The right face is at T_1.
(a) Find the temperature distribution in the rod material.
(b) Obtain a solution if the left face is insulated.

3.14 A long square rod $(L \times L)$ is insulated on two adjacent faces, one other face is subject to convection h to the surrounding at T_∞. The remaining face is at a uniform temperature $T_1 > T_\infty$.

(a) Derive an expression for the heat flow rate across the two uninsulated surfaces, in terms of T_1 and T_∞.

(b) Determine the temperature at the location where the insulated edge is in contact with the surface at T_1.

3.15 A long square rod has convection on two opposite faces, to an ambient fluid at T_∞. The other two faces are at T_1. What is the centre temperature of the rod ?

3.16 Find the steady temperature distribution for a rectangular bar $W \times H$, having the following boundary conditions:

At $x = 0$, $T = f(y)$; At $x = W$, $T = T_0$

At $y = 0$, $q'' =$ constant; At $y = H$, convection to an environment at T_∞.

3.17 Consider an infinitely long cylindrical shell of angular section Φ_0. The inner and outer radii of the shell are r_i and r_o, respectively. The outer surface receives a heat flux $q''(\Phi)$, while the inner surface is maintained at a uniform temperature T_0. The ends of the shell at $\Phi = 0$ and $\Phi = \Phi_0$ are insulated. Find the steady temperature distribution in the shell.

3.18 Consider a semi-infinite solid cylinder of radius R whose base is at temperature T_0 and whose periphery is exposed to a fluid at temperature T_∞ through a heat transfer coefficient h. Determine the steady two-dimensional temperature distribution in the cylinder.

Chapter 4

Unsteady-state Conduction

4.1 Introduction

Unsteady conduction problems are those where the temperature of the body in question varies with both space and time (as in a distributed system) or with only time (as in a lumped system). Time-dependent problems are of two types: transient and periodic problems.

A typical example of the transient case is the heating or cooling of ingots. Transient temperature calculations are very important for the purposes of melting, hot working, heat treatment, and so on, where from the variation of temperature the heat transfer analyst can predict the time required for a particular part to attain predetermined temperature levels. Other examples of industrial importance include starting up or shutting down of a nuclear reactor or a furnace, or of a turbine blade during the startup and shutdown of the turbine when it is subjected to sudden changes in gas temperature.

Periodic problems, on the other hand, are illustrated by the daily periodic variation of heat transfer from the sun to the earth's surface and the temperature fluctuations in the walls of internal combustion engines.

Before we discuss lumped and distributed systems, the concept of Biot number is introduced. We have seen that convective heat transfer to/from the boundaries is important in the formulation and solution of conduction problems. The Biot number, Bi, is defined as

$$\mathrm{Bi} = \frac{hL}{k} \qquad (4.1)$$

where h is the convective heat transfer coefficient, L is the characteristic dimension of the body, and k is the thermal conductivity of the body. Bi can also be rewritten as

$$\mathrm{Bi} = \frac{L/kA}{1/hA} = \frac{\text{conductive resistance}}{\text{convective resistance}}$$

When conductive (or internal) resistance is negligible, $k/L \to \infty$ and $\mathrm{Bi} \to 0$. This case corresponds to a small L or large k, and hence the spatial temperature distribution perpendicular to the boundary having this condition can be neglected. The aforesaid analysis is called lumped system analysis. When convective resistance (or external resistance) is negligible, which is the case for boiling, condensation, and highly turbulent flows, $h \to \infty$ and $\mathrm{Bi} \to \infty$. This implies that the boundary temperature approaches the ambient temperature. When the internal and external resistances are comparable, the general boundary condition cannot be simplified. This and the previous case (i.e., $h \to \infty$) require a distributed system approach.

In short, a lumped system approach assumes that temperatures at all points in the body are the same, whereas a distributed system implies that there is a temperature variation from point to point within a body. Typically, transient conduction in very small bodies or bodies of very high thermal conductivity can be modelled using the lumped system assumption.

The lumped model yields an initial-value problem (the governing equation is a first-order ordinary differential equation). The distributed model, on the contrary, results in an initial and boundary-value problem (the governing equation is a partial differential equation). The distributed system can be solved by the method of separation of variables, as will be seen later in this chapter.

4.2 Lumped System Transients

A small object of volume V, surface area A, density ρ, specific heat c, initially at temperature T_i, is suddenly exposed to an atmosphere at temperature T_∞ (Fig. 4.1). A good engineering application is the transient response of a thermocouple that is suddenly inserted into a flow system. The first step in the analysis is to define a system and indicate the important energy terms (Fig. 4.2). Note that in this case, the entire object is taken as the system rather than only a part of it, as spatial temperature variation is neglected. The only energy transfers that are considered are (1) the convection between the surroundings and the object and (2) the energy storage in the object. From Fig. 4.2, we see that

$$q_c = \dot{E}_s \tag{4.2}$$

where $q_c = hA\,(T_\infty - T)$ and $\dot{E}_s = \rho V c\,(dT/dt)$.

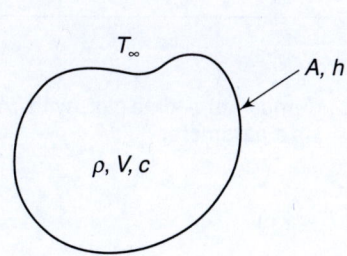

Fig. 4.1 A small object of arbitrary shape at T_i plunged into a fluid at T_∞

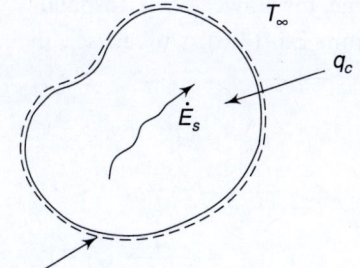

Fig. 4.2 Energy balance using lumped system model

Therefore, $hA\,(T_\infty - T) = \rho V c \dfrac{dT}{dt}$

or $\quad \dfrac{dT}{dt} + \dfrac{hA}{\rho V c}(T - T_\infty) = 0 \tag{4.3}$

Equation (4.3) is the governing equation for the lumped system analysis. It is a first-order ordinary differential equation. The initial condition is as follows:

$$\text{At } t = 0, \ T = T_i \tag{4.4}$$

The solution of Eq. (4.3) subject to the initial condition [Eq. (4.4)] is

$$\frac{T - T_\infty}{T_i - T_\infty} = \exp\left(-\frac{hA}{\rho c V}t\right) \tag{4.5}$$

The governing equation and its solution are valid for heating or cooling. The applicability of Eq. (4.5) is for very small Biot numbers, typically when $Bi = hL/k < 0.1$. L is the length scale corresponding to the maximum spatial temperature difference. While for a plane wall of thickness $2L$, the length scale to be used is L; for long cylinders or spheres, the length scale would be the radius r in a thermally and geometrically symmetric heating or cooling problem. For complex shapes, however, L may be replaced by the ratio of volume V to surface area A.

Using Eq. (4.5), T versus t can be plotted as shown in Fig. 4.3. The plot shows that as the quantity $hA/\rho cV$ decreases, the body takes a longer time to reach the steady-state temperature T_∞. The reciprocal of $hA/\rho Vc$ is called the time constant ϕ. Thus, in terms of ϕ, Eq. (4.5) can be expressed as

$$\frac{T - T_\infty}{T_i - T_\infty} = e^{-t/\phi} \tag{4.6}$$

where $\phi = \rho cV/hA$. The larger the value of ϕ, the slower the body is to respond to the change in temperature. For given thermal properties, the time constant ϕ is proportional to the ratio V/A. Therefore, the smaller the surface area of the body compared with its volume, the slower it will respond.

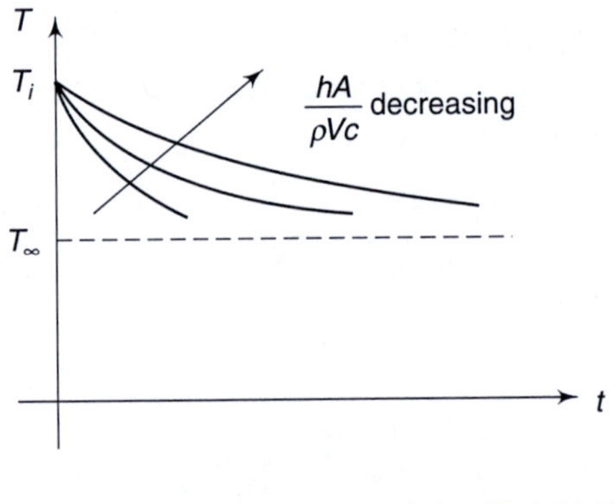

From Eq. (4.6), we can see that when $t = \phi$,

$$T - T_\infty = \frac{1}{e}(T_i - T_\infty)$$

Fig. 4.3 Temperature–time plots with $hA/\rho Vc$ as a parameter

Since $e = 2.7173$,

$$T - T_\infty = \frac{1}{2.7173}(T_i - T_\infty) = 0.368 \ (T_i - T_\infty)$$

or
$$\frac{T - T_\infty}{T_i - T_\infty} = 0.368 \tag{4.7}$$

The above equation shows that the response time (also called the e-folding time) is the time required for the temperature difference between the body and the surroundings to attain 36.8% of the temperature difference between the initial temperature of the body and the surroundings. The response time is nothing but the value of the time constant ϕ.

4.3 Electrical Network Analogy

The concept of a thermal circuit can also be used in the lumped system transient. Equation (4.3) can be also expressed as

$$\rho Vc \frac{dT}{dt} + \frac{T - T_\infty}{(1/hA)} = 0 \qquad (4.8)$$

In Eq. (4.8) the term $1/hA$ represents the entire thermal resistance of the problem, which has been entirely due to convection. The term ρVc represents the thermal capacitance of the problem. At initial time, that is, time zero, the switch is closed (the body is plunged into the fluid) and the capacitor discharges through the resistor to the ground state (i.e., the steady-state temperature of the fluid). The RC product is the time constant of the system. Figure 4.4 shows an equivalent electrical network.

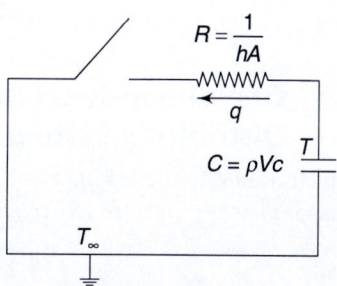

Fig. 4.4 Equivalent electrical network of the lumped system transient

---

Example 4.1   Lumped System Analysis of Transient Cooling of a Cylindrical Rod

During quenching, a cylindrical rod made of 1080 steel, 1 cm in diameter, and 20 cm in length is first heated to 750°C and then immersed in a water bath at 100°C. The heat transfer coefficient can be taken as 250 W/m² °C. The density, specific heat, and thermal conductivity of the steel are $\rho = 7801$ kg/m³, $c = 473$ J/kg °C, and $k = 43$ W/m °C, respectively. Calculate the time required for the rod to reach 300°C.

Solution

From the given data we have

$$\text{Bi} = \frac{hL}{k} = \frac{hr}{k}$$

$$= \frac{250(0.5 \times 10^{-2})}{43} = 0.029$$

Since, Bi < 0.1, we can use the lumped system analysis. From Eq. (4.6),

$$\frac{T - T_\infty}{T_i - T_\infty} = e^{-t/\phi}$$

where $\qquad \phi = \dfrac{\rho c V}{hA}$

$$= \frac{(7801)(473)}{250} \left(\frac{r}{2} \right)$$

$$= \frac{(7801)(473)(1 \times 10^{-2}/4)}{250} = 36.845 \text{ s}$$

Now, $\qquad \dfrac{T - T_\infty}{T_i - T_\infty} = \dfrac{300 - 100}{750 - 100} = 0.3077$

Therefore, $0.3077 = e^{-t/36.845}$

or $\quad \ln(0.3077) = -\dfrac{t}{36.845}$

or $\quad -1.17863 = -\dfrac{t}{36.845}$

or $\quad t = 43.48 \text{ s} \approx 43.5 \text{ s}$

Hence, the time required for the rod to reach 300°C is 43.5 s.

4.4 One-dimensional Transient Problems: Distributed System

Illustration 4.1 *An infinite plate of thickness 2L (Fig. 4.5) having the uniform initial temperature T_i is plunged into a bath at the constant temperature T_∞. The heat transfer coefficient is large. Assume constant k, ρ, c. Find the unsteady temperature of the plate.*

Solution

Case A: Large heat transfer coefficient

Taking $\theta = T - T_\infty$, the governing equation for this problem is

$$\frac{\partial \theta}{\partial t} = \alpha \frac{\partial^2 \theta}{\partial x^2} \qquad (4.9)$$

The initial condition is

$$\theta(x, 0) = \theta_i = T_i - T_\infty \qquad (4.10)$$

and the boundary conditions are

$$\text{BC-1: } \frac{\partial \theta}{\partial x}(0, t) = 0 \qquad (4.11)$$

$$\text{BC-2: } \theta(L, t) = 0 \qquad (4.12)$$

Note that the coordinate system is placed in the middle of the plate since the problem is geometrically and thermally symmetric. Also, at the surface of the plate, the temperature is assumed to be that of the fluid, as the heat transfer coefficient is large ($h \to \infty$).

Since the governing equation is linear and homogeneous, boundary conditions are homogeneous and the initial condition is non-homogenous, the method of separation of variables can be applied. Therefore, a product solution of the form $\theta(x,t) = X(x)\tau(t)$ can be assumed. This yields

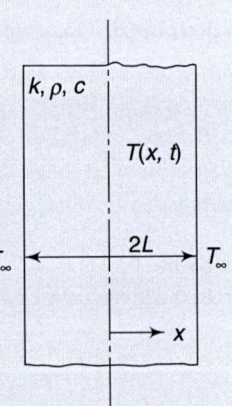

Fig. 4.5 Infinite plate of thickness 2L having unsteady temperature distribution

$$\frac{1}{X}\frac{d^2 X}{dx^2} = \frac{1}{\alpha\tau}\frac{d\tau}{dt} = \pm \lambda^2 \qquad (4.13)$$

$-\lambda^2$ is taken, as the x-direction gives a characteristic-value problem. Hence, we have

$$\frac{d^2 X}{dx^2} + \lambda^2 X = 0 \qquad (4.14\text{a})$$

with $\quad \dfrac{dX}{dx}(0) = 0 \qquad (4.14\text{b})$

$$X(L) = 0 \qquad (4.14\text{c})$$

and $\quad \dfrac{d\tau}{dt} + \alpha\lambda^2 \tau = 0 \qquad (4.15)$

Hence, as with steady problems, the non-homogeneous condition (initial condition) is left to the end of the problem. The general solution of Eq. (4.14a) is

$$X = A\cos \lambda x + B\sin \lambda x$$

$$\frac{dX}{dx} = -A\lambda \sin \lambda x + B\lambda \cos \lambda x$$

Applying Eq. (4.14b),

$$0 = B\lambda$$

The above equation is true only if $B = 0$. Therefore, $X = A\cos \lambda x$. Applying Eq. (4.14c),

$$0 = A\cos \lambda L$$

or

$$0 = \cos \lambda L = \cos (2n+1)\frac{\pi}{2}$$

Therefore, $\lambda_n L = (2n+1)\frac{\pi}{2}$

$$n = 0, 1, 2, 3, \dots$$

Therefore, $X_n(x) = A_n \cos \lambda_n x$ \hfill (4.16)

Now, from Eq. (4.15), we can write

$$\frac{d\tau}{\tau} = -\alpha\lambda^2 dt$$

Integrating, $\ln \tau = -\alpha\lambda^2 t + \ln C$

or

$$\ln\left(\frac{\tau}{C}\right) = -\alpha\lambda^2 t$$

Therefore, $\tau = Ce^{-\alpha\lambda^2 t}$

or

$$\tau_n = C_n e^{-\alpha\lambda_n^2 t}$$ \hfill (4.17)

Hence, the product solution becomes

$$\theta(x,t) = \sum_{n=0}^{\infty} a_n e^{-\alpha\lambda_n^2 t} \cos \lambda_n x$$ \hfill (4.18)

where $a_n = A_n C_n$. Finally, applying the initial condition [Eq. (4.10)],

$$\theta_i = \sum_{n=0}^{\infty} a_n \cos \lambda_n x$$ \hfill (4.19)

Equation (4.19) is a Fourier series expansion of θ_i over the interval $(0, L)$.

Therefore, $a_n = \dfrac{\displaystyle\int_0^L \theta_i \cos \lambda_n x\, dx}{\displaystyle\int_0^L \cos^2 \lambda_n x\, dx}$

$$= \frac{\theta_i \left.\dfrac{\sin \lambda_n x}{\lambda_n}\right|_0^L}{\dfrac{L}{2}}$$

$$= \frac{2\theta_i}{\lambda_n L}\left|\sin \lambda_n L\right|$$

$$= \frac{2\theta_i}{\lambda_n L} \sin(2n+1)\frac{\pi}{2}$$

$$= \frac{2\theta_i}{\lambda_n L}(-1)^n$$

Therefore, $\theta(x,t) = \sum_{n=0}^{\infty} \frac{2\theta_i}{\lambda_n L}(-1)^n e^{-\alpha\lambda_n^2 t} \cos \lambda_n x$

or $\qquad \frac{T(x,t) - T_\infty}{T_i - T_\infty} = 2\sum_{n=0}^{\infty} \frac{(-1)^n}{\lambda_n L} e^{-\alpha\lambda_n^2 t} \cos \lambda_n x \qquad (4.20)$

Case B: Moderate heat transfer coefficient

In this case, the only change will be in respect of the surface boundary condition, that is, BC-2:

$$\text{BC-2: } -k\left[\frac{\partial\theta(L,t)}{\partial x}\right] = h\theta(L,t) \qquad (4.21)$$

Recall the x-direction equation:

$$\frac{d^2 X}{dx^2} + \lambda^2 X = 0 \qquad (4.22a)$$

with $\qquad \frac{dX(0)}{dx} = 0 \qquad (4.22b)$

and $\qquad \frac{dX(L)}{dx} + \frac{h}{k}X(L) = 0 \qquad (4.22c)$

Compare Eq. (4.22c) with Eq. (4.14c) and note the difference. Now, we know the general solution of Eq. (4.22a) is $X = A\cos\lambda x + B\sin\lambda x$. Therefore,

$$\frac{dX}{dx} = -A\lambda\sin\lambda x + B\lambda\cos\lambda x$$

Applying Eq. (4.22b),

$$B = 0$$

Therefore, $X = A\cos\lambda x$. Applying Eq. (4.22c),

$$-A\lambda\sin\lambda L + \frac{h}{k}A\cos\lambda L = 0$$

or $\qquad \lambda\sin\lambda L = \frac{h}{k}\cos\lambda L \qquad (4.23)$

Note that Eq. (4.23) is a transcendental equation and the characteristic values are the roots of Eq. (4.23). Now, from Eq. (4.23), we can write

$$\tan\lambda_n L = \frac{h}{\lambda_n k} = \frac{hL/k}{\lambda_n L}$$

or $\qquad \tan\lambda_n L = \frac{Bi}{\lambda_n L} \qquad (4.24)$

when $\qquad Bi = \frac{hL}{k}$

The roots $(\lambda_n L)$ of Eq. (4.24) can be found either graphically as shown in Fig. 4.6 or by a standard numerical method such as the Newton–Raphson technique. Recall $\theta(x,t) = \sum_{n=1}^{\infty} a_n e^{-\alpha \lambda_n^2 t} \cos \lambda_n x$. Applying the initial condition,

$$\theta_i = \sum_{n=1}^{\infty} a_n \cos \lambda_n x$$

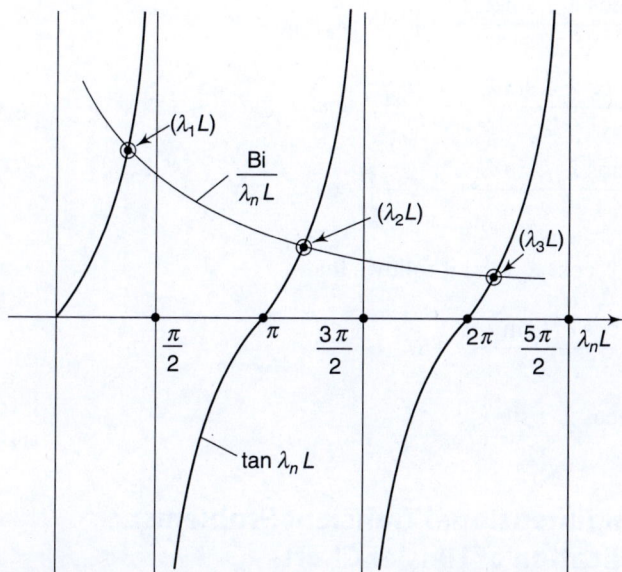

Fig. 4.6 Intersection of $\tan \lambda_n L$ and $\mathrm{Bi}/\lambda_n L$ curves showing the roots

where

$$a_n = \frac{\int_0^L \theta_i \cos \lambda_n x \, dx}{\int_0^L \cos^2 \lambda_n x \, dx} \tag{4.25}$$

$$= \frac{\theta_i \left. \dfrac{\sin \lambda_n x}{\lambda_n} \right|_0^L}{\dfrac{L}{2} + \dfrac{\sin \lambda_n L \cos \lambda_n L}{2\lambda_n}}$$

$$= \frac{2\theta_i \sin \lambda_n L}{\lambda_n L + \sin \lambda_n L \cos \lambda_n L}$$

Therefore,

$$\frac{T(x,t) - T_\infty}{T_i - T_\infty} = 2 \sum_{n=1}^{\infty} \left(\frac{\sin \lambda_n L}{\lambda_n L + \sin \lambda_n L \cos \lambda_n L} \right) e^{-\alpha \lambda_n^2 t} \cos \lambda_n x \tag{4.26}$$

Note that the denominator of Eq. (4.25) is evaluated by using the rule of integration by parts as follows:

$$I = \int_0^L \cos^2 \lambda_n x = \int_0^L \cos \lambda_n x \cos \lambda_n x\, dx$$

$$= \left| \cos \lambda_n x \int \cos \lambda_n x\, dx \right|_0^L - \int_0^L \left(-\lambda_n \sin \lambda_n x \int \cos \lambda_n x\, dx \right) dx$$

$$= \left| \cos \lambda_n x \left[\frac{\sin \lambda_n x}{\lambda_n} \right] \right|_0^L + \int_0^L \lambda_n \sin \lambda_n x \left| \frac{\sin \lambda_n x}{\lambda_n} \right| dx$$

$$= \frac{\cos \lambda_n L \sin \lambda_n L}{\lambda_n} + \int_0^L \sin^2 \lambda_n x\, dx$$

$$= \frac{\cos \lambda_n L \sin \lambda_n L}{\lambda_n} + \int_0^L (1 - \cos^2 \lambda_n x)\, dx$$

$$= \frac{\cos \lambda_n L \sin \lambda_n L}{\lambda_n} + L - \int_0^L \cos^2 \lambda_n x\, dx$$

Since, $I = \int_0^L \cos^2 \lambda_n x\, dx$, it follows that

$$2I = \frac{\cos \lambda_n L \sin \lambda_n L}{\lambda_n} + L$$

or $$I = \frac{\cos \lambda_n L \sin \lambda_n L}{2\lambda_n} + \frac{L}{2}$$

4.5 Multidimensional Transient Problems: Application of Heisler Charts

Let us now consider two-/three-dimensional transient heat transfer problems. To start with, we look at a two-dimensional problem illustrated below.

Illustration 4.2 *An infinitely long rod of rectangular cross-section (2L × 2l) having a uniform initial temperature T_i is plunged suddenly into a bath at constant temperature T_∞. The heat transfer coefficient is h. We wish to find the unsteady temperature of the rod. See Fig. 4.7 for the pictorial description of the problem.*

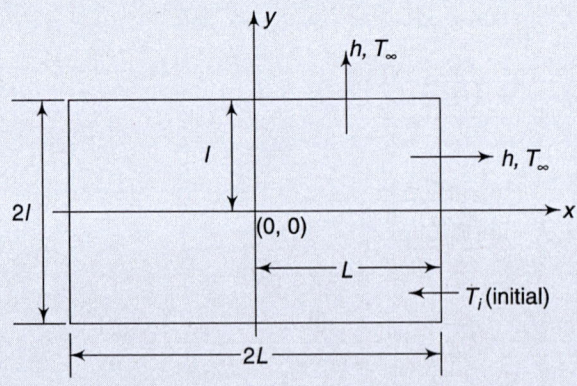

Fig. 4.7 An infinitely long bar of 2L × 2L cross-section plunged into a fluid at T_∞

Solution

The governing differential equation for the problem is

$$\frac{\partial^2 T}{\partial x^2} + \frac{\partial^2 T}{\partial y^2} = \frac{1}{\alpha}\frac{\partial T}{\partial t} \tag{4.27}$$

Let $\theta = (T - T_\infty)/(T_i - T_\infty)$. In terms of θ, the governing equation and the initial and boundary conditions are:

$$\text{GDE:}\quad \frac{\partial^2 \theta}{\partial x^2} + \frac{\partial^2 \theta}{\partial y^2} = \frac{1}{\alpha}\frac{\partial \theta}{\partial t} \tag{4.28}$$

Initial condition: $\theta(x, y, 0) = 1$ $\tag{4.29}$

$$\text{BC-1:}\quad \frac{\partial \theta}{\partial x}(0, y, t) = 0 \tag{4.30}$$

$$\text{BC-2:}\quad -k\frac{\partial \theta}{\partial x}(L, y, t) = h\theta(L, y, t) \tag{4.31}$$

$$\text{BC-3:}\quad \frac{\partial \theta}{\partial y}(x, 0, t) = 0 \tag{4.32}$$

$$\text{BC-4:}\quad -k\frac{\partial \theta}{\partial y}(x, l, t) = h\theta(x, l, t) \tag{4.33}$$

Note that the coordinate system is placed at the centre of the cross-section of the rod as the problem is thermally and geometrically symmetric. The problem could have been solved by the usual separation of variables approach by using a product solution of the form

$$\theta(x, y, t) = X(x)Y(y)\,\tau(t)$$

In the present discussion, however, a less restrictive form $\theta(x, y, t) = X(x, t)Y(y, t)$ will be assumed. If this method succeeds, then it is possible to express an unsteady two-dimensional problem as the product of two unsteady one-dimensional problems. Therefore, Eq. (4.28) becomes

$$\frac{1}{X}\left[\frac{\partial X}{\partial t} - \alpha\frac{\partial^2 X}{\partial x^2}\right] = -\frac{1}{Y}\left[\frac{\partial Y}{\partial t} - \alpha\frac{\partial^2 Y}{\partial y^2}\right] \tag{4.34}$$

Since x and y are independent variables, both sides of Eq. (4.34) must be independent of x and y, and equal to a parameter, say, $\pm \lambda^2(t)$, which can be a function of time. However, because of the geometric as well as thermal symmetry of the problem, the characteristic value problems in the x- and y-direction must be similar. This is only possible when $\lambda^2(t) = 0$. Therefore,

$$\frac{1}{X}\left[\frac{\partial X}{\partial t} - \alpha\frac{\partial^2 X}{\partial x^2}\right] = -\frac{1}{Y}\left[\frac{\partial Y}{\partial t} - \alpha\frac{\partial^2 Y}{\partial y^2}\right] = 0$$

which gives rise to

$$\frac{\partial X}{\partial t} = \alpha\frac{\partial^2 X}{\partial x^2} \tag{4.35a}$$

with $\quad X(x,0) = 1$ $\tag{4.35b}$

$$\frac{\partial X}{\partial x}(0,t) = 0 \tag{4.35c}$$

$$-k\frac{\partial X}{\partial x}(L,t) = hX(L,t) \tag{4.35d}$$

and
$$\frac{\partial Y}{\partial t} = \alpha \frac{\partial^2 Y}{\partial y^2} \tag{4.36a}$$

with
$$Y(y,0) = 1 \tag{4.36b}$$

$$\frac{\partial Y}{\partial y}(0,t) = 0 \tag{4.36c}$$

$$-k\frac{\partial Y}{\partial t}(l,t) = hY(l,t) \tag{4.36d}$$

Thus, the problem becomes expressible as a product of two one-dimensional transient problems.

Recall the solution of the one-dimensional transient heat condition problem with the moderate heat transfer coefficient h:

$$\frac{T - T_\infty}{T_i - T_\infty} = 2\sum_{n=1}^{\infty}\left(\frac{\sin \lambda_n L}{\lambda_n L + \sin \lambda_n L \cos \lambda_n L}\right) e^{-\alpha \lambda_n^2 t} \cos \lambda_n x \tag{4.37}$$

We now use non-dimensional variables such as

$$\xi = \frac{x}{L} \text{ or } \frac{y}{l}$$

$$\text{Fo} = \frac{\alpha t}{L^2} \text{ or } \frac{\alpha t}{l^2} \text{ (Fourier number)}$$

$$\text{Bi} = \frac{hL}{k} \text{ or } \frac{hl}{k}$$

The Fourier number (which is basically dimensionless time) is a measure of the rate at which heat is conducted through the thickness L of a body of volume L^3 (the heat conducting area being L^2) across a temperature difference of ΔT (i.e., $kL^2\Delta T/L$) relative to the rate of change of internal energy of the same body arising out of a temperature change of ΔT in time t (i.e., $\rho cL^3\Delta T/t$). Thus, a large value of the Fourier number indicates faster propagation of heat through the body.

Also, we represent $\lambda_n L$ or $\lambda_n l$ as μ_n, which are the roots of $\mu_n \sin \mu_n = \text{Bi} \cos \mu_n$. ξ is the dimensionless position coordinate and Fo is the dimensionless time. Equation (4.37) becomes

$$\left(\frac{T - T_\infty}{T_i - T_\infty}\right)_{\substack{2L \text{ or } 2l \\ \text{plate}}} = 2\sum_{n=1}^{\infty}\left(\frac{\sin \mu_n}{\mu_n + \sin \mu_n \cos \mu_n}\right) e^{-\mu_n^2 \text{Fo}} \cos \mu_n \xi$$

Therefore,
$$\left(\frac{T - T_\infty}{T_i - T_\infty}\right)_{\substack{2L, 2l \\ \text{plate}}} = \left(\frac{T - T_\infty}{T_i - T_\infty}\right)_{\substack{2L \\ \text{plate}}}\left(\frac{T - T_\infty}{T_i - T_\infty}\right)_{\substack{2l \\ \text{plate}}}$$

Figure 4.8 reproduces Heisler's chart (Heisler, 1947) for the history of temperature in the mid-plane of the plate, $T_c(t) = T(0, t)$. The lines drawn on the figure correspond to fixed values of the reciprocal of the Biot number, k/hL. It can be seen from the chart that for a fixed Bi, bodies with high thermal

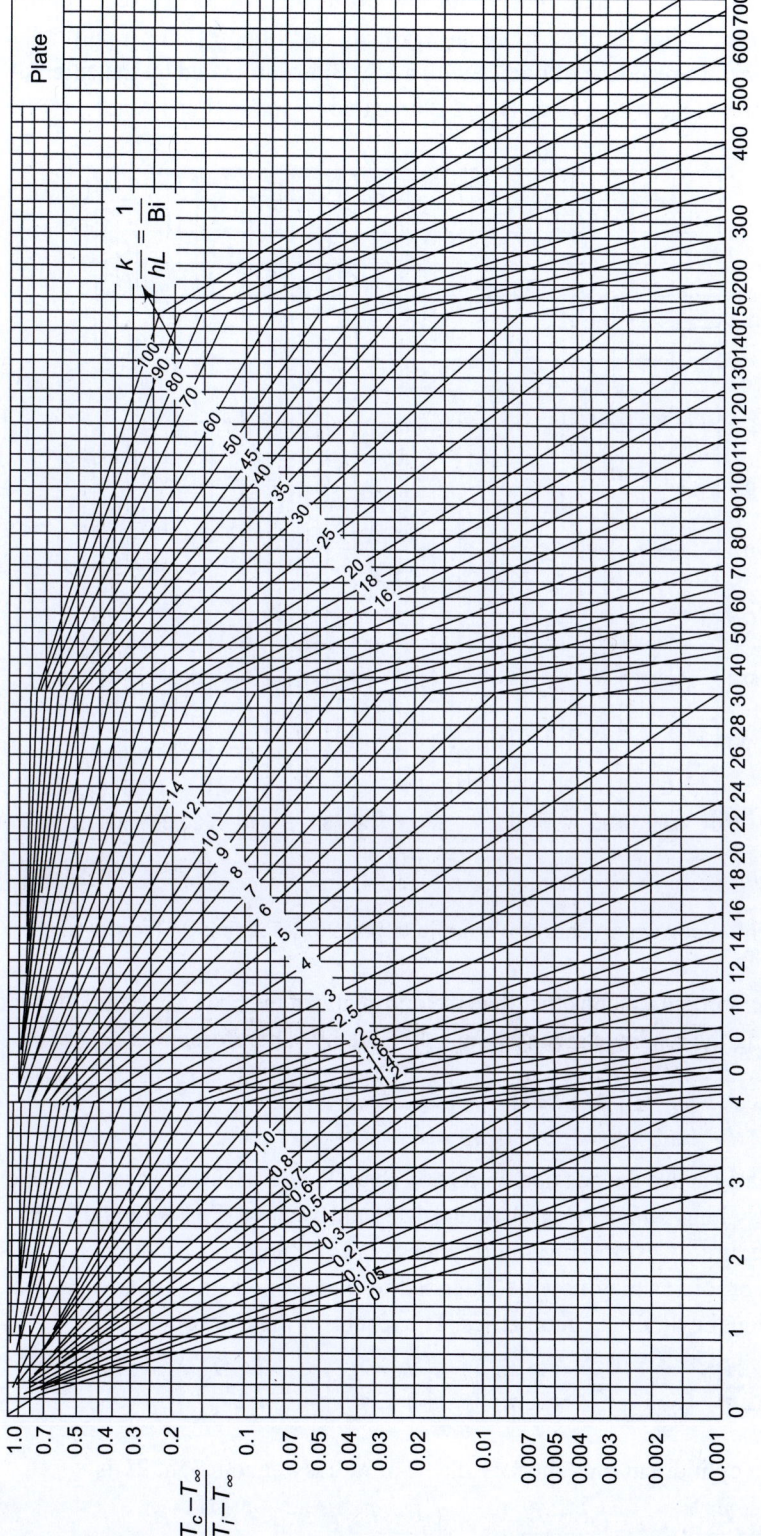

Fig. 4.8 Temperature-time history at mid-plane of an infinite plate of thickness 2 L (*Source*: Bejan, 1993)

diffusivity respond faster than those with low diffusivity, or for a fixed thermal diffusivity large bodies respond more slowly than small bodies. For a particular Fo, the temperature response of bodies with a low Bi is dominated by the external resistance while those with a large Bi are dominated by internal resistance.

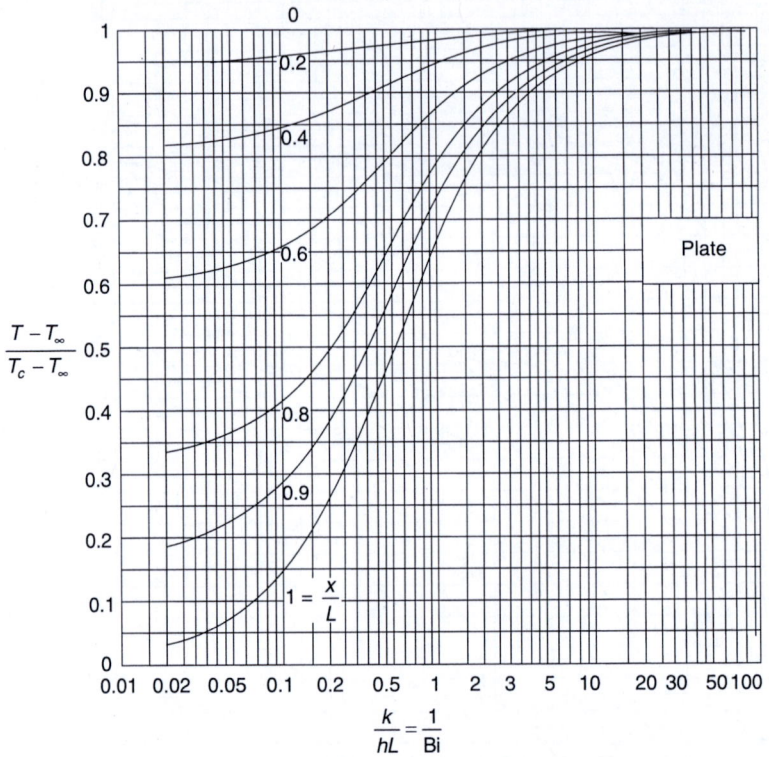

Fig. 4.9 Position-correction chart for an infinite plate of thickness 2L (*source:* Bejan, 1993)

The temperature in a plane other than the mid-plane of the plate can be calculated by multiplying the readings given by Figs 4.8 and 4.9. Therefore,

$$\left(\frac{T-T_\infty}{T_i-T_\infty}\right)_{\text{plate}} = \left(\frac{T-T_\infty}{T_c-T_\infty}\right)_{\text{Fig.4.9}} \left(\frac{T_c-T_\infty}{T_i-T_\infty}\right)_{\text{Fig.4.8}}$$

Figure 4.9 is also called the position-correction chart. Heisler charts are also available for an infinitely long cylinder (Figs 4.10 and 4.11) and a sphere (Figs 4.12 and 4.13).

The procedure of expressing multidimensional problems as products of one-dimensional problems may now be extended to three-dimensional Cartesian and two-dimensional cylindrical geometries. The result for the Cartesian case is

$$\left(\frac{T-T_\infty}{T_i-T_\infty}\right)_{\substack{2L,2l,2H \\ \text{plate}}} = \left(\frac{T-T_\infty}{T_i-T_\infty}\right)_{\substack{2L \\ \text{plate}}} \left(\frac{T-T_\infty}{T_i-T_\infty}\right)_{\substack{2l \\ \text{plate}}} \left(\frac{T-T_\infty}{T_i-T_\infty}\right)_{\substack{2H \\ \text{plate}}}$$

and that for a cylindrical rod (short cylinder) of radius r_0 and height 2L is

$$\left(\frac{T-T_\infty}{T_i-T_\infty}\right)_{\substack{2r_0,2L \\ \text{rod}}} = \left(\frac{T-T_\infty}{T_i-T_\infty}\right)_{\substack{\text{infinite} \\ 2r_0\text{rod}}} \left(\frac{T-T_\infty}{T_i-T_\infty}\right)_{\substack{2L \\ \text{plate}}}$$

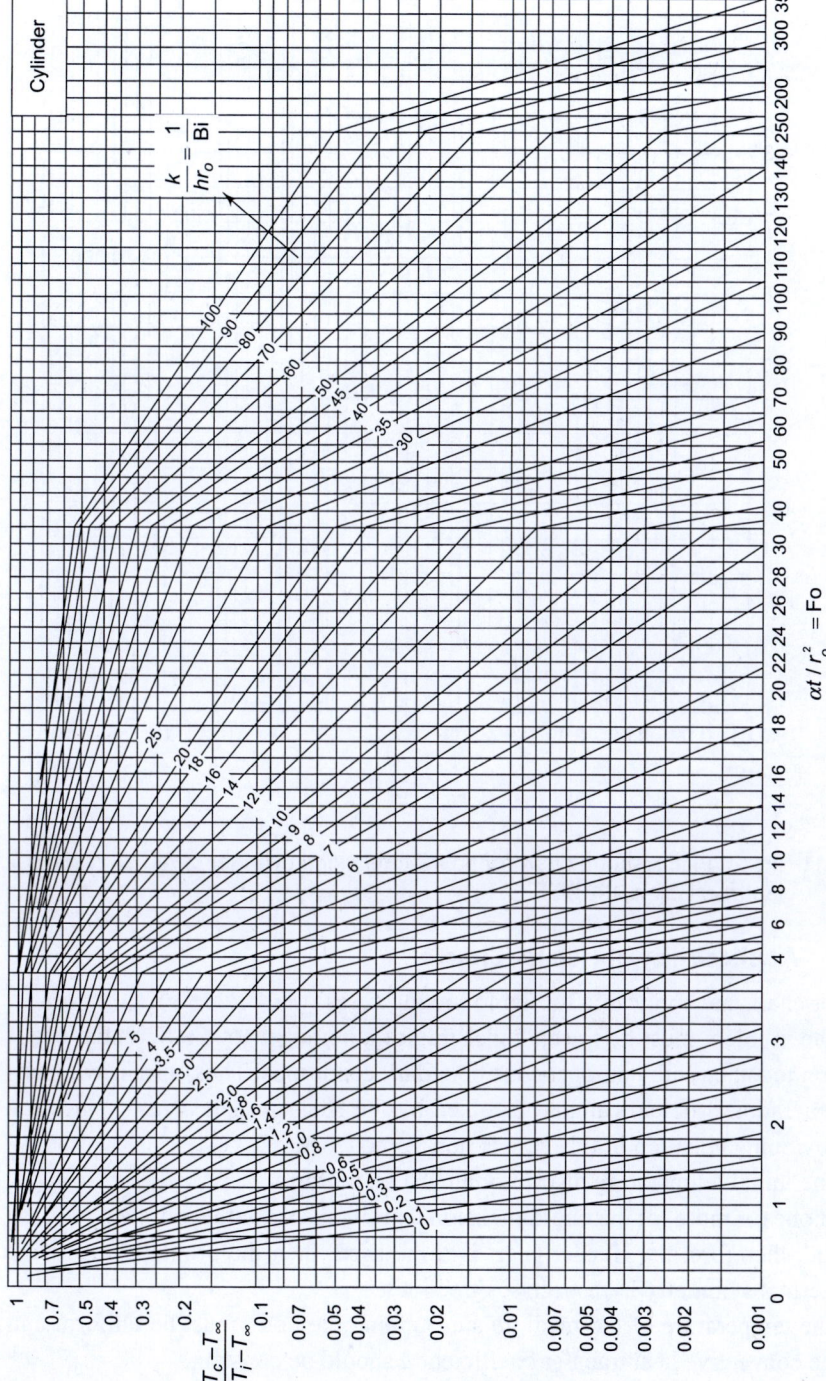

Fig. 4.10 Temperature-time history at the centre line of an infinitely long cylinder of radius r_0 (*Source:* Bejan, 1993)

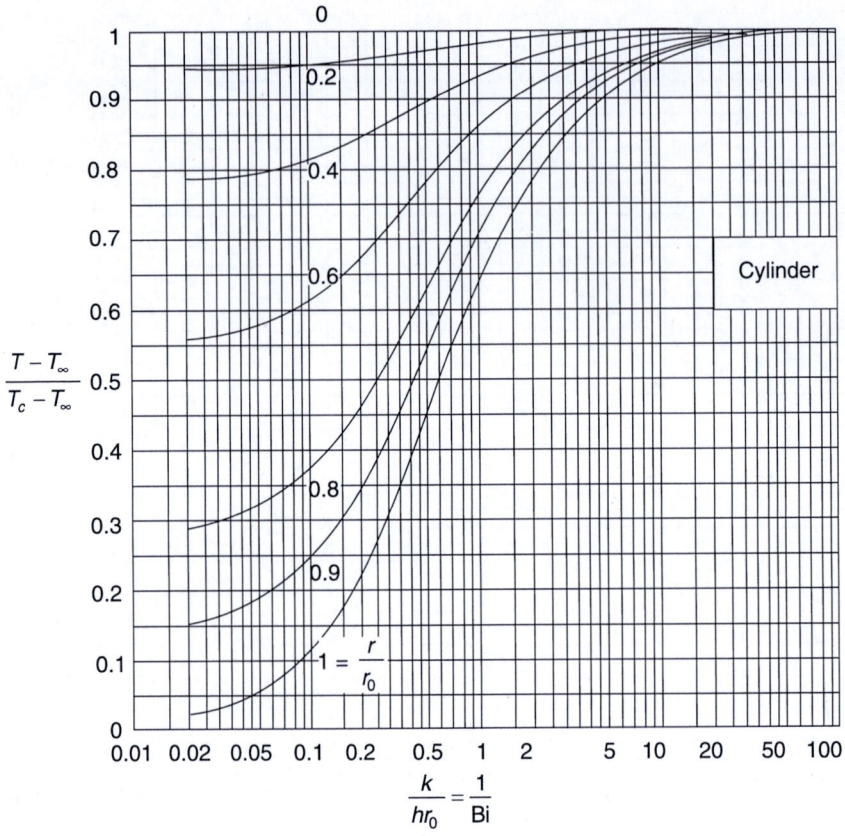

Fig. 4.11   Position-correction chart for an infinitely long cylinder of radius r_0 (*Source:* Bejan, 1993)

4.5.1   Applicability of Heisler Charts

Heisler charts are applicable for the following cases.

(a) The problem must be linear, consisting of a homogeneous differential equation together with homogeneous boundary conditions. Thus, a transient heat-generation problem cannot be solved by the Heisler chart method since the governing equation in this case is non-homogeneous.

(b) The initial temperature distribution must be uniform.

(c) If one (or more) of the one-dimensional problems is for the infinite plate case and, therefore, the Heisler chart is to be used, then the problem must have thermal symmetry in respective directions.

(d) The temperature of the medium surrounding the body must be uniform and the convective heat transfer coefficient h should be constant.

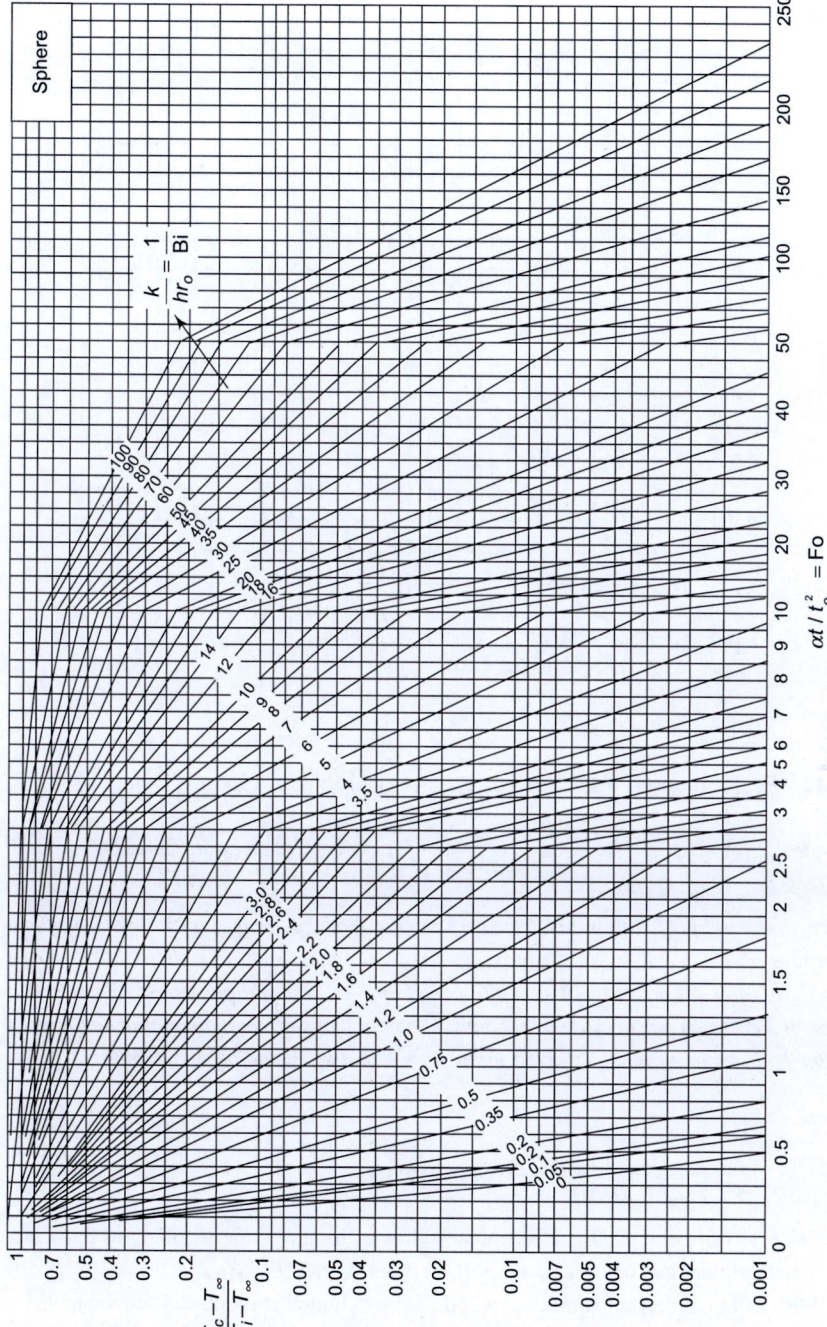

Fig. 4.12 Centre temperature in a solid sphere of radius r_0 (*Source*: Bejan, 1993)

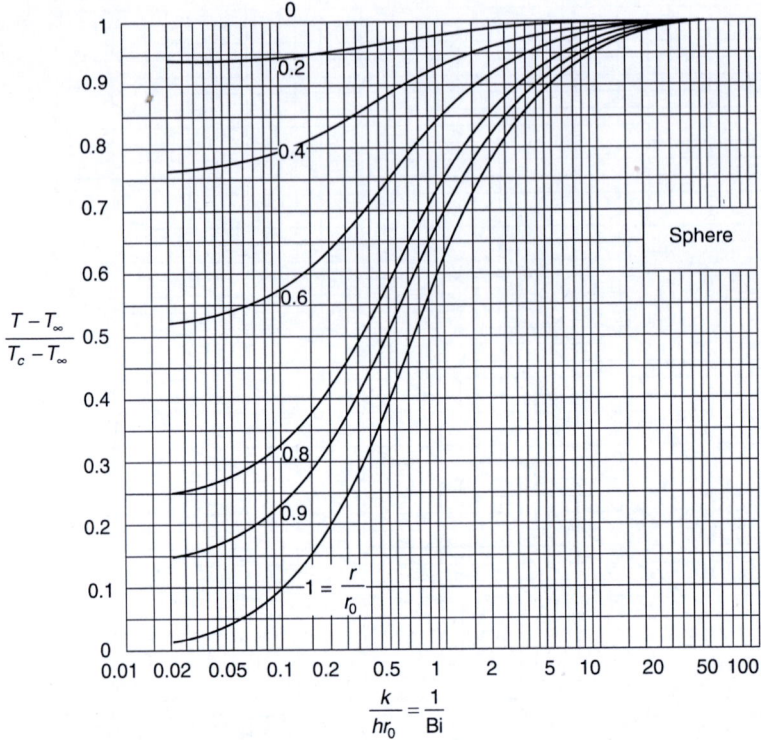

Fig. 4.13   Position-correction chart for a sphere of radius r_0 (*Source:* Bejan, 1993)

Example 4.2   Use of Heisler Charts to Obtain Transient Temperature Distribution in a Short Cylinder

A solid cylinder made of aluminium, 5 cm in diameter and 5 cm long, is initially at a uniform temperature of 300°C. It is suddenly immersed in a water bath at 20°C. Calculate the temperature, after it cools for 10 s, at the centre and at a radial position of 2 cm and a distance of 2 cm from one end of the cylinder. The heat transfer coefficient may be taken as 200 W/m²K. For aluminium, $\rho = 2707$ kg/m³, $c = 896$ J/kg°C, and $k = 204$ W/m°C.

Solution

$$\alpha = \frac{k}{\rho c} = \frac{204}{(2707)(896)} = 8.418 \times 10^{-5} \text{ m}^2 \text{/s}$$

$L/D = 5/5 = 1$. Typically, $L/D > 3$ is representative of long cylinders. Since, in this case, $L/D < 3$, the cylinder can be treated as short. In other words, $T = T(r, z)$.

For both a flat plate of thickness 5 cm and a long cylinder of radius 2.5 cm we have

$$\text{Fo} = \frac{\alpha t}{L^2} = \frac{8.418 \times 10^{-5} \times 10}{(0.025)^2} = 1.347$$

$$\frac{1}{\text{Bi}} = \frac{k}{hL} = \frac{204}{(200)(0.025)} = 40.8$$

Figure E4.2 shows the pictorial representation of the product solution of the infinite plate and infinite cylinder. From Fig. 4.8, for an infinite plate,

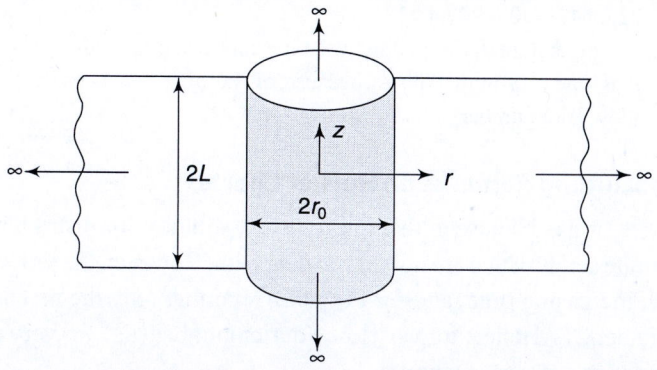

Fig. E4.2 Pictorial representation of the product solution of an infinite plate and an infinite cylinder

$$\left(\frac{T_C - T_\infty}{T_i - T_\infty}\right)_{2L \text{ plate}} = 0.9$$

From Fig. 4.10, for an infinite cylinder,

$$\left(\frac{T_C - T_\infty}{T_i - T_\infty}\right)_{2r_0 \text{ cylinder}} = 0.94$$

Hence,
$$\left(\frac{T_C - T_\infty}{T_i - T_\infty}\right)_{\text{finite cylinder}} = 0.9 \times 0.94 = 0.846$$

or
$$T_C = 0.846\,(T_i - T_\infty) + T_\infty$$
$$= 0.846\,(300 - 20) + 20 = 256.88\,°C$$

At the radial position

$$\frac{r}{r_0} = 0.8, \quad \frac{z}{L} = \frac{0.5}{2.5} = 0.2$$

$$\frac{1}{\text{Bi}} = \frac{k}{hr_0} = 40.8$$

From Fig. 4.9 (position-correction chart for an infinite plate)

$$\left(\frac{T_z - T_\infty}{T_C - T_\infty}\right)_{2L \text{ plate}} = 0.98$$

From Fig. 4.11 (position-correction chart for an infinite cylinder)

$$\left(\frac{T_r - T_\infty}{T_C - T_\infty}\right)_{\text{infinite cylinder}} = 0.98$$

Therefore,
$$\left(\frac{T - T_\infty}{T_C - T_\infty}\right)_{\text{finite cylinder}} = 0.98 \times 0.98 = 0.96$$

or $T - T_\infty = 0.96\,(T_C - T_\infty)$

$$= 0.96(256.88°C - 20) = 227.4°C$$

or $T = 227.4°C + 20 = 247.4°C$

We see that the temperature difference between the centre and the region near the surface of the cylinder is not very high. This is because of the high conductivity of aluminium (reflected by a low Biot number).

4.5.2   Concluding Remarks on Heisler Charts

To summarize, it has been seen that the transient solution for distributed systems involves infinite series which are not easy to deal with. However, the series converges rapidly with increasing time, and for Fo > 0.2, retaining only the first term results in an error under 2%. Hence, for very low Fourier number (i.e., for very early time) Heisler charts are not very accurate.

Heisler (1947) used this one-term approximation and obtained charts for plane wall, long cylinder, and sphere. Corresponding to each geometry there are two charts. The first chart (a semi-log plot with dimensionless centre-temperature on y-axis on a log scale) determines the centre temperature, T_c at a particular time t as a function of the reciprocal of Biot number ($1/\text{Bi} \to 0$ corresponds to $h \to \infty$). In this chart, scales are different in different regions of x-axis, which gives rise to changing gradients. The second chart (also called position-correction chart) determines temperature at locations other than the centre at the same time in terms of T_c. It is also a semi-log plot with $1/\text{Bi}$ on x-axis on a log scale. It may be noted that the second chart is for $t > 0$. One should not expect to retrieve the initial condition from the position-correction chart.

4.6   Semi-infinite Solid

Although all bodies have finite dimensions, a number of cases can be idealized as semi-infinite solids, in which there will be regions which still remain unaffected by a change of temperature on one of their surfaces. In other words, some parts of the body may still remain at the initial temperature even after a long time. Basically, a semi-infinite solid is an idealized body that has a single plane surface and extends to

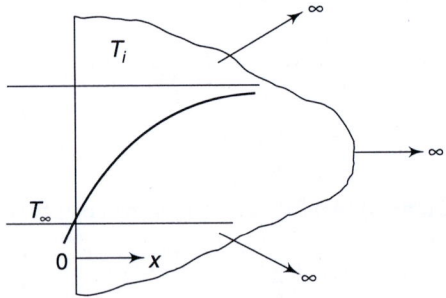

Fig. 4.14   Semi-infinite solid

infinity in all directions as shown in Fig. 4.14. The temperature change we are interested in is due to the imposed thermal condition on a single surface.

A thick plate can be considered as semi-infinite if the transient temperature response of the plate is to be examined for short periods of time after a temperature change on one of its surfaces. This is typical of many materials processing applications such as welding. After hot tea is poured into a porcelain tea cup, the wall may behave like a semi-infinite body at initial times, even though the cup may not be very thick. The earth, for example, can be treated as a semi-infinite medium in calculating the variation of temperature near its surface.

Consider a semi-infinite solid that is initially at temperature T_i. Assume that the surface temperature of the solid is suddenly changed to T_∞. We wish to find the unsteady temperature in the solid. See Fig. 4.14.

The formulation of the problem in terms of $\theta = T - T_\infty$ is as follows. The governing equation is

$$\frac{\partial \theta}{\partial t} = \alpha \frac{\partial^2 \theta}{\partial x^2} \tag{4.38}$$

The initial condition is

$$\theta(x,0) = T_i - T_\infty = \theta_i \tag{4.39}$$

The boundary conditions are

BC-1: $\theta(0, t) = 0$ (4.40)

BC-2: $\theta(\infty, t) = \theta_i$

It is interesting to see how the second boundary condition is expressed. Basically, it says that it will take an infinitely long time for the heat to penetrate to the other end. That is the reason why the temperature at $x \to \infty$ is specified as the initial temperature of the body. An alternative form of BC-2 is $\partial\theta/\partial x = 0$ since except in a small region near the surface the temperature in the rest of the body remains constant at θ_i. Therefore, the x-direction can be considered as the homogeneous direction.

Note that although the x-direction is homogeneous, the problem cannot be solved by the method of separation of variables (Fourier series technique) because in the x-direction, the solid is infinite. It may be recalled that the separation of variables requires finiteness in the homogeneous direction. Thus the Laplace transform, which readily yields a solution, becomes indispensable.

We now digress a little and introduce Laplace transform to the readers and return to the solution so that the readers can appreciate the solution technique.

Laplace transform   The direct Laplace transformation of a real piecewise[1] continuous function $f(t)$, denoted by $L\{f(t)\}$, is defined for positive t in terms of a new variable p (positive real or complex) as the integral

$$L\{f(t)\} = \bar{f}(p) = \int_0^\infty e^{-pt} f(t)dt \tag{4.41}$$

The new $\bar{f}$ function (p) is called the Laplace transform of $f(t)$ with respect to t. A transform is thus obtained simply by multiplying the known function $f(t)$ by e^{-pt} and integrating with respect to t from 0 to ∞. For example, if $f(t) = 1$, then,

$$\bar{f}(p) = \int_0^\infty e^{-pt} dt$$

$$= \left| \frac{-e^{-pt}}{p} \right|_0^\infty$$

---

[1]A function $f(t)$ is said to be piecewise continuous over a finite range if it is possible to divide that range into a finite number of intervals in each of which $f(t)$ is continuous.

$$= \left[0 - \left(-\frac{e^o}{p} \right) \right]$$

$$= \frac{1}{p}$$

The known function $f(t)$ is conversely called the inverse transform of $\bar{f}(p)$ and is denoted as

$$f(t) = L^{-1}\{\bar{f}(p)\}$$

Thus 1 is the inverse transform of $1/p$.

Solution procedure by Laplace transform

(a) Application of Laplace transform to the problem, that is, multiplication of its formulation by e^{-pt} and integration of the result with respect to t from 0 to ∞. The appropriate properties of Laplace transform are then employed to obtain the transform function.

(b) Inversion of the transform function by using either a table of transforms or the inversion theorem for Laplace transforms.

$$L\{1\} = \int_0^\infty e^{-pt} dt = \frac{1}{p}, \quad p > 0$$

$$L\{e^{-\alpha t}\} = \int_0^\infty e^{-(p+\alpha)t} dt = \frac{1}{p+\alpha}, \quad p+\alpha > 0$$

$$L\{\sin \alpha t\} = \int_0^\infty \sin \alpha t \, e^{-pt} dt$$

$$= \frac{\alpha}{p^2 + \alpha^2}, \quad p > 0$$

A typical Laplace transform table is shown in Table 4.1.

Table 4.1 A typical Laplace transform table

S.No.	Transform	Function
1.	$\dfrac{1}{p}$	1
2.	$\dfrac{1}{p+\alpha}$	$e^{-\alpha t}$
3.	$\dfrac{\alpha}{p^2 + \alpha^2}$	$\sin \alpha t$

An exhaustive Laplace transform table is given in Appendix A3.

Properties of Laplace transform

I. $L\{C_1 f(t) + C_2 g(t)\} = C_1 \bar{f}(p) + C_2 \bar{g}(p)$

II. $L\left\{\dfrac{df(t)}{dt}\right\} = p\bar{f}(p) - f(0)$

III. $L\left\{\dfrac{\partial^n f(x_i,t)}{\partial x_i^n}\right\} = \dfrac{\partial^n \overline{f}(x_i,p)}{\partial x_i^n}$

where x_i is a variable independent of t.

IV. $L\left\{\displaystyle\int_0^t f(\tau)d\tau\right\} = \dfrac{1}{p}\overline{f}(p)$

V. If α is a positive constant and $L\{f(t)\} = \overline{f}(p)$, then

$$L\{f(\alpha t)\} = \dfrac{1}{\alpha}\overline{f}\left(\dfrac{p}{\alpha}\right)$$

VI. If β is any constant and $L\{f(t)\} = \overline{f}(p)$, then

$$L\{e^{-\beta t}f(t)\} = \overline{f}(p+\beta)$$

VII. $L\left\{\dfrac{\partial^2}{\partial t^2}f(t)\right\} = p^2\overline{f}(p) - pf(0) - \dfrac{\partial}{\partial t}f(0)$

VIII. $L\left\{\displaystyle\int_a^t f(t)dt\right\} = \dfrac{1}{p}\overline{f}(p) - \dfrac{1}{p}\displaystyle\int_0^a f(t)\,dt$

IX. $L\left\{\displaystyle\int_0^t f_1(\tau)f_2(t-\tau)d\tau\right\} = \overline{f}_1(p)\overline{f}_2(p)$

τ is a dummy variable. The property IX is also called the Faltung or Borel theorem.

Solution of the semi-infinite body problem by the Laplace transform   Now that we know what the Laplace transform is and how it is applied, we return to our original semi-infinite body problem.

Taking the Laplace transform of the governing equation [Eq. (4.38)], we have

$$L\left\{\dfrac{\partial\theta}{\partial t}\right\} = \alpha L\left\{\dfrac{\partial^2\theta}{\partial x^2}\right\}$$

or $\qquad p\overline{\theta} - \theta(0) = \alpha\dfrac{\partial^2\overline{\theta}}{\partial x^2}$ $\qquad\qquad\qquad\qquad$ (4.42)

Note that properties II and III have been used to get the LHS and RHS, respectively, of Eq. (4.42). Note that Eq. (4.42) is an ordinary differential equation (ODE). Thus Laplace transform converts Eq. (4.38), which is originally a partial differential equation (PDE), into an ODE, which is easier to solve. The conversion of a difficult problem (which is not solvable directly) into an easy problem is typical of all transform methods. Rearranging Eq. (4.42),

$$\alpha\dfrac{d^2\overline{\theta}}{dx^2} - p\overline{\theta} = -\theta(0) = -\theta_i$$

or $\qquad \dfrac{d^2\theta}{dx^2} - \dfrac{p}{\alpha}\overline{\theta} = -\dfrac{\theta_i}{\alpha}$

Let $q^2 = p/\alpha$. Then,

$$\dfrac{d^2\overline{\theta}}{dx^2} - q^2\overline{\theta} = -\dfrac{\theta_i}{\alpha}$$ $\qquad\qquad\qquad\qquad$ (4.43)

Equation (4.43) is subject to the transforms of BC-1 and BC-2:

$$L\{\theta(0, t)\} = \bar{\theta}(0, p) = 0 \tag{4.44}$$

$$L\{\theta(\infty, t)\} = \bar{\theta}(\infty, p) = L\{\theta_i\} = \frac{\theta_i}{p} \tag{4.45}$$

The general solution of Eq. (4.43) is

$$\bar{\theta}(x, p) = Ae^{-qx} + Be^{qx} - \left(\frac{-\theta_i/\alpha}{q^2}\right)$$

$$= Ae^{-qx} + Be^{qx} + \frac{\theta_i}{p} \tag{4.46}$$

Applying Eq. (4.44),

$$0 = A + B + \frac{\theta_i}{p} \tag{4.47}$$

Applying Eq. (4.45)

$$\frac{\theta_i}{p} = Be^{\infty} + \frac{\theta_i}{p} \tag{4.48}$$

From Eqs (4.47) and (4.48)

$$B = 0$$

$$A = -\frac{\theta_i}{p}$$

Substituting A and B in Eq. (4.46), we have

$$\bar{\theta}(x, p) = -\frac{\theta_i}{p}e^{-qx} + \frac{\theta_i}{p} \tag{4.49}$$

or

$$\frac{\bar{\theta}(x, p)}{\theta_i} = \frac{1}{p} - \frac{1}{p}e^{-qx} \tag{4.50}$$

The next step is to invert the transformed solution that has been obtained in terms of $\bar{\theta}$ [Eq. (4.50)].

Taking an inverse transform of Eq. (4.50),

$$\frac{\theta(x, t)}{\theta_i} = 1 - \text{erfc}\left[\frac{x}{2(\alpha t)^{1/2}}\right] \tag{4.51}$$

The first and second terms in Eq. (4.51) are obtained by using Nos 1 and 27, respectively, from the table of Laplace transforms listed in Appendix A3.

Since, erfc(z) = 1– erf(z), where z is the argument of the function, Eq. (4.51) becomes

$$\frac{\theta(x, t)}{\theta_i} = \text{erf}\left[\frac{x}{2(\alpha t)^{1/2}}\right] \tag{4.52}$$

The error function table is listed in Appendix A4. erf(z) and erfc(z) are called the error function and the complementary error function, respectively:

$$erf(z) = \frac{2}{\sqrt{\pi}} \int_{\lambda=0}^{z} e^{-\lambda^2} d\lambda$$

$$\frac{d}{dz}[erf(z)] = \frac{2}{\sqrt{\pi}} e^{-z^2}$$

The surface heat flux at $x = 0$ is

$$q_s'' = k \frac{\partial T}{\partial x}\bigg|_{x=0}$$

In this case, $\partial T/\partial x|_{x=0}$ is positive as $T_i > T_\infty$. Now, from Eq. (4.52),

$$\frac{T - T_\infty}{T_i - T_\infty} = erf\left[\frac{x}{2(\alpha t)^{1/2}}\right]$$

or

$$T = T_\infty + (T_i - T_\infty)erf\left[\frac{x}{2(\alpha t)^{1/2}}\right]$$

or

$$\frac{\partial T}{\partial x} = (T_i - T_\infty)\frac{2}{\sqrt{\pi}}[e^{-x^2/4\alpha t}]\frac{1}{2(\alpha t)^{1/2}}$$

Therefore, $\dfrac{\partial T}{\partial x}\bigg|_{x=0} = (T_i - T_\infty)\dfrac{1}{\sqrt{\pi \alpha t}}$

$$q_s'' = k \frac{\partial T}{\partial x}\bigg|_{x=0} = \frac{k(T_i - T_\infty)}{\sqrt{\pi \alpha t}}$$

4.6.1 Other Surface Boundary Conditions

The boundary condition in the problem just discussed basically assumes constant surface temperature. However, in many practical circumstances, the transient is actually caused by a change in the environment temperature to T_∞, giving rise to convection heat transfer to/from the surface. Sometimes it may be due to a sudden thermal flux loading q'' at the surface, caused by radiation or by induction heating absorbed very near the surface. The solutions for such situations are given below.

Case A
Surface convection ($T_\infty > T_i$)

$$BC\text{-}1: \ -k\frac{\partial T}{\partial x}\bigg|_{x=0} = h[T_\infty - T(0, t)]$$

Solution

$$\frac{T(x, t) - T_i}{T_\infty - T_i} = erfc\left(\frac{x}{2\sqrt{\alpha t}}\right)$$

$$-\left[\exp\left(\frac{hx}{k} + \frac{h^2 \alpha t}{k^2}\right)\right]\left[erfc\left(\frac{x}{2\sqrt{\alpha t}} + \frac{h\sqrt{\alpha t}}{k}\right)\right]$$

Case B

Constant surface flux q″

BC-1: $q'' = -k\dfrac{\partial T}{\partial x}$

or $\dfrac{\partial T}{\partial x} = -\dfrac{q''}{k}$

Solution

$$T(x,t) - T_i = \frac{2q''(\alpha t/\pi)^{1/2}}{k}\exp\left(-\frac{x^2}{4\alpha t}\right) - \frac{q''x}{k}\operatorname{erfc}\left(\frac{x}{2\sqrt{\alpha t}}\right)$$

These solutions can be easily obtained by the Laplace transformation method.

4.6.2 Penetration Depth

The solution of the semi-infinite body problem with constant surface temperature T_∞ is plotted in Fig. 4.15 (T versus x for various t's). It is clear that the temperature gradient along each T versus x curve becomes smaller as time increases, that is, as the effect of having dropped the surface temperature from T_i to T_∞ diffuses into the semi-infinite solid. This brings us to the concept of *penetration depth* at a given time, which is defined as the distance up to which the temperature gradient exists and beyond which the body remains at the initial temperature. Mathematically, this is the distance at which $T_i - T$ is 1% of $T_i - T_\infty$. It is denoted by the symbol δ, and obviously it increases with time. The function of δ versus t can be found by using $\eta = x/2\sqrt{\alpha t}$ and within Eq. (4.52) as

$$\frac{\theta(x,t)}{\theta_i} = \frac{T - T_\infty}{T_i - T_\infty} = \operatorname{erf}(\eta)$$

$$(4.53)$$

The plot of θ/θ_i versus η is shown in Fig. 4.16. For $\eta = 1.8$, the value of erf(η) is 0.99 (Fig. 4.17), signifying that $T_i - T$ is only 1% of $T_i - T_\infty$.

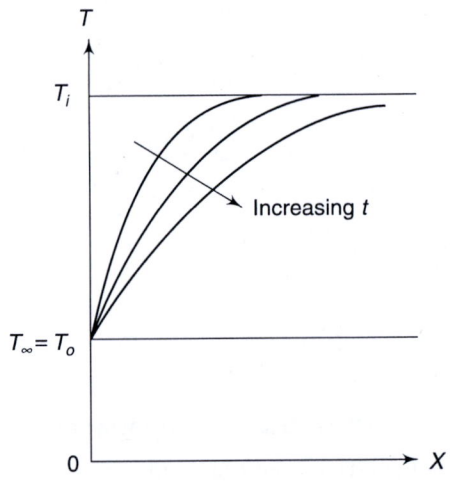

Fig. 4.15 Heat penetration into a semi-infinite solid with isothermal surface

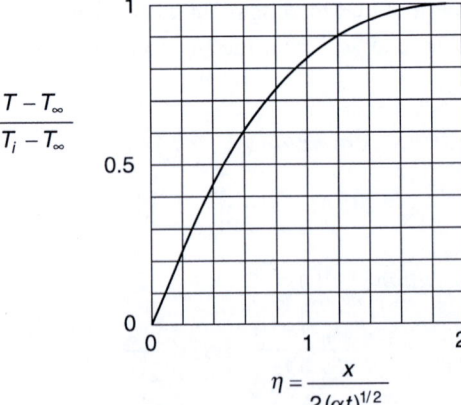

$$\eta = \frac{x}{2(\alpha t)^{1/2}}$$

Fig. 4.16 θ/θ_i versus η profile signifying merging of all T versus x curves into a single similarity profile

Beyond this value the semi-infinite body is assumed to be at the initial temperature. If $\delta(t)$ denotes the penetration depth at time t, then

$$\frac{\delta}{2\sqrt{\alpha t}} = 1.8$$

or $\qquad \delta = 3.6\sqrt{\alpha t}$ $\qquad$ (4.54)

The penetration depth is directly proportional to the square root of time for a given body. This also indicates that a low thermal diffusivity material closely satisfies the definition of a semi-infinite solid because at a given time the penetration depth will be very small.

Example 4.3 Finger Touching a Hot Wall : A Semi-infinite Body Conduction Model

Consider a finger touching a hot wall [Fig. E4.3(a)]. Initially both the finger and the hot wall (being made of low thermal diffusivity materials) will behave like semi-infinite solids. At the instant of contact, the finger–wall interface will assume an equilibrium temperature T_c, which will not change with time. Obtain an expression for T_c in terms of initial temperatures and the properties of the finger and the wall.

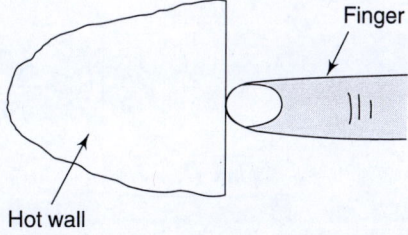

$$\eta = \frac{x}{2\sqrt{\alpha t}}$$

Fig. 4.17 Error function as listed in Appendix *A4* and the complementary error function

Solution
Basically, this is a problem of two semi-infinite bodies having different initial temperatures coming in intimate contact (Fig. E4.3(b)). Therefore, heat transfer is occurring from the body at a higher initial temperature (in this case, the hot wall) to that at a lower initial temperature (in this case, the finger). The temperature profiles with increasing time in the wall and the finger are also shown in Fig. E4.3(b).

Fig. E4.3(a) Finger touching a hot wall

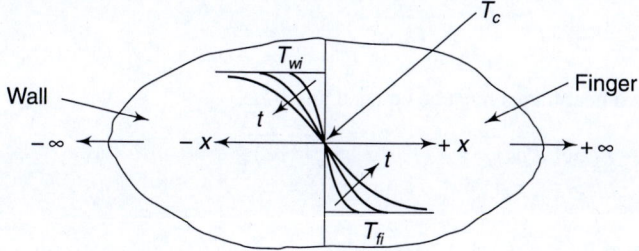

Fig. E4.3(b) Temperature transients in the finger and wall modelled as semi-infinite bodies

The governing differential equations (GDEs) and the initial (IC) and boundary conditions (BCs) for the finger and the wall are given below.

For finger

GDE: $\dfrac{\partial^2 T}{\partial x^2} = \dfrac{1}{\alpha}\dfrac{\partial T}{\partial t}$

IC: $T(0, x) = T_{f_i}$
BC-1: $T(0, t) = T_c$
BC-2: $T(\infty, t) = T_{f_i}$

For wall

GDE: $\dfrac{\partial^2 T}{\partial x^2} = \dfrac{1}{\alpha}\dfrac{\partial T}{\partial t}$

IC: $T(0, -x) = T_{w_i}$

BC-1: $T(0, t) = T_c$

BC-2: $T(-\infty, t) = T_{w_i}$

Note that here the unknown is T_c, which is to be obtained by satisfying the equality of heat fluxes at the interface of the finger and the wall. From the solution of a semi-infinite solid with isothermal surface temperature [Eq. (4.52)] we can write for this case,

$$T(x,t) = T_i + (T_c - T_i)\,\text{erfc}\,\dfrac{x}{2\sqrt{\alpha t}}$$

The heat flow per unit surface area into the solid is given by

$$q'' = -k\dfrac{\partial T}{\partial x}\bigg|_{x=0}$$

$$= -k(T_c - T_i)\left\{\dfrac{\partial}{\partial x}\left[\text{erfc}\,\dfrac{x}{2\sqrt{\alpha t}}\right]\right\}_{x=0}$$

$$= \dfrac{\sqrt{k\rho c}\,(T_c - T_i)}{\sqrt{\pi t}}$$

Consequently, the heat flow into the finger is given by

$$q''_f = \dfrac{1}{\sqrt{\pi t}}(T_c - T_{f_i})\sqrt{(k\rho c)_f}$$

and the heat flow from the wall is

$$q''_w = -\dfrac{1}{\sqrt{\pi t}}(T_c - T_{w_i})\sqrt{(k\rho c)_w}$$

Since these two heat fluxes must be equal at any time,

$$(T_c - T_{f_i})\sqrt{(k\rho c)_f} = -(T_c - T_{w_i})\sqrt{(k\rho c)_w}$$

which gives

$$T_c = \dfrac{T_{w_i}\sqrt{(k\rho c)_w} + T_{f_i}\sqrt{(k\rho c)_f}}{\sqrt{(k\rho c)_w} + \sqrt{(k\rho c)_f}}$$

Additional Examples

Example 4.4 Unsteady Heat Conduction in a Solid Sphere

A solid sphere of radius R having a uniform initial temperature T_0 is plunged suddenly into a bath at temperature T_∞. The heat transfer coefficient is large. Find the unsteady temperature of the sphere.

Solution

The formulation of the problem in terms of $\theta = T - T_\infty$ is as follows:

GDE: $\dfrac{1}{\alpha} \dfrac{\partial \theta}{\partial t} = \dfrac{1}{r^2} \dfrac{\partial}{\partial r} \left(r^2 \dfrac{\partial \theta}{\partial r} \right)$

IC: $\theta(r, 0) = \theta_0$

BC-1: $\theta(0, t) = $ finite

or $\qquad \dfrac{\partial \theta}{\partial r}(0, t) = 0$

BC-2: $\theta(R, t) = 0$

The problem can be solved by using a well-known transformation

$$\theta(r, t) = \frac{\psi(r, t)}{r}$$

Thus the transformed GDE, IC, and BCs are as follows:

GDE: $\qquad \dfrac{\partial \psi}{\partial t} = \alpha \dfrac{\partial^2 \psi}{\partial r^2}$

IC: $\qquad \psi(r, 0) = r\theta_0$

BC-1: $\psi(0, t) = 0$

BC-2: $\psi(R, t) = 0$

Using the product solution of the form $\psi(r, t) = \Re(r)\tau(t)$ yields

$$\frac{d^2 \Re}{dr^2} + \lambda^2 \Re = 0 \tag{A}$$

$$\Re(0) = 0 \tag{A1}$$
$$\Re(R) = 0 \tag{A2}$$

and $\qquad \dfrac{d\tau}{dt} = \alpha\lambda^2 \tau \tag{B}$

The solution of Eq. (A) subject to BCs (A1) and (A2) is

$$\Re_n(r) = A_n \phi_n(r)$$

where $\phi_n(r) = \sin \lambda_n r$; $\lambda_n R = n\pi$ and $n = 1, 2, 3, \ldots.$

The general solution of Eq. (B) is

$$\tau_n(t) = C_n e^{-\lambda_n^2 t}$$

Therefore, $\psi(r, t) = \displaystyle\sum_{n=1}^{\infty} a_n e^{-\alpha\lambda_n^2 t} \sin \lambda_n r \tag{C}$

where $a_n = A_n C_n$. Now, applying the IC, we obtain

$$r\theta_0 = \sum_{n=1}^{\infty} a_n \sin \lambda_n r$$

which is a sine Fourier series. Therefore,

$$a_n = \frac{\displaystyle\int_0^R (r\theta_0)\sin \lambda_n r\, dr}{\displaystyle\int_0^R \sin^2 \lambda_n r\, dr}$$

$$= \frac{\theta_0 \displaystyle\int_0^R r\sin \lambda_n r\, dr}{\dfrac{R}{2}}$$

$$= \frac{2\theta_0}{\lambda_n}(-1)^{n+1}$$

Finally, from Eq. (C) we can write

$$\frac{T(r,t) - T_\infty}{T_0 - T_\infty} = 2\sum_{n=1}^{\infty}(-1)^{n+1} e^{-\alpha\lambda_n^2 t} \frac{\sin \lambda_n r}{\lambda_n r}$$

Example 4.5   Cooling of Extruded Aluminium Bars

In an aluminium factory, aluminium bars of 2.5 cm × 5 cm cross-section are extruded at 500°C. An extruded bar passes through a water spray bath after it leaves the extrusion die. The water spray is available at a temperature of 25°C. The heat transfer coefficient between the cooling water and the surface of the bars is 5000 W/m² K. The properties of aluminium are: k = 230 W/mK, ρ = 2707 kg/m³, and c = 896 J/kgK.

(a) If the bars are extruded at a velocity of 0.5 m/s, determine the length L of the cooling tank required to reduce the centre-line temperature of the bars to 150°C.

(b) What is the maximum possible surface temperature that can be reached on the extruded bars after they leave the cooling tank?

Solution

(a) The aluminium bars (of $2L \times 2l$ cross-section) are initially at a temperature of 500°C. They are cooled by water at 25°C. The objective is to calculate the length of the cooling tank required to reduce the centre-line temperature of the bars to 150°C.

  If L (in m) is the length of the cooling tank, then the time t available for cooling is

$$t = \frac{L}{\text{speed of extrusion}} = \frac{L}{0.5}\, s$$

or $L = 0.5t$ (A)

Now, the problem boils down to finding t.

Given: $T_c = 150°C$, $T_i = 500°C$, $T_\infty = 25°C$

$$\left(\frac{T_c - T_\infty}{T_i - T_\infty}\right)_{\text{finite plate } (2L \times 2l)} = \left(\frac{T_c - T_\infty}{T_i - T_\infty}\right)_{\text{infinite } 2L \text{ plate}} \times \left(\frac{T_c - T_\infty}{T_i - T_\infty}\right)_{\text{infinite } 2l \text{ plate}}$$

$$= \frac{150 - 25}{500 - 25} = 0.263 \qquad (B)$$

Now, $\alpha = \dfrac{k}{\rho c} = \dfrac{230}{2707 \times 896} = 9.48 \times 10^{-5} \dfrac{m^2}{s}$

$h = 5000 \text{ W/m}^2 \text{ K}$

Also, for the $2l$ (i.e., 2.5 cm thick) infinite plate,

$$\dfrac{1}{Bi} = \dfrac{k}{hl} = \dfrac{230}{5000 \times 1.25 \times 10^{-2}} = 3.68$$

and $\quad Fo = \dfrac{\alpha t}{l^2} = \dfrac{(9.48 \times 10^{-5})t}{(1.25 \times 10^{-2})^2} = 0.6067\,t$

For the $2L$ (i.e., 5 cm thick) infinite plate,

$$\dfrac{1}{Bi} = 1.84$$

$$Fo = 0.1517\,t$$

In order to obtain t which will satisfy Eq. (B), $T_c - T_\infty/(T_i - T_\infty)$ has to be obtained iteratively (starting with, let us say, $t = 2$ sec) by the use of the Heisler chart (Fig. 4.8). It is seen that $t = 6.5$ sec nearly satisfies Eq. (B). Therefore, substituting $t = 6.5$ sec into Eq. (A) we get

$$L = 0.5 \times 6.5 = 3.25 \text{ m}$$

Hence, the length of the cooling tank is 3.25 m.

(b) It is obvious that the maximum surface temperature will occur at locations A (0, 1.25 cm) and B (0, −1.25 cm) of the bar cross-section which are at the least distance from the centre C (0, 0). Because of symmetry, $T_A = T_B$. So the determination of temperature at A is sufficient. Therefore, by using Heisler charts (Figs 4.8 and 4.9),

$$\left(\dfrac{T_A - T_\infty}{T_C - T_\infty} \right)_{\text{finite plate}} = \left(\dfrac{T_A - T_\infty}{T_C - T_\infty} \right)_{2.5 \text{ cm infinite plate}} \times \left(\dfrac{T_A - T_\infty}{T_C - T_\infty} \right)_{5 \text{ cm infinite plate}}$$

$$= 0.88 \times 1 = 0.88$$

from which we get

$T_A = 135\,°C$, which is the maximum surface temperature.

Example 4.6 Dimensionless Temperature in Lumped System as a Function of Biot Number and Fourier Number

Starting from Eqs (4.3) and (4.4) in lumped system transient analysis show the detailed solution methodology which led to the solution [Eq.(4.5)]. Also, show that

$$\dfrac{\theta}{\theta_i} = \dfrac{T - T_\infty}{T_i - T_\infty} = e^{-Bi \cdot Fo}$$

Solution

The governing differential equation [Eq.(4.3)] for lumped system transient is

$$\dfrac{dT}{dt} + \dfrac{hA}{\rho Vc}(T - T_\infty) = 0 \qquad\qquad\qquad (A)$$

Subject to the initial condition [Eq. (4.4)]

$\qquad$ At $t = 0$, $T = T_i$ $\qquad\qquad\qquad\qquad\qquad\qquad\qquad\qquad\qquad\qquad (B)$

Equation (A) is a non-homogeneous ordinary differential equation. Therefore, the following transformation is used to make Eq.(A) homogeneous.

$$\theta = T - T_\infty \tag{C}$$

Substituting Eq. (C) into Eq. (A), we get

$$\frac{d\theta}{dt} + \frac{hA}{\rho Vc}\theta = 0 \tag{D}$$

Equation (D) can also be written as

$$\frac{d\theta}{\theta} = -\frac{hA}{\rho Vc} dt \tag{E}$$

Integrating Eq. (E) and taking the limits from $\theta = \theta_i$ at $t = 0$ (which is the transformed Eq. (4.4)) to $\theta = \theta$ at $t = t$, we obtain

$$\ln\left(\frac{\theta}{\theta_i}\right) = -\frac{hA}{\rho Vc} t$$

or

$$\frac{\theta}{\theta_i} = e^{-\frac{hA}{\rho Vc}t} \tag{F}$$

which is same as Eq. (4.5).

Now, note that

$$\frac{hA}{\rho Vc} t = \frac{ht}{\rho c L_c} = \frac{hL_c}{k}\frac{k}{\rho c}\frac{t}{L_c^2} = \frac{hL_c}{k}\frac{\alpha t}{L_c^2} = \text{Bi.Fo} \tag{G}$$

where L_c = characteristic length = $\dfrac{V}{A}$

   $\text{Bi} = \text{Biot number} = \dfrac{hL_c}{k}$

   $\text{Fo} = \text{Fourier number} = \dfrac{\alpha t}{L_c^2}$

Thus, from Eqs (F) and (G), the final expression is

$$\frac{\theta}{\theta_i} = \frac{T - T_\infty}{T_i - T_\infty} = e^{-\text{Bi.Fo}}$$

Example 4.7   Biot Number Effect on Transient Temperature Profiles

Consider a plane wall of thickness 2L. Initially the wall is at a uniform temperature, T_i. Suddenly it is plunged into a bath of water at T_∞ ($T_i > T_\infty$). Draw qualitatively temperature profiles at an early time, an intermediate time and a large time for three cases, namely, Bi << 0.1, Bi = Moderate and Bi → ∞.

Solution
Case A: Bi << 0.1 (Lumped System)
Since Bi is very small and much below 0.1 (which is the upper limit for lumped system) the system can be treated as lumped. This means that conductivity of the wall is very high. Thus, temperature within the wall is almost uniform since the conductive resistance will be small as compared to the convective resistance. Figure E4.7(a) shows the transient temperature distributions. It can be seen that at an early time the core of the wall is still at the initial temperature since the effect of the boundary condition has not sunk in to the

interior. At an intermediate time the temperature at all points in the wall has dropped. At a large time (near steady state) the temperature of wall is very close to T_∞ which is its steady state temperature.

Case B: Bi = Moderate (Distributed System)

Bi = Moderate implies that conductive resistance is of the same order as the convective resistance and hence there will be temperature variation within the wall (Fig. E4.7(b)). Furthermore, there will be a finite temperature difference between the surface and the fluid, the difference being high at early times and low when the steady state is approached.

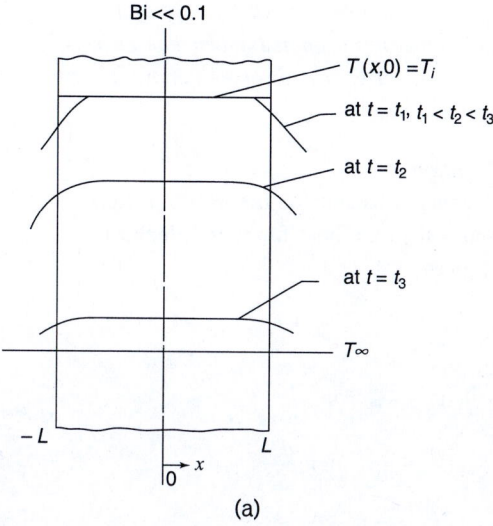

(a)

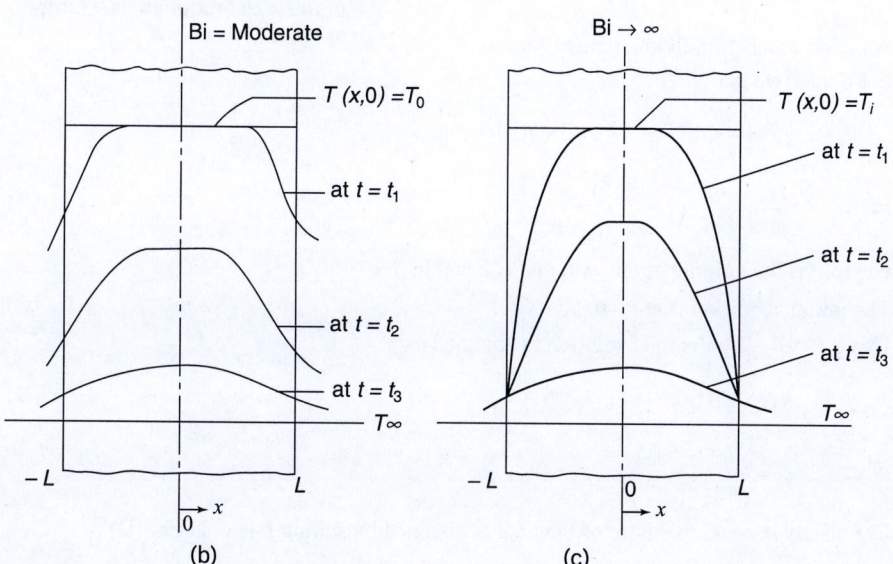

(b)   (c)

Fig. E4.7 Transient temperature profiles for (a) Bi ≪ 0.1, (b) Bi = Moderate, (c) Bi → ∞

Case C: Bi → ∞ (Distributed System)

Bi → ∞ means that 'h' is very large (e.g., boiling, condensation and highly turbulent flows) and hence the convective resistance is approaching zero. The implication of $h \to \infty$ is that in this situation the surface temperature will be very close to the ambient temperature at all times, i.e., at $t > 0$. The difference between surface and ambient temperatures is very small and same at all times. Figure E4.7(c) depicts the transient temperature profiles.

In all three cases the temperature profiles are symmetric about the mid-plane as the problem is thermally and geometrically symmetric.

Example 4.8 Lumped System Analysis of a Plane Wall Transient

A plane wall of thickness L, initially at temperature T_i is suddenly exposed to constant heat flux, q″ on one side and convective cooling on the other side (Fig. E4.8). Obtain the

*temperature variation in the wall as a func-
tion of time using lumped system approach.
Also, find an expression of the steady state
temperature.*

Solution

An energy balance on the wall of area A
(normal to the heat flow) and thickness
L gives

$$\dot{E}_{in} = \dot{E}_s + \dot{E}_{out} \qquad \text{(A)}$$

where $\dot{E}_{in} = q'' A$

$$\dot{E}_s = \rho c V \frac{dT}{dt}$$

$$\dot{E}_{out} = hA(T - T_\infty)$$

Substituting each individual energy term
in Eq. (A), we get

$$q'' A = \rho c V \frac{dT}{dt} + hA(T - T_\infty)$$

$$\Rightarrow \qquad \frac{dT}{dt} + \frac{hA}{\rho c V}(T - T_\infty) = \frac{q'' A}{\rho c V} \qquad \text{(B)}$$

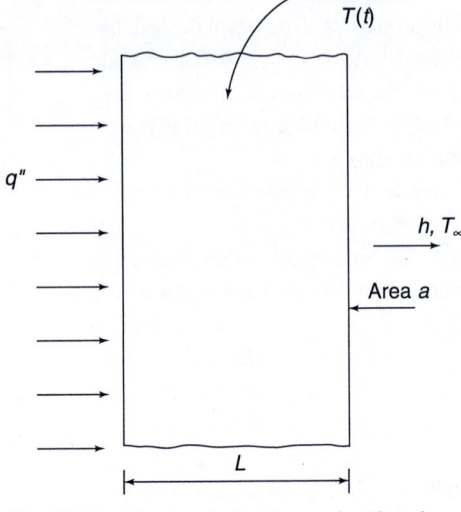

Fig. E4.8 Unsteady heat conduction in a plane wall modelled as a lumped system

where V is the volume of the wall and is equal to AL.

The initial condition is at $t = 0$, $T = T_1$ (C)

Using $\theta = T - T_\infty$, the final solution is obtained as

$$\theta(t) = \theta_i e^{-mt} + (1 - e^{-mt}) \frac{q''}{h} \qquad \text{(D)}$$

where $m = \dfrac{h}{\rho c L}$

The steady state temperature of the wall is obtained by setting $t \to \infty$ in Eq. (D).

$$\theta(\infty) = \frac{q''}{h}$$

Example 4.9   Proof of a Product Rule for the Transient Dimensionless Temperature
Distribution in a Rod of Cross-section $2L \times 2l$

*Prove the following expression (used in the solution of Example 4.5(b)) for a long rod of
cross-section $2L \times 2l$.*

$$\left(\frac{T - T_\infty}{T_c - T_\infty} \right)_{2L \times 2l} = \left(\frac{T - T_\infty}{T_c - T_\infty} \right)_{2L} \left(\frac{T - T_\infty}{T_c - T_\infty} \right)_{2l}$$

Solution

$$\left(\frac{T - T_\infty}{T_i - T_\infty} \right)_{2L \times 2L} = \left(\frac{T - T_\infty}{T_i - T_\infty} \right)_{2L} \left(\frac{T - T_\infty}{T_i - T_\infty} \right)_{2l} \qquad \text{(A)}$$

LHS of Eq. (A) can also be written as

$$\left(\frac{T-T_\infty}{T_i-T_\infty}\right)_{2L\times 2l}=\left(\frac{T_c-T_\infty}{T_i-T_\infty}\right)_{2L\times 2l}\left(\frac{T-T_\infty}{T_c-T_\infty}\right)_{2L\times 2l}$$ (B)

Similarly, RHS of Eq. (A) can be written as

$$\left(\frac{T-T_\infty}{T_i-T_\infty}\right)_{2L}\left(\frac{T-T_\infty}{T_i-T_\infty}\right)_{2l}=\left(\frac{T_c-T_\infty}{T_i-T_\infty}\right)_{2L}\left(\frac{T-T_\infty}{T_c-T_\infty}\right)_{2L}\left(\frac{T_c-T_\infty}{T_i-T_\infty}\right)_{2l}\left(\frac{T-T_\infty}{T_c-T_\infty}\right)_{2l}$$ (C)

Equating RHS of Eq. (B) and Eq. (C), we get

$$\left(\frac{T_c-T_\infty}{T_i-T_\infty}\right)_{2L\times 2l}\left(\frac{T-T_\infty}{T_c-T_\infty}\right)_{2L\times 2l}=\left(\frac{T_c-T_\infty}{T_i-T_\infty}\right)_{2L}\left(\frac{T-T_\infty}{T_c-T_\infty}\right)_{2L}\left(\frac{T_c-T_\infty}{T_i-T_\infty}\right)_{2l}\left(\frac{T-T_\infty}{T_c-T_\infty}\right)_{2l}$$ (D)

When $T=T_c$, Eq. (A) takes the form

$$\left(\frac{T_c-T_\infty}{T_i-T_\infty}\right)_{2L\times 2l}=\left(\frac{T_c-T_\infty}{T_i-T_\infty}\right)_{2L}\left(\frac{T_c-T_\infty}{T_i-T_\infty}\right)_{2l}$$ (E)

Substituting Eq. (E) into Eq. (D), we finally get

$$\left(\frac{T-T_\infty}{T_c-T_\infty}\right)_{2L\times 2l}=\left(\frac{T-T_\infty}{T_c-T_\infty}\right)_{2L}\left(\frac{T-T_\infty}{T_c-T_\infty}\right)_{2l}$$ (F)

Example 4.10 Plane Wall Transients: Use of Heisler Charts

Consider a 1.6 cm thick infinite slab of carbon steel (Fig. E4.10a) at the initial temperature $T_i=610°C$. This plate is suddenly plunged into a bath of water at the temperature $T_\infty=25°C$. The heat transfer coefficient h is 10^4 W/m^2 K. The properties of carbon steel are: k = 40 W/m K, and $\alpha=0.1$ cm^2/s. (a) Calculate the time t when the temperature in the mid-plane of the slab drops to $T_c=110$ °C (b) Determine also the corresponding temperature in a plane situated 0.2 cm from one of the cooled surfaces of the plate. Use Heisler charts.

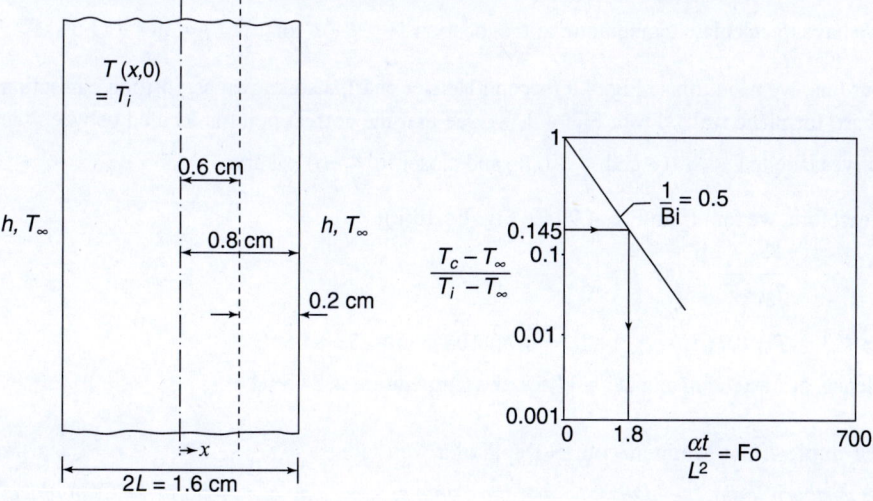

Fig. E4.10(a) Infinite slab of carbon steel

Fig. E4.10(b) Pictorial representation of the solution procedure by the use of Heisler chart shown in Fig. 4.8

Solution

(a) We have to use Fig. 4.8 (first Heisler chart for plane wall) in order to find the time when the centre-line temperature is 110°C. We need Biot number (Bi) and the dimensionless centre-line temperature.

$$Bi = \frac{hL}{k} = \frac{(10^4)(0.008)}{40} = 2$$

$$\frac{T_c - T_\infty}{T_i - T_\infty} = \frac{110 - 25}{610 - 25} = \frac{85}{585} = 0.145$$

Note that L is the half-thickness of the slab.

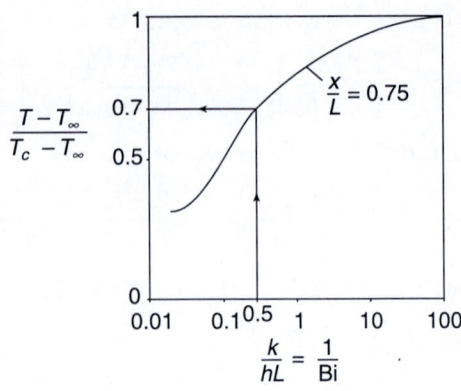

Fig. E4.10(c)   Pictorial representation of the solution procedure by the use of Heisler chart shown in Fig. 4.9

Next, we locate the line labelled $1/Bi = 1/2 = 0.5$ in Fig. 4.8 and we read the corresponding x-axis value [see Fig. E4.10(b)]

$$\frac{\alpha t}{L^2} = 1.8$$

which translates into

$$t = \frac{1.8\,L^2}{\alpha} = \frac{1.8(0.8)^2}{0.1} = 11.5\,\text{s}$$

Therefore, it will take 11.5 s for the mid-plane temperature to drop to 110°C from 610°C.

(b) $x = L - 0.2 = 0.8 - 0.2 = 0.6$ cm (Note that x is to be measured from the mid-plane.)

$$\frac{x}{L} = \frac{0.6}{0.8} = 0.75$$

$$\frac{1}{Bi} = 0.5$$

We have to calculate temperature at the location $\dfrac{x}{L} = 0.75$ for $\dfrac{1}{Bi} = 0.5$ at $t = 11.5$ s.

For this, we need to use Fig. 4.9 (second Heisler chart, also known as position correction chart, for plane wall). From Fig. 4.9, we see that the correct point is located between the curves labelled $\dfrac{x}{L} = 0.6$ and $\dfrac{x}{L} = 0.8$, and closer to $\dfrac{x}{L} = 0.8$ curve.

Therefore, we find from Fig. 4.9 [see Fig. E4.10(c)]

$$\frac{T - T_\infty}{T_c - T_\infty} = 0.7$$

$\Rightarrow$ $T = 0.7(T_c - T_\infty) + T_\infty = 0.7(110 - 25) + 25 = 84.5°\text{C}$

Hence, at $x = 0.6$ cm and at $t = 11.5$ s, the temperature is 84.5° C.

Example 4.11   Minimum Burial Depth of a Water Pipe

In northern India, the highest temperature on a summer day can go up to 45°C. In places where refrigeration facilities are not available, drinking such warm water or taking bath in it is very unpleasant. What minimum burial depth would you recommend to the company laying water pipelines so that even in summer one can get water at temperature not exceeding 25°C?

Assume that initially the soil is at 20°C and then it is subjected to a constant surface tem-perature of 40°C for 60 days. The thermal diffusivity of soil at 20°C is 0.138×10^{-6} m^2/s.

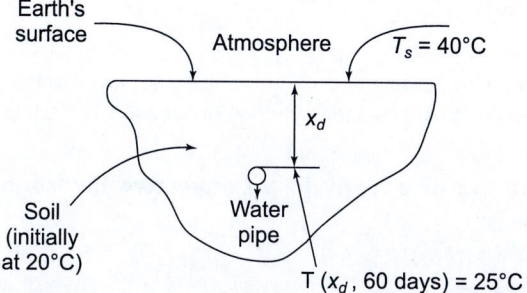

Fig. E4.11 Water pipe embedded in earth modelled as a semi-infinite solid

Solution

The earth can be modelled as a semi-infinite solid (Fig. E4.11) since it is very thick and the thermal diffusivity of earth is also low. Hence, it will take an infinitely long time for any change in the thermal condition on its surface to penetrate far into its interior. Therefore, the solution as given in Eq. (4.52) is valid for this problem.

Thus,

$$\frac{T - T_s}{T_i - T_s} = \text{erf}\left(\frac{x_d}{2\sqrt{\alpha t}}\right) \tag{A}$$

where $T_i = 20°C$

$T_s = 40°C$

$\alpha = 0.138 \times 10^{-6}$ m^2/s

$t = 60 \times 24 \times 3600 = 5.184 \times 10^6$ s

Substituting the above data into Eq. (A), we get

$$\frac{25 - 40}{20 - 40} = \text{erf}\left(\frac{x_d}{2\sqrt{0.138 \times 10^{-6} \times 5.184 \times 10^6}}\right) = \text{erf}\left(\frac{x_d}{2\sqrt{0.7154}}\right) = \text{erf}\left(\frac{x_d}{1.691}\right)$$

or

$$\text{erf}\left(\frac{x_d}{1.691}\right) = 0.75$$

From the error function table in Appendix A4, we can write approximately

$$\text{erf}\left(\frac{x_d}{1.691}\right) = 0.75$$

$\Rightarrow$ $x_d \approx 1.37$ m

Therefore, the recommended minimum burial depth of water pipes is 1.37 m.

Important Concepts and Formulae

Lumped system   Temperature of a body is uniform and only varies with time.

Distributed system   Temperature of a body is non-uniform and varies with both space and time.

Biot number (Bi)   Biot number is the ratio of conductive resistance to convective resistance.

$$\mathrm{Bi} = \frac{hL}{k}$$

In a thermally symmetric heating or cooling problem, L is taken as the half-thickness for a plane wall and, the radius for a long cylinder, or sphere. k is the thermal conductivity of the solid.

Heating or Cooling of a Body in a Convective Environment (Lumped System Transients)

$$\frac{T - T_\infty}{T_i - T_\infty} = e^{-\frac{hA}{\rho c V}t} = e^{-t/\phi}$$

where $\phi = \dfrac{\rho c V}{hA}$ which is known as time constant.

Time constant or response time or e-folding time is the time required for the temperature difference between the body and the surroundings to attain 36.8% of the temperature difference between the initial temperature of the body and the surroundings.

One-dimensional Transient Problems: Distributed Systems

Plane Wall (Large Heat Transfer Coefficient)

$$\frac{T(x,t) - T_\infty}{T_i - T_\infty} = 2 \sum_{n=0}^{\infty} \frac{(-1)^n}{\lambda_n L} e^{-\alpha_n^2 t} \cos \lambda_n x$$

where $\lambda_n L = (2n+1)\dfrac{\pi}{2}$

$n = 0, 1, 2, 3, ...$

Plane Wall (Moderate Heat Transfer Coefficient)

$$\frac{T(x,t) - T_\infty}{T_i - T_\infty} = 2 \sum_{n=1}^{\infty} \left(\frac{\sin \lambda_n L}{\lambda_n L + \sin \lambda_n L \cos \lambda_n L} \right) e^{-\alpha \lambda_n^2 t} \cos \lambda_n x$$

where the roots ($\lambda_n L$) can be found by solving the following transcendental equation graphically or numerically.

$$\tan \lambda_n L = \frac{\mathrm{Bi}}{\lambda_n L}$$

where $\mathrm{Bi} = \dfrac{hL}{k}$

The solution involves infinite series which are difficult to evaluate. However, the terms in the solutions converge rapidly with increasing time, and for Fo > 0.2, retaining only the first term and neglecting all other terms in the series results in an error of less than 2%.

Heisler Charts

In 1947, Heisler used this one-term approximation and obtained charts for plane wall, long cylinder, and sphere. There are two temperature charts associated with each geometry. The first chart determines the centre (or centre-line) temperature, T_C at a given time t. The second chart determines temperature at locations other than the centre (or centre-line) at the same time t in terms of T_C. Note that Heisler charts are applicable for problems involving symmetric heating or cooling.

Multidimensional Transient Problems

Multidimensional that is, 2D or 3D transient conduction problems can be solved by expressing their solutions as product of the solutions of 1D geometries whose intersection is the multi-dimensional body. Thus, Heisler charts can be used to obtain each of the 1D solutions and finally, the solution for 2D or 3D.

Semi-infinite Solid

A semi-infinite solid is an idealized body that has a single plane surface and extends to infinity in all directions. The temperature change we are interested in is due to the imposed thermal condition on a single surface.

Although all bodies have finite dimensions, a number of cases can be idealized as semi-infinite solids, in which there will be regions which still remain unaffected by a change of temperature on one of their surfaces. In other words, some parts of the body may still remain at the initial temperature even after a long time.

1D Transient Temperature Distribution

Surface Temperature Suddenly Changed to T_∞

$$\frac{\theta(x,t)}{\theta_i} = \mathrm{erf}\left[\frac{x}{2(\alpha t)^{1/2}}\right]$$

where $\quad \theta = T - T_\infty$

Surface Convection

$$\frac{T(x,t)-T_i}{T_\infty - T_i} = \mathrm{erfc}\left(\frac{x}{2\sqrt{\alpha t}}\right) - \left[\exp\left(\frac{hx}{k} + \frac{h^2 \alpha t}{k^2}\right)\right]\left[\mathrm{erfc}\left(\frac{x}{2\sqrt{\alpha t}}\right) + \frac{h\sqrt{\alpha t}}{k}\right]$$

Constant Surface Heat Flux q''

$$T(x,t)-T_i = \frac{2q''(\alpha t/\pi)^{1/2}}{k}\exp\left(-\frac{x^2}{4\alpha t}\right) - \frac{q''x}{k}\mathrm{erfc}\left(\frac{x}{2\sqrt{\alpha t}}\right)$$

Review Questions

4.1 Define Biot number and explain its physical significance.

4.2 What is the difference between lumped system and distributed system?

4.3 Define time constant.

4.4 Consider a cube of side r, a cylinder of diameter $2r$ and height $20r$, and a sphere of radius r. At $t = 0$, each one of them is at the same initial temperature, T_i. At $t > 0$, each body is immersed in a separate tank containing a fluid at T_∞ $(T_i > T_\infty)$. The heat transfer coefficient h is same for each. Each body is made of the same material. For each case, Bi < 0.1. Which one will cool fastest and why?

4.5 What is the difference between a large and moderate heat transfer coefficient?

4.6 Define Fourier number.

4.7 Under what conditions can the Heisler charts be used?

4.8 Define a semi-infinite solid. Can earth be considered a semi-infinite solid?

4.9 What is the basic method of solution in Laplace Transform?

4.10 What is penetration depth?

Problems

4.1 A thin wire (0.05 cm diameter) is initially heated to 200°C. It is suddenly exposed to an environment at 30°C. The heat transfer coefficient is 50 W/m²°C. (a) Can the lumped system approximation be made in this case? Justify. (b) Find the wire temperature after 10 sec if the wire is made of (i) copper, and (ii) aluminium.

Material properties

Copper: $\rho = 8954$ kg/m³, $c = 0.384$ kJ/kgK, $k = 398$ W/mK

Aluminium: $\rho = 2707$ kg/m³, $c = 0.896$ kJ/kgK, $k = 204$ W/mK

4.2 A plate 2.5 cm thick is made of chrome–nickel steel (15% Cr, 10% Ni) having $\rho = 7865$ kg/m³, $c = 0.46$ kJ/kgK, $k = 19$ W/mK. It is heated to some initial temperature and then exposed to air at 40°C. The heat transfer coefficient is 12 W/m²C. After justifying uniform interior temperature, find the initial temperature of the plate if its temperature after 10 min is 538°C.

4.3 Consider a long solid cylinder of outer radius R, which is heated initially to a known, axially symmetric, distribution of temperature, $f(r)$, and which is suddenly placed in contact with a convective fluid of constant temperature T_∞. The heat transfer coefficient is h. Obtain an expression for $T(r, t)$.

4.4 Stainless steel circular bars, each 12 cm in diameter, are to be quenched in a large oil bath maintained at 38°C. The initial temperature of the bars is 800°C. The maximum temperature within the bars at the end of the quenching process will have to be 200°C. How long must the bars be kept in the oil bath, if (a) the bars are infinitely long, (b) the length of the bars is twice the diameter? The properties of stainless steel are $k = 41$ W/mK, $\rho = 7865$ kg/m³, and $c = 460$ J/kgK. Take $h = 50$ W/m²K.

4.5 A long steel shaft of radius 10 cm ($\alpha = 1.6 \times 10^{-5}$ m²/s and $k = 61$ W/mK) is removed from a furnace at a uniform temperature of 500°C and immersed in a well-stirred large bath of coolant maintained at 20°C. The heat transfer coefficient between the shaft surface and coolant is 150 W/m²K. Calculate the time required for the shaft surface to reach 300°C.

4.6 A long 10 cm × 10 cm cross-section wood timber is initially at 40°C. It is suddenly exposed to flames at 600°C. The heat transfer coefficient is 20 W/m²°C. If the ignition temperature of wood is 482°C, how much time will elapse before any portion of the timber starts burning? The properties of wood are: $\rho = 800$ kg/m³, $c = 2.52$ kJ/kgK, $k = 0.346$ W/mK.

4.7 Hot tea having temperature 70°C is poured into a porcelain cup whose wall is initially at temperature $T_i = 25°C$. Assume that the surface of the porcelain wall instantly assumes the tea temperature $T_\infty = 70°C$. The thickness of the porcelain wall is 6 mm. The thermal diffusivity of porcelain is 0.004 cm²/s. Assuming the porcelain cup behaves like a semi-infinite body initially, estimate the time that passes until the wall temperature rises to 30°C at a point situated 2 mm from the wetted surface.

4.8 Obtain an expression for $T(x, t)$ for the semi-infinite body heat conduction problem for the case of constant surface heat flux q''.

4.9 A plate of thickness L initially has a sinusoidal temperature distribution varying from T_0 at $x = 0$ to T_m at $x = L/2$ and to T_0 at $x = L$. If the surfaces of the plate are held at T_0 for subsequent times, find the temperature distribution as a function of time.

4.10 A plate initially has a linear temperature distribution, from T_1 at the left face to T_2 at the right face. The two surface temperatures are suddenly changed to a constant value T_0 thereafter. Find the temperature as a function of the position within the plate material and time.

4.11 A steel cylinder of diameter 8 cm and length 12 cm is initially at 600°C. It is placed in oil at 20°C for rapid cooling, for case hardening of the surface layer. The value of h is 60 W/m²K.
   (a) Determine the time in which the maximum internal temperature level decreases to 500°C.
   (b) What is the minimum metal temperature at this time and where does it occur?
   (c) Plot the temperature distribution at that time along the cylinder axis.

4.12 A cubical region of 20 cm sides is initially at 100°C throughout. For $t > 0$ the surfaces are maintained at 50°C. The material conductivity and diffusivity are 237 W/mK and 97.1×10^{-6} m²/s, respectively. Calculate the time at which the maximum temperature in the region is reduced to 55°C.

4.13 Consider a solid body of volume V and surface area A, surrounded by a coolant at T_∞. The solid body is initially at temperature T_∞. For times $t \geq 0$, energy is generated in the solid at an exponential decay rate per unit volume to $q''' = q_0''' e^{-\beta t}$, where q_0''' and β are given as constants. Let the heat transfer coefficient h between the solid body and the coolant be constant. Assuming constant thermophysical properties and lumped capacitance, obtain an expression for the temperature of the solid body as a function of time for $t > 0$. What will be the maximum solid temperature and when is it reached?

4.14 A large stainless steel plate, 1.5 cm thick and initially at a uniform temperature of 24°C, is placed in a furnace maintained at 930°C. The combined convection and radiation heat transfer coefficient may be taken as 90 W/m²K. Estimate the time required for the mid-plane temperature to reach 540°C and the corresponding surface temperature.

4.15 A long rectangular aluminium rod, 10 cm × 20 cm in cross-section, is initially at a uniform temperature of 16°C. At time $t = 0$, the temperature of its surfaces is raised to 100°C by immersing it in boiling water and is subsequently maintained at this value. Calculate the temperature at the axis of the rod after 40 s have elapsed since the beginning of the cooling process.

4.16 The surface temperature of a very thick wall changes suddenly from 0°C to 1500°C and then remains constant at the new value. Initially the entire wall was at a temperature of 0°C. The wall is made of concrete with properties $k = 0.814$ W/mK, $c = 879$ J/kg K, and $\rho = 1906$ kg/m³.
   (a) Calculate the temperature at a depth of 20 cm from the surface after 8 h have elapsed.
   (b) How long will it take for the temperature at a depth of 40 cm to reach the value calculated in (a)?

4.17 Two flat plates made of the same material and with thickness a and b are initially at uniform temperatures T_1 and T_2, respectively. The plates are brought into contact at time $t = 0$. The external surfaces at $x = 0$ and $x = a + b$ are perfectly insulated. Assuming perfect thermal contact at the interface and constant thermophysical properties, obtain an expression for the temperature distribution in the system for $t > 0$. What is the steady temperature of the plates?

4.18 Solve Question 4.17 when the plates are made of different materials, all other conditions remaining the same.

4.19 In 1820, it was remarked by Fourier himself that the measured value of the geothermal gradient (rate of increase of temperature of the earth with depth) might be used to obtain a rough estimate of the time which has elapsed since the earth began to cool from an initially molten state. The surface is taken as the plane $x = 0$, and radiation takes place into a medium at absolute zero temperature. The temperature, when cooling began, taken at time $t = 0$, is constant and equal to T_0. He found that for large values of time t, the temperature gradient near the surface is approximately $T_0(\pi\alpha t)^{-1/2}$.

Lord Kelvin, in 1864, proposed a model by which he estimated the age of earth. He considered the earth as a semi-infinite solid bounded by the plane $x = 0$, the boundary being kept at 0 K and the initial temperature being T_0. He found out the geothermal gradient G at the surface: $G = (\partial T/\partial x)_{x=0} = T_0 (\pi\alpha t)^{-1/2}$ as in Fourier's problem. t in this equation corresponds to the age of earth.

Kelvin used the following data:

$$G = \frac{1}{2700} \ \text{C/cm}$$

$\alpha = 0.0118$ cm^2/s (which is the average thermal diffusivity of rock)

$T_0 = 3870°C$

Following Kelvin's treatment, calculate the age of earth.

Note that Kelvin estimated the age of earth to be 94×10^6 years (0.094 billion years) as compared to 4.7 billion years revealed by modern dating methods.

4.20 Consider a hot wall at 150°C and your finger at 37°C. Assuming the skin $k\rho c$ product is 41.8 J^2/sm^{4}K^2, calculate the contact temperatures for copper ($k\rho c = 13,340$), brick ($k\rho c = 19.37$), and asbestos ($k\rho c = 0.928$).

Chapter 5

Forced Convection Heat Transfer

5.1 Introduction

So far, we have discussed problems in conduction heat transfer. We considered convection only in relation to the boundary conditions imposed on a conduction problem. The main purpose of studying convective heat transfer is to predict the value of the convective heat transfer coefficient h. The subject of convection heat transfer requires an energy balance along with an analysis of fluid dynamics of the problem concerned.

Convection is not a separate mode of heat transfer. It describes a fluid system in motion, and heat transfer occurs by the mechanism of conduction alone. Obviously, we must allow for the motion of the fluid system in writing an energy balance, but there is no new basic mechanism of heat transfer involved.

An engineer often confronts the task of calculating heat transfer rates at the interface between a solid and a fluid, where the fluid may be visualized as moving relative to the stationary solid surface. If the fluid is at rest, the problem reduces to simple conduction where there are temperature gradients normal to the interface. However, if the fluid is in motion, heat is transported both by simple conduction and by the movement of the fluid itself. This complex transport process is referred to as convection. Thus the essential feature of convection heat transfer is the transport of energy to or from a surface by both molecular conduction processes and gross fluid motion.

If the fluid motion involved in the process is induced by some external means (pump, blower, wind, vehicle motion, etc.), the process is generally called *forced convection*. If the fluid motion arises due to external force fields, such as gravity, acting on density gradients induced by temperature gradients, the process is usually called *free convection* or *natural convection*.

Consider a fluid having velocity u_∞ and temperature T_∞ flowing over a surface of arbitrary shape and of area A [Fig. 5.1(a)]. The surface is maintained at temperature T_s ($> T_\infty$). The local heat flux q'' may be expressed as

$$q'' = h(T_s - T_\infty) \tag{5.1}$$

where h is the local convective heat transfer coefficient. The total heat transfer rate q may be obtained by integrating the local heat flux over the entire surface:

$$q = \int_A q'' dA \tag{5.2}$$

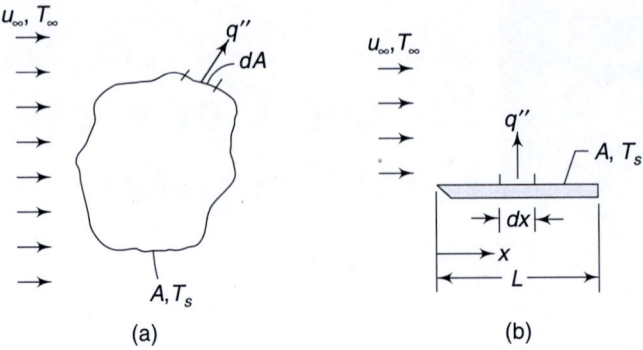

Fig. 5.1 Convection heat transfer from (a) a body of arbitrary shape (b) a horizontal flat plate

$$= (T_s - T_\infty) \int_A h \, dA \qquad (5.3)$$

Defining $\bar{h}$ as the average heat transfer coefficient which is expressed as

$$\bar{h} = \frac{1}{A} \int_A h \, dA \qquad (5.4)$$

Equation (5.3) can be written as

$$q = \bar{h} A (T_s - T_\infty) \qquad (5.5)$$

Note that for the special case of flow over a flat plate [Fig. 5.1(b)],

$$\bar{h} = \frac{1}{L} \int_0^L h \, dx \qquad (5.6)$$

5.2 Convection Boundary Layers

In this section the concepts of velocity (or momentum) and thermal boundary layers encountered in convection heat transfer to/from a flat plate are introduced.

5.2.1 Velocity (or Momentum) Boundary Layer

When fluids of small viscosity such as air or water move rapidly over a solid body (thus making Reynolds number of the flow very high), the friction between the fluid and the solid surface causes the movement of the fluid to be retarded within a thin region immediately adjacent to the solid surface. This thin region, in which large velocity gradients exist, is called the boundary layer (introduced by L. Prandtl in 1904). The region outside the boundary layer where the forces due to friction are small and may be neglected, and where the ideal or perfect fluid approximation is valid, is called the potential or inviscid flow region (Fig. 5.2). The salient features of the momentum boundary layer are as follows:

(a) The boundary layer thickness δ which is a function of x, the distance from the leading edge.

(b) The reason for boundary layer growth is that with increasing distance from the leading edge, the effect of viscosity penetrates further into the free stream.

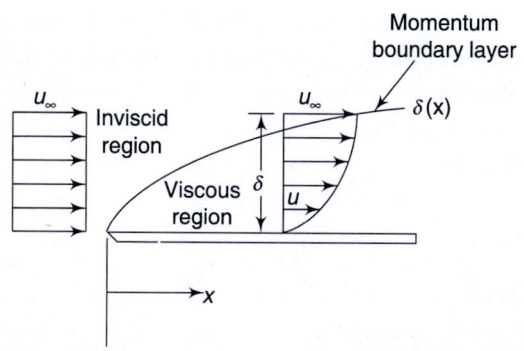

(c) The thickness $\delta(x)$ is conventionally defined as the distance normal to the wall at which $u/u_\infty = 0.99$.

Fig. 5.2 Momentum boundary layer over a flat plate

(d) Inside the boundary layer, except near the leading edge, the assumptions $u \gg v,\ \partial u/\partial y \gg \partial u/\partial x \approx \partial v/\partial y \gg \partial v/\partial x$ are valid. The large gradients normal to the wall imply that the streamwise diffusion of momentum is negligible.

(e) The pressure gradient across the boundary layer $\partial p/\partial y$ is negligible while that along the boundary layer $\partial p/\partial x$ is given by that of the inviscid free stream.

(f) An order of magnitude analysis reveals that the y-momentum can be neglected. The local skin friction coefficient C_f for flow over a flat plate is given as

$$C_f = \frac{\tau_s}{\frac{1}{2}\rho u_\infty^2} \tag{5.7}$$

where $\quad \tau_s = \mu \left.\frac{\partial u}{\partial y}\right|_{y=0} \tag{5.8}$

5.2.2 Thermal Boundary Layer

A thermal boundary layer develops when the free stream and surface temperatures differ. Due to no-slip condition at the plate surface, the stationary fluid particles have the same temperature as that of the plate surface after thermal equilibrium is reached. The fluid particles in contact with the plate exchange energy with those in the adjoining layer, and temperature gradients

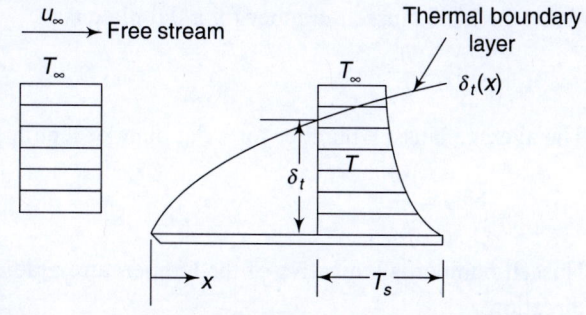

Fig. 5.3 Thermal boundary layer over a flat plate

develop in the fluid. The region of the fluid in which temperature gradients exist is the thermal boundary layer (Fig. 5.3), and its thickness δ_t is typically defined as the value of y for which the ratio $(T_s - T)/(T_s - T_\infty) = 0.99$.

As the distance from the leading edge increases, the effects of heat transfer penetrate further into the free stream and the thermal boundary layer grows. At the plate surface, since there is no fluid motion and heat transfer can only occur by conduction, we can apply Fourier's law to calculate the local surface heat flux as follows:

$$q_s'' = -k_f \left.\frac{\partial T}{\partial y}\right|_{y=0} \qquad (5.9)$$

Since

$$q_s'' = h(T_s - T_\infty) \qquad (5.10)$$

therefore, $h = \dfrac{-k_f \left.\dfrac{\partial T}{\partial y}\right|_{y=0}}{T_s - T_\infty}$ $\qquad (5.11)$

Equation (5.11) is the defining relation for the heat transfer coefficient. k_f is the thermal conductivity of the fluid. From Eq. (5.11) it is clear that the conditions in the thermal boundary layer, which strongly influence the plate surface temperature gradient $\partial T/\partial y|_{y=0}$, will determine the rate of heat transfer across the boundary layer. Note that, since $(T_s - T_\infty)$ is constant, independent of x, while δ_t increases with increasing x, temperature gradients in the boundary layer must decrease with increasing x. Accordingly, the magnitude of $\partial T/\partial y|_{y=0}$ decreases with increasing δ_t, and it follows that q_s'' and h decrease with increasing x. Since the thermal boundary layer thickness is zero at the leading edge, the heat transfer coefficient there is infinity.

5.3   Nusselt[1] Number

Frequently, the heat transfer coefficient is made non-dimensional by using a characteristic length L and defining the Nusselt number:

$$Nu = \frac{hL}{k_f} \qquad (5.12)$$

Thus, the local Nusselt number for a flat plate is

$$Nu_x = \frac{hx}{k_f} \qquad (5.13)$$

The average Nusselt number for a flat plate of length L is

$$\overline{Nu}_L = \frac{\overline{h}L}{k_f} \qquad (5.14)$$

Nusselt number is indicative of the temperature gradient at the wall in the normal direction.

---

[1]Wilhelm Nusselt (1882–1957) is well known for his works on dimensionless groups in modelling heat transfer and film condensation of steam. He was Professor at Karlsruhe and Technical University of Munich, Germany.

5.4 Prandtl[2] Number

Prandtl number is one of the most important dimensionless groups in heat transfer and is defined as

$$Pr = \frac{\mu c_p}{k_f} \qquad (5.15)$$

where μ, c_p, and k_f are the viscosity, specific heat, and conductivity of the fluid respectively. Pr can also be expressed as

$$Pr = \frac{(\mu/\rho)}{(k_f/\rho c_p)} = \frac{v}{\alpha} = \frac{\text{Kinematic viscosity}}{\text{Thermal diffusivity}} \qquad (5.16)$$

Kinematic viscosity is a diffusivity for momentum or for velocity, in the same sense that thermal diffusivity is a diffusivity for heat, or for temperature. Diffusivity is defined as the rate at which a particular effect is diffused through a medium. Both v and α have units of m^2/s.

Prandtl number also signifies the ratio of the momentum boundary layer thickness δ to the thermal boundary layer thickness δ_t. Thus,

$$Pr \sim \frac{\delta}{\delta_t} \qquad (5.17)$$

If $Pr = 1$, it means that velocity and thermal boundary layers grow together. If $Pr > 1$, it must follow that the velocity boundary layer develops faster than the thermal boundary layer. If $Pr < 1$, the opposite holds, that is, the thermal boundary layer develops more rapidly than the velocity boundary layer.

The Prandtl number spectrum of various fluids is shown in Fig. 5.4. Typical Prandtl numbers are: 0.7 for air, 7 for water at room temperature. Oils have very high values of Pr, and liquid metals have very low values of Pr. Prandtl number for any particular fluid generally varies somewhat with temperature.

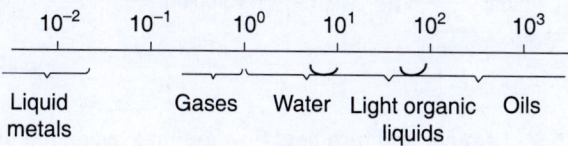

Fig. 5.4 Prandtl number spectrum of various fluids

5.5 Laminar and Turbulent Flows Over a Flat Plate

Initially, the flow in the boundary layer near the leading edge is laminar, regardless of the level of turbulence existing in the approaching free stream. The laminar flow can be visualized as having layers or laminates of fluid. Molecules may move from one lamina to another, carrying with them a momentum corresponding to the velocity of flow. There is a net momentum transport from regions of high velocity

---

[2]Ludwig Prandtl (1875–1953) was Professor at Göttingen University, Germany. Founder of modern fluid dynamics, he is best known for his boundary layer theory, wing theory, development of wind tunnel, and research on supersonic flow and turbulence.

to the regions of low velocity, thus creating a force in the direction of flow. This force manifests itself as viscous shear stress, $\tau_{yx} = \mu(\partial u/\partial y)$.

However, at some critical distance from the leading edge, small disturbances in the flow begin to amplify and a transition process takes place until the flow becomes turbulent. The turbulent flow regime may be visualized as a random churning action with chunks of fluid moving to and fro in all directions. Eddies or fluid packets of many sizes intermingle and fill the boundary layer. The transition from laminar to turbulent flow occurs when

$$\mathrm{Re}_{cr} = \frac{u_\infty x_{cr}}{v} = \frac{\rho u_\infty x_{cr}}{\mu} = 5 \times 10^5$$

The normal range for the beginning of transition is between 5×10^5 and 5×10^6. Actually the location of transition is not well defined since a transition zone occupies a finite length of the plate. Transition may also be affected by the surface roughness and the free stream turbulence level.

The laminar velocity profile is approximately parabolic, while the turbulent profile has a portion near the wall, which is nearly linear (Fig. 5.5). This region is called viscous sublayer and lies very close to the surface. In this region, velocity and temperature fluctuations are very small. Outside the sublayer the velocity profile is relatively flat in comparison with the laminar profile.

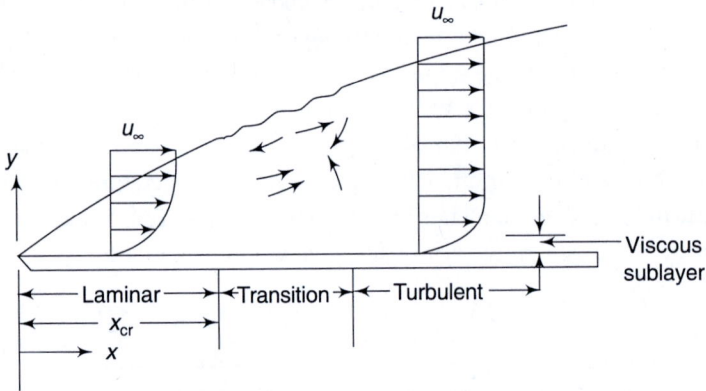

Fig. 5.5 Laminar and turbulent flow regimes over a flat plate

In the turbulent flow, one can imagine macroscopic chunks of fluid transporting energy and momentum instead of microscopic transport on the basis of individual molecules. Because of this, there is larger viscous shear force and heat transfer and better mixing which cause flat velocity as well as temperature profiles in turbulent flow.

5.6 Energy Equation in the Thermal Boundary Layer in Laminar Flow over a Flat Plate

We are now in a position to derive the energy equation in the thermal boundary layer in laminar flow over a flat plate using a differential control volume as shown in Fig. 5.6. The relevant assumptions are as follows.

(a) The flow is steady, incompressible, two-dimensional, and laminar.

(b) The fluid has constant viscosity, thermal conductivity, and specific heat.

(c) There is negligible heat conduction in the direction of flow (*x*-direction). This is because the thermal boundary layer is very thin. Hence the temperature gradient in the *y*-direction is quite large compared to that in the *x*-direction.

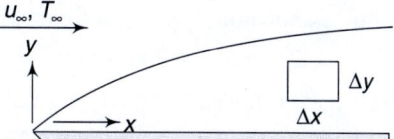

Fig. 5.6 Differential control volume in the thermal boundary layer

(d) Viscous work done at the *x*-faces of the control volume is negligible as $v \ll u$.

(e) There are no pressure gradients in *x* and *y* directions. $\partial p/\partial y = 0$ as the momentum boundary layer is very thin. $\partial p/\partial x = 0$ as u_∞ is constant for flow over a flat plate at zero incidence. The application of Bernoulli's equation reveals that pressure is constant in the *x*-direction in the free stream. Since, according to boundary layer theory, pressure outside the boundary layer can be impressed upon that in the boundary layer, $\partial p/\partial x = 0$ in the boundary layer equation.

(f) The *v*-velocity (i.e., the *y*-component of the fluid velocity) has negligible contribution to the kinetic energy of the fluid as $v \ll u$.

(g) $\mathrm{Pr} \geq 0.5$. This implies that the present analysis is applicable to most gases and liquids and $\delta_t < \delta$.

Figure 5.7 shows the expanded view of the control volume and the energy-in and energy-out terms. Note that thermal and kinetic energies are convected with bulk fluid motion across control surfaces. In addition, energy is also transferred across the control surface by conduction as well as by surface forces (pressure, viscous, etc.). Only for supersonic flows or the high-speed motion of lubricating oils, viscous dissipation may not be neglected.

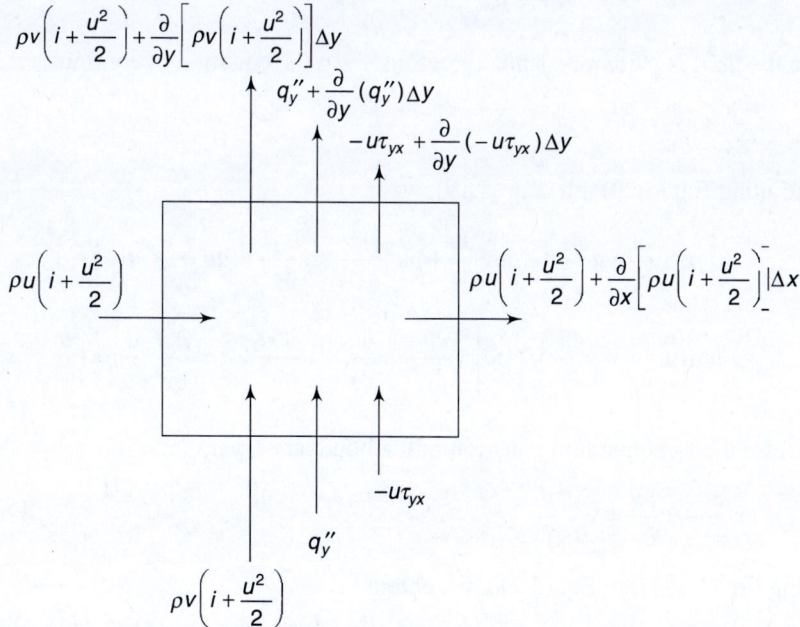

Fig. 5.7 Expanded view of the control volume with energy-in and energy-out terms

Applying the first law of thermodynamics on the energy terms on the control volume shown in Fig. 5.7, we can write

Energy-in = Energy-out

Note that i is the enthalpy (or thermal energy) per unit mass $(c_p T)$, $u^2/2$ is the kinetic energy per unit mass q_y'' is the conductive heat flux in the y-direction, and $-u\tau_{yx}$ is the viscous shear work in the x-direction. The negative sign on $u\tau_{yx}$ is given because viscous work is done on the control surface.

Writing all energy-in terms on the left-hand side and all energy-out terms on the right-hand side and cancelling terms, we have

$$0 = \frac{\partial}{\partial y}\left[\rho v\left(i + \frac{u^2}{2}\right)\right] + \frac{\partial}{\partial x}\left[\rho u\left(i + \frac{u^2}{2}\right)\right] + \frac{\partial}{\partial y}(q_y'') + \frac{\partial}{\partial y}(-u\tau_{yx}) \quad (5.18)$$

The first term on the right-hand side of Eq. (5.18) represents the net energy associated with fluid motion transferred into the control volume in the y-direction, that is, the net efflux of enthalpy i and kinetic energy. The second term represents the same in the x-direction. The third term is the net heat transfer by conduction in the y-direction and, finally, the last term is the net rate of work added to the control volume in the y-direction due to viscous dissipation. Substituting $q_y'' = -k_f(\partial T/\partial y)$ from Fourier's law and $\tau_{yx} = \mu(\partial u/\partial y)$ into Eq. (5.18), we get

$$0 = \rho v\left(\frac{\partial i}{\partial y} + u\frac{\partial u}{\partial y}\right) + \rho u\left(\frac{\partial i}{\partial x} + u\frac{\partial u}{\partial x}\right) + \rho\left(i + \frac{u^2}{2}\right)\left(\frac{\partial u}{\partial x} + \frac{\partial v}{\partial y}\right)$$
$$- k_f\frac{\partial^2 T}{\partial y^2} - u\frac{\partial}{\partial y}\left(\mu\frac{\partial u}{\partial y}\right) - \left(\mu\frac{\partial u}{\partial y}\right)\frac{\partial u}{\partial y} \quad (5.19)$$

Since the fluid is incompressible, therefore, from the equation of continuity,

$$\frac{\partial u}{\partial x} + \frac{\partial v}{\partial y} = 0 \quad (5.20)$$

Substituting Eq. (5.20) into Eq. (5.19), we get

$$0 = \rho\left(v\frac{\partial i}{\partial y} + u\frac{\partial i}{\partial x}\right) + \rho u v\frac{\partial u}{\partial y} + \rho u^2\frac{\partial u}{\partial x} - k_f\frac{\partial^2 T}{\partial y^2} - \mu u\frac{\partial^2 u}{\partial y^2} - \mu\left(\frac{\partial u}{\partial y}\right)^2$$

or
$$0 = \rho\left(u\frac{\partial i}{\partial x} + v\frac{\partial i}{\partial y}\right) + \rho u\left[u\frac{\partial u}{\partial y} + v\frac{\partial u}{\partial x} - v\frac{\partial^2 u}{\partial y^2}\right] - k_f\frac{\partial^2 T}{\partial y^2} - \mu\left(\frac{\partial u}{\partial y}\right)^2 \quad (5.21)$$

But, from the x-momentum equation of the boundary layer,

$$u\frac{\partial u}{\partial x} + v\frac{\partial u}{\partial y} = v\frac{\partial^2 u}{\partial y^2} \quad (5.22)$$

Putting Eq. (5.22) into Eq. (5.21), we obtain

$$0 = \rho\left(u\frac{\partial i}{\partial x} + v\frac{\partial i}{\partial y}\right) - k_f\frac{\partial^2 T}{\partial y^2} - \mu\left(\frac{\partial u}{\partial y}\right)^2 \quad (5.23)$$

But, $\partial i = c_p \partial T$ (5.24)

Substitution of Eq. (5.24) into Eq. (5.23) results in

$$O = \rho \left(uc_p \frac{\partial T}{\partial x} + vc_p \frac{\partial T}{\partial y} \right) - k_f \frac{\partial^2 T}{\partial y^2} - \mu \left(\frac{\partial u}{\partial y} \right)^2$$

or $$\rho c_p \left(u \frac{\partial T}{\partial x} + v \frac{\partial T}{\partial y} \right) = k_f \frac{\partial^2 T}{\partial y^2} + \mu \left(\frac{\partial u}{\partial y} \right)^2$$

or $$\underbrace{u \frac{\partial T}{\partial x} + v \frac{\partial T}{\partial y}}_{\text{convection}} = \underbrace{\alpha \frac{\partial^2 T}{\partial y^2}}_{\substack{\text{transversal} \\ \text{conduction}}} + \underbrace{\frac{\mu}{\rho c_p} \left(\frac{\partial u}{\partial y} \right)^2}_{\text{viscous dissipation}}$$ (5.25)

We see that the change in kinetic energy is balanced exactly by a portion of the work done by the viscous forces. The rest of the viscous work is dissipation.

The term $u(\partial T/\partial x) + v(\partial T/\partial y)$, which is on the left-hand side of Eq. (5.25), represents the net transport of energy into the control volume. In short, it is the convection term. The first term on the right-hand side, that is, $\alpha(\partial^2 T/\partial y^2)$ is the net heat conducted out of the control volume in the y-direction. It is the conduction term. The last term on the right-hand side, that is, $\mu/\rho c_p (\partial u/\partial y)^2$ is the net viscous work done on the element. It is called the viscous dissipation term. If the viscous dissipation is neglected (as in low-speed flows), Eq. (5.25) takes the form

$$u \frac{\partial T}{\partial x} + v \frac{\partial T}{\partial y} = \alpha \frac{\partial^2 T}{\partial y^2}$$ (5.26)

5.6.1 Importance of the Viscous Dissipation Term

The viscous dissipation term is of importance only for high-speed flows since its magnitude will be comparable to that of the conduction term. This may be shown with an order of magnitude analysis of the two terms on the right-hand side of the energy equation [Eq. (5.25)].

Recall Eq. (5.25),

$$u \frac{\partial T}{\partial x} + v \frac{\partial T}{\partial y} = \alpha \underbrace{\frac{\partial^2 T}{\partial y^2}}_{(1)} + \underbrace{\frac{\mu}{\rho c_p} \left(\frac{\partial u}{\partial y} \right)^2}_{(2)}$$

Now, u is of the order u_∞. y is of the order δ. $\partial T/\partial y$ is of the order $(T_s - T_\infty)/\delta$. Therefore,

$$\alpha \frac{\partial^2 T}{\partial y^2} \sim \alpha \frac{(T_s - T_\infty)}{\delta^2}$$

and $$\frac{\mu}{\rho c_p} \left(\frac{\partial u}{\partial y} \right)^2_{*} \sim \frac{\mu}{\rho c_p} \frac{u_\infty^2}{\delta^2}$$

If (2)/(1) is small, then viscous dissipation can be neglected. Now,

$$\frac{(2)}{(1)} = \frac{\dfrac{\mu}{\rho c_p}\left(\dfrac{u_\infty^2}{\delta^2}\right)}{\alpha \dfrac{(T_s - T_\infty)}{\delta^2}} = \frac{\mu}{\alpha \rho c_p} \frac{u_\infty^2}{T_s - T_\infty}$$

$$= \frac{Pr}{c_p} \frac{u_\infty^2}{T_s - T_\infty} = Pr\,Ec$$

where Ec = Eckert number

$$= \frac{u_\infty^2}{c_p(T_s - T_\infty)}$$

Therefore, if $Pr\,Ec \ll 1$, viscous dissipation can be neglected.

To illustrate this concept, consider the flow of air at $u_\infty = 5$ m/s, $T_\infty = 20°C$, $T_s = 60°C$, $p = 1$ atm. For these conditions, $c_p = 1005$ J/kg °C, Pr = 0.7. Therefore,

$$Pr\,Ec = Pr\frac{u_\infty^2}{c_p(T_s - T_\infty)} = (0.7)\frac{(5)^2}{(1005)(60 - 20)} = 0.000435$$

From the above, we see $Pr\,Ec \ll 1$ and hence in this case viscous dissipation is negligible. Thus, for low-speed incompressible flow, viscous dissipation can be neglected and Eq. (5.26) is valid.

5.6.2   Governing Equations and Boundary Conditions

From the foregoing discussion we see the governing differential equations for the thermal boundary layer over a flat plate at zero incidence are as follows:

$$\text{Continuity:} \qquad \frac{\partial u}{\partial x} + \frac{\partial v}{\partial y} = 0 \qquad\qquad (5.27)$$

$$x\text{-momentum:} \quad u\frac{\partial u}{\partial x} + v\frac{\partial u}{\partial y} = v\frac{\partial^2 u}{\partial y^2} \qquad\qquad (5.28)$$

$$\text{Energy:} \qquad u\frac{\partial T}{\partial x} + v\frac{\partial T}{\partial y} = \alpha\frac{\partial^2 T}{\partial y^2} \qquad\qquad (5.29)$$

For an isothermal flat plate, the boundary conditions are

$$\text{At } y = 0, \ u = 0, \ v = 0, \ T = T_s \qquad\qquad (5.30)$$

$$\text{At } y = \infty, \ u = u_\infty, \ T = T_\infty \qquad\qquad (5.31)$$

A close look at Eqs (5.28) and (5.29) reveals that they will have exactly the same form when $\alpha = v$. Thus, we should expect that the relative magnitudes of the thermal diffusivity and kinematic viscosity would have an important influence on convective heat transfer since these magnitudes relate the velocity distribution to the temperature distribution. Recall that $Pr = v/\alpha$. This explains why Prandtl number is so important in convective heat transfer studies.

5.6.3 Basic Solution Methodology

To solve a thermal boundary layer problem, the basic approach is to obtain the velocity field $u(x,y)$ and $v(x,y)$ by solving the continuity and x-momentum equations [Eqs (5.27) and (5.28)] and then substituting the velocity fields into the energy equation [Eq. (5.29)] to find the temperature distribution $T(x,y)$ which will enable one to compute the heat transfer. Of course, this decoupling of continuity and momentum equations from the energy equation is acceptable provided the fluid properties are not strong functions of temperature.

5.7 Solution of the Thermal Boundary Layer on an Isothermal Flat Plate

In this section the exact and approximate solution methodologies for the thermal boundary layer on an isothermal flat plate are presented.

5.7.1 Exact Solution: Similarity Analysis of Pohlhausen

Pohlhausen (1921) solved it by similarity method. By defining a stream function

$$\psi = \sqrt{u_\infty \nu x} f(\eta) \tag{5.32}$$

a similarity parameter

$$\eta = \sqrt{\frac{u_\infty}{\nu x}} y \tag{5.33}$$

and a non-dimensional temperature

$$\theta = \frac{T_s - T}{T_s - T_\infty} \tag{5.34}$$

the energy equation, Eq. (5.29) which is a partial differential equation, is transformed into an ordinary differential equation:

$$\frac{d^2\theta}{d\eta^2} + \frac{1}{2} \Pr f(\eta) \frac{d\theta}{d\eta} = 0 \tag{5.35}$$

Thus, θ is a function of η. Basically, at different x-locations, the temperature distribution will be similar with respect to δ_t. For an isothermal flat plate, the following boundary conditions apply:

$$\text{At } y = 0, \ T = T_s, \quad \text{or } \eta = 0, \ \theta = 0 \tag{5.36}$$

$$\text{At } y = \infty, \ T = T_\infty, \quad \text{or } \eta = \infty, \ \theta = 1 \tag{5.37}$$

Now, to integrate Eq. (5.35) directly,

$$\frac{d\theta'}{d\eta} + \frac{\Pr}{2} f\theta' = 0$$

or

$$\frac{d\theta'}{\theta'} + \frac{\Pr}{2} f d\eta = 0$$

or

$$\int_{\theta'(0)}^{\theta'(\eta)} \frac{d\theta'}{\theta'} = -\frac{\Pr}{2} \int_{\eta=0}^{\eta=\eta} f d\eta$$

or $\quad \theta'(\eta) = \theta'(0)\exp\left[-\dfrac{Pr}{2}\int_0^{\eta} f d\eta\right]$

or $\quad \theta(\eta) = \theta'(0)\left[\int_0^{\infty}\exp\left[-\dfrac{1}{2}Pr\int_0^{\eta} f d\eta\right]d\eta\right]$

$\theta'(0)$ is evaluated by using Eq. (5.37), that is, $\theta = 1$ and $\eta = \infty$.

$$1 = \theta(\infty) = \theta'(0)\left[\int_0^{\infty}\exp\left[-\dfrac{1}{2}Pr\int_0^{\eta} f d\eta\right]d\eta\right]$$

Therefore, $\quad \theta'(0) = \dfrac{1}{\displaystyle\int_0^{\infty}\exp\left[-\dfrac{1}{2}Pr\int_0^{\eta} f d\eta\right]d\eta}$ \hfill (5.38)

and $\quad \theta(\eta) = \dfrac{\displaystyle\int_0^{\eta}\exp\left[-\dfrac{1}{2}Pr\int_0^{\eta} f d\eta\right]d\eta}{\displaystyle\int_0^{\infty}\exp\left[-\dfrac{1}{2}Pr\int_0^{\eta} f d\eta\right]d\eta}$ \hfill (5.39)

From the Blasius similarity solution of the momentum boundary layer (Schlichting 1968), we know

$$f = \dfrac{\alpha\eta^2}{2!} - \dfrac{1}{2}\dfrac{\alpha^2\eta^5}{5!} + \dfrac{11}{4}\dfrac{\alpha^3\eta^8}{8!} - \dfrac{375}{8}\dfrac{\alpha^4\eta^{11}}{11!} + \cdots \tag{5.40}$$

where $\alpha = 0.332$ (do not confuse this α with thermal diffusivity). By substituting f from Eq. (5.40) into Eq. (5.38), Pohlhausen carried out the integration numerically. The following empirical equation expresses these results within a few percent agreement for $Pr > 0.5$:

$$\theta'(0) = \dfrac{d\theta(0)}{d\eta} = 0.332\,Pr^{0.343} \tag{5.41}$$

For simplicity, however, the exponent 0.343 is taken as 1/3. Similarly, Eq. (5.39) was evaluated using $f(\eta)$. The local heat transfer coefficient h is

$$h = \dfrac{-k_f\left(\dfrac{\partial T}{\partial y}\right)_{y=0}}{T_s - T_{\infty}}$$

$$= k_f\sqrt{\dfrac{u_{\infty}}{vx}}\left(\dfrac{d\theta}{d\eta}\right)_{\eta=0}$$

Now, $\quad \dfrac{d\theta}{d\eta}(0) = 0.332\,Pr^{1/3}$

Therefore, $h = k_f\sqrt{\dfrac{u_{\infty}}{vx}}\,0.332\,Pr^{1/3}$

$$Nu_x = \dfrac{hx}{k_f} = \sqrt{\dfrac{u_{\infty}x}{v}}\,0.332\,Pr^{1/3}$$

$$= \mathrm{Re}_x^{1/2}(0.332\,\mathrm{Pr}^{1/3})$$

$$= 0.332\,\mathrm{Re}_x^{1/2}\,\mathrm{Pr}^{1/3} \tag{5.42}$$

The average heat transfer coefficient,

$$\bar{h} = \frac{1}{L}\int_0^L h\,dx$$

$$= \frac{1}{L}\int_0^L k_f \sqrt{\frac{u_\infty}{vx}}\,0.332\,\mathrm{Pr}^{1/3}\,dx$$

$$= \frac{1}{L}(0.332)\,(2)k_f\,\mathrm{Pr}^{1/3}\sqrt{\frac{u_\infty L}{v}}$$

$$= \frac{1}{L}(0.664)k_f\,\mathrm{Pr}^{1/3}\,\mathrm{Re}_L^{1/2}$$

Therefore, $\overline{\mathrm{Nu}}_L = \dfrac{\bar{h}L}{k_f} = 0.664\,\mathrm{Re}_L^{1/2}\,\mathrm{Pr}^{1/3}$ (5.43)

$$= 2\,\mathrm{Nu}_{x=L} \tag{5.44}$$

Therefore, the average Nusselt number over a plate of length L is twice the local Nusselt number evaluated at $x = L$. All the above results are valid for $\mathrm{Pr} > 0.5$.

For liquid metals ($0.006 \le \mathrm{Pr} \le 0.03$), $\delta \ll \delta_t$ and hence the momentum boundary layer can be neglected. Thus, taking $u = u_\infty$, i.e., $f' = 1$, using Pohlhausen's method, one can show that

$$\mathrm{Nu}_x = 0.565\,\mathrm{Re}_x^{1/2}\,\mathrm{Pr}^{1/2} \tag{5.45}$$

The plot of θ versus η as a function of the Prandtl number is shown in Fig. 5.8. Note that for $\mathrm{Pr} = 1$, $\delta_t = \delta$, and the dimensionless temperature distribution is identical with dimensionless u-velocity or f'-distribution ($f' = u/u_\infty$). It is also seen that for higher Prandtl numbers, the thermal boundary layers are thinner and hence wall temperature gradients are steeper, thus resulting in higher heat transfer.

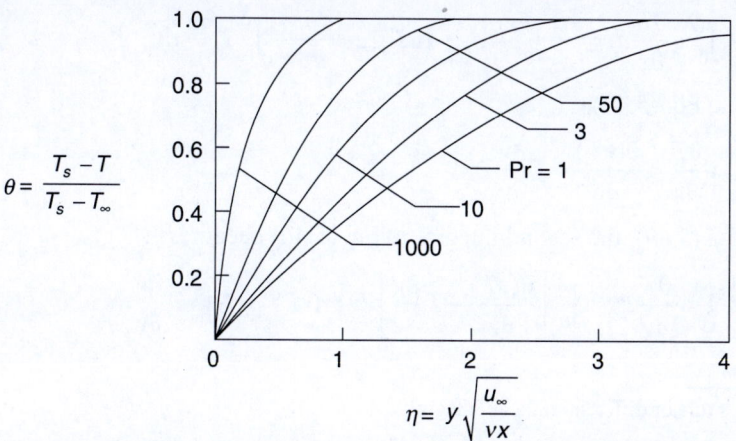

Fig. 5.8 Plot of θ versus η as a function of Prandtl number

5.7.2   Approximate Analysis: von Karman's Integral Method

Integration of the boundary layer equations is difficult, even for the case of constant properties. The basis of the technique developed by von Karman which reduces these difficulties is transformed boundary layer equations obtained by integrating the partial differential equations over the thickness of the boundary layer. The resulting momentum and energy equations are called the integral equations of the boundary layer.

The integral method is frequently termed approximate analysis of the boundary layer. It should be emphasized, however, that the integral equations themselves are exact within the boundary layer assumptions. The solutions of these equations are approximate only to the extent that the velocity and temperature profiles chosen are not exact. The integral technique is relatively simple to apply, and in many cases satisfactory results are obtained.

To derive the energy integral equation, the differential energy equation [Eq. (5.29)] is integrated over the thermal boundary layer thickness to obtain

$$\rho c_p \int_0^{\delta_t} \left(u \frac{\partial T}{\partial x} + v \frac{\partial T}{\partial y} \right) dy = k_f \left. \frac{\partial T}{\partial y} \right|_{y=0}^{y=\delta_t}$$

or
$$\rho c_p \left[\int_0^{\delta_t} u \frac{\partial T}{\partial x} dy + \int_0^{\delta_t} v \frac{\partial T}{\partial y} dy \right] = q_s'' \qquad (5.46)$$

where the relations $\partial T/\partial y = 0$ at $y = \delta_t$ and $q_s'' = -k_f(\partial T/\partial y)_s$ have been used. Recall that

$$\frac{\partial(uT)}{\partial x} = u \frac{\partial T}{\partial x} + T \frac{\partial u}{\partial x}$$

or
$$u \frac{\partial T}{\partial x} = \frac{\partial(uT)}{\partial x} - T \frac{\partial u}{\partial x} \qquad (5.47)$$

Using Eq. (5.47) and Leibniz's[3] rule, the first integral in Eq. (5.46) becomes

$$\int_0^{\delta_t} u \frac{\partial T}{\partial x} dy = \frac{d}{dx} \int_0^{\delta_t} uT dy - (uT)_{y=\delta_t} \frac{d\delta_t}{dx} - \int_0^{\delta_t} T \frac{\partial u}{\partial x} dy \qquad (5.48)$$

Similar to Eq. (5.47),

$$v \frac{\partial T}{\partial y} = \frac{\partial(vT)}{\partial y} - T \frac{\partial v}{\partial y} \qquad (5.49)$$

Using Eq. (5.49), the second integral in Eq. (5.46) becomes

$$\int_0^{\delta_t} v \frac{\partial T}{\partial y} dy = \int_0^{\delta_t} \left[\frac{\partial(vT)}{\partial y} - T \frac{\partial v}{\partial y} \right] dy = (vT)_{y=\delta_t} - \int_0^{\delta_t} T \frac{\partial v}{\partial y} dy \qquad (5.50)$$

---

[3]Leibniz's rule of differentiating an integral:
$$\frac{d}{dx} \int_{a(x)}^{b(x)} f(y, x) dy = f[b(x), x] b'(x) - f[a(x), x] a'(x) + \int_{a(x)}^{b(x)} \frac{\partial f(y, x)}{\partial x} dy$$

Since $v = 0$ at $y = 0$. Integrating the continuity equation over the thermal boundary layer thickness,

$$\int_0^{\delta_t} \frac{\partial u}{\partial x} dy + \int_0^{\delta_t} \frac{\partial v}{\partial y} dy = 0$$

or $\qquad \int_0^{\delta_t} \frac{\partial u}{\partial x} dy + (v)_{y=\delta_t} = 0$ $\hfill$ (5.51)

Using Leibniz's rule on the integrand in Eq. (5.51),

$$\int_0^{\delta_t} \frac{\partial u}{\partial x} dy = \frac{d}{dx} \int_0^{\delta_t} u\, dy - (u)_{y=\delta_t}\left(\frac{d\delta_t}{dx}\right) \tag{5.52}$$

Substituting Eq. (5.52) into Eq. (5.51), we obtain

$$(v)_{y=\delta_t} = -\frac{d}{dx}\int_0^{\delta_t} u\, dy - (u)_{y=\delta_t}\left(\frac{d\delta_t}{dx}\right) \tag{5.53}$$

Now multiplying Eq. (5.53) by T_∞: (note $T_\infty = T_{y=\delta_1}$),

$$(Tv)_{y=\delta_t} = -T_\infty \frac{d}{dx}\int_0^{\delta_t} u\, dy + (uT)_{y=\delta_t}\left(\frac{d\delta_t}{dx}\right) \tag{5.54}$$

and substituting Eq. (5.54) into Eq. (5.50),

$$\int_0^{\delta_t} v\frac{\partial T}{\partial y} dy = -T_\infty \frac{d}{dx}\int_0^{\delta_t} u\, dy + (uT)_{y=\delta_t}\left(\frac{d\delta_t}{dx}\right) - \int_0^{\delta_t} T\frac{\partial v}{\partial y} dy \tag{5.55}$$

Again, substituting Eqs (5.48) and (5.55) into Eq. (5.46),

$$\rho c_p \left[\frac{d}{dx}\int_0^{\delta_t} u(T - T_\infty)dy - \int_0^{\delta_t} T\left(\frac{\partial u}{\partial x} + \frac{\partial v}{\partial y}\right)dy \right] = q_s'' \tag{5.56}$$

But $\partial u/\partial x + \partial v/\partial y = 0$ from the continuity equation. So, Eq. (5.56) becomes

$$\rho c_p \frac{d}{dx}\int_0^{\delta_t} u(T - T_\infty)dy = q_s'' \tag{5.57}$$

which is the desired energy integral equation.

5.8 Procedure for Using Energy Integral Equation

(a) Assume velocity and temperature profiles consistent with boundary and compatibility requirements. This will result in $u = u(y, \delta)$ and $T = T(y, \delta_t)$. When we take

$$h = \frac{-k_f \left(\dfrac{\partial T}{\partial y}\right)_{y=0}}{T_s - T_\infty}$$

we find $\delta = \delta_t(x)$ is needed.

(b) Substitute u and T into the energy integral equation [Eq. (5.57)].
    (i) If $\delta_t < \delta$, we need to integrate only to δ_t.
    (ii) If $\delta_t = \delta$, integrate to δ_t, no conceptual problem.
    (iii) If $\delta_t > \delta$, integrate in two steps: 0–δ and δ–δ_t.
        The last one is a much more difficult problem. After performing integration, differentiate to get an ordinary differential equation, involving ratios of δ and δ_t.

(c) Solve the differential equation and obtain q''_s.

5.9 Application of Energy Integral Equation to the Thermal Boundary Layer over an Isothermal Flat Plate

We start by assuming that $\delta_t < \delta$, i.e., $\xi = \delta_t/\delta < 1$. We assume that the velocity function is a polynomial of the form

$$u = C_0 + C_1 y + C_2 y^2 + C_3 y^3 \tag{5.58}$$

Note that C's may be functions of x. Therefore, four boundary and compatibility conditions are required as follows:

$$y = 0, \qquad u = 0 \tag{5.59a}$$

$$y = 0, \qquad \frac{\partial^2 u}{\partial y^2} = 0 \tag{5.59b}$$

$$y = \delta, \qquad u = u_\infty \tag{5.59c}$$

$$y = \delta, \qquad \frac{\partial u}{\partial y} = 0 \tag{5.59d}$$

The second condition, i.e., $\partial^2 u/\partial y^2 = 0$ at $y = 0$ comes from the application of x-momentum equation at the wall. The application of the conditions (5.59a–d) gives

$$C_0 = 0 \tag{5.60a}$$

$$C_1 = \frac{3u_\infty}{2}\left(\frac{1}{\delta}\right) \tag{5.60b}$$

$$C_2 = 0 \tag{5.60c}$$

$$C_3 = -\frac{u_\infty}{2}\left(\frac{1}{\delta}\right)^3 \tag{5.60d}$$

Putting C_0, C_1, C_2, C_3 into Eq. (5.58) results in

$$\frac{u}{u_\infty} = \frac{3}{2}\frac{y}{\delta} - \frac{1}{2}\left(\frac{y}{\delta}\right)^3 \tag{5.61}$$

Similarly, we assume a temperature function:

$$T = a_0 + a_1 y + a_2 y^2 + a_3 y^3 \tag{5.62a}$$

with boundary and compatibility conditions:

$$y = 0, \quad T = T_s \tag{5.62b}$$

$$y = 0, \quad \frac{\partial^2 T}{\partial y^2} = 0 \tag{5.62c}$$

$$y = \delta_t, \quad T = T_\infty \tag{5.62d}$$

$$y = \delta_t, \quad \frac{\partial T}{\partial y} = 0 \tag{5.62e}$$

The second condition results from the application of energy equation at the wall. The application of conditions (5.62b–e) gives

$$a_0 = T_s \tag{5.63a}$$

$$a_1 = \frac{3}{2}(T_\infty - T_s)\left(\frac{1}{\delta_t}\right) \tag{5.63b}$$

$$a_2 = 0 \tag{5.63c}$$

$$a_3 = -\left(\frac{T_\infty - T_s}{2\delta_t^3}\right) \tag{5.63d}$$

Substituting a_0, a_1, a_2, a_3 into Eq. (5.61),

$$T = T_s + \left[\frac{3}{2}\frac{y}{\delta_t} - \frac{1}{2}\left(\frac{y}{\delta_t}\right)^3\right](T_\infty - T_s)$$

which can be written as

$$\frac{\theta}{\theta_\infty} = \frac{T - T_s}{T_\infty - T_s} = \left[\frac{3}{2}\left(\frac{y}{\delta_t}\right) - \frac{1}{2}\left(\frac{y}{\delta_t}\right)^3\right] \tag{5.64}$$

The energy integral equation can be written as

$$\frac{d}{dx}\int_0^{\delta_t} u(T - T_\infty)dy = \frac{q_s''}{\rho c_p} = -\alpha \left.\frac{\partial T}{\partial y}\right|_{y=0} \tag{5.65}$$

and the left-hand side is manipulated as follows:

$$\frac{d}{dx}\left\{\int_0^{\delta_t}[(T - T_s) - (T_\infty - T_s)]u\,dy\right\}$$

$$= \frac{d}{dx}\left\{\int_0^{\delta_t}[\theta - \theta_\infty]u\,dy\right\}$$

$$= \frac{d}{dx}\left\{\int_0^{\delta_t}\left(\frac{\theta}{\theta_\infty} - 1\right)u\theta_\infty\,dy\right\}$$

Equation (5.65) can now be written as

$$\frac{d}{dx}\left[\int_0^{\delta_t}\left(\frac{\theta}{\theta_\infty} - 1\right)u\theta_\infty\,dy\right] = -\alpha \left.\frac{\partial T}{\partial y}\right|_{y=0} \tag{5.66}$$

Now, substituting the temperature distribution [Eq. (5.64)] and the velocity distribution [Eq. (5.61)] into Eq. (5.65),

$$\frac{d}{dx}\left\{\int_0^{\delta_t}\left[\frac{3}{2}\frac{y}{\delta_t} - \frac{1}{2}\left(\frac{y}{\delta_t}\right)^3 - 1\right]\left[\frac{3}{2}\frac{y}{\delta} - \frac{1}{2}\left(\frac{y}{\delta}\right)^3\right]\theta_\infty u_\infty\,dy\right\}$$

$$= -\alpha\theta_\infty \left[\frac{3}{2}\left(\frac{1}{\delta_t}\right) - \frac{3}{2}\left(\frac{y}{\delta_t}\right)^2 \left(\frac{1}{\delta_t}\right) \right]_{y=0}$$

$$= -\frac{3\alpha\theta_\infty}{2\delta_t} \tag{5.67}$$

After carrying out the necessary integration on the left-hand side of Eq. (5.67), we obtain

$$\theta_\infty u_\infty \frac{d}{dx}\left[\delta\left(\frac{3}{20}\xi^2 - \frac{3}{280}\xi^4 \right) \right] = \frac{3\alpha\theta_\infty}{2\xi\delta}$$

where $\xi = \delta_t/\delta$. Since, we assumed $\xi < 1$, the term involving ξ^4 is small compared to the ξ^2 term so that we write

$$\frac{3}{20}\theta_\infty u_\infty \frac{d}{dx}(\delta\xi^2) = \frac{3\alpha\theta_\infty}{2\xi\delta} \tag{5.68}$$

or $$\frac{3}{20}\theta_\infty u_\infty \left(\xi^2 \frac{d\delta}{dx} + \delta\frac{d\xi^2}{dx} \right) = \frac{3\alpha\theta_\infty}{2\xi\delta}$$

Multiplying both sides by $20\xi\delta/3\theta_\infty$,

$$\xi^3 u_\infty \delta \frac{d\delta}{dx} + u_\infty \delta^2 2\xi^2 \frac{d\xi}{dx} = 10\alpha \tag{5.69}$$

From the momentum integral analysis (Rohsenow and Choi, 1961),

$$\delta\frac{d\delta}{dx} = \frac{140v}{13u_\infty} \tag{5.70}$$

and $$\delta^2 = 2\left(\frac{140}{13}\right)\frac{v}{u_\infty}x \tag{5.71}$$

Substituting Eqs (5.70) and (5.71) into Eq. (5.69) yields

$$\xi^3\left(\frac{140v}{13}\right) + \left(\frac{140v}{13}\right)4\xi^2 x \frac{d\xi}{dx} = 10\alpha$$

or $$\xi^3 + 4\xi^2 x \frac{d\xi}{dx} = \frac{13}{14\,Pr} \tag{5.72}$$

Recall $d\xi^3 = 3\xi^2 d\xi$. Substituting Eq. (5.73) into Eq. (5.72) gives   (5.73)

$$\xi^3 + \frac{4}{3}x\frac{d\xi^3}{dx} = \frac{13}{14\,Pr} \tag{5.74}$$

Let $\psi = \xi^3$ (5.75)

Then Eq. (5.74) becomes

$$\frac{d\psi}{dx} + \frac{3}{4}\frac{\psi}{x} = \frac{39}{56\,Pr\,x} \tag{5.76}$$

Equation (5.76) is a first-order ordinary linear differential equation. The solution is

$$\psi = Cx^{-3/4} + \frac{13}{14\,Pr} \tag{5.77}$$

Using the boundary condition

$$\delta_t = 0 \qquad \text{at} \qquad x = L$$

i.e., $\qquad \xi = 0 \qquad \text{at} \qquad x = L$

we obtain from Eq. (5.77)

$$\xi = \frac{\delta_t}{\delta} = \frac{1}{1.026\,Pr^{1/3}}\left[1-\left(\frac{L}{x}\right)^{3/4}\right]^{1/3} \tag{5.78}$$

While deriving Eq. (5.78) we assumed that the heating starts at $x = L$ (Fig. 5.9).

For no starting length, i.e., $L = 0$, Eq. (5.78) becomes

$$\xi = \frac{1}{1.026\,Pr^{1/3}} \tag{5.79}$$

To find the heat transfer coefficient, we start with

$$h = \frac{-k_f(\partial T/\partial y)_{y=0}}{T_s - T_\infty}$$

$$= \frac{-k_f(\partial \theta/\partial y)_{y=0}}{\theta_\infty}$$

$$= \frac{k_f\theta_\infty}{\theta_\infty}\left\{\frac{\partial}{\partial y}\left[\frac{3}{2}\frac{y}{\delta_t} - \frac{1}{2}\left(\frac{y}{\delta_t}\right)^3\right]\right\}_{y=0}$$

$$= \frac{3k_f}{2\delta_t} = \frac{3k_f}{2\xi\delta} \tag{5.80}$$

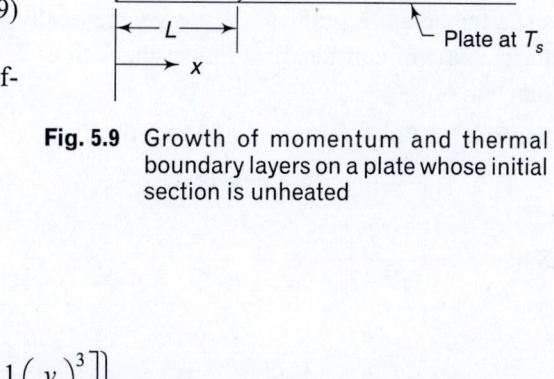

Fig. 5.9 Growth of momentum and thermal boundary layers on a plate whose initial section is unheated

L = Unheated length (Insulated section)

Substituting ξ from Eq. (5.78) and δ from Eq. (5.71), we get

$$h = \frac{3k_f}{2}\left(\frac{1.026}{4.64}\right)Pr^{1/3}\left[\frac{u_\infty}{vx}\right]^{1/2}\frac{1}{\left[1-\left(\frac{L}{x}\right)^{3/4}\right]^{1/3}}$$

or $\qquad Nu_x = 0.332\,Pr^{1/3}Re_x^{1/2}\dfrac{1}{\left[1-\left(\dfrac{L}{x}\right)^{3/4}\right]^{1/3}} \tag{5.81}$

Since we have taken $\xi \le 1$, so we require $Pr \ge 1$. For small Pr (0.006–0.03) Eq. (5.81) is not valid. For $L = 0$,

$$Nu_x = 0.332\,Pr^{1/3}\,Re_x^{1/2} \tag{5.82}$$

which is the same as obtained by using similarity analysis. For small Pr (liquid metals range), the integration must be performed on the energy integral equation in two parts:

$$0 < y < \delta, \ u \text{ as before}$$
$$\delta < y < \delta_t, \ u = u_\infty$$

This has been done by Eckert[4] for an entire plate heated, and the following result was obtained:

$$\text{Nu}_x = \frac{\text{Pe}_x^{1/2}}{1.55\,\text{Pr}^{1/2} + 3.09[0.372 - 0.15\,\text{Pr}]^{1/2}} \tag{5.83}$$

where Pe is Peclet[5] number and is equal to Re Pr. Peclet number signifies the ratio of the strength of convection to the strength of diffusion.

5.9.1   Energy Integral Solution for Uniform Heat Flux (q_s'' = constant) at the Wall

The correlations are obtained by integral analysis as no similarity solution is possible since temperature profiles are not geometrically similar at different x-locations for the case of constant heat flux at the wall as T_s is not a constant and will be a function of x.

For Pr ≥ 0.5 fluids (entire plate heated)

$$\text{Nu}_x = 0.453\,\text{Re}_x^{1/2}\,\text{Pr}^{1/3} \tag{5.84}$$

Since $h = \dfrac{q_s''}{T_s - T_\infty}$

$$\text{Nu}_x = \frac{hx}{k_f} = \frac{q_s'' x}{k_f(T_s - T_\infty)} \tag{5.85}$$

Substituting Eq. (5.84) into Eq. (5.85), we get

$$T_s(x) - T_\infty = \frac{q_s'' x}{0.453 k_f\,\text{Re}_x^{1/2}\,\text{Pr}^{1/3}} \tag{5.86}$$

It is clear from Eq. (5.86) the surface temperature will be increasing in the x-direction ($T_s \sim x^{1/2}$), that is, along the plate starting from the leading edge.

Now, $\overline{\text{Nu}}_L = \dfrac{\overline{h}_L L}{k_f}$

Also, $\overline{h}_L = \dfrac{q_s''}{T_{s,\text{avg}} - T_\infty}$

where $T_{s,\text{avg}} - T_\infty = \dfrac{1}{L}\displaystyle\int_0^L (T_s - T_\infty)\,dx$ \hfill (5.87)

---

[4] E.R.G. Eckert (1904–2004) was a legend in the field of heat transfer. The dimensionless group called Eckert number carries his name. He was Professor Emeritus of the Department of Mechanical Engineering at the University of Minnesota. He was the lead author of the classic textbook entitled *Heat and Mass Transfer*.

[5] Jean-Claude-Eugene Peclet (1793–1857) was a French physicist who wrote a remarkable treatise on heat transfer and its applications in 1829.

Substituting Eq. (5.86) into Eq. (5.87) and carrying out the integration, we obtain

$$T_{s,\text{avg}} - T_\infty = \frac{q_s''(L/k_f)}{0.6795\,\text{Re}_L^{1/2}\,\text{Pr}^{1/3}}$$

or, $$\frac{q_s''}{T_{s,\text{avg}} - T_\infty} = \left(\frac{k_f}{L}\right) 0.6795\,\text{Re}_L^{1/2}\,\text{Pr}^{1/3} \tag{5.88}$$

Hence, $$\overline{\text{Nu}}_L = \frac{\bar{h}_L L}{k_f} = \frac{q_s'' L}{\left(T_{s,\text{avg}} - T_\infty\right) k_f} \tag{5.89}$$

Using Eq. (5.88), Eq. (5.89) can be rewritten as

$$\overline{\text{Nu}}_L = 0.6795\,\text{Re}_L^{1/2}\,\text{Pr}^{1/3} \tag{5.90}$$

Comparing Eq. (5.90) with Eq. (5.91) we see that

$$\overline{\text{Nu}}_L = 1.5\,\text{Nu}\big|_{x=L} \tag{5.91}$$

For liquid metals, $0.006 \leq \text{Pr} \leq 0.03$ (entire plate heated)

$$\frac{\text{Nu}_x}{\sqrt{\text{Re}_x\,\text{Pr}}} = \frac{0.88}{1 + 1.317\sqrt{\text{Pr}}} \tag{5.92}$$

For $\text{Pr} \geq 0.5$ fluids (unheated starting length)

$$\text{Nu}_x = \frac{0.453\,\text{Re}_x^{1/2}\,\text{Pr}^{1/3}}{\left[1 - \left(\dfrac{L}{x}\right)^{3/4}\right]^{1/3}} \tag{5.93}$$

where L is the unheated starting length.

For liquid metals, $0.006 \leq \text{Pr} \leq 0.03$ (unheated starting length)

No energy integral solution is available in published literature.

5.10 Film Temperature

The results based on the thermal boundary layer analysis apply to constant property fluids. In real situations, fluid properties such as k_f, μ, ν, and α are not constant, as they depend primarily on the local temperature in the flow field. The assumption of constant property is valid provided the maximum temperature difference in the fluid $(T_s - T_\infty)$ is small relative to the absolute temperature level of the fluid (T_s or T_∞, expressed in kelvin). In such cases the properties needed for calculating the various dimensionless groups (Re_x, Pe_x, Pr, Nu_x) can be evaluated at the average temperature of the fluid in the thermal boundary layer,

$$T_f = \frac{1}{2}(T_s + T_\infty) \tag{5.94}$$

T_f is called the film temperature of the fluid, and is generally recommended for use in formulae based on the constant properties of the fluid.

For uniform heat flux boundary condition at the wall,

$$T_f = \frac{1}{2}\left(T_{s,\text{avg}} + T_\infty\right) \tag{5.95}$$

where $T_{s,\text{avg}}$ (also denoted as $\overline{T}_s$ in this book elsewhere) is the x-averaged wall temperature. However, since $T_{s,\text{avg}}$ is not known to start with, an iterative procedure is adopted. In the first iteration, T_∞ is used to evaluate the properties and using the correlation for average Nusselt number $T_{s,\text{avg}} - T_\infty$ is calculated. The newly obtained $T_{s,\text{avg}}$ is then used to evaluate properties at T_f and again $T_{s,\text{avg}} - T_\infty$ is calculated. Iteration continues till $\left| T_{s,\text{avg}}^{\text{old}} - T_{s,\text{avg}}^{\text{new}} \right| \le \varepsilon$ where ε is a small value such as 0.01, 0.1, 1 and so on. Example 5.1 illustrates this procedure.

Example 5.1   Flat Plate Local and Average Heat Transfer Coefficients

Calculate the average heat transfer coefficient and heat transfer at a distance of 10 cm from the leading edge of an entirely heated plate placed in an air stream. The air velocity is 10 m/s; its temperature $T_\infty = 30\,°C$. The surface temperature of the plate is $70\,°C$. The plate is 1 m wide.

Solution
The properties of air are evaluated at the film temperature

$$T_f = \frac{T_s + T_\infty}{2} = \frac{70 + 30}{2} = 50°C$$

At 50°C,
$$v = 18.02 \times 10^{-6}\ \text{m}^2/\text{s}$$
$$k_f = 0.02798\ \text{W/m°C}$$
$$\text{Pr} = 0.703$$

$$\text{Re}_x = \frac{u_\infty x}{v} = \frac{10 \times (10 \times 10^{-2})}{18.02 \times 10^{-6}}$$

$$= 5.55 \times 10^4$$
$$\text{Re}_{\text{crit}} = 5 \times 10^5$$

Since $\text{Re}_x < \text{Re}_{\text{crit}}$, the flow is laminar. Therefore,

$$\text{Nu}_x = 0.332\,\text{Re}_x^{1/2}\ \text{Pr}^{1/3}$$

$$= 0.332(5.55 \times 10^4)^{1/2}\,(0.703)^{1/3}$$
$$= 0.332\,(235.58)\,(0.889)$$
$$= 69.53$$

Therefore, $\text{Nu}_x = \dfrac{hx}{k_f} = 69.53$

$\Rightarrow$ $h = \dfrac{69.53\,k_f}{x}$

$$= \frac{(69.53)(0.02798)}{10 \times 10^{-2}}$$

$$= 19.45\ \text{W/m}^2°C$$

Therefore, $\bar{h}_L = 2h\big|_{x=L}$

$$= 2(19.45)$$
$$= 38.9 \text{ W/m}^2\,°C$$

Therefore, total heat transfer in a span of 10 cm from the leading edge of the plate is

$$q = \bar{h}_L\, A\,(T_s - T_\infty)$$
$$= (38.9)(10 \times 10^{-2} \times 1)(70 - 30)$$
$$= 155.6 \text{ W}$$

Example 5.2   Flat Plate Integral Thermal Boundary Layer Solution for Linear Velocity and Temperature Profiles

Assuming linear velocity and temperature profiles, carry out the integral analysis of the thermal boundary layer on an isothermal flat plate for $Pr \geq 1$ and obtain an expression for the local Nusselt number as a function of Reynolds number and Prandtl number.

For linear velocity profile: $(\delta/x)Re_x^{1/2} = 3.46$

Solution

Assuming a linear velocity profile,

$$u = C_0 + C_1 y \tag{A}$$
$$\text{BC-1: } y = 0, \ u = 0$$
$$\text{BC-2: } y = \delta, \ u = u_\infty$$

The application of BC-1 and BC-2 to Eq. (A) yields

$$C_0 = 0$$

$$C_1 = \frac{u_\infty}{\delta}$$

Therefore, $\dfrac{u}{u_\infty} = \dfrac{y}{\delta}$ $\tag{B}$

Assuming a linear temperature profile,

$$T = a_0 + a_1 y \tag{C}$$
$$\text{BC-1: } y = 0, \ T = T_s$$
$$\text{BC-2: } y = \delta_t, \ T = T_\infty$$

The application of BC-1 and BC-2 to Eq. (C) yields

$$T = T_s + \frac{T_\infty - T_s}{\delta_t} y$$

$$\Rightarrow \qquad \frac{\theta}{\theta_\infty} = \frac{T - T_s}{T_\infty - T_s} = \frac{y}{\delta_t} \tag{D}$$

The objective of the integral analysis is to evaluate h and Nu_x. Now,

$$h = \frac{-k_f\,(\partial T/\partial y)_{y=0}}{T_s - T_\infty}$$

$$= \frac{-k_f\,(T_\infty - T_s)}{(T_s - T_\infty)\delta_t}$$

$$= \frac{k_f}{\delta_t} \tag{E}$$

Recall the energy integral equation

$$\rho c_p \frac{d}{dx} \int_0^{\delta_t} u (T - T_\infty) \, dy = q_s''$$ (F)

Substituting u and $T - T_\infty$ from Eqs (B) and (D), respectively, we get

$$\frac{d}{dx} \int_0^{\delta_t} \frac{y}{\delta} u_\infty \left[(T - T_s) - (T_\infty - T_s) \right] dy = \frac{-k_f \left. \frac{\partial T}{\partial y} \right|_{y=0}}{\rho c_p}$$

or

$$\frac{d}{dx} \int_0^{\delta_t} \frac{y}{\delta} u_\infty \left(\frac{\theta}{\theta_\infty} - 1 \right) \theta_\infty \, dy = -\alpha \left. \frac{\partial \theta}{\partial y} \right|_{y=0}$$ (G)

The LHS of Eq. (G)

$$\frac{d}{dx} \left[\int_0^{\delta_t} \frac{u_\infty \theta_\infty}{\delta} \left(\frac{y^2}{\delta_t} - y \right) dy \right]$$

$$= \frac{d}{dx} \left(\frac{u_\infty \theta_\infty}{\delta} \right) \left[\frac{\delta_t^3}{3\delta_t} - \frac{\delta_t^2}{2} \right]$$

$$= -\frac{1}{6} u_\infty \theta_\infty \frac{d}{dx} \left(\frac{\delta_t^2}{\delta} \right)$$

The RHS of Eq. (G)

$$-\alpha \left. \frac{\partial \theta}{\partial y} \right|_{y=0} = -\alpha \frac{\theta_\infty}{\delta_t}$$

Therefore,

$$-\frac{u_\infty}{6} \theta_\infty \frac{d}{dx} \left(\frac{\delta_t^2}{\delta} \right) = -\alpha \frac{\theta_\infty}{\delta_t}$$

$$\Rightarrow \qquad -\frac{1}{6} u_\infty \frac{d}{dx} \left(\frac{\delta_t^2}{\delta} \right) = \frac{\alpha}{\delta_t}$$ (H)

Let $\xi = \delta_t / \delta$. Then, from Eq. (H),

$$\frac{u_\infty}{6} \frac{d}{dx} (\delta \xi^2) = \frac{\alpha}{\xi \delta}$$

$$\Rightarrow \qquad u_\infty \left[\delta \, 2\xi \frac{d\xi}{dx} + \xi^2 \frac{d\delta}{dx} \right] = \frac{6\alpha}{\xi \delta}$$

$$\Rightarrow \qquad u_\infty \left[2\xi^2 \delta^2 \frac{d\xi}{dx} + \xi^3 \delta \frac{d\delta}{dx} \right] = 6\alpha$$ (I)

Also,

$$\frac{\delta}{x} Re_x^{1/2} = 3.46$$

or

$$\delta = \frac{3.46 \, x}{Re_x^{1/2}}$$

$$\Rightarrow \qquad \frac{\delta}{x} \sqrt{\frac{u_\infty x}{\nu}} = 3.46$$

$$\Rightarrow \qquad \delta^2 = 11.97 \frac{v}{u_\infty} x \qquad\qquad\qquad\qquad \text{(J)}$$

$$\Rightarrow \qquad 2\delta \frac{d\delta}{dx} = 11.97 \frac{v}{u_\infty} \qquad\qquad\qquad\qquad \text{(K)}$$

Using Eqs (J) and (K) in Eq. (I), we obtain

$$2\xi^2 \,(11.97\,vx)\frac{d\xi}{dx} + \xi^3 \left(\frac{11.97}{2}\right) v = 6\alpha$$

or $\qquad \dfrac{2}{3}(11.97\,vx)\dfrac{d\xi^3}{dx} + \dfrac{11.97}{2} v\xi^3 = 6\alpha \qquad\qquad \text{(L)}$

Let $\psi = \xi^3$. Therefore, Eq. (L) becomes

$$\frac{d\psi}{dx} + \frac{3}{4}\frac{\psi}{x} = \frac{3}{4\,\text{Pr}\,x} \qquad\qquad\qquad\qquad \text{(M)}$$

Equation (M) is of the form

$$\frac{dy}{dx} + py = Q \qquad\qquad\qquad\qquad \text{(N)}$$

where p and Q are both functions of x alone or constants. The general solution of Eq. (N) is

$$y = \exp\left(-\int p\,dx\right)\left[\int Q\exp\int (p\,dx)\,dx + C\right] \qquad\qquad \text{(O)}$$

Comparing Eqs (M) and (N),

$$p = \frac{3}{4x}, \; Q = \frac{3}{4\,\text{Pr}\,x}, \; y = \psi$$

Therefore, from Eq. (O),

$$\psi = Cx^{-3/4} + \frac{1}{\text{Pr}}$$

$$\Rightarrow \qquad \xi^3 = \left(\frac{\delta_t}{\delta}\right)^3 = Cx^{-3/4} + \frac{1}{\text{Pr}} \qquad\qquad\qquad \text{(P)}$$

Now, at $x = L$, $\delta_t = 0$ or $\psi = 0$. Therefore,

$$0 = CL^{-3/4} + \frac{1}{\text{Pr}}$$

$$\Rightarrow \qquad C = \frac{-L^{3/4}}{\text{Pr}} \qquad\qquad\qquad\qquad \text{(Q)}$$

Substituting Eq. (Q) into Eq. (P),

$$\left(\frac{\delta_t}{\delta}\right)^3 = \frac{1}{\text{Pr}}\left[1 - \left(\frac{L}{x}\right)^{3/4}\right]$$

$$\Rightarrow \qquad \delta_t = \frac{\delta}{\text{Pr}^{1/3}}\left[1 - \left(\frac{L}{x}\right)^{3/4}\right]^{1/3} \qquad\qquad \text{(R)}$$

From Eq. (E),

$$h = \frac{k_f}{\delta_t}$$

$$= \frac{k_f}{\delta} \mathrm{Pr}^{1/3} \frac{1}{\left[1 - \left(\dfrac{L}{x}\right)^{3/4}\right]^{1/3}}$$

Also, $\qquad \delta = \dfrac{3.46\,x}{\mathrm{Re}_x^{1/2}}$

$\Rightarrow \qquad h = \dfrac{k_f}{3.46\,x/\mathrm{Re}_x^{1/2}} \mathrm{Pr}^{1/3} \dfrac{1}{\left[1 - \left(\dfrac{L}{x}\right)^{3/4}\right]^{1/3}}$

$\Rightarrow \qquad h = \dfrac{k_f}{3.46\,x} \mathrm{Re}_x^{1/2}\, \mathrm{Pr}^{1/3} \dfrac{1}{\left[1 - \left(\dfrac{L}{x}\right)^{3/4}\right]^{1/3}}$

$$\mathrm{Nu}_x = \frac{hx}{k_f} = \frac{1}{3.46} \mathrm{Re}_x^{1/2}\, \mathrm{Pr}^{1/3} \frac{1}{\left[1 - \left(\dfrac{L}{x}\right)^{3/4}\right]^{1/3}}$$

For no starting length, i.e., $L = 0$ (entire plate heated):

$$\mathrm{Nu}_x = \frac{1}{3.46} \mathrm{Re}_x^{1/2}\, \mathrm{Pr}^{1/3}$$

$$= 0.289\, \mathrm{Re}_x^{1/2}\, \mathrm{Pr}^{1/3}$$

5.11   Relationship Between Fluid Friction and Heat Transfer

It has been discussed earlier that the temperature and flow fields are related. In this section, we show how the heat transfer coefficient can be determined from the knowledge of the skin friction drag on a plate under conditions in which no heat transfer is involved.

$$\tau_s = C_{fx} \frac{\rho u_\infty^2}{2} \qquad\qquad (5.96)$$

where C_{fx} is the local skin friction coefficient. Also,

$$\tau_s = \mu \frac{\partial u}{\partial y}\bigg|_{y=0}$$

Using the velocity distribution

$$\frac{u}{u_\infty} = \frac{3}{2}\left(\frac{y}{\delta}\right) - \frac{1}{2}\left(\frac{y}{\delta}\right)^3$$

we have $\tau_s = \dfrac{3}{2} \dfrac{\mu u_\infty}{\delta}$

Also, $\qquad \delta = 4.64 \sqrt{\dfrac{vx}{u_\infty}}$

Therefore, $\tau_s = \dfrac{3}{2}\dfrac{\mu u_\infty}{4.64}\left(\dfrac{u_\infty}{vx}\right)^{1/2}$ (5.97)

Combining Eqs (5.96) and (5.97), we have

$$\frac{C_{fx}}{2} = \frac{3}{2}\frac{\mu u_\infty}{4.64}\left(\frac{u_\infty}{vx}\right)^{1/2}\frac{1}{\rho u_\infty^2}$$

$$= 0.323\,\mathrm{Re}_x^{-1/2}$$ (5.98)

The exact solution of the momentum boundary layer equation yields

$$\frac{C_{fx}}{2} = 0.332\,\mathrm{Re}_x^{-1/2}$$ (5.99)

Now, $\dfrac{\mathrm{Nu}_x}{\mathrm{Re}_x\,\mathrm{Pr}} = \dfrac{hx/k_f}{\left(\dfrac{u_\infty x}{v}\right)\left(\dfrac{\mu c_p}{k_f}\right)}$

$$= \frac{h}{\rho c_p u_\infty}$$

We also know $\quad \mathrm{Nu}_x = 0.332\,\mathrm{Re}_x^{1/2}\,\mathrm{Pr}^{1/3}$

or $\quad \dfrac{\mathrm{Nu}_x}{\mathrm{Re}_x\,\mathrm{Pr}} = 0.332\,\mathrm{Re}_x^{-1/2}\,\mathrm{Pr}^{-2/3}$

Therefore $\quad \dfrac{h}{\rho c_p u_\infty} = 0.332\,\mathrm{Re}_x^{-1/2}\,\mathrm{Pr}^{-2/3}$

$h/(\rho c_p u_\infty)$ *is called the Stanton[6] number. Therefore,*

$$\mathrm{St}_x\,\mathrm{Pr}^{2/3} = 0.332\,\mathrm{Re}_x^{-1/2}$$ (5.100)

Upon comparing Eqs (5.98) and (5.100), we note that the right-hand sides are identical except for a difference of about 3% in the constant, which is the result of the approximate nature of the boundary layer analysis. We recognize this approximation and write

$$\mathrm{St}_x\,\mathrm{Pr}^{2/3} = \frac{C_{fx}}{2}$$ (5.101)

Equation (5.101) is called the Reynolds[7]–Colburn[8] analogy, which expresses the relation between fluid friction and heat transfer for laminar flow on a flat plate. It turns out that Eq. (5.101) can also be applied to turbulent flow over a flat plate.

---

[6] Thomas Edward Stanton (1865–1931) was Professor of Engineering at Bristol University College, UK. His main research interests were the relationship between heat transfer and fluid flow, the aerodynamic loads on solid structures and the air cooling of IC engines.

[7] Osborne Reynolds (1842–1912) was Professor of Engineering at Owens College, Manchester, UK (later the University of Manchester). Best known for his work on turbulence and the discovery of the critical velocity for the laminar-turbulent transition in pipe flow (the critical Reynolds number), he also contributed to many other areas including the theory of lubrication.

[8] Allan Philip Colburn (1904–1955) was Professor of Chemical Engineering at the University of Delaware, USA. His main contributions were in the areas of the condensation of water vapour and the analogy between heat, mass, and momentum transfer.

Example 5.3   Flat Plate Heat Transfer using Reynolds-Colburn Analogy

Water at 20 °C and 1 atm flows over a flat plate at a speed of 0.5 m/s. The width of the plate is 1 m. The plate is entirely heated to a temperature of 60 °C. Calculate the heat transferred in the first 40 cm length of the plate using the Reynolds–Colburn analogy.

Solution

The film temperature T is

$$T_f = \frac{T_s + T_\infty}{2} = \frac{60 + 20}{2} = 40°C$$

Properties of water at 40°C:

$$\mu = 6.556 \times 10^{-4} \text{ kg/ms}$$
$$\rho = 992.04 \text{ kg/m}^3$$
$$k_f = 0.6328 \text{ W/m°C}$$
$$\text{Pr} = 4.334$$
$$c_p = 4.174 \text{ kJ/kg °C}$$

$$\text{Re}_x = \frac{\rho u_\infty x}{\mu}$$

$$= \frac{(992.04)(0.5)(0.4)}{6.556 \times 10^{-4}}$$

$$= 30.26 \times 10^4$$
$$= 3.026 \times 10^5$$

Hence, the flow is laminar since Re is less than 5×10^5. Now,

$$\frac{C_{fx}}{2} = 0.332 \, \text{Re}_x^{-1/2}$$

$$= 0.332 \, (3.026 \times 10^5)^{-1/2}$$
$$= 6.035 \times 10^{-4}$$

The Reynolds–Colburn analogy says

$$\text{St}_x \, \text{Pr}^{2/3} = \frac{C_{fx}}{2}$$

But, $$\text{St}_x = \frac{h}{\rho c_p u_\infty}$$

Therefore, $$\frac{h}{\rho c_p u_\infty} \text{Pr}^{2/3} = \frac{C_{fx}}{2}$$

or $$\frac{h}{(992.04)(4.174 \times 10^3)(0.5)} (4.334)^{2/3} = 6.035 \times 10^{-4}$$

or $$h = \frac{(6.035 \times 10^{-4})(992.04)(4.174 \times 10^3)(0.5)}{(4.334)^{2/3}}$$

$$= 470 \text{ W/m}^2 °C$$

The average heat transfer coefficient

$$\bar{h}_L = 2h\big|_{x=L} = 2(470) = 940 \text{ W/m}^2 °C$$

Therefore, the total heat transfer in the first 40 cm of the plate is

$$q = \bar{h}_L A (T_s - T_\infty) = 940 (0.4 \times 1)(60 - 20)$$

$$= 15,040 \text{ W} = 15.04 \text{ kW}$$

Check :

$$\overline{Nu}_L = 0.664 \, Re_L^{1/2} \, Pr^{1/3}$$

$$= 0.664 \, (3.026 \times 10^5)^{1/2} \, (4.334)^{1/3}$$

$$= 595.52$$

$$\Rightarrow \quad \frac{\bar{h}_L L}{k_f} = 595.52$$

$$\Rightarrow \quad \bar{h}_L = \frac{(595.52) k_f}{L}$$

$$= \frac{(595.52)(0.6328)}{0.4}$$

$$= 942.11 \text{ W/m}^2 \,^{\circ}\text{C}$$

From the Reynolds–Colburn analogy, we have

$$\bar{h}_L = 940 \quad \text{W/m}^2 \,^{\circ}\text{C}$$

Hence, the value of $\bar{h}_L$ matches very well with that obtained directly from the formula of thermal boundary layer analysis.

5.12   Turbulent Boundary Layer Over a Flat Plate

Turbulent flow is described as the motion in which an irregular fluctuation (mixing or Eddying motion) is superimposed on the main stream. The effects caused by the fluctuations are high mixing and as if the viscosity were increased by factors of one hundred, ten thousand, and even more. Mixing is responsible for the large resistance experienced by turbulent flows in pipes, for the drag encountered by ships and aeroplanes and for the losses in turbines and turbo compressors. However, we are forced to restrict ourselves to the considerations of time-averages of turbulent motion, because of tremendous complexity of the fluctuations.

5.12.1   Physical Aspects of Turbulent Boundary Layer

It has been observed from experiments that there are at least two regions, namely, (i) a predominantly viscous region very near the wall where momentum and heat transfer are occurring by the simple mechanism of viscous shear and molecular conduction; (ii) a fully turbulent region, comprising most of the boundary layer, where velocity is also a function of time. In this zone, "Eddy" motion is observed and heat and momentum are transported normally to the flow direction at rates that are much higher than those by viscous shear and molecular conduction alone.

In the fully turbulent region the velocity at any point seems to consist of a relatively large time-averaged velocity, on which is superimposed a smaller fluctuating velocity with instantaneous components in all three directions. This means that fluid is moving, at least temporarily, in the normal direction. This fluid carries momentum as well as thermal energy with it. Thus, there is a mean gradient of velocity and temperature in the normal direction. This is the primary mechanism by which momentum and heat are transported in the direction perpendicular to the plate.

The velocity fluctuations are the result of vorticity in the fluid. As a matter of fact, virtually every fluid particle is a part of fluid eddies that are turning over in various directions. In essence, this is what is meant by turbulent flow.

5.12.2   Time-averaged Equations

The time-averaging process is basically a smoothing out of the fluctuations of the turbulent flow field. It recognizes that a flow variable such as u is a function of spatial position and time, $u(x, y, z, t)$. At a fixed location in space, a typical u versus t plot is shown in Fig. 5.10. It is convenient to separate u into a mean motion, $\bar{u}$ and into a fluctuation or Eddying motion u' (Fig. 5.11). $\bar{u}$ is defined by the time-averaging operation

$$\bar{u} = \frac{1}{p} \int_0^p u \, dt \tag{5.102}$$

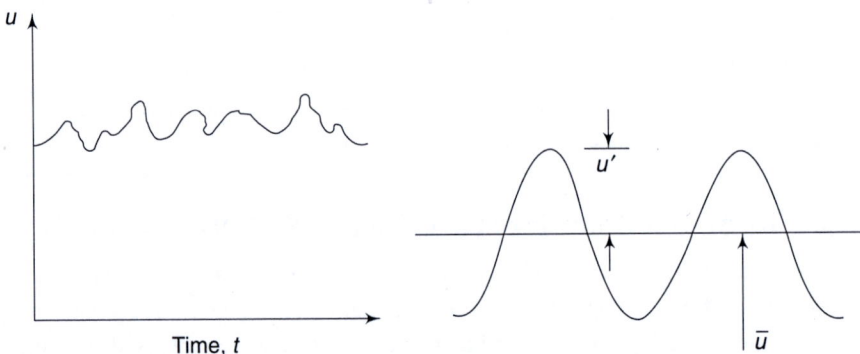

Fig. 5.10 u versus t plot for turbulent flow   **Fig. 5.11**   Separation of u into $\bar{u}$ and u'

The time-averaged value $\bar{u}$ is independent of time when the period p of the time-averaging operation exceeds the period of the slowest fluctuation exhibited by the actual variable u (Fig. 5.12). Thus,

$$u(x, y, z, t) = \bar{u}(x, y, z) + u'(x, y, z, t) \tag{5.103}$$

The same decomposition rule applies to the remaining variables of the flow field.

$$v = \bar{v} + v' \tag{5.104}$$

$$w = \bar{w} + w' \tag{5.105}$$

$$p = \bar{p} + p' \tag{5.106}$$

$$T = \bar{T} + T' \tag{5.107}$$

For compressible flow,

$$\rho = \bar{\rho} + \rho' \tag{5.108}$$

The next step is to time-average the continuity, momentum, and energy equations. This step consists of substituting u, v, w, p, T equations to the governing equations and then applying time-averaging operation to every term of the resulting equations. This analysis uses a special set of algebraic rules that follow from the time-averaging concept:

$$\bar{u}' = 0 \tag{5.109a}$$

$$\left(\frac{\partial u}{\partial x}\right) = \frac{\partial \bar{u}}{\partial x} \tag{5.109b}$$

$$\overline{u + v} = \bar{u} + \bar{v} \tag{5.109c}$$

$$\overline{\bar{u}u'} = 0 \tag{5.109d}$$

$$\frac{\partial \bar{u}}{\partial t} = 0 \tag{5.109e}$$

$$\overline{uv} = \bar{u}\,\bar{v} + \overline{u'v'} \tag{5.109f}$$

$$\left(\frac{\partial u}{\partial t}\right) = 0 \tag{5.109g}$$

$$\overline{u^2} = (\bar{u})^2 + \overline{(u')^2} \tag{5.109h}$$

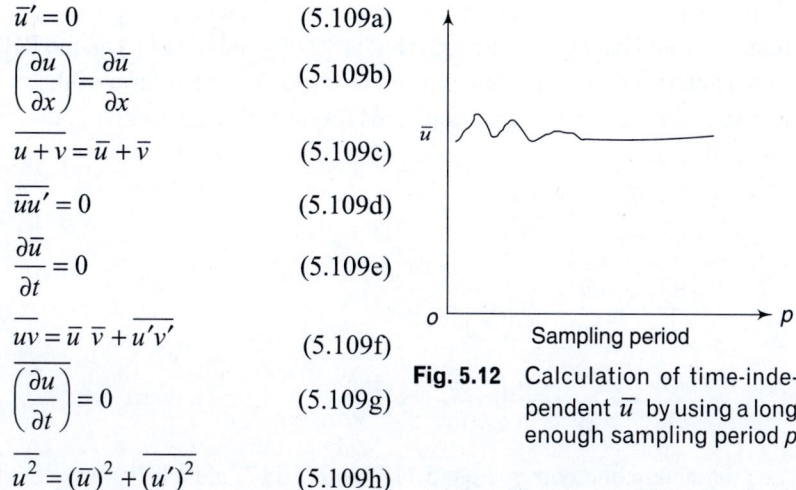

Fig. 5.12 Calculation of time-independent $\bar{u}$ by using a long enough sampling period p

It can be seen that the governing equations reduce to the following:

$$\frac{\partial \bar{u}}{\partial x} + \frac{\partial \bar{v}}{\partial y} + \frac{\partial \bar{w}}{\partial z} = 0 \tag{5.110}$$

$$\bar{u}\frac{\partial \bar{u}}{\partial x} + \bar{v}\frac{\partial \bar{u}}{\partial y} + \bar{w}\frac{\partial \bar{u}}{\partial z} = -\frac{1}{\rho}\frac{\partial \bar{p}}{\partial x} + \nu\nabla^2\bar{u} - \frac{\partial}{\partial x}(\overline{u'^2})$$

$$-\frac{\partial}{\partial y}(\overline{u'v'}) - \frac{\partial}{\partial z}(\overline{u'w'}) \tag{5.111}$$

$$\bar{u}\frac{\partial \bar{v}}{\partial x} + \bar{v}\frac{\partial \bar{v}}{\partial y} + \bar{w}\frac{\partial \bar{v}}{\partial z} = -\frac{1}{\rho}\frac{\partial \bar{p}}{\partial y} + \nu\nabla^2\bar{v} - \frac{\partial}{\partial x}(\overline{u'v'})$$

$$-\frac{\partial}{\partial y}(\overline{v'^2}) - \frac{\partial}{\partial z}(\overline{v'w'}) \tag{5.112}$$

$$\bar{u}\frac{\partial \bar{w}}{\partial x} + \bar{v}\frac{\partial \bar{w}}{\partial y} + \bar{w}\frac{\partial \bar{w}}{\partial z} = -\frac{1}{\rho}\frac{\partial \bar{p}}{\partial z} + \nu\nabla^2\bar{w}$$

$$-\frac{\partial}{\partial x}(\overline{u'w'}) - \frac{\partial}{\partial y}(\overline{v'w'}) - \frac{\partial}{\partial z}(\overline{w'^2}) \tag{5.113}$$

$$\bar{u}\frac{\partial \bar{T}}{\partial x} + \bar{v}\frac{\partial \bar{T}}{\partial y} + \bar{w}\frac{\partial \bar{T}}{\partial z} = \alpha\nabla^2\bar{T} - \frac{\partial}{\partial x}(\overline{u'T'})$$

$$-\frac{\partial}{\partial y}(\overline{v'T'}) - \frac{\partial}{\partial z}(\overline{w'T'}) \tag{5.114}$$

It may be noted that the turbulent flow field is three-dimensional (u, v, w) because by their very nature turbulent flows are instantaneously three-dimensional.

The turbulent boundary layer near a flat plate is two-dimensional only as a time-averaged flow field. That is, $\bar{w} = 0$ and $\partial/\partial z(\ ) = 0$. This requires, $\partial \bar{p}/\partial z = 0$, or in other words, $\bar{p} = \bar{p}(x, y)$, which means for a flat plate, $\bar{p} = p_\infty = $ constant, that is, $\partial \bar{p}/\partial x = 0$.

Furthermore, from the concept of boundary layer theory, streamwise diffusion of momentum and heat are neglected. That is $\partial/\partial x\ (\ )$ and $\partial^2/\partial x^2\ (\ )$ on the RHS of the x-momentum and energy equations are dropped. The final form of the boundary-layer-simplified time-averaged equations for the turbulent section of the boundary layer is therefore

$$\frac{\partial \bar{u}}{\partial x} + \frac{\partial \bar{v}}{\partial y} = 0 \tag{5.115}$$

$$\bar{u}\frac{\partial \bar{u}}{\partial x} + \bar{v}\frac{\partial \bar{u}}{\partial y} = \frac{\partial}{\partial y}\left(v\frac{\partial \bar{u}}{\partial y} - \overline{u'v'} \right) \tag{5.116}$$

$$\bar{u}\frac{\partial \bar{T}}{\partial x} + \bar{v}\frac{\partial \bar{T}}{\partial y} = \frac{\partial}{\partial y}\left(\alpha\frac{\partial \bar{T}}{\partial y} - \overline{v'T'} \right) \tag{5.117}$$

The momentum and energy Eqs (5.116) and (5.117) must be compared with their laminar counterparts to see that the products $\overline{u'v'}$ and $\overline{v'T'}$ survive the time-averaging operation. An inspection of Eqs (5.115) – (5.117) reveals that there are five unknowns $\bar{u}$, $\bar{v}$, $\bar{T}$, $\overline{u'v'}$, and $\overline{v'T'}$ in the three-equation system. The search for the two additional equations that are required to determine the five unknowns uniquely is recognized as turbulence modelling.

5.12.3   Eddy Diffusivities of Momentum and Heat

The time-averaged products $\overline{u'v'}$ and $\overline{v'T'}$ are replaced with the following expressions:

$$-\overline{u'v'} = \varepsilon \frac{\partial \bar{u}}{\partial y} \tag{5.118}$$

$$-\overline{v'T'} = \varepsilon_H \frac{\partial \bar{T}}{\partial y} \tag{5.119}$$

where ε (in m^2/s) is recognized as the Eddy diffusivity of momentum and ε_H (in m^2/s) is called the Eddy diffusivity of heat. Note that ε and ε_H are not the properties of the fluid in the sense v and α are.

The momentum and energy equations become

$$\bar{u}\frac{\partial \bar{u}}{\partial x} + \bar{v}\frac{\partial \bar{u}}{\partial y} = \frac{\partial}{\partial y}\left[(v + \varepsilon)\frac{\partial \bar{u}}{\partial y} \right] \tag{5.120}$$

$$\bar{u}\frac{\partial \bar{T}}{\partial x} + \bar{v}\frac{\partial \bar{T}}{\partial y} = \frac{\partial}{\partial y}\left[(\alpha + \varepsilon_H)\frac{\partial \bar{T}}{\partial y} \right] \tag{5.121}$$

ε and ε_H notations are given because on the RHS of Eqs (5.116) and (5.117) the time-averaged groups $(-\overline{u'v'})$ and $(-\overline{v'T'})$ are 'Eddy' contributions that enhance the effects of molecular diffusion which are represented, respectively, by $v(\partial \bar{u}/\partial y)$ and $\alpha(\partial \bar{T}/\partial y)$.

The RHS of Eq. (5.120) can be written as

$$v\frac{\partial \bar{u}}{\partial y} + \varepsilon\frac{\partial \bar{u}}{\partial y} = \frac{1}{\rho}\left(\underbrace{\mu\frac{\partial \bar{u}}{\partial y}}_{\tau_{mol}} + \underbrace{\rho\varepsilon\frac{\partial \bar{u}}{\partial y}}_{\tau_{Eddy}} \right) \tag{5.122}$$

where τ_{mol} and τ_{Eddy}, respectively, represent the usual molecular shear stress and the shear stress contribution made by the time-averaged effect of the eddies. The sum of the molecular and Eddy shear stresses is the apparent shear stress:

$$\tau_{app} = \tau_{mol} + \tau_{Eddy} \tag{5.123}$$

Similarly, the RHS of Eq. (5.121) can be written as

$$\alpha\frac{\partial \bar{T}}{\partial y} + \varepsilon_H\frac{\partial \bar{T}}{\partial y} = -\frac{1}{\rho c_p}\left[\underbrace{\left(-k_f\frac{\partial \bar{T}}{\partial y} \right)}_{q''_{mol}} + \underbrace{\left(-\rho c_p\varepsilon_H\frac{\partial \bar{T}}{\partial y} \right)}_{q''_{Eddy}} \right] \tag{5.124}$$

Both q''_{mol} and q''_{Eddy} are defined as positive when pointing in the positive y-direction, that is, away from the wall. Their sum represents the apparent heat flux:

$$q''_{app} = q''_{mol} + q''_{Eddy} \tag{5.125}$$

Therefore, Eqs (5.120) and (5.121) can be expressed as

$$\bar{u}\frac{\partial \bar{u}}{\partial x} + \bar{v}\frac{\partial \bar{u}}{\partial y} = \frac{\partial}{\partial y}\left[\frac{\tau_{app}}{\rho} \right] \tag{5.126}$$

$$\bar{u}\frac{\partial \bar{T}}{\partial x} + \bar{v}\frac{\partial \bar{T}}{\partial y} = \frac{\partial}{\partial y}\left[\frac{q''_{app}}{-\rho c_p} \right] \tag{5.127}$$

Thus, we see that to use Eqs (5.120) and (5.121) we need to have additional equations for ε and ε_H.

5.12.4 Prandtl's Mixing Length Hypothesis

Similar to the concept of mean free path in kinetic theory of gases, which is the average distance a particle travels between collisions, L. Prandtl in 1925 introduced a similar concept for describing turbulent flow phenomena. The Prandtl mixing length *l* is the distance travelled, on the average, by the turbulent lumps of a fluid in a direction normal to the mean flow without losing its identity. The distance *l* is of the same order as the Eddy diameter. Using this hypothesis, Prandtl wrote that

$$\tau_{Eddy} = \rho l^2 \left(\frac{\partial \bar{u}}{\partial y} \right)^2 \tag{5.128}$$

Comparing with

$$\tau_{Eddy} = \rho\varepsilon\frac{\partial \bar{u}}{\partial y}$$

$$\varepsilon = l^2\frac{\partial \bar{u}}{\partial y} \tag{5.129}$$

For more details, readers are referred to Schlichting (1968). Measurements of the $\bar{u}$ versus y profile suggest that the mixing length l is proportional to the distance from the wall:

$$l = \kappa y \qquad (5.130)$$

where $\kappa = 0.4$ is known as von Karman's constant. Therefore,

$$\varepsilon = \kappa^2 y^2 \frac{\partial \bar{u}}{\partial y} \qquad (5.131)$$

This is the simplest of the many Eddy diffusivity models that have been proposed.

5.12.5   Turbulent Prandtl Number

Regarding the thermal Eddy diffusivity ε_H, the simplest model is the assumption that ε_H is approximately the same as ε. By analogy with the definition of Prandtl number, $\mathrm{Pr} = v/\alpha$, the Eddy diffusivity ratio is called the turbulent Prandtl number. Therefore, the ε_H model consists in writing that Pr_t is a constant approximately equal to 1. Measurements of the temperature distribution in the turbulent boundary layers of $\mathrm{Pr} \geq 1$ fluids recommend the value $\mathrm{Pr}_t \cong 0.9$.

$$\frac{\varepsilon}{\varepsilon_H} = \mathrm{Pr}_t \qquad (5.132)$$

5.12.6   Wall Friction

Recall the momentum equation [Eq. (5.126)],

$$\bar{u} \frac{\partial \bar{u}}{\partial x} + \bar{v} \frac{\partial \bar{u}}{\partial y} = \frac{\partial}{\partial y} \left[\frac{\tau_{app}}{\rho} \right]$$

Sufficiently close to the wall, the inertia effect becomes negligible, and both sides of the above equation approach zero.

$$\frac{\partial}{\partial y} \left(\frac{\tau_{app}}{\rho} \right) = 0$$

$$\tau_{app} = \text{constant} = \tau_{s,\,x}$$

where $\tau_{s,\,x}$ is the value reached by τ_{app} right at the wall. Therefore, we conclude that in this layer τ_{app} is practically independent of y. This inner layer is recognized also as the constant-τ_{app} region of the boundary layer. In the outer layer, also called the wake region, τ_{app} decreases to zero (the inertia of the flow is finite and nega-tive) as y approaches the boundary layer thickness (Fig. 5.13).

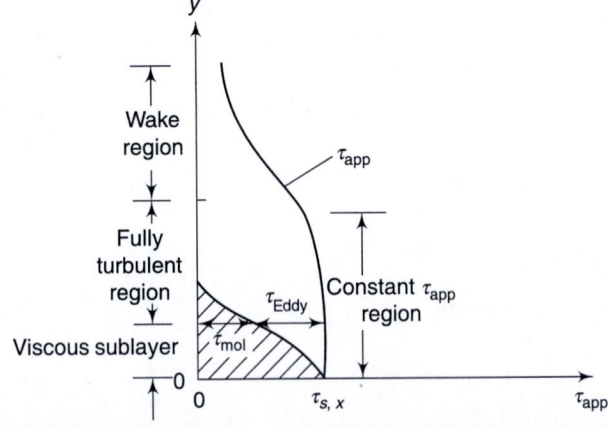

Fig. 5.13 τ_{app} versus y plot for turbulent flow over a flat plate

Prandtl's mixing length hypothesis does not work well for the outer layer where the model predicts $\varepsilon \approx 0$ because $\partial \overline{u}/\partial y$ approaches zero. On the contrary, ε should be large there. However, Prandtl's hypothesis turns out to be useful for the near-wall region, that is for the viscous sublayer and fully turbulent region.

By using the momentum equation and the mixing length model, the velocity distribution that is obtained for the constant τ_{app} region is as follows:

$$u^+ = \begin{cases} y^+ & \text{(viscous sublayer, } \varepsilon \ll v) & (5.133) \\ \dfrac{1}{\kappa}\ln y^+ + B & \text{(fully turbulent region, } \varepsilon \gg v) & (5.134) \end{cases}$$

in which u^+ and y^+ are the dimensionless 'wall coordinates' defined by

$$u^+ = \frac{\overline{u}}{(\tau_{s,x}/\rho)^{1/2}} \tag{5.135}$$

$$y^+ = \frac{y}{v}\left(\frac{\tau_{s,x}}{\rho}\right)^{1/2} \tag{5.136}$$

Up to $y^+ = 40$, Prandtl's $u^+(y^+)$ distribution matches well with the experimental data. Experimental measurements of $u^+(y^+)$ distribution indicate $B = 5.5$. The interface between the viscous sublayer and the fully turbulent region is located at $y^+ = 11.6$.

A simpler empirical $u^+(y^+)$ expression that approximates most of the curve represented by Eqs (5.133) and (5.134) is the so-called Prandtl's 1/7th power law:
$$u^+ = 8.7(y^+)^{1/7} \tag{5.137}$$

The power law expression fits the logarithmic form fairly well out to at least $y^+ = 1500$.

Note that the profile does not hold in the immediate vicinity of the wall, since at the wall the profile predicts $\dfrac{d\overline{u}}{dy} = \infty$.

Using the momentum integral equation

$$\frac{d}{dx}\int_0^\delta \overline{u}(u_\infty - \overline{u})\,dy = \frac{\tau_{s,x}}{\rho} \tag{5.138}$$

and inserting 1/7th power law for $\overline{u}$ in the integrand, we obtain

$$\frac{\delta}{x} = 0.37\left(\frac{u_\infty x}{v}\right)^{-1/5} \tag{5.139}$$

Equation (5.139) implies that turbulent boundary layer originates at $x = 0$, without a preceding laminar boundary layer and transition region. Thus $x = 0$ is a fictitious virtual origin of the turbulent boundary layer, provided that the same turbulent transport mechanisms were applicable down to zero Re. We will use Eq. (5.139) with the underlying assumption that x is the distance from the virtual origin of the turbulent boundary layer. Practically speaking, if the turbulent region on the plate is sufficiently long, the difference between the real and virtual origins is often negligibly small so that little error is introduced if the preceding laminar boundary layer is ignored.

$$\frac{\tau_{s,x}}{\rho u_\infty^2} = \frac{1}{2}C_{f,x} = 0.0296\left(\frac{u_\infty x}{v}\right)^{-1/5}.$$

(5.140)

$$\overline{\tau_{s,L}} = 0.037\rho u_\infty^2\,Re_L^{-1/5}$$

(5.141)

The $C_{f,x}$ formula is valid up to $Re_x = 10^8$. $\tau_{s,x}$ decreases as $x^{-1/5}$ in the downstream, that is, at a much slower rate than in the laminar section (Fig. 5.14). δ increases as $x^{4/5}$, almost linearly. The increase is considerably steeper than in the leading laminar section, where δ increases as $x^{1/2}$ (Fig. 5.15).

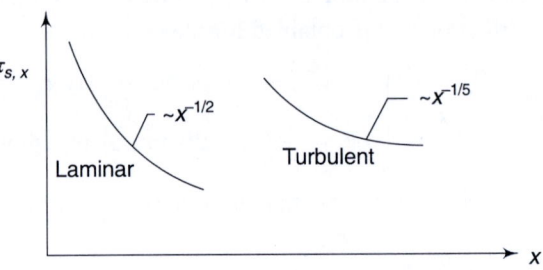

Fig. 5.14   The behaviour of wall shear stress in the laminar and turbulent sections of the boundary layer over a flat plate

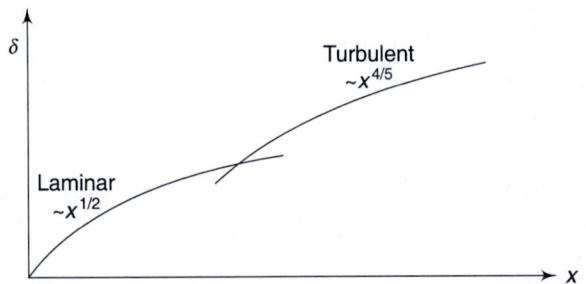

Fig. 5.15   The growth of a boundary layer in the laminar and turbulent sections of the boundary layer over a flat plate

It may be noted that von Karman introduced the concept of buffer layer which lies between the viscous sub-layer and the fully turbulent region. In the buffer layer, $\varepsilon \approx v$. von Karman's universal velocity profile is represented by the following equations:

Viscous sub-layer:	$0 < y^+ < 5,$	$u^+ = y^+$	(5.142)
Buffer layer:	$5 < y^+ < 30,$	$u^+ = 5.0 + 5\ln(y^+/5)$	(5.143)
Turbulent region:	$30 < y^+ < 400,$	$u^+ = 2.5\ln y^+ + 5.5$	(5.144)

5.12.7   Basic Approach in Solving Turbulent Heat Transfer on a Flat Plate

In turbulent flow, momentum and thermal boundary layers have the same thickness. For a laminar flow, this is not the case, the relative thickness depending primarily on the Prandtl number. Since for the turbulent boundary layer the momentum layer provides the primary transport mechanism (the Eddy diffusivity) in the outer region, it is not possible for the thermal layer to have a thickness that is significantly different, except for a low Prandtl number fluid or for a case where the virtual origins of the two boundary layers are very different. In analysis that follows we will not be considering a very low Prandl number fluid or a case where the virtual origins of the two boundary layers are very different.

5.12.8 Heat Transfer

For Pr ≥ 1 fluids, the nature of the time-averaged temperature distribution resembles that of the longitudinal velocity distribution. Figure 5.16 shows that q''_{app} is constant in the near-wall region. The thickness of the constant-τ_{app} region is more or less the same as that of the constant-q''_{app} region. In it, both sides of the energy equation (5.121) or (5.127) are zero. Therefore, integrating in y we obtain

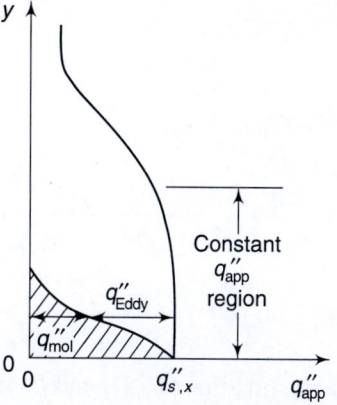

Fig. 5.16 q''_{app} versus y plot for turbulent heat transfer over a flat plate

$$q''_{app} = \text{constant} = q''_{s,x} \qquad (5.145)$$

Right at the wall, ε_H and q''_{Eddy} are zero. Now, the constant-q''_{app} can be rewritten as

$$-(k_f + \rho c_p \varepsilon_H)\frac{\partial \overline{T}}{\partial y} = q''_{s,x} \qquad (5.146)$$

Also, the constant-τ_{app} condition can be written as

$$(\mu + \rho\varepsilon)\frac{\partial \overline{u}}{\partial y} = \tau_{s,x} \qquad (5.147)$$

Dividing Eq. (5.147) by Eq. (5.146), we get

$$\frac{\rho(v + \varepsilon)}{\rho c_p(\alpha + \varepsilon_H)}\frac{\partial \overline{u}}{\partial \overline{T}} = -\frac{\tau_{s,x}}{q''_{s,x}} \qquad (5.148)$$

and assuming that $v = \alpha$ and $\varepsilon = \varepsilon_H$ (i.e., Pr = 1 and Pr$_t$ = 1), we obtain

$$\frac{1}{c_p}\frac{d\overline{u}}{d\overline{T}} = -\frac{\tau_{s,x}}{q''_{s,x}} \qquad (5.149)$$

Integrating from the wall ($\overline{u} = 0$, $\overline{T} = T_s$, at $y = 0$) to a large y where $\overline{u} = u_\infty$ and $\overline{T} = T_\infty$,

$$\frac{u_\infty}{c_p(T_\infty - T_s)} = -\frac{\tau_{s,x}}{q''_{s,x}} \qquad (5.150)$$

Now, the local heat transfer coefficient is

$$h = \frac{q''_{s,x}}{(T_s - T_\infty)}$$

This implies h is proportional to the local shear stress $\tau_{s,x}$:

$$St_x = \frac{h}{\rho c_p u_\infty} = \frac{q''_{s,x}}{\rho c_p u_\infty(T_s - T_\infty)} \qquad (5.151)$$

$$= \frac{Nu_x}{Re_x\, Pr}$$

Now, recall

$$C_{f,x} = \frac{\tau_{s,x}}{\frac{1}{2}\rho u_\infty^2} \tag{5.152}$$

From Eq. (5.150),

$$\frac{u_\infty}{c_p(T_\infty - T_s)} = \frac{-\frac{1}{2}\rho u_\infty^2 C_{f,x}}{q_{s,x}''}$$

$$\Rightarrow \qquad \frac{1}{2}C_{f,x} = \frac{q_{s,x}''}{\rho c_p u_\infty (T_s - T_\infty)} \tag{5.153}$$

Comparing Eqs (5.151) and (5.153)

$$St_x = \frac{1}{2}C_{f,x} \tag{5.154}$$

Equation (5.154) is valid for $Pr = Pr_t = 1$ and is called the Reynolds analogy between wall friction and heat transfer. For fluids with $Pr \neq 1$, Colburn proposed the following empirical relation that shows $Pr^{2/3}$ as a factor:

$$St_x \, Pr^{2/3} = \frac{1}{2}C_{f,x} \quad (Pr \geq 0.5) \tag{5.155}$$

Equation (5.155) is called the Reynolds–Colburn analogy. Since

$$St_x = \frac{Nu_x}{Re_x \, Pr}$$

and $\qquad \dfrac{1}{2}C_{f,x} = 0.0296 \, Re_x^{-1/5}$

therefore, from Eq. (5.155), we get

$$Nu_x = 0.0296 \, Re_x^{4/5} \, Pr^{1/3} \quad (Pr \geq 0.5) \tag{5.156}$$

Equation (5.156) also works well for a uniform wall flux. Thus, in turbulent flow heat transfer coefficient is not sensitive to thermal boundary condition at the wall. On the contrary, the local heat transfer coefficient is 36% higher for constant heat flux boundary condition at the wall than that for constant wall temperature in laminar flow. This is because in turbulent flow heat transfer resistance is primarily in the sublayer and since sublayer thickness is independent of the wall boundary condition, there is not much change in h. Note that when the wall flux is uniform, the local Nusselt number is defined as

$$Nu_x = \frac{q_s''x}{k_f[T_x(x) - T_\infty]} \tag{5.157}$$

The properties are evaluated at the film temperature $T_f = (\bar{T}_s + T_\infty)/2$, where $\bar{T}_s$ is the x-averaged wall temperature.

From Eq. (5.156) it can be inferred that in the turbulent section, h decreases as $x^{-1/5}$. This behaviour differs significantly from the $h \sim x^{-1/2}$ decrease in the laminar portion of the boundary layer (Fig. 5.17).

In the turbulent section of the plate, $\partial h / \partial x$ is small because the rate of increase of sublayer boundary layer thickness is very small with x.

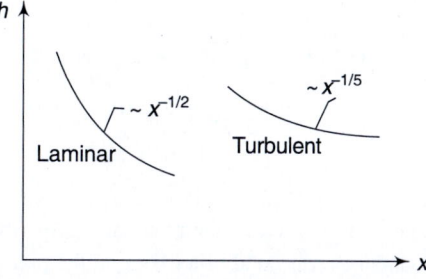

Fig. 5.17 The variation of local heat transfer coefficient in the laminar and turbulent sections of the boundary layer

The average heat transfer coefficient $\overline{h}_L$ for the mixed boundary layer condition on an isothermal wall is evaluated from

$$\overline{h}_L = \frac{1}{L}\left(\int_0^{x_{\text{crit}}} h_{\text{laminar}}\, dx + \int_{x_{\text{crit}}}^{L} h_{\text{turbulent}}\, dx\right) \tag{5.158}$$

where h_{laminar} and $h_{\text{turbulent}}$ are substituted from Eqs (5.82) and (5.156), respectively. x_{crit} is calculated from

$$\frac{u_\infty x_{\text{crit}}}{v} = 5 \times 10^5$$

Finally, the following expression for $\overline{\text{Nu}}_L$ is obtained:

$$\overline{\text{Nu}}_L = \frac{\overline{h}_L L}{k_f} = 0.037\,\text{Pr}^{1/3}(\text{Re}_L^{4/5} - 23{,}550) \tag{5.159}$$

Equation (5.159) is valid for $\text{Pr} \geq 0.5$ and $5 \times 10^5 < \text{Re}_L < 10^8$. It should be noted that if a critical Reynolds number different from 5×10^5 is used, Eq. (5.159) will change accordingly.

Example 5.4 Total Heat Transfer from an Isothermal Plate

Air at 20°C and 1 atm flows over a flat plate at 50 m/s. The plate is 100 cm long and is maintained at 60°C. The width of the plate is 2 m. Calculate the total heat transfer from the plate.

Solution

The properties are evaluated at the film temperature T_f,

$$T_f = \frac{20 + 60}{2} = 40°\text{C} = 313\,\text{K}$$

$$\rho = \frac{p}{RT} = \frac{1.0132 \times 10^5\ \text{N/m}^2}{(287\ \text{Nm/kgK})(313\ \text{K})} = 1.128\ \text{kg/m}^3$$

$\text{Pr} = 0.7$, $k_f = 0.02723$ W/m°C

$c_p = 1.007$ KJ/kg°C

$\mu = 1.906 \times 10^{-5}$ kg/ms

$$\mathrm{Re}_L = \frac{\rho u_\infty L}{\mu}$$

$$= \frac{(1.128)(50)(100 \times 10^{-2})}{1.906 \times 10^{-5}}$$

$$= 2.96 \times 10^6$$

Since $\mathrm{Re}_L > 5 \times 10^5$, the boundary layer is turbulent beyond $x = x_{crit}$. Therefore, a mixed boundary layer condition exists on the plate, and Eq. (5.159) can be used to calculate the average heat transfer coefficient over the plate.

$$\overline{\mathrm{Nu}_L} = \frac{\overline{h}_L L}{k_f} = 0.037\,\mathrm{Pr}^{1/3}\,(\mathrm{Re}_L^{4/5} - 23{,}550)$$

$$= 0.037(0.7)^{1/3}\,[2.96 \times 10^6)^{4/5} - 23{,}550]$$
$$= 0.037(0.888)\,[150{,}325 - 23{,}550]$$
$$= 4165$$

$$\overline{h}_L = \overline{\mathrm{Nu}_L}\,\frac{k_f}{L}$$

$$= \frac{(4165)(0.02723)}{100 \times 10^{-2}}$$

$$= 113.4\ \mathrm{W/m^2\ ^\circ C}$$

$$q = \overline{h}_L\,A\,(T_s - T_\infty)$$
$$= (113.4)(1 \times 2)\,(60 - 20)$$
$$= 4536\ \mathrm{W}$$

Note if Re_L were smaller than 5×10^5, the entire length L is covered by laminar boundary layer flow, and the $\overline{\mathrm{Nu}_L}$ formula given by Eq. (5.43) will be applicable.

5.13 Heat Transfer in Laminar Tube Flow

We now consider a forced convection configuration in which the flow is 'internal', that is, surrounded by the wall of a tube or a duct at a temperature different from that of the fluid. We assume laminar flow ($\mathrm{Re}_D \le 2300$).

Several problems can be defined for this general category with major differences in the ease or even the possibility of solution depending upon the region of the tube considered and boundary conditions.

1. Consider the following configuration shown in Fig. 5.18.

   Assumptions
   (a) steady, 2D laminar flow and heat transfer (r, z)
   (b) incompressible, constant properties
   (c) $v_r = 0$
   (d) no swirl, $v_\theta = 0$
   (e) θ-symmetry
   (f) no viscous dissipation

2. For μ, ρ, and k_f constant, the conservation equations in

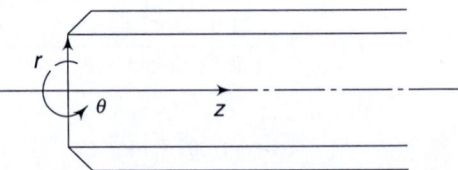

Fig. 5.18 Tube coordinate system

cylindrical coordinates are as follows:

Continuity:

$$\frac{1}{r}\frac{\partial}{\partial r}(rv_r) + \frac{1}{r}\frac{\partial v_\theta}{\partial \theta} + \frac{\partial v_z}{\partial z} = 0 \tag{5.160}$$

z-direction momentum:

$$\rho\left[\frac{\partial v_z}{\partial t} + v_r\frac{\partial v_z}{\partial r} + \frac{v_\theta}{r}\frac{\partial v_z}{\partial \theta} + v_z\frac{\partial v_z}{\partial z}\right]$$

$$= F_z - \frac{\partial p}{\partial z} + \mu\left[\frac{\partial^2 v_z}{\partial r^2} + \frac{1}{r}\frac{\partial v_z}{\partial r} + \frac{1}{r^2}\frac{\partial^2 v_z}{\partial \theta^2} + \frac{\partial^2 v_z}{\partial z^2}\right] \tag{5.161}$$

Energy equation:

$$\rho c_p\left[\frac{\partial T}{\partial t} + v_r\frac{\partial T}{\partial r} + \frac{v_\theta}{r}\frac{\partial T}{\partial \theta} + v_z\frac{\partial T}{\partial z}\right]$$

$$= k_f\left[\frac{\partial^2 T}{\partial r^2} + \frac{1}{r}\frac{\partial T}{\partial r} + \frac{1}{r^2}\frac{\partial^2 T}{\partial \theta^2} + \frac{\partial^2 T}{\partial z^2}\right]$$

$$+ \beta T\left[\frac{\partial p}{\partial t} + v_r\frac{\partial p}{\partial r} + \frac{v_\theta}{r}\frac{\partial p}{\partial \theta} + v_z\frac{\partial p}{\partial z}\right] +$$

$$\mu\left[2\left\{\left(\frac{\partial v_r}{\partial r}\right)^2 + \left(\frac{1}{r}\frac{\partial v_\theta}{\partial \theta} + \frac{v_r}{r}\right)^2 + \left(\frac{\partial v_z}{\partial z}\right)^2\right\}\right.$$

$$+ \left(\frac{1}{r}\frac{\partial v_z}{\partial \theta} + \frac{\partial v_\theta}{\partial z}\right)^2 + \left(\frac{\partial v_r}{\partial z} + \frac{\partial v_z}{\partial r}\right)^2 + \left(\frac{1}{r}\frac{\partial v_r}{\partial \theta} + \frac{\partial v_\theta}{\partial r} + \frac{v_\theta}{r}\right)^2$$

$$\left. - \frac{2}{3}\left(\frac{\partial v_r}{\partial r} + \frac{1}{r}\frac{\partial v_\theta}{\partial \theta} + \frac{v_r}{r} + \frac{\partial v_z}{\partial z}\right)^2\right] \tag{5.162}$$

Note that F_z is the body force per unit volume in the z-direction and β is the volumetric thermal expansion coefficient $[=-1/\rho(\partial\rho/\partial T)_p]$.

3. By our assumptions, the above simplify to the following:
Continuity:

$$\frac{\partial v_z}{\partial z} = 0 \tag{5.163}$$

z-direction momentum:

$$0 = -\frac{\partial p}{\partial z} + \mu\left[\frac{\partial^2 v_z}{\partial r^2} + \frac{1}{r}\frac{\partial v_z}{\partial r}\right] \tag{5.164}$$

Energy:

$$\rho c_p v_z\frac{\partial T}{\partial z} = k_f\left(\frac{\partial^2 T}{\partial r^2} + \frac{1}{r}\frac{\partial T}{\partial r} + \frac{\partial^2 T}{\partial z^2}\right)$$

$$= k_f \left\{ \frac{1}{r} \frac{\partial}{\partial r} \left(r \frac{\partial T}{\partial r} \right) + \frac{\partial^2 T}{\partial z^2} \right\} \qquad (5.165)$$

If the conduction in the axial direction is small relative to the axial energy transport by the bulk movement of the fluid,

$$\frac{\partial^2 T}{\partial z^2} \rightarrow 0 \qquad (5.166)$$

It may be recalled that the significant dimensionless number which consider the influence of axil conduction is the Peclet number (RePr) which indicates the ratio of strength of convection to strength of conduction.

For $Re_D Pr > 100$, axial conduction is usually neglected. Then the energy equation becomes

$$\rho c_p v_z \frac{\partial T}{\partial z} = \frac{k_f}{r} \frac{\partial}{\partial r} \left(r \frac{\partial T}{\partial r} \right) \qquad (5.167)$$

4. In general, the ease or possibility of solution of the above equations depends upon boundary conditions and how far down the tube we start our solution. In order of increasing difficulty, our distance down the tube at the start of solution would be:
   (a) 'Far down the tube', that is, fully developed flow and heat transfer.
   (b) Fully developed flow, thermal entry length solutions.
   (c) Tube entrance, i.e., developing velocity and temperature profiles.
5. For each of the above, from the infinite number of boundary conditions, the most frequently studied are
   (a) Constant wall flux
   (b) Constant wall temperature
   (c) Wall temperature a function of z

   Hydrodynamic:
   (a) Parabolic profile
   (b) Slug flow (Navier–Stokes equation not used or applicable, suitable for liquid metal flow).
6. While a fully developed velocity profile implies $v_r = 0$, $\partial v_z/\partial z = 0$, fully developed temperature profile criteria and consequences have to be understood. First, we define mixing cup or bulk or mixed mean temperature. 'Mixing cup' temperature is the temperature the fluid would assume if placed in a mixing chamber and allowed to come to equilibrium. The convected energy rate in the z-direction is

$$\dot{m} c_p T_m = (A v_m \rho) c_p T_m$$

$$= \int_A v_z \rho c_p T \, dA$$

where A is the cross-sectional area of the tube. Therefore, the mixing cup temperature T_m is

$$T_m = \frac{1}{Av_m} \int_A v_z T \, dA \tag{5.168}$$

where v_m is the mean velocity of the flow:

$$v_m = \frac{1}{A} \int_A v_z \, dA \tag{5.169}$$

Substituting Eq. (5.169) into Eq. (5.168), we have

$$T_m = \frac{\int_A v_z T \, dA}{\int_A v_z \, dA} \tag{5.170}$$

For a circular tube,

$$T_m = \frac{\int_0^{r_0} v_z T 2\pi r \, dr}{\int_0^{r_0} v_z 2\pi r \, dr}$$

$$= \frac{\int_0^{r_0} v_z T r \, dr}{\int_0^{r_0} v_z r \, dr} \tag{5.171}$$

Now we define a dimensionless temperature in terms of the wall temperature T_s, which for 'thermally' fully developed flow is a function of r only. In dimensionless form:

$$\frac{T_s - T}{T_s - T_m} = f\left(\frac{r}{r_0}\right) \tag{5.172}$$

The statement that this profile is invariant with z implies two constraints:
(a) Since we are assuming steady state, at any point along the wall,

$$\left[\frac{\partial}{\partial r}\left(\frac{T_s - T}{T_s - T_m}\right)\right]_{r=r_0} = \text{constant} = \frac{-\frac{\partial T}{\partial r}\Big|_{r=r_0}}{T_s - T_m}$$

Let $q'' = h(T_s - T_m)$. Also

$$q'' = +k_f \left(\frac{\partial T}{\partial r}\right)_{r=r_0} \quad \text{(because T is increasing with r for heated tube)}$$

Therefore, $\dfrac{\dfrac{q''}{k_f}}{\dfrac{q''}{h}} = \dfrac{h}{k_f} = \dfrac{+\dfrac{\partial T}{\partial r}\Big|_{r=r_0}}{T_s - T_m} = \text{constant}$

Thus, h is a constant and does not vary with z.

Experimentally, also this is true far down the tube. Thus, in thermally fully developed flow the non-dimensional temperature profile is invariant with tube length.

(b) Since the shape of the temperature profile does not change, the slope at r is the same for all z, or

$$\left[\frac{\partial}{\partial z}\left(\frac{T_s - T}{T_s - T_m} \right) \right]_{r=\text{constant}} = 0$$

Performing the indicated differentiation and solving for $\partial T/\partial z$,

$$\frac{\partial T}{\partial z} = \frac{dT_s}{dz} - \left(\frac{T_s - T}{T_s - T_m} \right)\frac{dT_s}{dz} + \left(\frac{T_s - T}{T_s - T_m} \right)\frac{dT_m}{dz}$$

(i) For constant heat flux (electric resistance heating, radiant heating, nuclear heating, etc.):

$$q'' = h(T_s - T_n) = \text{constant}$$

with q'' and h both constant

$$\frac{\partial}{\partial z}\left(\frac{q''}{h} \right) = 0 = \frac{\partial T_s}{\partial z} - \frac{\partial T_m}{\partial z}$$

or

$$\frac{dT_s}{dz} = \frac{dT_m}{dz} = \frac{\partial T}{\partial z} = \text{constant}$$

To find the value of this constant an energy balance on a fluid element of size $\Delta z \times \dfrac{\pi D^2}{4}$ is carried out as shown in Fig.5.19.

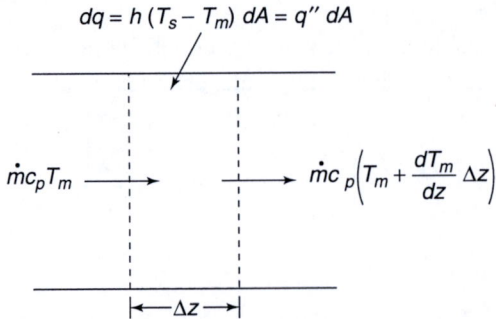

$$dq = h\,(T_s - T_m)\,dA = q''\,dA$$

$\dot{m}c_p T_m \longrightarrow$ $\longrightarrow \dot{m}c_p\left(T_m + \dfrac{dT_m}{dz}\Delta z\right)$

$\longleftarrow \Delta z \longrightarrow$

Fig. 5.19   Energy balance on a volume element in a tube flow

Hence,

$$\dot{m}c_p T_m + q''(p\Delta z) = \dot{m}c_p\left(T_m + \frac{dT_m}{dz}\Delta z\right)$$

$$\Rightarrow \qquad \frac{dT_m}{dz} = \frac{q''p}{\dot{m}c_p} \tag{5.173}$$

where p is the perimeter of the tube. Since for constant heat flux wall boundary

condition q'' constant, the RHS of Eq. (5.173) is also constant.
Thus,

$$\frac{\partial T}{\partial z} = \frac{dT_s}{dz} = \frac{dT_m}{dz} = \frac{q''p}{\dot{m}c_p} = \text{constant} \tag{5.174}$$

For a circular tube,

$$p = 2\pi r_0 \tag{5.175}$$

$$\dot{m} = \rho v_m A = \rho v_m \pi r_o^2 \tag{5.176}$$

Substituting Eqs (5.175) and (5.176) into Eq. (5.174), we finally get

$$\frac{\partial T}{\partial z} = \frac{dT_s}{dz} = \frac{dT_m}{dz} = \frac{2q''}{\rho v_m c_p r_o} = \text{constant} \tag{5.177}$$

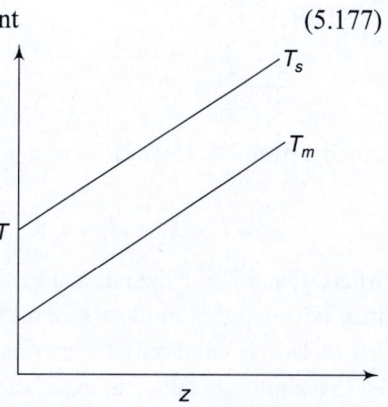

The above analysis is also applicable to turbulent heat transfer. The only difference is that fluid temperature is a time-averaged quantity.

T_s, T_m, versus z plots are shown in Fig. 5.20.

(ii) For constant wall temperature (evaporators, condensers, etc.)

$$\frac{dT_s}{dz} = 0$$

and therefore,

$$\frac{\partial T}{\partial z} = \left(\frac{T_s - T}{T_s - T_m}\right)\frac{dT_m}{dz}$$

Fig. 5.20 T_s, T_m versus z plots for the constant wall flux case

T_s, T_m, versus z plots are shown in Fig. 5.21.

An expression for T_m as a function of z in the region of thermally fully developed flow for constant wall temperature can be obtained by doing an energy balance on a control volume as shown in Fig. 5.19.
On energy balance,

$$\dot{m}c_p T_m + h(T_s - T_m)dA = \dot{m}c_p\left(T_m + \frac{dT_m}{dz}\Delta z\right)$$

$$\Rightarrow \qquad \dot{m}c_p\frac{dT_m}{dz}\Delta z = h(T_s - T_m)dA \tag{5.178}$$

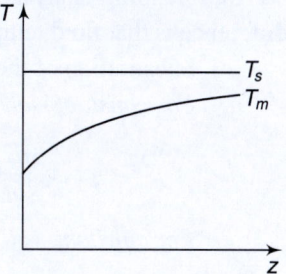

Fig. 5.21 T_s, T_m versus z plots for the constant wall temperature case

Now, $\quad dA = p\Delta z \tag{5.179}$

Since T_s is constant

$$dT_m = - d(T_s - T_m) \tag{5.180}$$

Putting Eqs (5.179) and (5.180) in Eq. (5.178), we obtain

$$\dot{m}c_p\frac{\{-d(T_s - T_m)\}}{dz}\Delta z = h(T_s - T_m)p\Delta z$$

$$\Rightarrow \qquad \frac{d(T_s - T_m)}{T_s - T_m} = -\frac{hp}{\dot{m}c_p} dz$$

Integrating,
$$\int_{T_m=T_i}^{T_m=T_e} \frac{d(T_s - T_m)}{T_s - T_m} = -\frac{hp}{\dot{m}c_p} \int_{z=0}^{z=L} dz$$

$$\Rightarrow \qquad \ln \frac{T_s - T_e}{T_s - T_i} = -\frac{hpL}{\dot{m}c_p} = -\frac{hA}{\dot{m}c_p} \qquad (5.181)$$

Note that
$$A = pL = \pi DL.$$

From Eq. (5.181),

$$\frac{T_s - T_e}{T_s - T_i} = e^{-\frac{h\pi DL}{\dot{m}c_p}} \qquad (5.182)$$

Finally, from Eq. (5.182) we obtain

$$T_e = T_s - (T_s - T_i)e^{-\frac{h\pi DL}{\dot{m}c_p}}$$

where T_i and T_e are the inlet and exit bulk temperature of the fluid, respectively. It may be noted that in the above derivation 'h' is treated as constant and therefore, Eq. (5.182) is valid only for thermally fully developed flow.

In a similar manner, an expression for T_m as a function of z can be written as

$$T_m = T_s - (T_s - T_i)e^{-\frac{h\pi Dz}{\dot{m}c_p}} \qquad (5.183)$$

The foregoing analysis is also applicable to turbulent heat transfer. The only difference is that fluid temperature is a time-averaged quantity.

(c) Substitution of these values into the energy equation yields:

For constant wall flux

$$\rho c_p v_z \frac{dT_m}{dz} = \frac{k_f}{r} \frac{\partial}{\partial r}\left(r \frac{\partial T}{\partial r}\right) \qquad (5.184)$$

For constant wall temperature

$$\rho c_p v_z \left(\frac{T_s - T}{T_s - T_m}\right) \frac{dT_m}{dz} = \frac{k_f}{r} \frac{\partial}{\partial r}\left(r \frac{\partial T}{\partial r}\right) \qquad (5.185)$$

(d) Note that in both Eqs (5.184) and (5.185) the independent variables (z, r) have been separated so that it is possible to hold z constant and integrate with respect to r to determine the temperature profile.

7. Now let us add to our assumptions:

(e) Fully developed flow and heat transfer
(f) Constant heat flux

Starting with the Navier–Stokes equation and separated variables:

$$\frac{\mu}{r} \frac{d}{dr}\left(r \frac{dv_z}{dr}\right) = \frac{dp}{dz} = C_0 \qquad (5.186)$$

Integrating twice yields

$$v_z = \frac{C_0 r^2}{4\mu} + C_1 \ln r + C_2 \tag{5.187}$$

Boundary conditions:

$$r = r_0, \quad v_z = 0 \ (\text{no slip}) \tag{5.188a}$$

$$r = 0, \quad \frac{dv_z}{dr} = 0 \ (\text{axisymmetry}) \tag{5.188b}$$

Constants are solved to obtain

$$v_z = \frac{C_0}{4\mu}(r^2 - r_0^2)$$

In terms of the mean velocity, $v_m = - C_0 r_0^2 / 8\mu$, the expression for velocity becomes

$$v_z = 2v_m \left[1 - \left(\frac{r}{r_0}\right)^2 \right] \tag{5.189}$$

Substituting Eq. (5.189) into the energy equation Eq. (5.184):

$$\rho c_p (2v_m) \left[1 - \left(\frac{r}{r_0}\right)^2 \right] \frac{dT_m}{dz} = \frac{k_f}{r} \frac{\partial}{\partial r}\left(r \frac{\partial T}{\partial r} \right) \tag{5.190}$$

Integrating twice with respect to r, we obtain

$$T = \frac{dT_m}{dz} \left[\frac{2\rho c_p v_m}{k_f} \right] \left[\frac{r^2}{4} - \left(\frac{1}{r_0}\right)^2 \left(\frac{r^4}{16}\right) \right] + C_3 \ln r + C_4 \tag{5.191}$$

Boundary conditions are

$$r = 0, \quad \frac{dT}{dr} = 0 \ (\text{axisymmetry}) \tag{5.192a}$$

$$r = r_0, \quad T = T_s \tag{5.192b}$$

which yield values of constants

$$C_3 = 0$$

$$C_4 = T_s - \frac{dT_m}{dz} \left[\frac{2\rho c_p v_m}{k_f} \right] \frac{3r_0^2}{16}$$

Therefore, $\quad T_s - T = \frac{\rho c_p v_m}{8 r_0^2 k_f} \left(\frac{dT_m}{dz} \right)(3r_0^4 - 4r_0^2 r^2 + r^4) \tag{5.193}$

Using the above temperature profile and the velocity profile v_z [Eq. (5.189)] in Eq. (5.171), we get

$$T_m = T_s - \frac{11}{96} \left(\frac{2v_m}{\alpha} \right) \left(\frac{dT_m}{dz} \right) r_0^2$$

where $\alpha = k_f / \rho c_p$.

Since, $q_s'' = h(T_s - T_m) = k_f \left(\dfrac{\partial T}{\partial r} \right)_{r=r_0}$

therefore, $h \left(\dfrac{11}{96} \right) \left(\dfrac{2v_m}{\alpha} \right) \left(\dfrac{dT_m}{dz} \right) r_0^2 = \dfrac{r_0 v_m \rho c_p}{2} \dfrac{dT_m}{dz}$

$\Rightarrow \qquad h = \dfrac{48}{11} \dfrac{k_f}{D} = 4.364 \dfrac{k_f}{D}$

$\Rightarrow \qquad \mathrm{Nu}_D = \dfrac{hD}{k_f} = 4.364$

$\Rightarrow \qquad \mathrm{Nu}_D = 4.364 \qquad\qquad\qquad\qquad\qquad\qquad (5.194)$

8. If assumption 7 had been constant wall temperature, we might have solved the energy equation, Eq. (5.185), by assuming a trial temperature profile, substituting into Eq. (5.185), and solving. The T thus obtained then becomes the trial T and the entire procedure is repeated until the difference between two successive T's is very small. For constant wall temperature,
$$\mathrm{Nu}_D = 3.658$$

5.13.1   Effect of Axial Conduction in the Fluid in Laminar Tube Flow

In the previous analysis the axial fluid conduction term $\dfrac{\partial^2 T}{\partial z^2}$ was neglected. In the fully developed heat transfer situation, with constant heat flux at the wall this term is always zero since $\dfrac{\partial T}{\partial z}$ is constant. Hence, Nusselt number is not influenced by axial conduction. But for the case of constant wall temperature axial conduction can be of significance at low values of Re and/or Pr. Michelsen and Villadsen (1974) recommended the following equations for such case based on their analytical solution.

$$\mathrm{Nu} = 4.180654 - 0.183460\,\mathrm{Pe}, \quad \text{for Pe} < 1.5 \qquad\qquad (5.195)$$

$$= 3.656794 + \dfrac{4.487}{\mathrm{Pe}^2}, \quad \text{for Pe} > 5 \qquad\qquad (5.196)$$

where Pe (= RePr) is the Peclet number.

It can be seen from Eqs (5.195) and (5.196) that the effect of axial fluid conduction is negligible for Pe = 100, and is quite insignificant even for Pe =10. Therefore, axial conduction becomes important for laminar flow of liquid metals which have very low Prandtl numbers. For gases axial conduction can be considered only at extremely low Reynolds number, and for most liquids it can be dropped.

In conclusion it can be said that axial conduction is absent for all fluids if the outer wall of the tube is exposed to constant heat flux boundary condition in thermally fully developed flow. Otherwise, it should be considered; however, its importance depends on the product of Reynolds number and Prandtl number.

5.14 Hydrodynamic and Thermal Entry Lengths

Hydrodynamic entry length is the distance from the entrance of a tube beyond which the flow becomes fully developed. For laminar flow,

$$\left(\frac{z_{fd,h}}{D}\right)_{lam} = 0.05\,\mathrm{Re}_D \tag{5.197}$$

where $\mathrm{Re}_D = \rho v_m D/\mu$. Although there is no satisfactory general expression for the entry length in turbulent flow ($\mathrm{Re}_D > 2300$), as a first approximation

$$10 \le \left(\frac{z_{fd,h}}{D}\right)_{turb} \le 60 \tag{5.198}$$

Just like a flat plate, a boundary layer develops as the fluid makes contact with the wall. The distance from the entrance at which the boundary layers merge at the centre line is termed the hydrodynamic entry length $z_{fd,h}$ (Fig. 5.22). The developing flow that takes place in the entrance region has two regions—the annular boundary layer and the core that surrounds the centre line. The fluid that once entered the tube with the velocity v_∞ is slowed down by the wall in the boundary layer ($v_z < v_m$). The conservation of the flow rate $\rho\pi r_0^2 v_\infty$ (or $\rho\pi r_0^2 v_m$) in each cross-section requires $v_z > v_m$ in the core region. The maximum boundary layer thickness δ_{max} is equal to the radius of the tube, r_0.

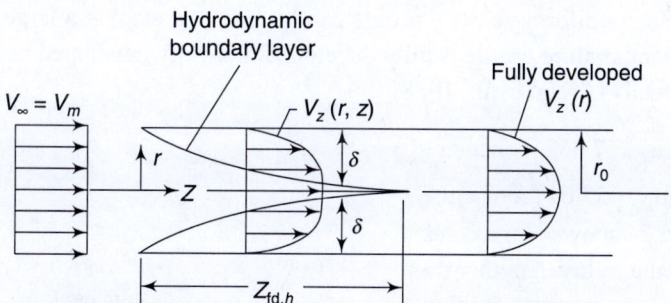

Fig. 5.22 Hydrodynamic entrance region and hydrodynamically fully developed region

If the fluid enters the tube at a uniform temperature that differs from the surface temperature, convection heat transfer occurs and a thermal boundary layer develops (Fig. 5.23). In this region, known as the thermal entrance region ($0 < z < z_{fd,t}$), the shape of the temperature profile develops, that is, T-profile changes from one profile to the next. In the thermally fully developed region ($z > z_{fd,t}$) the shape of the temperature profile is preserved. For laminar flow, the thermal entry length is expressed as

$$\left(\frac{z_{fd,t}}{D}\right)_{lam} \approx 0.05\,\mathrm{Re}_D\mathrm{Pr} \tag{5.199}$$

Comparing Eqs (5.197) and (5.199),

$$\frac{z_{fd,h}}{z_{fd,t}} = \frac{1}{\mathrm{Pr}} \tag{5.200}$$

So, if Pr > 1, the hydrodynamic boundary layer develops more rapidly than the thermal boundary layer, while it is inverse for Pr < 1.

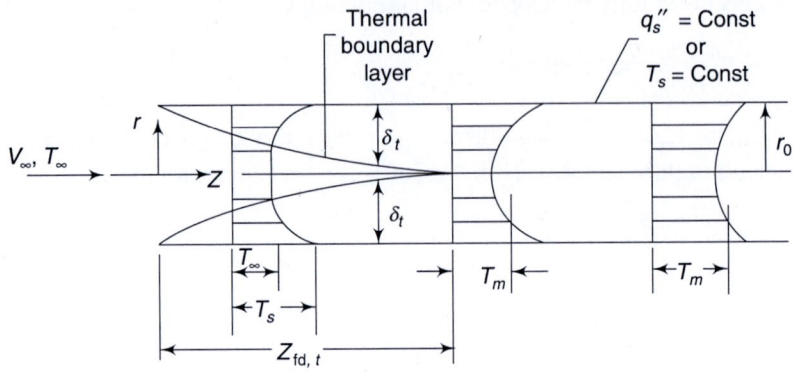

Fig. 5.23 Thermal entrance region and thermally fully developed region

Moreover, for very high Pr fluids, such as oils (Pr ≥ 100), $z_{fd,h}$ is very much smaller than $z_{fd,t}$ and it is reasonable to assume a fully developed velocity profile throughout the thermal entry region. On the other hand, for very low Pr fluids (liquid metals, $0.006 \le Pr \le 0.03$), $z_{fd,h} \gg z_{fd,t}$ and $v_z = v_m = v_\infty$ over the cross-section like the velocity distribution of a solid slug. So, there is no need to solve the Navier–Stokes equation. So, a uniform velocity profile can be assumed even at a large enough z where the temperature profile is fully developed. For fully developed heat transfer in liquid metals (Asako et al., 1988),

$$\mathrm{Nu}_D = 7.962 \ (\text{constant } q_s'') \qquad (5.201)$$

$$\mathrm{Nu}_D = 5.769 \ (\text{constant } T_s) \qquad (5.202)$$

Figure 5.24 shows Nu_D versus z plots in the entire length of a tube subject to constant heat flux and constant wall temperature. For the constant wall flux case, a closed form expression that covers both the entrance and the fully developed region was developed by Churchill and Ozoe (1973):

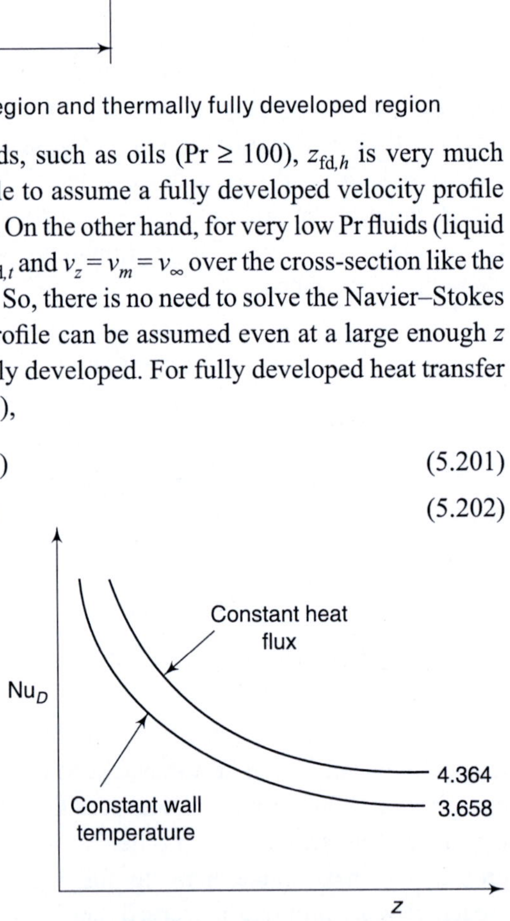

Fig. 5.24 Nu_D versus z plots for constant heat flux and constant wall temperature for a circular tube for Pr ≥ 0.5 fluids

$$\frac{\mathrm{Nu}_D}{4.364 \left[1 + \left(\dfrac{\mathrm{Gz}}{29.6} \right)^2 \right]^{1/6}} =$$

$$\left[1 + \left(\frac{\dfrac{\mathrm{Gz}}{19.04}}{\left[1 + \left(\dfrac{\mathrm{Pr}}{0.0207} \right)^{2/3} \right]^{1/2} \left[1 + \left(\dfrac{\mathrm{Gz}}{29.6} \right)^2 \right]^{1/3}} \right)^{3/2} \right]^{1/3} \qquad (5.203)$$

where the new dimensionless group Gz is called the Graetz[9] number:

$$Gz = \frac{\pi D^2 v_m}{4\alpha z} = \frac{\pi}{4}\left(\frac{z/D}{\text{Re}_D \text{ Pr}}\right)^{-1} \tag{5.204}$$

Example 5.5 Heat Transfer in a Water Pipe Subjected to Constant Heat Flux

Water at 25 °C enters a pipe with constant wall heat flux $q_s'' = 1$ kW/m². The flow is hydrodynamically and thermally fully developed. The mass flow rate of water is $\dot{m} = 10$ g/s and the pipe radius $r_0 = 1$ cm. Calculate (a) Reynolds number, (b) the heat transfer coefficient, and (c) the difference between the local wall temperature and the local mean (bulk) temperature.

Solution

(a) The properties of water at 25 °C are:

$$\mu = 8.96 \times 10^{-4} \text{ kg/m s}$$
$$k_f = 0.6109 \text{ W/m°C}$$

$$\text{Re}_D = \frac{\rho v_m D}{\mu}$$

Now, $$\rho v_m = \frac{\dot{m}}{A}$$

$$= \frac{10}{\pi (1)^2} = 3.18 \text{ g/cm}^2 \text{ s}$$

Therefore, $$\text{Re}_D = \frac{(3.18)(2)}{\dfrac{8.96 \times 10^{-4} \times 1000}{100}} = 709$$

Since, $\text{Re}_D < 2300$, the flow is laminar.

(b) For constant heat flux,

$$\text{Nu}_D = 4.364$$

or $$\frac{hD}{k_f} = 4.364$$

or $$h = \text{Nu}_D \frac{k_f}{D}$$

$$= \frac{(4.364)(0.6109)}{2 \times 10^{-2}} = 133.3 \text{ W/m}^2\text{K}$$

(c) Since, $q_s'' = h(T_s - T_m)$, therefore,

$$T_s - T_m = \frac{q_s''}{h} = \frac{1000}{133.3}$$

$$= 7.5 °C \text{ or } 7.5 \text{ K}$$

---

[9] Leo Graetz (1856–1941) was Professor of Physics at the University of Munich. He worked on a wide range of topics covering heat transfer to mechanics and electromagnetism.

Example 5.6   Nuclear Reactor Cooling by Liquid Sodium Flow in Circular Pipes

Power generation in a nuclear reactor is limited principally by the ability to transfer heat in a reactor. A solid-fuel reactor is cooled by liquid sodium flowing inside small-diameter stainless steel tubes. (a) Develop an expression for Nusselt number for this case with suitable assumptions. (b) Would water have been a better coolant? Take $k_{liq.\,sodium}/k_{water} = 117.3$, $Pr_{water} = 10$.

Solution

(a)   For liquid sodium flowing in circular tubes, because of its very low Prandtl number (~ 0.005), the velocity boundary layer develops much less rapidly than the thermal boundary layer; consequently, the temperature profile becomes fully developed, while the velocity profile is still uniform across the cross-section. This situation is called 'slug flow.'

We assume that the flow is steady and laminar and heat transfer is fully developed. The energy equation is

$$\rho c_p v_z \frac{\partial T}{\partial z} = \frac{k_f}{r} \frac{\partial}{\partial r}\left(r \frac{\partial T}{\partial r}\right)$$

Boundary conditions:

$$r = 0, \qquad \frac{\partial T}{\partial r} = 0$$

$$r = r_0, \qquad T = T_s$$
$$v_z = \text{constant (slug flow)}$$

For constant wall flux,

$$\frac{\partial T}{\partial z} = \frac{dT_s}{dz} = \frac{dT_m}{dz} = \text{constant}$$

Therefore, the energy equation can be written as

$$\frac{k_f}{\rho c_p}\left[\frac{1}{r}\frac{\partial}{\partial r}\left(r\frac{\partial T}{\partial r}\right)\right] = v_z \frac{dT_m}{dz}$$

or

$$\frac{1}{r}\frac{\partial}{\partial r}\left(r\frac{\partial T}{\partial r}\right) = \left(\frac{\rho c_p}{k_f}\right)v_z \frac{\partial T_m}{\partial z} = \text{constant} = C$$

Therefore, $\dfrac{\partial}{\partial r}\left(r\dfrac{\partial T}{\partial r}\right) = Cr$

Integrating twice,

$$T = C\frac{r^2}{4} + C_1 \ln r + C_2$$

Applying the boundary conditions,

$$C_1 = 0$$

$$C_2 = T_s - \frac{Cr_0^2}{4}$$

Therefore, $T - T_s = C/4\,(r^2 - r_0^2)$.

Now, $\qquad T_m = \dfrac{\displaystyle\int_0^{r_0} v_z\, T r\, dr}{\displaystyle\int_0^{r_0} v_z\, r\, dr}$

Since, $v_z =$ constant

$$T_m = \frac{\int_0^{r_0} Tr\,dr}{\int_0^{r_0} r\,dr} = \frac{2}{r_0^2}\int_0^{r_0} Tr\,dr$$

$$= \frac{2}{r_0^2}\int_0^{r_0}\left\{T_s - \frac{C}{4}(r_0^2 - r^2)\right\}r\,dr$$

$$= T_s - \frac{C}{4}\frac{r_0^4}{4}\frac{2}{r_0^2}$$

Therefore, $T_s - T_m = \dfrac{C}{8}r_0^2$

Now, $h = \dfrac{k_f \left.\dfrac{\partial T}{\partial r}\right|_{r=r_0}}{T_s - T_m}$

$$= \frac{k_f\left[\dfrac{C}{2}r_0\right]}{\dfrac{C}{8}r_0^2} = \frac{4k_f}{r_0}$$

Therefore, $\mathrm{Nu}_D = \dfrac{h(2r_0)}{k_f} = 8$

(b) Since Pr for water is 10, the velocity profile would be parabolic (fully developed) and $\mathrm{Nu}_D = 4.364$. Therefore,

$$\frac{h_{\text{liq. sodium}}}{h_{\text{water}}} = \frac{8\,k_{\text{liq. sodium}}}{4.36\,k_{\text{water}}}$$

$$= 1.83\frac{k_{\text{liq.sodium}}}{k_{\text{water}}}$$

$$= (1.83)(114.3) = 209$$

Therefore, h_{water} is much less than $h_{\text{liq. sodium}}$. So, water would have been a worse coolant.

5.15 Heat Transfer in Turbulent Tube Flow

Experimental observations indicate that the fully developed laminar flow becomes unstable when $\mathrm{Re}_D \approx 2000$. The turbulent regime takes over when $\mathrm{Re}_D > 2300$. It may be noted that the laminar regime can survive even at a considerably higher Reynolds number if the pipe wall is exceptionally smooth and the disturbance in the fluid entering the tube is very small. Actually, Reynolds showed that for smooth entrance conditions used in his original experiment, the critical values of Reynolds number were in the range of 11,800–14,300. The onset of turbulence was visualized by dye injection method. Reynolds described the appearance of turbulence as follows: '... all at once, the colour band appeared to expand and mix with the water. On viewing the tube by the light of an electric spark, the mass of colour resolved itself into a mass of more or less distinct curls, showing eddies.'

In terms of the time-averaged flow variables discussed earlier, the turbulent pipe flow is described by the mean axial velocity component $\bar{v}_z$, the radial component $\bar{v}_r$, the pressure $\bar{p}$, and the temperature $\bar{T}$. In fully developed flow, $\bar{v}_r = 0$, $\partial\bar{v}_z/\partial z = 0$ and therefore, $\bar{v}_z = \bar{v}_z(r)$, $\bar{p} = \bar{p}(z)$ and the momentum and energy equations reduce to

$$0 = -\frac{1}{\rho}\frac{d\bar{p}}{dz} + \frac{1}{r}\frac{\partial}{\partial r}\left[(v+\varepsilon)r\frac{\partial\bar{v}_z}{\partial r}\right] \tag{5.205}$$

$$\bar{v}_z\frac{\partial\bar{T}}{\partial z} = \frac{1}{r}\frac{\partial}{\partial r}\left[(\alpha+\varepsilon_H)r\frac{\partial\bar{T}}{\partial r}\right] \tag{5.206}$$

τ_{app} and q''_{app} are written as

$$\tau_{app} = -\mu\frac{\partial\bar{v}_z}{\partial r} - \rho\varepsilon\frac{\partial\bar{v}_z}{\partial r} \tag{5.207}$$

$$q''_{app} = \underbrace{k_f\frac{\partial\bar{T}}{\partial r}}_{\text{molecular}} + \underbrace{\rho c_p\varepsilon_H\frac{\partial\bar{T}}{\partial r}}_{\text{Eddy}} \tag{5.208}$$

Note that $\partial\bar{v}_z/\partial r$ is negative and $\partial\bar{T}/\partial r$ is positive as r increases. Therefore, for the fully developed flow, Eqs (5.205) and (5.206) are written in terms of Eqs (5.207) and (5.208) as

$$0 = -\frac{1}{\rho}\frac{d\bar{p}}{dz} - \frac{1}{\rho r}\frac{\partial}{\partial r}(r\tau_{app}) \tag{5.209}$$

$$\bar{v}_z\frac{\partial\bar{T}}{\partial z} = \frac{1}{\rho c_p r}\frac{\partial}{\partial r}(rq''_{app}) \tag{5.210}$$

The axial velocity distribution $\bar{v}_z$ obeys the universal law $u^+ = u^+(y^+)$ presented earlier, where

$$u^+ = \frac{\bar{v}_z}{\left(\dfrac{\tau_s}{\rho}\right)^{1/2}} \tag{5.211}$$

$$y^+ = \frac{y}{v}\left(\frac{\tau_s}{\rho}\right)^{1/2} \tag{5.212}$$

and $y = r_0 - r$ is the distance measured away from the wall. The velocity profile is remarkably flat over most of the central portion of the cross-section (Fig. 5.25) and obeys Prandtl's 1/7th power law:

$$u^+ = 8.7(y^+)^{1/7} \tag{5.213}$$

Just like in a flat plate, τ_{app} and q''_{app} are

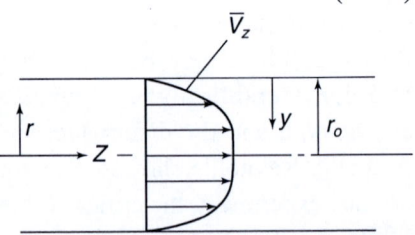

Fig. 5.25   Axial velocity profile in turbulent pipe flow

nearly constant in the vicinity of the wall. It can be shown that

$$\frac{\tau_{app}}{\tau_s} = \frac{r}{r_0} \tag{5.214}$$

and $\quad \dfrac{q''_{app}}{q''_s} = \dfrac{r}{r_0} \tag{5.215}$

Therefore, in a sufficiently thin region close to the wall,

$$\tau_{app} = \tau_s \tag{5.216}$$

and $\quad q''_{app} = q''_s \tag{5.217}$

The foregoing equations [Eqs (5.214) and (5.215)] form the basis for the analysis that led to the Stanton number formula, Eq. (5.226).
In terms of y,

$$\frac{q''_{app}}{\rho c_p} = (\alpha + \varepsilon_H)\frac{d\bar{T}}{dy} \tag{5.218}$$

$$\frac{\tau_{app}}{\rho} = (v + \varepsilon)\frac{d\bar{v}_z}{dy} \tag{5.219}$$

Dividing Eq. (5.218) by Eq. (5.219), assuming $\varepsilon = \varepsilon_H$ and $v = \alpha$, or $\text{Pr} = 1$,

$$\frac{q''_{app}}{c_p \tau_{app}} d\bar{v}_z = -d\bar{T} \tag{5.220}$$

An additional assumption that falls out from Eqs (5.214) and (5.215) is

$$\frac{q''_{app}}{\tau_{app}} = \text{constant} = \frac{q''_s}{\tau_s} \tag{5.221}$$

Integrating Eq. (5.220) between wall conditions and mean bulk conditions gives

$$\frac{q''_s}{c_p \tau_s} \int_{\bar{v}_z=0}^{\bar{v}_z=v_m} d\bar{v}_z = \int_{T_s}^{T_m} -d\bar{T}$$

or $\quad \dfrac{q''_s v_m}{c_p \tau_s} = T_s - T_m \tag{5.222}$

But the heat transfer at the wall may be expressed as

$$q''_s = h(T_s - T_m) \tag{5.223}$$

and the shear stress at the wall may be calculated from the force balance on a fluid element filling the entire cross-section of the tube as

$$\tau_s = \frac{\Delta p(\pi D^2)}{4\pi DL} = \frac{\Delta p}{4}\frac{D}{L}$$

The pressure drop may be expressed in terms of a friction factor as

$$\Delta p = f\frac{L}{D}\rho\frac{v_m^2}{2} \tag{5.224}$$

so that $\quad \tau_s = \dfrac{f}{8}\rho v_m^2$ $\hspace{6cm}$ (5.225)

Substituting the expressions for τ_s and q_s'' into Eq. (5.222), we obtain

$$St = \frac{h}{\rho c_p v_m} = \frac{Nu_D}{Re_D\,Pr} = \frac{f}{8} \hspace{4cm} (5.226)$$

Equation (5.226) is called the Reynolds analogy for the tube flow.

An empirical formula for the turbulent friction factor up to $Re_D = 2 \times 10^5$ for flow in smooth tubes is

$$f = \frac{0.316}{Re_D^{1/4}} \hspace{5cm} (5.227)$$

Inserting this in Eq. (5.226),

$$\frac{Nu_D}{Re_D\,Pr} = 0.0395\,Re_D^{-1/4}$$

or $\quad Nu_D = 0.0395\,Re_D^{3/4}$ $\hspace{5cm}$ (5.228)

The above relation is highly restrictive because it is applicable for $Pr \approx 1$. It turns out that for other Prandtl number fluids ($Pr \geq 0.5$), the following relation works well:

$$St\,Pr^{2/3} = \frac{f}{8} \hspace{5cm} (5.229)$$

$$Nu_D = 0.0395\,Re_D^{3/4}\,Pr^{1/3} \hspace{4cm} (5.230)$$

For calculation purposes, a more correct relation to use for the turbulent flow in a smooth tube is given below:

$$Nu_D = 0.023\,Re_D^{0.8}\,Pr^n \hspace{4cm} (5.231)$$

The above is valid for $0.7 \leq Pr \leq 120$, $2500 \leq Re_D \leq 1.24 \times 10^5$, and $L/D > 60$. The Prandtl number exponent is $n = 0.4$ when the fluid is being heated ($T_s > T_m$) and $n = 0.3$ when the fluid is being cooled ($T_s < T_m$). All the physical properties needed for the calculation of Nu_D, Re_D, and Pr are to be evaluated at the bulk temperature T_m. Equation (5.231) is known as the Dittus–Boelter (1930) correlation.

The condition in the Dittus–Boelter equation regarding the direction of heat transfer (heating or cooling) is probably a variable-properties correction. Equation (5.231) tends to overpredict the Nusselt number for gases by at least 20% and underpredict the Nusselt number for the higher Prandtl number fluids by 7–10%. Therefore, this correlation should be used with caution.

Kakac et al. (1987) recommended the following correlation in the range of $0.5 \leq Pr < 2000$ and $2300 \leq Re_D \leq 5 \times 10^6$:

$$Nu_D = \frac{(Re_D - 1000)\,Pr\,C_f/2}{1.0 + 12.7\sqrt{C_f/2}\,(Pr^{2/3} - 1.0)} \hspace{3cm} (5.232)$$

Sleicher and Rouse (1975) suggest a somewhat more convenient empirical formulation:

$$Nu_D = 5 + 0.015 \, Re_D^a \, Pr^b \qquad (5.233)$$

where $\quad a = 0.88 - \dfrac{0.24}{4 + Pr}$

$b = 0.333 + 0.5 e^{-0.6 \, Pr}$

The range of validity of this equation is stated to be

$$0.1 \le Pr \le 10^4, \quad 10^4 \le Re_D \le 10^6$$

It may be noted that for $Pr \ge 0.5$ fluids, Nusselt number is virtually independent of the constant heat flux and constant surface temperature conditions. Recall that this is certainly not true for laminar flows. For liquid metals ($0.006 \le Pr \le 0.03$) the correlation proposed by Notter and Sleicher (1972) fits fairly well the experimental data:

$$\text{Constant heat flux:} \quad Nu_D = 6.3 + 0.0167 \, Re_D^{0.85} \, Pr^{0.93} \qquad (5.234)$$

$$\text{Constant surface temperature:} \quad Nu_D = 4.8 + 0.0156 \, Re_D^{0.85} \, Pr^{0.93} \qquad (5.235)$$

All the properties in Eqs (5.234) and (5.235) are evaluated at the mean temperature $T_m = 1/2(T_{in} + T_{out})$.

5.15.1 Salient Features of Liquid Metal Heat Transfer in Turbulent Tube Flow

Turbulent heat transfer in liquid metals flow in a pipe is quite different from that in common fluids. Therefore, it is worthwhile remembering the distinguishing aspects which are stated below point by point.

1. $\alpha \gg \varepsilon_H$ for liquid metals. That is, molecular conduction is the dominant diffusion mechanism even in the fully turbulent part of the flow field.
2. As $Pr \to 0$, turbulent Eddy diffusion becomes insignificant. This means that liquid metal heat transfer characteristics in turbulent flow have many similarities with laminar flow, while at the same time its turbulent momentum characteristics are no different from those of other common liquids.
3. The thermal entry length tends to be large, similar to laminar flow.
4. Outer surface thermal boundary condition (constant heat flux or constant wall temperature) has a substantial effect on the convection heat transfer coefficient.
5. Although Nusselt number is low for a liquid metal, the heat transfer coefficient is very high because of its high thermal conductivity. This is the reason of its widespread use in cooling of reactor core in nuclear power industries.
6. Longitudinal conduction (also called axial conduction) is an important factor in liquid metal heat transfer except in the case of constant heat flux boundary condition at the outer wall of the tube.

5.16 External Flows over Cylinders, Spheres, and Banks of Tubes

In this section, Nusselt number correlations for heat transfer in external flows over cylinders, spheres, and banks of tubes are given.

5.16.1 Single Cylinder in Crossflow

Consider the heat transfer between a long cylinder placed across a fluid stream of uniform velocity u_∞ and temperature T_∞. The cylinder surface has a uniform temperature T_s. A very reliable correlation based on data from many independent sources was developed by Churchill and Bernstein (1977), which is given below:

$$\overline{Nu}_D = 0.3 + \frac{0.62\,Re_D^{1/2}\,Pr^{1/3}}{\left[1+\left(\dfrac{0.4}{Pr}\right)^{2/3}\right]^{1/4}}\left[1+\left(\frac{Re_D}{282,000}\right)^{5/8}\right]^{4/5} \tag{5.236}$$

where $\overline{Nu}_D = \bar{h}D/k$. Equation (5.236) holds for all values of Re_D and Pr, provided the Peclet number Pe_D is greater than 0.2. The physical properties needed for calculating Nu_D, Pr, and Re_D are evaluated at the film temperature, $(T_s + T_\infty)/2$. Equation (5.236) applies also to a cylinder with constant heat flux, in which case the average heat transfer coefficient is based on the perimeter-averaged temperature difference between the cylinder surface and free stream.

5.16.2 Sphere

Whitaker (1972) proposed the following correlation for the heat transfer between an isothermal free stream (u_∞, T_∞) and an isothermal spherical surface (T_s):

$$\overline{Nu}_D = 2 + (0.4\,Re_D^{1/2} + 0.06\,Re_D^{2/3})\,Pr^{0.4}\left(\frac{\mu_\infty}{\mu_w}\right)^{1/4} \tag{5.237}$$

The relation is valid for $0.71 < Pr < 380$, $3.5 < Re_D < 7.6 \times 10^4$, and $1 < (\mu_\infty/\mu_w) < 3.2$. All the physical properties in Eq. (5.237) are evaluated at the free-stream temperature T_∞, except $\mu_w = \mu(T_w)$. Equation (5.237) also applies to spherical surfaces with constant heat flux, with $\overline{Nu}_D$ based on the surface-averaged temperature difference between the sphere and the surrounding stream.

5.16.3 Bank of Tubes in Crossflow

A bank of tubes in crossflow has application in heat exchangers. The geometry is represented by the cylinder diameter D, the longitudinal spacing of two consecutive rows (longitudinal pitch X_l), and the transverse spacing of the two consecutive cylinders (transverse pitch X_t). A large body of literature is available on the heat transfer performance of banks of cylindrical tubes in crossflow, the most comprehensive review being that by Zukauskas (1987). For aligned arrays of cylinders, the array-averaged Nusselt number is anticipated within ± 15% by

$$\overline{Nu}_D = 0.9C_n Re_D^{0.4} Pr^{0.36}\left(\frac{Pr}{Pr_w}\right)^{1/4}, \quad Re_D = 1\text{--}10^2$$

$$\overline{Nu}_D = 0.52C_n Re_D^{0.5} Pr^{0.36}\left(\frac{Pr}{Pr_w}\right)^{1/4}, \quad Re_D = 10^2\text{--}10^3 \tag{5.238}$$

$$\overline{Nu}_D = 0.27C_n Re_D^{0.63} Pr^{0.36}\left(\frac{Pr}{Pr_w}\right)^{1/4}, \quad Re_D = 10^3\text{--}(2\times10^5)$$

$$\overline{Nu}_D = 0.033C_n Re_D^{0.8} Pr^{0.4}\left(\frac{Pr}{Pr_w}\right)^{1/4}, \quad Re_D = (2\times10^5)\text{--}(2\times10^6)$$

where C_n is a function of the total number of rows in the array. The Reynolds number Re_D is based on the average velocity through the narrowest cross-section formed by the array, that is, the maximum average velocity u_{max},

$$Re_D = \frac{u_{max}D}{v} \tag{5.239}$$

The narrowest flow cross-section forms in the plane that contains the centres of all the cylinders of one row. The conservation of mass through such a plane requires that

$$u_\infty X_t = u_{max}(X_t - D) \tag{5.240}$$

$\overline{Nu}_D$ is based on $\overline{h}$, which is the heat transfer coefficient averaged over all the cylindrical surfaces in the array. The total area of these surfaces is $nm\pi DL$, where n is the number of rows, m is the number of cylinders in each row (across the flow direction), and L is the length of the array in the direction perpendicular to the plane of the paper.

All the physical properties except Pr_w in Eq. (5.238) are evaluated at the mean temperature of the fluid that flows through the spaces formed between the cylinders. Pr_w is evaluated at the temperature of the cylindrical surface.

Additional Examples

Example 5.7 Similarity Solution for High Prandtl Number Fluid Flow Over an Isothermal Plate

Using Pohlhausen's similarity approach show that for a high Prandtl number fluid flowing over an entirely heated plate maintained at constant temperature, the local Nusselt number can be expressed as $Nu_x = 0.339\,Re_x^{1/2}\,Pr^{1/3}$.

Solution
When the Pr is very high, the velocity boundary layer will be very much thicker than the thermal boundary layer. If we assume that the thermal boundary layer is entirely within the part of the velocity boundary layer in which the velocity profile is linear, we can get a simple solution for the Nusselt number.

From the Blasius similarity solution for flow over a flat plate we know that $f''(0) = 0.3321$. Then letting $f'' = f''(0)$ in the region of interest, integrating twice with respect to η, and noting that $f(0) = 0$ and $f'(0) = 0$, we get

$$f = \frac{0.3321}{2}\eta^2 \qquad \text{(A)}$$

Now, we know from Eq. (5.38)

$$\theta'(0) = \frac{1}{\int_0^\infty \exp\left[-\frac{1}{2}\Pr\int_0^\eta f\,d\eta\right]d\eta} \qquad \text{(B)}$$

Substituting Eq. (A) into Eq. (B) and carrying out the integration, we finally obtain

$$\mathrm{Nu}_x = 0.339\,\mathrm{Re}_x^{1/2}\,\mathrm{Pr}^{1/3} \qquad \text{(C)}$$

Equation (C) corresponds almost precisely to the exact solution for $\Pr \geq 10$.

Example 5.8 Integral Solution for Very Low Prandtl Number Fluid Flow Over an Isothermal Plate

Using von Karman's integral analysis, show that for a liquid metal flowing over a flat plate maintained at constant temperature the local Nusselt number is related to the Reynolds number and the Prandtl number as $Nu_x = 0.53Re_x^{1/2}\,Pr^{1/2}$.

Solution
Since, for liquid metals (very low Pr) δ/δ_t is small, we take $u = u_\infty$ over the entire thermal boundary layer thickness δ_t as an approximation. For temperature distribution, we use a cubic parabola as before.

Energy integral equation [Eq. (5.65)]:

$$\frac{d}{dx}\left[\int_0^{\delta_t}(T_\infty - T)u\,dy\right] = \alpha\left(\frac{\partial T}{\partial y}\right)_{y=0} \qquad \text{(A)}$$

Also, from Eq. (5.64),

$$\frac{\theta}{\theta_\infty} = \frac{T - T_s}{T_\infty - T_s} = \frac{3}{2}\left(\frac{y}{\delta_t}\right) - \frac{1}{2}\left(\frac{y}{\delta_t}\right)^3 \qquad \text{(B)}$$

and $u = u_\infty$ (C)

Inserting Eqs (B) and (C) in Eq. (A) and carrying out the necessary integration and differentiation, we obtain

$$2\delta_t\,d\delta_t = \frac{8\alpha}{u_\infty}dx \qquad \text{(D)}$$

Using the condition $\delta_t = 0$ at $x = 0$, since the entire plate is heated, the solution of Eq (D) is

$$\delta_t = \sqrt{\frac{8\alpha x}{u_\infty}} \qquad \text{(E)}$$

Now,

$$h_x = \frac{-k\left(\dfrac{\partial T}{\partial y}\right)_{y=0}}{T_s - T_\infty} = \frac{3k}{2\delta_t} \qquad \text{(F)}$$

Substituting Eq. (E) in Eq. (F) we get

$$h_x = \frac{3\sqrt{2}}{8}k\sqrt{\frac{u_\infty}{\alpha x}}$$

$$\text{Therefore, Nu}_x = \frac{h_x x}{k} = \frac{\dfrac{3\sqrt{2}}{8} k \sqrt{\dfrac{u_\infty}{\alpha x}} x}{k} = \frac{3\sqrt{2}}{8} \sqrt{\frac{u_\infty}{vx}} x \sqrt{\frac{v}{\alpha}}$$

$$= 0.53 \sqrt{\frac{u_\infty x}{v}} \sqrt{\frac{v}{\alpha}} = 0.53 \text{Re}_x^{1/2} \, \text{Pr}^{1/2}$$

Example 5.9 Forced Convection over a Building Wall

Wind is blowing at 5 km/h over a building wall of size 5 m × 10 m as shown in Fig.E5.9. Air temperature is 0°C and the wall surface temperature is 10°C. Calculate (a) the average forced convection heat transfer coefficient over the 10 m length of the wall; (b) the local forced convection heat transfer coefficient at a location of 10 cm from the leading edge of the wall. For property values see Table A1.4 of the book.

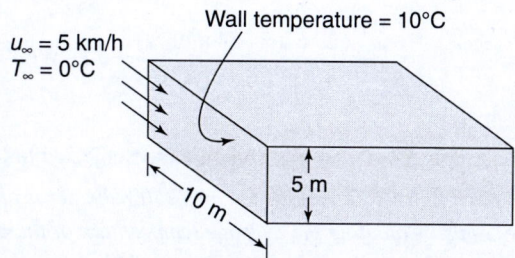

Fig. E5.9 Forced convection over a wall of a house

Solution

(a) The properties will have to be evaluated at the film temperature,

$$T_f = \frac{T_s + T_\infty}{2} = \frac{10 + 0}{2} = 5°C = 5 + 273 = 278 \text{ K}$$

By linear interpolation of the properties between 250 K and 300 K, we get the following at 278 K.

$$v = 13.932 \times 10^{-6} \text{ m}^2/\text{s}$$

$$\text{Pr} = 0.713$$

$$k_f = 24.54 \times 10^{-6} \text{ W/m K}$$

Now, $\qquad u_\infty = \dfrac{5 \times 1000}{3600} = 1.39 \text{ m/s}$

$$\text{Re}_L = \frac{u_\infty L}{v} = \frac{(1.39)(10)}{13.932 \times 10^{-6}} = 9.97 \times 10^5$$

Since $\text{Re}_L > \text{Re}_{cr}$ (= 5×10^5) , the flow is turbulent at $x = L = 10$ m. Therefore, a mixed boundary layer flow exists on the wall surface. Hence,

$$\overline{\text{Nu}}_L = 0.037 \, \text{Pr}^{1/3} \left(\text{Re}_L^{4/5} - 23550 \right)$$

$$= 0.037 (0.713)^{1/3} \, [(9.97 \times 10^5)^{4/5} - 23550] = 1301.5$$

Therefore, $\overline{\text{Nu}}_L = \dfrac{\overline{h}_L L}{k_f} = 1301.5$

$\Rightarrow \qquad \overline{h}_L = \dfrac{\overline{\text{Nu}}_L \, k_f}{L} = \dfrac{(1301.5)(24.54 \times 10^{-3})}{10} = 3.19 \text{ W/m}^2\text{K}$

(b) At $x = 10$ cm $= 0.1$ m

$$\text{Re}_x = \frac{u_\infty x}{v} = \frac{(1.39)(0.1)}{13.932 \times 10^{-6}} = 9.97 \times 10^3$$

Since $\text{Re}_x < \text{Re}_{cr} (= 5 \times 10^5)$ the flow is laminar at $x = 0.1$ m.

Therefore, $\text{Nu}_x = 0.332 \text{Re}_x^{1/2} \text{ Pr}^{1/3} = 0.332(9.97 \times 10^3)^{1/2} (0.713)^{1/3} = 29.6$

Hence,

$$\text{Nu}_x = \frac{hx}{k_f} = 29.6$$

$\Rightarrow$

$$h = \frac{(29.6)(24.54 \times 10^{-3})}{0.1} = 7.26 \text{ W/m}^2 \text{ K}$$

Example 5.10   Airflow Over a Plate Subjected to Constant Heat Flux

A plate (0.6 m × 0.6 m) heated by a 250 W heater is placed in an airflow at 27°C, 1 atm with = 5 m/s. Calculate the average temperature of the plate and the local temperature of the plate at the trailing edge. Assume steady state.

Solution

The properties should be evaluated at the film temperature, that is, $\dfrac{T_{s,\text{avg}} + T_\infty}{2}$. But since we do not know the plate temperature at the beginning, we take the properties at the free-stream temperature ($T_\infty = 27°C = 300$ K) in the first iteration.

Iteration # 1

From Table A1.4 of the book

$$v = 15.89 \times 10^{-6} \text{ m}^2/\text{s}$$
$$\text{Pr} = 0.707$$
$$k_f = 26.3 \times 10^{-3} \text{ W/m K}$$

$$\text{Re}_L = \frac{u_\infty L}{v} = \frac{(5)(0.6)}{15.89 \times 10^{-6}} = 1.88 \times 10^5$$

Since $\text{Re}_L < \text{Re}_{cr} (= 5 \times 10^5)$, the flow is laminar over the entire plate.

The relevant correlation for the constant heat flux boundary condition at the wall is

$$T_{s,\text{avg}} - T_\infty = \frac{q_s'' L}{0.6795 k_f \text{ Re}_L^{1/2} \text{ Pr}^{1/3}}$$

Thus, $T_{s,\text{avg}} - T_\infty = \dfrac{\left(\dfrac{250}{(0.6)^2}\right)(0.6)}{0.6795(0.0263)(1.88 \times 10^5)^{1/2} (0.707)^{1/3}} = 60.42°C$

Hence, $T_{s,\text{avg}} = 60.42 + T_\infty = 60.42 + 27 = 87.42°$ C. Now, we go back and evaluate properties at $(87.42 + 27)/2 = 57.21°C$ and begin the second iteration.

Iteration #2

At $T_f = 57.21°C = 330.21$ K ≈ 330 K, 1 atm the properties of air are

$$v = 18.908 \times 10^{-6} \text{ m}^2/\text{s}$$
$$\text{Pr} = 0.7028$$
$$k_f = 28.52 \times 10^{-3} \text{ W/m K}$$

$$\text{Re}_L = \frac{(5)(0.6)}{18.908 \times 10^{-6}} = 1.58 \times 10^5 \ (< 5 \times 10^5) \text{ and hence the flow is laminar.}$$

$$= 60.84°C$$

Hence,
$$T_{s,\text{avg}} - T_\infty = \frac{\left(\dfrac{250}{(0.6)^2}\right)(0.6)}{0.6795(0.02852)(1.58 \times 10^5)^{1/2}(0.7028)^{1/3}} = 60.84°C$$

or,
$$T_{s,\text{avg}} = 60.84 + T_\infty = 60.84 + 27 = 87.84°C$$

We see that the absolute difference between the first and the second iteration values is 0.42, which is quite small. Therefore, the iterations stop at this point and we finally take $T_{s,\text{avg}} = 87.84°C$. However, if more accuracy is desired the iterations may continue till a pre-specified tolerance limit (such as 0.1, 0.01, 0.001 and so on) is reached. That is, at the end of each iteration the absolute difference between successive iteration values of $T_{s,\text{avg}}$ will have to be computed and compared with the tolerance limit.

Second Part

$$T_s(x) - T_\infty = \frac{q_s'' x}{0.453 k_f \text{Re}_x^{1/2} \text{Pr}^{1/3}}$$

Using the property values at $T_f = 330$ K in the above expression, we obtain,

$$T_s \big|_{x=0.6\,m} - T_\infty = \frac{\left(\dfrac{250}{(0.6)^2}\right)(0.6)}{0.453(0.02852)(1.58 \times 10^5)^{1/2}(0.7028)^{1/3}} = 91.27°C$$

Therefore, at $x = 0.6$ m, $T_s = 91.27 + 27 = 118.27°$ C

Example 5.11   Mixed Boundary Layer Flow over an Isothermal Plate

Water at 1 atm pressure flows with the velocity = 0.2 m/s at 10°C over a plate which is maintained at 30°C. At x = 6 m, a probe is inserted in the viscous sublayer to the position represented by $y^+ = 2.7$ (Fig. E5.11). (a) Calculate the actual spacing y (mm) between the probe and the wall. (b) Obtain the heat transfer coefficient averaged over the length x.

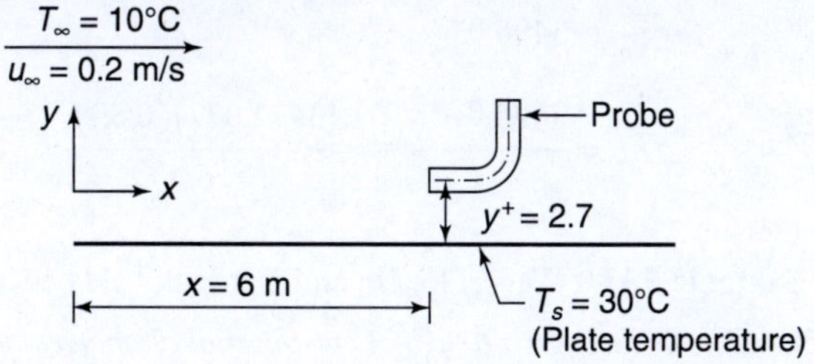

Fig. E5.11   Forced convection over a heated plate with a probe at $x = 6$ m

Solution

(a) At $T_f = \dfrac{30 + 10}{2} = 20°C$, 1 atm, the properties of water are

$$\rho = 998.2 \ \text{kg/m}^3$$

$$v = 0.01004 \times 10^{-4} \ \text{m}^2/\text{s}$$

$$\mu = 0.001002 \ \text{kg/m s}$$

$$k_f = 0.59 \ \text{W/m K}$$

$$\text{Pr} = 7.07$$

Given: $u_\infty = 0.2 \ \text{m/s}, x = 6 \ \text{m}$

Now, $$\text{Re} = \frac{\rho u_\infty x}{\mu} = \frac{(998.2)(0.2)(6)}{0.001002} = 1.195 \times 10^6$$

Since $\text{Re} > \text{Re}_{cr}$ where $\text{Re}_{cr} = 5 \times 10^5$ the flow at $x = 6$ m is turbulent.
Therefore,

$$\tau_{s,x} = 0.0296 \, \rho u_\infty^2 \left(\text{Re}_x\right)^{-1/5} = (0.0296)(998.2)(0.2)^2 \ (1.195 \times 10^6)^{-1/5} = 0.07197 \ \text{N/m}^2$$

Recall

$$y^+ = \frac{y}{v}\left(\frac{\tau_{s,x}}{\rho}\right)^{1/2}$$

$$\Rightarrow \qquad y^{+^2} = \frac{y^2}{v^2}\frac{\tau_{s,x}}{\rho}$$

$$\Rightarrow \qquad y^2 = \frac{y^{+^2} v^2 \rho}{\tau_{s,x}} = \frac{(2.7)^2 \ (0.01004 \times 10^{-4})^2 \ (998.2)}{0.07197} = (101918.76)(10^{-12})$$

$$\Rightarrow \qquad y = 319.247 \times 10^{-6} \approx 0.32 \times 10^{-3} \ \text{m}$$

Therefore, $y \approx 0.32$ mm

(b) Since $\text{Re}_x > \text{Re}_{cr}$, a mixed boundary layer flow exists on the plate.

$$\overline{\text{Nu}_L} = \frac{\overline{h_L} L}{k_f} = 0.037 \, \text{Pr}^{1/3} \ (\text{Re}_L^{4/5} - 23550)$$

$$= 0.037 (7.07)^{1/3} \ [(1.195 \times 10^6)^{4/5} - 23550]$$

$$\overline{\text{Nu}_L} = 3494.077$$

Thus, $$\overline{h_L} = \frac{3494.077 \, k_f}{L} = \frac{(3494.077)(0.59)}{6} = 343.58 \ \text{W/m}^2\text{K}$$

Example 5.12   Water Flow in an Electrically Heated Tube

Water flows through a tube of 3 cm-internal diameter and 20 m length. The outside surface of the tube is heated electrically so that it is subjected to uniform heat flux circumferentially and axially (Fig.E5.12). The inlet and exit temperatures of the water are 10°C and 70°C, respectively. The mass flow rate of the water is 720 kg/h. Disregard the thermal resistance of the tube wall. Assume steady flow and heat transfer. Estimate the inner surface temperature at the tube exit.

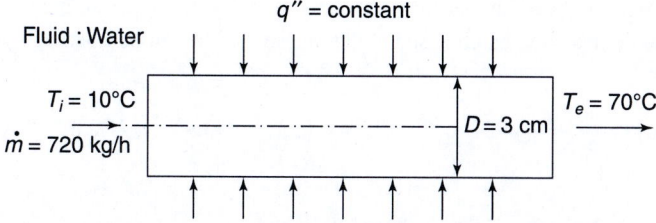

Fig. E5.12 Heating of water flowing in an electrically heated circular tube

Solution

The properties of water at the bulk mean temperature of $T_b = \dfrac{T_i + T_e}{2} = \dfrac{10 + 70}{2} = 40°C$ are

$\rho = 992.1 \text{ kg/m}^3$

$c_p = 4179 \text{ J/kg K}$

$k_f = 0.631 \text{ W/m K}$

$\text{Pr} = 4.32$

$v = 0.658 \times 10^{-6} \text{ m}^2/\text{s}$

Given:

$$\dot{m} = 720 \text{ kg/h} = \frac{720}{3600} = 0.2 \text{ kg/s}$$

$D = 3 \text{ cm} = 0.03 \text{ m}$

$T_i = 10° \text{ C}$

$T_e = 70° \text{ C}$

$L = 20 \text{ m}$

It can be seen from the energy balance on the entire tube that the rate at which heat must be supplied to the water is

$$q = \dot{m} c_p \, (T_e - T_i) = (0.2)(4179)(70 - 10) = 50148 \text{ W}$$

Hence, $q'' = \dfrac{q}{A_s} = \dfrac{q}{\pi D L} = \dfrac{50148}{\pi(0.03)(20)} = 26605.125 \approx 26605 \text{ W/m}^2$

Also, $q'' = h(T_s = T_m)$

Therefore, the surface temperature T_s at any location can be determined from

$$T_s = T_m + \frac{q''}{h}$$

To obtain 'h' we need to first obtain Reynolds number and check whether the flow is laminar or turbulent.

$$v_m = \frac{\dot{m}}{\rho A_c} = \frac{\dot{m}}{\rho(\frac{\pi}{4} D^2)} = \frac{0.2}{992.1(\frac{\pi}{4}(0.03)^2)} = \frac{0.2}{992.1(7.069 \times 10^{-4})} = 0.285 \, \text{m/s}$$

$$\text{Re}_D = \frac{v_m D}{v} = \frac{(0.285)(0.03)}{0.658 \times 10^{-6}} = 12993.9 \text{ which is greater than 2300.}$$

Therefore, the flow is turbulent. The entry length is approximately, $z_{fd,h} \approx z_{fd,t} = 10D = (10)$ $(0.03) = 0.3$ m which is much less than the length of the pipe, that is, 20 m.

Hence, the flow and heat transfer may be treated as fully developed in the entire pipe and Dittus–Boelter correlation can be used to determine the Nusselt number.

For heating:

$$\text{Nu}_D = \frac{hD}{k_f} = 0.023\,\text{Re}_D^{0.8}\;\text{Pr}^{0.4} = 0.023\,(12993.9)^{0.8}\;(4.32)^{0.4} = 80.68$$

Then,
$$h = \frac{k_f}{D}\,\text{Nu}_D = \frac{0.631}{0.03}(80.68) = 1697\,\text{W/m}^2\text{K}$$

Hence, the inner surface temperature of the tube at the exit is

$$T_s = T_e + \frac{q''}{h} = 70 + \frac{26605}{1697} = 70 + 15.67 = 85.67°\,\text{C} = 85.7°\text{C}$$

Important Concepts and Formulae

Basic Terms for Convective Heat Transfer on a Flat Plate

$$h = \frac{-k_f\left.\dfrac{\partial T}{\partial y}\right|_{y=0}}{T_s - T_\infty}$$

$$\bar{h} = \frac{1}{L}\int_0^L h\,dx$$

$$q = \bar{h}A\left(T_s - T_\infty\right)$$

$$\text{Nu}_x = \frac{hx}{k_f}$$

$$\overline{\text{Nu}_L} = \frac{\bar{h}L}{k_f}$$

Heat Transfer Correlations for Laminar Flow over a Flat Plate
Isothermal Plate (T_s = constant)
Similarity Solution

$$\text{Nu}_x = 0.332\,\text{Re}_x^{1/2}\;\text{Pr}^{1/3} \quad (\text{Valid for Re} \le 5\times10^5 \;\text{Pr} \ge 0.05)$$

where

$$\text{Re}_x = \frac{u_\infty x}{\nu} \quad (\text{Note that: Re}_{cr} = 5\times10^5)$$

$$\text{Pr} = \frac{\nu}{\alpha} = \frac{\mu c_p}{k_f}$$

$$\overline{\text{Nu}_L} = 0.664\,\text{Re}_L^{1/2}\;\text{Pr}^{1/3} \quad (\text{Valid for Re} \le 5\times10^5,\; \text{Pr} \ge 0.5)$$

For liquid metals ($\text{Re} \le 5\times10^5,\, \text{Pr} \ge 0.5$)

$$\text{Nu}_x = 0.565\,\text{Re}_x^{1/2}\;\text{Pr}^{1/2}$$

Integral Solution (for T_s = constant in the heated section of the plate having an insulated starting section of length L)

For $\text{Pr} \geq 0.5$

$$\text{Nu}_x = 0.332 \text{Re}_x^{1/2}\ \text{Pr}^{1/3}\ \frac{1}{\left[1-\left(\dfrac{L}{x}\right)^{3/4}\right]^{1/3}}$$

For $L = 0$

$$\text{Nu}_x = 0.332 \text{Re}_x^{1/2}\ \text{Pr}^{1/3}$$

For $0.006 \leq \text{Pr} \leq 0.03$ *(Entire Plate at T_c constant)*

$$\text{Nu}_x = \frac{\text{Pe}_x^{1/2}}{1.55\,\text{Pr}^{1/2} + 3.09\left[0.372 - 0.15\,\text{Pr}\right]^{1/2}}$$

where Pe = RePr

For Plate Subjected to Constant Heat Flux (q_s'' = constant):
Similarilty Solution
Not possible for this case.
Integral Solution

For Pr ≥ 0.5 fluids (entire plate heated)

$$\text{Nu}_x = 0.453 \text{Re}_x^{1/2}\ \text{Pr}^{1/3}$$

$$T_s\,(x) - T_\infty = \frac{q''\,x}{0.453\,k_f\ \text{Re}_x^{1/2}\ \text{Pr}^{1/3}}$$

$$\overline{\text{Nu}_L} = \frac{\bar{h}_L\,L}{k_f}$$

Also,

$$\bar{h}_L = \frac{q_s''}{T_{s,\text{avg}} - T_\infty}$$

where

$$T_{s,\text{avg}} - T_\infty = \frac{1}{L}\int_0^L (T_s - T_\infty)\,dx$$

$$T_{s,\text{avg}} - T_\infty = \frac{q_s''\,(L/k_f)}{0.6795\,\text{Re}_L^{1/2}\ \text{Pr}^{1/3}}$$

$$\overline{\text{Nu}_L} = 0.6795\,\text{Re}_L^{1/2}\ \text{Pr}^{1/3}$$

For liquid metals, 0.006 ≤ Pr ≤ 0.03 (entire plate heated)

$$\frac{\text{Nu}_x}{\sqrt{\text{Re}_x\,\text{Pr}}} = \frac{0.88}{1 + 1.317\sqrt{\text{Pr}}}$$

For Pr ≥ 0.5 fluids (unheated starting length)

$$\text{Nu}_x = \frac{0.453 \text{Re}_x^{1/2}\ \text{Pr}^{1/3}}{\left[1-\left(\dfrac{L}{x}\right)^{3/4}\right]^{1/3}}$$

where L is the unheated starting length.

For liquid metals, $0.006 \leq Pr \leq 0.03$ (unheated starting length)

No energy integral solution is available in published literature.

Heat Transfer Correlations for Turbulent Flow over a Flat Plate

For turbulent flow over the entire plate (either T_s = constant or q''_s = constant)

For $Pr \geq 0.5$ and $Re > 5 \times 10^5$

$$Nu_x = 0.0296 \, Re_x^{4/5} \, Pr^{1/3}$$

For Mixed Boundary Layer Flow Over a Flat Plate (T_s = constant)

$$\overline{Nu}_l = 0.037 \, Pr^{1/3} \, (Re_L^{4/5} - 23{,}550) \text{ valid for } Pr \geq 0.5 \text{ and } 5 \times 10^5 < Re_L < 10^8$$

All the physical properties are to be evaluated at the film temperature.

Heat Transfer in Tube Flow

Basic Terms

$$Nu_D = \frac{hD}{k_f}$$

$$Re_D = \frac{v_m D}{v} \text{ (Note that } Re_{cr} = 2300)$$

For heating

$$h = \frac{+k_f \left.\dfrac{\partial T}{\partial r}\right|_{r=r_0}}{T_s - T_m}$$

$$T_m = \frac{\displaystyle\int_0^{r_o} v_z T r \, dr}{\displaystyle\int_o^{r_o} v_z r \, dr}$$

$$q'' = h(T_s - T_m)$$

For laminar flow

$$\left(\frac{z_{fd,h}}{D}\right)_{lam} = 0.05 \, Re_D$$

$$\left(\frac{z_{fd,t}}{D}\right)_{lam} = 0.05 \, Re_D \, Pr$$

For turbulent flow

$$10 \leq \left(\frac{z_{fd,h}}{D}\right)_{turb} \leq 60$$

$$z_{fd,h} = z_{fd,t}$$

Laminar Flow

Fully Developed Flow and Heat Transfer

Constant Wall Heat Flux (q''_s = constant)

$$Nu_D = 4.364$$

Constant Wall Temperature (T_s = constant)

$$Nu_D = 3.658$$

Slug Flow and Fully Developed Heat Transfer in Liquid Metals ($0.006 \leq Pr \leq 0.03$)
Constant Wall Heat Flux (q''_s = constant)

$$Nu_D = 7.962$$

Constant Wall Temperature (T_s = constant)

$$Nu_D = 5.769$$

Turbulent Flow

Fully Developed Flow and Heat Transfer

Dittus-Boelter Correlation (Valid for $0.7 \leq Pr \leq 120, 2500 \leq Re_D \leq 1.24 \times 10^5$)

Constant Wall Heat Flux and Constant Wall Temperature

$$Nu_D = 0.023 Re_D^{0.8} Pr^n$$

$n = 0.4$ when the fluid is being heated ($T_s > T_m$)

$n = 0.3$ when the fluid is being cooled ($T_s < T_m$)

Liquid Metals ($0.006 \leq Pr \leq 0.03$)

Notter and Sleicher (1972) correlation

Constant Wall Heat Flux

$$Nu_D = 6.3 + 0.0167 Re_D^{0.85} Pr^{0.93}$$

Constant Wall Temperature

$$Nu_D = 4.8 + 0.0156 Re_D^{0.85} Pr^{0.93}$$

All the physical properties are to be calculated at the mean temperature
($(T_{in} + T_{out})/2$).

Review Questions

5.1 What do you mean by velocity and thermal boundary layers on a flat plate?

5.2 What is the difference between laminar and turbulent flow on a flat plate?

5.3 Define Nusselt number and explain its physical significance.

5.4 Define Prandtl number and explain its physical significance.

5.5 In what kind of fluids and flow viscous dissipation may not be neglected?

5.6 What is the advantage of integral analysis over the similarity technique in solving boundary layer problems?

5.7 What is Reynolds-Colburn analogy?

5.8 Define turbulent flow.

5.9 Why are turbulent shear stress and heat transfer much higher than their laminar counterparts?

5.10 Define Eddy diffusivities of momentum and heat.

5.11 What is Prandtl's mixing length hypothesis?

5.12 Can Prandtl's mixing length hypothesis be used to predict Eddy diffusivity of momentum in the outer layer, that is, wake region?

5.13 For $Pr \geq 1$ fluids, are the thicknesses of momentum and thermal boundary layers same or different? Explain.

5.14 What is the importance of viscous sub-layer?

5.15 Define fully developed flow and heat transfer in a tube.

5.16 What do you mean by hydrodynamic and thermal entry lengths?

5.17 For laminar fully developed flow and heat transfer in a tube subject to constant heat flux or constant wall temperature, Nusselt number is constant and is not a function of Reynolds number. Why?

5.18 For $Pr \geq 0.5$ fluids, Nusselt number is virtually independent of the constant heat flux and constant surface temperature in turbulent flow in a tube. Why is this so?

5.19 What is the basic difference in the heat transfer mechanism in liquid metals as compared to other fluids in turbulent flow in a tube?

5.20 Plot qualitatively Nu vs. z covering entrance and fully developed regions for laminar as well as turbulent tube flow for both constant heat flux and constant wall temperature boundary conditions.

Problems

5.1 If the local heat transfer coefficient for the thermal boundary over a flat plate has a power-law dependence on x,

$$h_x = Cx^n$$

where C is a constant, then show that the quantity averaged from $x = 0$ to x (in this case, $h_{x, avg}$) is simply

$$h_{x, avg} = \frac{h_x}{1 + n}$$

Apply this to show that for the laminar thermal boundary layer over a flat plate, the average heat transfer coefficient over a distance x is twice its local value.

5.2 Starting with the Pohlhausen's similarity equation show that for a very low Prandtl number fluid (liquid metal) flowing over a flat plate maintained at constant temperature, the local Nusselt number can be expressed as

$$Nu_x = 0.564 Re_x^{1/2} Pr^{1/2}$$

5.3 Mercury at 70°C flows with a velocity of 0.1 m/s over a 0.4-m-long flat plate maintained at 130°C. Determine the average heat transfer coefficient over the length of the plate. The physical properties of the fluid taken at $T_f = (70 + 130)/2 = 100$°C are
$$k_f = 10.51 \text{ W/m°C}, \quad v = 0.0928 \times 10^{-6} \text{ m}^2\text{/s}, \quad Pr = 0.0162.$$
Use the Nusselt number formula given in Question 5.2.

5.4 Air at 1 atm and 40°C flows with a velocity of 40 m/s past a flat plate 1 m long and 0.5 m wide. The surface is maintained at 300°C.
(a) Should the viscous dissipation be considered? (b) Find the total heat transfer and the total drag force on one side of the plate. (c) What error in part (b) results if the boundary layer is assumed to be turbulent from the leading edge?

5.5 Liquid sodium at 370°C flows at a velocity of 0.6 m/s over a flat plate, 7.5 cm long in the flow direction. The plate temperature is uniform at 200°C. Calculate the average heat flux to the plate.

5.6 Air at atmospheric pressure and 30°C enters a 2.5 cm pipe at 3 m/s. If the pipe wall temperature is maintained at 80°C, how long must the pipe be to increase the bulk temperature of the air to (a) 60°C, (b) 80°C.

5.7 A flow of 45 kg/h molasses at 37°C is pumped through a 5-cm-inner- diameter pipe. After a very long unheated length, the fluid passes through a 1 m-long heated section where the tube wall is maintained at 80°C by condensing steam. Calculate the mean temperature of molasses leaving the heated section. Molasses properties: $\rho = 1120 \text{ kg/m}^3$, $\mu = 0.062$ kg/m s, $c_p = 1.672$ kJ/kg°C, and $k_f = 0.865$ W/m°C.

5.8 Power generation in a nuclear reactor is limited principally by the ability to transfer heat in the reactor. A solid-fuel reactor is cooled by a fluid flowing inside 0.625 cm diameter stainless steel tubes. If the tube wall temperature is 300°C, compare the relative merits of using water or liquid sodium as the coolant. In each case, the velocity is 4.6 m/s and the fluid inlet temperature is 200°C.

5.9 Water flows in a 3-cm-diameter pipe so that $Re_D = 1500$ (laminar flow). The arithmetic mean of the inlet and outlet temperatures is 35°C (the average bulk temperature). (a) Calculate the maximum water velocity in the tube. (b) What would be the heat transfer coefficient for such a system if the tube wall is subjected to a constant heat flux and the velocity and temperature profiles are fully developed? Evaluate properties at the average bulk temperature.

5.10 Wind blows at 0.5 m/s parallel to the short side of a flat roof with the rectangular area 10 m × 20 m. The roof temperature is 45°C, and the air temperature is 30°C. Calculate the total force experienced by the roof. Estimate also the total heat transfer rate by laminar forced convection from the roof to the atmosphere.

5.11 An unspecified fluid with Pr = 17.8 flows parallel to a flat isothermal wall and develops a laminar boundary layer over it. At a certain location x along the wall the local skin friction coefficient is 0.008. Calculate the value of the local Nusselt number at that location.

5.12 Evaluate the average heat transfer coefficient for a flat plate of length L, with laminar and turbulent flow on it. Show that in Pr ≥ 0.5 fluids the average Nusselt number is given by Eq. (5.159).

5.13 Air at 20°C flows with a uniform velocity u_∞ over a thin metallic flat plate in a direction parallel to its width, which is 2 cm. The plate is considerably longer in the direction normal to its thickness; and, therefore, the boundary layer flow that develops is two-dimensional. A temperature sensor mounted on the trailing edge of the plate reads $T_s = 30°C$. The plate is heated volumetrically by an electric current so that the power dissipated by it is 300 W/m². Calculate the stream velocity u_∞.

5.14 Air flows with a velocity of 3.3 m/s over the top surface of a flat iceberg. The air temperature outside the boundary layer is 40°C, while the ice surface temperature is 0°C. The length of the iceberg in the direction of air flow is $L = 100$ m. The latent heat of melting for ice is 333.4 kJ/kg. Calculate (a) the heat flux into the iceberg averaged over length L; (b) the rate of melting caused by this heat flux (in mm/h).

5.15 Water flows with a mean velocity of 6 cm/s through a 2.5-cm diameter pipe whose wall is kept at a constant temperature of 10°C. The inlet water temperature is 30°C. The total length of the pipe is $L = 20$ m. Calculate the exit temperature of the fluid by assuming that the flow is fully developed. The water properties are to be evaluated at the film temperature at the inlet [that is, (30 + 10)/2 = 20°C]. Verify that the flow is in the laminar regime.

5.16 A stream of water is heated with constant heat flux in a pipe with a diameter of 2 cm and length of 3 m. The stream enters the pipe with a mean velocity of 10 cm/s and a temperature of 10°C. The mean temperature at the pipe exit is 20°C. The water properties can be evaluated at the average bulk temperature, that is, $(T_{inlet} + T_{exit})/2$.
   (a) Calculate the thermal entry length and compare it with the length of the pipe. Is the stream fully developed at the pipe exit?
   (b) Estimate the local Nusselt number at $z = 3$ m by assuming that both the temperature profile and the velocity profile are fully developed.

(c) Calculate the wall heat flux and local wall temperature at the pipe outlet.

5.17   Water flows with a mean velocity of 10 cm/s through a copper tube of inner diameter $D = 0.5$ cm and total length $L = 4$ m. The tube wall is isothermal with a temperature of 30 °C. The inlet temperature of the tube is 20 °C. The physical properties of water can be evaluated at 25 °C.

(a) Verify that the flow is hydrodynamically fully developed over most of the tube length.

(b) Verify also that the flow is thermally fully developed along most of the tube.

(c) Calculate the mean outlet temperature of the water stream, and the total heat transfer rate to the stream from the tube wall.

5.18   A highly viscous fluid is forced through a straight circular pipe of inner radius r_0. Due to viscous heat generation the fluid tends to warm up as it flows through the pipe. To offset this effect, cooling is provided all along the pipe wall, which is isothermal (T_s = constant). The flow is hydro-dynamically and thermally fully developed. The energy equation for the fluid with constant properties reduces in this case to

$$k \frac{1}{r} \frac{d}{dr} \left(r \frac{dT}{dr} \right) + \mu \phi = 0$$

where ϕ is the viscous dissipation term: $[\phi = (dv_z/dr)^2]$, where $v_z(r)$ is the fully developed velocity profile.

(a) Determine the temperature distribution $T(r)$ in the fluid.

(b) Calculate the total heat transfer rate.

5.19   Water flows at a rate of 0.5 kg/s through a 10-m-long pipe with an inner diameter of 2 cm. It is being heated with a constant wall heat flux of 50 kW/m². The water properties can be evaluated at 20 °C. Assume that the velocity and temperature fields are fully developed. Calculate:

(a) The heat transfer coefficient based on the Dittus–Boelter correlation, Eq. (5.231).

(b) The difference between the wall temperature and the local mean water temperature.

(c) The temperature rise of water from the inlet to the outlet.

Natural Convection Heat Transfer

6.1 Introduction

Natural or free convection flow arises when a heated object is placed in a quiescent fluid (i.e., a fluid at rest), the density of which varies with temperature. The density in the neighbourhood of the surface of the object decreases, which in a normal fluid is associated with a temperature increase. This causes the layers near the surface to rise, causing the surrounding cold fluid to take its place. This, in turn, creates a free convection flow which now transports the heat away from the object. The cause of such a flow is a body force, which in this case is the gravitational force. Basically, in free convection, fluid motion is due to buoyancy effects. Buoyancy force is due to the combined presence of a density gradient within the fluid and a body force which is proportional to fluid density.

Free convection flows have been studied most extensively because they are found frequently in nature as well as in engineering applications. Flows can be caused by other body forces, such as, centrifugal,[1] Coriolis,[2] electric, magnetic forces. The flow of cooling air through passages in the rotating blades of gas turbines is an example of flow under the influence of centrifugal and Coriolis forces. In hypersonic flights (Mach number > 5) of missiles, the surrounding air temperatures may become so high that the air is ionized, which means that the atoms and molecules carry electrical charges. In this case electric or magnetic forces may arise which influence the flow as body forces.

Typical applications of natural convection are as follows: (i) heat transfer from pipes and transmission lines as well as from various electronic devices; (ii) dissipation of heat from the coil of a refrigeration unit to the surrounding air; (iii) heat transfer from a heater to room air; (iv) heat transfer from nuclear fuel rods to the surrounding coolant; (v) heated and cooled enclosures; (vi) quenching, wire-drawing, and extrusion; (vii) atmospheric and oceanic circulation.

In this chapter, we deal with only the simplest free convection solutions such as vertical plate geometry. For others, only empirical correlations will be provided.

---

[1] The term $\rho v_\theta^2/r$ is the centrifugal force. It gives the effective force in the r-direction resulting from fluid motion in the θ-direction.

[2] The term $\rho v_r v_\theta/r$ is the Coriolis force. It is an effective force in the θ-direction when there is flow both in the r- and θ-directions.

6.1.1 Physical Mechanism of Natural Convection

Figure 6.1(a) shows a cup of hot tea (whose typical temperature is 70°C) exposed to cold air at 25°C. The temperature of the air adjacent to the cup is higher, and thus its density decreases. So, light air is surrounded by high-density or heavier air, and according to *Archimedes' law* (see the review below) the light air rises. The space vacated by the warmer air in the proximity of the hot tea is replaced by the neighbouring cooler air which then removes the heat. The process of rising of hot air and its replenishment by cold air continues till the cup of tea is cooled to the temperature of the ambient, that is, 25°C. This phenomenon is called natural convection.

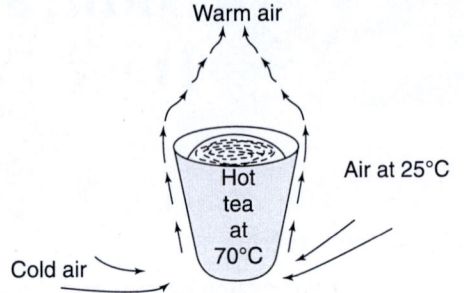

Fig. 6.1(a)   Cooling of hot tea

Reversed natural convection also takes place when a cold object is placed in a hot environment. Figure 6.1(b) depicts a cold drink bottle at 10°C being warmed by air at 25°C. Note the opposite direction of fluid motion as compared to that in Fig. 6.1(a).

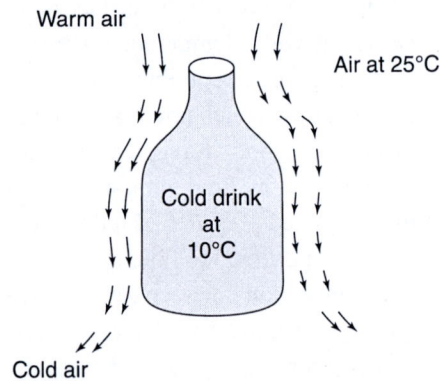

Fig. 6.1(b)   Warming up of cold drink

Review of Archimedes' Law

The upward force exerted by a fluid on a body completely or partially submerged in it is called the buoyancy force. Archimedes' law states that the magnitude of the buoyancy force is equal to the weight of the fluid displaced by the body. That is,

$$F_{\text{buoyancy}} = \rho_{\text{fluid}}\, g V_{\text{body}} \qquad (6.1)$$

where V_{body} = Volume of the portion of the body immersed in the fluid (for bodies totally immersed in the fluid, the full volume of the body must be considered) Thus,

$$F_{\text{net}} = F_{\text{buoyancy}} - W = \rho_{\text{fluid}}\, g V_{\text{body}} - \rho_{\text{body}}\, g V_{\text{body}} = (\rho_{\text{fluid}} - \rho_{\text{body}})\, g V_{\text{body}} \quad (6.2)$$

where W is the weight of the body.

From Eqs (6.1) and (6.2) it is clear that without gravity there can be no buoyancy and hence, no natural convection.

6.2   Free Convection from a Vertical Plate

Let us now consider the natural convection heat transfer from a heated vertical plate placed in an extensive quiescent medium (Fig. 6.2). Since $T_s > T_\infty$, the fluid adjacent to the vertical surface gets heated, becomes lighter, and rises. The fluid from the neighbouring areas rushes in to take the place of this rising fluid. Eventually, a flow of the form shown in Fig. 6.2, known as boundary layer flow, develops adjacent to the vertical surface. The analysis and study of such a flow give the desired information on heat transfer rates, flow field, temperature field, etc.

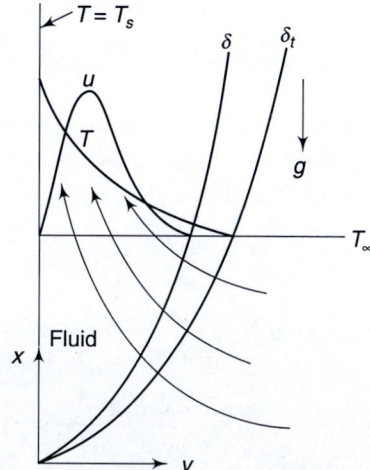

The u-velocity is positive and increases upwards. The v-velocity is negative as the fluid is drawn towards the plate and decreases upwards. The u-velocity is zero at the plate and at the edge of the boundary layer (contrast this with forced convection where the velocity is maximum at the edge of the boundary layer). The u-velocity reaches a maximum in the boundary layer. The temperature decreases monotonically as shown in Fig. 6.2.

Fig. 6.2 Momentum and thermal boundary layers for natural convection on a vertical plate

6.2.1 Analysis

To analyse the boundary layer flow shown in Fig. 6.2, we make the following assumptions.

1. The flow is steady, laminar, and two-dimensional.
2. The temperature difference between the plate and the fluid is small to moderate, in which case the fluid may be treated as having constant properties.
3. Also, with one exception, the fluid is incompressible. The exception involves accounting for the effect of variable density in the buoyancy force, since it is this variation that induces fluid motion. This is known as the Boussinesq[3] approximation.
4. Boundary layer approximations are valid.

6.2.2 Governing Equations

The x-momentum equation reduces to

$$u\frac{\partial u}{\partial x} + v\frac{\partial u}{\partial y} = -\frac{1}{\rho}\frac{\partial p}{\partial x} + v\frac{\partial^2 u}{\partial y^2} + \frac{X}{\rho} \qquad (6.3)$$

The body force per unit volume is

$$X = -\rho g \qquad (6.4)$$

Therefore, Eq. (6.3) becomes

$$u\frac{\partial u}{\partial x} + v\frac{\partial u}{\partial y} = -\frac{1}{\rho}\frac{\partial p}{\partial x} - g + v\frac{\partial^2 u}{\partial y^2} \qquad (6.5)$$

---

[3] Joseph Boussinesq (1842–1929) was Professor of Physics and Experimental Mechanics at the Faculty of Sciences in Paris. Apart from his contributions to the theory of elasticity and magnetism in 1877, he pioneered the use of the time-averaged Navier–Stokes equations in the analytical study of turbulent flows.

Invoking boundary layer approximations, we can write

$$\frac{\partial p}{\partial y} = 0 \tag{6.6}$$

Furthermore, the x-pressure gradient at any point in the boundary layer must be equal to the pressure gradient in the quiescent region outside the boundary layer. However, in this region, $u = v = 0$. Therefore,

$$\frac{\partial p}{\partial x} = -\rho_\infty g \tag{6.7}$$

Hence, $\quad u\frac{\partial u}{\partial x} + v\frac{\partial u}{\partial y} = -\frac{1}{\rho}(-\rho_\infty g) - g + v\frac{\partial^2 u}{\partial y^2}$

or $\quad u\frac{\partial u}{\partial x} + v\frac{\partial u}{\partial y} = \frac{g}{\rho}(\rho_\infty - \rho) + v\frac{\partial^2 u}{\partial y^2} \tag{6.8}$

Equation (6.8) must apply at every point in the momentum boundary layer. The first term on the right-hand side of Eq. (6.8) is the buoyancy force, and the flow originates because the density ρ is a variable. This is related to the temperature difference by using the definition of the volumetric thermal expansion coefficient β, which is

$$\beta = -\frac{1}{\rho}\left(\frac{\partial \rho}{\partial T}\right)_p \tag{6.9}$$

This thermodynamic property of the fluid provides a measure of the amount by which the density changes in response to a change in temperature at constant pressure. It is expressed in the following approximate form

$$\beta = -\frac{1}{\rho}\left(\frac{\rho_\infty - \rho}{T_\infty - T}\right) \tag{6.10}$$

p is the reference pressure level, that is, the pressure at the bottom of the plate. For most natural convection flows, the pressure correction is not required.

Substituting Eq. (6.10) into Eq. (6.8), we obtain

$$\underbrace{u\frac{\partial u}{\partial x} + v\frac{\partial u}{\partial y}}_{\text{inertia}} = \underbrace{g\beta(T - T_\infty)}_{\text{buoyancy}} + \underbrace{v\frac{\partial^2 u}{\partial y^2}}_{\text{friction}} \tag{6.11}$$

The presence of temperature in the buoyancy term of the momentum Eq. (6.11) couples the flow to the temperature field and vice versa. The overall mass and energy conservation equations remain unchanged. So, finally the set of governing equations may be expressed as follows:

Continuity: $\quad \dfrac{\partial u}{\partial x} + \dfrac{\partial v}{\partial y} = 0 \tag{6.12}$

x-momentum: $\quad u\dfrac{\partial u}{\partial x} + v\dfrac{\partial u}{\partial y} = g\beta(T - T_\infty) + v\dfrac{\partial^2 u}{\partial y^2} \tag{6.13}$

Energy: $\quad u\dfrac{\partial T}{\partial x} + v\dfrac{\partial T}{\partial y} = \alpha\dfrac{\partial^2 T}{\partial y^2} \tag{6.14}$

Note that the viscous dissipation term has been neglected in the energy equation because of the small velocities associated with free convection. Free convection effects obviously depend on the expansion coefficient β. The manner in which β is obtained depends on the nature of the fluid. For a perfect gas, $\rho = p/RT$ and hence $\beta = -1/\rho \, (\partial \rho / \partial T)_p = 1/T$ where T is the absolute temperature (in kelvin). T is taken as $(T_s + T_\infty)/2$, the film temperature.

6.2.3 Non-dimensionalization

Using the following dimensionless parameters, let us now non-dimensionalize the momentum and energy equations:

$$x^* = \frac{x}{L} \tag{6.15a}$$

$$y^* = \frac{y}{L} \tag{6.15b}$$

$$u^* = \frac{u}{u_0} \tag{6.15c}$$

$$v^* = \frac{v}{u_0} \tag{6.15d}$$

$$T^* = \frac{T - T_\infty}{T_s - T_\infty} \tag{6.15e}$$

L is a characteristic length (in this case, the length of the plate) and u_0 is an arbitrary reference velocity.

$$x\text{-momentum: } u^* \frac{\partial u^*}{\partial x^*} + v^* \frac{\partial u^*}{\partial y^*} = \frac{g\beta(T_s - T_\infty)\, L}{u_0^2} T^* + \frac{1}{\mathrm{Re}_L} \frac{\partial^2 u^*}{\partial y^{*2}} \tag{6.16}$$

$$\text{Energy: } u^* \frac{\partial T^*}{\partial x^*} + v^* \frac{\partial T^*}{\partial y^*} = \frac{1}{\mathrm{Re}_L \, \mathrm{Pr}} \frac{\partial^2 T^*}{\partial y^{*2}} \tag{6.17}$$

The dimensionless parameter in the first term on the right-hand side of the x-momentum equation is a direct fallout of the buoyancy force. The coefficient of T^* in Eq. (6.16) can be expressed as

$$\frac{g\beta(T_s - T_\infty)L}{u_0^2} = \frac{\mathrm{Gr}_L}{\mathrm{Re}_L^2}$$

where $\mathrm{Re}_L = u_0 L/v$ is the Reynolds number based on the arbitrary reference velocity and

$$\mathrm{Gr}_L = \frac{g\beta(T_s - T_\infty)L^3}{v^2} \text{ is the Grashof[4] number}$$

$$\mathrm{Re} = \text{inertia force/viscous force}$$

$$\mathrm{Gr} = \text{buoyancy force/viscous force}$$

---

[4] Franz Grashof (1826–1893) was Professor of Mechanical Engineering at the University of Karlsruhe, Germany.

Gr_L/Re_L^2 basically represents the ratio of buoyancy force to inertia force. There-fore, in a flow where both forced and free convection effects are significant (mixed convection),

$$Nu_L = f(Re_L, Gr_L, Pr)$$

$$Nu_L = f(Re_L, Gr_L, Pr) \Rightarrow \frac{Gr_L}{Re_L^2} = 1 \text{ (mixed convection)}$$

$$Nu_L = f(Re_L, Pr) \Rightarrow \frac{Gr_L}{Re_L^2} \ll 1 \text{ (forced convection)}$$

$$Nu_L = f(Gr_L, Pr) \Rightarrow \frac{Gr_L}{Re_L^2} \gg 1 \text{ (free convection)}$$

6.2.4 Genesis of the Physical Meaning of Gr, Re, and Gr/Re² from Dimensional Analysis

The inertia force on a surface of a cube of fluid having side L and density ρ in a stream of density ρ_∞ which is flowing at normal velocity V can be approximately expressed as:

$$F_i \sim ma = \rho L^3 \frac{V}{t} = \rho L^3 \frac{V}{L/V} = \rho V^2 L^2$$

The net buoyancy force on the cube is [see Eq.(6.2)]

$$F_b \sim g\Delta\rho L^3$$

where $\Delta\rho = \rho_\infty - \rho$

The viscous force on a side L oriented parallel to the stream is

$$F_\mu \sim \tau A = \mu \frac{du}{dy} A = \mu \frac{V}{L} L^2 = \mu VL$$

Now,

Reynolds number = Inertia Force/Viscous Force = $\dfrac{F_i}{F_\mu} = \dfrac{\rho V^2 L^2}{\mu VL} = \dfrac{\rho VL}{\mu} = \dfrac{VL}{v}$

where $v = \dfrac{\mu}{\rho}$

Grashof number = Buoyancy force/Viscous force

$$= \frac{F_b}{F_\mu} = \frac{g\Delta\rho L^3}{\mu VL} = \frac{g\Delta\rho L^3}{\rho v VL} = \frac{g \dfrac{\Delta\rho}{\rho} L^3}{v^2}$$

(Note that since both v and VL have the unit of m²/s, the product vVL is written as v^2 in the denominator of the above expression of Grashof number.)

Thus, $\dfrac{F_b}{F_i}$ = Buoyancy force/Inertia force = $\dfrac{g\Delta\rho L^3}{\rho V^2 L^2} = \dfrac{g \dfrac{\Delta\rho}{\rho} L^3/v^2}{\left(\dfrac{VL}{v}\right)^2} = \dfrac{Gr}{Re^2}$

6.3 Flow Regimes in Free Convection over a Vertical Plate

It is well known that when Navier–Stokes equations are non-dimensionalized, a length Reynolds number appears as a dimensionless parameter. For very low Reynolds number (Re $\rightarrow$ 0) the inertia force is much smaller than viscous force and the flow is termed "Creeping flow". At moderate Reynolds number, the flow is still laminar but both inertia and viscous forces are important. At high Re, the flow is turbulent. Similar flow regimes are also observed in free convection. Then the question arises: what is the length of Reynolds number for free convection over a vertical plate? Note that to define a Reynolds number a characteristic velocity is needed to describe the inertia force within the boundary layer. Following Kays and Crawford (1993), consider the region of the boundary layer where the velocity reaches its maximum. This is one-third of the total boundary layer thickness (shown later in the integral analysis). Neglecting the effect of momentum diffusion $\left(v\dfrac{\partial^2 u}{\partial y^2} \right)$ and cross-stream momentum convection $\left(v\dfrac{\partial u}{\partial y} \right)$ in this region and assuming that T is more or less equal to T_s, the x-momentum equation [Eq. (6.13)] reduces to

$$u\frac{du}{dx} = g\beta(T_s - T_\infty) \tag{6.18}$$

The LHS of Eq. (6.18) is convective acceleration term and the RHS is the buoyancy force term. Integrating Eq. (6.18),

$$\int_0^u u\,du = g\beta(T_s - T_\infty)\int_0^x dx$$

$$\Rightarrow \qquad \frac{u^2}{2} = g\beta(T_s - T_\infty)x$$

$$\Rightarrow \qquad u \propto \sqrt{g\beta(T_s - T_\infty)x}$$

A Reynolds number for this flow then can be written as

$$\text{Re}_x = \frac{ux}{v} = \frac{\sqrt{g\beta(T_s - T_\infty)x}\,x}{v} = \sqrt{\frac{g\beta(T_s - T_\infty)x^3}{v^2}}$$

$$\Rightarrow \qquad \text{Re}_x = \text{Gr}_x^{1/2}$$

From this approximate analysis, it is clearly seen that the Grashof number in free convection from vertical flat surfaces plays the same role as the length Reynolds number in forced convection.

It may be noted that creeping flow is found below $\text{Gr}_x = 10^4$. In the creeping flow regime boundary layer assumptions are no longer valid. Transition from laminar to turbulent flow occurs at $\text{Gr}_x = 10^9$.

6.4 Basic Solution Methodology

Since the momentum and energy equations are coupled through the buoyancy force term, natural convection flows are generally more difficult to solve as compared to the corresponding forced flow circumstances. The basic solution methodology is to solve continuity, momentum, and energy equations simultaneously to obtain the velocity and temperature fields and hence heat transfer.

6.4.1 Similarity Solution

Equations (6.12)–(6.14) are solved subject to the following boundary conditions:

$$y = 0, \ u = v = 0, \ T = T_s \tag{6.19a}$$

$$y \to \infty, \ u \to 0, \ T \to T_\infty \tag{6.19b}$$

A similarity solution has been obtained by Ostrach (1952) using the similarity parameter η, non-dimensional stream function $f(\eta)$, and non-dimensional temperature T^*, which are defined as follows:

$$\eta = \frac{y}{x}\left(\frac{Gr_x}{4}\right)^{1/4} \tag{6.20}$$

$$\psi(x, y) = f(\eta)\left[4v\left(\frac{Gr_x}{4}\right)^{1/4}\right] \tag{6.21}$$

$$T^* = \frac{T - T_\infty}{T_s - T_\infty} \tag{6.22}$$

The transformed x-momentum equation [Eq. (6.13)] and energy equation [Eq. (6.14)] are

$$x\text{-momentum:} \quad f''' + 3ff'' - 2(f')^2 + T^* = 0 \tag{6.23}$$

$$\text{Energy:} \qquad T^{*''} + 3\text{Pr} \, f T^{*'} = 0 \tag{6.24}$$

The continuity equation is identically satisfied. The transformed boundary conditions are

$$\eta = 0, \ f = f' = 0, \ T^* = 1 \tag{6.25}$$

$$\eta = \infty, f' \to 0, \ T^* \to 0 \tag{6.26}$$

The coupled ordinary differential equations were solved numerically by Ostrach (1952).

Figure 6.3 shows the resulting similarity profiles for the vertical velocity and temperature near the isothermal wall. Figure 6.3(b) may also be used to obtain the appropriate from of the heat transfer correlation.

A low Prandtl number (liquid metals) implies that viscous effects are small, and a high Prandtl number (heavy oils) implies that inertia effects are small (creeping motion). This is also revealed in Fig. 6.3(a), which shows a decrease of the maximum vertical velocity with an increase of the Prandtl number. Figure 6.3(b) indicates that the thermal boundary layer thickness decreases as the Prandtl number increases, implying higher heat transfer. It is also seen that for $\text{Pr} \leq 1$, $\delta_t \simeq \delta$ and for $\text{Pr} > 1$, $\delta_t < \delta$.

Now, to calculate the local Nusselt number, we have to evaluate h, which is expressed as

$$h = \frac{-k_f \left.\dfrac{\partial T}{\partial y}\right|_{y=0}}{T_s - T_\infty} \tag{6.27}$$

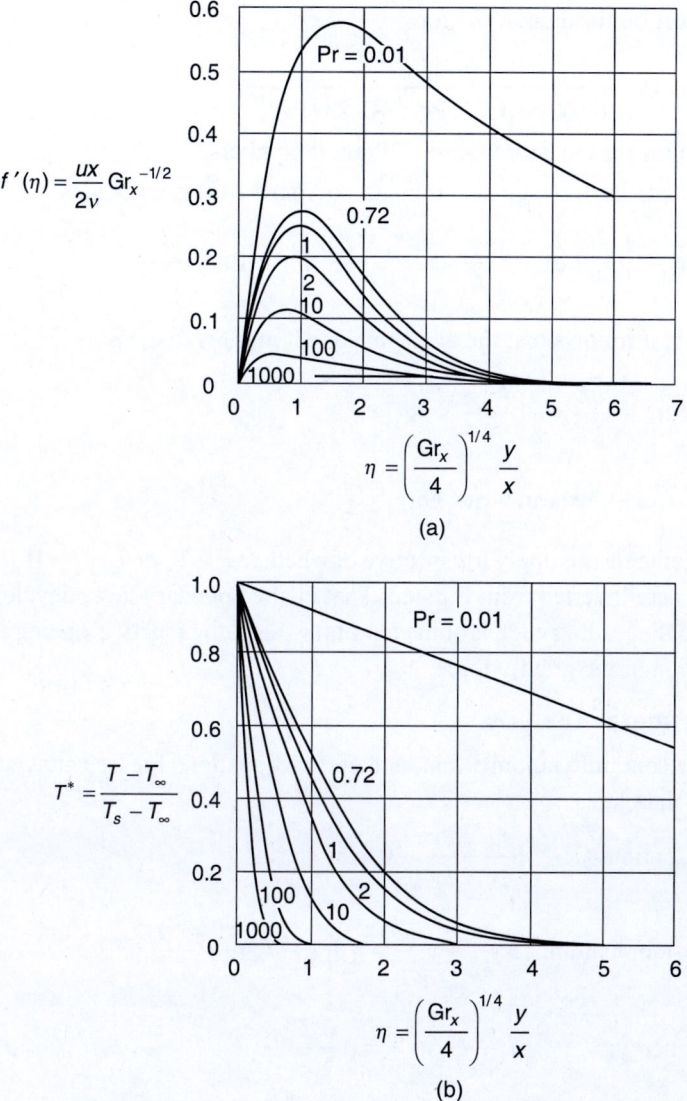

Fig. 6.3(a,b) Laminar, free convection boundary layer conditions on an isothermal vertical surface for various Prandtl number fluids: (a) vertical velocity profiles, (b) temperature profiles

In terms of dimensionless variables,

$$-k_f \frac{\partial T}{\partial y}\bigg|_{y=0} = -\frac{k_f}{x}(T_s - T_\infty)\left(\frac{Gr_x}{4}\right)^{1/4} \frac{\partial T^*}{\partial \eta}\bigg|_{\eta=0} \qquad (6.28)$$

Using Eqs (6.27) and (6.28), we obtain

$$Nu_x = \frac{hx}{k_f} = -\left(\frac{Gr_x}{4}\right)^{1/4} \frac{\partial T^*}{\partial \eta}\bigg|_{\eta=0} = \left(\frac{Gr_x}{4}\right)^{1/4} g(Pr) \qquad (6.29)$$

$g(\mathrm{Pr})$ can be obtained by using Fig. 6.3(b). The results have been correlated with an interpolation formula of the form

$$g(\mathrm{Pr}) = \frac{0.75\,\mathrm{Pr}^{1/2}}{(0.609 + 1.221\,\mathrm{Pr}^{1/2} + 1.238\,\mathrm{Pr})^{1/4}} \tag{6.30}$$

which applies for the entire range of Prandtl numbers.

To evaluate the average heat transfer coefficient, $\bar{h}$ is written as

$$\bar{h} = \frac{1}{L}\int_0^L h\,dx = \frac{k_f}{L}\left[\frac{g\beta(T_s - T_\infty)}{4\nu^2}\right]^{1/4} g(\mathrm{Pr})\int_0^L \frac{dx}{x^{1/4}}$$

Integrating, it follows that the average Nusselt number $\overline{\mathrm{Nu}}_L$ is

$$\overline{\mathrm{Nu}}_L = \frac{\bar{h}L}{k_f} = \frac{4}{3}\left(\frac{\mathrm{Gr}_L}{4}\right)^{1/4} g(\mathrm{Pr})$$

Using Eq. (6.29), we can write $\overline{\mathrm{Nu}}_L = \dfrac{4}{3}\mathrm{Nu}_L$. $\tag{6.31}$

The foregoing results apply irrespective of whether $T_s > T_\infty$ or $T_s < T_\infty$. If $T_s < T_\infty$, the conditions are inverted from Fig. 6.2. That is, the boundary layer develops downwards and the leading edge is at the top of the plate, the positive x being defined in the direction of the gravity force.

6.4.2 Integral Analysis

Recall the continuity, momentum, and energy equations for free convection over a vertical flat plate.

$$\text{Continuity}: \quad \frac{\partial u}{\partial x} + \frac{\partial v}{\partial y} = 0 \tag{6.32}$$

$$x\text{-momentum}: \quad u\frac{\partial u}{\partial x} + v\frac{\partial u}{\partial y} = g\beta(T - T_\infty) + \nu\frac{\partial^2 u}{\partial y^2} \tag{6.33}$$

$$\text{Energy}: \quad u\frac{\partial T}{\partial x} + v\frac{\partial T}{\partial y} = \alpha\frac{\partial^2 T}{\partial y^2} \tag{6.34}$$

As in the forced convection case, integration is carried out over the momentum and thermal boundary layer thicknesses to obtain the following.
Integral momentum:

$$\frac{d}{dx}\left[\int_0^\delta \rho u^2\,dy\right] = -\tau_s + \int_0^\delta \rho g\beta(T - T_\infty)\,dy$$

$$= -\mu\frac{\partial u}{\partial y}\bigg|_{y=0} + \int_0^\delta \rho g\beta(T - T_\infty)\,dy \tag{6.35}$$

Integral energy:

$$\frac{d}{dx}\left[\int_0^{\delta_t} u(T - T_\infty)\,dy\right] = -\alpha\frac{\partial T}{\partial y}\bigg|_{y=0} \tag{6.36}$$

Note that the functional form of both velocity and temperature distributions must be known in order to arrive at the solution.

The following conditions apply for the temperature distribution:

$$T = a_0 + a_1 y + a_2 y^2 \tag{6.37}$$

with boundary and compatibility conditions

$$y = 0, \quad T = T_s \tag{6.38a}$$

$$y = \delta_t, \quad T = T_\infty \tag{6.38b}$$

$$y = \delta_t, \quad \frac{\partial T}{\partial y} = 0 \tag{6.38c}$$

so that we obtain for the temperature distribution,

$$\frac{T - T_\infty}{T_s - T_\infty} = \left(1 - \frac{y}{\delta_t}\right)^2 \tag{6.39}$$

The following conditions apply for the vertical velocity profile:

$$\frac{u}{u_0} = C_0 + C_1 y + C_2 y^2 + C_3 y^3 \tag{6.40}$$

where u_0 is a fictitious velocity, which is a function of x. The boundary and compatibility conditions are:

$$y = 0, \; u = 0 \tag{6.41a}$$

$$y = \delta, \; u = 0 \tag{6.41b}$$

$$y = \delta, \; \frac{\partial u}{\partial y} = 0 \tag{6.41c}$$

$$y = 0, \; \frac{\partial^2 u}{\partial y^2} = -g\beta\left(\frac{T_s - T_\infty}{\nu}\right) \tag{6.41d}$$

The velocity profile is

$$\frac{u}{u_0} = \frac{\beta \delta^2 g(T_s - T_\infty)}{4 u_0 \nu} \frac{y}{\delta}\left(1 - \frac{y}{\delta}\right)^2 \tag{6.42}$$

The term involving the temperature difference $(T_s - T_\infty)$, δ^2, and u_0 may be incorporated into the function u_0 so that the final relation to be assumed for the velocity profile is

$$\frac{u}{u_0} = \frac{y}{\delta}\left(1 - \frac{y}{\delta}\right)^2 \tag{6.43}$$

u_{max} occurs according to Eq. (6.43) at a distance $y = \delta/3$ from the wall, and $u_{max} = (4/27)u_0$. Substituting Eqs (6.39) and (6.43) into Eq. (6.35) and carrying out the integrations and differentiations yields

$$\frac{1}{105} \frac{d}{dx}(u_0^2 \delta) = \frac{1}{3} g\beta(T_s - T_\infty)\delta - \nu\frac{u_0}{\delta} \tag{6.44}$$

Substituting Eqs (6.39) and (6.43) into Eq. (6.36) and carrying out the necessary operations,

$$\frac{1}{30}(T_s - T_\infty)\frac{d}{dx}(u_0 \delta_t) = 2\alpha\frac{(T_s - T_\infty)}{\delta_t} \tag{6.45}$$

We assume that $\delta = \delta_t$. This assumption is justified because of the small computational work it involves and because the results of the calculation performed with this value agree quite well with the experimental results of Touloukian et al. (1948). Sparrow and Gregg (1956) performed an analysis for $\delta \neq \delta_t$. The results deviated only moderately from those presented here. Therefore, Eq. (6.45) becomes

$$\frac{1}{30}(T_s - T_\infty)\frac{d}{dx}(u_0\delta) = 2\alpha\frac{(T_s - T_\infty)}{\delta} \tag{6.46}$$

We assume solutions of the form

$$u_0 = Ax^m \tag{6.47}$$
$$\delta = Bx^n \tag{6.48}$$

Substituting Eqs (6.47) and (6.48) into Eqs (6.44) and (6.46), we get

$$\frac{1}{60}AB(m+n)x^{m+n-1} = \frac{\alpha}{B}x^{-n} \tag{6.49}$$

and $\quad \dfrac{1}{105}A^2B(2m+n)x^{2m+n-1} = \dfrac{1}{3}g\beta(T_s - T_\infty)Bx^n - v\dfrac{A}{B}x^{m-n}$ (6.50a)

or $\quad \dfrac{1}{105}A^2B(2m+n)x^{2m+n-1} = x^n\left[\dfrac{1}{3}g\beta(T_s - T_\infty)B - v\dfrac{A}{B}x^{m-2n}\right]$ (6.50b)

Since Eqs (6.49) and (6.50a) are satisfied for all values of x, the exponents of the individual terms in each must be equal. Therefore, from Eq. (6.50a),

$$2m + n - 1 = n = m - n \tag{6.51}$$

and, from Eq. (6.49),

$$m + n - 1 = -n \tag{6.52}$$

Hence, solving Eqs (6.51) and (6.52), we get

$$m = \frac{1}{2}$$

$$n = \frac{1}{4}$$

Substituting the values of m and n into Eqs (6.49) and (6.50b), we get

$$\frac{1}{80}AB^2 = \alpha \tag{6.53}$$

and $\quad \dfrac{5}{420}A^2B = \dfrac{1}{3}g\beta(T_s - T_\infty)B - v\dfrac{A}{B}$ (6.54)

Solving Eqs (6.53) and (6.54) simultaneously,

$$A = 5.17v\left[\frac{20}{21} + \frac{v}{\alpha}\right]^{-1/2}\left(\frac{g\beta(T_s - T_\infty)}{v^2}\right)^{1/2} \tag{6.55}$$

and $\quad B = 3.93\left(\dfrac{\alpha}{v}\right)^{1/2}\left[\dfrac{20}{21} + \dfrac{v}{\alpha}\right]^{1/4}\left(\dfrac{g\beta(T_s - T_\infty)}{v^2}\right)^{-1/4}$ (6.56)

Putting B and n into Eq. (6.48), the expression for the boundary layer thickness is obtained as

$$\frac{\delta}{x} = 3.93\,\mathrm{Pr}^{-1/2}(0.952 + \mathrm{Pr})^{1/4}\,\mathrm{Gr}_x^{-1/4} \tag{6.57}$$

The heat transfer rate is calculated as

$$\frac{q}{A} = -k_f \left(\frac{\partial T}{\partial y}\right)_{y=0} = hA(T_s - T_\infty)$$

Using $(T - T_\infty)/(T_s - T_\infty) = (1 - y/\delta)^2$, the expression for h is

$$h = \frac{2k_f}{\delta}$$

or

$$\frac{hx}{k_f} = \mathrm{Nu}_x = \frac{2x}{\delta} \tag{6.58}$$

Substituting Eq. (6.57) into Eq. (6.58),

$$\mathrm{Nu}_x = 0.508\,\mathrm{Pr}^{1/2}(0.952 + \mathrm{Pr})^{-1/4}\,\mathrm{Gr}_x^{1/4} \tag{6.59}$$

The above agrees well with Nu_x obtained from the similarity solution:

$$\overline{\mathrm{Nu}_L} = \frac{4}{3}\mathrm{Nu}_L \tag{6.60}$$

For air, with $\mathrm{Pr} = 0.714$,

$$\mathrm{Nu}_x = 0.378(\mathrm{Gr}_x)^{1/4}$$

which agrees well with the experimental result where the coefficient of $(\mathrm{Gr}_x)^{1/4}$ is 0.360 instead.

6.4.3 Turbulent Processes

The laminar boundary layer flow over a flat plate may turn into turbulent flow if the Rayleigh[5] number $(\mathrm{Gr}_x\mathrm{Pr})$ exceeds a certain critical value. For a long time, it was thought that the critical value of the Rayleigh number is 10^9, regardless of the value of the Prandtl number. This established view was challenged by Bejan and Lage (1990) who showed that it is the Grashof number that should be used to determine whether the free convection flow is laminar or turbulent. Therefore,

$$\mathrm{Gr}_{x_{\text{critical}}} = 10^9 \tag{6.61}$$

The universal transition criterion can be expressed also in terms of the Rayleigh number:

$$\mathrm{Ra}_{x_{\text{critical}}} = 10^9\,\mathrm{Pr} \tag{6.62}$$

It is supported very well by experimental observations. The traditional Rayleigh number criterion $(\mathrm{Ra}_x \sim 10^9)$ is true for fluids whose Prandtl numbers are of order 1 (e.g., air). However, for liquid metals $(\mathrm{Pr} \sim 10^{-3}\text{–}10^{-2})$, the Grashof number criterion means that the actual critical Rayleigh number is of order $10^6\text{–}10^7$, which is well below the often-mentioned critical Rayleigh number of 10^9.

When $\mathrm{Ra}_x > 10^9\mathrm{Pr}$, the Nusselt number can be calculated with the help of the following empirical correlation of Churchill and Chu (1975) for constant surface temperature conditions:

---

[5] Lord Rayleigh (John William Strutt, 1842–1919) was Professor of Natural Philosophy at the Royal Institution of Great Britain. He is known for his work on hydrodynamic stability and cellular convection in a fluid layer heated from below (Benard convection).

$$\overline{\mathrm{Nu}}_L = \left\{ 0.825 + \frac{0.387\,\mathrm{Ra}_L^{1/6}}{\left[1+\left(\dfrac{0.492}{\mathrm{Pr}}\right)^{9/16}\right]^{8/27}} \right\}^2 \tag{6.63}$$

This correlation is valid for $10^{-1} < \mathrm{Ra}_x < 10^{12}$ and for all Prandtl numbers. Equation (6.63) can also be applied for a constant wall flux with 0.492 replaced by 0.437. For the case of constant wall flux, Ra_L is based on the L-averaged temperature difference, namely, $\overline{T}_s - T_\infty$. The physical properties used in the definition of $\overline{\mathrm{Nu}}_L$, Ra_L, and Pr are evaluated at the film temperature $[(T_s + T_\infty)/2]$.

In the laminar range, $\mathrm{Gr}_x < 10^9$, a correlation that represents the experimental data more accurately is another correlation by Churchill and Chu (1975) for both constant wall flux and constant surface temperature:

$$\overline{\mathrm{Nu}}_L = 0.68 + \frac{0.67\,\mathrm{Ra}_L^{1/4}}{\left[1+\left(\dfrac{0.492}{\mathrm{Pr}}\right)^{9/16}\right]^{4/9}} \tag{6.64}$$

Equations (6.63) and (6.64) are alternatives to the Nusselt number relations developed using similarity and integral analyses when the Rayleigh number is low (i.e., when boundary layer approximations are no longer valid).

The integral analysis of turbulent free convection by Eckert and Jackson (1950) gives the following local Nusselt number expression:

$$\mathrm{Nu}_x = 0.0295 \left[\frac{\mathrm{Pr}^7}{(1+0.494\,\mathrm{Pr}^{2/3})^6} \right]^{1/5} \mathrm{Gr}_x^{2/5} \tag{6.65}$$

The average Nusselt number is

$$\overline{\mathrm{Nu}}_L = 0.834\,\mathrm{Nu}_L \tag{6.66}$$

Example 6.1   Natural Convection from an Isothermal Vertical Glass Plate

A 0.5-m high flat plate of glass at 93 °C is removed from an annealing furnace and hung vertically in the air at 28 °C, 1 atm. Calculate the initial rate of heat transfer to the air. The plate is 1 m wide.

Solution
At the film temperature $T_f = (93 + 28)/2 = 60.5\,°C$, the properties of air are

$\mu = 1.999 \times 10^{-5}$ kg/m s
$k_f = 0.02874$ W/m °C
$c_p = 1.0078$ kJ/kg °C
$\rho = 1.1164$ kg/m^3
$\beta = 1/T_f = 1/(60.5 + 273) = 2.998 \times 10^{-3}$ K^{-1}

$$\mathrm{Gr}_L = \frac{g\beta\,(T_s - T_\infty)\,L^3}{\nu^2} = \frac{g\beta\,(T_s - T_\infty)\,L^3}{\left(\dfrac{\mu}{\rho}\right)^2}$$

$$= \frac{(9.8)(2.998 \times 10^{-3})(93 - 28)(0.5)^3}{\left(\dfrac{1.999 \times 10^{-5}}{1.1164}\right)^2}$$

$$= 0.2387/(3.206 \times 10^{-10}) = 7.44 \times 10^8$$

Since $Gr_L < 10^9$, the flow is laminar. Now,

$$Pr = \mu c_p/k_f = (1.999 \times 10^{-5})(1.0078 \times 10^3)/0.02874$$

$$= 0.701$$

Using the Nusselt number relation obtained from the similarity analysis (Eq. 6.31),

$$\overline{Nu_L} = \frac{4}{3} Nu_L$$

where $\quad Nu_L = \left(\dfrac{Gr_L}{4}\right)^{1/4} g\,(Pr)$

where $\quad g(Pr) = \dfrac{0.75\,(Pr)^{1/2}}{(0.609 + 1.221\,Pr^{1/2} + 1.238\,Pr)^{1/4}}$

Now, $\quad g(Pr) = \dfrac{0.75\,(0.701)^{1/2}}{[0.609 + 1.221\,(0.701)^{1/2} + 1.238\,(0.701)]^{1/4}}$

$$= \frac{0.75\,(0.837)}{[0.609 + 1.221\,(0.837) + 1.238\,(0.701)]^{1/4}}$$

$$= \frac{0.62775}{(0.609 + 1.021977 + 0.8678)^{1/4}}$$

$$= 0.62775/(2.498)^{1/4} = 0.62775/1.257$$

$$= 0.499$$

$$Nu_L = \left(\frac{7.44 \times 10^8}{4}\right)^{1/4} (0.499)$$

$$= (1.1678)\,(10^2)(0.499)$$

$$= 58.27$$

$$\overline{Nu_L} = \frac{4}{3} Nu_L = \frac{4}{3}(58.27) = 77.69$$

$$\bar{h}_L = 77.69 \frac{k_f}{L} = \frac{77.69\,(0.02874)}{0.5}$$

$$= 4.465 \; W/m^2\,K$$

$$q = \bar{h}_L\,A(T_s - T_\infty)$$

$$= (4.465)(0.5 \times 1)(93 - 28)$$

$$= 145.1 \; W$$

In addition to this heat transfer, there will be heat transfer by radiation as well.

6.5   Free Convection from Other Geometries

In this section we will discuss free convection from some more geometries.

6.5.1   Inclined Plate

For many natural convection flows, the heated surface is inclined to the direction of gravity. As a result, the buoyancy force causing the motion has a component in the flow in both tangential and normal directions. If θ is the angle of inclination from the vertical with $+\theta$ representing an upward-facing heated surface (unstable with a buoyancy component away from the surface) and $-\theta$ is downward-facing (stable), then by replacing g with $g\cos\theta$ in the Grashof number, the vertical-surface Nusselt number correlations can be used in the range $+20°$ to $-60°$ relative to the vertical. This applies to laminar flow on constant-surface temperature and constant-heat-flux surfaces for the case where the surface span is wide enough to neglect edge effects. The Nusselt number is independent of the angle for turbulent flow with $0° < \theta < +30°$. For the stable condition, $0° < \theta < -80°$, the Grashof number should be computed using $g\cos^2\theta$.

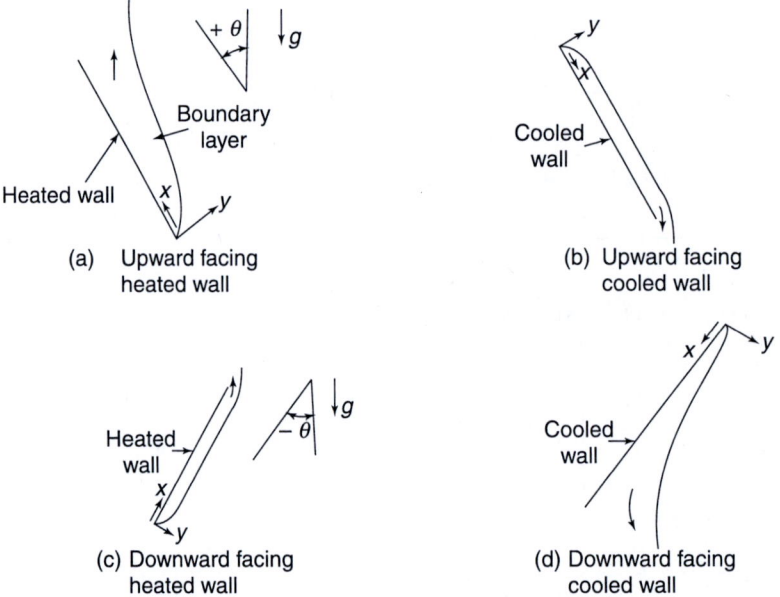

(a)   Upward facing heated wall

(b)   Upward facing cooled wall

(c)   Downward facing heated wall

(d)   Downward facing cooled wall

Fig. 6.4   Free convection over one side of heated and cooled walls inclined with respect to the vertical direction.

Figure 6.4 shows four possible orientations of inclined walls assumed to be thermally interacting with the ambient fluid from one side only. When a plate is inclined with respect to the gravitational vector, the buoyancy force has a component normal as well as along the plate. Since there is a reduction in the buoyancy force along the wall, there is a decline in the fluid velocities and hence, subsequent fall in convection heat transfer. In Figs 6.4(a), (d), for the cases of heated wall tilted upward and cooled wall tilted downward, the normal component of the buoyancy force will tend to push the flow outward resulting in the thickening of the tail end of the boundary layer and a tendency of the wall jet to separate from the wall. The opposite effect is observed in Figs 6.4(b), (c), i.e., for the cases of upward facing cooled wall and downward facing heated wall. Here, the wall jet is pressed against the wall (as the normal component of the buoyancy force acts in a direction towards the wall) until it flows over the trailing edge.

6.5.2 Horizontal Surfaces

Figure 6.5 shows free convection flow patterns when a hot plate faces upward or downward, or when a cold plate faces downward or upward. For a horizontal plate, buoyancy force acts exclusively perpendicular to the plate. In Figs 6.5(a), (b), that is, for upward facing hot plate or downward facing cold plate the natural convection flow leaves the boundary layer as a vertical plume originating at the central portion of the plate. For the case (a), if the temperature difference between the plate and the fluid is sufficiently large, the heated fluid ascends from the entire surface (Bejan, 1993). Since the law of conservation of mass dictates that colder fluid is replaced by the warmer fluid and vice-versa from the ambient, heat transfer is very effective.

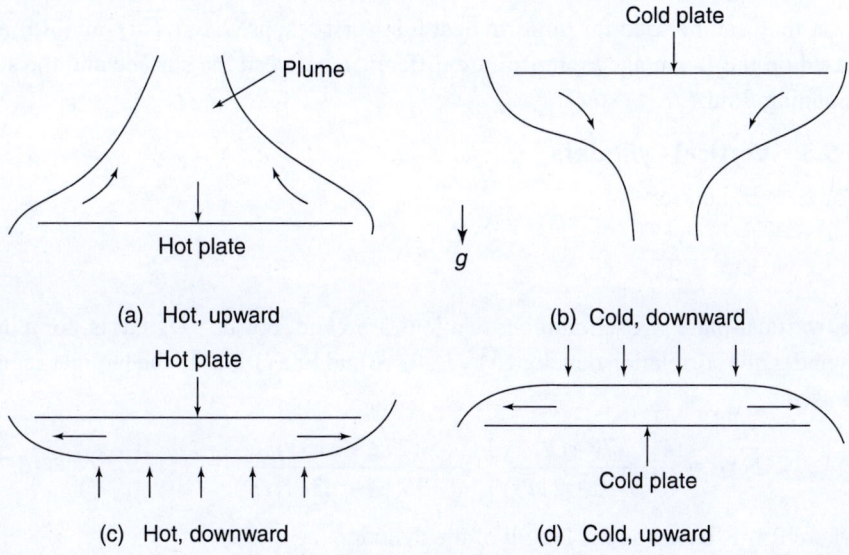

Fig. 6.5 Free convection over or below horizontal surface with central plume [(a), (b)] and without plume flow [(c), (d)]

In contrast, in Figs 6.5(c), (d), that is, for the cases of downward facing hot plate and upward facing cold plate, the tendency of the fluid to rise or fall is restricted by the plate. The flow must bifurcate horizontally before it can ascend or spill from the edges, and hence the heat transfer is much less effective. Here, the boundary layer covers the entire surface.

In Fig. 6.5, it is assumed that only one side of the plate is thermally interacting with the ambient.

Case A: Hot surface facing upward or cold surface facing downward
In this case, the dense layer will be above the light layer; the buoyancy force will be away from the surface and heat transfer will be more effective. The correlations are

$$\overline{Nu}_L = 0.54 \, Ra_L^{1/4} \tag{6.67}$$

$$(10^4 \le Ra_L \le 10^7)$$

$$\overline{Nu}_L = 0.15 Ra_L^{1/3} \tag{6.68}$$

$$(10^7 \le Ra_L \le 10^{11})$$

Case B: Cold surface facing upward or hot surface facing downward

In this case, the light layer will be above the dense layer; the buoyancy force will be towards the surface and heat transfer will be less effective. The correlations are

$$\overline{Nu}_L = 0.27 \, Ra_L^{1/4} \tag{6.69}$$

$$(10^5 \le Ra_L \le 10^9)$$

Here, $L = A_s/P =$ surface area/perimeter. For a disc of diameter D,
$$L = D/4$$

The foregoing correlations apply for isothermal surfaces and for $Pr \ge 0.5$. However, they can be used for uniform heat flux surfaces, provided $\overline{Nu}_L$ and Ra_L are based on the L-averaged temperature difference between the surface and the surrounding fluid.

6.5.3   Vertical Cylinders

For $\delta_t \ll D$, i.e., for

$$\frac{D}{L} \ge \frac{35}{Gr_L^{1/4}}$$

the vertical-plate Nusselt number relations are valid. For $\delta_t > D$, that is, for a thin cylinder, the correlation developed by Lefevre and Ede (1956) in the laminar region is used:

$$\overline{Nu}_L = \frac{4}{3}\left[\frac{7 Ra_L Pr}{5(20 + 21 Pr)}\right]^{1/4} + \frac{4(272 + 315 Pr)L}{35(64 + 63 Pr)D} \tag{6.70}$$

where Ra_L is based on the length of the cylinder.

6.5.4   Horizontal Cylinders

For average Nusselt numbers on isothermal horizontal cylinders, the following correlations are valid:

Laminar

$$\overline{Nu}_D = 0.518\left[1 + \left(\frac{0.599}{Pr}\right)^{3/5}\right]^{-5/12} (Gr_D \, Pr)^{1/4} \tag{6.71}$$

Turbulent $(Gr_D Pr > 10^9)$

$$\overline{Nu}_D = 0.1(Gr_D \, Pr)^{1/3} \tag{6.72}$$

To account for curvature effects or low Gr_D when the boundary layer is thick compared to the cylinder diameter (thin cylinder),

$$\overline{Nu}_{D_{\text{thin cylinder}}} = \cfrac{2}{\ln\left[1 + \cfrac{2}{\overline{Nu}_{D_{\text{from Eq. (6.71) or Eq. (6.72)}}}}\right]} \tag{6.73}$$

6.5.5 Enclosed Space Between Infinite Parallel Plates

Case A: Upper plate hotter than the lower plate

In this case, the light layers are above the dense layers and the system is in stable equilibrium. There is no fluid motion by free convection. Heat transfer is by pure conduction if radiation is neglected. The temperature profile between the plates is linear in the steady state.

Case B: Lower plate hotter than the upper plate

For values of $Gr_\delta Pr < 1708$ (when δ is the separation distance), pure conduction is still observed and $\overline{Nu}_\delta = 1.0$. For $Gr_\delta Pr > 1708$, free convection occurs in a distinctive hexagonal cellular flow pattern (known as Benard cells, in honour of H. Benard who reported the first investigation of this phenomenon in 1900). Heated air rises in the interior of these cells, and dense air moves down at the rims. At even higher Rayleigh numbers, the cells multiply (become narrower) and eventually the flow becomes oscillatory and turbulence sets in at $Gr_\delta Pr = 50{,}000$ and destroys the cellular pattern.

Experimental measurements in the range $3 \times 10^5 < Gr_\delta Pr < 7 \times 10^9$ support the correlation

$$\overline{Nu}_\delta = 0.069 Ra_\delta^{1/3} Pr^{0.074} \tag{6.74}$$

where $Ra_\delta = Gr_\delta Pr$ and $\overline{Nu}_\delta = \overline{h}\,\delta/k_f$.

The physical properties needed for calculating $\overline{Nu}_\delta$, Ra_δ, and Pr are evaluated at the average fluid temperature $(T_h + T_c)/2$, where T_h and T_c are the hot and cold surface temperatures, respectively. Equation (6.74) holds when the horizontal layer is sufficiently wide so that the effect of the short vertical sides is minimal. A typical example of Benard convection is the heating of water in a tea kettle.

6.5.6 Enclosed Space Between Vertical Parallel Plates

Flows induced in enclosed spaces that are subjected to temperature differences in the horizontal direction are an important class of internal natural convection problems. A typical example is the circulation of air in the slot of a double-pane window. The relevant correlations are

$$\overline{Nu}_H = 0.22\left(\frac{Pr}{0.2 + Pr}Ra_H\right)^{0.28}\left(\frac{L}{H}\right)^{0.09} \tag{6.75}$$

valid for

$$2 < \frac{H}{L} < 10, \;\; Pr < 10^5, \;\; Ra_H < 10^{13}$$

and

$$\overline{Nu}_H = 0.18\left(\frac{Pr}{0.2 + Pr}Ra_H\right)^{0.29}\left(\frac{L}{H}\right)^{-0.13} \tag{6.76}$$

valid for

$$1 < \frac{H}{L} < 2, \ 10^{-3} < \text{Pr} < 10^5, \ 10^3 < \frac{\text{Pr}}{0.2 + \text{Pr}} \, \text{Ra}_H \left(\frac{L}{H}\right)^3$$

where $\overline{\text{Nu}}_H = \dfrac{\overline{h}H}{k_f}$

$$\text{Ra}_H = \frac{g\beta(T_h - T_c)H^3}{\alpha v}$$

where T_h is the hot side temperature, T_c is the cold side temperature, L is the distance between the hot plate and the cold plate, and H is the height of the plates. Equation (6.76) is valid in the 'wide' cavity limit, that is, when $\delta_t < L$, or $L/H > \text{Ra}_H^{-1/4}$.

6.6 Correlations for Free Convection over a Vertical Plate Subjected to Uniform Heat Flux

For the case of uniformly heated plate ($q_s'' = \text{Constant}$) Fujii and Fujii (1976) obtained the following correlation (valid for $0.01 < \text{Pr} < 1000$) based on the similarity solution of Yang (1960) and Sparrow and Gregg (1958) as given in Kays and Crawford (1993) for laminar free convection flow over a vertical plate.

$$\text{Nu}_x = \left(\frac{\text{Pr}}{4 + 9\,\text{Pr}^{1/2} + 10\,\text{Pr}}\right)^{1/5} (\text{Gr}_x^* \, \text{Pr})^{1/5} \tag{6.77}$$

where $\text{Gr}_x^* = \text{Gr}_x \text{Nu}_x = \dfrac{g\beta q_s'' x^4}{k_f v^2}$ is the modified Grashof number.

Note that $\text{Gr}_x = \dfrac{g\beta(T_s - T_\infty)x^3}{v^2}$

$$\text{Nu}_x = \frac{q_s'' x}{(T_s - T_\infty)k_f}$$

It can be easily shown that

$$\overline{\text{Nu}}_L = \frac{5}{4}\text{Nu}\big|_{x=L} \tag{6.78}$$

In the fluids of air–water Prandtl number range (Bejan, 1993)

$$\text{Ra}_{x,cr}^* = 10^{13} \tag{6.79}$$

where $\text{Ra}_x^* = \text{Gr}_x^* \text{Pr}$ is the modified Rayleigh number.

Vliet and Liu (1969) recommend the following correlations for the local and average Nusselt numbers for air-water Prandtl number range.

Laminar Flow $\left(10^5 < \text{Ra}_x^* < 10^{13}\right)$

$$\text{Nu}_x = 0.6\text{Ra}_x^{*1/5} \tag{6.80a}$$

$$\overline{\text{Nu}}_L = 0.75\text{Ra}_x^{*1/5} \tag{6.80b}$$

Turbulent Flow $(10^{13} < \mathrm{Ra}_x^* < 10^{16})$

$$\mathrm{Nu}_x = 0.568\mathrm{Ra}_x^{*0.22} \qquad\qquad (6.81\mathrm{a})$$

$$\overline{\mathrm{Nu}_L} = 0.645\mathrm{Ra}_x^{*0.22} \qquad\qquad (6.81\mathrm{b})$$

Remember that $\quad \overline{\mathrm{Nu}_L} = \dfrac{\overline{h}L}{k_f}$

where $\quad \overline{h} = \dfrac{q_s''}{T_{s,\mathrm{avg}} - T_\infty}$

As in forced convection the solution methodology for free convection with constant wall flux boundary condition is iterative with property values evaluated at T_∞ in the first iteration and at $T_f = \dfrac{T_{s,\mathrm{avg}} + T_\infty}{2}$ in the subsequent iterations.

6.7 Mixed Convection

When both forced and free convection effects are significant, mixed convection occurs. The effect of buoyancy is to alter the Nusselt number and friction coefficient for the forced convection case. If a vertical surface (or a vertical tube) is heated such that $T_s > T_\infty$, the resulting buoyancy aids the forced convective motion, which would result in the enhancement of the Nusselt number corresponding to the forced convection value at that Reynolds number. The opposite is true when $T_s < T_\infty$. A typical example of mixed convection is the heat transfer from an extruded plate or a rod as it emerges from the die vertically.

For flow over a horizontal surface, a similar increase or decrease of the Nusselt number occurs. Forced convection in heated horizontal tubes can result in enhanced Nusselt numbers if the buoyancy sets up a secondary flow.

Additional Examples

Example 6.2 Natural Convection from a Hot Oven Door

The door of a hot oven is 0.5 m high and is at 200 °C. The outer surface is exposed to atmospheric pressure air at 20 °C. Estimate the average heat transfer coefficient at the outer surface of the door.

Solution

$$T_f = \frac{200 + 20}{2} = 110°\mathrm{C}$$

Table A1.4 in Appendix A1 gives

$$v = 24.10 \times 10^{-6} \ \mathrm{m^2/s}$$

$$k = 0.03194 \ \mathrm{W/m\,K}$$

$$\mathrm{Pr} = 0.704$$

At $T_f = 110°\mathrm{C}$, $\beta = \dfrac{1}{383} = 0.00261 \,\mathrm{K}^{-1}$

$$L = 0.5 \text{ m}$$

$$\Delta T = T_s - T_\infty = 200 - 20 = 180°C$$

$$g = 9.8 \text{ m/s}^2$$

$$Gr_L = \frac{g\beta L^3 \Delta T}{v^2} = \frac{(9.8)(0.00261)(0.5)^3 (180)}{(24.10 \times 10^{-6})^2} = 9.91 \times 10^8$$

Since $Gr_L < 10^9$, the correlation of Churchill and Chu (1975), Eq. (6.64), is valid as it represents the experimental data more accurately. Now,

$$Ra_L = Gr_L \, Pr = 9.91 \times 10^8 \times 0.704 = 6.98 \times 10^8$$

$$\overline{Nu_L} = \overline{h}\frac{L}{k} = 0.68 + \frac{0.67 \, Ra_L^{1/4}}{\left[1 + \left(\frac{0.492}{Pr}\right)^{9/16}\right]^{4/9}}$$

$$= 0.68 + \frac{0.67 \, (6.98 \times 10^8)^{1/4}}{\left[1 + \left(\frac{0.492}{0.704}\right)^{9/16}\right]^{4/9}} = 0.68 + \frac{109}{1.3} = 84.52$$

Therefore,

$$\overline{h} = \overline{Nu_L}\frac{k}{L} = \frac{(84.52)(0.03194)}{(0.5)} = 5.399 \text{ W/m}^2 \text{ K}$$

Example 6.3   Minimum Air Velocity for Ignoring Buoyancy Effect

If the oven door described in Example 6.2 is subjected to an upward forced flow of air, find the minimum free stream air velocity for which free convection effects may be neglected.

Solution

The following table (Chapman 1989) shows the maximum values of Gr_x / Re_x^2 for neglection of buoyancy on forced convection past a vertical plate with less than 5% error.

Pr	100	10	0.72	0.03–0.003
Gr_x / Re_x^2	0.24	0.13	0.08	0.056–0.05

The calculations of Example 6.2 showed that

$$Gr_L = 9.91 \times 10^8$$

For the air temperature in question, $Pr = 0.704$. At this value of Pr, the above table shows that less than 5% error will result by treating the flow as pure forced convection if

$$\frac{Gr_L}{Re_L^2} < 0.08$$

$$Re_L > 1.13 \times 10^5$$

Thus,

$$\frac{u_\infty L}{v} > 1.13 \times 10^5$$

or

$$\frac{u_\infty (0.5)}{24.1 \times 10^{-6}} > 1.13 \times 10^5$$

or

$$u_\infty > 5.5 \text{ m/s}$$

Example 6.4   Free Convection from a Hot Metal Cube

A hot metal cube of dimensions 10 cm × 10 cm × 10 cm (Fig. E6.4) with its five faces at 520°C is hanging vertically in air. The bottom face is insulated. Calculate the rate of heat loss from the sides of the cube by free convection into the surrounding quiescent air at atmospheric pressure and at 30°C. The properties of air at the film temperature (548 K) are listed below.

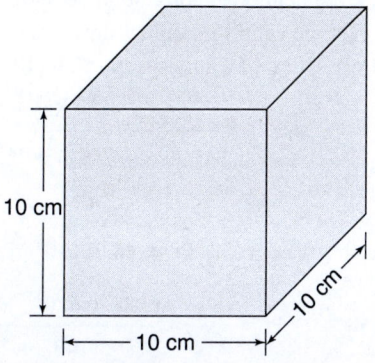

$$\rho = 0.6329 \text{ kg/m}^3$$

$$c_p = 1.04 \text{ kJ/kg K}$$

$$\mu = 288.4 \times 10^{-7} \text{ Ns/m}^2$$

$$k_f = 43.9 \times 10^{-3} \text{ W/mK}$$

Fig. E6.4   Free convection from a hot metal cube

Solution

The total heat loss is the sum of the heat loss from the four vertical sides and the top face of the cube. There is no heat loss from the bottom face since it is insulated.

$$q_{total} = q''_{vertical} (4)(10 \times 10^{-2})(10 \times 10^{-2}) + q''_{top}(10 \times 10^{-2})(10 \times 10^{-2})$$

$$= \left(10^2 \times 10^{-4}\right)\{4 h_{avg,vertical} + h_{avg,top}\}(T_s - T_\infty)$$

$$= (10^{-2})(520 - 30)\left\{4 h_{avg,vertical} + h_{avg,top}\right\}$$

$$= 4.9 \left\{4 h_{avg,vertical} + h_{avg,top}\right\}$$

Now, $$\beta = \frac{1}{T_f} = \frac{1}{548} = 1.82 \times 10^{-3} \text{ K}^{-1}$$

Heat Transfer from the Vertical Faces

$$Gr_{vertical} = \frac{g\beta(T_s - T_\infty)L^3}{v^2}$$

where $$T_s = 520°C$$

$$T_\infty = 30°C$$

$$L = 10 \times 10^{-2} = 0.1 \text{ m}$$

$$v = \frac{\mu}{\rho} = \frac{288.4 \times 10^{-7}}{0.6329} = 4.56 \times 10^{-5} \text{ m}^2/s$$

$$\beta = 1.82 \times 10^{-3} \text{K}^{-1}$$

Hence, $$Gr_{vertical} = \frac{(9.81)(1.82 \times 10^{-3})(520 - 30)(0.1)^3}{(4.56 \times 10^{-5})^2} = 4.2 \times 10^6$$

Since $Gr_{vertical} < 10^9$, the free convection flow is laminar.

Therefore, the correlation of Churchill and Chu (1975) based on experimental data can be used to calculate the average Nusselt number and average heat transfer coefficient. The aforementioned correlation being relatively more accurate is chosen. However, other correlations valid for laminar flow over a vertical plate can also be used.

Correlation of Churchill and Chu (1975):

$$\overline{Nu}_L = 0.68 + \frac{0.67\,Ra_L^{1/4}}{\left[1+\left(\frac{0.492}{Pr}\right)^{9/16}\right]^{4/9}}$$

Now, $\quad Ra_L = Gr_L\,Pr = (4.2 \times 10^6\,)Pr$

where $\quad Pr = \dfrac{\mu c_p}{k_f} = \dfrac{(288.4 \times 10^{-7}\,)(1.04 \times 10^3\,)}{43.9 \times 10^{-3}} = 0.683$

Therefore, $Ra_L = (4.2 \times 10^6)(0.683) = 2.87 \times 10^6$

Thus, $\quad \overline{Nu}_L = 0.68 + \dfrac{0.67\,(2.87 \times 10^6\,)^{1/4}}{\left[1+\left(\dfrac{0.492}{0.683}\right)^{9/16}\right]^{4/9}} = 21.89$

Hence, $\quad \dfrac{\overline{h}_L L}{k_f} = 21.89$

$\Rightarrow \qquad \overline{h}_L = \dfrac{21.89\,k_f}{L} = \dfrac{(21.89)(43.9 \times 10^{-3}\,)}{0.1} = 9.61 \text{ W/m}^2\text{K}$

Heat Transfer from the Top Horizontal Face

$$L = \frac{A_s}{P} = \frac{10 \times 10}{40} = 2.5 \text{ cm} = 2.5 \times 10^{-2} \text{ m}$$

$$Gr_L = \frac{g\beta(T_s - T_\infty\,)L^3}{v^2} = \frac{(9.81)(1.82 \times 10^{-3}\,)(520-30)(0.025)^3}{(4.56 \times 10^{-5}\,)^2} = 6.56 \times 10^4$$

Hence, $\quad Ra_L = Gr_L\,Pr = (6.56 \times 10^4\,)(0.683) = 4.48 \times 10^4$

For a hot surface facing upwards the following correlation is valid for $10^4 \le Ra_L \le 10^7$ and $Pr \ge 0.5$.

$$\overline{Nu}_L = 0.54\,Ra_L^{1/4}$$

Since the present problem satisfies the above requirements, therefore,

$$\overline{Nu}_L = 0.54(4.48 \times 10^4\,)^{1/4} = 7.83$$

Hence, $\quad \dfrac{\overline{h}_L L}{k_f} = 7.83$

$\Rightarrow \qquad \overline{h}_L = \dfrac{7.83\,k_f}{L} = \dfrac{(7.83)(43.9 \times 10^{-3}\,)}{0.025} = 13.75 \text{ W/m}^2\text{K}$

Finally, $q_{total} = 4.9\{4(9.61)+13.75\} = 255.7 \text{ W}$

Example 6.5 Free Convection from a Long Horizontal Cylinder

A horizontal cylinder has a diameter of D = 5 cm and length L =50 cm. Its surface is maintained at 35ºC while the surrounding air is at 20ºC. Neglecting radiation calculate the rate of heat transfer from the cylindrical surface at the steady state. The properties of air at the film temperature (approximately, 300 K) are

$$v = 15.89 \times 10^{-6} \text{ m}^2/\text{s}$$

$$k_f = 26.3 \times 10^{-3} \text{ W/mK}$$

$$\text{Pr} = 0.707$$

Solution

$$D = 5 \text{ cm} = 0.05 \text{ m}$$

$$L = 50 \text{ cm} = 0.5 \text{ m}$$

$$T_s = 35º \text{ C}$$

$$T_\infty = 20º \text{ C}$$

$$\beta = \frac{1}{T_f} = \frac{1}{300} = 3.33 \times 10^{-3} \text{ K}^{-1}$$

$$\text{Gr}_D = \frac{g\beta(T_s - T_\infty)D^3}{v^2} = \frac{(9.81)(3.33 \times 10^{-3})(35 - 20)(0.05)^3}{(15.89 \times 10^{-6})^2} = 2.425 \times 10^6$$

$$\text{Gr}_D \text{ Pr} = 2.425 \times 10^6 \times 0.707 = 1.714 \times 10^6$$

Since $\text{Gr}_D \text{ Pr} < 10^9$, the flow is laminar.

The relevant correlation is

$$\overline{\text{Nu}}_D = 0.518 \left[1 + \left(\frac{0.599}{\text{Pr}} \right)^{3/5} \right]^{-5/12} (\text{Gr}_D \text{ Pr})^{1/4}$$

Substituting the values of $\text{Gr}_D \text{ Pr}$ and Pr in the above expression, we get

$$\overline{\text{Nu}}_D = 14.34$$

Hence, $$\overline{h}_D = \frac{14.34 k_f}{D} = \frac{14.34(26.3 \times 10^{-3})}{0.05} = 7.54 \text{ W/m}^2\text{K}$$

Thus, $$q = \overline{h}_D A(T_s - T_\infty) = \overline{h}_D (\pi DL)(T_s - T_\infty)$$

$$= (7.54)(3.1415 \times 0.05 \times 0.5)(35 - 20) = 8.88 \text{ W}$$

So, at the steady state the heat transfer from the cylindrical surface is 8.88 W.

Important Concepts and Formulae

Free Convection from a Vertical Plate

Boussinesq Approximation   In the present analysis of free convection, the fluid is treated as incompressible. The exception involves accounting for the effect of variable density in the buoyancy force, since it is this variation that induces fluid motion. This is known as the Boussinesq approximation.

Grashof Number

$$Gr_L = \frac{g\beta(T_s - T_\infty)L^3}{v^2}$$

Grashof number signifies the ratio of buoyancy force to viscous force.

Reynolds Number

$$Re_L = \frac{u_o L}{v}$$

where u_o is an arbitrary reference velocity.

Richardson Number

$$\frac{Gr_L}{Re_L^2} = \frac{g\beta(T_s - T_\infty)L^3}{u_o^2}$$

Mixed Convection

$$Nu_L = f(Re_L, Gr_L, Pr) \Rightarrow \frac{Gr_L}{Re_L^2} = 0(1)$$

Forced Convection

$$Nu_L = f(Re_L, Pr) \Rightarrow \frac{Gr_L}{Re_L^2} \ll 1$$

Free Convection

$$Nu_L = f(Gr_L, Pr) \Rightarrow \frac{Gr_L}{Re_L^2} \gg 1$$

Rayleigh Number

$$Ra_L = Gr_L \, Pr$$

Constant Wall Temperature (T_s = constant)
Laminar Flow

$$Gr_{cr} = 10^9$$

Similarity Analysis

$$Nu_x = \left(\frac{Gr_x}{4}\right)^{1/4} g(Pr)$$

where $\quad g(Pr) = \dfrac{0.75 \, Pr^{1/2}}{(0.609 + 1.221 \, Pr^{1/2} + 1.238 \, Pr)^{1/4}}$

which applies for the entire range of Prandtl numbers and $10^4 \le Gr_x \le 10^9$.

$$\overline{Nu_L} = \frac{4}{3} Nu_L$$

Integral Analysis

$$\overline{Nu_L} = 0.508 \, Pr^{1/2} (0.952 + Pr)^{-1/4} \, Gr_x^{1/4}$$

$$\overline{Nu_L} = \frac{4}{3} Nu_L$$

Experimental Correlation of Churchill and Chu (1975)
(valid for $Gr_x < 10^9$):

$$\overline{Nu}_L = 0.68 + \frac{0.67\,Ra_L^{1/4}}{\left[1+\left(\dfrac{0.492}{Pr}\right)^{9/16}\right]^{4/9}}$$

Turbulent Flow
Integral Analysis of Eckert and Jackson (1950):

$$Nu_x = 0.0295\left[\frac{Pr^7}{(1+0.494\,Pr^{2/3}\,)^6}\right]^{1/5} Gr_x^{2/5}$$

$$\overline{Nu}_L = 0.834\,Nu_L$$

Experimental correlation of Churchill and Chu (1975) (valid for $10^{-1} < Ra_x < 10^{12}$ and for all Prandtl numbers):

$$\overline{Nu}_L = \left\{0.825 + \frac{0.387\,Ra_L^{1/6}}{\left[1+\left(\dfrac{0.492}{Pr}\right)^{9/16}\right]^{8/27}}\right\}^2$$

Constant Wall Heat Flux (q_s'' = constant)
For the case of uniformly heated plate (q_s'' = constant) Fujii and Fujii (1976) obtained the following correlation (valid for $0.01 < Pr < 1000$) based on the similarity solution of Yang (1960) and Sparrow and Gregg (1958) as given in Kays and Crawford (1993) for laminar free convection flow over a vertical plate.

$$Nu_x = \left(\frac{Pr}{4+9\,Pr^{1/2}+10\,Pr}\right)^{1/5} (Gr_x^* \, Pr)^{1/5}$$

where $Gr_x^* = Gr_x\,Nu_x = \dfrac{g\beta q_s'' x^4}{k_f \nu^2}$ is the modified Grashof number.

Note that

$$Gr_x = \frac{g\beta(T_s - T_\infty)x^3}{\nu^2}$$

$$Nu_x = \frac{q_s'' x}{(T_s - T_\infty)k_f}$$

It can be easily shown that

$$\overline{Nu}_L = \frac{5}{4}Nu\Big|_{x=L}$$

In the fluids of air–water Prandtl number range (Bejan, 1993)

$$Ra_{x,cr}^* = 10^{13}$$

where $Ra_x^* = Gr_x^*\,Pr$ is the modified Rayleigh number.

Vliet and Liu (1969) recommend the following correlations for the local and average Nusselt numbers for air–water Prandtl number range.

Laminar Flow ($10^5 < Ra_x^* < 10^{13}$)

$$Nu_x = 0.6\,Ra_x^{*1/5}$$

$$\overline{Nu}_L = 0.75\,Ra_x^{*1/5}$$

Turbulent Flow ($10^{13} < Ra_x^* < 10^{16}$)

$$Nu_x = 0.568\,Ra_x^{*0.22}$$

$$\overline{Nu}_L = 0.645\,Ra_x^{*0.22}$$

Remember that

$$\overline{Nu}_L = \frac{\overline{h}L}{k_f}$$

where $\quad \overline{h} = \dfrac{q_s''}{T_{s,avg} - T_\infty}$

As in forced convection the solution methodology for free convection with constant wall flux boundary condition is iterative with property values evaluated at T_∞ in the first iteration and at $T_f = \dfrac{T_{s,avg} + T_\infty}{2}$ in the subsequent iterations.

Inclined Plates

When a plate is inclined with respect to the gravitational vector, the buoyancy force has a component normal as well as along the plate. Since there is a reduction in the buoyancy force along the wall, there is a decline in the fluid velocities and hence, subsequent fall in convection heat transfer. For the cases of heated wall tilted upward and cooled wall tilted downward, the normal component of the buoyancy force will tend to push the flow outward resulting in the thickening of the tail end of the boundary layer and a tendency of the wall jet to separate from the wall. The opposite effect is observed in the cases of upward facing cooled wall and downward facing heated wall. Here, the wall jet is pressed against the wall (as the normal component of the buoyancy force acts in a direction towards the wall) until it flows over the trailing edge.

Horizontal Surfaces

Isothermal Surface

Case A: Hot surface facing upward or cold surface facing downward

$\overline{Nu}_L = 0.54\,Ra_L^{1/4}$ (Valid for $10^4 \le Ra_L \le 10^7$ and for $Pr \ge 0.5$)

$\overline{Nu}_L = 0.15\,Ra_L^{1/3}$ (Valid for $10^7 \le Ra_L \le 10^{11}$ and for $Pr \ge 0.5$)

Case B: Cold surface facing upward or hot surface facing downward

$\overline{Nu}_L = 0.27\,Ra_L^{1/4}$ (Valid for $10^5 \le Ra_L \le 10^9$)

where $\quad L = \dfrac{A_s}{P}$ surface area /perimeter.

The foregoing correlations can be used also for uniform heat flux surfaces, provided $\overline{Nu}_L$ and Ra_L are based on the L-averaged temperature difference between the surface and the surrounding fluid.

Vertical Cylinders

Isothermal Surface

For $\dfrac{D}{L} \geq \dfrac{35}{Gr_L^{1/4}}$ (that is, $\delta_t \leq D$)

The vertical-plate Nusselt number relations are valid.

For $\delta_t > D$, that is, for a thin cylinder, the following correlation of Lefvre and Ede (1956) in the laminar region is used.

$$\overline{Nu}_L = \frac{4}{3}\left[\frac{7\,Ra_L\,Pr}{5(20+21\,Pr)}\right]^{1/4} + \frac{4(272+315\,Pr)\,L}{35(64+63\,Pr)\,D}$$

where Ra_L is based on the length of the cylinder.

Horizontal Cylinders

Isothermal Surface

Laminar ($Gr_D\,Pr \leq 10^9$)

$$\overline{Nu}_D = 0.518\left[1+\left(\frac{0.599}{Pr}\right)^{3/5}\right]^{-5/12}\left(Gr_D\,Pr\right)^{1/4}$$

Turbulent ($Gr_D\,Pr > 10^9$)

$$\overline{Nu}_D = 0.1\left(Gr_D\,Pr\right)^{1/3}$$

Enclosed Space between Infinite Horizontal Parallel Plates

Case A: Upper plate hotter than the lower plate
In this case, the light layers are above the dense layers and the system is in stable equilibrium. There is no fluid motion by free convection. Heat transfer is by pure conduction if radiation is neglected. The temperature profile between the plates is linear in the steady state.

Case B: Lower plate hotter than the upper plate
For values of $Gr_\delta\,Pr < 1708$ (where δ is the separation distance), pure conduction is still observed. For $Gr_\delta\,Pr > 1708$, free convection occurs in a distinctive hexagonal cellular flow pattern (known as Benard cells). Heated air rises in the interior of these cells, and dense air moves down at the rims. At even higher Rayleigh numbers, the cells multiply (become narrower) and eventually the flow becomes oscillatory and turbulence sets in at $Gr_\delta\,Pr = 50000$ and destroys the cellular pattern.

Enclosed Space between Vertical Parallel Plates

Flows induced in enclosed spaces that are subjected to temperature differences in the horizontal direction are an important class of internal natural convection problems. A typical example is the circulation of air in the slot of a double pane window or in a hollow wall.

Mixed Convection

When both forced and free convection effects are significant, mixed convection occurs. The effect of buoyancy is to alter the Nusselt number and friction coefficient for the forced convection case.

Review Questions

6.1 Explain the physical mechanism of natural convection. How does it differ from forced convection?

6.2 What is Boussinesq approximation?

6.3 Define Grashof number and explain its physical significance.

6.4 Consider $Gr/Re^2 = 0.01$ for a vertical flat plate. Does it represent mixed convection, forced convection or free convection?

6.5 Define Rayleigh number.

6.6 Discuss the natural convection flow physics for (i) hot surface facing upward and (ii) hot surface facing downward.

6.7 What is Benard convection?

6.8 Give an example of mixed convection.

6.9 Qualitatively plot non-dimensional vertical velocity and temperature vs. η for different Pr numbers for an isothermal vertical plate and physically explain the trends of the graphs.

6.10 Show by dimensional analysis that Gr/Re^2 represents the ratio of buoyancy and inertia forces.

Problems

6.1 A vertical wall at 65°C is exposed to air at 25°C. Find the Grashof number at 0.6 m from the lower edge.

6.2 At what speed must air be blown over the wall in Question 6.1 so that the mixed convection effect becomes important? The plate is 1 m long.

6.3 A flat ribbon heat strip 1 m wide and 3 m long is held vertically on an insulating substrate. Its energy dissipation is 0.25 kW/m^2 to air at 20 °C. What are the average heat transfer coefficient and surface temperature of the ribbon? At what distance from the lower edge will the transition to turbulent flow occur?

6.4 Develop the momentum integral equation [Eq. (6.35)] for the laminar free convection over an isothermal vertical flat plate.

6.5 Consider free convection cooling of a thick, square plate of aluminium with one surface exposed to air and the other surfaces insulated. The temperature of air is 25 °C and that of aluminium is 45 °C. The side of the square is 10 cm. Compare the average heat transfer coefficients for three exposed face orientations: vertical face, face inclined 45 ° to the vertical (downward-facing heated surface), and horizontal face (heated surface facing upward).

6.6 Consider a copper wire of diameter 0.00040 cm and length 600 times its diameter. Let the wire temperature be 250 °C and the air temperature be 25 °C. Compare the heat transfer coefficients for the wire placed in the horizontal and vertical positions.

6.7 A sheet of paper of size 0.3 m × 0.3 m is suspended by a calibrated spacing in still air at 20 °C. The initial weight of the paper is 20 g. When the sun shines on the paper so that it is uniformly heated to 60 °C, do you expect the weight of the paper to increase or decrease?

6.8 Two 50-cm horizontal square plates are separated by a distance of 1 cm. The lower plate is maintained at a constant temperature of 50 °C and the upper plate is at 20 °C. Water at atmospheric pressure occupies the space between the plates. Calculate the heat lost by the lower plate. Compare the results if the space between the plates is occupied by air.

6.9 The door of a household refrigerator is 0.9 m tall and 0.6 m wide. The temperature of its external surface is 20 °C while the room air temperature is 30 °C. Verify that the natural convection boundary layer that descends along the door is laminar, and calculate the heat transfer from air to the door.

6.10 An immersion heater for water consists of a thin vertical plate of height 8 cm and length 15 cm. The plate is heated electrically and maintained at 60 °C while the water is at 20 °C. Show that the natural convection flow upwards along the plate is laminar. Also, calculate the total heat transfer from the heater to the water pool.

6.11 In order to simulate in the laboratory the natural convection of air in a 4-m-high room, an experiment is to be conducted by using a small box filled with water. The researcher must select a certain vertical dimension (height H) for the water enclosure so that the Rayleigh number of the water flow matches the Rayleigh number of the air flow in the actual room. In the water experiment the average water temperature and the cold-wall temperature are 25 °C and 15 °C, respectively. What is the required height H of the water apparatus?

6.12 A horizontal disc of 9 cm diameter, which is at a temperature of 45 °C, is immersed facing upward in a pool of 25 °C water. Calculate the total heat transfer from the surface to the water.

6.13 A bare 10-cm-diameter horizontal steam pipe is 100 m long and has a surface temperature of 150 °C. Estimate the total heat loss by free convection to atmospheric air at 25 °C.

6.14 A 5-cm-outer-diameter tube of 40 cm length, maintained at a uniform temperature of 100 °C, is placed vertically in quiescent air at atmospheric pressure and at a temperature of 27 °C. Calculate the average natural convection heat transfer coefficient and the rate of heat loss from the tube to the air. What will be the total heat loss taking radiation into account? Assume the outer surface of the tube is black.

6.15 A double-glass window consists of two vertical parallel plates of glass, each 1 m × 1 m in size, separated by a 1.5-cm air gap at atmospheric pressure. Calculate the free convection heat transfer coefficient for the air space for a temperature difference of 25 °C. Assume the mean temperature of air to be 27 °C.

6.16 Water is contained between two parallel horizontal plates separated by a distance of 2 cm. The lower plate is at a uniform temperature of 130 °C, and the upper plate is at 30 °C. Determine the natural convection heat transfer per unit area between the plates.

6.17 Solve Question 6.16 if the lower plate is at 30 °C while the upper plate is at 130 °C. Will the heat transfer rate be the same? If not, why? Neglect radiation heat transfer between the plates.

6.18 A horizontal plate 0.2 m × 0.2 m in size at a uniform temperature of 150 °C is exposed to atmospheric air at 20 °C. Calculate the free convection heat transfer coefficient for the heated surface (a) facing up and (b) facing down.

6.19 A hot iron block of dimensions 10 cm × 15 cm × 20 cm at 400 °C is placed on an asbestos insulation with its 10 cm side oriented vertically. Calculate the total rate of heat loss from the block by free convection into the surrounding quiescent air at atmospheric pressure and at 25 °C.

6.20 An uninsulated horizontal duct of diameter $D = 20$ cm transporting cold air at 10 °C is exposed to quiescent atmospheric air at 35 °C. Determine the heat gain by free convection per meter length of the duct.

Chapter 7

Boiling and Condensation

7.1 Boiling

Many engineering problems involve boiling and condensation. For example, both processes are essential to all closed-loop power and refrigeration cycles. In a thermal power cycle, pressurized liquid is converted to vapour in a boiler. After expansion in a turbine, the vapour is transformed back to its liquid state in a condenser, whereupon it is pumped to the boiler to repeat the cycle. Today, boiling is also used to cool electronic equipment, such as supercomputers, laptop computers, and mobile phones. Evaporators and condensers are also essential components in vapour compression refrigeration cycles. To design such components requires that the associated phase change processes be well understood.

The large change in the density associated with the phase change process can give rise to vigorous natural convection. For example, heating or cooling air at atmospheric pressure through the temperature range 90–100°C induces a 2.7% density change, whereas heating water under the same conditions induces a much larger density change of 99.9%.

The formation of vapour from a liquid is called boiling when bubbles are formed; the formation of liquid from vapour is called condensation. Because of their importance in industry, boiling and condensation have been studied extensively by researchers.

7.1.1 Evaporation

If the liquid at a heated wall is only slightly superheated above the saturation temperature, then only a few or even no vapour bubbles are formed. In a vessel filled with liquid and heated from below, the temperature profile develops as depicted in Fig. 7.1.

Above the heated base with the temperature T_w, there forms a boundary layer of the order of magnitude of 1 mm with a strong temperature drop, while in the core the liquid temperature is almost constant above the height z (mean value T_L). At the free surface the temperature in a thin layer drops to the value T_0, which is slightly above the saturation temperature T_{sat}. However, for technical calculations this liquid superheating at the surface can be disregarded. Therefore, in the following, the saturation temperature $T_0 = T_{sat}$ is always ascribed to a vapour forming surface.

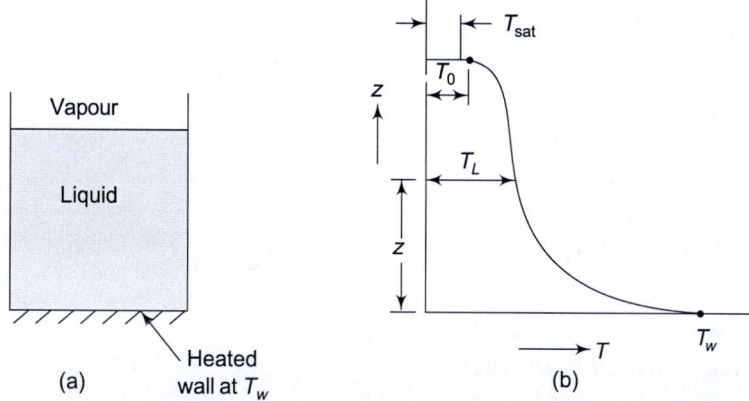

Fig. 7.1 (a) Evaporation of a liquid at a heated wall, (b) temperature profile in the liquid during surface evaporation

In the thin layer near the wall, the temperature drops steeply, as shown in Fig. 7.1. In this layer, conduction dominates. In the liquid below, heat transport is provided by the ascending and descending convective flows. They produce the uniform temperature in the core of the liquid. The two boundary layers above and below are differentiated from one another by the fact that the free surface is movable because of the vapour formation and that finite parallel velocities can also occur there, in contrast to the liquid at the wall.

Because vapourization or evaporation occurs at the free surface, one speaks of 'quiet boiling'. This process, by its nature, belongs to the phenomenon of free convection in closed spaces. The heat transfer coefficient from the heating surface to the liquid can be defined as

$$h = \frac{q''}{T_w - T_{sat}} \tag{7.1}$$

$T_w - T_{sat}$ instead of $T_w - T_L$ is used in Eq. (7.1) because T_L is not known in advance and, as explained above, deviates only slightly from the saturation temperature. For evaporation, the laws of heat transfer by free convection apply. Thus,

$$h = c_1 (\Delta T)^{1/4} \quad \text{for laminar flow} \tag{7.2}$$

$$h = c_2 (\Delta T)^{1/3} \quad \text{for turbulent flow} \tag{7.3}$$

Since, $q'' = h\Delta T$,

$$h = c_1' (q'')^{1/5} \quad \text{for laminar flow} \tag{7.4}$$

$$h = c_2' (q'')^{1/4} \quad \text{for turbulent flow} \tag{7.5}$$

For free convection in horizontal enclosed spaces, where the bottom is hotter than the top, the critical Rayleigh number is 50,000.

$$Ra_\delta = Gr_\delta Pr \tag{7.6}$$

where δ is the vertical distance between the bottom and the top and k is the conductivity of the liquid.

$$Gr_\delta = \frac{g\beta(\Delta T)\delta^3}{v^2} \tag{7.7}$$

where β is the volumetric thermal expansion coefficient.

7.1.2 Nucleate Boiling

If one raises the wall temperature by increasing the heat input, vapour bubbles form when a definite wall temperature is reached. As observation shows, these arise only at certain points on the heating surface. The number of vapour bubbles formed increases with the heat input. This type of heat transfer is characterized by nucleate boiling. Figure 7.2 shows a typical temperature pattern over a horizontal plate.

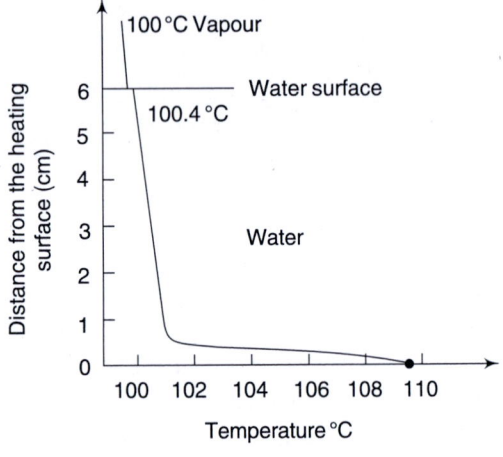

Fig. 7.2 Temperature profile above a heating surface during nucleate boiling (*source*: Stephan, 1992)

In contrast to Fig. 7.1, one notices that the temperature difference $T_w - T_L$ is substantially greater and that $T_L - T_{sat}$ is much smaller. The bubble movement at the surface permits no exact measurement of the boundary layer. Once again, h is calculated on the basis of $T_w - T_{sat}$ as for evaporation.

The heat transfer is much better than during evaporation and

$$h = c_3 (\Delta T)^3 \tag{7.8}$$

$$h = c_3'(q'')^{3/4} \tag{7.9}$$

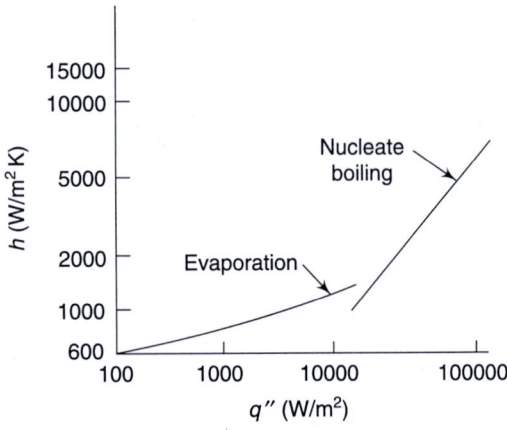

Fig. 7.3 Heat transfer coefficient (h) versus wall heat flux (q'') for boiling water at 100 °C (*source*: Stephan, 1992)

Figure 7.3 shows h versus q'' on log-log scale. The graph clearly indicates two distinct straight lines, one for evaporation and another for nucleate boiling.

7.2 Review of Phase Change Processes of Pure Substances

Before the mechanism of nucleate boiling is discussed, it is worthwhile to review equilibrium phase-change diagram of pure substances. It may be noted that a pure substance consists of a single chemical species and has a fixed chemical composition throughout.

7.2.1 *p-v-T* surface

The state of a pure compressible substance is determined by any two independent intensive properties. Recall that intensive properties are those that are independent of the size of the system such as T, p, ρ, and v. On the other hand, extensive properties are those that depend on the size or extent of the system. Mass m, volume V, and total energy E represent extensive properties.

Knowing that $z = z(x, y)$ represents a surface in space, $p-v-T$ behaviour of a substance can be represented as a surface in space (Fig. 7.4). It can be seen that T and v are the independent variables (the base) and p is the dependent variable (the height).

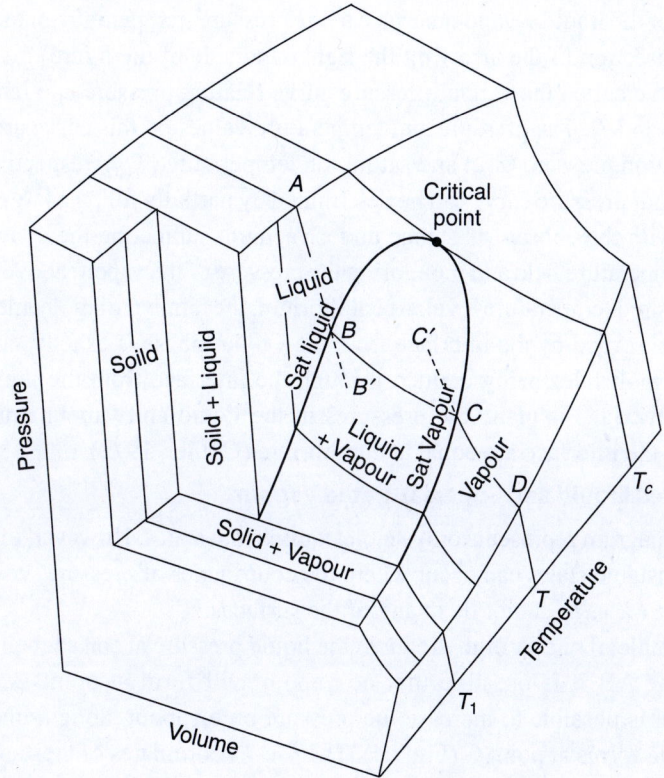

Fig. 7.4 *p-v-T* surface for a pure substance

A $p-v$ diagram is simply a projection of $p-v-T$ surface on the $p-v$ plane, and a $T-v$ diagram is nothing but the top view of this surface.

The various stable equilibrium phase states are shown in Fig. 7.4. The stable regions are liquid alone, liquid and vapour, and vapour alone. This sets the temperature range of interest to T_1 (the triple point) at the lower limit and T_c (the critical point) at the upper limit. For example, for water, triple point temperature and pressure are 0.01°C and 0.6113 kPa, respectively. This means all three phases of water (solid, liquid and vapour) will exist in equilibrium only if T and p have exactly these values. No substance can exist in liquid phase in stable equilibrium at pressures below the triple point pressure. The critical point is defined as the point at which the saturated liquid and saturated vapour states cannot be distinguished. The critical pressure and temperature of water are 22.09 MPa and 374.14°C, respectively (Cengel and Boles, 2003). At $p > p_c$ there is no distinct liquid-to-vapour phase change process, that is, it has all vapour phase. For a constant temperature T within the aforesaid range the pressure and volume vary along a line such as ABCD. Along line AB only liquid exists where vapour state exists along the line CD. Liquid–vapour mixture exists along the line BC. The saturation curve is the locus of the points such as B and C. The pressure remains invariant along the line BC while the specific volume, v (volume per unit mass) varies and is determined by its enthalpy. The projection of the liquid + vapour surface in the pressure–temperature plane (viewed along the direction of the arrow on the right hand side of the figure) gives rise to a single curve called the vapour pressure curve relating pressure and temperature between T_1 and T_c. The pressure and temperature values on this curve are referred to as saturation pressure (p_{sat}) and saturation temperature (T_{sat}), respectively.

The vapour pressure curve can be determined by partially filling a large diameter container with the chosen substance under vacuum, subjecting the container to a uniform temperature field and monitoring the pressure of the vapour above the liquid for each desired temperature level. At equilibrium, the number of molecules striking and being absorbed by the interface from the vapour phase is exactly same as the number of molecules being emitted through the interface from the liquid phase. Since the interface is plane the pressures in the liquid and vapour immediately adjacent to the interface are equal at equilibrium (Collier, 1972).

Superheated Liquid and Supersaturated Vapour

The p-v-T diagram represents only stable equilibrium states. However, other metastable or unstable states can occur where the coordinates of pressure, volume, and temperature (p, v, T) do not lie in any of the surfaces.

For example, if one carefully reduces the liquid pressure at constant temperature along a line AB it is possible that no vapour will form at point B (Fig.7.4). Similarly, it is possible to increase the pressure on a vapour along a line DC and yet no liquid forms at point C (Fig.7.4). The p-v-T coordinates of these metastable states lie along an extrapolation of AB to B$'$ and DC to C$'$ (Collier, 1972). Point B$'$ may also be obtained by carefully heating the liquid above saturation temperature corresponding to the imposed static pressure on the liquid. This process is referred to as *superheating* and the metastable liquid state is called superheated liquid. The equivalent cooling process leading to non-formation of liquid along the extrapolated line CC$'$ is called *supersaturation* and the metastable vapour state is called supersaturated vapour.

Requirement of Liquid Superheating for Vapour Bubble Formation

Vapour and liquid phases can coexist in unstable equilibrium along lines BB$'$ and CC$'$. In this case, unlike in stable equilibrium the liquid and vapour pressures in the vicinity of the interface are no longer equal. If the interface is concave with the centre of curvature in the vapour phase then the vapour pressure p_v and the liquid pressure p_l will be related by the following equation (Collier, 1972):

$$p_v - p_l = \sigma \left(\frac{1}{r_1} + \frac{1}{r_2} \right) \qquad (7.10)$$

where σ is the surface tension of the liquid–vapour interface, and r_1 and r_2 are the main radii.

For a perfectly spherical bubble, $r_1 = r_2 = r$ and Eq. (7.10) will be reduced to

$$p_v - p_1 = \frac{2\sigma}{r} \qquad (7.11)$$

The above expression can be derived by balancing the pressure difference between the inside and outside of the bubble and the forces of surface tension, as indicated in Fig.7.5. The pressure force is $(p_v - p_l)\pi r^2$ and the surface tension force is $2\pi r\sigma$. On equating these two forces, the expression given in Eq. (7.11) will result.

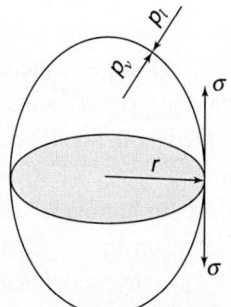

Since the vapour pressure is greater than liquid pressure the number of molecules striking and being absorbed by the interface from the vapour phase is greater than when the interface is planar. To maintain equilibrium the number of molecules emitted through the interface from the liquid phase must increase correspondingly (Collier, 1972). This

Fig. 7.5 Force balance on a spherical vapour bubble

enhancement can only be possible by increasing the temperature of the system above that necessary for equilibrium with planar interface, that is, the saturation temperature (T_{sat}) corresponding to the liquid pressure (p_1). Therefore, the liquid adjacent to the curved interface is superheated with respect to the imposed liquid pressure.

To conclude it can be said that some amount of liquid superheating is necessary for vapour bubbles to form.

7.3 Formation of Vapour Bubbles

This section explains how vapour bubbles form on a hot surface during boiling of a liquid.

It is well known that no bubbles form in a completely pure, carefully degassed liquid, unless the liquid is extremely superheated. Furthermore, it has been observed that the bubbles reappear at the same place on the heater surface with a varying frequency. Obviously, there must be some relation of this phenomenon with highly active centres that catalyse the transformation from unstable superheated liquid to stable vapour. These centres are the remains of gas or vapour in cavities on the surface. Figure 7.6 shows a highly expanded view of a heater surface. Most machining operations score tiny grooves or cavities on the surface. When a surface is wetted, liquid is prevented by surface tension from completely filling these holes and therefore, small gas or vapour pockets are formed. These little pockets serve as bubble nucleation sites.

Gas
pockets
serving
as nucleation sites

Fig. 7.6 Expanded view of a heater surface

When the surface is heated the gas or vapour remnants expand until a critical size corresponding to the size of a viable bubble is reached. Then a vapour bubble can grow further as a result of the superheating of the liquid, until finally buoyancy and dynamic forces become larger than the adhesion (such as surface tension force) and the bubble detaches from the heated surface. Subsequently, the bubble lifts off leaving a small remnant of gas or vapour in the cavity. This remnant vapour will be cooled by the cold liquid coming from the interior of the fluid to the wall, and then heated by the addition of the heat from the wall. A new nucleus for the growth of a vapour bubble is born. This explains why the surface roughness is an important parameter in boiling heart transfer.

The nucleation mechanism explained above is termed *heterogeneous nucleation*. Vapour bubbles almost always form on the solid surfaces or on suspended particles. The homogeneous nuclei formation, with bubbles formed by "themselves" within the liquid as a result of natural fluctuation of local molecular density and energy plays a very minor role and becomes important only when the liquid temperature is much above its saturation temperature corresponding to its pressure. For example, the homogeneous nucleation temperature of water at 1 bar is 300°C.

7.4 Bubble Departure Diameter and Frequency of Bubble Release

The bubble departure diameter (d_d) is the diameter of a bubble at the time of its detachment from the heated surface. The frequency of bubble release (f) is defined as

$$f = \frac{1}{\tau} \tag{7.12}$$

where τ is the time period associated with the growth of each bubble and is equal to the sum of the waiting period and the time required for the bubble to grow to its departure diameter.

The frequency of bubble release is directly related to how large the bubble must become for release to occur, and, as a result, on the rate at which the bubble can grow to this size.

The bubble departure diameter is determined by the net effect of the adhesive force which tries to hold the bubble in place on the surface and the buoyancy force (for an upward facing horizontal surface) and dynamic forces (if the liquid adjacent to the surface has a bulk motion associated with it) which tend to dislodge the growing

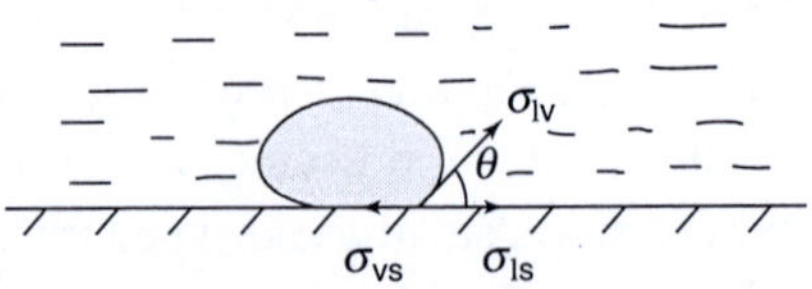

Fig. 7.7 Vapour bubble on a surface

bubble from the surface. Surface tension acting along the contact line invariably plays a key role in keeping the bubble attached to the surface. Figure 7.7 shows a bubble attached to a surface and three surface tension forces acting on it. σ_{lv} is the surface tension between the liquid and vapour and is determined by the properties of the vapour and the liquid and is acting in the liquid–vapour interface. σ_{ls} is the surface tension between the liquid and the solid surface and is determined by the properties of the solid surface and the liquid and is acting along the solid surface on the liquid side. Finally, σ_{vs} is the surface tension between the vapour and the solid and is determined by the properties of the vapour and the solid surface and is acting along the solid surface on the vapour side. The equilibrium in the horizontal direction requires

$$\sigma_{vs} - \sigma_{ls} = \sigma_{lv} \cos\theta \tag{7.13}$$

where θ is the bubble contact angle in degrees and depends on the relative magnitude of different surface tensions. It may be smaller or larger than 90°. When $\theta < 90°$, the liquid is said to wet the surface. When $\theta > 90°$, the surface is considered to be non-wetted. For water at atmospheric pressure, $\theta = 40 - 45°$. For refrigerants of halogenated hydrocarbons $\theta = 35°$.

7.4.1 Departure Diameter Correlations

The departure diameter was typically obtained from high-speed movies of the boiling process. Based on this kind of data, various investigators have proposed correlations which can predict the departure diameter of the bubbles during nucleate boiling. Some of them are given below.
Fritz (1935)

$$\text{Bo}_d^{1/2} = 0.0208\theta \tag{7.14}$$

where θ is the contact angle in degrees.

Equation (7.14) was obtained based on a simple balance of surface tension forces and buoyancy at the time of departure. The effect of contact angle has been taken into account in an empirical manner.

$$\text{Bo}_d^{1/2} = \frac{g(\rho_l - \rho_v)d_d^2}{\sigma} \tag{7.15}$$

Bo_d is called Bond number.
Zuber (1959)

$$\text{Bo}_d^{1/2} = \left[\frac{\sigma}{g(\rho_l - \rho_v)} \right]^{-1/6} \left[\frac{6k_l(T_w - T_{sat})}{q''} \right]^{1/3} \tag{7.16}$$

The ratio $\dfrac{k_l(T_w - T_{sat})}{q''}$ is a length scale which represents the superheated thermal boundary layer thickness adjacent to the surface. Thus surface tension, buoyancy, and the size of the bubble relation to this boundary layer thickness are represented in this correlation.

Cole (1967)

$$\text{Bo}_d^{1/2} = 0.04 Ja \tag{7.17}$$

where $Ja = \dfrac{\rho_l c_{pl} \left[T_w - T_{\text{sat}}(p_l) \right]}{\rho_v h_{fg}}$ is called Jacob number.

Thus Eq. (7.17) indicates a simple functional dependence of Bond number on Jakob number.

7.4.2   Frequency of Bubble Release Correlations

The frequency of bubble release depends on the bubble diameter at the time of release and on the rate of bubble growth to departure diameter. Hence the frequency of release is a function of the departure diameter and all the conditions and fluid properties affecting the waiting time and the bubble growth rate. The size and nature of each cavity affects the nucleation and waiting time behaviour. Nucleation sites with rather small cavity diameters emit bubbles with higher frequency as compared to the sites having larger cavity diameters. Although different cavities emit bubbles at different frequencies, it is useful to take the mean bubbling frequency f associated with the boiling process.

A number of researchers have obtained correlations predicting frequency of bubble release. Two of them are listed below.

Jacob and Fritz (1931)

$$fd_d = 0.078 \text{ m/s} \tag{7.18}$$

Equation (7.18) is valid for hydrogen and water vapour bubbles.

Zuber (1963)

$$fd_d = 0.59 \left[\frac{\sigma g (\rho_l - \rho_v)}{\rho_l^2} \right]^{1/4} \tag{7.19}$$

The above correlation is based on an analogy between the bubble release process and natural convection.

7.5   Boiling Modes

Boiling is characterized by the formation of vapour bubbles, which grow and subsequently detach from the surface. Vapour bubble growth and dynamics depend, in a complicated manner, on the excess temperature or wall superheat, $\Delta T_w \simeq T_w - T_{\text{sat}}$, the nature of the surface, and thermophysical properties of the fluid, such as its surface tension, viscosity, and the latent heat of vapourization.

Two types of boiling are normally encountered, namely, pool boiling and flow boiling. In pool boiling the liquid is quiescent and its motion near the surface is due to free convection and mixing induced by the bubble growth and detachment. In contrast, in flow boiling, fluid motion is induced by externally imposed pressure differences, as well as by free convection and bubble-induced mixing. Boiling may be classified according to whether it is sub-cooled or saturated. In *sub-cooled boiling*, the temperature of the liquid is below the saturation temperature and bubbles formed at the surface may condense in the liquid. On the other hand, the temperature of the liquid slightly exceeds the saturation temperature in *saturated boiling*. Bubbles formed at the surface are then propelled through the liquid by buoyancy forces. In this chapter both saturated pool boiling and flow boiling will be discussed.

7.5.1 Saturated Pool Boiling

Many researchers have studied saturated pool boiling. In pool boiling flow results only by means of the buoyancy of the produced bubbles and by the differences in the density. An appreciation of the physical mechanism of boiling may be obtained by examining the boiling curve.

7.5.2 Boiling Curve

Nukiyama (1934) was the first to identify different regimes of pool boiling using the apparatus shown in Fig. 7.8. By calibrating the resistance of a nichrome wire as a function of temperature before the experiment, he was able to obtain both the heat flux and the temperature using the observed current and voltage (Fig. 7.9).

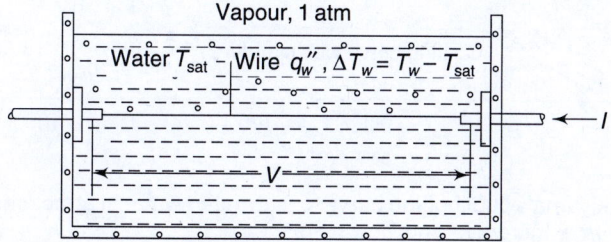

Vapour, 1 atm

Water T_{sat} — Wire q_w'', $\Delta T_w = T_w - T_{sat}$

Fig. 7.8 Nukiyama experiment for pool boiling

The heat flux from a horizontal nichrome wire to saturated water was determined by measuring the current flow I and the voltage drop V. The wire temperature was determined from the knowledge of the manner in which its electrical resistance varied with temperature. This arrangement is called *power-controlled heating*, wherein the wire temperature T_w (and hence the excess temperature or wall superheat ΔT_w) is the dependent variable and the power setting (and hence the heat flux q_w'') is the independent variable. Following the arrows

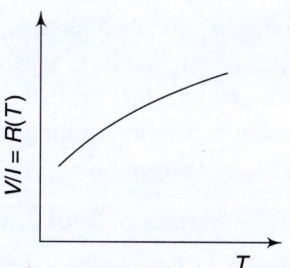

Fig. 7.9 Calibration curve to determine wire temperature from voltage drop and current flow

of the heating curve of Fig. 7.10, it is evident that as power is applied, the heat flux increases, at first slowly and then very rapidly with increase in wall superheat.

Nukiyama observed that boiling did not begin until $\Delta T_w \simeq 5\,°C$. As the imposed heat flux increased slightly above the critical value q_{max}'', the wire temperature jumped to the melting point and burnout occurred. However, repeating the experiment with a platinum wire having a higher melting point (2045 K versus 1500 K), Nukiyama was able to maintain heat fluxes above the maximum value q_{max}'' without burnout. When he subsequently ran the experiment in reverse, that is, by decreasing the heat flux, the variation of ΔT_w with q_w'' followed the cooling curve of Fig. 7.10. When the heat flux is lowered slightly below q''_{min}, the vapour film collapses, isolated bubbles form, and the wire temperature drops to the low level associated with the nucleate boiling regime.

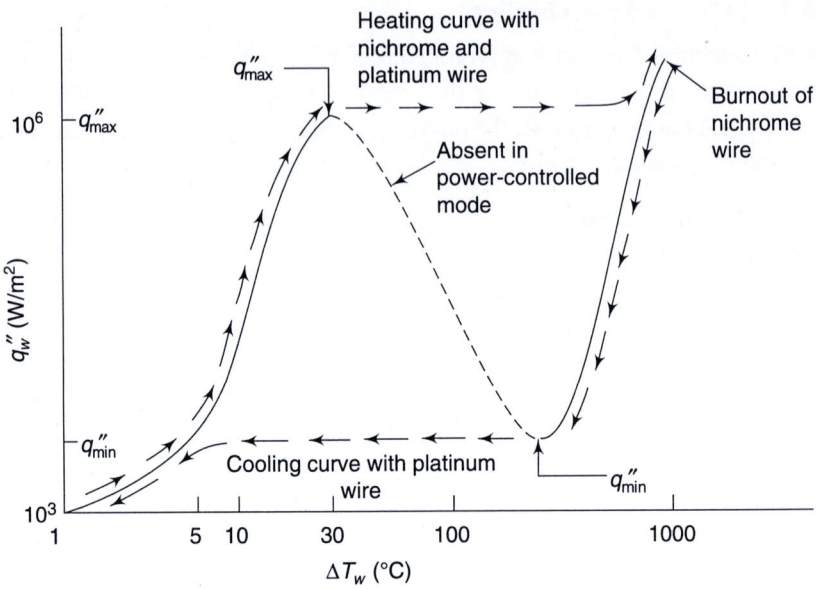

Fig. 7.10 Nukiyama's pool boiling curve for saturated water at atmospheric pressure (*source:* Incropera and Dewitt, 1998)

To sum up, during heat-flux-controlled boiling the transition boiling regime is inaccessible, and certain portions of the boiling curve can be obtained only by varying q''_w in one direction. It is said that heat flux-controlled boiling has an hysteresis effect, that is, it depends not only on the imposed condition (q''_w) but also on its previous history, in this case the previous value of q''_w. The existence of the missing transitional boiling regime of the curve was obtained by Drew and Mueller (1937) by temperature (T_w)-controlled experiments.

7.5.3 Modes of Pool Boiling

Figure 7.11 depicts the individual modes or regimes that occur during boiling of water at atmospheric pressure. The heat flux q''_w is plotted as a function of the wall superheat $\Delta T_w = T_w - T_{sat}$ on a log-log scale. Similar trends of the curve have been observed for other fluids. In the left ascending branch of the curve, the region up to point A represents free convection evaporation (i.e., no bubble is seen at this stage). In this regime ($\Delta T_w \leq 5°C$), the fluid motion is determined principally by free convection effects known as Benard convection.

The portion ABC of the curve represents the region of *developed nucleate boiling*. The heat flux increases with the wall temperature (or wall superheat). In region A–B, isolated bubbles form at nucleation sites and detach from the surface. The formation, growth, and departure of vapour bubbles produce strong local turbulent flows close to the wall, resulting in considerable fluid mixing near the surface. This substantially increases h and q''_w. In this regime, most of the convective heat exchange is through direct transfer from the surface to the liquid in motion at the surface, and not through the vapour bubbles rising from the surface. With increase in the wall superheat (also called the excess temperature), more and more nucleation sites are activated. This increased bubble formation causes bubble interference and coalescence. In region *B–C*, the vapour escapes in the form of jets or columns, which subsequently coalesce into slugs of vapour. Because the vapour isolates the heating surface more and more from the liquid, the heat transfer coefficient falls as more vapour is formed. Point *P* of Fig. 7.11 is an inflection point in the boiling curve, at which the heat transfer coefficient is maximum. After this point, h begins to decline with increasing ΔT_w, although q_w'' continues to increase. With further increase in the wall temperature, the isolating effect of the vapour finally predominates and the heat flux decreases again in spite of the rising water temperature. The point after which the heat flux shows a fall is called the *maximum heat flux* or *critical heat flux point*, $q_{w,\,max}''$ (or $q_{w,\,c}''$). For water, the value of $q_{w,\,max}''$ is around 1 MW/m^2 at atmospheric pressure. The approximate magnitude of h is around 10^4 W/m^{2}K, which is several orders of magnitude larger than the corresponding value in single-phase convection.

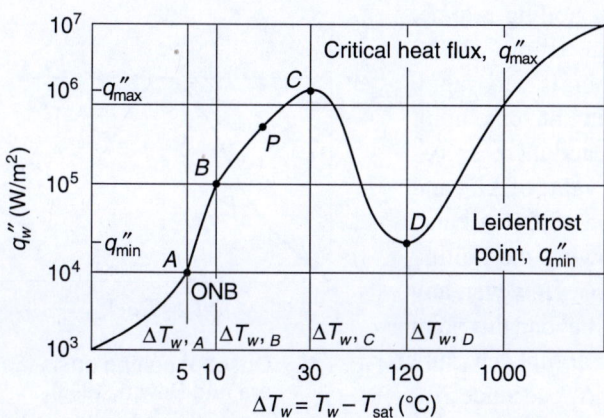

Fig. 7.11 Typical boiling curve for water at 1 atm (q_w''versus ΔT_w) (*source:* Incropera and Dewitt, 1998)

To conclude, nucleate boiling exists in the range $\Delta T_{w,\,A} \leq \Delta T_w \leq T_{w,\,C}$, where $\Delta T_{w,\,C} = 30\,°C$. Point *A* is referred to as the onset of nucleate boiling, ONB. Point *C* is called the departure from nucleate boiling, DNB.

After reaching a maximum, the transferred heat flux decreases in spite of the rising wall temperature. This falling region of the boiling curve is designated as *partial film boiling*, because the heating surface is covered partially by vapour. This region, corresponding to $\Delta T_{w,\,C} \leq \Delta T_w \leq \Delta T_{w,\,D}$ (where $\Delta T_{w,\,D} \approx 120\,°C$), is also termed *transitional boiling* or *unstable film boiling*. At any point on the surface, conditions may alternate between film and nucleate boiling, but the fraction of the total surface covered by the vapour film increases with increasing ΔT_w. Because of low vapour conductivity as compared to liquid, h (and q_w'') must decrease with increasing ΔT_w.

After surpassing a minimum heat flux, the heat flux increases again with the wall temperature. The right ascending branch of the boiling curve is characterized as the region of *film boiling*, because the heating surface there is completely covered by a vapour film. This region corresponds to $\Delta T_w \geq \Delta T_{w,D}$. The point D at which the heat flux is minimum, $q''_{w,D} = q''_{w,min}$, is referred to as the *Leidenfrost point*. In 1756, Leidenfrost observed that water droplets supported by a vapour film slowly boil away as they move about a hot surface. Because the rate of vapour generation just equals the rate at which the vapour leaves the film in the form of detaching bubbles, the film is steadily present. As the surface temperature increases, radiation becomes the dominant heat transfer mode and the heat flux increases with increasing ΔT_w.

The foregoing discussion shows that the famous N-shaped pool boiling curve, which at first seems rather strange, is physically plausible.

7.5.4 Importance of Critical Heat Flux

The knowledge of the value of the critical heat flux (CHF) is extremely important in the design of boiling equipment. Although the discussion of the boiling curve assumes that the T_w-controlled experiment is possible, but in most practical situations, such as in a nuclear reactor or in an electric resistance heating device or in a boiler in a thermal power plant, boiling is q''_w-controlled, which does not go through the transitional boiling regime, and the change from CHF to the burnout point is very rapid. Consider starting at some point P in Fig. 7.12, and increase q''_w gradually. The value of ΔT_w, and hence the value of T_w, will also increase, following the boiling curve to point C. However, any increase in q''_w beyond this point will result in an abrupt departure from $\Delta T_{w,C}$ to $\Delta T_{w,E}$. Since $T_{w,E}$

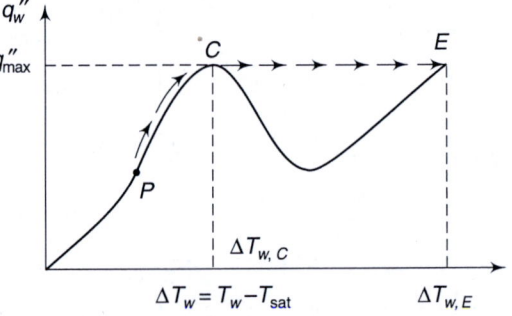

Fig. 7.12 Onset of boiling crisis (*Source:* Incropera and Dewitt, 1998)

may exceed the melting point of the material of the heating surface, destruction or failure of the system may occur. For this reason, point C is often less accurately termed the *burnout point* or the *boiling crisis*. Hence, the accurate knowledge of the CHF, $q''_{w,C} \equiv q''_{w,max}$, is important. An engineer may want to operate a boiling equipment close to this value, but would never risk exceeding it.

7.5.5 T_w versus q''_w Curve

Traditionally, the boiling curve is plotted as a q''_w versus ΔT_w curve. But one may like to know the effect of the wall heat flux on the wall temperature directly. In that case, the T_w versus q''_w plot would be more convenient.

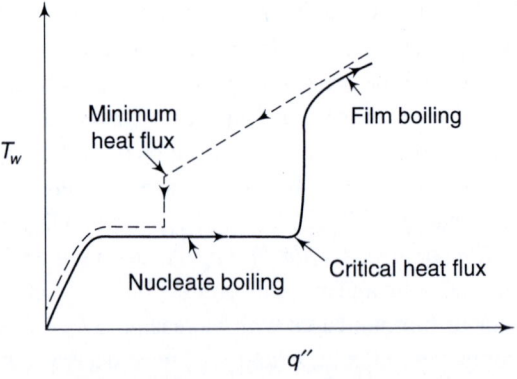

Fig. 7.13 T_w versus q'' curve for power controlled situation

Figure 7.13 shows the T_w versus q_w'' curve for a heat-flux-controlled situation. The same form of the curve is observed on increasing the heat flux: the system passes through the free convection and nucleate boiling regimes. When the maximum heat flux is reached, there is an excursion into film boiling, as shown, with a rapid (and often catastrophic) increase in the wall temperature. When the heat flux is decreased, the wall temperature decreases until the minimum heat flux is reached, after which there is an excursion back into nucleate boiling.

7.6 Heat Transfer Mechanism in Nucleate Boiling: Rohsenow's Model and its Basis

Rohsenow (1952) postulated that due to formation, growth and departure of vapour bubbles, strong local flows close to the heater surface are produced. The flow is turbulent in nature and turbulence is induced by the bubbles. The bubble stirring action is the reason for high heat transfer coefficient associated with nucleate boiling. Essentially, heat flows from the surface first to the adjacent liquid, as in any single-phase convection process (Carey, 2008).

The above hypothesis suggests that it may be possible to adapt a single-phase forced convection heat transfer correlation to nucleate pool boiling, if the appropriate length and velocity scales are specified (Carey, 2008). Thus,

$$\mathrm{Nu}_b = \frac{hL_b}{k_l} = A\,\mathrm{Re}_b^n\,\mathrm{Pr}_l^m \tag{7.20}$$

where L_b is an appropriate bubble length scale. The Reynolds number Re_b is given by

$$\mathrm{Re}_b = \frac{\rho_v u_b L_b}{\mu_l} \tag{7.21}$$

Equation (7.21) indicates a ratio of vapour inertia force to liquid viscous force, u_b is an appropriate velocity scale.

Rohsenow took the length scale and the velocity scale to be the bubble departure diameter d_d and the vapour superficial velocity, respectively. They are defined as

$$L_b = d_d = C_b\theta \left[\frac{2\sigma}{g(\rho_l - \rho_v)}\right]^{1/2} \tag{7.22}$$

$$u_b = \frac{q''}{\rho_v h_{fg}} \tag{7.23}$$

where u_b is the average velocity of vapour receding from the surface.

In Eq. (7.22) θ is the contact angle and C_b is a constant specific to the surface–liquid combination. Equation (7.22) is equivalent to the departure diameter correlation of Fritz (1935) [see Eq. (7.14)]. Now, we know

$$h = \frac{q''}{T_w - T_{\mathrm{sat}}(p_l)} \tag{7.24}$$

The relationship of Nu with Re and Pr is postulated to be of the form

$$\mathrm{Nu}_b = A\,\mathrm{Re}_b^{(1-r)}\,\mathrm{Pr}_l^{(1-s)} \tag{7.25}$$

Substituting Eqs (7.21)–(7.24) into Eq. (7.25) and rearranging yields

$$\frac{q''}{\mu_l h_{fg}}\left[\frac{\sigma}{g(\rho_l-\rho_v)}\right]^{1/2}=\left(\frac{1}{C_{s,f}}\right)^{1/r}\mathrm{Pr}_l^{-s/r}\left[\frac{c_{pl}(T_w-T_{sat}(p_l))}{h_{fg}}\right]^{1/r} \tag{7.26}$$

where $\quad C_{s,f}=\dfrac{\left(\sqrt{2}C_b\theta\right)^r}{A}$ (7.27)

Equation (7.27) is of the form of the well-known Rohsenow correlation for pool boiling heat transfer given in the next section. Originally, values of $r = 0.33$ and $s = 1.7$ were recommended for Eq. (7.26).

Subsequently, Rohsenow recommended that for water only be changed to 1. $C_{s,f}$ depends on the surface–liquid combinations and are listed in Table 7.1. The values of $C_{s,f}$ were obtained based on fits to pool boiling data available in literature. For those surface–liquid combinations not listed in Table 7.1 or in other published literature it is recommended that an experiment be conducted to determine the appropriate value of $C_{s,f}$ for the particular surface–liquid combination of interest. If this is not possible, a value of $C_{s,f} = 0.013$ is suggested as a first approximation.

7.7 Empirical Correlations and Application Equations

In this section empirical correlation for heat flux in the nucleate pool boiling regime of the pool boiling curve and the application equation for critical heat flux (CHF) are given.

7.7.1 Correlation of Rohsenow in the Nucleate Pool Boiling Regime

This is the first and most useful correlation for nucleate boiling:

$$q_w''=\mu_l\,h_{fg}\,[g(\rho_l-\rho_v)/\sigma]^{1/2}\,(c_{p,l}\,\Delta T_w/\,C_{s,f}h_{fg}\,\mathrm{Pr}_l^s)^3 \tag{7.28}$$

The appearance of the surface tension σ (N/m) follows from the significant effect this fluid has on bubble formation and development. The coefficient $C_{s,f}$ and the exponent s depend on the surface–liquid combination, and representative values are presented in Table 7.1.

Table 7.1   Values of $C_{s,f}$ and s for various liquid–surface combinations

Liquid–surface combination	$C_{s,f}$	s
Water–copper		
Scored	0.0068	1.0
Polished	0.0130	1.0
Water–stainless steel		
Chemically etched	0.0130	1.0
Mechanically polished	0.0130	1.0
Ground and polished	0.0060	1.0
Water–brass	0.0060	1.0
Water–nickel	0.006	1.0
Water–platinum	0.0130	1.0

(Contd.)

(Contd.)

n-pentane–copper		
Polished	0.0154	1.7
Lapped	0.0049	1.7
Benzene–chromium	0.101	1.7
Ethyl alcohol–chromium	0.0027	1.7

The Rohsenow correlation (1952) applies only for clean surfaces. When it is used to estimate the heat flux, errors can amount to $\pm 100\%$. However, since $\Delta T_w \, (q_w'')^{1/3}$, this error is reduced by a factor of 3 when the expression is used to estimate ΔT_w from the knowledge of q_w''. Also, since $q_w'' \propto (h_{fg})^{-2}$ and h_{fg} decreases markedly with increasing saturation pressure (temperature), the nucleate boiling heat flux will increase significantly as the liquid is pressurized.

Note that $c_{p,l}$ is the specific heat of saturated liquid (J/kg K), $\Delta T_w = T_w - T_{sat}$ is the temperature excess (°C), h_{fg} is the enthalpy of vapourization (J/kg), Pr_l is the Prandtl number of saturated liquid, q_w'' is the wall heat flux (W/m^2), μ_l is the liquid viscosity (kg/ms), σ is the surface tension of the liquid–vapour interface (N/m), g is the gravitational acceleration (9.81 m/s^2), ρ_l is the density of saturated liquid (kg/m^3), ρ_v is the density of saturated vapour (kg/m^3), $C_{s,f}$ is a constant (determined from experimental data), and $s = 1.0$ for water and 1.7 for other liquids.

7.7.2   Critical Heat Flux for Nucleate Pool Boiling

Kutateladze (1948) and Zuber (1958) obtained an expression of the form

$$q_{w,max}'' = (\pi/24) \, h_{fg} \rho_v \, [\sigma g(\rho_l - \rho_v)/ \rho_v^2 \,]^{1/4} \{1 + (\rho_v /\rho_l)\}^{1/2} \tag{7.29}$$

which, as a first approximation, is independent of the surface material and only weakly dependent on geometry. Replacing the Zuber constant $(\pi/24)$ = 0.131 by an experimental value of 0.149 and approximating the last term in the parentheses to unity, the above equation becomes

$$q_{w,max}'' = 0.149 \, h_{fg} \, \rho_v \, [\sigma \, g(\rho_l - \rho_v)/ \rho_v^2 \,]^{1/4} \tag{7.30}$$

In principle, since this expression applies to a horizontal heated surface of infinite extent, there is no characteristic length; in practice, however, the expression is applicable if the characteristic length is large compared to the mean bubble diameter.

The critical heat flux depends strongly on pressure, mainly through the pressure dependence of the surface tension and the enthalpy of vapourization. With increasing pressure and at constant wall superheating, more vapour bubbles are formed. The heat transfer, therefore, increases with pressure. However, at sufficiently high pressure, a more or less complete vapour film is formed, which in turn leads to a reduction in heat transfer.

7.8   Heat Transfer in the Vicinity of Ambient Pressure

The measurement on water in the range of fully developed nucleate boiling for heat fluxes between 10^4 W/m^2 and 10^6 W/m^2 and saturation pressure between 0.5 bar and 20 bars (1 bar = 10^5 N/m^2) can be represented well, according to the investigations by Fritz (1963), by a simple empirical equation in which h is in W/m^{2}K, q_w'' is in W/m^2, and p is in bars.

$$h = 1.95(q_w'')^{0.72}(p)^{0.24} \tag{7.31}$$

7.9 Minimum Heat-flux Expression

The minimum heat flux q_{min}'' occurs at the lowest heater temperature where the film is continuous and stable.

Zuber (1958) used the theory of stability of the vapour–liquid interface of the film to derive the following expression for the minimum heat flux from a large horizontal plate. The unstable wavy configuration that can be assumed by the horizontal interface between a heavy fluid (above) and a lighter fluid (below) is called the Taylor instability

$$q_w'' = c\rho_v h_{fg}\left[\frac{g\sigma(\rho_l - \rho_v)}{(\rho_l + \rho_v)^2}\right]^{1/4} \tag{7.32}$$

The constant, $c = 0.09$, has been experimentally determined by Berenson (1961). The result is accurate to approximately 50% for most fluids at moderate pressures but provides poor estimates at higher pressures. A similar result has also been obtained for horizontal cylinders.

7.10 Film Boiling Correlations

For the ascending portion of the film boiling curve, the correlations have forms similar to those obtained in the study of film condensation (see Section 7.11). One such result, which applies to film boiling on a cylinder or sphere of diameter D, is of the form

$$\overline{Nu}_D = \frac{\overline{h}_{conv}D}{k_v} = c\left[\frac{g(\rho_l - \rho_v)h_{fg}'D^3}{v_v k_v(T_w - T_{sat})}\right]^{1/4} \tag{7.33}$$

The correlation constant $c = 0.62$ for horizontal cylinders and 0.67 for spheres. The corrected latent heat h_{fg}' accounts for the sensible energy required to maintain temperatures within the vapour blanket above the saturation temperature. Although it may be approximated as $h_{fg}' = h_{fg} + 0.8c_{p,v}(T_w - T_{sat})$, it is known to depend weakly on the Prandtl number of the vapour. Vapour properties are calculated at the film temperature $T_f = (T_w + T_{sat})/2$, and the liquid density is evaluated at the saturation temperature.

At elevated surface temperatures ($T_w \geq 300\,^\circ$C), the radiation heat transfer across the vapour film becomes significant. Bromley (1950) investigated film boiling from the outer surface of horizontal tubes and suggested calculating the total heat transfer coefficient from a transcendental equation of the form

$$\overline{h}^{4/3} = \overline{h}_{conv}^{4/3} + \overline{h}_{rad}\overline{h}^{1/3} \tag{7.34}$$

where $\quad \overline{h}_{rad} = \dfrac{\varepsilon\sigma(T_w^4 - T_{sat}^4)}{T_w - T_{sat}} \tag{7.35}$

Note that in Eq. (7.34) ε is the emissivity of the heater surface, σ is the Stefan–Boltzmann constant $(5.667 \times 10^{-8} \text{ W/m}^2\text{K}^4)$, and T is in Kelvins.

Example 7.1 Boiling of Water in a Saucepan

In a saucepan 1 l of water at atmospheric pressure (1 bar) is to be boiled on an electric heater. The power of the heater is q = 3 kW. The diameter of the heater is the same as that of the saucepan, i.e., 0.3 m.

(a) *How long does it take for the water to start boiling if the initial temperature is 20°C? The heat losses to the surroundings amount to 30% of the heat input.*

(b) *What is the temperature at the bottom of the saucepan when the water begins to boil?*

(c) *Estimate the time required for complete vapourization of all the water.*

(d) *Calculate the maximum heat flux.*

Solution

The following data are given:

$$T_{sat} = 100°C \text{ at } 1.013 \text{ bars}, \quad T_{initial} = 20°C, \quad q = 3 \text{ kW}$$

The properties of saturated water, liquid (100°C):

$$\rho_l = 958.1 \text{ kg/m}^3, \quad c_{p,l} = 4.216 \text{ kJ/kg K}, \quad \rho_v = 0.5974 \text{ kg/m}^3$$

$$\sigma = 58.92 \times 10^{-3} \text{ N/m}, \quad h_{fg} = 2257.3 \text{ kJ/kg}$$

(a) Until the boiling point is reached, the following amount of heat must be supplied to the heater.

$$Q = mc_{p,l}(T_{sat} - T_{initial}) = \rho_v V c_{p,l}(T_{sat} - T_{initial})$$

$$= (958.1)(10^{-3})(4.216)(100 - 20) = 323 \text{ kJ}$$

Since 30% of the heat supplied is lost to the surroundings, we can write

$$(q - 0.3q)\, t = Q$$

or $t = Q/0.7q = 323/(0.7)(3) = 154 \text{ s} = 2.56 \text{ min}$

(b) The heat transfer coefficient in boiling is obtained from Eq. (7.31) as follows:

$$h = 1.95 (q_w'')^{0.72} \, p^{0.24}$$

with $q_w'' = (0.7)(3)/((\pi/4)(0.3)^2 = 2.971 \times 10^4 \text{ W/m}^2$

So, $h = 1.95 (2.97 \times 10^4)^{0.72} (1.013)^{0.24} = 3250 \text{ W/m}^2\text{K}$

and $\Delta T_w = \dfrac{q_w''}{h} = \dfrac{2.97 \times 10^4}{3250} = 9.1°C$

The temperature at the bottom of the pan is

$$T_w = T_{sat} + \Delta T_w = 100 + 9.1 = 109.1°C$$

(c) In order to completely vapourize the entire mass of water, the following amount of heat is required

$$Q = \rho_l V h_{fg} = (958.1)(10^{-3})(2257.3) = 2163 \text{ kJ}$$

Now, 70% of the heat supplied is fed to the water. Therefore, the time required to completely vapourize all the water is

$$t = \frac{Q}{q_w} = \frac{2163}{(0.7)(3)} = 1030 \text{ s} = 17.2 \text{ min}$$

(d) The maximum heat flux is found from Eq. (7.30):

$$q_{w,max}'' = 0.149\, h_{fg}\, \rho_v \left[\frac{\sigma g (\rho_l - \rho_v)}{\rho_v^2} \right]^{1/4}$$

$$= 1.26 \times 10^6 \text{ W/m}^2 = 1.26 \text{ MW/m}^2$$

Example 7.2   Film Boiling on a Horizontal Cylinder

A metal-clad heating element of 6 mm diameter and emissivity ε = 1 is horizontally immersed in a bath of water. The surface temperature of the metal is 255 °C under steady-state boiling conditions. Calculate the power dissipation per unit length of the heater.

Solution

Given data:

Saturated water, liquid (100 °C), $\rho_l = 957.9$ kg/m³, $h_{fg} = 2257$ kJ/kg

Saturated water vapour [$T_f = (T_w + T_{sat})/2 = (255 + 100)/2 + 273 = 450.5$ K], $\rho_v = 4.808$ kg/m³, $c_{p,v} = 2.56$ kJ/kgK, $k_v = 0.0331$ W/m K, $\mu_v = 14.85 \times 10^{-6}$ N s/m²

The wall superheat is

$$\Delta T_w = T_w - T_{sat} = 255 - 100 = 155 °C$$

According to the boiling curve (Fig. 7.10), stable film boiling conditions exist, in which case heat transfer is due to both convection and radiation.

$$q_w' = q_w'' \pi D = \bar{h} \pi D \Delta T_w$$

From Eqs (7.33) and (7.34)

$$\bar{h}_{conv} = 0.62 \left[\frac{\rho_v k_v^3 (\rho_l - \rho_v)(h_{fg} + 0.8 c_{p,v} \Delta T_w) g}{\mu_v D \Delta T_w} \right]^{1/4}$$

Substituting all the values in the above equation, we get

$$\bar{h}_{conv} = 460 \text{ W/m}^2 \text{ K}$$

Now from Eq. (7.35), we obtain

$$\bar{h}_{rad} = 21.3 \text{ W/m}^2 \text{ K}$$

Solving Eq. (7.34), which is a transcendental equation, by trial and error, it follows that

$$\bar{h} = 476 \text{ W/m}^2 \text{ K}$$

Hence, $q_w' = \bar{h} \pi D \Delta T_w = 476 (\pi)(6 \times 10^{-3})(155) = 1390$ W/m $= 1.39$ kW/m

7.11   Condensation

Condensation occurs when the temperature of a vapour is reduced below its saturation temperature. In industrial equipment, the process commonly occurs from contact between the vapour and a cooled surface [Figs 7.14(a), (b)]. The enthalpy of condensation is released, heat is transferred to the surface, and the condensate is formed, which as a result of its contact with the wall surface is sub-cooled, so that further vapour can condense upon the previously formed condensate.

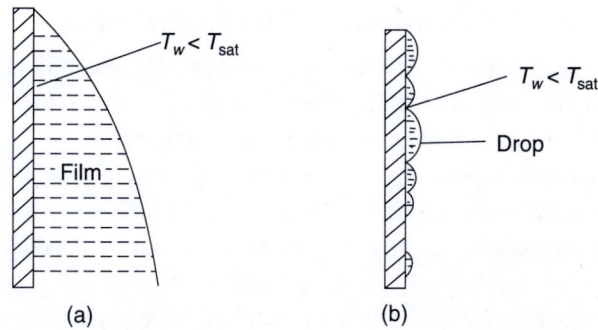

Fig. 7.14 (a) Film condensation, (b) dropwise condensation

As shown in Figs 7.14(a) and (b), condensation may occur in one of two ways, depending on the condition of the surface. The dominant form of condensation is one in which a liquid film covers the entire condensing surface, and under the action of gravity the film flows continuously from the surface. Film condensation is generally characteristic of clean, uncontaminated surfaces. However, if a surface is coated with a substance that inhibits wetting, it is possible to maintain dropwise condensation. The drops form in cracks, pits, and cavities on the surface and may grow and coalesce through condensation. Typically, more than 90% of the surface is covered by drops, ranging from a few micrometres in diameter to agglomerations visible to the naked eye. The droplets flow from the surface due to the action of gravity.

Examples of equipment in which condensation occurs on a cooled surface are condensers for Rankine power generation cycles and vapour compression refrigeration cycles, dehumidifiers for air conditioners, and heat pipes.

Regardless of whether it is in the form of a film or droplets, the condensate provides resistance to heat transfer between the vapour and the surface. Because this resistance increases with condensate thickness, which increases in the flow direction, it is desirable to use short vertical surfaces or horizontal cylinders in situations involving film condensation. Most condensers therefore consist of horizontal tube bundles through which a liquid coolant flows and around which the vapour to be condensed is circulated. In terms of maintaining high condensation and heat transfer rates, droplet formation is superior to film formation. It is therefore common practice to use surface coatings such as silicones, teflon, and an assortment of waxes and fatty acids that inhibit wetting and hence stimulate dropwise condensation. However, such coatings may not stay for long due to the oxidation, fouling, or outright removal, and film condensation eventually occurs. For this reason, condenser design calculations are often based on the assumption of film condensation.

The process of condensation of a vapour flowing inside a horizontal or vertical tube is one of importance in refrigeration and air-conditioning. The details can be found in Akers et al. (1958). In this chapter, however, only the theory of film condensation on a vertical plate and horizontal cylinders will be discussed.

7.11.1 Laminar Film Condensation on a Vertical Plate

Nusselt (1916) developed the first theory of film condensation by making simple assumptions leading to useful results. The assumptions are as follows:

1. The liquid film has laminar flow and constant properties.
2. The gas is a pure vapour and at a uniform temperature of T_{sat}. With no temperature gradient in the vapour, heat transfer to the liquid–vapour interface can occur only by condensation at the interface and not by conduction from the vapour.
3. The shear stress at the liquid–vapour interface is assumed to be negligible, in which case $\partial u/\partial y|_{y=\delta} = 0$.
4. Momentum and energy transfer by convection in the condensate film are negligible. This is because of low velocities associated with the film. Therefore, heat transfer across the film occurs only by conduction, in which case the liquid temperature distribution is linear.

Figure 7.15 shows the film conditions resulting from the aforementioned assumptions. From assumption 4, momentum convection terms may be neglected, and the x-momentum equation may be expressed as

$$\frac{\partial^2 u}{\partial y^2} = \frac{1}{\mu_l}\frac{\partial p}{\partial x} - \frac{X}{\mu_l} \tag{7.36}$$

where X is the body force per unit volume. Now,

$$X = \rho_l g$$

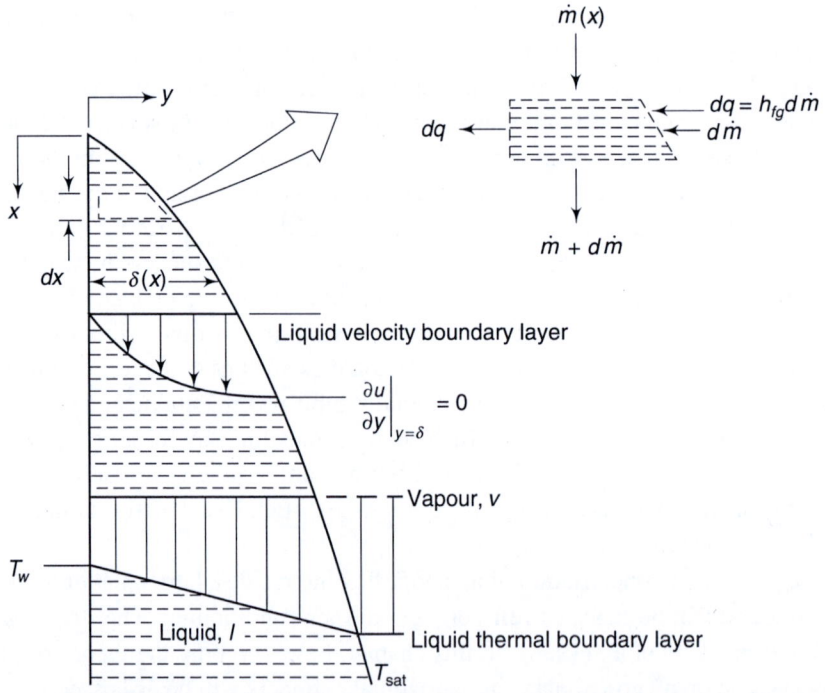

Fig. 7.15 Boundary layer conditions for Nusselt's analysis of film condensation on a vertical plate of width b

Invoking the boundary layer approximation in the liquid film, we can write

$$\frac{\partial p}{\partial x} = \rho_v g$$

$$\frac{\partial p}{\partial y} = 0$$

Therefore, from Eq. (7.36) it follows

$$\frac{\partial^2 u}{\partial y^2} = -\frac{g}{\mu_l}(\rho_l - \rho_v) \tag{7.37}$$

Equation (7.37) is subject to the boundary conditions

$$u(0) = 0$$

$$\left[\frac{\partial u}{\partial y}\right]_{y=\delta} = 0$$

Solving Eq. (7.37), the velocity profile in the film is obtained as

$$u(y) = \frac{g(\rho_l - \rho_v)\delta^2}{\mu_l}\left[\frac{y}{\delta} - \frac{1}{2}\left(\frac{y}{\delta}\right)^2\right] \tag{7.38}$$

From this result the condensate mass flow rate per unit width, $\Gamma(x)$, may be expressed as

$$\Gamma(x) = \frac{\dot{m}(x)}{b} = \int_0^{\delta(x)} \rho_l u(y)dy = \frac{g\rho_l(\rho_l - \rho_v)\delta^3}{3\mu_l} \tag{7.39}$$

Evaluation of δ   From an energy balance of a differential element of the film as shown in the expanded view of Fig. 7.15 we have

$$dq = h_{fg}\,d\dot{m} = q_w''(b\,dx)$$

But $$q_w'' = \frac{k_l(T_{sat} - T_w)}{\delta}$$

Therefore, we can write

$$\frac{d\Gamma}{dx} = \frac{k_l(T_{sat} - T_w)}{\delta h_{fg}} \tag{7.40}$$

Also from Eq. (7.39), we get

$$\frac{d\Gamma}{dx} = \frac{g\rho_l(\rho_l - \rho_v)\delta^2}{\mu_l}\frac{d\delta}{dx} \tag{7.41}$$

Therefore, from Eqs (7.40) and (7.41),

$$\delta^3 d\delta = \left[\frac{k_l\mu_l(T_{sat} - T_w)}{g\rho_l(\rho_l - \rho_v)h_{fg}}\right]dx$$

At $x = 0$, $\delta = 0$. Therefore,

$$\delta(x) = \left[\frac{4k_l\mu_l(T_{sat} - T_w)x}{g\rho_l(\rho_l - \rho_v)h_{fg}}\right]^{1/4} \tag{7.42}$$

Equation (7.42) may be substituted in Eq. (7.39) to obtain $\Gamma(x)$.

Calculation of the average heat transfer coefficient and the average Nusselt number

$$q_w'' = h_x(T_{sat} - T_w) = \frac{k_l(T_{sat} - T_w)}{\delta}$$

$$\Rightarrow \qquad h_x = \frac{k_l}{\delta}$$

Substituting δ from Eq. (7.42) in the above equation gives

$$h_x = \left[\frac{g\rho_l(\rho_l - \rho_v)k_l^3 h_{fg}}{4\mu_l(T_{sat} - T_w)x} \right]^{1/4}$$

$$\bar{h}_L = \frac{1}{L}\int_0^L h_x\,dx = \frac{4}{3}h_L$$

$$\bar{h}_L = 0.943 \left[\frac{g\rho_l(\rho_l - \rho_v)k_l^3 h_{fg}}{\mu_l(T_{sat} - T_w)L} \right]^{1/4} \tag{7.43}$$

$$\overline{Nu}_L = \bar{h}_L\frac{L}{k_l} = 0.943 \left[\frac{\rho_l g(\rho_l - \rho_v)h_{fg}L^3}{\mu_l k_l(T_{sat} - T_w)} \right]^{1/4} \tag{7.44}$$

In using Eqs (7.43) and (7.44) all liquid properties should be evaluated at $T_f = (T_{sat} + T_w)/2$ and h_{fg} should be evaluated at T_{sat}.

Calculation of the mass flow rate of the condensate

$$\dot{m} = \frac{q}{h_{fg}} = \bar{h}_L\frac{A(T_{sat} - T_w)}{h_{fg}} \tag{7.45}$$

7.11.2   Laminar Film Condensation on Inclined Plates

Equation (7.43) can be used by replacing g by $g\cos\theta$, where θ is the angle between the vertical and the surface. However, it must be used with caution for large values of θ and does not apply if $\theta = \pi/2$.

7.11.3   Laminar Film Condensation on the Inner or Outer Surface of a Vertical Tube

The vertical plate correlation [Eq. (7.43)] can be used if $R \gg \delta$.

7.12   Turbulent Film Condensation

As in the case of other convection phenomena, turbulent flow conditions may exist in film condensation. Consider the vertical surface of Fig. 7.16(a). The transition criterion may be expressed in terms of a Reynolds number defined as

$$Re_\delta = \frac{\rho_l u_m D_h}{\mu_l} = \frac{\rho_l u_m(4\delta)}{\mu_l} \tag{7.46}$$

where D_h is the hydraulic diameter [= 4 × cross-sectional area/wetted perimeter = $4(b\delta)/b = 4\delta$], u_m is the average velocity in the film, and δ is the film thickness (which is the characteristic length in this case). Equation (7.46) may also be reframed as

$$\text{Re}_\delta = \frac{4\Gamma}{\mu_l} \tag{7.47}$$

where Γ is the mass flow rate per unit width of the plate.

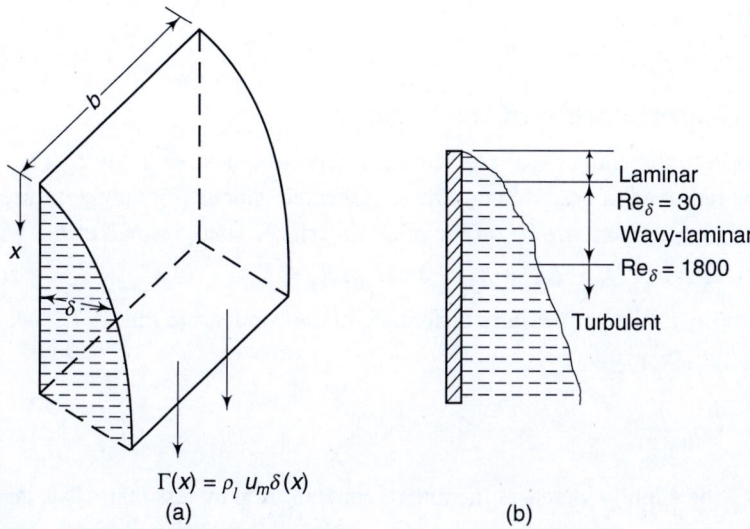

$$\Gamma(x) = \rho_l \, u_m \delta(x)$$
(a)   (b)

Fig. 7.16   Film condensation on a vertical plate: (a) Mass rate of condensation for a plate of width b, (b) various flow regimes

Figure 7.16(b) shows various flow regimes of the condensate film. For $\text{Re}_\delta \le 30$, the film is laminar and wave-free; for $30 \le \text{Re}_\delta \le 1800$, ripples or waves form on the condensate film; $\text{Re}_\delta = 1800$ is the critical Reynolds number; $\text{Re}_\delta > 1800$ indicates turbulent flow.

In the wavy-laminar region, Kutateladze recommends a correlation of the form

$$\frac{\bar{h}_L (v_l^2/g)^{1/3}}{k_l} = \frac{\text{Re}_\delta}{1.08\,\text{Re}_\delta^{1.22} - 5.2} \tag{7.48}$$

For the turbulent region, Labunstov recommends

$$\frac{\bar{h}_L}{k_l}(v_l^2/g)^{1/3} = \frac{\text{Re}_\delta}{8750 + 58\,\text{Pr}^{-0.5}(\text{Re}_\delta^{0.75} - 253)} \tag{7.49}$$

v_l and Pr are the kinematic viscosity and Prandtl number of the condensate, respectively.

7.13   Sub-cooling of Condensate

As the wall temperature is lower than the saturation temperature, not only the condensation enthalpy is released at the wall but also heat flow from the sub-cooling of the condensate exists. Therefore,

$$\dot{m}h'_{fg} = \dot{m}h_{fg} + \int_0^\delta \rho_l u b c_{p,l}(T_{\text{sat}} - T)\,dy \tag{7.50}$$

Substituting the u and T profiles (note that the T-profile is linear) and performing the integration

$$h'_{fg} = h_{fg} + \frac{3}{8}c_{p,l}(T_{sat} - T_w) \tag{7.51a}$$

Hence, for more accurate calculations, h_{fg} in Eq. (7.43) should be replaced by h'_{fg} as computed from Eq. (7.51a). Using a more accurate temperature profile in Eq. (7.50), one obtains

$$h'_{fg} = h_{fg} + 0.68c_{p,l}(T_{sat} - T_w) \tag{7.51b}$$

7.14 Superheating of the Vapour

In addition to the enthalpy of condensation, the superheat enthalpy $c_{p,g}(T_g - T_{sat})$ has to be removed in order to cool the superheated vapour from temperature T_g to the saturation temperature T_{sat} at the phase interface. Then, instead of h_{fg}, we use

$$h''_{fg} = c_{p,g}(T_g - T_{sat}) + h_{fg} + 0.68c_{p,l}(T_{sat} - T_w) \tag{7.52}$$

Furthermore, as the temperature difference in the condensate film is $T_{sat} - T_w$,

$$q'' = \bar{h}_L(T_{sat} - T_w)$$

and

$$\frac{\dot{m}}{A} = \frac{q''}{h''_{fg}}$$

Superheating slightly increases the rate of condensation by less than 10%, an effect that is appreciable only if the wall and superheat temperatures differ by less than about 3 °C.

7.15 Laminar Film Condensation on Horizontal Tubes (Nusselt's Approach)

Figure 7.17 shows a condensate flowing over a horizontal tube. An energy balance on the condensate film between θ and $\theta + d\theta$ per unit length of the cylinder gives

$$h_{fg}d\dot{m} = q''_w(1)dx \tag{7.53}$$

Neglecting the curvature effect, from Eq. (7.53), substituting for q''_w results in

$$h_{fg}\frac{d\dot{m}}{dx} = \frac{k_l(T_{sat} - T_w)}{\delta} \tag{7.54}$$

Noting that $\theta = x/R$, we get from Eq. (7.54)

$$\frac{h_{fg}}{R}\frac{d\dot{m}}{d\theta} = \frac{k_l(T_{sat} - T_w)}{\delta} \tag{7.55}$$

The component of gravity acting tangentially to the tube surface is $g \sin\theta$. Thus, with the parabolic velocity distribution used for a vertical flat plate [Eq. (7.38)],

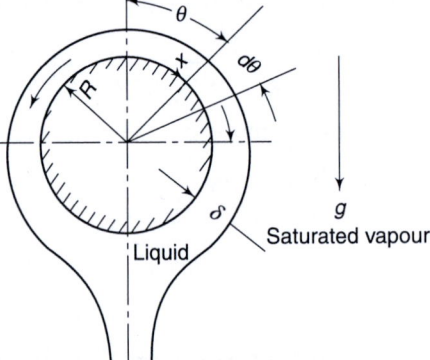

Fig. 7.17 Film condensation on a horizontal tube

replacing g by $g \sin\theta$, the condensate mass flow rate is given by

$$\dot{m} = \rho_l \int_0^\delta u\,dy$$

u is to be substituted in the above expression from Eq. (7.38). Therefore,

$$\dot{m} = \rho_l(\rho_l - \rho_v)\delta^3 g \frac{\sin\theta}{3\mu_l} \tag{7.56}$$

Also, $\qquad \delta = \left[\dfrac{4k_l\mu_l(T_{sat} - T_w)x}{g\sin\theta\rho_l(\rho_l - \rho_v)h_{fg}}\right]^{1/4}$

Now $x = R\theta$; hence,

$$\delta = \left[\frac{4k_l\mu_l(T_{sat} - T_w)R\theta}{g\sin\theta\rho_l(\rho_l - \rho_v)h_{fg}}\right]^{1/4} \tag{7.57}$$

Combining Eqs (7.55) – (7.56), we get

$$\dot{m}^{1/3}\,d\dot{m} = \frac{Rk_l(T_{sat} - T_w)}{h_{fg}}\left[\frac{g(\rho_l - \rho_v)}{3v_l}\right]^{1/3}\sin^{1/3}\theta\,d\theta \tag{7.58}$$

Integrating Eq. (7.58) from $\theta = 0$ to $\theta = \pi$ gives the condensate production from one side:

$$\dot{m} = 1.924\left[\frac{R^3k_l^3(T_{sat} - T_w)^3 g(\rho_l - \rho_v)}{h_{fg}^3 v_l}\right]^{1/4} \tag{7.59}$$

An energy balance on the entire tube yields

$$2h_{fg}\dot{m} = \bar{h}_D 2\pi R(1)(T_{sat} - T_w) \tag{7.60}$$

Therefore, from Eqs (7.59) and (7.60), we obtain

$$\bar{h}_D = 0.728\left[\frac{g\rho_l(\rho_l - \rho_v)k_l^3 h_{fg}}{\mu_l(T_{sat} - T_w)D}\right]^{1/4} \tag{7.61}$$

7.16   Vertical Tier of *n* Horizontal Tubes

Some condensers have banks of horizontal tubes in vertical rows, as depicted in Fig. 7.18. In this situation, first analysed by Nusselt, the condensate from the topmost tube drips onto the second tube and so on. The thicker condensate film on lower tubes renders them less effective condensing surfaces than the upper tubes. For each tube, the analysis leading to Eq. (7.58) applies. Hence, for each tube

$$\dot{m}_{bottom,\,n}^{4/3} = \dot{m}_{top,\,n}^{4/3} + B \tag{7.62}$$

where $\quad B^{3/4} = 1.924\left[\dfrac{R^3 k_l^3(T_{sat} - T_w)^3 g(\rho_l - \rho_v)}{h_{fg}^3 v_l}\right]^{1/4} \qquad (7.63)$

For tube 1,

$$\dot{m}_{b,1}^{4/3} = B$$

For tube 2,

$$\dot{m}_{b,2}^{4/3} = \dot{m}_{b,1}^{4/3} + B = 2B$$

Fig. 7.18   Vertical tier of *n* tubes

Extending the above results to the nth tube, we get

$$\dot{m}_{b,n}^{4/3} = nB \qquad (7.64)$$

The average heat transfer coefficient is found from an energy balance as follows:

$$2h_{fg}\dot{m}_{b,n} = 2\pi Rn\overline{h}_D(T_{sat} - T_w) \qquad (7.65)$$

Thus, from Eqs (7.64) and (7.65), we have

$$\overline{h}_D(nD) = \frac{2h_{fg}}{\pi(T_{sat} - T_w)}(nB)^{3/4} \qquad (7.66)$$

Equation (7.66) reveals that the result for a single horizontal tube can be used directly if D is replaced by nD in Eq. (7.61). For n tubes in a vertical row, the average heat transfer coefficient for a single tube in the array is

$$\overline{h}_{D,n} = 0.728\left[\frac{g\rho_l(\rho_l - \rho_v)k_l^3 h_{fg}}{n\mu_l(T_{sat} - T_w)D}\right]^{1/4} \qquad (7.67)$$

That is, $\overline{h}_{D,n} = \overline{h}_D n^{-1/4}$ $\qquad (7.68)$

7.16.1 Chen's Modification of Nusselt's Correlation

It has been found that the average heat transfer coefficient of a single tube in a vertical tier of n horizontal tubes as predicted by Nusselt's theory [see Eq. (7.67)] is much less than the experimental value.

Chen (1961) modified the Nusselt's correlation [Eq. (7.67)] by considering the additional condensation on the subcooled liquid layer between the tubes. He provided the following result:

$$\overline{h}_{D,n} = 0.728\left[1 + 0.2\frac{c_{pl}(T_{sat} - T_w)}{h_{fg}}(n-1)\right]\left[\frac{g\rho_l(\rho_l - \rho_v)k_l^3 h_{fg}'}{nD\mu_l(T_{sat} - T_w)}\right]^{1/4} \qquad (7.69)$$

which is a good approximation provided

$$\frac{c_{pl}(T_{sat} - T_w)}{h_{fg}} < 2$$

Equation (7.69) agrees well with most of the available experimental data for condensation on banks of tubes. However, Eq. (7.69) does not take into account external vibrations causing ripples in the liquid surface, splashing off tubes in a bank, or uneven run off due to bowing or inclination of tubes (Rohsenow, 1973).

7.17 Staggered Tube Arrangement

It is evident from Eq. (7.68) that for a vertical stack of n tubes the average heat transfer coefficient is reduced, as expected from the fact that the film thickness is greater for the lower tubes, increasing the resistance to heat transfer.

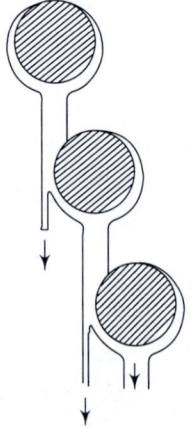

Fig. 7.19 Staggered tube arrangement

Thus, improved performance in a condenser can be achieved by using a staggered arrangement as illustrated in Fig. 7.19.

Example 7.3 Film Condensation on the Outer Surface of a Vertical Tube

The outer surface of a vertical tube, which is 1 m long and has an outer diameter of 80 mm, is exposed to saturated steam at 1.0133 bars and is maintained at 50°C by the flow of cool water through the tube. Calculate the rate of heat transfer to the coolant and the rate of condensation of steam at the outer surface.

Solution

Assumption: Laminar film condensation on a vertical plate.

Given data:

Saturated steam $(p = 1.0133$ bars), $T_{sat} = 100°C$, $\rho_v = 0.596$ kg/m^3, $h_{fg} = 2257$ kJ/kg

Saturated water $(T_f = 75°C)$, $\rho_l = 975$ kg/m^3, $\mu_l = 375 \times 10^{-6}$ N s/m^2, $k_l = 0.668$ W/mK, $c_{p,l} = 4193$ J/kgK

Recall

$$\bar{h}_L = 0.943 \left[\frac{g\rho_l \, (\rho_l - \rho_v) \, k_l^3 \, h'_{fg}}{\mu_l \, (T_{sat} - T_w) \, L} \right]^{1/4}$$

Obtaining h'_{fg} from Eq. (7.51b) and using the given data in the above equation, we have

$$\bar{h}_L = 4094 \text{ W/m}^2 \text{ K}$$

Therefore, $q = \bar{h}_L \, (\pi DL)(T_{sat} - T_w) = 4094 \, (\pi)(0.08)(1)(100 - 50) = 51,446$ W

Hence, $\dot{m} = \dfrac{q}{h'_{fg}} = \dfrac{51,446}{2.4 \times 10^6} = 0.0214$ kg/s

Validity check of the laminar flow assumption

$$\text{Re}_\delta = \frac{4\dot{m}}{\mu_l b} = \frac{4(0.0214)}{375 \times 10^{-6} \, (\pi)(0.08)} = 908$$

Since $30 < \text{Re}_\delta < 1800$, a significant portion of the condensate is in the wavy-laminar region. Hence the wave-free laminar assumption may be a poor one.

$$\delta(L) = \left[\frac{4 k_l \mu_l \, (T_{sat} - T_w) L}{g\rho_l \, (\rho_l - \rho_v) h'_{fg}} \right]^{1/4} = 2.18 \times 10^{-4} \text{ m} = 0.218 \text{ mm}$$

Since $\delta(L) \ll D/2$, the use of vertical plate correlation for a vertical cylinder is justified. **A more accurate approach** The solution using Kutateladze's correlation Eq. (7.48) for the wavy-laminar region will give a more accurate result.

Example 7.4 Film Condensation on a Horizontal Tube

Saturated Freon-12 at 50°C condenses on a 3-cm-diameter horizontal tube, the outer surface of which is kept at 40°C. Find the average condensation heat transfer coefficient.

Solution

Given data:

$$T_{sat} = 50°C, \ T_f = 45°C, \ T_{sat} - T_w = 10°C$$

At 50°C, $h_{fg} = 121.43$ kJ/kg

At $T_f = 45°C$, $\rho_l = 1236.55$ kg/m^3, $c_{p,l} = 1.01175$ kJ/kg°C, $k_l = 0.068$ W/m K,
$$\mu_l = 2.362 \times 10^{-4} \, N\,s/m^2$$

Also, $D = 0.03$ m. Thus, Eq. (7.61) gives (neglecting ρ_v as compared to ρ_l and replacing h_{fg} by h_{fg}')

$$\bar{h}_D = 0.728 \left[\frac{g\rho_l^2 \, k_l^3 \, h_{fg}'}{\mu_l \, (T_{sat} - T_w)\,D} \right]$$

Substituting the values from the given data, we get

$$\bar{h}_D = 1244 \, W/m^2 K$$

7.18   Flow Boiling

In this section the fundamentals of flow boiling are introduced to the readers.

7.18.1   Introduction

Boiling in forced flow is called flow boiling or convective boiling. Generally, in most vapour-producing equipment vapourization occurs under forced convection. The flow conditions are influenced to a great extent by the pressure gradient along the heating surface. The vapour content increases along the path of flow up to the point of complete vapourization. In general, a liquid enters in a sub-cooled condition into a heated channel. Vapour bubbles formed at the wall due to nucleation condense again in the colder core of the liquid. If the liquid in the core is heated up to saturation temperature, then saturated nucleate boiling results.

Figure 7.20 shows the various regimes in succession in the flow boiling in a vertical heated tube. After the *bubbly* flow characterizing nucleate boiling the individual bubbles grow together into large bubbles, that is, they develop into a *slug* flow. With increasing vapour content the large bubbles also grow together. So that at first there is *semi-annular* flow and, subsequently, there forms at the tube wall a liquid film and a vapour core with liquid drops

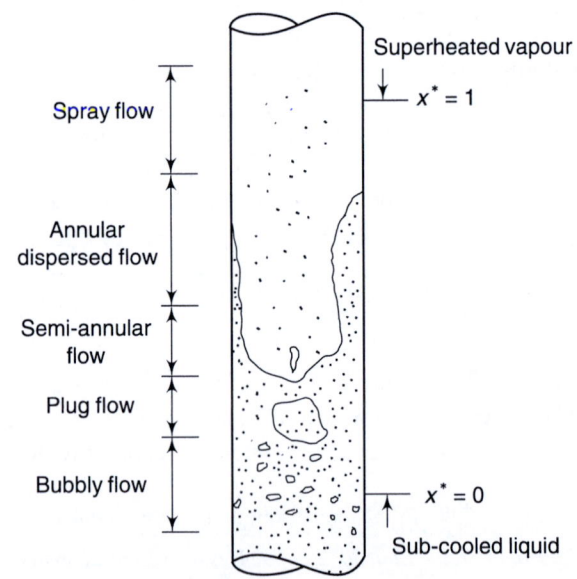

Fig. 7.20   Various regimes of flow boiling in a vertical heated tube

which is called *annular-dispersed* flow. With further addition of heat, the liquid film disappears downstream and in this part the two-phase flow contains vapour with liquid drops. This regime is called *spray* or *mist* flow. In most technical applications, annular-dispersed flow occurs frequently. Slug flow occurs if the flow velocity is small.

We already know that in pool boiling the flow field and heat transfer are determined by the difference between the hot surface and saturation temperatures, as well as by the properties of the fluid and the heating surface. On the other hand, in boiling in forced flow the velocities of the vapour and liquid phases and the distribution of phases are additional factors that influence the flow field and heat transfer. Therefore, in flow boiling the empirical heat transfer coefficient relationships are of the form $h = c(q'')^n (m'')^s f(x^*)$, where m'' and x^* represent the mass flux of the fluid and the vapour quality (to be defined in Section 7.18.2). In nucleate boiling the heat transfer coefficient is chiefly dependent on the heat flux and practically not at all on the flow velocity. On the contrary, in convective boiling the heat transfer coefficient is primarily influenced by the velocity of the flow or by the mass flux but hardly by the heat flux. In flow boiling, $n \approx 0$, s lies between 0.6 and 0.8, while in the nucleate boiling portion, $n \approx 0.75$ and s lies between 0.1 and 0.3.

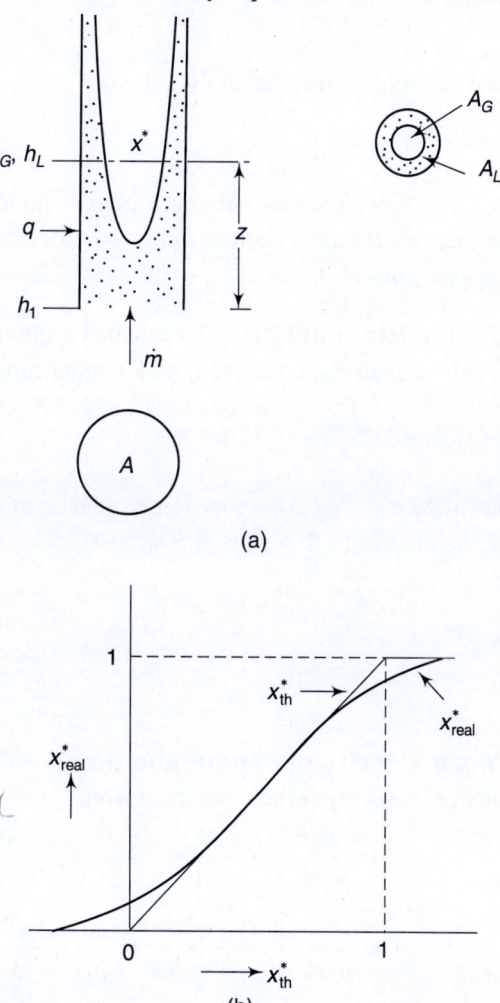

(a)

(b)

Fig. 7.21 (a) Axial section of a tube in which gas and liquid are flowing, (b) Qualitative plot of the real quality against the thermodynamic quality

7.18.2 Definitions of Some Basic Terms

Before we go into the analysis of the two-phase flow, some basic terms and definitions are introduced. Figure 7.21(a) shows the section of a channel in which a mixture of gas and liquid is flowing. An annular flow is shown in the figure for the sake of simplicity. The following terms are also valid for other flow patterns.

Void fraction ε It is defined as the ratio of the cross-sectional area A_G, filled by the gas at any location along the tube length to the total cross-sectional area A of the pipe:

$$\varepsilon = \frac{A_G}{A} \qquad (7.70)$$

For the liquid phase the corresponding fraction is

$$1 - \varepsilon = \frac{A_L}{A} \tag{7.71}$$

because

$$A_G + A_L = A \tag{7.72}$$

These fractions do not change within a sufficiently small tube section Δz. Thus,

$$\varepsilon = \frac{A_G \Delta z}{A \Delta z} \tag{7.73}$$

Therefore, the *volume fraction* (which is the same as the void fraction) of the vapour in the tube section under consideration is then

$$\varepsilon = \frac{V_G}{V} \tag{7.74}$$

and the volume fraction in the liquid is

$$1 - \varepsilon = \frac{V_L}{V} \tag{7.75}$$

V_G, V_L, V are the gas volume, liquid volume, and the total volume of the gas mixture in the tube section under consideration. ε is also called the *volumetric vapour content*.

Volumetric quality $\mathcal{E}$ The *volumetric quality* is the ratio of the volumetric flow rate of vapour to that of the liquid–vapour mixture:

$$\varepsilon^* = \frac{\dot{V}_G}{\dot{V}} \tag{7.76}$$

Quality x^*   Quality x^* represents the ratio of the mass flow rate $\dot{m}_G$ of the vapour to the total mass flow rate $\dot{m} = \dot{m}_G + \dot{m}_L$:

$$x^* = \frac{\dot{m}_G}{\dot{m}} \tag{7.77}$$

$$1 - x^* = \frac{\dot{m}_L}{\dot{m}} \tag{7.78}$$

Mean velocity of vapour and liquid   The mean velocity of both phases in any cross-section can be obtained from

$$v_G = \frac{\dot{m}_G}{\rho_G A_G} = \frac{x^* \dot{m}}{\rho_G \varepsilon A} = \frac{x^* \dot{m}''}{\rho_G \varepsilon} \tag{7.79}$$

$$v_L = \frac{\dot{m}_L}{\rho_L A_L} = \frac{(1 - x^*)\dot{m}}{\rho_L (1 - \varepsilon) A} = \frac{(1 - x^*)\dot{m}''}{\rho_L (1 - \varepsilon)} \tag{7.80}$$

Slip factor s   The ratio of vapour and liquid velocities is defined as *slip* or *slip factor*. Thus,

$$s = \frac{v_G}{v_L} = \frac{x^*}{1 - x^*} \frac{1 - \varepsilon}{\varepsilon} \frac{\rho_L}{\rho_G} \tag{7.81}$$

Relationship between quality x^* and volumetric quality ε^*

$$x^* = \frac{\dot{m}_G}{\dot{m}_G + \dot{m}_L} = \frac{\dot{V}_G \rho_G}{\dot{V}_G \rho_G + \dot{V}_L \rho_L}$$

$$= \frac{\varepsilon^* \dot{V} \rho_G}{\varepsilon^* \dot{V} \rho_G + (1-\varepsilon^*)\dot{V}\rho_L} = \frac{\varepsilon^*}{\varepsilon^* + (1-\varepsilon^*)\rho_L/\rho_G} \qquad (7.82)$$

Relationship between volumetric quality ε^* and void fraction ε, when $s = 1$

For $s = 1$, that is, when $v_G = v_L$,

$$\varepsilon^* = \frac{\dot{V}_G}{\dot{V}} = \frac{v_G A_G}{v_G A_G + v_L A_L} = \frac{A_G}{A} = \varepsilon \qquad (7.83)$$

In general, for flow in unheated channels, the mass flow rates of individual phases are known and, hence, the quality can be easily calculated. In a heated channel the quality is yielded from an energy balance as shown in the next section.

Example 7.5 Void Fraction in an Annular Flow Pattern

In a vertical annular flow pattern the thickness of the liquid film on the tube wall is δ. The tube diameter is D. If $\delta \ll D$, what is the expression for the void fraction ε ?

Solution

$$\varepsilon = \frac{A_G}{A}$$

$$= \frac{\frac{\pi}{4}(D - 2\delta)^2}{\frac{\pi}{4}D^2} = \left(1 - 2\frac{\delta}{D}\right)^2 = 1 - 4\frac{\delta}{D}\left[1 - \frac{\delta}{D}\right]$$

Since $\delta/D \ll 1$, therefore, $1 - \delta/D \approx 1$.

Hence, $\varepsilon = 1 - \dfrac{4\delta}{D}$

Example 7.6 Slip Factor

Derive an expression for the slip s in terms of void fraction ε and volumetric quality ε^*.

Solution
From Eq. (7.81)

$$s = \frac{x^*}{1 - x^*} \frac{1 - \varepsilon}{\varepsilon} \frac{\rho_L}{\rho_G}$$

But, we know from Eq. (7.82) that

$$x^* = \frac{\varepsilon^*}{\varepsilon^* + (1 - \varepsilon^*)\dfrac{\rho_L}{\rho_G}}$$

Therefore, $\dfrac{x^*}{1 - x^*} = \left(\dfrac{\varepsilon^*}{1 - \varepsilon^*}\right)\left(\dfrac{\rho_G}{\rho_L}\right)$

Hence, $s = \left(\dfrac{\varepsilon^*}{1-\varepsilon^*}\right)\left(\dfrac{\rho_G}{\rho_L}\right)\left(\dfrac{1-\varepsilon}{\varepsilon}\right)\left(\dfrac{\rho_L}{\rho_G}\right) = \left(\dfrac{\varepsilon^*}{1-\varepsilon^*}\right)\left(\dfrac{1-\varepsilon}{\varepsilon}\right)$

7.19   Calculation of x^* in a Heated Channel

Consider Fig. 7.21(a), it is assumed that the liquid enters the evaporator tube in a sub-cooled state. Its specific enthalpy (kJ/kg) at the inlet is h_1. The heat transfer rate q on the outer surface of the tube heats the liquid to the saturation temperature, when it begins to evaporate. The energy balance on the control volume of length z, disregarding kinetic and potential energies, gives rise to

$$\dot{m}h_1 + q = \dot{m}_G h_G + \dot{m}_L h_L = \dot{m}[x^* h_G + (1-x^*)h_L] \tag{7.84}$$

If it is further assumed that the vapour and liquid are in thermal equilibrium in the cross-section at the location z, then their specific enthalpies in the saturated state have to be calculated at pressure $p(z)$. This gives

$$h_G = h_v \tag{7.85}$$
$$h_L = h_l \tag{7.86}$$

and $h_v - h_l = h_{fg}$ \hfill (7.87)

h_v, h_l, and h_{fg} are the saturation vapour enthalpy, liquid enthalpy, and enthalpy of vapourization, respectively. Since thermodynamic equilibrium has been assumed here, the quality is indicated by x^*_{th} and is called *thermodynamic quality*. Thus, from Eq. (7.84), we can write

$$\dot{m}h_1 + q = \dot{m}[x^*_{th}(h_v - h_l) + h_l] \tag{7.88}$$

Therefore, from Eqs (7.87) and (7.88), we can write

$$x^*_{th} = \frac{1}{h_{fg}}\left(\frac{q}{\dot{m}} + h_1 - h_l\right) \tag{7.89}$$

Note that the assumption of thermodynamic equilibrium implies that the vapour and liquid at a cross-section are at the same temperature.

7.19.1   Cases of Failure of Eq. (7.89)

Case I: Low vapour content

Equation (7.89) fails at the tube inlet where the quality x^* is still low. In the vicinity of the inlet, vapour bubbles can form on the hot wall even when the core flow is still sub-cooled. So the vapour and liquid are at different temperatures. The real quality is thus positive, whereas according to Eq. (7.89) a negative value of the thermodynamic quality is predicted because the liquid is still sub-cooled, and, therefore, $q + \dot{m}h_1 < \dot{m}h_l$.

Case II: High vapour content

At high vapour content, spray flow develops. Heat is primarily transferred to the vapour which is in a superheated state, although fluid drops (which evaporate slowly) are still present in the core flow. The real quality is, therefore, lower than 1 even though the thermodynamic quality has already reached a value of 1.

7.19.2 Applicability of Eq. (7.89)

In the regions of intermediate quality where neither sub-cooling nor spray flow occurs, Eq. (7.89) predicts exact values. Figure 7.21(b) shows a qualitative plot of the real quality x_{real}^* against the thermodynamic quality x_{th}^*.

7.20 Pressure Drop in a Two-phase Flow

In a two-phase flow, the boiling temperature (that is, saturation vapour temperature) falls in the direction of flow as a result of the pressure drop. This results in a change in the driving temperature drop which is responsible for heat transfer along the flow path. Calculation of heat transfer without simultaneous computation of the pressure drop is, therefore, impossible.

In the following derivation of the expression for the pressure drop let us consider a channel inclined at angle γ to the horizontal, through which a two-phase fluid flows (see Fig. 7.22). The force balance in the z-direction on the differential control volume of length Δz, assuming a steady one-dimensional flow, gives

$$\left[p - \left(p + \frac{dp}{dz}\Delta z\right)\right]A - \tau_0 c\Delta z - \rho g A \sin \gamma \Delta z = m\frac{Dv}{Dt}$$

$$= \rho A\Delta z\left[\frac{\partial v}{\partial t} + v\frac{\partial v}{\partial z}\right] \tag{7.90}$$

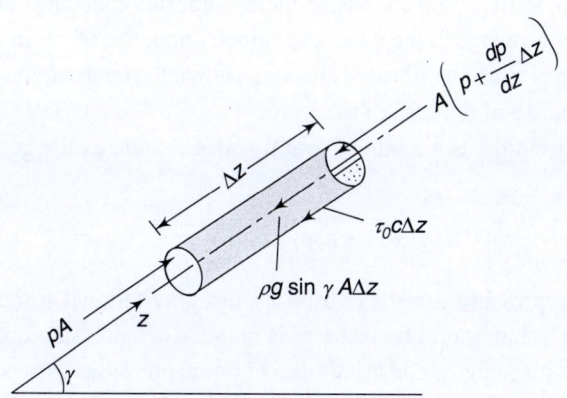

Fig. 7.22 Pressure drop in a two-phase flow in an inclined tube

Since the flow is steady, $\partial v/\partial t = 0$ and $\dot{m} = \rho Av = $ constant, Eq. (7.89) can be rewritten as

$$-\frac{dp}{dz} = \tau_0 \frac{c}{A} + \rho g \sin \gamma + \frac{1}{A}\frac{d}{dz}(\dot{m}v) \tag{7.91}$$

where τ_0 is the shear stress at the channel wall and c is the circumference of the channel. The density of a two-phase mixture is calculated from

$$\rho = \varepsilon \rho_G + (1 - \varepsilon)\rho_L \tag{7.92}$$

The flow momentum is composed of that of the gas and the liquid, according to

$$\dot{m}v = \dot{m}_G v_G + \dot{m}_L v_L \tag{7.93}$$

Using Eqs (7.79) and (7.80), Eq. (7.92) can be written as

$$\dot{m}v = \frac{\dot{m}^2}{A}\left[\frac{(x^*)^2}{\varepsilon\rho_G} + \frac{(1-x^*)^2}{(1-\varepsilon)\rho_L}\right] \tag{7.94}$$

Thus, putting Eqs (7.92) and (7.94) into Eq. (7.91) we get

$$-\frac{dp}{dz} = \tau_0\frac{c}{A} + \frac{\dot{m}^2}{A^2}\frac{d}{dz}\left[\frac{(x^*)^2}{\varepsilon\rho_G} + \frac{(1-x^*)^2}{(1-\varepsilon)\rho_L}\right]$$

$$+ [\varepsilon\rho_G + (1-\varepsilon)\rho_L]g\sin\gamma \tag{7.95}$$

The total pressure drop consists of the following three parts.

(i) The pressure drop due to friction:

$$-\left(\frac{dp}{dz}\right)_f = \tau_0\frac{c}{A} \tag{7.96}$$

The pressure drop due to friction exists because of the shear stress between the fluid flow and the channel wall.

(ii) The acceleration pressure drop:

$$-\left(\frac{dp}{dz}\right)_a = \frac{\dot{m}^2}{A^2}\frac{d}{dz}\left[\frac{(x^*)^2}{\varepsilon\rho_G} + \frac{(1-x^*)^2}{(1-\varepsilon)\rho_L}\right] \tag{7.97}$$

The acceleration pressure drop arises due to the change in momentum in both phases such as in heated channels where evaporation occurs due to heat input. This causes a change in the mass and velocity and, therefore, in the momentum flux of both phases. In unheated channels, however, evaporation due to flashing occurs because of the loss of pressure.

(iii) The pressure drop as a result of gravity, also known as the geodetic pressure drop:

$$-\left(\frac{dp}{dz}\right)_g = [\varepsilon\rho_G + (1-\varepsilon)\rho_L]g\sin\gamma \tag{7.98}$$

The geodetic pressure drop is caused by the gravitational force acting on the fluid. For flows in horizontal tubes there is no geodetic pressure drop.

In channels with bends or constrictions additional pressure drops occur. However, in the present discussion they will not be considered.

Geodetic and acceleration pressure drops are often negligible in comparison with frictional pressure drop. However, in heated channels with large heat and mass fluxes the acceleration pressure drop can be considerable and can no longer be neglected. The acceleration pressure drop can be determined from

$$(p_1 - p_2)_a = \frac{\dot{m}^2}{A^2}\left[\frac{(x_2^*)^2}{\varepsilon_2\rho_{G_2}} - \frac{(x_1^*)^2}{\varepsilon_1\rho_{G_1}} + \frac{(1-x_2^*)^2}{(1-\varepsilon_2)\rho_{L_2}} - \frac{(1-x_1^*)^2}{(1-\varepsilon_1)\rho_{L_1}}\right] \tag{7.99}$$

In Eq. (7.99) the subscript 1 denotes the inlet of the channel and the subscript 2 denotes the outlet of the channel. If the slip factor is set to $s = 1$ (homogeneous flow), then from Eq. (7.81) by putting $s = 1$, we get

$$\varepsilon = \left(\frac{\rho_G}{\rho_L}\frac{1-x^*}{x^*} + 1\right)^{-1} \tag{7.100}$$

In complete vapourization with a change in the quality from $x_1^* = 0$ to $x_2^* = 1$, $\varepsilon_1 = 0$ and $\varepsilon_2 = 1$, the acceleration pressure drop will be

$$(p_1 - p_2)_a = \frac{\dot{m}^2}{A^2}\left[\frac{1}{\rho_{G_2}} - \frac{1}{\rho_{L_1}}\right] \tag{7.101}$$

The frictional pressure drop usually constitutes the largest fraction of the total pressure drop. However, only empirical methods are available for its calculation. It includes not only the momentum transfer between the fluid and the wall but also the momentum transfer between the individual phases. These two processes cannot be estimated separately except for simple flows. Thus, only sketchy ideas of the influence of the momentum transfer between the phases exist.

7.21 Determination of Frictional Pressure Drop: Lockhart and Martinelli Approach

Methods for the determination of the frictional pressure usually start with simple models. In most cases, either homogeneous flow (homogeneous distribution of phases, i.e., $s \rightarrow 1$) or heterogeneous flow (heterogeneous distribution of phases, i.e., $s > 1$) is assumed.

In the computation of frictional pressure drop it is advantageous to define a few parameters that are suitable for the representation of the two-phase frictional pressure drop and the volumetric quality. The frictional pressure drop is often reduced to the pressure drop or single-phase flow, using the definition of Lockhart and Martinelli (1949) as follows:

$$\left(\frac{dp}{dz}\right)_f = \phi_L^2 \left(\frac{dp}{dz}\right)_L \tag{7.102}$$

or

$$\left(\frac{dp}{dz}\right)_f = \phi_G^2 \left(\frac{dp}{dz}\right)_G \tag{7.103}$$

The subscript f stands for two-phase flow. The subscripts L and G represent liquid-alone and gas-alone flows, respectively. $(dp/dz)_L$ is the frictional pressure drop in the liquid, and $(dp/dz)_G$ is that of vapour, under the assumption that each of the two phases is flowing by itself through the tube. While ϕ_L^2 is the two-phase frictional multiplier based on the pressure gradient for the liquid-alone flow, ϕ_G^2 is the same based on the pressure gradient for the gas-alone flow. Their definitions result from Eqs (7.102) and (7.103). If these factors are known, then only the pressure drop in the individual phases has to be determined, in order to calculate the frictional pressure drop for the two-phase flow.

For the frictional pressure drop in a liquid flow through a pipe, it is well known that

$$\left(\frac{dp}{dz}\right)_L = -f_L \frac{1}{d}\frac{\rho_L v_L^2}{2} = -f_L \frac{(\dot{m}_L'')^2}{2d\rho_L} \tag{7.104}$$

where $\dot{m}''_L = \dot{m}(1 - x^*)$, d is the diameter of the pipe, and f is the friction factor:

$$f_L = \frac{c_1}{\text{Re}_L^n} \tag{7.105}$$

where c_1 and n are functions of the flow, that is, whether laminar or turbulent, and the roughness of the pipe.

$$\text{Re}_L = \frac{v_L \rho_L d}{\mu_L} = \frac{\dot{m}''_L d}{\mu_L} = \frac{\dot{m}''(1 - x^*)d}{\mu_L} \tag{7.106}$$

For laminar flow in a smooth or rough pipe, $c_1 = 64$, $n = 1$, whereas for turbulent flow in a smooth pipe ($2300 \le \text{Re}_L \le 10^5$), $c_1 = 0.3164$, and $n = 0.25$ (from Blasius correlation) in Eq. (7.105). For turbulent flow in a rough pipe, Colbrook's relation Eq. (7.107) is valid.

$$\frac{1}{f_L^{0.5}} = -2.0\log\left(\frac{e/D}{3.7} + \frac{2.51}{\text{Re}_L \, f_L^{0.5}}\right) \tag{7.107}$$

where e is the roughness height. Equation (7.107) is a transcendental equation and is to be solved iteratively to obtain f_L.

7.21.1 Homogeneous Model

In flows with a high proportion of small bubbles, $x^* \to 0$, in spray or drop flow, $x^* \to 1$, or in flows with small density differences such as those near the critical point, the frictional pressure drop can well be represented by the homogeneous model. The basic principles upon which the model is based are the assumptions of:

(i) equal vapour and liquid velocities,
(ii) the attainment of thermodynamic equilibrium between the phases, and
(iii) the use of a suitably defined friction factor for a two-phase flow.

In the homogeneous model, calculations are similar to those for a single-phase flow, although they involve suitably defined mean property values. The following is applicable for the frictional drop in a homogeneous two-phase flow:

$$\left(\frac{dp}{dz}\right)_f = -f\frac{1}{d}\frac{\rho v^2}{2} = -f\frac{1}{d}\frac{(\dot{m}'')^2}{2\rho} \tag{7.108}$$

With the assumption of equal liquid and vapour velocities, the volumetric vapour content is obtained from

$$\varepsilon = \left(\frac{\rho_G}{\rho_L}\frac{1-x^*}{x^*} + 1\right)^{-1} \tag{7.109}$$

Using Eq. (7.109) the density of the homogeneous flow is

$$\rho = \varepsilon\rho_G + (1-\varepsilon)\rho_L = \left(\frac{x^*}{\rho_G} + \frac{1-x^*}{\rho_L}\right)^{-1} \tag{7.110}$$

The friction factor is calculated in the same way as for the single-phase flow.

$$f = \frac{c_1}{\text{Re}^n} \tag{7.111}$$

where $\quad Re = \dfrac{\dot{m}''d}{\mu}$ (7.112)

The mixture viscosity μ can be calculated from any of the following three relations, the most commonly used being that of McAdam and his co-workers.

The relation of McAdam and co-workers:

$$\frac{1}{\mu} = \frac{x^*}{\mu_G} + \frac{1-x^*}{\mu_L}$$ (7.113)

The relation of Cicchitti and co-workers:

$$\mu = x^*\mu_G + (1-x^*)\mu_L$$ (7.114)

The relation of Dukler and co-workers:

$$\mu = \rho\left[x^*\frac{\mu_G}{\rho_G} + (1-x^*)\frac{\mu_L}{\rho_L} \right]$$ (7.115)

Therefore, from Eqs (7.108) and (7.110),

$$\left(\frac{dp}{dz}\right)_f = -f\frac{1}{d}\frac{(\dot{m}'')^2}{2}\left(\frac{x^*}{\rho_G} + \frac{1-x^*}{\rho_L}\right)$$ (7.116)

On the other hand, the pressure drop, under the assumption that only liquid is flowing through the tube is

$$\left(\frac{dp}{dz}\right)_L = -f_L\frac{1}{d}\frac{(\dot{m}'')^2(1-x^*)^2}{2\rho_L}$$ (7.117)

where $\quad f_L = \dfrac{c}{Re_L^n}$ (7.118)

and $\quad Re_L = \dfrac{\dot{m}''(1-x^*)d}{\mu_L}$ (7.119)

Therefore, substituting Eqs (7.111)–(7.113) and (7.116)–(7.119) into Eq. (7.102), we obtain

$$\phi_L^2 = \frac{c_1}{c}(1-x^*)^{n-2}\frac{\left[1 + x^*\left(\dfrac{\rho_L}{\rho_G} - 1\right)\right]}{\left[1 + x^*\left(\dfrac{\mu_L}{\mu_G} - 1\right)\right]^n}$$ (7.120)

Example 7.7 Two-phase Frictional Multiplier

Calculate the two-phase frictional multiplier ϕ_L^2 for evaluating the frictional pressure gradient in a smooth evaporator tube for the following conditions. Assume turbulent liquid-alone flow and turbulent two-phase flow.

Fluid: Steam-water

Pressure: 179.7 bars

Quality: 0.1825

Solution

$$n = 0.25 \text{ (from the Blasius correlation)}$$

$$x^* = 0.1825$$

Properties at $p = 179.7$ bars (see Table A1.6 in Appendix A1):

$$\rho_L = \frac{1}{1.856 \times 10^{-3}} = 538.8 \text{ kg/m}^3$$

$$\rho_G = \frac{1}{0.0075} = 133.3 \text{ kg/m}^3$$

$$\mu_L = 67 \times 10^{-6} \text{ Ns/m}^2$$

$$\mu_G = 28 \times 10^{-6} \text{ Ns/m}^2$$

Therefore, substituting the above values in Eq. (7.120) we obtain

$$\phi_L^2 = 2.0906$$

7.21.2   Heterogeneous Model

In this model the two phases flow separately and have different velocities, so that a slip exists between the phases. The basic idea of Lockhart–Martinelli is that the frictional pressure drop in a two-phase flow can be determined, by the use of a correction factor, from the frictional pressure drop in individual phases. The correction factor takes into account the momentum transfer between the phases. Thus, we write

$$X^2 = \frac{\phi_G^2}{\phi_L^2} = \frac{\left(\dfrac{dp}{dz}\right)_L}{\left(\dfrac{dp}{dz}\right)_G} \tag{7.121}$$

where $\left(\dfrac{dp}{dz}\right)_L = -f_L \dfrac{1}{d} \dfrac{(\dot{m}'')^2 (1-x^*)^2}{2\rho_L}$

and $\left(\dfrac{dp}{dz}\right)_G = -f_G \dfrac{1}{d} \dfrac{(\dot{m}'')^2 x^{*2}}{2\rho_G}$

Therefore, $X^2 = \dfrac{\left(\dfrac{dp}{dz}\right)_L}{\left(\dfrac{dp}{dz}\right)_G} = \dfrac{f_L}{f_G} \left(\dfrac{1-x^*}{x^*}\right)^2 \dfrac{\rho_G}{\rho_L}$ (7.122)

In general, X is called the Lockhart–Martinelli parameter. X assumes different values depending on the type of flow for two phases, whether laminar or turbulent. Lockhart and Martinelli assumed that each of the two factors ϕ_G and ϕ_L can be represented as a function of parameter X.

$$\phi_L^2 = 1 + \frac{C}{X} + \frac{1}{X^2} \tag{7.123}$$

$$\phi_G^2 = 1 + CX + X^2 \tag{7.124}$$

For laminar vapour-laminar liquid flow C is 5. For a laminar vapour-turbulent liquid, turbulent vapour-laminar liquid, and turbulent vapour-turbulent liquid C takes on a value of 10, 12, and 20 respectively.

The flow of a phase can be assumed to be laminar when Re < 1000 and turbulent when Re > 2000. The transition region 1000 < Re < 2000 is more difficult to predict, but, to be on the safe side, the values for a turbulent flow should be taken.

The accuracy of the Lockhart–Martinelli procedure is within ±50%. Larger deviations are to be expected for tube diameters greater than 0.1 m.

7.22 Various Heat Transfer Regimes in a Two-phase Flow

Let us consider a sub-cooled liquid fed into the bottom of a vertical evaporator tube that is uniformly heated along its entire length. The heat flux q'' is assumed to be low, and the tube is long enough that the liquid can be completely evaporated. As long as the wall temperature is below the saturation temperature, heat will be transferred by single-phase, forced flow. If the wall temperature exceeds the saturation temperature, vapour bubbles can form even though the core liquid is still sub-cooled. In this area the wall temperature is virtually constant and is slightly above the saturation temperature. The transition to nucleate boiling starts when the liquid reaches the saturation temperature at its centre.

In the nucleate boiling regime heat transfer is primarily determined by the formation of vapour bubbles and only to a small extent by convection. This regime covers the bubble, plug, churn, and a part of the annular flow regimes. The vapour content continuously increases downstream, and at a sufficiently high vapour content, the churn flow converts into an annular flow, with a liquid film at the wall and vapour, with liquid droplets in the core. The entire nucleate boiling regime is dominated by the formation of vapour bubbles at the wall.

However, in annular flow, the liquid film downstream is so thin and its resistance to heat transfer is so low that the liquid close to the wall is no longer superheated enough to sustain nucleation at the wall. Heat is here mainly conducted by the liquid that is evaporating at its surface. In other words, heat is transferred by convective evaporation. As soon as the liquid film at the wall is completely evaporated, the temperature of the wall subject to a constant heat flux rises because of low vapour conductivity. This transition iss known as *dryout*. The spray flow region now commences, followed by a region where all the liquid droplets being carried along the vapour are completely evaporated, in which heat is transferred by convection to the vapour.

7.23 Methodology of Calculation of the Heat Transfer Coefficient in a Two-phase Flow: The Chen Approach

In flow boiling, it is assumed that the heat transfer coefficient is a combination of two parts which are independent of each other. h_B' is associated with bubble formation while h_c' is associated with convection. Thus, the two-phase heat transfer coefficient is

$$h_{2\text{-ph}} = h_B' + h_c' \tag{7.125}$$

The h'_B part is based on the heat transfer coefficient in nucleate pool boiling. However, in a forced flow more heat is transferred as compared to that in nucleate pool boiling, and as a result the bubble formation is partially suppressed. Chen (1966) accounted for this effect with a suppression factor $S \le 1$, such that h'_B is written as

$$h'_B = Sh_B \tag{7.126}$$

The factor S approaches 1 for a vanishing small mass flux, the heat transfer coefficient h'_B being then the same as that in pool boiling. It approaches zero at a large mass flux because then the heat transfer is exclusively determined by the convective part h'_c, which may be written as

$$h'_c = Fh_c \tag{7.127}$$

h_c is the single phase heat transfer coefficient. F is called the enhancement factor and is ≥ 1. The basic idea is that the rapidly flowing vapour and vapour bubbles enhance the heat transfer coefficient of the single-phase forced liquid flow. Factor F is principally determined by the shear stress exerted by the vapour on the liquid and as Chen showed, may be expressed by the Lockhart–Martinelli parameter, X_{tt}, that is, for turbulent vapour-turbulent liquid flow. Therefore,

$$h_{2\text{-ph}} = Sh_B + Fh_c \tag{7.128}$$

In the limiting case of $S = F = 1$, this equation corresponds to that for saturated nucleate boiling. For vertical tubes, S and F are calculated from

$$S = (1 + 1.15 \times 10^{-6} F^2 \text{Re}_l^{1.17})^{-1} \tag{7.129}$$

$$F = 1 + 2.4 \times 10^4 B_0^{1.16} + 1.37 X_{tt}^{-0.86} \tag{7.130}$$

where $\text{Re}_l = [\dot{m}''(1 - x^*)]d/\mu_l$ is the liquid Reynolds number and $B_0 = q''/\dot{m}''h_{fg}$ is called the boiling number. The summary of the procedure of the calculation of the two-phase heat transfer coefficient is as follows.

1. Determine X_{tt}, the Lockhart–Martinelli parameter.
2. Determine h_B from an appropriate nucleate pool boiling correlation.
3. Determine h_c from Dittus–Boelter or any other suitable correlation for heat transfer in a single-phase turbulent flow in tubes (see Chapter 5).
4. Evaluate S and F.
5. Calculate $h_{2\text{-ph}}$.

7.24 Critical Boiling States

There are two fundamental types of boiling crisis in flow boiling.

Case I: Small volumetric vapour content
In this case film boiling occurs. Once a critical heat flux is reached, a vapour film forms at the wall that separates the liquid from the hot wall. The higher thermal resistance of the vapour film leads to a drop in the heat flux if the wall temperature is specified, or to an increase in the wall temperature if the heat flux is imposed on the wall. In the film boiling region, the critical heat flux decreases approximately linearly with the quality.

Case II: Large volumetric vapour content
In this case, annular flow develops. At the wall there is liquid, and in the core there exists vapour. Once the critical heat flux is reached, the liquid film disappears at the wall, which then becomes covered with vapour. This is known as *dryout*. If the vapour content is sufficiently high the heated surface dries out at a very low heat flux. In dryout the critical heat flux drops sharply with the quality. However, after the dryout, small liquid droplets can reach the heated surface from the core flow, and because of a low heat flux they are only partially vapourized. This type of dryout with subsequent spray cooling of the wall with liquid droplets is called *deposition controlled burn out*. In this region the critical heat flux falls weakly with quality.

7.25 Condensation of Flowing Vapour in Tubes

Condensation of flowing vapour in horizontal or vertical tubes generally occurs in refrigeration and air-conditioning systems. The flow regime is complicated and depends on the velocity of the vapour flowing in the tube. There are usually two kinds of flow: (a) stratified and (b) annular.

Stratified flow   Stratified flow exists if the vapour velocity is small and the condensation occurs in the manner depicted in Fig. 7.23. That is, the condensate flow is from the upper portion of the tube to the bottom, from where it flows in a longitudinal direction with the vapour. This type of flow is also characterized by low interfacial shear forces. In these circumstances the Nusselt equation for condensation on the outer surface of a horizontal tube can be applied in a modified form. It is assumed that a laminar film of condensate runs down the inner surface of the tube and collects as a stratified layer of liquid in the lower part of the tube. Small vapour velocity implies that

$$\text{Re}_{v_i} = \left(\frac{\rho_v u_{m,v} D}{\mu_v} \right)_i < 35{,}000$$

where i refers to the tube inlet.

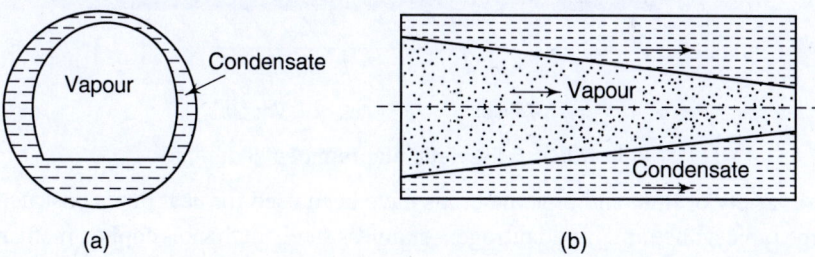

(a)   (b)

Fig. 7.23   (a) Cross-section of condensate flow for low vapour velocities: Stratified flow, (b) longitudinal section of condensate flow for large vapour velocities: annular flow

Chato (1962) recommends the following correlation for the average heat transfer coefficient:

$$\bar{h}_D = 0.557 \left[\frac{g\rho_l(\rho_l - \rho_v)k_l^3 h_{fg}'}{\mu_l(T_{sat} - T_w)D} \right]^{1/4} \tag{7.131}$$

where $\quad h_{fg}' = h_{fg} + \dfrac{3}{8}c_{p,l}(T_{sat} - T_w)$ (7.132)

Annular flow At higher vapour velocities the two-phase flow becomes annular. The vapour occupies the core of the annulus, diminishing in diameter as the thickness of the outer condensate layer increases in the flow direction [Fig. 7.23(b)].

7.26 Heat Pipe

The heat pipe is a novel device that allows the transfer of huge amounts of heat through small surface areas. In other words, it is a high heat flux removal equipment. The basic configuration of the device is shown in Fig. 7.24. A circular pipe has a layer of wicking material covering the inner surface, with a hollow core in the centre. A condensable fluid is also contained in the pipe, and the liquid permeates the wicking material by capillary action. When heat is supplied to one end of the pipe (the evaporator), liquid is vapourized in the wick and the vapour moves to the central core due to the resulting pressure difference. At the other end of the pipe, heat is removed (the condenser) and the vapour condenses back into the wick. Liquid is replenished in the evaporator section by capillary action.

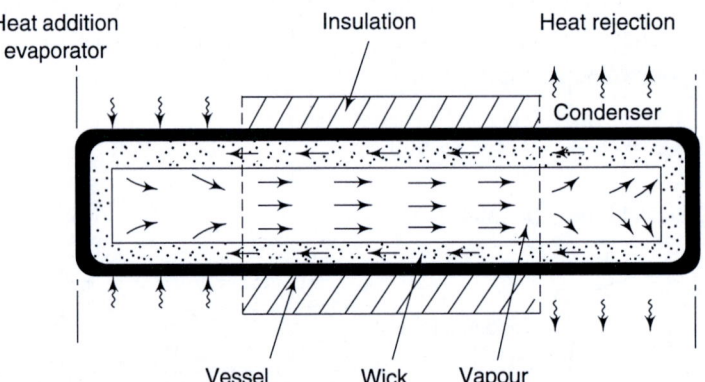

Fig. 7.24 Schematic diagram of a heat pipe

A variety of fluid and pipe materials have been used for heat pipe construction. Some typical ones are: liquid nitrogen–stainless steel, methanol–copper, methanol–nickel, methanol–stainless steel, water–copper, water–nickel.

Heat pipes have been used for cooling of electronic systems such as transistors, laptop computers, cell phones, because in such systems heat generation must be dissipated from very small surface areas and the performance of the electronic devices is strongly temperature dependent. The heat pipe facilitates transfer of heat from a small area to a larger area where it can be dissipated more easily, for example, by use of cooling fins.

Additional Examples

Example 7.8 Nucleate Boiling on an Immersion Heater

An immersion heater for a coffee mug (of volume 250 ml) is to be made of stainless steel. It must heat water from the room temperature of 20°C to its saturation temperature in 5 min and, to avoid possible early material failure, the heater must operate in the beginning of fully developed nucleating boiling (jets and columns regime, that is, corresponding to an excess temperature of 10°C). Estimate the heater area and the wattage of the heater. Take $C_{s,f}$ for stainless steel–water combination as 0.013.

Solution

Properties at $p = 1.0133$ bars (see Table A1.6 in Appendix A1):

$$\mu_l = 279 \times 10^{-6} \, Ns/m^2, \, h_{fg} = 2257 \, kJ/kg,$$

$$\rho_l = \frac{1}{1.044 \times 10^{-3}} = 958 \, kg/m^3$$

$$\rho_v = \frac{1}{1.679} = 0.5955 \, kg/m^3, \, \sigma = 58.9 \times 10^{-3} \, N/m,$$

$$c_{p,l} = 4.217 \, kJ/kgK, \, Pr_l = 1.76$$

Other data:

$$g = 9.81 \, m/s^2, \, s = 1.0, \, \Delta T_w = 10K, \, C_{s,f} = 0.013$$

From Rohsenow's nucleate pool boiling correlation [Eq. (7.10)], using the aforementioned data, we get

$$q_w'' = 137167 \, W/m^2$$

The rate of heat input required to raise the water temperature to the saturation temperature is

$$q = \frac{\rho_l V c_{p,l} \, (T_{sat} - T_{init})}{\Delta t}$$

$$= \frac{(958)(250 \times 10^{-6})(4.217 \times 10^3)(100 - 20)}{(5 \times 60)} = 269.3 \, W$$

where V is the volume of the coffee mug (in m^3) and Δt is the time taken (in s) for raising the water temperature by 80°C.

Now, $\quad q_w'' = \dfrac{q}{A}$

Therefore, $A = \dfrac{q}{q_w''} = \dfrac{269.3}{137167} = 0.00196 \, m^2 = 19.6 \, cm^2$

That is, heater area = 19.6 cm² and heater power = 269.3 W.

Example 7.9 Minimum Heat Flux for a Horizontal Plate

A horizontal plate is exposed to a stagnant pool of water at 100°C. Determine the minimum heat flux.

Solution

Using saturated water and steam properties at 100°C (see the values in Example 7.8) and from Eq. (7.14), we obtain

$$q_w'' = C\rho_v h_{fg} \left[\frac{g\sigma (\rho_l - \rho_v)}{(\rho_l + \rho_v)^2} \right]^{1/4}$$

$$= 0.09\,(0.5955)(2257\times10^3)\left[\frac{9.81\,(58.9\times10^{-3}\,)(958-0.5955)}{(958+0.5955)^2}\right]^{0.25}$$

$$= 18{,}942.95 \text{ W/m}^2$$

Example 7.10   Bubble Departure Diameter and Bubble Frequency

(a) *Estimate the bubble departure diameter and bubble frequency for boiling of saturated liquid water at atmospheric pressure with a wall superheat of 15°C.*

(b) *A very small amount of a surfactant is added to the water, which reduces the surface tension to 20% of its original value. Other properties for the water are unchanged. What effect does it have on the bubble frequency and departure diameter?*

For water at atmospheric pressure, T_{sat} = 100°C, v_l = 0.00104 m^3/kg, v_v = 1.673 m^3/kg, h_{fg} = 2257 kJ/kg, σ = 0.0588 N/m, c_{pl} = 4.22 kJ/kg K.

Solutions

(a) The bubble departure diameter is obtained by the use of the correlation of Cole (1967) given below.

Bubble Departure Diameter

$$\text{Bo}_d^{1/2} = 0.04\,\text{Ja}$$

where $$\text{Bo}_d = \frac{g\left(\rho_l - \rho_v\right)d_d^2}{\sigma}$$

$$\text{Ja} = \frac{\rho_l c_{pl}\left[T_w - T_{sat}\,(p_l\,)\right]}{\rho_v h_{fg}}$$

Now, $$\text{Ja} = \frac{\left(\dfrac{1}{0.00104}\right)(4.22)\,x\,10^3\,(15)}{\left(\dfrac{1}{1.673}\right)(2257)(10^3\,)} = 45.12$$

$$\text{Bo}_d^{1/2} = 0.04\,\text{Ja} = (0.04)\,(45.12) = 1.8048$$

Also, $$\text{Bo}_d = \frac{g\,(\rho_l - \rho_v\,)d_d^2}{\sigma}$$

Hence, $$d_d = \left[\frac{\sigma\text{Bo}_d}{g\,(\rho_l - \rho_v\,)}\right]^{1/2}$$

$$= \left[\frac{\sigma}{g\,(\rho_l - \rho_l\,)}\right]^{1/2}\left(\text{Bo}_d\,\right)^{1/2}$$

$$= \left[\frac{(0.0588}{9.8\left(\dfrac{1}{0.00104} - \dfrac{1}{1.673}\right)}\right]^{1/2}(1.8048)$$

$$= 4.508\times10^{-3}\text{ m} \approx 4.51\text{ mm}$$

The bubble frequency is determined by the following correlation of Zuber (1963).
Bubble Frequency

$$fd_d = 0.59 \left[\frac{\sigma g (\rho_l - \rho_v)}{\rho_l^2} \right]^{1/4}$$

Hence, $\quad fd_d = 0.59 \left[\dfrac{0.0588(9.8)\left(\dfrac{1}{0.00104} - \dfrac{1}{1.673} \right)}{\left(\dfrac{1}{0.00104} \right)^2} \right]^{1/4} = 0.0923$ m/s

Now, $\quad d_d = 4.51 \times 10^{-3}$ m

Therefore, $f = \dfrac{0.0923}{4.51 \times 10^{-3}} = 20.46$ s$^{-1} \approx 20.5$ s^{-1}

(b) $\quad d_{d_1} = \left[\dfrac{\sigma_1 \mathrm{Bo}_{d_1}}{g (\rho_l - \rho_v)} \right]^{1/2}$

$$d_{d_2} = \left[\frac{\sigma_2 \mathrm{Bo}_{d_2}}{g (\rho_l - \rho_v)} \right]^{1/2}$$

$\Rightarrow \quad \dfrac{d_{d_1}}{d_{d_2}} = \left[\dfrac{\sigma_1 \, \mathrm{Bo}_{d_1}}{\sigma_2 \, \mathrm{Bo}_{d_2}} \right]^{1/2}$

The subscripts '1' and '2' refer to the original departure diameter and the departure diameter after surface tension of the liquid has been reduced to 20% of the original value. Since all other properties are constant, $\mathrm{Bo}_{d_1} = \mathrm{Bo}_{d_2}$ as Ja is not a function of σ.

Hence, $\quad \dfrac{d_{d_1}}{d_{d_2}} = \left[\dfrac{\sigma_1}{\sigma_2} \right]^{1/2}$

$\Rightarrow \quad d_{d_2} = d_{d_1} \left[\dfrac{\sigma_2}{\sigma_1} \right]^{1/2}$

Given: $\quad \sigma_2 = 0.2 \sigma_1$

$\Rightarrow \quad \dfrac{\sigma_2}{\sigma_1} = 0.2$

Therefore, $d_{d_2} = (4.51)(0.2)^{1/2} = 2.017 = 2.02$ mm

Now, $\quad f_1 d_{d_1} = 0.59 \left[\dfrac{\sigma_1 g (\rho_l - \rho_v)}{\rho_l^2} \right]^{1/4}$

$$f_2 d_{d_2} = 0.59 \left[\frac{\sigma_2 g (\rho_l - \rho_v)}{\rho_l^2} \right]^{1/4}$$

$$\frac{f_1 d_{d_1}}{f_2 d_{d_2}} = \left[\frac{\sigma_1}{\sigma_2} \right]^{1/4}$$

$$\Rightarrow \qquad f_2 = f_1 \left(\frac{d_{d_1}}{d_{d_2}}\right)\left[\frac{\sigma_2}{\sigma_1}\right]^{1/4}$$

$$= (20.5)\left(\frac{4.51}{2.02}\right)(0.2)^{1/4} = 30.62 \text{ s}^{-1}$$

This example problem demonstrates that reduction of surface tension of a liquid reduces the bubble departure diameter and increases the bubble frequency during boiling.

Example 7.11   Effect of Gravity on Critical Heat Flux for Saturated Pool Boiling

(a) *Compare the CHF (in kW/m^2) of saturated pool boiling of liquid nitrogen at 101.3 kPa on a large horizontal plate (L/Lb > 30) carried out on the surfaces of the earth and the moon. Take $g_{moon} = \dfrac{g_{earth}}{6}$ where $g_{earth} = 9.8 \ m/s^2$.*

(b) *Using the results of 3(a) and the Rohsenow correlation with $C_{sf} = 0.013$, determine ΔT_w at $q'' = 0.1 q''_{max}$ for boiling of liquid nitrogen on the surface of the earth and the moon, respectively.*

   The saturation property values of nitrogen at 101.3 kPa are given below. The property values are not functions of acceleration due to gravity.

$$\rho_l = 807.1 \ kg/m^3$$
$$\rho_v = 4.621 \ kg/m^3$$
$$h_{fg} = 197.6 \ kJ/kg$$
$$c_{pl} = 2.064 \ kJ/kg \ K$$
$$\mu_l = 163 \times 10^{-6} \ Ns/m^2$$
$$Pr_l = 2.46$$
$$\sigma = 8.85 \times 10^{-3} \ N/m$$

Solution

(a) For an infinite flat plate, Zuber (1958) correlation for CHF is

$$q''_{max} = 0.149 \, \rho_v h_{fg}\left[\frac{\sigma(\rho_l - \rho_v)g}{\rho_v^2}\right]^{1/4}$$

Since liquid and vapour properties are invariant with gravity the above expression can be written after substitution of property values as

$$q''_{max} = 103.55 \, g^{1/4}$$

Earth

$$g = 9.8 \text{ m/s}^2$$

Hence, $q''_{max,earth} = 103.55(9.8)^{1/4} = 183.18 \text{kW/m}^2$

Moon

$$g_{moon} = \frac{g_{earth}}{6} = \frac{9.8}{6} = 1.633 \text{ m/s}^2$$

Hence, $q''_{max,moon} = (103.55)(1.633)^{1/4} = 117.01 \text{ kW/m}^2$

(b) From Rohsenow (1952) correlation we can write after substituting $s = 1.7$ (for fluids other water in this case):

$$q'' = \mu_l h_{fg} \left[\frac{g(\rho_l - \rho_v)}{\sigma} \right]^{1/2} \left[\frac{c_{p_l} \Delta T_w}{C_{s,f} h_{fg}} \right]^{3.0} \mathrm{Pr}_l^{-5.15}$$

Substituting property values and $C_{s,f}$ 0.013 in the above expression, we get

$$\Delta T_w = \left[\frac{q''}{(48.77)g^{1/2}} \right]^{1/3}$$

Given $\quad q'' = 0.1 q''_{max}$

Earth

$$q''_{max} = 183.18 \text{ kW/m}^2$$
$$g = 9.8 \text{ m/s}^2$$

Hence, $(\Delta T_w)_{earth} = \left[\dfrac{(0.1)(183.18)(10^3)}{(48.77)(9.8)^{1/2}} \right]^{1/3}$ = 4.85 K (or °C)

Moon

$$q''_{max} = 117.01 \text{ kW/m}^2$$
$$g = 1.633 \text{ m/s}^2$$

Hence, $(\Delta T_w)_{moon} = \left[\dfrac{(0.1)(117.01)(10^3)}{(48.77)(1.633)^{1/2}} \right]^{1/3}$ = 5.63K (or °C)

Example 7.12 Thermodynamic Quality in Flow Boiling

A horizontal tubular test section is to be installed in a flow boiling experiment. The tube is having 10.16 mm internal diameter and is 3.66 m long and is heated uniformly over its length. The water enters the test section at 204°C in a subcooled condition and exits at 68.9 bar in saturated condition. The mass flow rate of water is 0.108 kg/s and the power applied to the tube is 100 kW. (a) What is the length of the tube required to preheat water to the saturation temperature? (b) Obtain the thermodynamic quality at the tube exit. Assume negligible pressure drop across the tube length.

Data: Specific enthalpy of water at the inlet temperature of 204°C = 0.872 MJ/kg. Specific enthalpy of the saturated water at 68.9 bar = 1.26 MJ/kg. Latent heat of vapourization at 68.9 bar = 1.51 MJ/kg.

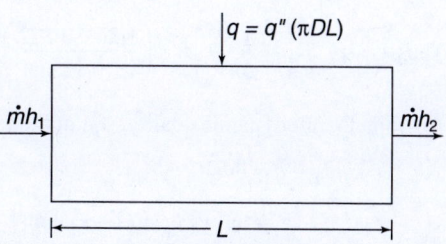

Fig. E7.12(a) Energy balance on the entire tube of length *L*

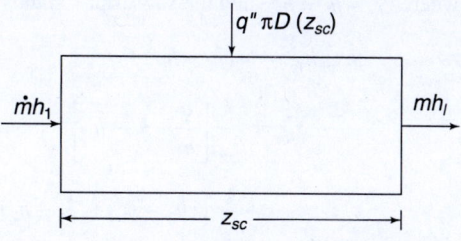

Fig. E7.12(b) Energy balance on the part of tube length (z_{sc}) up to which subcooled condition exists

Solution

(a) Figure E7.12(a) shows the energy balance on the entire tube of length L.

Hence, $\dot{m}h_1 + q = \dot{m}h_2$

$\Rightarrow$ $\dot{m}(h_2 - h_1) = q$

$\Rightarrow$ $\Delta h = \dfrac{q}{\dot{m}} = \dfrac{100 \times 10^3}{(0.108)} = 0.925 \times 10^6$ J/kg $= 0.925$ MJ/kg

Note that h is the specific enthalpy (enthalpy per unit mass, J/kg). The subscripts 1 and 2 represent inlet and exit, respectively of the tube.

Figure E7.12(b) shows the energy balance on a part of the tube length (z_{sc}). 'sc' stands for subcooled condition. At $z = z_{sc}$, the water attains the saturation temperature corresponding to $p = 68.9$ bar.

On energy balance,

$$\dot{m}(h_l - h_1) = q'' \pi D(z_{sc}) \tag{A}$$

h_l is the specific saturation enthalpy of water at 68.9 bar.

Also, from Fig. E7.12(a) it can be seen that

$$\dot{m}\Delta h = q''(\pi DL) \tag{B}$$

Dividing Eq. (A) by Eq. (B), we get

$$\frac{h_l - h_1}{\Delta h} = \frac{z_{sc}}{L} \tag{C}$$

The subscript l stands for saturated liquid.

Thus, $z_{sc} = \left(\dfrac{h_l - h_1}{\Delta h}\right) L = \left(\dfrac{1.26 - 0.872}{0.925}\right)(3.66) = 1.535$ m

(b) The thermodynamic quality, x_{th}^* at the tube exit is given by

$$\dot{m}h_1 + q = \dot{m}h_2$$

$$= \dot{m}\left[x_{th}^* h_v + (1 - x_{th}^*)h_l\right]$$

$$= \dot{m}\left[x_{th}^* h_{fg} + h_l\right]$$

where $h_v - h_l = h_{fg}$ and the subscript v stands for saturated vapour.

$\Rightarrow$ $\dot{m}x_{th}^* h_{fg} = \dot{m}(h_1 - h_l) + q$

$\Rightarrow$ $x_{th}^* = \dfrac{1}{h_{fg}}\left[\dfrac{q}{\dot{m}} + h_1 - h_l\right]$

$$= \frac{1}{h_{fg}}\left[(h_2 - h_1) + (h_1 - h_l)\right]$$

$$= \frac{1}{1.51}(0.925 + 0.872 - 1.26) = 0.356$$

Example 7.13 Film Condensation on a Square Array of Tubes

A square array of nine hundred horizontal tubes having 1.25 cm outer diameter is used to condense steam at atmospheric pressure (101 kPa). The tube walls are maintained at 80°C by a coolant flowing inside the tubes. Calculate the total amount of steam condensed per hour per unit length of the tubes using (a) Nusselt's correlation, (b) Chen's correlation. Data: Saturated Water (101 kPa):

$$T_{sat} = 100°C, \; \rho_v = 0.597 \; kg/m^3, \; h_{fg} = 2257 \; kJ/kg$$

The liquid properties are evaluated at $T_f = \dfrac{T_w + T_{sat}}{2} = \dfrac{80+100}{2} = 90°C = 363 \; K$

Liquid Properties at 363 K:

$\rho_l = 965.3 \; kg/m^3$

$\mu_l = 314.4 \times 10^{-6} \; Ns/m^2$

$k_l = 675.3 \times 10^{-3} \; W/mK$

$c_{pt} = 4.21 \; kJ/kg \; K$

Solution

Assumptions

1. Laminar film condensation is occurring on the tubes.
2. Condensation heat transfer on a tube in a vertical tier is not influenced by the presence of tubes in the neighbouring tiers.

Since the condenser consists of a square array of 900 tubes there are 30 rows and 30 columns of tubes. In each vertical tier there are 30 tubes. The average condensation rate per unit length for a single tube in a tier may be obtained by

$$\dot{m}'_1 = \frac{q'_1}{h'_{fg}} = \bar{h}_{D,n} \, (\pi D)(T_{sat} - T_w)/h'_{fg} \tag{A}$$

where $h'_{fg} = h_{fg} \left[1 + 0.68 \, Ja\right]$

where $Ja = \dfrac{c_{pl}(T_{sat} - T_w)}{h_{fg}} = \dfrac{4.21 \times 10^3 \, (100 - 80)}{2257 \times 10^3} = 0.0373$

$h'_{fg} = 2257 \left[1 + 0.68(0.0373)\right] = 2314.2 \; kJ/kg$

(a) From Nusselt's theory of laminar film condensation on a vertical tier of *n* horizontal tubes, we can write

$$\bar{h}_{D,n} = 0.728 \left[\frac{g\rho_l \, (\rho_l - \rho_v) k_l^3 \, h'_{fg}}{n\mu_l \, (T_{sat} - T_w) D} \right]^{1/4} \tag{B}$$

Here *n* = 30, *D* = 1.25 × 10⁻² m

From Eq. (B)

$$\bar{h}_{D,n} = 0.728 \left[\frac{9.8(965.3)(965.3 - 0.597)(675.3 \times 10^{-3})^3 \, (2314.2 \times 10^3)}{(30)(314.4 \times 10^{-6})(100 - 80)(1.25 \times 10^{-2})} \right]^{1/4}$$

$$= 5240 \; W/m^2K$$

Hence, from Eq. (A)

$$\dot{m}'_1 = (5240)(\pi)(1.25 \times 10^{-2})(20)/2314.2 \times 10^3 = 177.8 \times 10^{-5} \; kg/s \; m$$

For the complete array,

$$m' = n^2 \, \dot{m}_1' = (900)(177.8 \times 10^{-5}) = 1.6 \text{ kg/s m} = 5760 \text{ kg/h m}$$

(b) Using Chen (1961) correlation which takes into account the additional film condensation on the subcooled liquid layer between the tubes in a vertical tier in the array

$$\overline{h}_{D,n} = \overline{h}_{D,\,n,\,\text{Nusselt}} \left[1 + 0.2 \frac{c_{p_l} (T_{\text{sat}} - T_w)}{h_{fg}} (n-1) \right]$$

$$= (5240) \left[1 + 0.2 \frac{4.21 \times 10^3 (20)}{2257 \times 10^3} (30 - 1) \right] = 6373.8 \text{ W/m}^2 \text{ K}$$

Hence, from Eq. (A)

$$\dot{m}_1' = \frac{(6373.8)(\pi)(1.25 \times 10^{-2})(20)}{2314.2 \times 10^3}$$

$$= 216.3 \times 10^{-5} \text{ kg/s m}$$

For the complete array,

$$\dot{m}' = n^2 \, \dot{m}_1'$$

$$= (900)(216.3 \times 10^{-5}) = 1.947 \text{ kg/s m} = 7009.2 \text{ kg/h m}$$

Thus, it is clearly seen that a considerable amount of film condensation is taking place on the liquid layer between the successive tubes in each vertical tier of the square array and should not be neglected.

Important Concepts and Formulae

Boiling

The formation of vapour from a liquid with appearance of bubbles is called boiling.

Bubble Departure Diameter and Frequency of Bubble Release

The bubble departure diameter (d_d) is the diameter of a bubble at the time of its detachment from the heated surface. The frequency of bubble release (f) is defined as

$$f = \frac{1}{\tau}$$

where τ is the time period associated with the growth of each bubble and is equal to the sum of the waiting period and the time required for the bubble to grow to its departure diameter.

The frequency of bubble release thus is directly related to how large the bubble must become for release to occur, and, as a result, on the rate at which the bubble can grow to this size.

Departure Diameter Correlations

Fritz (1935)

$$\text{Bo}_d^{1/2} = 0.0208 \, \theta$$

where θ is the contact angle in degrees.

$$\text{Bo}_d^{1/2} = \frac{g(\rho_l - \rho_v) d_d^2}{\sigma}$$

Bo_d is called Bond number.

Zuber (1959)

$$\text{Bo}_d^{1/2} = \left[\frac{\sigma}{g(\rho_l - \rho_v)}\right]^{-1/6} \left[\frac{6k_l(T_w - T_{\text{sat}})}{q''}\right]^{1/3}$$

Cole (1967)

$$\text{Bo}_d^{1/2} = 0.04\, Ja$$

where $Ja = \dfrac{\rho_l c_{pl}\left[T_w - T_{\text{sat}}(p_l)\right]}{\rho_v h_{fg}}$ is called Jacob number.

Frequency of Bubble Release Correlations

Jacob and Fritz (1931)

$$fd_d = 0.078 \text{ m/s}$$

The above correlation is valid for hydrogen and water vapour bubbles.

Zuber (1963)

$$fd_d = 0.59\left[\frac{\sigma g(\rho_l - \rho_v)}{\rho_l^2}\right]^{1/4}$$

Saturated Pool Boiling Correlations

Correlation of Rohsenow (1952) in the Nucleate Pool Boiling Regime

$$q_w'' = \mu_l h_{fg}\left[g(\rho_l - \rho_v)/\sigma\right]^{1/2}\left(c_{p,l}\Delta T_w / C_{s,f} h_{fg}\, \text{Pr}_l^s\right)^3$$

$$s = 1.0 \text{ for water}$$
$$= 1.7 \text{ for other liquids}$$
$$\Delta T_w = T_w - T_{\text{sat}}$$

$C_{s,f}$ depends on the surface-liquid combinations and are listed in Table 7.1. The properties are to be evaluated at the saturation temperature.

Critical Heat Flux (CHF) for Nucleate Pool Boiling

Correlation of Zuber (1958)

$$q_{w\,\text{max}}'' = 0.149\, h_{fg}\,\rho_v\left[\sigma g(\rho_l - \rho_v)/\rho_v^2\right]^{1/4}$$

The above expression applies to a horizontal surface of infinite extent. It is independent of surface material and is weakly dependent on geometry.

Minimum Heat Flux for Nucleate Pool Boiling

Correlation for a Large Horizontal Plate, Berenson (1961)

$$q_{w,\text{min}}'' = c\rho_v h_{fg}\left[\frac{g\sigma(\rho_l - \rho_v)}{(\rho_l + \rho_v)^2}\right]^{1/4}$$

$$c = 0.09$$

Film Boiling Correlations

$$\overline{\text{Nu}}_D = \bar{h}_{\text{conv}}\frac{D}{k_v} = c\left[\frac{g(\rho_l - \rho_v)h_{fg}' D^3}{v_v k_v(T_w - T_{\text{sat}})}\right]^{1/4}$$

$$c = 0.62 \text{ for horizontal cylinder}$$
$$= 0.67 \text{ for sphere}$$

$$h'_{fg} = h_{fg} + 0.8 c_{p,v} (T_w - T_{sat})$$

Vapour properties are calculated at the film temperature $T_f = \left(\dfrac{T_w + T_{sat}}{2} \right)$, and the liquid density is evaluated at the saturation temperature.

At the elevated surface temperatures ($T_w \geq 300\ C$), the radiation heat transfer across the vapour film becomes significant. Bromley (1950) investigated film boiling from the outer surface of horizontal tubes and suggested calculating the total heat transfer coefficient from a transcendental equation of the form

$$\bar{h}^{-4/3} = \bar{h}_{conv}^{-4/3} + \bar{h}_{rad}\, \bar{h}^{-1/3}$$

where $\bar{h}_{rad} = \dfrac{\varepsilon \sigma (T_w^4 - T_{sat}^4)}{T_w - T_{sat}}$

ε = emissivity of the heater surface

σ = Stefan-Boltzmann constant = 5.667×10^{-8} W/m^2 K^4

Flow Boiling

Boiling in forced flow is called flow boiling or convective boiling.

Some basic Terms

Void Fraction, ε : It is defined as the ratio of the cross-sectional area A_G, filled with gas (or vapour) at any location along the tube length to the total cross-sectional area A of the pipe.

$$\varepsilon = \frac{A_G}{A}$$

These fractions do not change within a sufficiently small tube section Δz.

Thus, $\varepsilon = \dfrac{A_G \Delta z}{A \Delta z}$

Therefore, the volume fraction (which is same as the void fraction) of the vapour in the tube section under consideration is then

$$\varepsilon = \frac{V_G}{V}$$

Quality, x^* : It represents the ratio of the mass flow rate m_g of the gas (or vapour) and the total mass flow rate $\dot{m} = \dot{m}_G + \dot{m}_L$. Hence,

$$x^* = \frac{\dot{m}_G}{\dot{m}}$$

Mean velocity of vapour and liquid: The mean velocity of both phases in any cross-section can be obtained from

$$v_G = \frac{\dot{m}_G}{\rho_G A_G} = \frac{x^* \dot{m}}{\rho_G \varepsilon A}$$

$$v_L = \frac{\dot{m}_L}{\rho_L A_L} = \frac{(1 - x^*) \dot{m}}{\rho_L (1 - \varepsilon) A}$$

Slip Factor, s: The ratio of vapour and liquid velocities is defined as slip or slip factor.

Thus, $s = \dfrac{v_G}{v_L} = \dfrac{x^*}{1 - x^*} \dfrac{1 - \varepsilon}{\varepsilon} \dfrac{\rho_L}{\rho_G}$

Thermodynamic Quality in a Heated Channel

$$x_{th}^* = \frac{1}{h_{fg}}\left(\frac{q}{\dot{m}} + h_1 - h_l\right)$$

where h_1 is the specific enthalpy at the inlet (sub-cooled state)

h_l is the specific saturation liquid enthalpy

Homogeneous Flow (homogeneous distribution of phases): $s \to 1$

Example: In flows with high proportion of small bubbles, $x^* \to 0$, in spray or drop flow, $x^* \to 1$, or in flows with small density differences such as those near the critical point, the homogeneous model can be used to calculate the frictional pressure drop.

Heterogeneous Flow (heterogeneous distribution of phases): $s > 1$

Methodology of Calculation of the Heat Transfer Coefficient in a Two-Phase Flow: The Approach of Chen (1966)

In flow boiling, it is assumed that the heat transfer coefficient is a combination of two parts which are independent of each other. h_B is associated with bubble formation while h_c is associated with convection. Thus, the two-phase heat transfer coefficient is

$$h_{2-ph} = h'_B + h'_c$$

Condensation

The formation of liquid from vapour is called condensation.

Laminar Film Condensation on a Vertical Plate

Correlation of Nusselt (1916)

$$\overline{Nu}_L = \overline{h}_L \frac{L}{k_f} = 0.943 \left[\frac{\rho_l g (\rho_l - \rho_v) h_{fg} L^3}{\mu_l k_l (T_{sat} - T_w)} \right]^{1/4}$$

All liquid properties should be evaluated at $T_f = \dfrac{T_{sat} + T_w}{2}$ and h_{fg} should be evaluated at T_{sat}.

Calculation of the Mass Flow Rate of the Condensate

$$\dot{m} = \frac{q}{h_{fg}} = \overline{h}_L \frac{A(T_{sat} - T_w)}{h_{fg}}$$

Turbulent Film Condensation on a Vertical Plate

$$Re_\delta = \frac{4\Gamma}{\mu_l}$$

where Γ is the mass flow rate per unit width of the plate.

For $Re_\delta \leq 30$, the film is laminar and wave-free and the correlation of Nusselt (1916) is applicable.

For $30 \leq Re_\delta \leq 1800$, ripples or waves form on the condensate film. This is called wavy-laminar regime for which Kutateladze recommends a correlation of the form

$$\overline{h}_L \frac{\left(v_l^2 / g\right)^{1/3}}{k_l} = \frac{Re_\delta}{1.08 Re_\delta^{1.22} - 5.2}$$

$Re_\delta = 1800$ is the critical Reynolds number.

$Re_\delta > 1800$, the flow is turbulent. For turbulent regime Labunstov recommends

$$\bar{h_l}\frac{\left(v_l^2/g\right)^{1/3}}{k_l}=\frac{\text{Re}_\delta}{8750+58\,\text{Pr}^{-0.5}\left(\text{Re}_\delta^{0.75}-253\right)}$$

Laminar Film Condensation on a Single Horizontal Tube
Nusselt's Correlation

$$\bar{h}_D=0.728\left[\frac{g\rho_l\left(\rho_l-\rho_v\right)k_l^3 h_{fg}}{\mu_l\left(T_{\text{sat}}-T_w\right)D}\right]^{1/4}$$

Laminar Film Condensation on a Vertical Tier of n Horizontal Tubes
Nusselt's Correlation

$$\bar{h}_{D,n}=0.728\left[\frac{g\rho_l\left(\rho_l-\rho_v\right)k_l^3 h_{fg}}{n\mu_l\left(T_{\text{sat}}-T_w\right)D}\right]^{1/4}$$

In all the above correlations for film condensation, in order to get more accurate results the effect of subcooling of the condensate should be considered. This is achieved by changing h_{fg} by h'_{fg} given in the expression below.

$$h'_{fg}=h_{fg}+0.68c_{p,l}\left(T_{\text{sat}}-T_w\right)$$

Modification of Nusselt's Correlation by Chen (1961)
It has been found that the average heat transfer coefficient of a single tube in a vertical tier of n horizontal tubes as predicted by Nusselt's theory is much less than the experimental value.

Chen (1961) modified the Nusselt's correlation for laminar film condensation on a vertical tier of n horizontal tubes by considering the additional condensation on the subcooled liquid layer between the tubes. He provided the following result:

$$\bar{h}_{D,n}=0.728\left[1+0.2\frac{c_{pl}\left(T_{\text{sat}}-T_w\right)}{h_{fg}}(n-1)\right]\left[\frac{g\rho_l\left(\rho_l-\rho_v\right)k_l^3 h'_{fg}}{nD\mu_l\left(T_{\text{sat}}-T_w\right)}\right]^{1/4}$$

which is a good approximation provided $\dfrac{c_{pl}\left(T_{\text{sat}}-T_w\right)}{h_{fg}}<2$.

Condensation of Flowing Vapour in Tubes
Condensation of flowing vapour in horizontal or vertical tubes generally occurs in refrigeration and air-conditioning systems. The flow boiling is complicated and depends on the velocity of the vapour flowing in the tube. There are usually two kinds of flow: (a) stratified and (b) annular.

Stratified Flow: It occurs if the vapour velocity is small. The condensate flow is from the upper portion of the tube to the bottom, from where it flows in the longitudinal direction with the vapour.

$$\bar{h}_D=0.557\left[\frac{g\rho_l\left(\rho_l-\rho_v\right)k_l^3 h'_{fg}}{\mu_l\left(T_{\text{sat}}-T_w\right)D}\right]^{1/4}$$

Chato (1962) recommends the following correlation for the average heat transfer coefficient:

where $\quad h'_{fg}=h_{fg}+\dfrac{3}{8}c_{p,l}\left(T_{\text{sat}}-T_w\right)$

Annular Flow: At higher vapour velocities the two-phase flow becomes annular. The vapour occupies the core of the annulus, diminishing in diameter as the thickness of the outer condensate layer increases in the flow direction.

Review Questions

7.1 What is the difference between boiling and evaporation?

7.2 Define heat transfer coefficient for boiling.

7.3 Define superheated liquid and supersaturated vapour.

7.4 How is a vapor bubble formed in a surface cavity?

7.5 Define departure diameter and bubble frequency.

7.6 Draw temperature-controlled saturated pool boiling curve for a liquid and explain its various regimes.

7.7 How will the pool boiling curve differ if the boiling is heat flux controlled?

7.8 Define critical heat flux (CHF) in pool boiling. Why is it important to know CHF?

7.9 Define flow boiling. Show various regimes of flow boiling in a vertical tube.

7.10 Define void fraction, quality, and slip factor.

7.11 Differentiate between homogeneous and heterogeneous models of flow boiling.

7.12 Discuss boiling crisis in flow boiling.

7.13 Distinguish between film condensation and dropwise condensation.

7.14 Why are condensers generally designed on the basis of film condensation?

7.15 State all assumptions in Nusselt's theory of laminar film condensation on a vertical flat plate.

7.16 Define heat transfer coefficient in film condensation.

7.17 Why is staggered tube arrangement used in condensers?

7.18 What is the difference between stratified and annular flow in condensation inside a tube?

7.19 What is a heat pipe?

7.20 Mention some applications of heat pipe.

Problems

7.1 A heated copper plate (polished) is submerged in a container of water at atmospheric pressure. The plate temperature is 220°C. Calculate the heat transfer per unit area of the plate.

7.2 Water at 5 atm is boiled on a 1-mm diameter platinum wire which is at 10°C above the saturation temperature. Estimate the heat transfer on a 1 m length of the tube.

7.3 It is desired to boil 2 kg/h of water at atmospheric pressure in a kettle with a flat bottom of 30 cm diameter. At what temperature must the bottom surface of the kettle be maintained to accomplish this?

7.4 Calculate the peak heat flux for boiling water at atmospheric pressure on a brass pan. [Hint: Use Eq. (7.30)]

7.5 Heat transfer coefficients for boiling are usually large compared with those for single phase convection. Estimate the flow velocity that would be required to produce a value of h for forced convection through a smooth 6-mm-diameter brass tube comparable with that which could be obtained by pool boiling with $\Delta T_w = 16°C$, $p = 7$ bars, and water as the fluid.

7.6 A horizontal tube having 1.2 cm outer diameter is submerged in water at 1 atm and 100°C. Calculate the heat flux for surface temperatures of (a) 520°C, (b) 630°C, and (c) 810°C. Assume $\varepsilon = 0.9$.

7.7 A square array of four hundred 1.25 cm OD horizontal tubes is used to condense steam at atmospheric pressure. The tube walls are maintained at 80°C by a coolant flowing inside the tubes. Calculate the total amount of steam condensed per hour per unit length of the tubes.

7.8 A vertical plate 25 cm wide and 1.1 m high is maintained at 60°C and exposed to saturated steam at 1 atm. Calculate the heat transfer and the total mass of steam condensed per hour.

7.9 A circular tube of 1.6 cm outer diameter and 1.8 m length has a surface temperature of 40°C. Saturated steam at 55°C is condensing on its surface. Find the mass of steam condensed per hour, if the tube is (a) horizontal and (b) vertical.

7.10 A circular horizontal tube of outer radius R is exposed to saturated steam at T_{sat}. The outer tube wall is maintained at $T_w (< T_{sat})$ by passing a liquid metal at high velocity through the tube. Suppose this system is inside a satellite in space under zero gravity conditions. Show that the rate at which the condensate layer grows on the outer surface of the tube is governed by the following expression:

$$\left(\frac{R_\delta}{R}\right)^2 \left[\ln\left(\frac{R_\delta}{R}\right) - \frac{1}{2}\right] + \frac{1}{2} = \frac{2k\Delta T}{\rho h_{fg} R^2}t$$

where $R_\delta - R$ is the thickness of the condensate layer at any time t, k is the thermal conductivity of the condensate, $\Delta T = T_{sat} - T_w$, ρ is the density of the condensate, and h_{fg} is the enthalpy of condensation.

7.11 A platinum wire is submerged in water at a pressure of 500 kN/m². If the temperature excess is 15°C, find the heat flux from the wire.

7.12 An electrical heating element consists of a horizontal rod of 0.5 cm in diameter. It is maintained at a surface temperature of 240°C in a pool of water at atmospheric pressure. Assuming that the element has a surface emissivity of 0.8, estimate the power required to operate the heater per metre of length.

7.13 Compare the critical heat flux for water boiling at (a) 1 atm pressure and (b) 5 atm pressure.

7.14 Calculate the heat flux and the heat transfer coefficient for nucleate pool boiling of water at atmospheric pressure on a chemically etched stainless steel surface when the surface temperature is (a) 110°C and (b) 118°C.

7.15 Repeat Question 7.14 if the surface is nickel.

7.16 From experience it is known that the risk of burning foods with non-stick pan (Teflon coated) is less than that with ordinary uncoated utensils. Can you explain, why is this so?

7.17 A vertical plate 30 cm wide and 1 m high is maintained at 50°C and is exposed to saturated steam at 1 atm pressure. Find the total heat transferred and the amount of steam condensed per hour from both sides of the plate.

7.18 For the data of Question 7.17, find the thickness of the condensate film and the maximum film velocity at the bottom edge of the plate and at locations halfway and three-fourths down the plate.

7.19 Saturated Freon-12 at 310 K condenses on the outer surface of a horizontal 10-cm diameter pipe of 1 m length. The pipe has a uniform surface temperature of 12°C. Determine the total rate of condensation in kg/h.

7.20 For Question 7.19 what would be the mass rate of condensation if Freon-12 were condensing on a vertical tier of five horizontal tubes?

Chapter 8

Radiation Heat Transfer

8.1 Introduction

Thermal radiation, commonly known as radiation heat transfer, is distinctly different from conduction and convection. While, for most applications, conductive and convective heat transfer rates are linearly proportional to the temperature differences, the radiation heat transfer is proportional to the differences of the individual absolute temperatures of the bodies each raised to the fourth power. Thus, it is evident that the importance of radiation becomes intensified at high absolute temperature levels. Consequently, radiation contributes substantially in combustion applications such as fires, furnaces, IC engines, in nuclear reactions such as in the sun or in nuclear explosions and so on.

The other distinguishing feature of radiative heat transfer is that while both conduction and convection require the presence of a medium for the transfer of energy, thermal radiation is by electromagnetic waves, or photons, which may travel a long distance without interacting with a medium. Thus thermal radiation is of great importance in vacuum and space applications. Some common examples are heat leakage through the evacuated walls of a Thermos flask, or the heat dissipation from the filament of a vacuum tube. Radiation is used to reject waste heat from a power plant operating in space.

8.2 Physical Mechanism of Energy Transport in Thermal Radiation

Thermal radiation is one of the many types of electromagnetic radiation travelling at the speed of light. Alternatively, from a quantum point of view, energy is transported by photons, all of which travels at the speed of light. There is, however, a distribution of energy among the photons. The energy associated with each photon is $h\nu$, where h is Planck's constant and ν is the frequency of radiation.

Three parameters may be employed in characterizing radiation. They are (i) the frequency ν, (ii) wavelength λ, and (iii) the wave or photon speed c. Of these only two are independent, since they are related by

$$c = \lambda\nu \tag{8.1}$$

The speed of light, c, within a given medium is related to that in a vacuum, c_0, by

$$c = \frac{c_0}{n} \tag{8.2}$$

where n is the index of refraction.

Thermal radiation is defined as the radiant energy emitted by a medium, which is due solely to the temperature of the medium which governs the emission of thermal radiation. The wavelength range encompassed by thermal radiation is approximately 0.3–50 µm, where 1 µm $= 10^{-6}$ m. In turn, this wavelength range includes three sub-ranges: the ultraviolet, visible, and infrared ranges. These sub-ranges are illustrated schematically in Fig. 8.1.

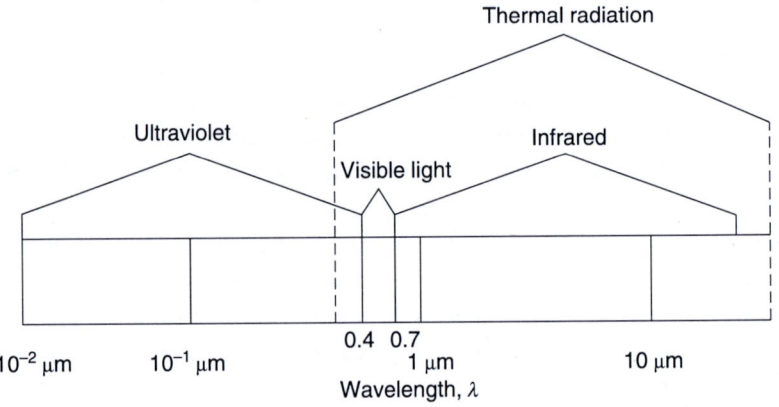

Fig. 8.1 Classification of radiation

8.3 Laws of Radiation

In this section we will discuss three basic laws of radiation. These are Planck's law, Wien's displacement law, and Stefan–Boltzmann law.

8.3.1 Planck's Law

It can be shown by the application of the second law of thermodynamics that there is a maximum amount of radiant energy that can be emitted at a given temperature and at a given wavelength. The emitter of such radiation is called a black body. The energy emission per unit time and per unit area from a black body in a frequency range dv is $E_{bv}dv$, where E_{bv} is called the spectral (or monochromatic) emissive power of the black body radiation. Furthermore, the black body emissive power is also a function of the absolute temperature T of the black body. The explicit form of $E_{bv}(T)$ is given by Planck's[1] law as

$$E_{bv}(T) = \frac{2\pi h v^3 n^2}{c_0^2 [\exp(h v / kT) - 1]} \tag{8.3}$$

In Eq. (8.3), h and k are Planck's and Boltzmann's constants, respectively. The index of refraction, n, refers to the medium bounding the black body.

---

[1] Max Planck (1858–1947), German physicist, was awarded the Nobel Prize in Physics in 1918 for his development of the quantum theory.

Equation (8.3) is sometimes recast in terms of wavelength. This is, however, useful only when the index of refraction of the bounding medium is independent of frequency, for example, for the case of vacuum ($n = 1$) and gases ($n \approx 1$).

Assuming that n is independent of frequency (or wavelength), applying Eqs (8.1) and (8.2) in Eq. (8.3), the following form of Planck's law is obtained:

$$E_{b\lambda}(T) = \frac{C_1}{n^2 \lambda^5 (e^{C_2/n\lambda T} - 1)} \qquad (8.4)$$

where $C_1 = 2\pi h c_0^2$ and $C_2 = hc_0/k$. Equation (8.4) can be further written as

$$\frac{E_{b\lambda}}{\sigma n^3 T^5} = \frac{C_1/\sigma}{(n\lambda T)^5 (e^{C_2/n\lambda T} - 1)} \qquad (8.5)$$

where σ is the Stefan[2]–Boltzmann[3]constant. The values of k, h, c_0, σ, C_1, and C_2 are listed in Table 8.1. The normalized black body emissive power spectrum is shown in Fig. 8.2.

Table 8.1 Black body radiation constants

Boltzmann's constant $k = 1.380 \times 10^{-16}$ erg/K
Planck's constant $h = 6.625 \times 10^{-27}$ erg sec
Speed of light $c_0 = 2.998 \times 10^{10}$ cm/sec
Stefan–Boltzmann constant $\sigma = 5.668 \times 10^{-8}$ W/m^{2}K^4
$C_1 = 3.740 \times 10^{-5}$ erg cm^2/sec
$C_2 = 1.4387$ cm K
[1 erg = 10^{-7} J]

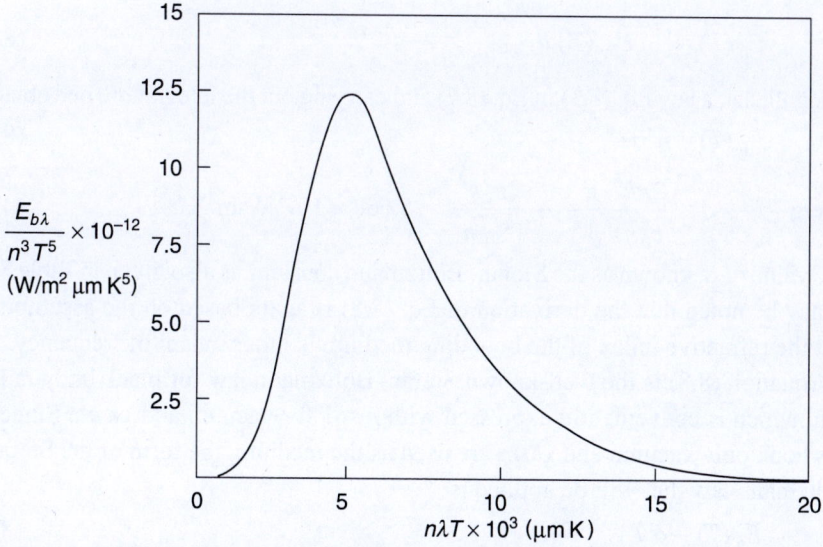

Fig. 8.2 Normalized black body emissive power spectrum

[2] Josef Stefan (1835–1893), Austrian physicist and Professor at the University of Vienna, in 1879, based on his experiments, determined that black body emission was proportional to the absolute temperature to the fourth power.

[3] Ludwig Erhard Boltzmann (1844–1906), Austrian physicist, contributed tremendously in the area of statistical mechanics. Boltzmann derived the fourth-power law from thermodynamic considerations in 1889.

8.3.2 Wien's Displacement Law

A careful look at Eq. (8.5) reveals that the quantity $E_{b\lambda}/\sigma n^3 T^5$, appearing on the left side, is a function solely of $n\lambda T$. The maximum value of the emissive power occurs at

$$(n\lambda T)_{max} = 0.2898 \text{ cm K} = 2898 \text{ μm K} \tag{8.6}$$

Equation (8.6) is known as Wien's[4] displacement law since it was developed independently by Wilhelm Wien in 1891 (i.e., much earlier than the publication of Planck's law).

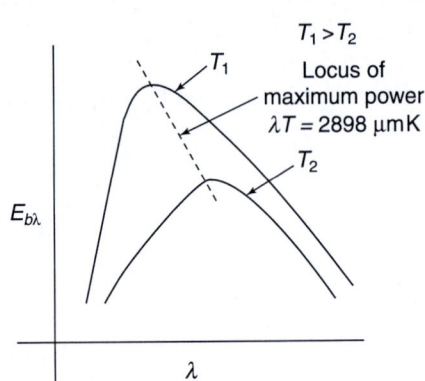

Figure 8.3 shows a graphical representation of Wien's displacement law. Equation (8.6) can also be derived by differentiating Eq. (8.4) with respect to λ while holding T constant and setting the result equal to zero. Equation (8.6) is basically the locus of the maxima of the radiation emission curves as is clearly seen in Fig. 8.3.

Fig. 8.3 Graphical representation of Wien's displacement law

8.3.3 Stefan–Boltzmann Law

The total black body emissive power E_b represents the energy emitted per unit time and area by a black body over all frequencies (or wavelengths). Thus,

$$E_b(T) = \int_0^\infty E_{b\nu}(T)\, d\nu \tag{8.7}$$

Using Planck's law Eq. (8.3) in Eq. (8.7) and carrying out the integration, one obtains

$$E_b(T) = n^2 \sigma T^4 \tag{8.8}$$

where $\sigma = \dfrac{2\pi k^4}{c_0^2 h^3}\dfrac{\pi^4}{15} = \dfrac{2\pi^5 k^4}{15 c_0^2 h^3} = 5.668 \times 10^{-8} \text{ W/m}^2\text{K}^4$

The value of σ, known as the Stefan–Boltzmann constant, is also given in Table 8.1. It may be noted that the derivation of Eq. (8.8) is again based on the assumption that the refractive index of the bounding medium is independent of frequency.

Equation (8.8) is the well-known Stefan–Boltzmann law for black body radiation, which is conveniently expressed with $n = 1$ for vacuum and gases. Since in this book only vacuum and gases are used as the medium, the form of the Stefan–Boltzmann law that will be applied is

$$E_b(T) = \sigma T^4 \tag{8.9}$$

Note that T is in kelvin and E_b is in W/m².

---

[4]Wilhelm Wien (1864–1928), German physicist and Professor at the University of Giessen and later at the University of Munich, was awarded the Nobel Prize in Physics in 1911 for the discovery of displacement law.

Example 8.1   Application of Wien's Displacement Law

What wavelengths correspond to maximum emissive powers of the sun and earth? Take
T_{sun} *= 5762 K and* T_{earth} *= 290 K.*

Solution

From Wien's displacement law [Eq. (8.6)], with the sun's surface at 5762 K and bounded
by vacuum ($n = 1$), it follows that

$$\lambda_{max,\ sun} = \frac{2898\ \mu m\ K}{5762\ K} = 0.5\ \mu m$$

Note that 0.5 μm is near the centre of the visible region (Fig. 8.1) and is situated in that part
of the electromagnetic spectrum where maximum daylight is available.

$$\lambda_{max,\ earth} = \frac{2898\ \mu m\ K}{290\ K} = 10\ \mu m$$

Thus, earth's maximum emission occurs in the immediate infrared, requiring infrared cameras
and detectors for night vision.

8.3.4   Explanation for Change in Colour of a Body when it is Heated

From Wien's displacement law we see that maximum spectral emissive power is
displaced to shorter wavelengths with increasing temperature. This fact explains the
change in the colour of a body as it is heated. Since the band of the wavelength visible to the eye lies between 0.4 and 0.7 μm, only a very small portion of the radiant
energy spectrum at lower temperatures is detected by the eye. As the body is heated,
the maximum emission occurs at shorter wavelengths. The first visible sign of the
increase in the temperature of a body is a dark-red colour. As the temperature is raised
further, the colour appears as bright red, then bright yellow, and finally white since in
the last case a larger portion of the total radiation falls within the visible range.

8.4   Intensity of Radiation

The concept of the
intensity of radiation is
introduced to treat the
directional effects of
surface radiation. Radia-
tion emitted by a surface
propagates in all pos-
sible directions. Also,
radiation incident on a
surface may come from
different directions. A
radiation analyst might
be interested in knowing
the directional distribu-
tion of the emission and/
or the manner in which

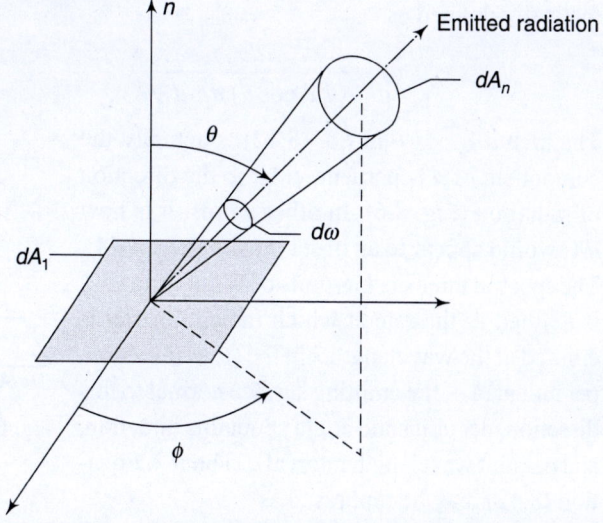

Fig. 8.4   Emission of radiation from differential area dA_1

the surface responds to the incoming radiation from different directions.

Consider emission in a particular direction from an element of area dA_1, as shown in Fig. 8.4. This direction is specified in terms of a spherical coordinate system. Furthermore, a differentially small surface dA_n in space, through which this radiation passes, subtends a solid angle $d\omega$ when viewed from a point on dA_1, where

$$d\omega = \frac{dA_n}{r^2} \tag{8.10}$$

Note that dA_n represents an area that is normal to the (θ, ϕ) direction. Therefore, from Fig. 8.5, and using Eq. (8.10),

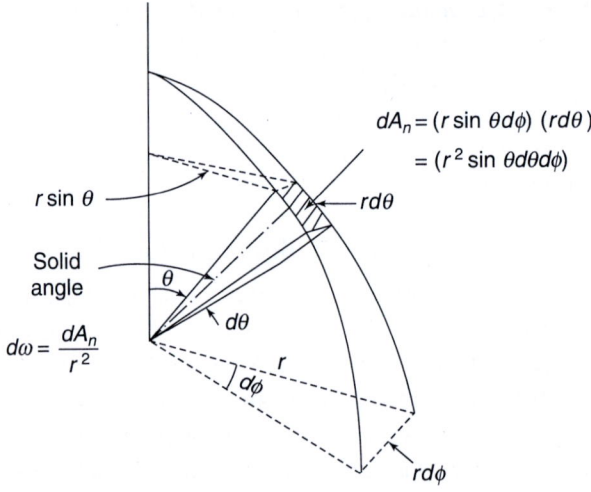

Fig. 8.5   Differential solid angle $d\omega$

$$d\omega = \frac{r^2 \sin \theta \, d\theta \, d\phi}{r^2} = \sin \theta d\theta d\phi \tag{8.11}$$

The unit of solid angle is steradian (Sr). The spectral intensity of radiation is mathematically defined as

$$I_{\lambda, e} (\lambda, \theta, \phi) = \frac{dq}{(dA_1 \cos \theta) \, d\omega \, d\lambda} \tag{8.12}$$

The area $dA_1 \cos \theta$ in Eq. (8.12) is actually the component of dA_1 perpendicular to the direction of radiation (Fig. 8.6). In other words, it is how dA_1 would appear to an observer situated on dA_n. The spectral intensity has units of W/m² Srµm. $I_{\lambda,e}$ is defined as the rate at which radiant energy is emitted at the wavelength λ in the (θ, ϕ) direction, per unit area of the emitting surface normal to this direction, per unit solid angle about this direction, and per unit wavelength interval $d\lambda$ about λ. Equation (8.12) may be rephrased as

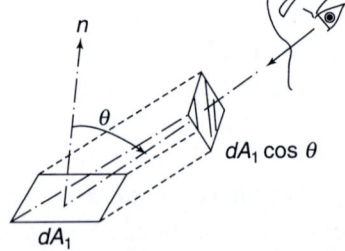

Fig. 8.6   The projection of dA_1 normal to the direction of radiation

$$dq_\lambda = I_{\lambda, e} (\lambda, \theta, \phi) \, dA_1 \cos \theta \, d\omega \tag{8.13}$$

where $dq_\lambda = dq/d\lambda$ and has units of W/μm. From Eq. (8.13) we can write after substituting $d\omega$ from Eq. (8.11),

$$\frac{dq_\lambda}{dA_1} = dq''_\lambda = I_{\lambda,e}(\lambda, \theta, \phi)\cos\theta\sin\theta\, d\theta\, d\phi$$

Integrating,

$$q''_\lambda = \int_0^{2\pi}\int_0^{\pi/2} I_{\lambda,e}(\lambda, \theta, \phi)\cos\theta\sin\theta\, d\theta\, d\phi$$

$$= E_\lambda = \text{spectral emissive power}$$

The total heat flux associated with emission over all directions and at all wavelengths is then

$$q'' = \int_0^\infty q''_\lambda(\lambda)\, d\lambda = E = \text{total emissive power} \tag{8.14}$$

For the special case of a diffuse emitter for which the emitted radiation is independent of direction, we can write

$$I_{\lambda,e}(\lambda, \theta, \phi) = I_{\lambda,e}(\lambda)$$

Therefore,

$$E_\lambda(\lambda) = I_{\lambda,e}(\lambda)\int_0^{2\pi}\int_0^{\pi/2}\cos\theta\sin\theta\, d\theta\, d\phi$$

$$= I_{\lambda,e}(\lambda)\pi$$

Therefore, $\qquad E_\lambda(\lambda) = \pi I_{\lambda,e}(\lambda)$ $\hfill$ (8.15)

Similarly, $\qquad\qquad E = \pi I$ $\hfill$ (8.16)

where I is the total intensity of the emitted radiation. Note that the constant π has units of Sr.

8.4.1 Relation to Irradiation

Irradiation G refers to radiation from all directions incident on a surface. The spectral irradiation G_λ (W/m$^2\mu$m) is defined as the rate at which radiation is incident on a surface per unit area of the surface at the wavelength λ, per unit wavelength interval $d\lambda$ about λ. Thus,

$$G_\lambda(\lambda) = \int_0^{2\pi}\int_0^{\pi/2} I_{\lambda,i}(\lambda, \theta, \phi)\cos\theta\sin\theta\, d\theta\, d\phi \tag{8.17}$$

Total irradiation G is

$$G = \int_0^\infty G_\lambda(\lambda)\, d\lambda \tag{8.18}$$

If the incident radiation is diffuse, then

$$G_\lambda(\lambda) = \pi I_{\lambda,i}(\lambda) \tag{8.19}$$

and $\qquad G = \pi I_i$ $\hfill$ (8.20)

8.4.2 Relation to Radiosity

Radiosity J accounts for all of the radiant energy leaving a surface. Thus radiosity constitutes the reflected portion of the irradiation, as well as direct emission. The spectral radiosity J_λ (W/m$^2\mu$m) represents the rate at which radiation leaves a unit area of the surface at wavelength λ, per unit wavelength interval $d\lambda$ about λ.

Thus, $J_\lambda (\lambda) = \int_0^{2\pi} \int_0^{\pi/2} I_{\lambda,e+r} (\lambda, \theta, \phi) \cos\theta \sin\theta \, d\theta \, d\phi$ (8.21)

$$J = \int_0^\infty J_\lambda(\lambda) \, d\lambda \qquad (8.22)$$

The subscripts e and r of I_λ in Eq. (8.21) represent emission and reflection, respectively. If the surface is both a diffuse reflector and an emitter, $I_{\lambda,e+r}$ is independent of θ and ϕ. Therefore,

$$J_\lambda (\lambda) = \pi I_{\lambda, e+r} \qquad (8.23)$$
$$J = \pi I_{e+r} \qquad (8.24)$$

8.4.3   Relation between Radiosity and Irradiation

Figure 8.7 depicts pictorially the relationship between radiosity (J) and irradiation (G). It is evident from the figure that

$$J = \varepsilon E_b + \rho G \qquad (8.25)$$

In the above expression ρ is the reflectivity of the surface (defined in Section 8.6) and ε is the emissivity of the surface (defined in Section 8.8).

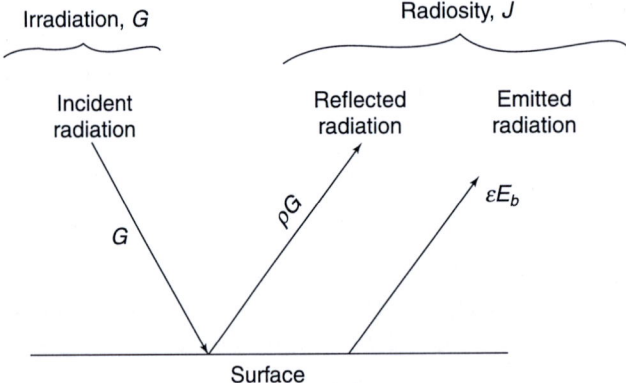

Fig. 8.7   Relationship between radiosity and irradiation

8.5   Diffuse Surface and Specular Surface

A surface is termed *diffuse* or *specular* depending on its response to incident radiation. If the angle of incidence is equal to the angle of reflection, the reflection is called specular [Fig. 8.8(a)]. On the other hand, when an incident beam is distributed uniformly in all directions after reflection, the reflection is called diffuse [Fig. 8.8(b)]. No real surface is completely specular or completely diffuse. An ordinary mirror is quite specular for visible light, but would not be necessarily specular over the entire wavelength of thermal radiation. Normally, a rough surface exhibits diffuse behaviour much more than a highly polished surface. Similarly, a polished surface is more specular than a rough surface.

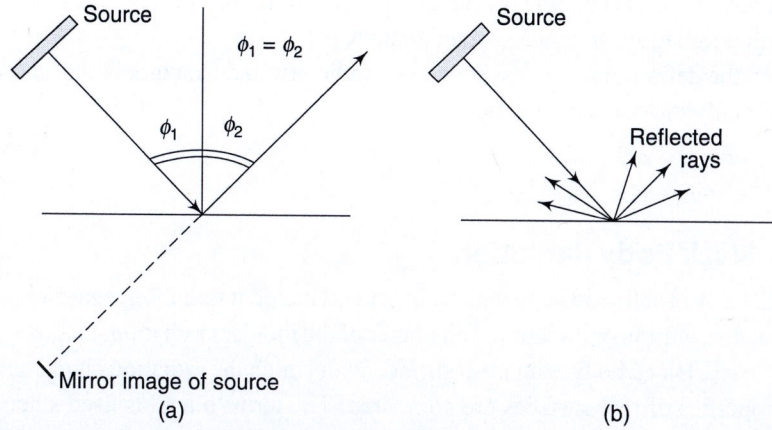

Fig. 8.8 (a) Specular surface (b) diffuse surface

8.6 Absorptivity, Reflectivity, and Transmissivity

When a radiation beam strikes a surface, a part of it may be reflected away from the surface, a portion may be absorbed by the surface, while the rest may be transmitted through the surface (Fig. 8.9). These fractions of reflected, absorbed, and transmitted energy are called reflectivity ρ, absorptivity α, and transmissivity τ, respectively. Thus,

$$\rho_\lambda + \alpha_\lambda + \tau_\lambda = 1 \qquad (8.26)$$

or
$$\rho + \alpha + \tau = 1 \qquad (8.27)$$

In general, both the spectral (or monochromatic) and total surface properties are dependent on the surface composition, roughness, temperature, etc. The monochromatic properties are dependent on the wavelength of the incident radiation, and the total properties are dependent on the spectral distribution of the incident energy.

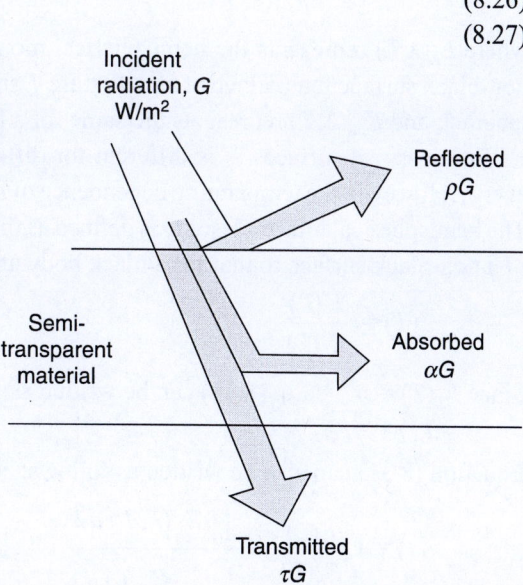

Fig. 8.9 Pictorial representation of absorptivity, reflectivity, and transmissivity

For gases, the aforementioned surface properties are also dependent on the geometrical size and shape of the gas bulk through which the radiation passes. Most gases have high transmissivities τ and low absorptivities α and reflectivities ρ. For example, air at atmospheric pressure is virtually transparent to thermal radiation, so that $\alpha = \rho = 0$ and $\tau = 1$. Other gases, especially water vapour and carbon dioxide, may be highly absorptive to thermal radiation—at least at certain wavelengths. Most solids, except for glass, are opaque to thermal radiation, so that $\tau = 0$ and, hence, $\rho + \alpha = 1$.

From the definitions of emissive power, radiosity, and irradiation, one can write for thermally opaque solid surfaces,

$$J = E + \rho G$$

or $\quad J = E + (1 - \alpha)G$ (8.28)

8.7   Black Body Radiation

A black body is defined as one which absorbs all incident radiation regardless of the spectral distribution or directional character of the incident radiation, i.e., $\alpha_\lambda = \alpha = 1$ or $\rho_\lambda = \rho = 0$. Black body is an ideal surface with which the radiation characteristics and properties of real surfaces are compared. The term 'black' is used since dark surfaces normally show high values of absorptivity. The intensity of black body radiation, I_b, is uniform. Thus, black body radiation is diffuse and

$$E_b = \pi I_b$$ (8.29)

8.8   Radiation Characteristics of Non-black Surfaces: Monochromatic and Total Emissivity

The monochromatic emissivity is defined as

$$\varepsilon_\lambda (\lambda, T) = \frac{E_\lambda(\lambda, T)}{E_{b\lambda}(\lambda, T)}$$ (8.30)

where $E_\lambda(\lambda, T)$ represents the hemispherical monochromatic emissive power of a non-black surface maintained at temperature T and measured at a particular wavelength λ, and $E_{b\lambda}(\lambda, T)$ represents the same for a black body.

For most real surfaces ε_λ is different for different wavelengths of the emitted energy. However, the temperature dependence of ε_λ may be small and often ignored. The hemispherical total emissivity is defined as the ratio of the total emissive power of a non-black surface to that for a black body at the same temperature:

$$\varepsilon (T) = \frac{E(T)}{E_b(T)}$$ (8.31)

Since $E_b(T) = \sigma T^4$, Eq. (8.31) can be written as

$$E(T) = \varepsilon \sigma T^4$$ (8.32)

Equation (8.31) can also be written as follows:

$$\varepsilon(T) = \frac{E(T)}{E_b(T)} = \frac{\int_0^\infty E_\lambda (\lambda, T)\, d\lambda}{\int_0^\infty E_{b\lambda}(\lambda, T)\, d\lambda}$$

or $\quad \varepsilon (T) = \dfrac{\int_0^\infty \varepsilon_\lambda (\lambda, T)\, E_{b\lambda}(\lambda, T)\, d\lambda}{\int_0^\infty E_{b\lambda}(\lambda, T)\, d\lambda} = \dfrac{\int_0^\infty \varepsilon_\lambda (\lambda, T)\, E_{b\lambda}(\lambda, T)\, d\lambda}{\sigma T^4}$ (8.33)

Since $\varepsilon_\lambda(\lambda, T) = f(\lambda, T)$, Eq. (8.33) shows that $\varepsilon = f(T)$ only.

An ideal gray body, which is a special type of non-black surfaces, is defined as the one for which the monochromatic emissivity is independent of wavelengths, i.e., the ratio of E_λ to $E_{b\lambda}$ is the same for all wavelengths of emitted energy at a given temperature. Thus, for the ideal gray surface,

$$\varepsilon(T) = \varepsilon_\lambda(T) \tag{8.34}$$

Comparisons of the emissivity and emissive power of a real surface with those of a gray surface and a black body at the same temperature are shown in Fig. 8.10 and Fig. 8.11, respectively. It may be noted that the radiation emission from a real surface generally deviates from that predicted by Planck's law, and the emissive curve may have undulations as shown in Fig. 8.11. A gray surface should emit the same amount of radiation as the real surface it approximates at the same temperature. Therefore, the area under the curve of a gray surface must be equal to that under the curve of the corresponding real surface.

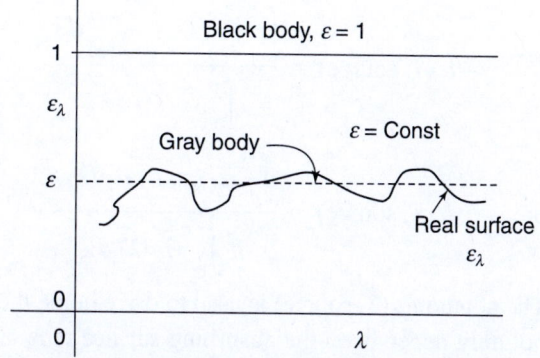

Fig. 8.10 Comparison of the emissivity of a real surface with that of a gray surface and a black body at the same temperature

Typical emissivity values of various materials are listed in Table A1.8. Metals usually have low emissivities and non-metals such as ceramics and organic materials have high ones. The emissivity of metals increases with temperature. Furthermore, oxidation causes significant increase in the emissivity of metals. This is because, with oxidation the metallic surface looks somewhat dull and hence its reflectivity decreases.

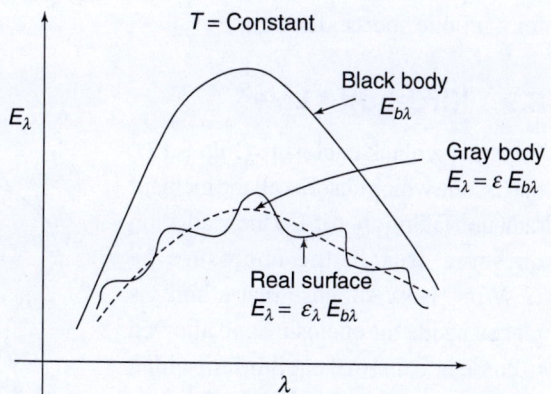

Fig. 8.11 Comparison of the emissive power of a real surface with that of a gray surface and a black body at the same temperature

8.8.1 Monochromatic and Total Absorptivities

Unlike a black body, real surfaces do not generally absorb all incident energy. Ignoring directional preferences by considering only the hemispherical irradiation G_λ incident upon a surface, the hemispherical monochromatic absorptivity is defined as

$$\alpha_\lambda(\lambda, T) = \frac{[G_\lambda(\lambda)]_{\text{absorbed}}}{G_\lambda(\lambda)} \qquad (8.35)$$

where G_λ is the irradiation, from all directions, incident on the surface at a particular wavelength and can be expressed as

$$G_\lambda(\lambda) = \int_0^{2\pi} \int_0^{\pi/2} I_{\lambda,i}(\lambda, \theta, \phi) \cos\theta \sin\theta \, d\theta \, d\phi \qquad (8.36)$$

The hemispherical total absorptivity is defined as the sum of the absorbed irradiation over all wavelengths divided by the total incident energy. Thus,

$$\alpha(T, \text{source}) = \frac{\int_0^\infty [G_\lambda(\lambda)]_{\text{absorbed}} \, d\lambda}{\int_0^\infty G_\lambda(\lambda) \, d\lambda} \qquad (8.37)$$

or

$$\alpha(T, \text{source}) = \frac{\int_0^\infty \alpha_\lambda(\pi, T) \, G_\lambda(\lambda) \, d\lambda}{\int_0^\infty G_\lambda(\lambda) \, d\lambda} \qquad (8.38)$$

The notation $\alpha(T, \text{source})$ is used to drive home the point that the total absorptivity not only depends on the absorbing surface temperature but also on the source of the incident radiation. In other words, the spectral and spatial characteristics of the incident radiation also influence the total absorptivity value. Thus, the total absorptivity is not a simple surface property that can be tabulated like emissivity values. A separate table has to be made for each source. This is not done, except for a unique source, the sun.

8.9 Kirchhoff's Law[5]

Consider a black enclosure (Fig. 8.12), that is, one which absorbs all the incident radiation falling upon it. Let the irradiation on some area in the enclosure be G W/m^2. Now suppose that a body is placed inside the enclosure and allowed to come into thermal equilibrium with it (i.e., the temperature of the body is the same as that of the wall). At equilibrium the energy absorbed by the body must be equal to the energy emitted so that

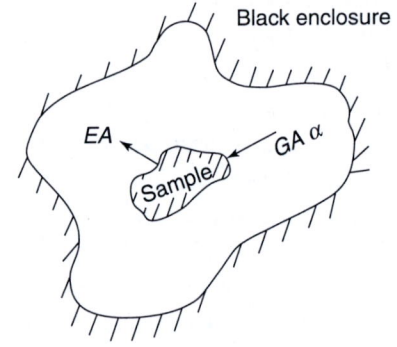

Fig. 8.12 Model for deriving Kirchhoff's law

$$EA = GA\alpha \qquad (8.39)$$

If the body in the enclosure is replaced by a black body of the same size and shape and allowed to come to equilibrium with the enclosure at the same temperature, then

$$E_b A = GA(1) \qquad (8.40)$$

---

[5] Gustav Robert Kirchhoff (1824–1887), German physicist, served as Professor at the University of Heidelberg for 22 years and later at the University of Berlin. Together with Robert Bunsen he established the theory of spectrum analysis.

since $\alpha = 1$ for a black body. Dividing Eq. (8.39) by Eq. (8.40),

$$\frac{EA}{E_b A} = \frac{GA\alpha}{GA}$$

or
$$\frac{EA}{E_b A} = \alpha \qquad (8.41)$$

Now, since $\varepsilon = E/E_b$, we can write

$$\varepsilon = \alpha \qquad (8.42)$$

Equation (8.42) is known as Kirchhoff's identity. Since $\alpha \leq 1$, then $\varepsilon \leq 1$; that is, a black body, a perfect absorber is also a perfect emitter.

8.9.1 Restrictions of Kirchhoff's Law

It should be noted that Kirchhoff's law is subject to thermal equilibrium in an isothermal enclosure and diffuse radiation. In the case of monochromatic radiation, the form of Kirchhoff's law is similar, that is,

$$\alpha_\lambda (\lambda, T) = \varepsilon_\lambda (\lambda, T) \qquad (8.43)$$

Equation (8.43) is valid provided the incident radiation is diffuse or the surface is such that α_λ and ε_λ have no directional dependencies. Now,

$$\varepsilon (T) = \frac{\int_0^\infty \varepsilon_\lambda (\lambda, T)\, E_{b,\lambda}(\lambda, T)\, d\lambda}{\int_0^\infty E_{b\lambda}(\lambda, T)\, d\lambda} \qquad (8.44)$$

$$\alpha (T, \text{source}) = \frac{\int_0^\infty \alpha_\lambda (\lambda, T)\, G_\lambda(\lambda)\, d\lambda}{\int_0^\infty G_\lambda(\lambda)\, d\lambda} \qquad (8.45)$$

Comparing Eqs (8.44) and (8.45) it is observed that Kirchhoff's law, i.e., $\alpha(T) = \varepsilon(T)$ occurs when either of the following two conditions is satisfied.

1. If the incident irradiation on the receiving surface has a spectral distribution the same as that of emission from a black body at the same temperature, i.e., $G_\lambda (\lambda) = E_{b\lambda} (\lambda)$.
2. If the receiving surface is an ideal gray surface, i.e., $\varepsilon_\lambda \neq f(\lambda)$ so that it may be moved outside the integral in Eq. (8.44).

Condition 1 is a restriction on the source of irradiation, which must be that of a black body at the same temperature as the receiver. This was the case of the isothermal enclosure which gave rise to $\varepsilon = \alpha$, but seldom thermal equilibrium is realized in heat transfer applications.

Condition 2, that of grayness, puts a restriction on the nature of the surface rather than on incident radiation. While very few surfaces meet the definition of grayness, there are many instances in which this idealization may be made. In such cases the total form of Kirchhoff's law can be used even though thermal equilibrium does not exist.

8.9.2 Note on a Gray Body

In spite of the fact that for most of the engineering materials ε_λ and α_λ are not constant over the entire range of wavelengths, many materials do qualify as gray bodies, at least approximately. For example, suppose that non-negligible values of $E_{b\lambda}$ and G_λ occur in the same finite range of wavelengths. ε_λ and α_λ are essentially constant in that finite range, the gray body definition is met, and $\alpha = \varepsilon$. The gray body condition is likely to be violated when $E_{b\lambda}$ and G_λ lie in different wavelength ranges, for instance, when $E_{b\lambda}$ represents infrared emission and G_λ represents irradiation from a high temperature source such as the sun.

8.10 View Factor

To compute the magnitude of radiation exchange between any two surfaces, the knowledge of view factor (also called configuration factor or shape factor) is a must. The view factor is a quantity that depends on the surface geometry and orientations. The view factor F_{ij} (or F_{i-j}) is defined as the fraction of the radiation leaving surface i, which is directly intercepted by surface j. Thus

$$F_{ij} = \frac{q_{i \to j}}{A_i J_i} \tag{8.46}$$

where $q_{i \to j}$ is the amount of radiation leaving surface i and intercepted by surface j and $A_i J_i$ is the total radiation leaving surface i which may be due to both emission and reflection.

8.10.1 View Factor Integral

To derive an expression for F_{ij}, the arbitrarily oriented surfaces A_i and A_j are considered as shown in Fig. 8.13(a). Elemental areas on each surface (dA_i and dA_j) are connected by a line of length R, which forms the polar angles θ_i and θ_j, respectively, with the surface normals n_i and n_j. The magnitudes of R, θ_i, and θ_j are functions of the position of the elemental areas on A_i and A_j.

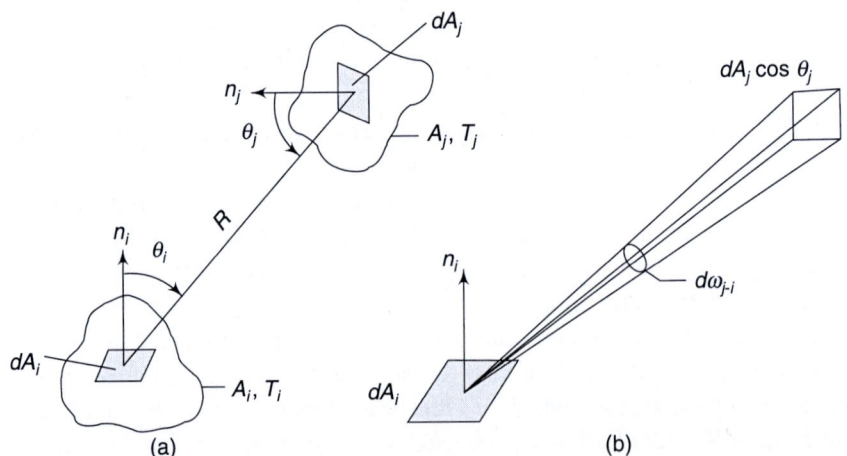

Fig. 8.13 (a) View factor geometrical parameters, (b) solid angle $d\omega_{j-i}$ subtended by dA_j when viewed from dA_i

From the definition of radiation intensity, it follows:

$$I_i = \frac{dq_{i \to j}}{dA_i \cos \theta_i \, d\omega_{j\text{-}i}}$$

or $\qquad dq_{i \to j} = I_i \cos \theta_i \, dA_i \, d\omega_{j\text{-}i}$ $\hfill$ (8.47)

where $dq_{i \to j}$ is the rate at which radiation leaves dA_i and is intercepted by dA_j, and $d\omega_{j\text{-}i}$ is the solid angle subtended by dA_j when viewed from dA_i [Fig. 8.13(b)].

Now, $\qquad d\omega_{j\text{-}i} = \dfrac{\cos \theta_j dA_j}{R^2}$ $\hfill$ (8.48)

Putting Eq. (8.48) into Eq. (8.47), we have

$$dq_{i \to j} = I_i \frac{\cos \theta_i \cos \theta_j}{R^2} \, dA_i \, dA_j \qquad (8.49)$$

Assuming that surface i emits and reflects diffusely and using $J_i = \pi I_i$ in Eq. (8.49), we get

$$dq_{i \to j} = J_i \frac{\cos \theta_i \cos \theta_j}{\pi R^2} \, dA_i \, dA_j \qquad (8.50)$$

Thus, $\qquad q_{i \to j} = J_i \displaystyle\int_{A_i} \int_{A_j} \frac{\cos \theta_i \cos \theta_j}{\pi R^2} dA_i \, dA_j$ $\hfill$ (8.51)

Note that the radiosity J_i could be moved outside the integral in Eq. (8.51) as it is uniform over surface A_i because of the fact that surface A_i is diffuse.

From the definition of the view factor [see Eq. (8.46)] it follows that

$$F_{ij} = \frac{1}{A_i} \int_{A_i} \int_{A_j} \frac{\cos \theta_i \cos \theta_j}{\pi R^2} dA_i \, dA_j \qquad (8.52)$$

Similarly,

$$F_{ji} = \frac{q_{j \to i}}{A_j J_i} \qquad (8.53)$$

and $\qquad F_{ji} = \dfrac{1}{A_j} \displaystyle\int_{A_i} \int_{A_j} \frac{\cos \theta_i \cos \theta_j}{\pi R^2} dA_i dA_j$ $\hfill$ (8.54)

Equations (8.52) and (8.54) are called view factor integrals. Either of these equations may be used to determine the view factor associated with any two surfaces that are diffuse emitters and reflectors and have uniform radiosity.

8.10.2 View Factor Relations

It is obvious from Eqs (8.52) and (8.54) that

$$A_i F_{ij} = A_j F_{ji} \qquad (8.55)$$

Equation (8.55) is called the reciprocity relation and is useful in determining one view factor from the knowledge of the other.

Another important view factor relation applies to the surfaces of an enclosure. From the definition of view factor it follows that for each of the N surfaces of an enclosure the following summation rule is valid:

$$\sum_{j=1}^{N} F_{ij} = 1 \qquad\qquad (8.56)$$

Equation (8.56) originates from the conservation requirement that all of the radiation leaving surface i must be intercepted by the enclosure surfaces. It may be noted that

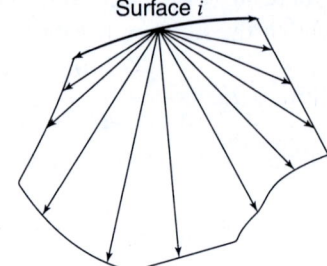

$$F_{ii} = 0, \ \text{if the surface } i \text{ is plane or convex}$$

$$F_{ii} \neq 0, \ \text{if the surface } i \text{ is concave}$$

The term F_{ii} represents the fraction of the radiation that leaves surface i and is directly intercepted by surface i itself.

Figure 8.14 demonstrates the summation rule. Figure 8.15 shows how F_{ii} change for plane, convex, and concave surfaces.

Fig. 8.14 Pictorial demonstration of summation rule

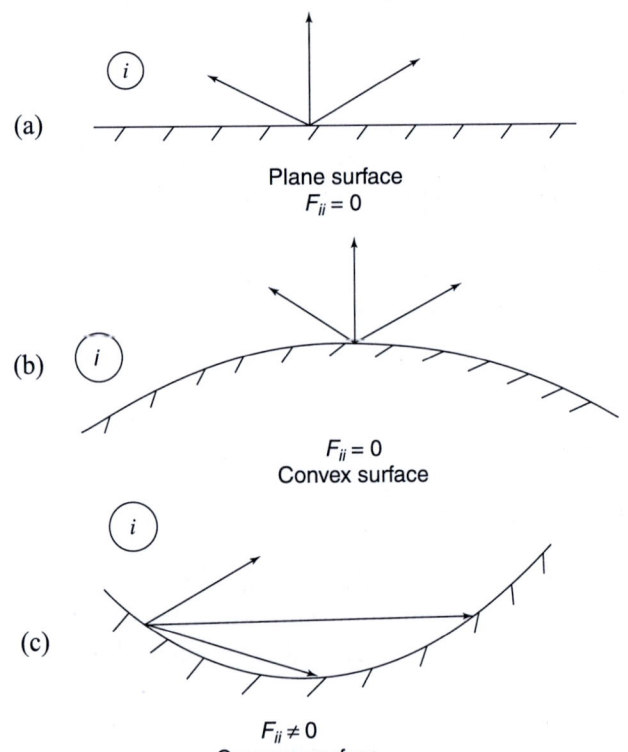

Fig. 8.15 View factor from a surface to itself (Fii) for plane, convex, and concave surfaces

---

Example 8.2 View Factor from View Factor Integral

Consider a circular disc of diameter D and area A_j above a plane surface of area A_i ($<< A_j$). The surfaces are parallel to each other, and A_i is located at a distance L from the centre

of A_j. Obtain an expression for the view factor F_{ij}. See Fig. E8.2.

Solution

Recall

$$F_{ij} = \frac{1}{A_j} \int_{A_i} \int_{A_j} \frac{\cos\theta_i \cos\theta_j}{\pi R^2} dA_i \, dA_j$$

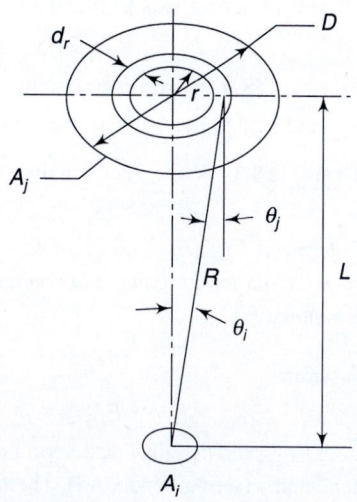

Note that θ_i, θ_j, and R are approximately independent of the position on A_i. Thus, the above expression reduces to

$$F_{ji} = \int_{A_j} \frac{\cos\theta_i \cos\theta_j}{\pi R^2} dA_j$$

Now, from Fig. E8.2, we see that

$$\theta_i = \theta_j = \theta$$

Therefore, $F_{ij} = \displaystyle\int_{A_j} \frac{\cos^2\theta}{\pi R^2} dA_j$

Furthermore, $R^2 = r^2 + L^2$

$$\cos\theta = \frac{L}{R}$$

$$dA_j = 2\pi r \, dr$$

Fig. E8.2 View factor parameters for radiation from a small area element to a circular disc

Hence, $\quad F_{ij} = \displaystyle\int_0^{D/2} \frac{\left(\dfrac{L}{R}\right)^2}{\pi R^2} \, 2\pi r \, dr$

$$= L^2 \int_0^{D/2} \frac{2r \, dr}{(r^2 + L^2)^2} = \frac{D^2}{D^2 + 4L^2}$$

8.10.3 View Factor Algebra

Suppose the view factor between surface A_3 to the combined area $A_{1,2}$ is desired (Fig. 8.16). The view factor $F_{3\text{-}1,2}$ can be expressed as

$$F_{3\text{-}1,2} = F_{3\text{-}1} + F_{3\text{-}2} \qquad (8.57)$$

In other words, the total view factor is the sum of its parts. Now, multiplying both sides of Eq. (8.57) by A_3, we have

$$A_3 F_{3\text{-}1,2} = A_3 F_{3\text{-}1} + A_3 F_{3\text{-}2} \qquad (8.58)$$

From the reciprocity relations, we can write

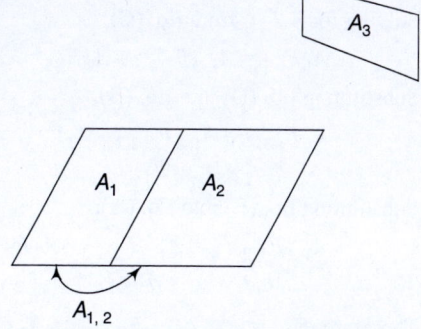

Fig. 8.16 View factor between A_3 and composite area $A_{1,2}$

$$A_3 F_{3\text{-}1,2} = A_{1,2} F_{1,2\text{-}3} \qquad (8.59)$$

$$A_3 F_{3\text{-}1} = A_1 F_{1\text{-}3} \qquad (8.60)$$

$$A_3 F_{3\text{-}2} = A_2 F_{2\text{-}3} \qquad (8.61)$$

Therefore, using Eqs (8.58)–(8.61), the following can be obtained:

$$A_{1,2}F_{1,2\text{-}3} = A_1 F_{1\text{-}3} + A_2 F_{2\text{-}3} \qquad (8.62)$$

Equation (8.62) simply states that the total radiation arriving at surface 3 from the $A_{1,2}$ surface is the sum of the radiations from surfaces 1 and 2.

Example 8.3   View Factor using View Factor Algebra

Determine $F_{1\text{-}3}$ (Fig. E8.3) in terms of known view factors for perpendicular rectangles with a common edge.

Solution

$$F_{1\text{-}2,3} = F_{1\text{-}2} + F_{1\text{-}3}$$

Both $F_{1\text{-}2,3}$ and $F_{1\text{-}2}$ may be determined from standard charts (see Appendix A5). Therefore,

$$F_{1\text{-}3} = F_{1\text{-}2,3} - F_{1\text{-}2}$$

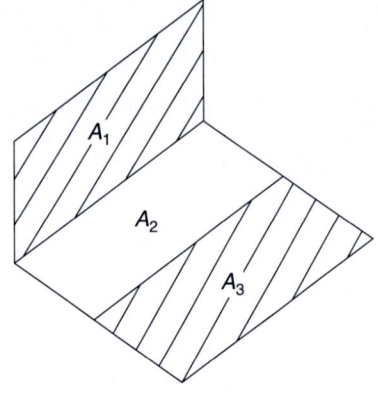

Fig. E8.3   View factor between perpendicular rectangles of areas A_1 and A_3

Example 8.4   Additional Example of View Factor Algebra

Determine $F_{1\text{-}4}$ (Fig. E8.4) in terms of known view factors for perpendicular rectangles with a common edge.

Solution

$$A_{1,2}F_{1,2\text{-}3,4} = A_1 F_{1\text{-}3,4} + A_2 F_{2\text{-}3,4} \quad (\text{A})$$

Also, $A_1 F_{1\text{-}3,4} = A_1 F_{1\text{-}3} + A_1 F_{1\text{-}4}$ (B)

Furthermore,

$$A_{1,2}F_{1,2\text{-}3} = A_1 F_{1\text{-}3} + A_2 F_{2\text{-}3} \quad (\text{C})$$

Solving for $A_1 F_{1\text{-}3}$ from Eq. (C),

$$A_1 F_{1\text{-}3} = A_{1,2} F_{1,2\text{-}3} - A_2 F_{2\text{-}3} \quad (\text{D})$$

Substituting Eq. (D) into Eq. (B),

$$\begin{aligned} A_1 F_{1\text{-}3,4} &= A_{1,2} F_{1,2\text{-}3} - A_2 F_{2\text{-}3} \\ &\quad + A_1 F_{1\text{-}4} \end{aligned} \quad (\text{E})$$

Substituting Eq. (E) into Eq. (A),

$$\begin{aligned} A_{1,2} F_{1,2\text{-}3,4} &= A_{1,2} F_{1,2\text{-}3} - A_2 F_{2\text{-}3} \\ &\quad + A_1 F_{1\text{-}4} + A_2 F_{2\text{-}3,4} \end{aligned}$$

Therefore, $F_{1\text{-}4} = \dfrac{1}{A_1} (A_{1,2} F_{1,2\text{-}3,4} + A_2 F_{2\text{-}3}$

$$- A_{1,2} F_{1,2\text{-}3} - A_2 F_{2\text{-}3,4})$$

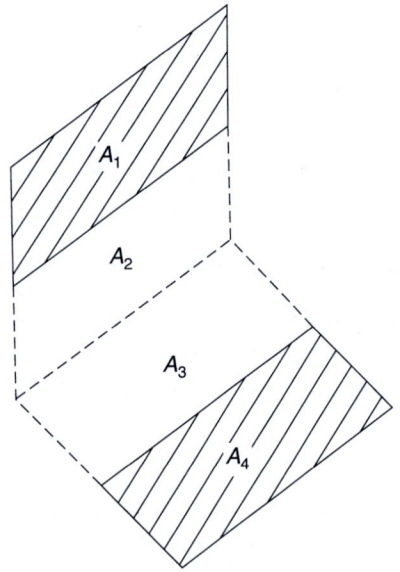

Fig. E8.4   View factor between perpendicular rectangles of areas A_1 and A_4

Note: In the foregoing solutions the tacit assumption has been made that various bodies do not see themselves, i.e., $F_{1\text{-}1} = F_{2\text{-}2} = F_{3\text{-}3} = 0$.

Example 8.5 View Factors in an Enclosure of Triangular Cross-section

An enclosure of triangular cross-section is made up of three plane plates, each of finite width and infinite length (thus forming an infinitely long triangular prism). Derive an expression for the configuration factor (or view factor) between any two of the plate widths L_1, L_2, L_3 (see Fig. E8.5).

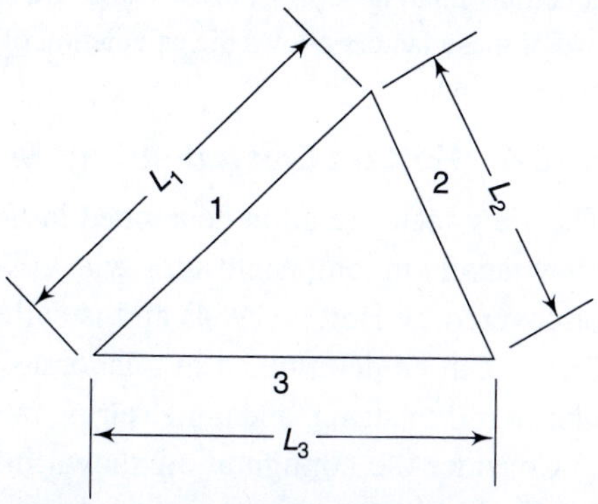

Solution

For plate 1, from the summation rule for an enclosure,

$$F_{1\text{-}1} + F_{1\text{-}2} + F_{1\text{-}3} = 1 \qquad \text{(A)}$$

Since $F_{1\text{-}1} = 0$, Eq. (A) reduces to

$$F_{1\text{-}2} + F_{1\text{-}3} = 1 \qquad \text{(B)}$$

Fig. E8.5 A triangular enclosure having sides of finite width and infinite length

Using similar relations for other two plates, we get

$$F_{2\text{-}1} + F_{2\text{-}3} = 1 \qquad \text{(C)}$$
$$F_{3\text{-}1} + F_{3\text{-}2} = 1 \qquad \text{(D)}$$

Multiplication of Eqs (B)–(D) by the respective plate areas A_1, A_2, and A_3, respectively, results in

$$A_1 F_{1\text{-}2} + A_1 F_{1\text{-}3} = A_1 \qquad \text{(E)}$$
$$A_2 F_{2\text{-}1} + A_2 F_{2\text{-}3} = A_2 \qquad \text{(F)}$$
$$A_3 F_{3\text{-}1} + A_3 F_{3\text{-}2} = A_3 \qquad \text{(G)}$$

By applying reciprocity relations, we obtain

$$A_1 F_{1\text{-}2} + A_1 F_{1\text{-}3} = A_1 \qquad \text{(H)}$$
$$A_1 F_{1\text{-}2} + A_2 F_{2\text{-}3} = A_2 \qquad \text{(I)}$$
$$A_1 F_{1\text{-}3} + A_2 F_{2\text{-}3} = A_3 \qquad \text{(J)}$$

Thus Eqs (H)–(J) give three equations for three unknown view factors. Solving Eqs (H)–(J) simultaneously, we get

$$F_{1\text{-}2} = \frac{A_1 + A_2 - A_3}{2 A_1}$$

or

$$F_{1\text{-}2} = \frac{L_1 + L_2 - L_3}{2 L_1} \qquad \text{(K)}$$

On a careful look at the three simultaneous Eqs (H)–(J), it is observed that (i) the first equation [Eq. (H)] involves two unknowns $F_{1\text{-}2}$ and $F_{1\text{-}3}$, (ii) the second equation [Eq. (I)] has one additional unknown $F_{2\text{-}3}$, and (iii) the final equation [Eq. (J)] has no additional unknowns. Generalizing the procedure to any N-sided enclosure made up of plane or convex surfaces shows that of N simultaneous equations, the first would involve $N-1$ unknowns, the second $N-2$ unknowns, and so forth. The total number of unknowns, U, is then

$$U = (N-1) + (N-2) + \cdots + 1$$

which is an arithmetic series having $N-1$ terms. Therefore,

$$U = \frac{N-1}{2} \{ 2(N-1) + (N-2)\,(-1) \}$$

$$= \frac{N\,(N-1)}{2}$$

Thus, for a four-sided enclosure made up of planar or convex surfaces of known area, four equations relating $4(4-1)/2$ or 6 unknown view factors can be written. Specifying any two of these factors allows the calculation of the rest by solving the set of four simultaneous equations.

8.10.4 Hottel's Crossed-strings Method

The view factor relation developed in Example 8.5 may be used to determine all view factors in long enclosures with constant cross-section. The method was first discovered by Hottel (1954) and is called crossed-strings method since the view factors can be determined in a laboratory by a person. All he/she needs are four pins, a roll of string, and a measuring tape.

Consider the configuration shown in Fig. 8.17. Our objective is to determine $F_{1\text{-}2}$. The surfaces shown are rather irregular (partly convex, partly concave) and the view between them may be obstructed. One can imagine how difficult it can be to obtain the view factor by integration.

In the crossed-strings method, four pins are placed at the two ends of each surface, as indicated by a, b, c, and d in Fig. 8.17. Now the points a and c and b and d are connected by tight strings. Similarly, a and b, c and d, a and d, and b and c are connected tightly. Now, assuming the strings to be imaginary surfaces A_{ab}, A_{ac}, and A_{bc} for the triangle abc, we can write

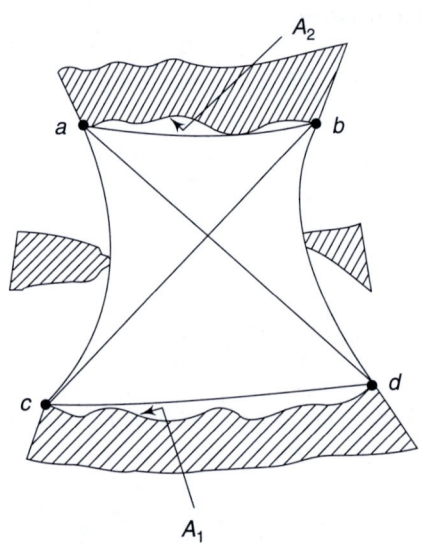

Fig. 8.17 Hottel's crossed-strings method

$$F_{ab\text{-}ac} = \frac{A_{ab} + A_{ac} - A_{bc}}{2A_{ab}} \quad (8.63)$$

It should be noted here that since a tightened string always forms a convex surface, the rule of triangle developed in Example 8.5 can be used. Similarly, for Δabd,

$$F_{ab\text{-}bd} = \frac{A_{ab} + A_{bd} - A_{ad}}{2A_{ab}} \quad (8.64)$$

From the summation rule,

$$F_{ab\text{-}ac} + F_{ab\text{-}bd} + F_{ab\text{-}cd} = 1 \quad (8.65)$$

Thus, using Eqs (8.63) – (8.65),

$$F_{ab\text{-}cd} = \frac{(A_{bc} + A_{ad}) - (A_{ac} + A_{bd})}{2A_{ab}} \quad (8.66)$$

It is also observed from Fig. 8.17 that all radiation leaving A_{ab} travelling to A_{cd} will be received by the surface A_1. At the same time all radiation from A_{ab} going to A_1

must pass through A_{cd}. Therefore,

$$F_{ab-cd} = F_{ab-1} \tag{8.67}$$

Using the reciprocity relation and repeating the argument for surfaces A_{ab} and A_2, we see

$$F_{ab\text{-}cd} = F_{ab-1}$$

$$= \frac{A_1}{A_{ab}} F_{1\text{-}ab}$$

$$= \frac{A_1}{A_{ab}} F_{1\text{-}2} \tag{8.68}$$

Substituting Eq. (8.68) into Eq. (8.66), we get

$$F_{1-2} = \frac{(A_{bc} + A_{ad}) - (A_{ac} + A_{bd})}{2 A_1} \tag{8.69}$$

or

$$F_{1\text{-}2} = \frac{\text{diagonals} - \text{sides}}{2 \times (\text{originating area})}$$

$$= \frac{\left(\begin{array}{c}\text{sum of lengths of}\\\text{crossed strings}\end{array}\right) - \left(\begin{array}{c}\text{sum of lengths of}\\\text{uncrossed strings}\end{array}\right)}{2 \times (\text{originating length})} \tag{8.70}$$

Example 8.6   View Factor using Hottel's Crossed-strings Method

Two infinitely long semi-cylindrical surfaces of radius R are separated by a minimum distance D as shown in Fig. E8.6. Derive the view factor $F_{1\text{-}2}$ by Hottel's crossed-strings method.

Solution
The length of crossed string *abcde* is denoted as L_1, and of uncrossed string *ef* as L_2. From the symmetry of the problem,

$$F_{1\text{-}2} = \frac{2L_1 - 2L_2}{2L_{A_1}} = \frac{2L_1 - 2L_2}{2(\pi R)}$$

$$= \frac{L_1 - L_2}{\pi R}$$

The length L_2 is given by

$$L_2 = D + 2R$$

The length of $L_1 = 2$ (length of *cde*). The segment of L_1 from *c* to *d* is found from the right-angled triangle *ocd* to be

$$L_{1,\,c\text{-}d} = \left[\left(\frac{D+2R}{2}\right)^2 - R^2\right]^{1/2}$$

$$= \left[D\left(\frac{D}{4} + R\right)\right]^{1/2}$$

and the segment of length L_1 from *d* to *e* is

$$L_{1,\,d\text{-}e} = R\theta$$

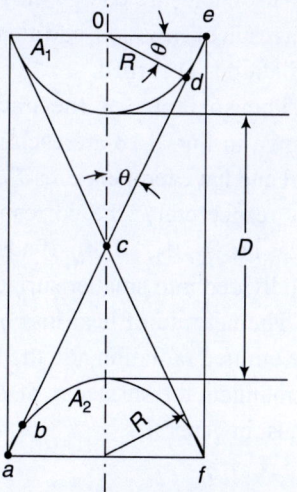

Fig. E8.6   View factor between two infinite semi-cylindrical surfaces by Hottel's crossed-strings method

From Δocd, the angle θ is given by

$$\theta = \sin^{-1} \frac{R}{\dfrac{D}{2} + R}$$

Therefore, $F_{1\text{-}2} = \dfrac{L_1 - L_2}{\pi R}$

$$= \frac{2\,(L_{1,c\text{-}d} + L_{1,d\text{-}e}) - L_2}{\pi R}$$

or $\quad F_{1\text{-}2} = \dfrac{\left[4D\left(\dfrac{D}{4} + R\right)\right]^{1/2} + 2R\sin^{-1}\left[\dfrac{R}{\dfrac{D}{2} + R}\right] - D - 2R}{\pi R}$

8.11 Radiation Exchange in a Black Enclosure

Consider an enclosure having N black surfaces (Fig. 8.18). In general, radiation may leave a surface of an enclosure due to both reflection and emission, and on reaching another surface, experience reflection as well as absorption. However, since for black bodies there is no reflection, in a black enclosure energy only leaves as a result of emission, and all incident radiation is absorbed.

The surfaces of the enclosure shown in Fig. 8.18 are each isothermal and have temperatures $T_1, T_2, ..., T_N$, respectively; the corresponding

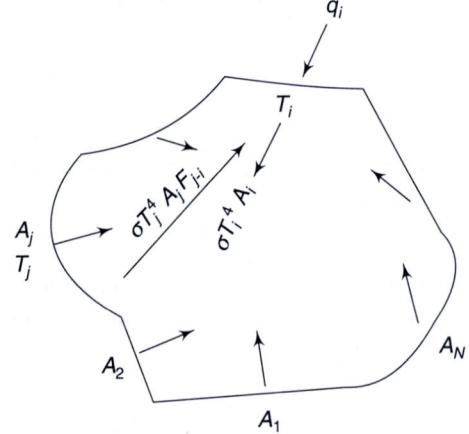

Fig. 8.18 A black enclosure of N surfaces

areas are $A_1, A_2, ..., A_N$. If a given physical surface is not actually isothermal, it is subdivided into smaller surfaces, each of which is essentially isothermal.

The net rate of heat loss q_i from a typical surface i is the difference between the emitted radiation and the absorbed portion of the incident radiation. Note that to maintain the surface A_i at T_i, q_i must be supplied from outside to the surface A_i. Thus, in general,

$$\frac{q_i}{A_i} = \varepsilon_i \sigma T_i^4 - \alpha_i G_i \tag{8.71}$$

where G_i is the irradiation on surface i. For a black enclosure, $\varepsilon_i = \alpha_i = 1$. Therefore, Eq. (8.71) transforms to

$$\frac{q_i}{A_i} = \sigma T_i^4 - G_i \tag{8.72}$$

The radiant heat flux incident on surface i comes from the other surfaces of the enclosure. Consider the radiation (i.e., emission in this case) coming from any other surface j. Since, surface j is a black body emitter, radiant energy $\sigma T_j^4 A_j$ emanates from j in all directions. The energy that arrives at surface i is $\sigma T_j^4 A_j F_{j-i}$ or $\sigma T_j^4 A_i F_{i-j}$ (by reciprocity rule). Contributions such as this arrive at A_i from all surfaces of the enclosure. Thus,

$$G_i A_i = \sum_{j=1}^{N} \sigma T_j^4 A_i F_{i-j}$$

or $\qquad G_i = \sum_{j=1}^{N} \sigma T_j^4 F_{i-j}$ $\hfill$ (8.73)

Substituting Eq. (8.73) into Eq. (8.72), we get

$$\frac{q_i}{A_i} = \sigma T_i^4 - \sum_{j=1}^{N} \sigma T_j^4 F_{i-j} \hfill (8.74)$$

Now, we know, from the summation rule for an enclosure,

$$\sum_{j=1}^{N} F_{i-j} = 1$$

Therefore, $\qquad \sigma T_i^4 = \sum_{j=1}^{N} \sigma T_i^4 F_{i-j}$ $\hfill$ (8.75)

Putting Eq. (8.75) into Eq. (8.74),

$$\frac{q_i}{A_i} = \sum_{j=1}^{N} \sigma T_i^4 F_{i-j} - \sum_{j=1}^{N} \sigma T_j^4 F_{i-j}$$

or $\qquad \dfrac{q_i}{A_i} = \sum_{\substack{j=1 \\ j \neq i}}^{N} \sigma (T_i^4 - T_j^4) F_{i-j}$ $\hfill$ (8.76)

or $\qquad q_i = \sum_{\substack{j=1 \\ j \neq i}}^{N} A_i F_{i-j} \sigma (T_i^4 - T_j^4)$ $\hfill$ (8.77)

8.12 Radiation Exchange in a Gray Enclosure

The difference between radiation in a black enclosure and gray enclosure is the surface reflection. In a gray enclosure, radiation may experience multiple reflections off all surfaces, with partial absorption occurring at each.

It may be recalled from the discussion in Section 8.11 that the heat loss at a surface of a enclosure has the general form

$$\frac{q_i}{A_i} = \varepsilon_i \sigma T_i^4 - \alpha_i G_i \hfill (8.78)$$

Now, we know, at a gray surface (Fig. 8.19)

$$J = \varepsilon \sigma T^4 + \rho G$$

Therefore, at a surface i,

$$J_i = \varepsilon_i \sigma T_i^4 + \rho_i G_i$$

But, since $\rho_i = 1 - \varepsilon_i$,

$$J_i = \varepsilon_i \sigma T_i^4 + (1 - \varepsilon_i) G_i$$

or

$$G_i = \frac{J_i - \varepsilon_i \sigma T_i^4}{1 - \varepsilon_i}$$

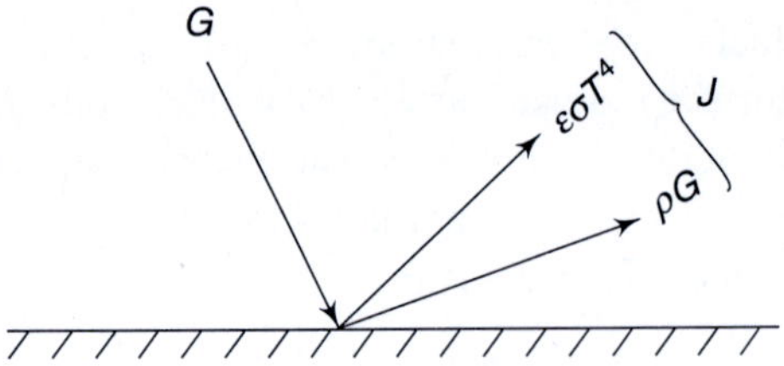

Hence, from Eq. (8.78),

Fig. 8.19   Irradiation G and radiosity J at a surface

$$\frac{q_i}{A_i} = \varepsilon_i \sigma T_i^4 - \frac{\varepsilon_i}{1 - \varepsilon_i}(J_i - \varepsilon_i \sigma T_i^4) \quad (\text{since } \alpha_i = \varepsilon_i \text{ from Kirchhoff's law})$$

Simplifying further, we can now write for a gray surface,

$$\frac{q_i}{A_i} = \frac{\varepsilon_i}{1 - \varepsilon_i}(\sigma T_i^4 - J_i) \tag{8.79}$$

Thus, to determine the heat transfer rate at a surface having a prescribed temperature, it is only necessary to determine the radiosity J. Now,

$$J_i = \varepsilon_i \sigma T_i^4 + (1 - \varepsilon_i) G_i$$

But,

$$G_i = \sum_{j=1}^{N} J_j F_{i-j}$$

Therefore, $J_i = \varepsilon_i \sigma T_i^4 + (1 - \varepsilon_i) \sum_{j=1}^{N} J_j F_{i-j} \quad (1 \leq i \leq N)$

Now, we get N linear, non-homogeneous algebraic equations for N unknown radiosities $J_1, J_2, \ldots, J_N$. Consequently, these equations are solved simultaneously and J_i values obtained, and with these q_i can be determined from Eq. (8.79).

8.13   Electric Circuit Analogy

Equation (8.79) can be rephrased in the following form:

$$q_i = \frac{\sigma T_i^4 - J_i}{\dfrac{1 - \varepsilon_i}{A_i \varepsilon_i}} \tag{8.80}$$

Now, since $E_{bi} = \varepsilon T_i^4$, Eq. (8.80) can be written as

$$q_i = \frac{E_{bi} - J_i}{\dfrac{1 - \varepsilon_i}{A_i \varepsilon_i}} \tag{8.81}$$

Equation (8.81) is analogous to the electric current flow representation by Ohm's law. The radiative transfer q_i is associated with the driving potential $(E_{bi} - J_i)$ and a surface radiative resistance of the form $(1 - \varepsilon_i)/A_i\varepsilon_i$. The positive sign of q_i indicates that there is a net radiative heat transfer from the surface, while the negative sign signifies net transfer to the surface.

It may be noted that

(a) the surface resistance of a black body is zero since $\varepsilon_i = 1$. Thus, for a black body, $J_i = E_{bi}$;

(b) for an adiabatic surface, $q_i = 0$ and hence, $J_i = E_{bi} = \sigma T_j^4$. This shows that the temperature of an *adiabatic* surface (also called *reradiating* surface) is independent of its emissivity. Furthermore, the surface resistance of such a surface is disregarded as there is no net heat transfer through it. This is analogous to the case when a resistance is not considered in an electrical circuit if no current is flowing through it.

In practical applications the surface whose back sides are well insulated are modelled as adiabatic surfaces. That is, the net heat transfer through such a surface is zero. Neglecting the convection effects on the front (heat transfer) side of such a surface and assuming steady state conditions, the net heat lost by the surface must equal the net gain to it. Hence, $q_i = 0$. In such instances, the surface is considered to reradiate all the radiation incident upon it, and such a surface is designated as a *re-radiating* surface

Now, consider the radiant exchange by two surfaces A_i and A_j. The amount that is intercepted by surface j from the radiation leaving surface i is $J_i A_i F_{i-j}$ and of that total energy leaving surface j that reaches surface i is $J_j A_j F_{j-i}$. The net exchange between the two surfaces is

$$q_{i-j} = J_i A_i F_{i-j} - J_j A_j F_{j-i}$$

Now, from the reciprocity rule,

$$A_i F_{i-j} = A_j F_{j-i}$$

Therefore, $q_{i-j} = (J_i - J_j) A_i F_{i-j} = (J_i - J_j) A_j F_{j-i}$

$$= \frac{J_i - J_j}{\dfrac{1}{A_i F_{i-j}}} \tag{8.82}$$

The denominator on the right-hand side of Eq. (8.82), that is, $1/A_i F_{i-j}$ is called space resistance. Hence,

$$q_i = \sum_{j=1}^{N} q_{i-j} = \sum_{j=1}^{N} A_i F_{i-j} (J_i - J_j) \tag{8.83}$$

This result suggests that the net rate of radiation transfer from surface i, that is, q_i may be represented as a sum of the components, that is, q_{i-j}, related to the radiation exchange with other surfaces. Each component may be represented as a network element for which $(J_i - J_j)$ is the driving potential and $1/A_i F_{i-j}$ is the space or geometric resistance. So, from Eqs (8.81) and (8.83) we can write

$$\frac{E_{bi} - J_i}{(1 - \varepsilon_i)/A_i \varepsilon_i} = \sum_{j=1}^{N} \frac{J_i - J_j}{(1/A_i F_{i-j})} \tag{8.84}$$

The equivalent electric circuit is depicted in Fig. 8.20.

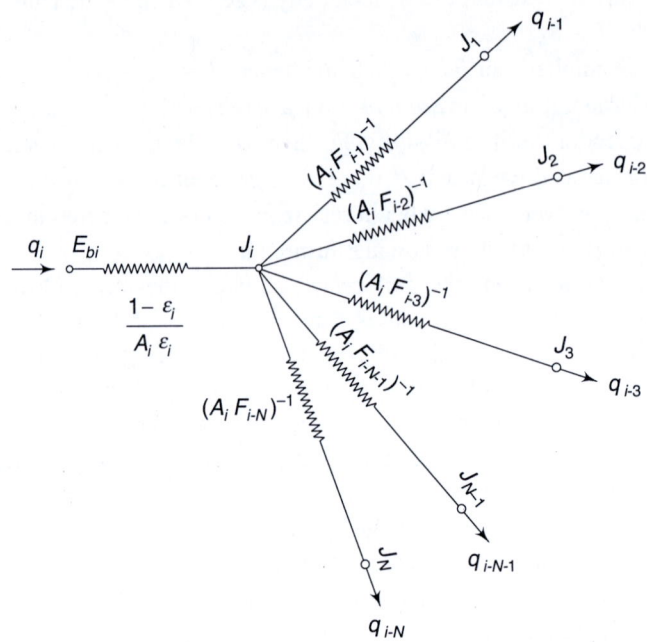

Fig. 8.20 Equivalent electric circuit for an *N*-surface enclosure

It may be noted that Eq. (8.84) is useful when the surface temperature T_i (and hence E_{bi}) is known. However, there may be cases when the net radiative transfer rate at the surface, q_i, rather than T_i is known. In such cases, the preferred form is

$$q_i = \sum_{j=1}^{N} \frac{J_i - J_j}{(1/A_i F_{i-j})} \tag{8.85}$$

To solve this problem, Eq. (8.84) is written for each surface at which T_i is known, and Eq. (8.85) is written for each surface for which q_i is known. The resulting set of N linear, algebraic equations are then solved for N unknowns, $J_1, J_2, ..., J_N$. Thus Eq. (8.83) may be used to determine the net radiative transfer rate q_i at each surface of known T_i or the value of T_i at each surface of known q_i. The equations may be solved by using standard numerical methods such as matrix inversion, Gaussian elimination or Gauss–Jordon elimination, and so on.

8.14 Three-surface Enclosure

The electrical network analogy concept described in Section 8.13 can be easily extended to determine radiation exchange among the surfaces of a three or more surface enclosure. Recall the surface resistance at surface A_i is given by

$$R_i = \frac{1 - \varepsilon_i}{A_i \varepsilon_i} \tag{8.86}$$

And the space resistance between two surfaces i and j is given by

$$R_{ij} = \frac{1}{A_i F_{i-j}} = \frac{1}{A_j F_{j-i}} \quad (8.87)$$

Figure 8.21(a) shows a three-surface enclosure each side of which has a fixed temperature. Figure 8.21(b) depicts the corresponding analogous electrical network. To solve this problem, the algebraic sum of the currents at the nodes J_1, J_2, J_3 is set equal to zero (Kirchhoff's current law).

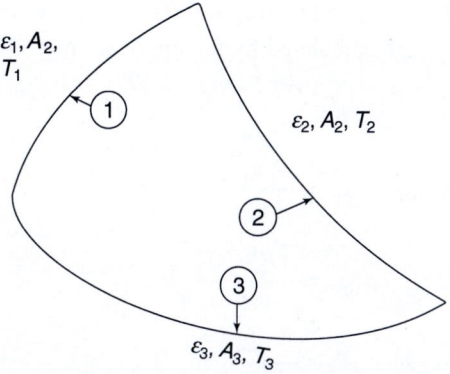

Fig. 8.21 (a) A three-surface enclosure

This leads to three simultaneous algebraic equations [Eqs (8.88), (8.89), and (8.90)] for the determination of three unknown radiosities J_1, J_2, and J_3.

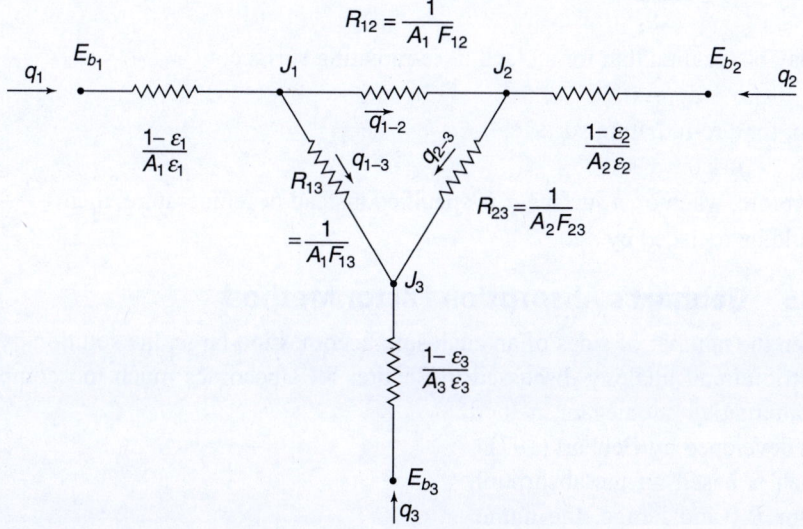

Fig. 8.21 (b) Corresponding analogous electrical network

At Node J_1,

$$\frac{E_{b_1} - J_1}{R_1} = \frac{J_1 - J_2}{R_{12}} + \frac{J_1 - J_3}{R_{13}} \quad (8.88)$$

At Node J_2,

$$\frac{E_{b_2} - J_2}{R_2} = \frac{J_2 - J_1}{R_{12}} + \frac{J_2 - J_3}{R_{23}} \quad (8.89)$$

At Node J_3,

$$\frac{E_{b_3} - J_3}{R_3} = \frac{J_3 - J_1}{R_{13}} + \frac{J_3 - J_2}{R_{23}} \quad (8.90)$$

Solving Eqs (8.88), (8.89), and (8.90) simultaneously J_1, J_2, and J_3 are obtained.

Once the radiosities are known, the net heat loss/gain at a surface (i.e., q_1, q_2 and q_3) are calculated by the application of Ohm's law as follows:

$$q_1 = \frac{\sigma T_1^4 - J_1}{R_1} \tag{8.91}$$

where $R_1 = \dfrac{1-\varepsilon_1}{A_1 \varepsilon_1}$

$$q_2 = \frac{\sigma T_2^4 - J_2}{R_2} \tag{8.92}$$

where $R_2 = \dfrac{1-\varepsilon_2}{A_2 \varepsilon_2}$

and $$q_3 = \frac{\sigma T_3^4 - J_3}{R_3} \tag{8.93}$$

where $R_3 = \dfrac{1-\varepsilon_3}{A_3 \varepsilon_3}$

It may be recalled that for a black or re-radiating surface,

$$J_i = E_{b_i} = \sigma T_i^4$$

Also, for a re-radiating surface,

$$q_i = 0$$

Therefore, when on a surface q_i is specified instead of temperature, then $\dfrac{E_{b_i} - J_i}{R_i}$ should be replaced by q_i.

8.15   Gebhart's Absorption Factor Method

When the number of sides of an enclosure becomes too large, the solution by the electric circuit analogy discussed in Section 8.13 becomes much too complex. An alternative but elegant method was developed by Gebhart (1971), which is based on the absorption factor $B_{i\text{-}j}$ and hence, the name, absorption factor method.

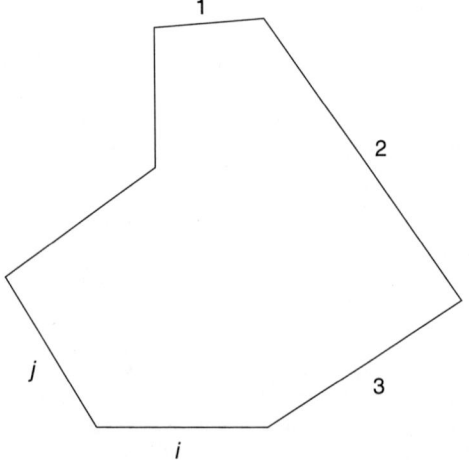

The absorption factor $B_{i\text{-}j}$ is defined as the fraction of the emitted energy from surface i, that is, $E_i A_i$, which is absorbed by surface j (Fig. 8.22) taking into account all paths whereby this radiant energy may reach surface j for absorption. The intervening medium is assumed to be non-participating. Therefore, the net rate of energy loss/gain from/at A_j is

Fig. 8.22   An enclosure of N surfaces

$$q_j = E_j A_j - \sum_{i=1}^{N} B_{i\text{-}j} E_i A_i \tag{8.94}$$

where E_i or E_j is $\varepsilon_i\,\sigma T_i^4$ or $\varepsilon_i\,\sigma_i T_i^4$, respectively. An equation similar to Eq. (8.94) can be written for each surface. In each equation there are N^2 values of $B_{i\text{-}j}$, and the following matrix can be written:

$$\bar{B} = \begin{bmatrix} B_{1\text{-}1} & B_{1\text{-}2} & \cdots & B_{1\text{-}N} \\ B_{2\text{-}1} & \cdots & \cdots & \cdots \\ \vdots & \vdots & \vdots & \vdots \\ B_{N\text{-}1} & \cdots & \cdots & B_{N\text{-}N} \end{bmatrix} \tag{8.95}$$

Knowing the view factors and the values of the reflectivity and the emissivity, the values of the $B_{i\text{-}j}$'s can be determined numerically. The following equations can be written for each of the N surfaces:

$$B_{1\text{-}j} = F_{1\text{-}j}\varepsilon_j + F_{1\text{-}1}\rho_1 B_{1\text{-}j} + F_{1\text{-}2}\rho_2 B_{2\text{-}j} + \cdots + F_{1\text{-}N}\,\rho_N B_{N\text{-}j}$$

$$B_{2\text{-}j} = F_{2\text{-}j}\varepsilon_j + F_{2\text{-}1}\rho_1 B_{1\text{-}j} + F_{2\text{-}2}\rho_2 B_{2\text{-}j} + \cdots + F_{2\text{-}N}\,\rho_N B_{N\text{-}j}$$

$$\vdots$$

$$B_{N\text{-}j} = F_{N\text{-}j}\varepsilon_j + F_{N\text{-}1}\rho_1 B_{1\text{-}j} + F_{N\text{-}2}\rho_2 B_{2\text{-}j} + \cdots + F_{N\text{-}N}\,\rho_N B_{N\text{-}j} \tag{8.96}$$

where ε_i and ρ_i $(i = 1, \ldots, N)$ represent the emissivity and reflectivity of the ith surface. These N linear equations can be transposed and rearranged to obtain

$$(F_{1\text{-}1}\rho_1 - 1)\,B_{1\text{-}j} + F_{1\text{-}2}\rho_2 B_{2\text{-}j} + \cdots + F_{1\text{-}N}\,\rho_N B_{N\text{-}j} + F_{1\text{-}j}\varepsilon_j = 0$$

$$F_{2\text{-}1}\rho_1 B_{1\text{-}j} + (F_{2\text{-}2}\rho_2 - 1)\,B_{2\text{-}j} + \cdots + F_{2\text{-}N}\,\rho_N B_{N\text{-}j} + F_{2\text{-}j}\varepsilon_j = 0$$

$$\vdots$$

$$F_{N\text{-}1}\rho_1 B_{1\text{-}j} + F_{N\text{-}2}\rho_2 B_{2\text{-}j} + \cdots + (F_{N\text{-}N}\,\rho_N - 1)\,B_{N\text{-}j} + F_{N\text{-}j}\varepsilon_j = 0 \tag{8.97}$$

Equation (8.97) can be solved for the N unknowns $B_{1\text{-}j}, B_{2\text{-}j}, \ldots, B_{N\text{-}j}$. Equation (8.97) is valid for any choice of j, that is, $1, 2, \ldots, N$.

With the values of $B_{i\text{-}j}$ known, the radiant heat loss or gain q_j is calculated using Eq. (8.94). Some special points of importance are noted below.

1. The view factor $F_{i\text{-}j}$ represents the fraction of the radiant energy leaving the area A_i and arriving directly at the area A_j, while the absorption factor $B_{i\text{-}j}$ represents the fraction of the radiant energy leaving the area A_i and absorbed by the area A_j following multiple reflections before it finally arrives at the area A_j.

2. In the special case of black surfaces: $B_{i\text{-}j} = F_{i\text{-}j}$.

3. Since the energy emitted by each surface is absorbed by the collection of N surfaces which form the enclosure, we have

$$\sum_{j=1}^{N} B_{i\text{-}j} = 1$$

4. Gebhart also proved a reciprocity relationship between the absorption factors. For diffuse radiation and reflection:

$$\varepsilon_i B_{i\text{-}j} A_i = \varepsilon_j B_{j\text{-}i} A_j$$

5. Absorption factors depend only on the geometry, through view factors and the properties ε and ρ of the surfaces comprising the enclosure. They do not

depend on the surface temperature and the heat input. This is particularly advantageous because B_{i-j} values are obtained independent of the surface temperature and the heat input. Therefore, in many complicated problems of the thermal design of industrial systems, the absorption factor method is more suitable as compared to other methods since the numerical solution for B_{i-j} is carried out conveniently.

Example 8.7　Use of Electrical Network Analogy and Gebhart's Absorption Factor Methods

Two parallel discs 60 cm in diameter are spaced 30 cm apart with one disc located directly above the other disc. The upper disc is maintained at 500°C while the lower one is at 227°C. The emissivities of the discs are 0.2 and 0.4, respectively. The discs are located in a very large space whose walls are maintained at 60°C. Determine the rate of heat loss by radiation from the inner surfaces of each disc. Solve the problem by (a) electrical network analogy method and (b) Gebhart's absorption factor method (see Fig. E8.7).

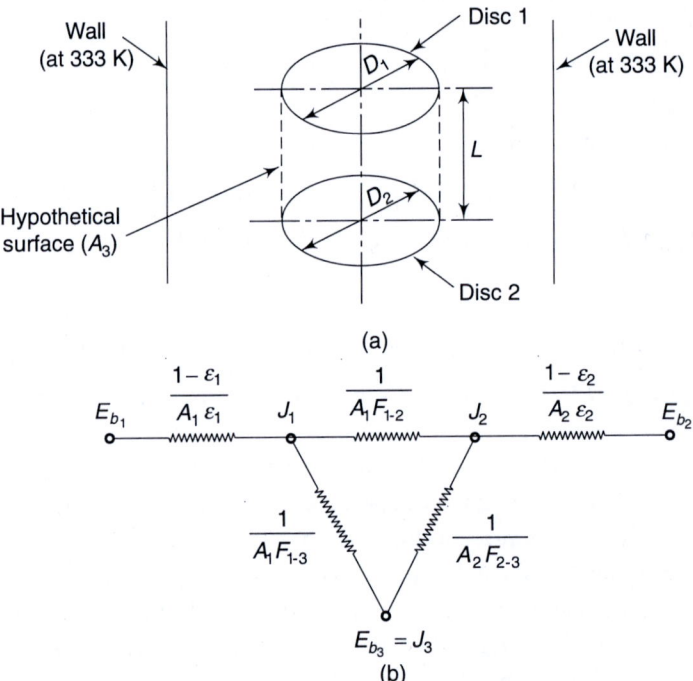

Fig. E8.7　Radiation heat transfer between two parallel discs placed in a large room

Solution

(a)　By electrical network analogy method

The surroundings may be represented by a hypothetical surface A_3, which completes the enclosure and which is approximated as a black body, the temperature of which is 333 K (that is, the same as that of the walls). Referring to Fig. E8.7,

$$T_1 = 773 \text{ K}, \ T_2 = 500 \text{ K}, \ T_3 = 333 \text{ K}$$
$$\varepsilon_1 = 0.2, \ \varepsilon_2 = 0.4, \ L = 30 \text{ cm}$$
$$D_1 = D_2 = 60 \text{ cm}$$

The problem may be viewed as a three-surface enclosure problem for which we are interested to obtain the net rate of heat loss/gain at disc 1 and disc 2. We know,

$$q_1 = \frac{E_{b1} - J_1}{(1 - \varepsilon_1)/A_1 \varepsilon_1}$$

$$q_2 = \frac{E_{b2} - J_2}{(1 - \varepsilon_2)/A_2 \varepsilon_2}$$

Therefore, to evaluate q_1 and q_2, J_1 and J_2 are to be determined since everything else is known. The equivalent electrical network representation is shown in Fig. E8.7(b). The view factors are obtained from the chart (see Appendix A5).

$$F_{1-2} = F_{2-1} = 0.38$$
$$F_{1-3} = 1 - F_{1-2} = 0.62$$
$$F_{2-3} = 1 - F_{2-1} = 0.62$$

Also, $A_1 = A_2 = \pi r_1^2 = \pi (0.3)^2 = 0.283 \text{ m}^2$

$$\frac{1 - \varepsilon_1}{A_1 \varepsilon_1} = \frac{1 - 0.2}{(0.283)(0.2)} = 14.1 \text{ m}^{-2}$$

$$\frac{1 - \varepsilon_2}{A_2 \varepsilon_2} = \frac{1 - 0.4}{(0.283)(0.4)} = 5.3 \text{ m}^{-2}$$

$$\frac{1}{A_1 F_{1-2}} = \frac{1}{(0.283)(0.38)} = 9.3 \text{ m}^{-2}$$

$$\frac{1}{A_1 F_{1-3}} = \frac{1}{A_2 F_{2-3}} = \frac{1}{(0.283)(0.62)} = 5.7 \text{ m}^{-2}$$

$$E_{b1} = \sigma T_1^4 = (5.668 \times 10^{-8})(773)^4 = 20{,}200 \text{ W/m}^2$$
$$E_{b2} = \sigma T_2^4 = (5.668 \times 10^{-8})(500)^4 = 3540 \text{ W/m}^2$$
$$J_3 = E_{b3} = \sigma T_3^4 = (5.668 \times 10^{-8})(333)^4 = 695 \text{ W/m}^2$$

Using Kirchhoff's law at nodes 1 and 2:

$$\frac{20{,}200 - J_1}{14.1} + \frac{J_2 - J_1}{9.3} + \frac{695 - J_1}{5.7} = 0$$

$$\frac{3540 - J_2}{5.3} + \frac{J_1 - J_2}{9.3} + \frac{695 - J_2}{5.7} = 0$$

Solving the above equations simultaneously,

$$J_1 = 5266 \text{ W/m}^2$$
$$J_2 = 2876 \text{ W/m}^2$$

Therefore, $q_1 = \dfrac{E_{b1} - J_1}{\dfrac{1 - \varepsilon_1}{A_1 \varepsilon_1}}$

$$= \frac{20{,}200 - 5266}{(1 - 0.2)/(0.283)(0.2)} = 1059 \text{ W}$$

and $q_2 = \dfrac{E_{b2} - J_2}{\dfrac{1 - \varepsilon_2}{\varepsilon_2 A_2}}$

$$= \frac{3540 - 2876}{(1 - 0.4)/(0.283)(0.4)} = 125 \text{ W}$$

(b) By Gebhart's absorption factor method

$$q_j = E_j A_j - \sum_{i=1}^{N} B_{i\text{-}j} E_i A_i$$

Therefore, $\quad q_1 = E_1 A_1 - (B_{1\text{-}1}E_1 A_1 + B_{2\text{-}1}E_2 A_2 + B_{3\text{-}1}E_3 A_3)$

and $\quad\quad\quad q_2 = E_2 A_2 - (B_{1\text{-}2}E_1 A_1 + B_{2\text{-}2}E_2 A_2 + B_{3\text{-}2}E_3 A_3)$

where $E_1 = \varepsilon_1 E_{b2}$, $E_2 = \varepsilon_2 E_{b2}$, and $E_3 = \varepsilon_3 E_{b3}$. In order to evaluate q_1, we need to know $B_{1\text{-}1}$, $B_{2\text{-}1}$, and $B_{3\text{-}1}$ which can be found often by solving the following sets of equations simultaneously. Note that the equations are obtained by substituting $j = 1$ and $N = 3$ in Eq. (8.97).

$$(\eta_{1\text{-}1} - 1)B_{1\text{-}1} + \eta_{1\text{-}2}B_{2\text{-}1} + \eta_{1\text{-}3}B_{3\text{-}1} = -F_{1\text{-}1}\varepsilon_1$$
$$\eta_{2\text{-}1} B_{1\text{-}1} + (\eta_{2\text{-}2} - 1)B_{2\text{-}1} + \eta_{2\text{-}3}B_{3\text{-}1} = -F_{2\text{-}1}\varepsilon_1$$
$$\eta_{3\text{-}1} B_{1\text{-}1} + \eta_{3\text{-}2} B_{2\text{-}1} + (\eta_{3\text{-}3} - 1)B_{3\text{-}1} = -F_{3\text{-}1}\varepsilon_1$$

where $\quad \eta_{1\text{-}1} = F_{1\text{-}1}\rho_1 = (0)\rho_1 = 0$

$$\eta_{2\text{-}1} = F_{2\text{-}1}\rho_1 = F_{2\text{-}1}(1 - \varepsilon_1)$$
$$= 0.38(1 - 0.2) = 0.304$$
$$\eta_{3\text{-}1} = F_{3\text{-}1}\rho_1 = F_{3\text{-}1}(1 - \varepsilon_1)$$
$$= \frac{A_1 F_{1\text{-}3}}{A_3}(1 - \varepsilon_1)$$
$$= \frac{\pi(0.3)^2 \, (0.62)(1 - 0.2)}{\pi(0.6)(0.3)} = 0.248$$

$$\eta_{1\text{-}2} = F_{1\text{-}2}\rho_2 = 0.38 \, (1 - \varepsilon_2)$$
$$= 0.38(1 - 0.4) = 0.228$$
$$\eta_{2\text{-}2} = F_{2\text{-}2}\rho_2 = (0)\rho_2 = 0$$
$$\eta_{3\text{-}2} = F_{3\text{-}2}\rho_2 = \frac{A_2 F_{2\text{-}3}}{A_3}(1 - \varepsilon_2)$$
$$= \frac{\pi(0.3)^2 \, (0.62)(1 - 0.4)}{\pi(0.6)(0.3)} = 0.186$$

$$\eta_{1\text{-}3} = F_{1\text{-}3}\rho_3 = F_{1\text{-}3}(0) = 0$$
$$\eta_{2\text{-}3} = F_{2\text{-}3}\rho_3 = F_{2\text{-}3}(0) = 0$$
$$\eta_{3\text{-}3} = F_{3\text{-}3}\rho_3 = F_{3\text{-}3}(0) = 0 \quad \text{(since } \rho_3 = 1 - \varepsilon_3 = 1 - 1 = 0)$$

Substituting all these values, we get

$$-B_{1\text{-}1} + 0.228 B_{2\text{-}1} = 0$$
$$0.304 B_{1\text{-}1} - B_{2\text{-}1} = -0.076$$
$$0.248 B_{1\text{-}1} + 0.186 B_{2\text{-}1} - B_{3\text{-}1} = -0.062$$

Solving the above equations simultaneously,

$$B_{1\text{-}1} = 0.0186$$
$$B_{2\text{-}1} = 0.08166$$
$$B_{3\text{-}1} = 0.08178$$

Therefore, $q_1 = E_1 A_1 - (B_{1\text{-}1}A_1 + B_{2\text{-}1}E_2 A_2 + B_{3\text{-}1}E_3 A_3)$
$$= \varepsilon_1 E_{b1} A_1 - (B_{1\text{-}1}\varepsilon_1 E_{b1}A_1 + B_{2\text{-}1}\varepsilon_2 E_{b2}A_2 + B_{3\text{-}1}\varepsilon_3 E_{b3}A_3)$$
$$= 1142 - (21.24 + 32.69 + 32.14)$$
$$= 1055.93 \approx 1056 \text{ W}$$

Therefore $q_1 = 1056$ W, which matches very closely the value of $q_i = 1059$ W obtained by the electrical network analogy method. For the calculation of q_2, we need $B_{1\text{-}2}$, $B_{2\text{-}2}$, $B_{3\text{-}2}$. Using the same approach the result can be obtained.

8.16 Two-surface Enclosure

Radiation in two-surface enclosure is the simplest of the enclosure problems. In this case there are two surfaces that exchange radiation only with each other. Such an enclosure is depicted in Fig. 8.23(a).Since there are only two surfaces, the net rate of heat loss from surface 1 must equal the net rate of heat gain by surface 2, and both quantities must equal the net rate at which radiation is exchanged between surface 1 and surface 2. It is assumed that $T_1 > T_2$. Thus,

$$q_1 = -q_2 = q_{1-2} \qquad (8.98)$$

A negative sign before q_2 is given to make q_2 positive since according to the sign convention used q_2 is negative as heat is gained by surface 2.

The radiation heat transfer rate can be easily obtained by transforming the problem into an analogous electrical network problem. Such a network is shown in Fig. 8.23(b) from which it can be seen that the total resistance to radiation exchange between sides 1 and 2 is composed of two surface resistances and one space resistance. Hence, the net radiation exchange between two surfaces may be expressed as

$$q = q_1 = -q_2 = q_{1-2} = \dfrac{E_{b1} - E_{b2}}{\dfrac{1 - \varepsilon_1}{A_1 \varepsilon_1} + \dfrac{1}{A_1 F_{1-2}} + \dfrac{1 - \varepsilon_2}{A_2 \varepsilon_2}} \qquad (8.99)$$

where $E_{b1} = \sigma T_1^4$ and $E_{b2} = \sigma T_2^4$.

Equation (8.99) can be used for any two diffuse, gray surfaces that form an enclosure. Some special cases are discussed in Section 8.17.

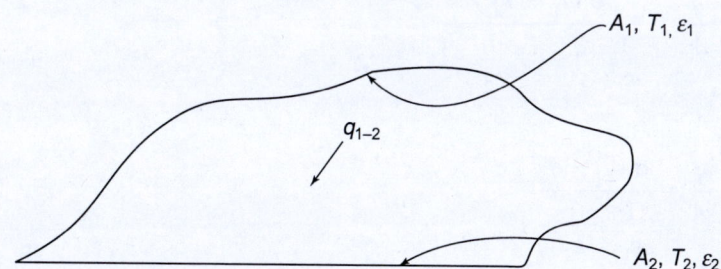

Fig. 8.23(a)   A two-surface enclosure

Fig. 8.23(b)   Analogous electrical network

8.17 Infinite Parallel Planes

When two infinite parallel planes are considered (Fig. 8.24), A_1 and A_2 are equal and the radiation view factor (or shape factor) is unity since all radiation leaving one plane reaches the other. The equivalent electrical circuit is shown in Fig. 8.25.

Thus, $q = \dfrac{E_{b1} - E_{b2}}{\dfrac{1 - \varepsilon_1}{A_1 \varepsilon_1} + \dfrac{1}{A_1 F_{1\text{-}2}} + \dfrac{1 - \varepsilon_2}{A_2 \varepsilon_2}}$ (8.100)

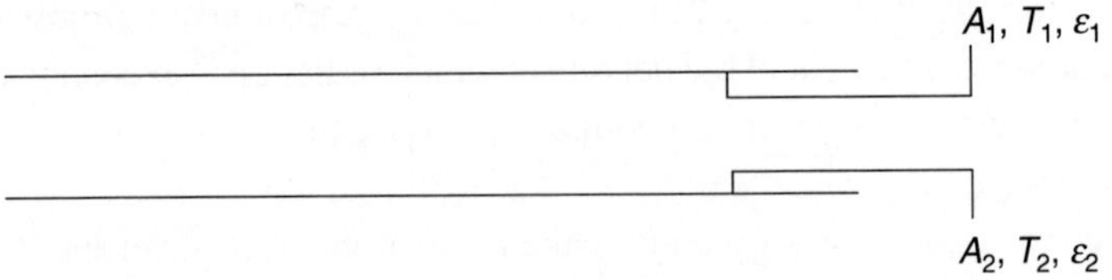

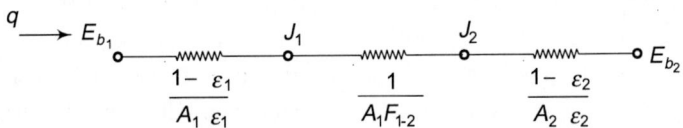

A_1, T_1, ε_1

A_2, T_2, ε_2

Fig. 8.24   Infinite parallel planes

$q \longrightarrow E_{b_1} \quad\text{—wwww—}\quad J_1 \quad\text{—wwww—}\quad J_2 \quad\text{—wwww—}\quad E_{b_2}$

$\dfrac{1 - \varepsilon_1}{A_1 \varepsilon_1} \qquad\qquad \dfrac{1}{A_1 F_{1\text{-}2}} \qquad\qquad \dfrac{1 - \varepsilon_2}{A_2 \varepsilon_2}$

Fig. 8.25   Equivalent electrical circuit for radiation between infinite parallel planes

Letting $A_1 = A_2$, $F_{1\text{-}2} = 1$, Eq. (8.100) transforms to

$$q = \dfrac{\sigma\,(T_1^4 - T_2^4)}{\dfrac{1 - \varepsilon_1}{\varepsilon_1 A_1} + \dfrac{1}{A_1} + \dfrac{1 - \varepsilon_2}{\varepsilon_2 A_1}}$$

or $$\dfrac{q}{A_1} = \dfrac{\sigma\,(T_1^4 - T_2^4)}{\dfrac{1}{\varepsilon_1} - 1 + 1 + \dfrac{1}{\varepsilon_2} - 1}$$

Putting $A_1 = A$,

$$\dfrac{q}{A} = \dfrac{\sigma\,(T_1^4 - T_2^4)}{\dfrac{1}{\varepsilon_1} + \dfrac{1}{\varepsilon_2} - 1}$$ (8.101)

When two long concentric cylinders exchange heat, $F_{1\text{-}2} = 1$ (Fig. 8.27). Therefore,

$$q = \dfrac{\sigma\,(T_1^4 - T_2^4)}{\dfrac{1 - \varepsilon_1}{\varepsilon_1 A_1} + \dfrac{1}{A_1} + \dfrac{1 - \varepsilon_2}{\varepsilon_2 A_2}}$$

$$= \dfrac{\sigma A_1\,(T_1^4 - T_2^4)}{\dfrac{1 - \varepsilon_1}{\varepsilon_1} + 1 + \dfrac{1 - \varepsilon_2}{\varepsilon_2}\dfrac{A_1}{A_2}}$$

$$= \dfrac{\sigma A_1\,(T_1^4 - T_2^4)}{\dfrac{1}{\varepsilon_1} + \left(\dfrac{1}{\varepsilon_2} - 1\right)\dfrac{A_1}{A_2}}$$ (8.102)

Equation (8.102) is particularly important when applied to the limiting case of a convex object completely surrounded by a very large concave surface (Fig. 8.26). In this instance, $A_1/A_2 \to 0$ and Eq. (8.102) becomes

$$q = \sigma \varepsilon_1 A_1 (T_1^4 - T_2^4) \quad (8.103)$$

Equation (8.103) is also used to calculate the radiation loss from a hot object at T in a large room at T_∞. In this case,

$$q = \sigma \varepsilon A (T^4 - T_\infty^4) \quad (8.104)$$

where A is the surface area of the hot object.

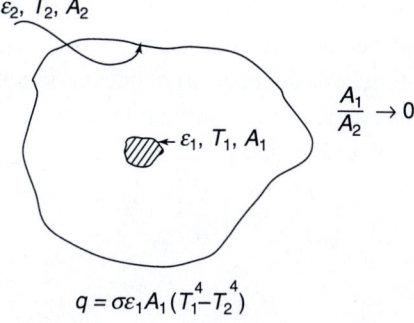

$$q = \sigma \varepsilon_1 A_1 (T_1^4 - T_2^4)$$

Fig. 8.26 Convex object surrounded by a very large concave surface

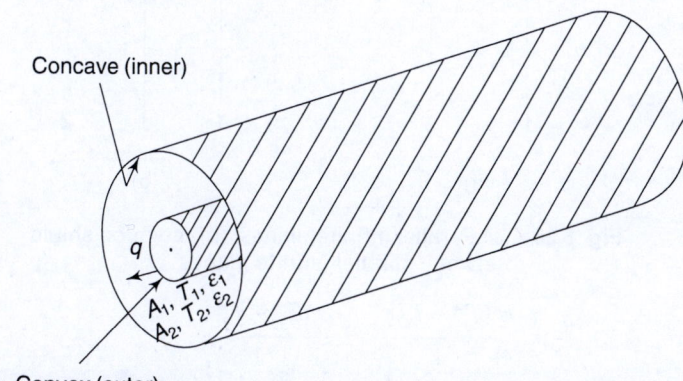

Fig. 8.27 Radiation between two long concentric cylinders

8.18 Radiation Shields

Radiation shields (constructed from highly reflective materials) are used to reduce the radiation heat transfer between two particular surfaces. These shields do not deliver or remove any heat from the overall system. They only place another resistance in the heat flow path so that the overall heat transfer decreases. Typical applications of radiation shields are in cryogenics and space. Also, a temperature sensor used for measuring the temperature of fluids is placed in a radiation shield to reduce the error caused by the radiation effect when the temperature sensor (e.g., thermometer, thermocouple) is exposed to surfaces that are much hotter or colder than the fluid itself.

Consider two parallel infinite planes as shown in Fig. 8.28(a). It has been demonstrated in Section 8.17 that the heat transfer between these surfaces may be calculated as

$$\frac{q}{A} = \frac{\sigma (T_1^4 - T_2^4)}{\dfrac{1}{\varepsilon_1} + \dfrac{1}{\varepsilon_2} - 1} \quad (8.105)$$

Now, we would like to see what the heat transfer rate will be if a radiation shield is placed between the same two planes [Fig. 8.28(b)].

Since the shield does not deliver or remove heat from the system, the heat transfer between plate 1 and the shield must be exactly the same as that between the shield and plate 2. Thus,

$$\left(\frac{q}{A}\right)_{1\text{-}3} = \left(\frac{q}{A}\right)_{3\text{-}2} = \frac{q}{A}$$

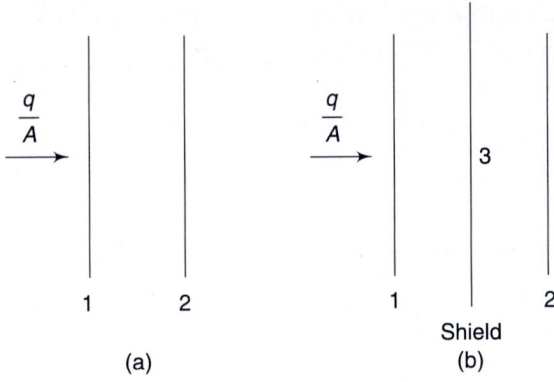

Fig. 8.28 (a) Parallel infinite planes, (b) radiation shield between parallel infinite planes

Therefore,
$$\frac{q}{A} = \frac{\sigma\,(T_1^4 - T_3^4)}{\dfrac{1}{\varepsilon_1} + \dfrac{1}{\varepsilon_3} - 1} = \frac{\sigma\,(T_3^4 - T_2^4)}{\dfrac{1}{\varepsilon_3} + \dfrac{1}{\varepsilon_2} - 1} \qquad (8.106)$$

If we assume $\varepsilon_1 = \varepsilon_2 = \varepsilon_3$, then from Eq. (8.106) we obtain

$$T_1^4 - T_3^4 = T_3^4 - T_2^4$$

or
$$T_3^4 = \frac{1}{2}\,(T_1^4 + T_2^4) \qquad (8.107)$$

Substituting T_3^4 from Eq. (8.107) into Eq. (8.106), we get

$$\frac{q}{A} = \frac{\dfrac{1}{2}\,\sigma\,(T_1^4 - T_2^4)}{\dfrac{1}{\varepsilon_1} + \dfrac{1}{\varepsilon_3} - 1} \qquad (8.108)$$

But, since $\varepsilon_3 = \varepsilon_2$, we see that

$$\frac{q}{A} = \frac{\dfrac{1}{2}\,\sigma\,(T_1^4 - T_2^4)}{\dfrac{1}{\varepsilon_1} + \dfrac{1}{\varepsilon_2} - 1} \qquad (8.109)$$

Comparing Eq. (8.109) with Eq. (8.105) it is observed that this heat flow is just one-half of that which would be experienced if there were no shield present. In general, it can be shown that for N number of shields, in the special case for which all emissivities are equal,

$$\left(\frac{q}{A}\right)_{\text{with shields}} = \frac{1}{N+1}\left(\frac{q}{A}\right)_{\text{without shields}} \tag{8.110}$$

Thus, Eq. (8.109) results when $N = 1$ in Eq. (8.110).

It can also be seen from Eq. (8.108) that when the emissivity of the shield (ε_3) is low, that is, when the reflectivity of the shield is high, the heat transfer decreases significantly. That is why radiation shields are made of highly reflective materials.

8.19 Radiation Heat Transfer Coefficient

The concept of radiation heat transfer coefficient is useful when the total heat transfer by both convection and radiation is the objective of an analysis. Thus,

$$q_{\text{rad}} = h_r A_1 (T_1 - T_2) \tag{8.111}$$

where T_1 and T_2 are the temperatures of the two bodies exchanging heat by radiation. The value of h_r, corresponding to Eq. (8.105), can be calculated from

$$\frac{q}{A_1} = \frac{\sigma\,(T_1^4 - T_2^4)}{\dfrac{1}{\varepsilon_1} + \dfrac{A_1}{A_2}\left(\dfrac{1}{\varepsilon_2} - 1\right)} = h_r\,(T_1 - T_2)$$

or

$$h_r = \frac{\sigma\,(T_1^2 + T_2^2)\,(T_1 + T_2)}{\dfrac{1}{\varepsilon_1} + \left(\dfrac{A_1}{A_2}\right)\left(\dfrac{1}{\varepsilon_2} - 1\right)} \tag{8.112}$$

Obviously, the radiation heat transfer coefficient is a very strong function of temperature. For example, the heat loss by free convection and radiation from a hot object in a large room whose walls are at the same temperature as that of the fluid can be obtained from

$$q = (h_c + h_r)\,A_1 (T_s - T_\infty) \tag{8.113}$$

8.20 Gas Radiation

So far, in this chapter we have focused our attention on the characteristics of solid surfaces and the exchange of radiation between solid surfaces separated by non-participating media. The solid surfaces considered were taken as opaque ($\tau = 0$) while the media between the surfaces were assumed to be transparent and non-emitting ($\tau = 1$, $\varepsilon = \alpha = 0$).

Elementary gases with symmetrical molecules such as nitrogen and oxygen are, indeed, transparent to thermal radiation. Thus, air can be considered as a non-participating medium. Many gases with more complex molecules such as carbon dioxide, water vapour, ammonia, and most hydrocarbon gases do emit and absorb thermal radiation—at least in certain wavelength bands. Figure 8.29 shows spectral absorptivity of CO_2 at 830 K and 10 atm for a path length of 38.8 cm. Such gases are called participating media. In this chapter, however, we will consider only CO_2 and H_2O (vapour) as these are commonly encountered in practice (combustion products in furnaces and combustion chambers). The basic principle can be extended to other gases.

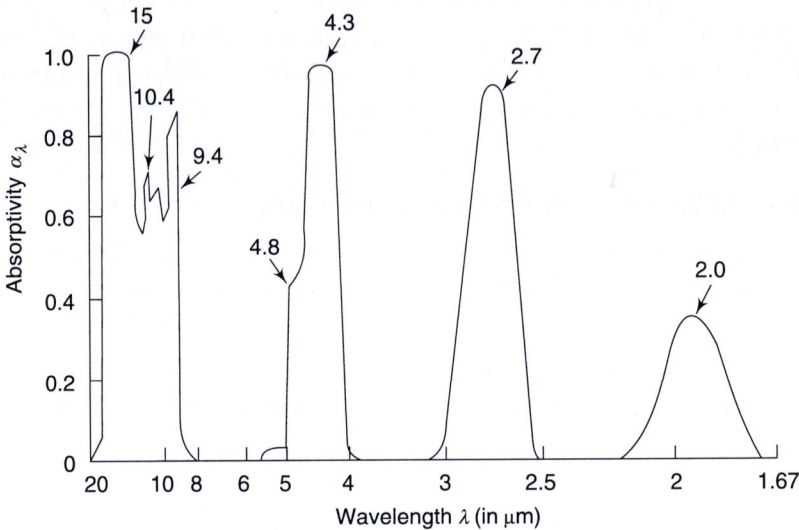

Band designation λ (in μm)

Fig. 8.29 Spectral absorptivity of CO_2 at 830 K and 10 atm for a path length of 38.8 cm (from Cengel 2003)

8.20.1 Participating Medium

The following are the main characteristics of a participating medium.

- It emits and absorbs radiation throughout its entire volume. That is, gaseous radiation is a volumetric phenomenon, and thus depends on the size and shape of the body.
- Such gases emit and absorb at a number of narrow-wavelength bands. Therefore, the assumption of a gray body may not always be appropriate for such gases even when enclosing surfaces are gray.
- The emission and absorption characteristics of the constituents of a gas mixture also depend on the temperature, pressure, and composition of the gas mixture.

8.20.2 Beer's Law

Consider a monochromatic beam of radiation having intensity I_λ impinging on the gas layer of thickness dx (Fig. 8.30). The attenuation in the intensity resulting from absorption in the layer is assumed to be proportional to the thickness of the layer and the intensity of radiation at that point. Thus,

$$dI_\lambda = -\kappa_\lambda I_\lambda dx \tag{8.114}$$

where κ_λ is called the monochromatic absorption coefficient. Integrating Eq. (8.114), we obtain

$$\int_{I_{\lambda_0}}^{I_{\lambda_x}} \frac{dI_\lambda}{I_\lambda} = \int_0^x -\kappa_\lambda \, dx$$

or
$$\frac{I_{\lambda_x}}{I_{\lambda_0}} = e^{-\kappa_\lambda x} \qquad (8.115)$$

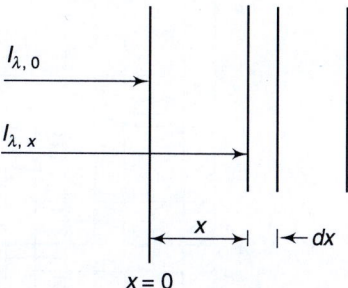

Equation (8.115) is called *Beer's Law.*

The foregoing result may be used to define the spectral transmissivity τ_λ and spectral absorptivity α_λ of the medium as follows:

$$\tau_\lambda = \frac{I_{\lambda,x}}{I_{\lambda,0}} = e^{-k_\lambda x} \qquad (8.116)$$

Fig. 8.30 Absorption in a gas layer

$$\alpha_\lambda = 1 - \tau_\lambda = 1 - e^{-k_\lambda x} \qquad (8.117)$$

Equation (8.117) is based on the assumption of $\rho_\lambda = 0$. If Kirchhoff's law is assumed to be valid, then $\alpha_\lambda = \varepsilon_\lambda$, and hence

$$\varepsilon_\lambda = 1 - e^{-k_\lambda x} \qquad (8.118)$$

For an optically thick medium (i.e., a medium with high $\kappa_\lambda x$) $\varepsilon_\lambda \approx \alpha_\lambda \approx 1$. For example, for $\kappa_\lambda x = 5$, $\varepsilon_\lambda = \alpha_\lambda = 0.993$.

8.20.3 Mean Beam Length

Equations (8.115) and (8.117) describe the change of intensity and absorptivity with distance in a gas layer of thickness x. However, these are the values one might measure in a laboratory experiment with radiation passing normal to the layer. But in a practical problem of a gas contained between two large parallel plates which emit radiation diffusely, the radiant energy transmitted through the gas travels many distances. While the energy transmitted normal to the plate traverses a path equal to the spacing between the plates, that emitted at shallow angles is absorbed in the gas over a much larger distance, and so on. Hottel and Egbert (1942) presented the gas emissivities for carbon dioxide and water vapour as shown in Figs 8.31 and 8.32, respectively. In these figures, L_e is the mean beam length, which is the radius of an equivalent hemisphere. Table 8.2, according to Hottel (1954) and Eckert and Drake (1972) presents mean beam lengths for specific geometries of the enclosure containing the gas. In the absence of mean beam length information for a particular geometry, a satisfactory approximation can be obtained from

$$L_e = 3.6 \frac{V}{A} \qquad (8.119)$$

where V is the total volume of the gas and A is the total surface area.

For $p <$ or > 1 atm correction factors are provided. Figures 8.33 and 8.34 show the correction factor plots for carbon dioxide and water vapour, respectively. When both carbon dioxide and water vapour are present in the gas mixture an addition correction $\Delta\varepsilon$ is subtracted from the total of the emissivities of the two components. Thus, the total gas emissivity ε_g of the mixture is expressed as

$$\varepsilon_g = C_c\varepsilon_c + C_w\varepsilon_w - \Delta\varepsilon \qquad (8.120)$$

To obtain $\Delta\varepsilon$, Fig. 8.35 is used.

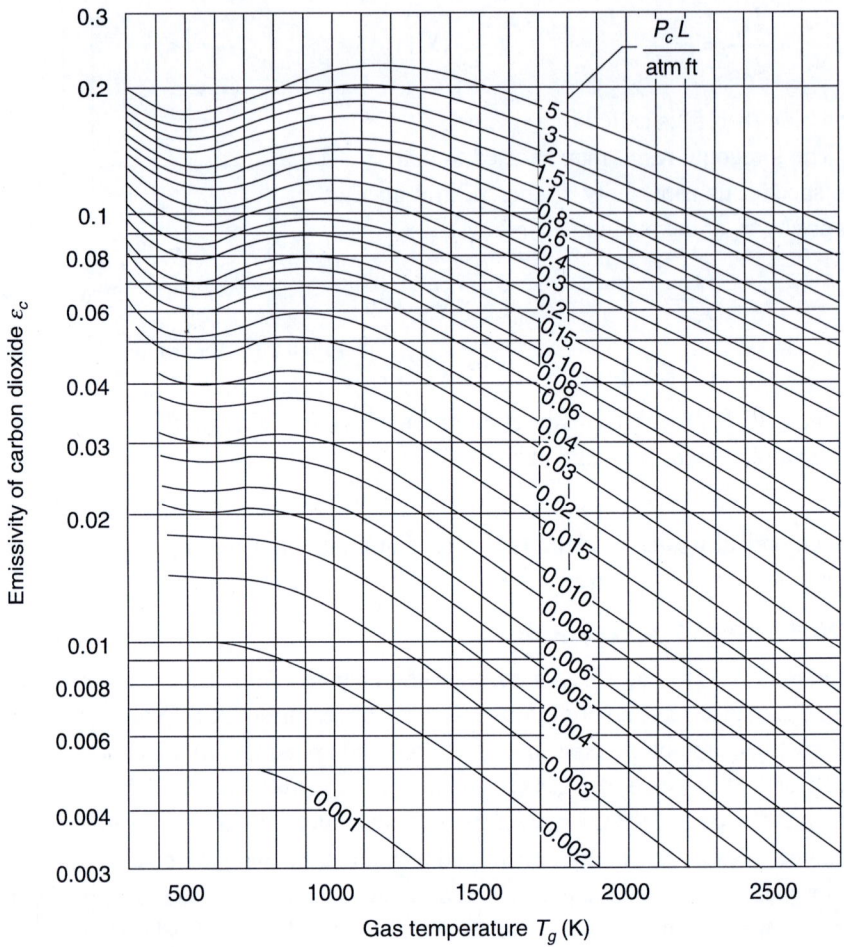

Fig. 8.31 Emissivity of carbon dioxide in a mixture with non-participating gases at a mixture pressure of 1 atm (from Bejan 1993)

8.20.4 Heat Exchange Between Gas Volume and Black Enclosure

Consider a gas volume at uniform temperature T_g enclosed by a black surface of arbitrary geometry at T_w. Since the bounding surface is black, the surface will emit radiation to the gas without reflecting any amount of it, and the gas will absorb this radiation at a rate of $\alpha_g A \sigma T_w^4$, where A is the bounding surface area. Therefore, the net heat transfer per unit area from the gas to the enclosure is

$$q'' = \text{(energy per unit time per unit area emitted by the gas)}$$

$$- \text{(energy per unit time per unit area absorbed by the gas)}$$

$$= \varepsilon_g(T_g)\sigma T_g^4 - \alpha_g(T_w)\,\sigma T_w^4 \tag{8.121}$$

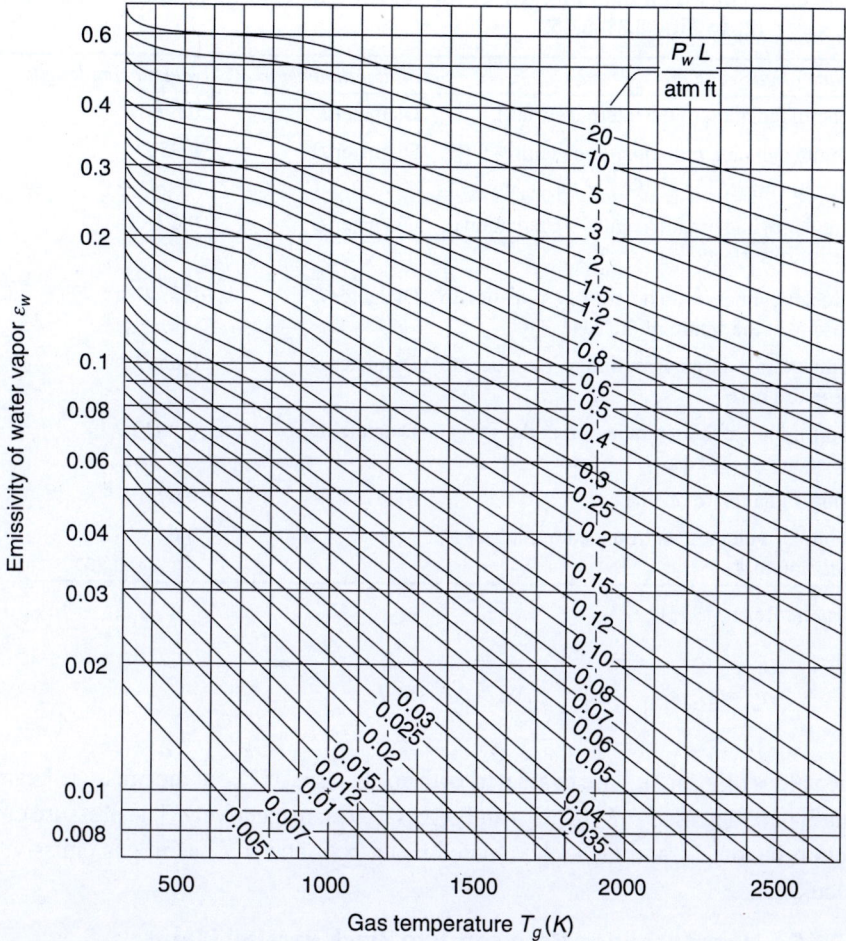

Fig. 8.32 Emissivity of water vapour in a mixture with non-participating gases at a mixture pressure of 1 atm (*Source*: Bejan, 1993)

where $\varepsilon_g(T_g)$ is the gas emissivity at T_g and $\alpha_g(T_w)$ is the gas absorptivity for radiation from the black enclosure at T_w and is a function of both T_w and T_g. For a mixture of CO_2 and H_2O (vapour) an empirical relation for α_g is

$$\alpha_g(T_w) = \alpha_c + \alpha_w - \Delta\alpha \qquad (8.122)$$

where

$$\alpha_c = C_c \varepsilon_c' \left(\frac{T_g}{T_w}\right)^{0.65}$$

Table 8.2   The mean beam length L_e for several gas volume shapes (L replaces L_e in Figs 8.31–8.35)

Shape of gas volume	Actual dimension	Mean beam length
Sphere, radiation to the internal surface	Diameter D	$0.6\,D$
Infinite cylinder, radiation to the entire internal surface	Diameter D	$0.95\,D$
Circular cylinder with height $= D$, radiation to the entire surface	Diameter D	$0.6\,D$
Circular cylinder with height $= D$, radiation to the spot at the centre of the base	Diameter D	$0.77\,D$
Semi-infinite circular cylinder, radiation to the entire base	Diameter D	$0.65\,D$
Semi-infinite cylinder, radiation to the spot at the centre of the base	Diameter D	$0.9\,D$
Cube, radiation to one face	Side a	$0.67\,a$
Arbitrary volume V surrounded by surface A, radiation to A	V, A	3.6

(*Source*: Bejan, 1993)

$$\alpha_w = C_w \varepsilon_w' \left(\frac{T_g}{T_w}\right)^{0.45} \quad ; \Delta\alpha = \Delta\varepsilon \quad \text{at } T_w$$

The values of ε_c' and ε_w' are evaluated with an abscissa of T_w but the pressure-beam-length parameters of $p_c L_e (T_w/T_g)$ and $p_w L_e (T_w/T_g)$, respectively. The pressure correction factors C_c and C_w are evaluated using $p_c L_e$ and $p_w L_e$ as in gas emissivity calculations.

8.20.5   Heat Exchange Between Two Black Parallel Plates

Let us consider two black parallel plates at different temperatures T_1 and T_2 enclosing a gas volume. For plate 1, the net rate of energy gain is

$$q_1 = G_1 A_1 - E_{b_1} A_1 \tag{8.123}$$

Similarly for plate 2,

$$q_2 = G_2 A_2 - E_{b_2} A_2 \tag{8.124}$$

$G_1 A_1$ and $G_2 A_2$ are irradiation on surface 1 and surface 2, respectively. Thus,

$$G_1 A_1 = A_g F_{g1} \varepsilon_g (T_g) E_{bg} + A_2 F_{21} \tau_g (T_2) E_{b_2} \tag{8.125}$$

where $\tau_g(T_2) = 1 - \alpha_g(T_2)$.

Similarly, $G_2 A_2 = A_g F_{g2} \varepsilon_g (T_g) E_{bg} + A_1 F_{12} \tau_g (T_1) E_{b_1}$ (8.126)

Note that, by reciprocity rule

$$A_g F_{g1} = A_1 F_{1g}$$
$$A_g F_{g2} = A_2 F_{2g}$$

Also note that since $F_{1g} = F_{2g} = 1$ and $A_g = A_1 = A_2$, we have $F_{g1} = F_{g2} = 1$.

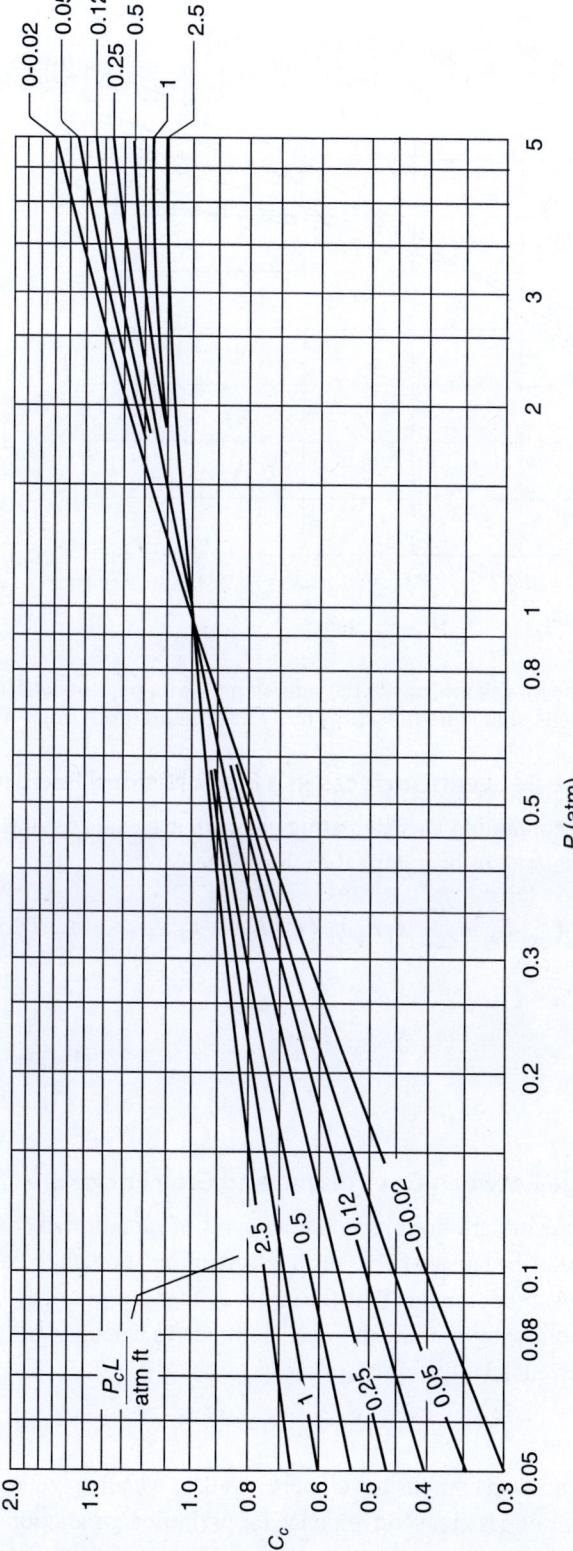

Fig. 8.33 Correction factor for the emissivity of carbon dioxide in a mixture with non-participating gases at mixture pressures other than 1 atm (Bejan 1993)

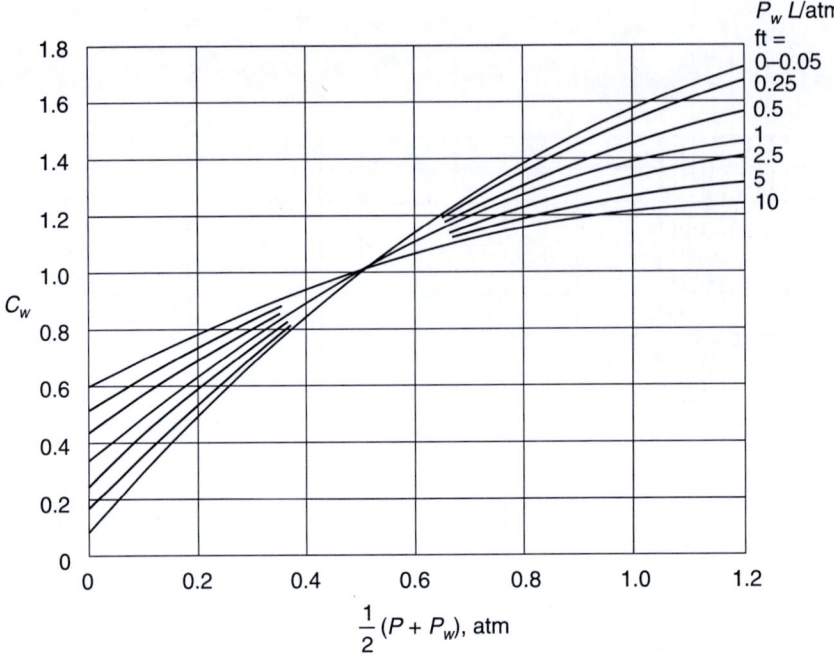

P_w L/atm
ft =
0–0.05
0.25
0.5
1
2.5
5
10

$\frac{1}{2}(P + P_w)$, atm

Fig. 8.34   Correction factor for the emissivity of water vapour in a mixture with non-participating gases at mixture pressures other than 1 atm (from Bejan 1993)

8.20.6   Heat Exchange Between Surfaces in a Black N-sided Enclosure

For an N-sided enclosure containing a participating gas with each side having different temperatures, the net rate of heat gain at each surface is

$$q_i = A_g F_{gi} \, \varepsilon_g(T_g) \, E_{b_g} + \sum_{j=1}^{N} A_j \, F_{ji} \, \tau_g \, (T_j) \, E_{bj} - E_{b_i} A_i \qquad (8.127)$$

$$A_g F_{gi} = A_i F_{ig}, \quad i = 1, ..., N$$

Since $F_{ig} = 1$, therefore,

$$F_{gi} = \frac{A_i}{A_g}$$

8.20.7   Heat Exchange Between Gas Volume and Gray Enclosure

If the surfaces of the enclosure are not black, the analysis of radiation exchange becomes more complicated because of the surface reflection. For engineering calculations, Hottel (1954) recommends that for surfaces that are gray but have emissivity $\varepsilon_w > 0.8$, the following modification can be made to calculate the heat gain at each surface of the enclosure:

$$q_{gray} = \frac{\varepsilon_w + 1}{2} \, q_{black} \qquad (8.128)$$

The emissivity of the furnace and combustion chamber walls is usually greater than 0.8, and thus Eq. (8.128) provides great convenience for preliminary radiation heat transfer calculations.

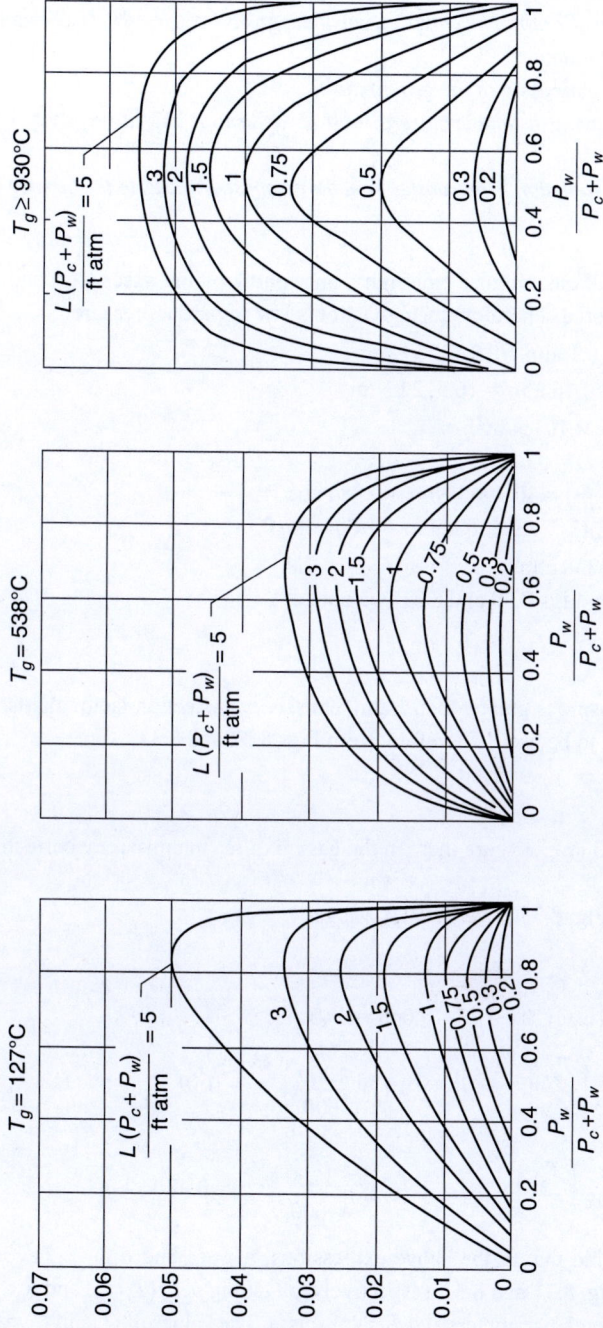

Fig. 8.35 Correction for gas emissivity when carbon dioxide and water vapour are present simultaneously in a mixture with non-participating gases (Bejan 1993)

Example 8.8   Gas Radiation in a Cylindrical Furnace

A cylindrical furnace whose height and diameter are 5 m contains combustion gases at 1200 K at a total pressure of 2 atm. The composition of the gas is 80% N_2, 8% H_2O (vapour), 7% O_2, and 5% CO_2 by volume.

(i) Determine the total emissivity of the gas mixture.

(ii) For a black wall having a temperature of 600 K, determine the absorptivity of the combustion gases.

(iii) Calculate the rate of radiation heat transfer from the combustion gases to the furnace wall.

Solution

(i) To calculate the total gas mixture emissivity, only participating gases CO_2 and H_2O (vapour) are considered. Therefore, using Dalton's law of partial pressure

$$p_c = 0.05(2) = 0.1 \text{ atm} = 10.132 \text{ kN/m}^2$$
$$p_w = 0.08(2) = 0.16 \text{ atm} = 16.512 \text{ kN/m}^2$$

Note that 1 atm $= 1.0132 \times 10^5$ N/m².

From Table 8.2, $L_e = 0.6 \, D = 0.6(5) = 3$ m.

Hence, $p_c L_e = (10.132)(3) = 30.396$ kN/m $= 0.984$ atm ft

$ p_w L_e = (16.512)(3) = 49.536$ kN/m $= 1.604$ atm ft

Note that 1 kN/m $= 0.03238$ atm ft

Now, $T_g = 1200$ K. Using Figs 8.30 and 8.31, we obtain

$$\varepsilon_{c, \, 1 \text{ atm}} = 0.16$$
$$\varepsilon_{w, \, 1 \text{ atm}} = 0.23$$

Now, since the total pressure is greater than 1 atm, emissivity correction factor charts, Figs 8.32 and 8.33, will have to be used. Therefore, from Figs 8.32 and 8.33,

$$C_c = 1.1$$
$$C_w = 1.4$$

Since both CO_2 and H_2O (vapour) are there in the gas mixture, an emissivity correction is required.

From the third chart of Fig. 8.35 ($T_g \geq 930\,°C$), we get

$$\Delta \varepsilon = 0.048$$

Therefore, $\varepsilon_g = C_c \varepsilon_{c, \, 1 \text{ atm}} + C_w \varepsilon_{w, \, 1 \text{ atm}} - \Delta \varepsilon$

$$= 1.1 \times 0.16 + 1.4 \times 0.23 - 0.048 = 0.45$$

(ii) $\alpha_c = C_c \varepsilon'_{c, \, 1 \text{ atm}} \left(\dfrac{T_g}{T_w} \right)^{0.65} = (1.1)(0.11) \left(\dfrac{1200}{600} \right)^{0.65} = 0.19$

$ \alpha_w = C_w \varepsilon'_{w, \, 1 \text{ atm}} \left(\dfrac{T_g}{T_w} \right)^{0.45} = (1.4)(0.25) \left(\dfrac{1200}{600} \right)^{0.45} = 0.48$

It may be kept in mind that in the above expressions, $\varepsilon'_{c, \, 1\text{atm}}$ and $E'_{w, \, 1\text{atm}}$ have been calculated by reading Fig. 8.31 and 8.32, respectively, by taking $p_c L_e (T_w/T_g)$ and $p_w L_e (T_w/T_g)$ as pressure-beam-length parameters and T_w as abscissa. The values of C_c and C_w remain unchanged. Also, $\Delta \alpha = \Delta \varepsilon$ at $T_w = 600$ K. But there is no chart for 600 K in Fig. 8.35. However, we can read $\Delta \varepsilon$ values at 400 K and 800 K, and take their arithmetic mean.

At $p_w/(p_w + p_c) = 0.615$ and $p_c L_e + p_w L_e = 2.6$, we read from Fig. 8.35

$$(\Delta \varepsilon)_{400 \text{ K}} = 0.025$$
$$(\Delta \varepsilon)_{800 \text{ K}} = 0.029$$

Therefore, $\Delta\varepsilon = \dfrac{0.025 + 0.029}{2} = 0.027$

Hence, $\quad \alpha_g = \alpha_c + \alpha_w - \Delta\alpha = 0.19 + 0.48 - 0.027 = 0.64$

(iii) The total surface area of the cylinder is

$$A = \pi DH + 2\frac{\pi D^2}{4} = \pi(5)(5) + 2\frac{\pi(5)^2}{4} = 118 \text{ m}^2$$

Therefore, $q = A\sigma[(\varepsilon_g(T_g)\, T_g^4 - \alpha_g\,(T_w)\, T_w^4)]$

$\quad\quad\quad = (118)(5.67 \times 10^{-8})\,[0.45(1200)^4 - 0.64(600)^4]$

$\quad\quad\quad = 569 \times 10^4 \text{ W} = 5.69 \text{ MW}$

8.21 Solar Radiation

The sun is our primary source of energy. It is a nearly spherical body having a diameter of 1.39×10^9 m and a mass of 2×10^{30} kg and is situated at a mean distance of 1.5×10^{11} m from the earth. It emits radiant energy continuously at the rate of 3.8×10^{26} W. A very small fraction of this energy (about 1.7×10^{17} W) reaches the earth. The sun is essentially a nuclear reactor producing energy by fusion reaction during which two hydrogen atoms fuse to form one atom of helium. The core temperature of the sun is 40,000,000 K. However, the temperature drops to about 5780 K in the outer region of the sun.

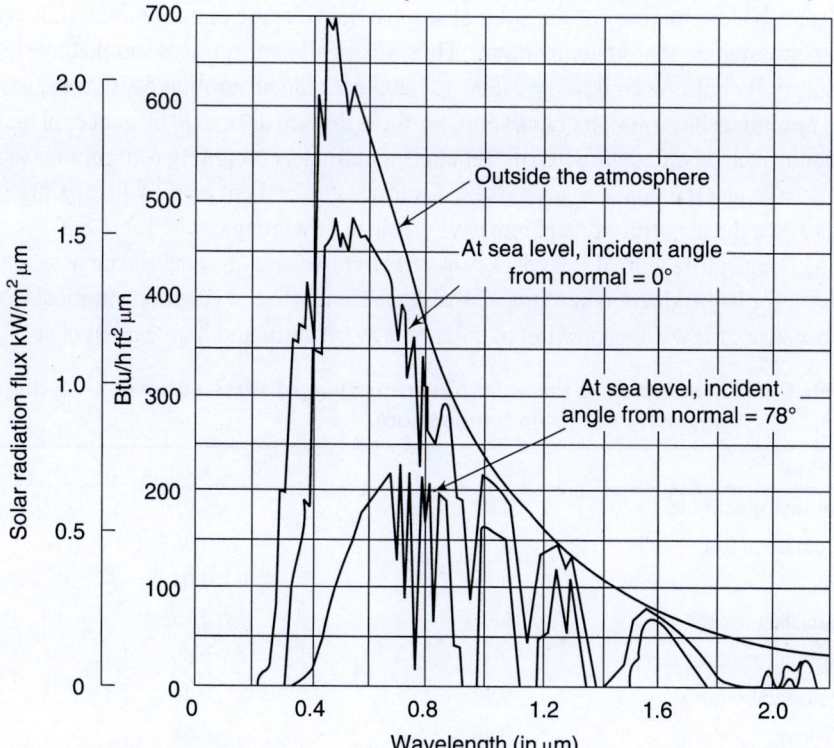

Fig. 8.36 Spectral distribution of solar radiation as a function of atmospheric conditions and angle of incidence (Holman, 1997)

The solar energy reaching the earth's atmosphere is called the total solar irradiance, $G_s = 1373$ W/m^2. It is also called *solar constant*. Because of strong absorption by carbon dioxide and water vapour in the atmosphere, not all the energy expressed by the solar constant reaches the surface of the earth. The dust and other pollutants in the atmosphere also affect the incident solar radiation on earth's surface. Figure 8.36 shows the atmospheric absorption effects for a sea-level location on clear days in a moderately dusty atmosphere with moderate vapour content. It is clear from the figure that maximum solar energy reaches the earth's surface when the rays are directly incident on the earth's surface; the reasons being a larger view area to the incoming solar flux and the shortest distance of travel by the solar rays so that there is less absorption than there would be for an incident angle tilted from the normal. It is also observed from the figure that solar radiation which reaches the surface of the earth does not behave like the radiation from an ideal gray body, while outside the atmosphere the distribution follows more or less an ideal pattern. The solar spectrum is similar to a black body at 5780 K, and hence the sun can be treated as a black body.

The spectral distribution of the incident solar radiation is quite different from the spectral distribution of emitted radiation by the surfaces, since the former is concentrated in the short-wavelength region while the latter is concentrated in the infrared region. Therefore, the radiation properties of surfaces will be quite different for the incident and emitted radiation, and the surfaces cannot be assumed gray. Table 8.3 shows a comparison of the solar absorptivities of some common materials with their emissivities at room temperature. Thus, solar collector surfaces should have high absorptivity but low emissivity values to maximize the absorption of solar radiation and minimize the emission of radiation. Surfaces that are desired to be kept cool under the sun, such as outer surfaces of fuel tanks and trucks containing refrigerated food, must have just the opposite properties. A surface can be kept cool by just painting it white (see the absorptivity and emissivity values of white paint in Table 8.3).

Although utilization of solar energy is attractive because one does not have to spend money for its production and it is available in abundance, it is not economical to do so because of low concentration of solar energy on earth and high capital cost.

Table 8.3   Comparison of the solar absorptivity a_s of some surfaces with their emissivity ε at room temperature

Surface	α_s	ε
Polished aluminium	0.09	0.03
Aluminium foil	0.15	0.05
Polished copper	0.18	0.03
Tarnished copper	0.65	0.75
Polished stainless steel	0.37	0.60
Dull stainless steel	0.50	0.21
Concrete	0.60	0.88
White marble	0.46	0.95

(Contd)

(Contd)

Red brick	0.63	0.93
Asphalt	0.90	0.90
Black paint	0.97	0.97
White paint	0.14	0.93
Snow	0.28	0.97

(Source: Cengel, 2003)

Example 8.9 Radiation Equilibrium Temperature

Calculate the radiation equilibrium temperature for the roof of a house exposed to a solar flux of 300 W/m² and a surrounding temperature of 25 °C if the roof is (a) made of white marble or (b) coated with black paint. Neglect convection.

Solution

At radiation equilibrium the net energy absorbed from the sun must equal the long-wavelength exchange with the surroundings. That is,

$$\left(\frac{q}{A}\right)_{solar} \alpha_s = \varepsilon\sigma\,(T^4 - T_\infty^4) \tag{a}$$

(a) For white marble, we obtain from Table 8.3,

$\alpha_s = 0.46$ and $\varepsilon = 0.95$.

Therefore, from Eq. (a) we have

$(300)(0.46) = (0.95)(5.67 \times 10^{-8})(T^4 - 298^4)$

or $T = 319.7$ K $= 46.7$°C

(b) For black paint, we obtain from Table 8.3,

$\alpha_s = 0.97$, $\varepsilon = 0.97$

Therefore, Eq. (a) becomes

$(300)(0.97) = (0.97)(5.67 \times 10^{-8})(T^4 - 298^4)$

or $T = 338.8$ K $= 65.8$°C

From this example we see what we have expected from the start, that white surfaces are cooler than black surfaces in the sunlight.

8.22 Greenhouse Effect

It is common experience that when one leaves one's car under direct sunlight on a sunny day, the interior of the car gets much warmer than the environment. This can be explained by observing the spectral transmissivity curve of glass as shown in Fig. 8.37. It is seen that glass at thicknesses encountered in practice transmits over 90% radiation in the visible range and is practically opaque to radiation in the longer-wavelength infrared regions of the electromagnetic spectrum (approximately $\lambda > 3$ μm). Thus the glass windows of the car allow the solar radiation to come in but do not let the infrared radiation from the interior surfaces to escape. This creates the familiar *heat trap* effect and hence a rise in temperature inside the car. This heating effect is known as the *greenhouse effect*, since it is utilized primarily in greenhouses (Fig. 8.38), that is, glasshouses for delicate plants.

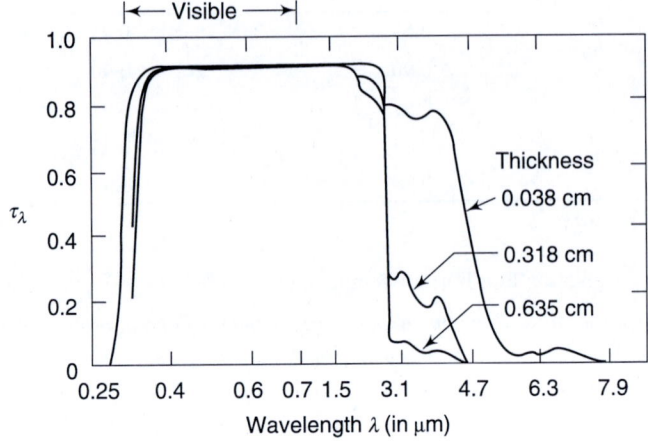

Fig. 8.37 Spectral transmissivity of low-iron glass at room temperature for different thicknesses (Cengel, 2003)

The greenhouse effect is also experienced on a large scale on earth due to pollution caused by the exhaust gases from the automobile engines. The combustion products such as carbon dioxide and water vapour in the atmosphere transmit the bulk of the solar radiation but absorb the infrared radiation emitted by the surface of the earth.

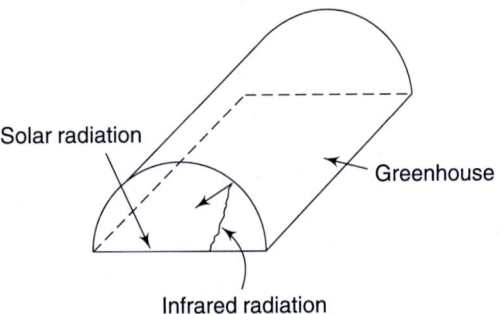

Fig. 8.38 A greenhouse

We know that the earth's surface warms up during the daytime due to solar heating and cools down at night by radiating its energy into outer space as infrared radiation. Thus, there is concern that the heat trap effect on earth will eventually cause global warming and thus drastic changes in climate.

In coastal areas, where humidity is high, there is not much difference between daytime and night-time temperatures because water vapour acts as a barrier on the path of infrared radiation emanating from the earth, and thus slows down the cooling process at night. On the other hand, in deserts, where the air is dry and the skies are clear, there is a large swing between daytime and night-time temperatures because of the absence of the water vapour barrier for infrared radiation.

Example 8.10 Derivation of the Expression of Planck's Law as given in Eq. (8.4)

Starting from Eq. (8.3) and using Eq. (8.1) and Eq. (8.2) obtain the expression of Planck's law as given in Eq. (8.4).

Solution
Equation (8.3) is

$$E_{bv}(T) = \frac{2\pi h v^3 n^2}{c_0^2 \left[\exp(hv/kT) - 1\right]} \tag{A}$$

From Eqs (8.1) and (8.2),

$$v = \frac{c_0}{n\lambda} \tag{B}$$

Assuming n is independent of frequency, differentiating Eq. (B)

$$dv = -\frac{c_0}{n\lambda^2} d\lambda \tag{C}$$

Also, $$E_b(T) = \int_0^\infty E_{bv}\, dv \tag{D}$$

Substituting Eqs (A) and (B) into Eq. (D) we see that Eq. (A) can be recast in terms of wavelength.

Thus, $$E_b(T) = -\int_0^\infty \frac{2\pi h v^3 n^2}{c_0^2 \left[\exp(hv/kT)-1\right]} \left(-\frac{c_0}{n\lambda^2}\right) d\lambda$$

Using Eq. (B),

$$E_b(T) = -\int_0^\infty \frac{2\pi h \left(\dfrac{c_0}{n\lambda}\right)^3 n^2}{c_0^2 \left[\exp(hc_0/n\lambda kT)-1\right]} \left(-\frac{c_0}{n\lambda^2}\right) d\lambda$$

$$= \int_0^\infty \frac{2\pi h c_0^2}{n^2 \lambda^5 \left[\exp(hc_0/n\lambda kT)-1\right]} d\lambda \tag{E}$$

We also know that

$$E_b(T) = \int_0^\infty E_{b\lambda}\, d\lambda \tag{F}$$

Therefore, comparing Eq. (E) with Eq. (F) we can write

$$E_{b\lambda} = \frac{2\pi h c_0^2}{n^2 \lambda^5 \left[\exp(hc_0/n\lambda kT)-1\right]} \tag{G}$$

Using $C_1 = 2\pi h c_0^2$ and $C_2 = hc_0/k$, Eq. (G) can be finally expressed as

$$E_{b\lambda} = \frac{C_1}{n^2 \lambda^5 \left(e^{\frac{C_2}{n\lambda T}} - 1\right)} \tag{H}$$

Equation (H) is Eq. (8.4).

Example 8.11   View Factor for an Infinitely Long Wedge-shaped Groove

Obtain a general expression for F_{1-2} for the infinitely long wedge-shaped groove as shown in E8.11. Side 1 is of length a, and Side 2 is of length b. What would be the expression if c = d = 0?

Solution

Since the configuration is infinitely long in the direction perpendicular to the plane of the paper Hottel's crossed-strings method can be used to find the view factor. Hence,

$$F_{1-2} = \frac{\text{(Sum of lengths of crossed strings)} - \text{(Sum of the lengths of uncrossed strings)}}{2\,(\text{Original length})}$$

In Fig. E8.11,

$$s_1 = \sqrt{(c-d\cos\alpha)^2 + d^2\sin^2\alpha} = \sqrt{c^2 + d^2 - 2cd\cos\alpha}$$

$$s_2 = \sqrt{(a+c)^2 + (b+d)^2 - 2(a+c)(b+d)\cos\alpha}$$

$$d_1 = \sqrt{c^2 + (b+d)^2 - 2c(b+d)\cos\alpha}$$

$$d_2 = \sqrt{(a+c)^2 + d^2 - 2(a+c)d\cos\alpha}$$

Sum of the lengths of crossed strings=
sum of the lengths of diagonals $= d_1 + d_2$
Sum of the lengths of uncrossed strings
= sum of the lengths of sides $= s_1 + s_2$.
Originating length = Length of side
1 = a

Therefore,

$$F_{1-2} = \frac{(d_1 + d_2) - (s_1 + s_2)}{2a}$$

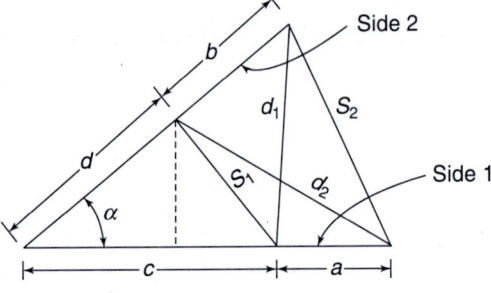

(A) **Fig. E8.11** Cross-section of an infinitely long wedge-shaped groove

If $c = d = 0$, then

$$s_1 = 0$$

$$s_2 = \sqrt{a^2 + b^2 - 2ab\cos\alpha}$$

$$d_1 = b$$

$$d_2 = a$$

Hence, Eq. (A) reduces to

$$F_{1-2} = \frac{a+b-\sqrt{a^2 + b^2 - 2ab\cos\alpha}}{2a}$$

Example 8.12   Radiation Exchange in a Black Rectangular Furnace

A very long furnace is shown in Fig. E8.12. The furnace is 50 cm × 40 cm in cross-section, and all surfaces are black. The top and bottom walls are maintained at temperature $T_1 = T_3 = 1200\ K$, while the side walls are at temperature $T_2 = T_4 = 800\ K$. Determine the net radiative heat loss or gain (per unit furnace length) on each surface.

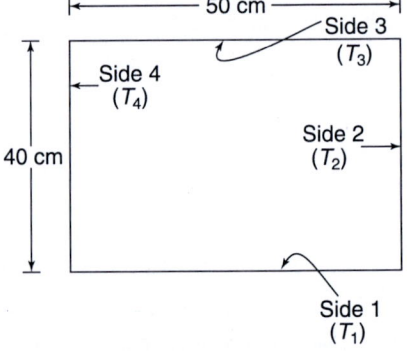

Fig. E8.12   Cross-section of a long furnace

Solution

From the theory of radiation exchange in a black enclosure (Section 8.11) the net heat loss/gain at a surface can be expressed as

$$q_i = \sum_{j=1}^{N} A_i F_{i-j} \sigma\left(T_i^4 - T_j^4\right) \tag{A}$$

Note that in Eq. (A) $j \neq i$.

Now, expanding Eq. (A) we can write

$$q_1 = A_1 F_{1-2}\sigma(T_1^4 - T_2^4) + A_1 F_{1-3}\sigma(T_1^4 - T_3^4) + A_1 F_{1-4}\sigma(T_1^4 - T_4^4)$$

or

$$q_1'' = \sigma\{F_{1-2}(T_1^4 - T_2^4) + F_{1-3}(T_1^4 - T_3^4) + F_{1-4}(T_1^4 - T_4^4)\} \tag{B}$$

Now, since $T_1 = T_3$ and $T_2 = T_4$, and from symmetry $F_{1-2} = F_{1-4}$, Eq. (B) reduces to

$$q_1'' = 2\sigma F_{1-2}(T_1^4 - T_2^4) \tag{C}$$

Again, from symmetry of the problem

$$q_1'' = q_3''$$

From Eq.(A) after simplification and using symmetry

$$q_2'' = q_4'' = 2\sigma F_{2-1}(T_2^4 - T_1^4) \tag{D}$$

The view factors can be calculated by Hottel's crossed-strings method since the furnace is very long. Hence,

$$F_{1-2} = \frac{40 + 50 - \left(\sqrt{40^2 + 50^2} + 0\right)}{2(50)} = 0.26$$

From the rule of reciprocity,

$$F_{2-1} = \frac{A_1}{A_2}F_{1-2} = \frac{50}{40}(0.26) = 0.325$$

Therefore, from Eq. (C), on per unit length basis,

$$q_1' = q_3' = 2l_1\sigma F_{1-2}(T_1^4 - T_2^4)$$

$$= 2(0.5)(5.67 \times 10^{-8})(0.26)(1200^4 - 800^4) = 24530 \text{ W/m}$$

Similarly, from Eq. (D),

$$q_2' = q_4' = 2l_2\sigma F_{2-1}(T_2^4 - T_1^4)$$

$$= 2(0.4)(5.67 \times 10^{-8})(0.325)(800^4 - 1200^4) = -24530 \text{ W/m}$$

It is clearly seen from the results that sum of all surface heat losses/gains is equal to zero. This is because of the conservation of energy which states that the total heat transfer into the enclosure (i.e., heat transfer rates summed over all the surfaces) must be equal to the rate of change of radiative energy within the enclosure. Since radiation travels at the speed of light, steady state is reached almost instantly, so that the rate of change of radiative energy is negligible (Modest, 1993). Hence, in an enclosure $\sum_{i=1}^{N} q_i = 0$. The aforesaid relationship is useful to check the correctness of surface heat loss/gain calculations for an enclosure.

Important Concepts and Formulae

Thermal Radiation

Thermal radiation is one of the many types of electromagnetic radiation travelling in straight line at the speed of light. Alternatively, from a quantum point of view, energy is transported by photons, all of which travels at the speed of light. There is, however, a distribution of energy among the photons. The energy associated with each photon is hv, where h is Planck's constant and v is the frequency of radiation.

Three parameters may be employed in characterizing radiation. They are (i) the frequency, v, (ii) wavelength, λ, and (iii) the wave or photon speed c. Of these only two are independent, since they are related by

$$c = \lambda v$$

The speed of light, c, within a given medium is related to that in a vacuum, c_0, by

$$c = \frac{c_0}{n}$$

where n is the index of refraction of the bounding medium.

Thermal radiation is defined as the radiant energy emitted by a medium, which is due solely to the temperature of the medium which governs the emission of thermal radiation. The wavelength range encompassed by the thermal radiation is approximately 0.3–50 μm.

Laws of Radiation

Planck's Law

There is a maximum amount of radiant energy that can be emitted at a given temperature and at a given wavelength. The emitter of such a radiation is called a black body. The energy emission per unit time and per unit area from a black body in a frequency range dv is $E_{bv}dv$, where E_{bv} is called the spectral (or monochromatic) emissive power of the black body radiation. The explicit form of E_{bv} (T) is given by Planck's law as

$$E_{bv}\ (T) = \frac{2\pi h v^3\ n^2}{c_o^2\left[\exp\left(hv/kT\right)-1\right]}$$

where h and k are Planck's and Boltzmann's constants, respectively. The index of refraction, n, refers to the medium bounding the black body. T is in kelvin.

Wien's Displacement Law

The law states that the maximum value of the emissive power occurs at

$$(n\lambda T)_{max} = 2989 \ \mu m \ K$$

Stefan-Boltzmann Law

The law states that the total black body emissive power E_b (unit: W/m^2) is given by

$$E_b\ (T) = n^2\sigma T^4$$

where σ is known as Stefan-Boltzmann constant which as a value of 5.668×10^{-8} W/m^2 K^4. When $n = 1$ (for vacuum and gases) the law takes the form

$$E_b = \sigma T^4$$

Note that T is in kelvin.

Intensity of Radiation

The concept of the intensity of radiation is introduced to treat the directional effects of radiation. Radiation emitted by a surface propagates in all possible directions. Also, radiation incident on a surface may come from different directions. A radiation analyst might be interested in knowing the directional distribution of the emission and/or the manner in which the surface responds to the incoming radiation from different directions.

The spectral intensity (unit: W/m^2 Sr μm), $I_{\lambda,e}$ is defined as the rate at which radiant energy is emitted at the wavelength λ in the (θ, ϕ) direction, per unit area of the emitting surface normal to this direction, per unit solid angle about this direction, and per unit wavelength interval $d\lambda$ about λ.

$$I_{\lambda,e}\ (\lambda,\theta,\varphi) = \frac{dq}{(dA_1\ \cos\theta)\ d\omega d\lambda}$$

For the special case of diffuse emitter for which the emitted radiation is independent of direction, we can write

$$I_{\lambda,\varepsilon}\ (\lambda,\ \theta,\ \phi) = I_{\lambda,\varepsilon}\ (\lambda)$$

The spectral emissive power, $E_\lambda (\lambda)$ is

$$E_\lambda(\lambda) = \pi I_{\lambda,\varepsilon} (\lambda)$$

Similarly, the total emissive power, E is

$$E = \pi I$$

where I is the total intensity of the emitted radiation. Note that the constant π has the unit of steradian (Sr).

Irradiation

Irradiation G refers to the radiation from all directions incident on a surface. If the incident radiation is diffuse, then

$$G_\lambda(\lambda) = \pi I_{\lambda,i} (\lambda)$$

and

$$G = \pi I$$

Radiosity

Radiosity J accounts for all the radiant energy leaving a surface. Thus radiosity constitutes the reflected portion of the irradiation, as well as direct emission. If the surface is both a diffuse reflector and an emitter, then

$$J_\lambda(\lambda) = \pi I_{l,e+r}$$
$$J = \pi I_{e+r}$$

Relation between Radiosity and Irradiation

$$J = \varepsilon E_b + \rho G$$

In the above expression ρ is the reflectivity of the surface and ε is the emissivity of the surface.

Diffuse Surface and Specular Surface

A surface is termed diffuse or specular depending on its response to incident radiation. If the angle of incidence is equal to the angle of reflection, the reflection is called specular. On the other hand, when an incident beam is distributed uniformly in all directions after reflection, the reflection is called diffuse.

Absorptivity, Reflectivity, and Transmissivity

When a radiation beam strikes a surface, a part of it may be reflected away from the surface, a portion may be absorbed by the surface, while the rest may be transmitted through the surface. These fractions of reflected, absorbed, and transmitted energy are called reflectivity p, absorptivity α, and transmissivity τ, respectively. Thus,

$$\rho_\lambda + \alpha_\lambda + \tau_\lambda = 1$$

or

$$\rho + \alpha + \tau = 1$$

Black Body Radiation

A black body is defined as one which absorbs all incident radiation regardless of the spectral distribution or directional character of the incident radiation, i.e., $\alpha_\lambda = \alpha = 1$ or $\rho_\lambda = \rho = 0$. The term 'black' is used since dark surfaces normally show high values of absorptivity. The intensity of black body radiation, I_b is uniform. Thus, black body radiation is diffuse and

$$E_b = \pi I_b$$

Monochromatic and Total Emissivity of Non-black Surfaces

The monochromatic emissivity of a surface is defined as

$$\varepsilon_\lambda (\lambda, T) = \frac{E_\lambda (\lambda, T)}{E_{b\lambda} (\lambda, T)}$$

The total emissivity of a surface is defined as

$$\varepsilon(T) = \frac{E(T)}{E_b(T)}$$

For a black surface, $\varepsilon = \varepsilon_\lambda = 1$.

Gray Body

An ideal gray body, which is a special type of non-black surfaces, is defined as the one for which the monochromatic emissivity is independent of wavelengths, i.e., the ratio of E_λ to $E_{b\lambda}$ is the same for all wavelengths of emitted energy at a given temperature. Thus, for the ideal gray surface,

$$\varepsilon(T) = \varepsilon_\lambda(T)$$

Monochromatic and Total Absorptivities

The monochromatic absorptivity is defined as

$$\alpha_\lambda(\lambda, T) = \frac{[G_\lambda(\lambda)]_{absorbed}}{G_\lambda(\lambda)}$$

The total absorptivity is defined as the sum of the absorbed radiation over all wavelengths divided by the total incident energy. Thus,

$$\alpha(T, source) = \frac{\int_0^\infty [G_\lambda(\lambda)]_{absorbed}\, d\lambda}{\int_0^\infty G_\lambda(\lambda)\, d\lambda}$$

The notation $\alpha(T, source)$ is used to drive home the point that the total absorptivity not only depends on the absorbing surface temperature but also on the source of the incident radiation.

Kirchhoff's Law

For thermal equilibrium in an isothermal enclosure and diffuse radiation,

$$\varepsilon(\lambda, T) = \varepsilon_\lambda(T)$$
$$\varepsilon = \alpha$$

Since $\alpha = 1$ for a black body, $\varepsilon = 1$.

Hence, a black body, a perfect absorber is also a perfect emitter.

View Factor

The view factor F_{ij} (or F_{i-j}) is defined as the fraction of the radiation leaving surface i, which is directly intercepted by surface j. Thus,

$$F_{i-j} = \frac{q_{i\rightarrow j}}{A_i J_i}$$

where $q_{i\rightarrow j}$ is the radiation leaving surface i and intercepted by surface j and $A_i J_i$ is the total radiation leaving surface i which may be due to both emission and reflection.

View Factor Integral

$$F_{ij} = \frac{1}{A_i} \int_{A_i} \int_{A_j} \frac{\cos\theta_i \cos\theta_j}{\pi R^2} dA_i\, dA_j$$

The above integral may be used to determine the view factor associated with any two surfaces that are diffuse emitters and reflectors and have uniform radiosity.

View Factor Relations

Reciprocity Relation

$$A_i F_{ij} = A_j F_{ji}$$

The reciprocity relation is useful in determining one view factor from the knowledge of the other.

Summation Rule

$$\sum_{j=1}^{N} F_{ij} = 1$$

The summation rule originates from the conservation requirement that all of the radiation leaving surface i must be intercepted by the enclosure surfaces including the surface i itself.

It may be noted that

$F_{ii} = 0$, if the surface i is plane or convex

$F_{ii} \neq 0$, if the surface i is concave

The term F_{ii} represents the fraction of the radiation that leaves surface i and is directly intercepted by surface i itself.

Radiation Exchange among Surfaces in an Enclosure

In this analysis it is assumed that the solid surfaces are opaque ($\tau = 0$) while the medium enclosed by the surfaces is transparent and non-emitting ($\tau = 1$, $\varepsilon = \alpha = 0$). Hence, this kind of medium is called non-participating. Elementary gases with symmetrical molecules such as nitrogen and oxygen are, indeed, transparent to thermal radiation. Thus, air can be considered as a non-participating medium.

Radiation Exchange in a Black Enclosure

The net rate of heat loss q_i from a typical surface i is the difference between the emitted radiation and the absorbed portion of the incident radiation. Thus, in general,

$$\frac{q_i}{A_i} = \varepsilon_i \sigma T_i^4 - \alpha_i G_i$$

For a black enclosure, $\varepsilon_i = \alpha_i = 1$.

Therefore, $q_i = \sigma T_i^4 - G_i$

For an N-surface black enclosure

$$G_i A_i = \sum_{j=1}^{N} \sigma T_j^4 A_i F_{i-j}$$

Applying the summation rule, we finally get

$$q_i = \sum_{j=1}^{N} A_i F_{i-j} \sigma (T_i^4 - T_j^4)$$

Note that in the summation expression, $j \neq i$.

Radiation Exchange in a Gray Enclosure

The difference between radiation in a black enclosure and gray enclosure is the surface reflection. In a gray enclosure, radiation may experience multiple reflections off all surfaces, with partial absorption occurring at each.

The net rate of heat loss from a gray surface is

$$q_i = \frac{A_i \varepsilon_i}{1 - \varepsilon_i} \left(\sigma T_i^4 - J_i \right)$$

where $J_i = \varepsilon_i \sigma T_i^4 + (1 - \varepsilon_i) \sum_{j=1}^{N} J_j F_{i-j}$ $(i \le i \le N)$

Now, we get N linear, non-homogeneous algebraic equations for N unknown radiosities $J_1, J_2, ..., J_N$. Consequently, these equations are solved simultaneously and J_i values obtained, and with these q_i can be determined.

Electric Circuit Analogy

q_i can be written as

$$q_i = \frac{E_{bi} - J_i}{\dfrac{1 - \varepsilon_i}{A_i \varepsilon_i}}$$

The above equation is analogous to the electric current flow representation by Ohm's law. The radiative transfer q_i is associated with the driving potential $(E_{bi} - J_i)$ and a surface resistance of the form $(1 - \varepsilon_i)/A_i \varepsilon_i$. The positive sign of q_i indicates that there is a net radiative heat transfer from the surface, while the negative sign signifies net transfer to the surface.

Now, consider the radiant exchange by two surfaces A_i and A_j. The amount that is intercepted by surface j from the radiation leaving surface i is $J_i A_i F_{i-j}$ and of the total energy leaving surface j that reaches surface i is $J_j A_j F_{j-i}$. The net exchange between the two surfaces is

$$q_{i-j} = J_i A_i F_{i-j} - J_j A_j F_{j-i}$$

Now, from the reciprocity rule,

$$A_i F_{i-j} = A_j F_{j-i}$$

Therefore, $q_{i-j} = (J_i - J_j) A_i F_{i-j} = \dfrac{J_i - J_j}{\dfrac{1}{A_i F_{i-j}}}$

$\dfrac{1}{A_i F_{i-j}}$ is called space resistance.

Hence, $q_i = \sum_{j=1}^{N} q_{i-j} = \sum_{j=1}^{N} A_i F_{i-j} (J_i - J_j)$

This result suggests that the net rate of radiation transfer from surface i, that is, q_i may be represented as a sum of the components, that is, q_{i-j}, related to the radiation exchange with other surfaces. Each component may be represented as a network element for which $(J_i - J_j)$ is the driving potential and $1/A_i F_{i-j}$ is the space or geometric resistance. So, finally we can write

$$\frac{E_{bi} - J_i}{\dfrac{(1 - \varepsilon_i)}{A_i \varepsilon_i}} = \sum_{j=1}^{N} \frac{J_i - J_j}{(1 / A_i F_{i-j})}$$

It may be noted that

(a) the surface resistance of a black body is zero since $\varepsilon_i = 1$. Thus, for a black body, $j_i = E_{bi}$;

(b) for an adiabatic surface, $q_i = 0$ and hence, $j_i = \varepsilon_{bi} = \sigma T_i^4$. This shows that the temperature of an *adiabatic* surface (also called *reradiating* surface) is independent of its emissivity. Furthermore, the surface resistance of such a surface is disregarded as there is no net heat transfer through it. This is analogous to the case when a resistance is not considered in an electrical circuit if no current is flowing through it.

In practical applications, the surfaces whose back sides are well insulated are modeled as adiabatic surfaces. That is, the net heat transfer through such a surface is zero. Neglecting the convection effects on the front (heat transfer) side of such a surface and assuming steady state conditions, the net heat lost by the surface must equal the net gain to it. Hence, $q_i = 0$. In such instances, the surface is considered to reradiate all the radiation incident upon it, and such a surface is designated as a *reradiating* surface.

Three-surface Enclosure

The electrical network analogy concept can be easily extended to determine radiation exchange among the surfaces of a three or more surface enclosure.

Two-surface Enclosure

Radiation in two surface enclosure is the simplest of the enclosure problems. In this case there are two surfaces that exchange radiation only with each other.

The radiation heat transfer rate can be easily obtained by transforming the problem into an analogous electrical network problem. It can be seen that the total resistance to radiation exchange between sides 1 and 2 is composed of two surface resistances and one space resistance. Hence, the net radiation exchange between two surfaces may be expressed as

$$q = q_1 = -q_2 = q_{1-2} = \frac{E_{b_1} - E_{b_2}}{\dfrac{1-\varepsilon_1}{A_1 \varepsilon_1} + \dfrac{1}{A_1 F_{1-2}} + \dfrac{1-\varepsilon_2}{A_2 \varepsilon_2}}$$

where $E_{b_i} = \sigma T_1^4$ and $E_{b_i} = \sigma T_2^4$

The expression for q can be used for any two diffuse, gray surfaces that form an enclosure.

Special Cases of Two-surface Enclosure

Case A: Two Infinite Parallel Planes

$$q = \frac{\sigma A (T_1^4 - T_2^4)}{\dfrac{1}{\varepsilon_1} + \dfrac{1}{\varepsilon_2} - 1}$$

Case B: Two Concentric Long Cylinders

$$q = \frac{\sigma A_1 (T_1^4 - T_2^4)}{\dfrac{1}{\varepsilon_1} + \left(\dfrac{1}{\varepsilon_2} - 1\right)\dfrac{A_1}{A_2}}$$

Case C: A Convex Object Surrounded by a Very Large Concave Surface

$$q = \sigma \varepsilon_1 A_1 (T_1^4 - T_2^4)$$

Case D: Radiation Loss from a Hot Object of Surface Area A at T in a Large Room at T_∞

$$q = \sigma \varepsilon A (T^4 - T_\infty^4)$$

Gas Radiation

Gases such as carbon dioxide, water vapour, ammonia, and most hydrocarbon gases do emit and absorb thermal radiation—at least in certain wavelength bands. The methodology of radiation exchange calculations in an enclosure containing CO_2 and H_2O (vapour) which are commonly encountered in practice (combustion products in furnaces and combustion chambers) has been discussed in this section. The basic principle can be extended to other gases. The medium in which such gases are present is called participating medium.

The following are the main characteristics of a participating medium.

(a) It emits and absorbs radiation throughout its entire volume. That is, gaseous radiation is a volumetric phenomenon, and thus depends on the size and shape of the body.

(b) Such gases emit and absorb at a number of narrow-wavelength bands. Therefore, the assumption of a gray body may not always be appropriate for such gases even when enclosing surfaces are gray.

(c) The emission and absorption characteristics of the constituents of a gas mixture also depend on the temperature, pressure, and composition of the gas mixture.

Review Questions

8.1 What are the distinguishing features of thermal radiation? Why is it distinctly different from conduction and convection?

8.2 What is the physical mechanism of energy transport in thermal radiation?

8.3 What is a black body?

8.4 State Stefan-Boltzmann law. How is it obtained from Planck's law?

8.5 State Wien's displacement law.

8.6 Define irradiation and radiosity.

8.7 What is the difference between a diffuse surface and specular surface?

8.8 What is the relation among absorptivity, reflectivity and transmissivity?

8.9 Define monochromatic and total emissivity and absorptivity.

8.10 State Kirchhoff's law of radiation. What are the restrictions of Kirchhoff's law?

8.11 What is a gray body? How does it differ from a real body and black body?

8.12 Define view factor.

8.13 When can Hottel's crossed-strings method be used?

8.14 Define surface resistance and space resistance.

8.15 What is the advantage of Gebhart's absorption factor method over electrical network analogy method?

8.16 What is a radiation shield? Where is it used?

8.17 Define radiation heat transfer coefficient.

8.18 What are the important features of a participating medium?

8.19 Define mean beam length and explain its importance in gas radiation.

8.20 What is solar constant? Explain what you mean by greenhouse effect.

Problems

8.1 Determine the solid angle at which the sun is seen from the earth. Note that the image of the sun that we see from the earth is a circular disc of radius $R_s = 6.96 \times 10^8$ m at a distance S of approximately 1.496×10^{11} m. See Fig. Q8.1.

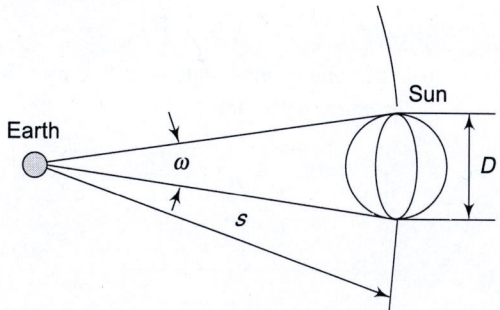

Fig. Q8.1

8.2 A solar collector mounted on a satellite orbiting earth is directed at the sun (i.e., normal to the sun's rays). Determine the total solar heat flux incident on the collector. Note that the sun is closest to the definition of a black body. Take the effective temperature of the sun as 5762 K. [Hint: Place an imaginary spherical shell (which includes the solar collector) around the sun at a distance $S = 1.496 \times 10^{11}$ m from earth (see Fig. Q8.2).]

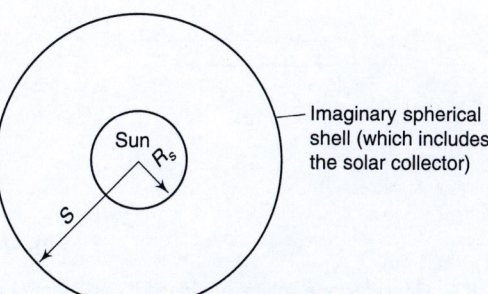

Fig. Q8.2

8.3 Consider the very long isosceles triangular duct shown in Fig. Q8.3. Determine $F_{1\text{-}2}$. Note that A_3 is the area of the base of the duct.

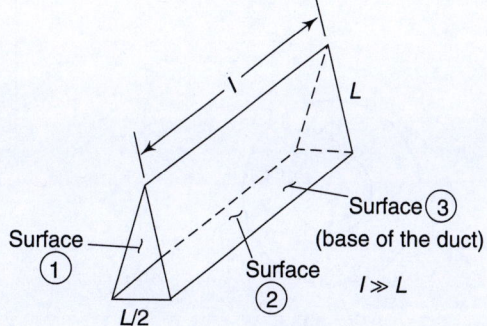

Fig. Q8.3

8.4 For the configuration shown in Fig. Q8.4, calculate $F_{1\text{-}2}$ by Hottel's crossed string method.

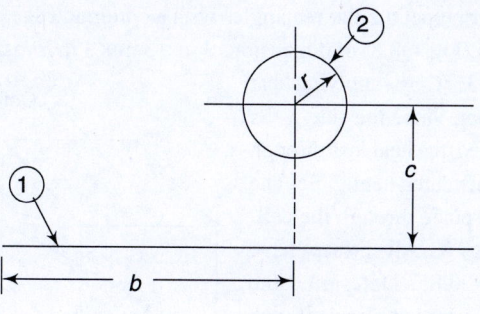

Fig. Q8.4

8.5 (a) Show the view factor $F_{1\text{-}2}$ between two infinite parallel plates just one above the other [Fig. Q8.5(a)] is

$$F_{1-2} = \sqrt{1 + \left(\frac{D}{L}\right)^2} - \frac{D}{L}$$

(b) When one parallel plate is shifted by a distance L [Fig. Q8.5(b)] with respect to the other, show that

$$F_{1-2} = \sqrt{1 + \left(\frac{D}{2L}\right)^2} + \frac{D}{2L} - \sqrt{1 + \left(\frac{D}{L}\right)^2}$$

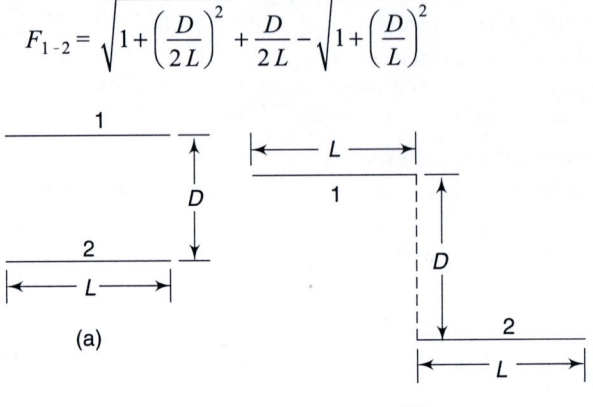

(a)

(b)

Fig. Q8.5

8.6 Consider one sphere enclosed by an other (Fig. Q8.6). Find F_{1-1}, F_{1-2}, F_{2-1}, and F_{2-2}.

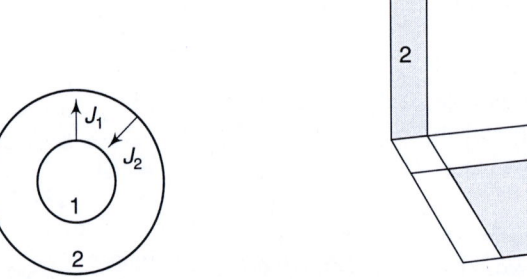

Fig. Q8.6   **Fig. Q8.7**

8.7 Find the shape factor F_{1-2} for the configuration shown in Fig. Q8.7 in terms of the shape factor for perpendicular rectangles with a common edge.

8.8 A cubical room (Fig. Q8.8) of dimensions $3\ m \times 3\ m \times 3\ m$ is maintained at a uniform temperature of $37\,°C$ by supplying heat through the floor. Since the side walls are well insulated, the heat loss through them can be considered negligible. The heat loss takes place through the ceiling, which is at $7\,°C$. All surfaces have emissivity $\varepsilon = 0.85$. Determine the rate of heat loss by radiation through the ceiling. Solve the problem by both the electrical network analogy and Gebhart's absorption factor method.

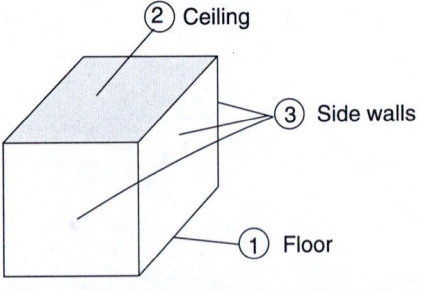

Fig. Q8.8

8.9 Figure Q8.9 shows a furnace which is 5 m high, 10 m wide, and 20 m long. The floor of the furnace, A_1, acts as a black plane at $T_1 = 200°C$, the left side wall acts as a gray plane, A_2, with $T_2 = 400°C$ and $\varepsilon_2 = 0.4$. The two 5 m × 10 m ends act as a single insulated surface A_3 and the remaining 10 m × 20 m ceiling and 5 m × 20 m right-side wall act

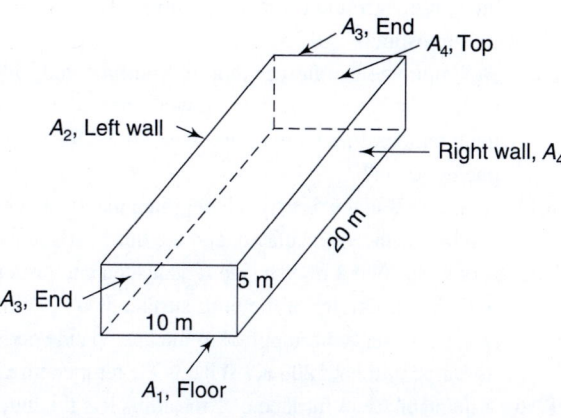

Fig. Q8.9

as a second ins lated surface A_4. Find the heat flow at the two active surfaces (A_1 and A_2) and the temperature of the adiabatic (i.e., insulated) surfaces (A_3 and A_4).

8.10 In Darjeeling, a shoe store with a display window in the front is to be heated by making the floor a black, radiant heating panel at 45°C. The glass window acts as a black plane at 10°C and the other walls and ceiling act as black planes at 25°C. (a) Find the net heat lost by the floor. (b) What difference

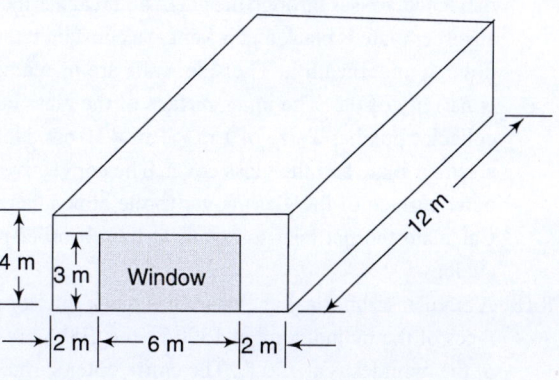

Fig. Q8.10

will result if the ceiling height is raised to 4.5 m, the other dimensions remaining the same? See Fig. Q8.10.

Given: $F_{1-2} = 0.058$ where the floor is designated as A_1 and the window as A_2.

8.11 Two parallel, infinite planes directly opposed to one another are maintained at 300°C and 400°C, respectively. (a) What is the net heat flux between the two planes if one has an emissivity of 0.6 and the other an emissivity of 0.7? Does it matter which plane has which emissivity? (b) What is the net heat flux between the planes if they are black? (c) Repeat part (a) if the temperature of the 400°C plane is raised to 500°C.

8.12 Consider two large, opposed parallel plates, one at $T_1 = 400°C$ with emissivity $\varepsilon_1 = 0.8$ and the other at 300°C with emissivity $\varepsilon_2 = 0.4$. An aluminium radiation shield with emissivity $\varepsilon_3 = 0.05$ is placed between the plates. Compare heat transfer rates with and without the radiation shield.

8.13 A 20-megaton nuclear bomb is detonated at a height of 10 km above the ground level and dissipates its energy uniformly over a period of 4 s. Assume that the radiation leaves the fireball in a spherically symmetric manner and that 50% of the total energy is dissipated as thermal radiation. Calculate the radiant energy flux directly below the burst at the ground level. 1 megaton = 10^{15} cal and the average transmissivity (τ) of the atmosphere is 0.35.

8.14 An employee in a flower shop in Shimla noted for two seasons that water collecting in the plastic coverings over flowers formed 0.25 inch thick ice at night when the official temperature reading was well above freezing. How do you explain this strange phenomenon?

8.15 A furnace is in the form of a long, triangular duct in which one surface is kept at 1200 K, another surface is insulated, and the third surface is maintained at 500 K. The triangle is of width W = 1 m on a side. The heated and insulated surfaces have an emissivity of 0.8. The emissivity of the third surface is 0.4. During steady-state operation, at what rate must energy be supplied to the heated side per unit length of the duct to maintain its temperature at 1200 K? What is the temperature of the insulated surface?

8.16 A hemispherical furnace of 1 m radius has the inner surface ($\varepsilon = 1$) of its roof maintained at 800 K while its floor ($\varepsilon = 0.5$) is kept at 600 K. Calculate the net radiative heat transfer from the roof to the floor.

8.17 A solar collector consists of a glass cover plate, a collector plate, and side walls. The glass is totally transparent to solar irradiation which falls on the glass cover at normal incidence, passes through the glass, and reaches the absorber plate at 1000 W/m². The absorber plate is black and is kept at a constant temperature of 77°C by heating water flowing underneath it. The side walls are insulated and made of a material with an emissivity of 0.5. The inner surface of the glass cover has an emissivity of 0.9. The collector box has a size of 1 m × 1 m × 10 cm. Neglect free convection between the absorber plate and the glass cover. The convective heat transfer coefficient from the outer surface of the glass cover to the atmosphere, which is at 17°C, is 5 W/m²K. Calculate the net heat loss/gain at the absorber plate using the electrical network analogy.

8.18 A circular cylindrical enclosure has black interior surfaces. The top and bottom surfaces of the cylinder are at 1500 K and 500 K, respectively. The peripheral surface of the cylinder is at 750 K. The entire outer surface of the cylinder is insulated such that this surface does not radiate to the surroundings. What rate of heat per unit area (W/m²) is supplied to each area as a result of the internal radiation exchange?

8.19 An enclosure with black interior surfaces has one side open to an environment at temperature T_∞. The sides of the enclosure are maintained at uniform temperatures T_1, T_2, T_3, ... How are the heat inputs to the sides q_1, q_2, q_3, ..., influenced by the value of T_∞?

8.20 A frustum of a cone of 7.5 cm bottom and 5 cm top radii and 7.5 cm height has its base exposed to a constant heat flux of 6000 W/m². The top is maintained at 1800 K while the side is perfectly insulated. All surfaces are diffuse-gray. The emissivities of the top, bottom, and side are 0.5, 0.6, and 0.8 respectively. What is the temperature achieved by the bottom surface as a result of radiation exchange within the enclosure?

8.21 What is the ratio of heat transfer with shield to heat transfer without shield if a single, thin radiation shield is inserted between two concentric spheres, the outer surface of the inner sphere (of radius R_1) and the inner surface of the outer sphere (of radius R_2) being at temperatures T_1 and T_2 ($T_1 > T_2$), respectively? The outer surface of the outer sphere is insulated. Assume the sphere and radiation shield surfaces are diffuse-gray, with emissivities independent of temperature. Both sides of the shield have the same emissivity ε_s, and the inner and outer spheres have respective emissivities ε_1 and ε_2.

8.22 A gas turbine combustion chamber is 0.3 m in diameter and the walls (which may be considered as black) are maintained at 505 °C. The products of combustion are at 1005 °C, a pressure of 1 atm, and contain 15% of CO_2 and 15% of H_2O (vapour) by volume. Assuming the combustion chamber to be very long (that is, an infinite cylinder), determine the net radiant heat exchange between the gases and the combustion chamber wall.

Chapter 9

Heat Exchangers

9.1 Introduction

A heat exchanger is an equipment that is used to transfer heat from one fluid to another, usually through a separating wall. Heat exchangers are employed in a variety of applications, such as, steam power plants, chemical processing plants, building heating, air conditioning, refrigeration systems, and mobile power plants for automotive, marine, and aerospace vehicles. In this chapter our discussion will be limited to heat exchangers, where the primary modes of heat transfer are conduction and convection. However, radiation is also important in the heat exchanger design, for example, in power plants operating in spacecraft.

The present chapter deals with the analysis of three types of heat exchangers, e.g., double-pipe, shell-and-tube, and crossflow heat exchangers.

9.2 Classification of Heat Exchangers

Heat exchangers are classified according to the fluid flow arrangement and types of applications. The details are given as under.

9.2.1 Fluid Flow Arrangement

Most heat exchangers may be classified according to the fluid flow arrangement. The four most common types of flow-path configuration are shown schematically in Fig. 9.1. In parallel-flow or cocurrent-flow units [Fig. 9.1(a)] two fluid streams enter together at one end, flow in same direction, and exit at the other end; whereas in countercurrent or counter-flow units, two fluid streams move in opposite directions [Fig. 9.1(b)]. In single-pass crossflow systems the flow path of one fluid cuts at right angle to that of the other fluid [Fig. 9.1(c)]. In multi-pass crossflow systems one fluid stream zigzags across the flow path of the other fluid stream, usually giving a crossflow approximation to counter-flow [Fig. 9.1(d)]. Amongst the four basic types, counter-flow requires the least heat transfer surface area to produce a given temperature rise from a given inlet temperature difference.

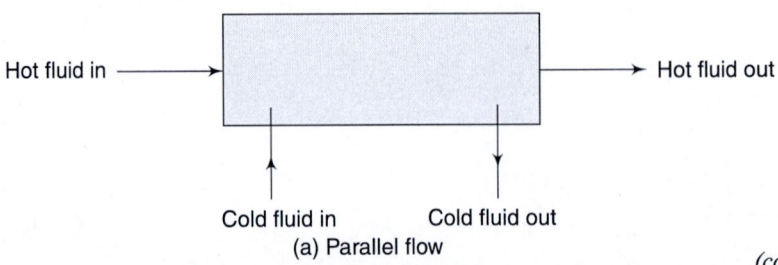

(a) Parallel flow

(contd)

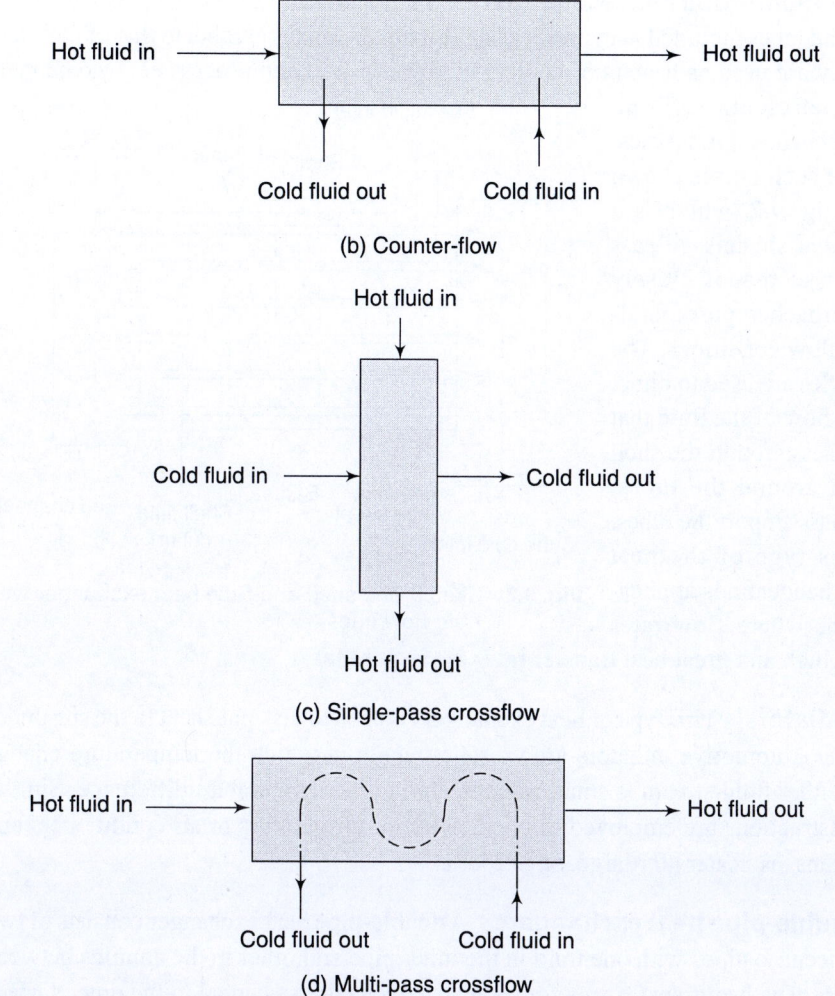

(b) Counter-flow

(c) Single-pass crossflow

(d) Multi-pass crossflow

Fig. 9.1 Types of flow-path configuration through heat exchangers

9.2.2 Types of Application

Heat exchangers are also classified according to the application for which they are designed. Some of these applications are discussed below.

Boilers Steam boilers have been used to generate power for over two centuries. These are one of the earliest heat exchange equipment. The types of boilers range from many small, relatively simple, units used for space heating units (during winter in cold countries) to the huge, complex, and expensive boilers in modern power plants in which the heat source is the hot products of combustion.

Condensers Its application mainly lies in steam power plants and refrigeration and air-conditioning systems. A single-pass condenser in a typical modern 300-MW steam power plant employs nearly 1,000,000 ft. of tubing in a matrix of around 20,000 tubes in which cold water flows and over which steam exiting from the turbine condenses.

Shell-and-tube heat exchangers Shell-and-tube heat exchangers are built of round tubes mounted in cylindrical shells with their axes parallel to that of the shell. They are used as heaters or coolers in power plants and process heat exchangers in petroleum-refining and chemical industries. One such unit is shown in Fig. 9.2, which is a baffled, single-shell-pass unit so that it closely approaches pure parallel flow conditions. The baffles are used to direct the flow of the fluid that passes through the shell and around the tubes, and to support the tubes. This type of the heat exchanger finds applications where flow rates are high and great heat transfer rates are required.

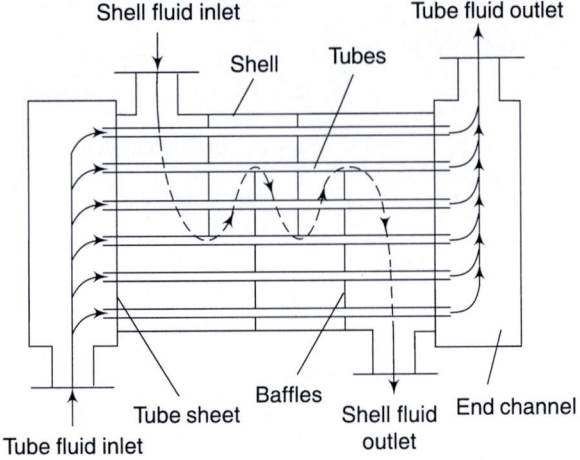

Fig. 9.2 Sketch of a shell-and-tube heat exchanger with fluid flow lines

Radiators This type of heat exchangers is used to dissipate heat to the surroundings. Automotive radiators are crossflow units in which the temperature change in either fluid stream is small as compared to the temperature difference. Similar constructions are employed as condensers in refrigerators or air conditioners and in fans, as heaters for large, open rooms.

Double-pipe heat exchangers A double-pipe heat exchanger consists of two concentric pipes with one fluid in the inner pipe and other in the annulus between them. The heat transfer area for such heat exchangers is equal to the outer surface area of the inner tube. The flow and heat transfer rates in this type are moderate because the equipment is relatively small. Both parallel-flow and counter-flow arrangements are available in these exchangers. The schematic diagram of a double-pipe heat exchanger is shown in Fig. 9.3.

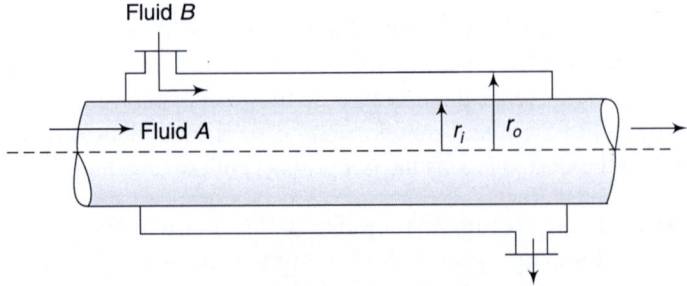

Fig. 9.3 Schematic diagram of a double-pipe heat exchanger

Other types of heat exchangers These include cooling towers, regenerators and recuperators, and immersion heaters and coolers.

9.3 Overall Heat Transfer Coefficient

The overall heat transfer coefficient U has already been defined in Section 2.9. The heat transfer through a plane wall bounded by hot and cold fluids at T_{∞_1} and T_{∞_2} respectively, is expressed as

$$q = UA(T_{\infty_1} - T_{\infty_2}) = UA\,\Delta T_{\text{overall}} \tag{9.1}$$

where
$$\left[U = \frac{1}{h_1} + \frac{L}{k} + \frac{1}{h_2} \right]^{-1} \tag{9.2}$$

However, a plane wall in a heat exchanger is a rarity. Most frequently, a cylindrical wall as in double-pipe heat exchangers (Fig. 9.3) is the surface through which heat transfer takes place. q in this case is expressed as

$$q = \frac{T_{\infty_1} - T_{\infty_2}}{\dfrac{1}{h_1 A_1} + \dfrac{\ln \dfrac{r_o}{r_i}}{2\pi kL} + \dfrac{1}{h_o A_o}} \tag{9.3}$$

Note that in the cylindrical wall case, A is not a constant, but varies from $2\pi r_i L$ to $2\pi r_o L$. Therefore, the definition of U in this case depends on the area selected. If the inner surface area is taken as the basis, then

$$U_i = \frac{1}{\dfrac{1}{h_i} + \dfrac{r_i}{k} \ln \dfrac{r_o}{r_i} + \dfrac{r_i}{h_o r_o}} \tag{9.4}$$

If the outer surface area is used, then

$$U_o = \frac{1}{\dfrac{r_o}{r_i} \dfrac{1}{h_i} + \dfrac{r_o \ln \left(\dfrac{r_o}{r_i} \right)}{k} + \dfrac{1}{h_o}} \tag{9.5}$$

The value of U is dictated in many cases by only one of the convection heat transfer coefficients. In most practical problems the conduction resistance is small as compared to the convective resistances.

9.4 Fouling Factor

Prolonged operation of a heat exchanger may result in the heat transfer surfaces being coated with various deposits in the flow systems. Furthermore, the surfaces may become corroded as a result of the interaction between the fluids and the material used for the construction of the heat exchanger. In either case, this coating gives rise to an additional resistance to the heat flow, and thus results in low performance. The overall effect is usually represented by a quantity called 'fouling factor' or fouling resistance, R_f, which must be included along with the other thermal resistances making up the overall heat transfer coefficient.

Fouling factors are obtained experimentally by determining the values of U for both clean and dirty conditions in the heat exchanger. The fouling factor is defined as

$$R_f = \frac{1}{U_{\text{dirty}}} - \frac{1}{U_{\text{clean}}} \tag{9.6}$$

U_{clean} can be obtained from either Eq. (9.4) or Eq. (9.5); and from the knowledge of the fouling factor (see Table 9.1), U_{dirty} can be calculated. The value of U_{dirty} should be used in the design of heat exchangers.

Table 9.1   Normal fouling factors

Type of fluid	Fouling factor ($m^2\ ^\circ C/W$)
Sea water, below 125 °F[*]	0.00009
Above 125 °F[*]	0.0002
Treated boiler feed water above 125 °F[*]	0.0002
Fuel oil	0.0009
Quenching oil	0.0007
Alcohol vapours	0.00009
Steam, non-oil-bearing	0.00009
Industrial air	0.0004
Refrigerating liquid	0.0002

[*]$C/5 = (F - 32)/9$, where C represents Celsius and F Fahrenheit.
(*Source*: Holman, 1997)

9.5   Typical Temperature Distributions

Figure 9.4 shows typical temperature distributions for a number of idealized cases. Note that the temperature distribution is plotted as a function of the distance from the cold fluid inlet end of the heat exchanger. In all cases, the heat transfer surface area per unit length is assumed to be constant throughout the exchanger and heat transfer coefficients independent of the axial position, that is, the local fluid temperature.

Figures 9.4(a)–(d) indicate varying slopes of the fluid temperature curves with the distance from the inlet. This effect is particularly pronounced in Figs 9.4(c) and (d), for which the temperature on one side of the heat exchanger is constant irrespective of the distance from the fluid inlet.

In general, the temperature distribution in an idealized parallel-flow or counter-flow heat exchanger is as depicted in Fig. 9.4(a) or (b) if there is no change of phase in either fluid.

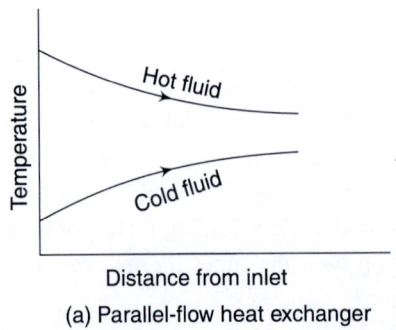

(a) Parallel-flow heat exchanger

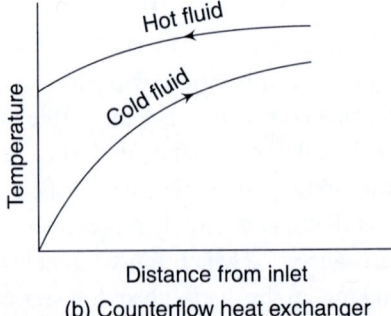

(b) Counterflow heat exchanger

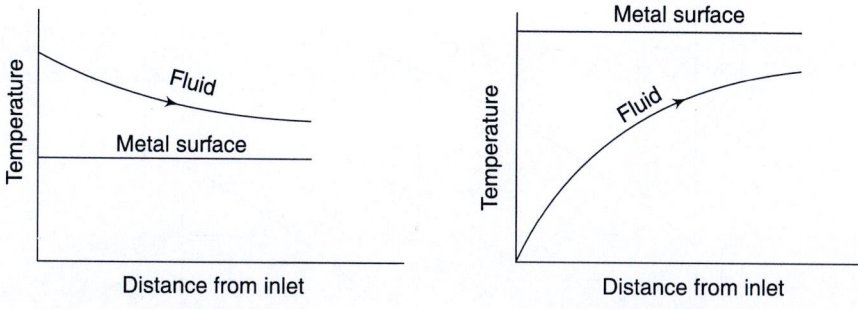

(c) Uniform surface temperature
   (as in a gas-heated boiler)

(d) Uniform surface temperature
   (as in a water-cooled condenser)

Fig. 9.4 Typical axial temperature distributions in heat exchangers

9.6 Temperature Distribution in Counter-flow Heat Exchangers

Consider a differential length dx at a distance x from the cold fluid inlet as shown in Fig. 9.5(b). Heat added to the cold fluid will result in a temperature rise dT [Fig. 9.5(a)]. This can be equated to the heat transferred through the increment of the surface area in the length dx as

$$\dot{m}_c c_c dT_c = U\, dA\, \Delta T \tag{9.7}$$

Now,

$$dA = A\frac{dx}{L} \tag{9.8}$$

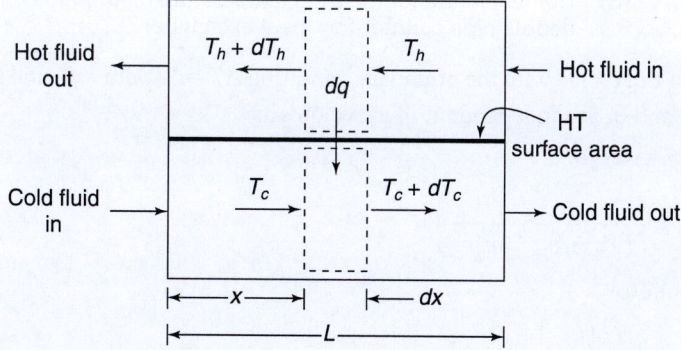

Fig. 9.5(a) Energy balance in a counter-flow heat exchanger

Substituting Eq. (9.8) into Eq. (9.7), we get

$$\dot{m}_c c_c dT_c = \left(\frac{UA}{L}\right)\Delta T\, dx \tag{9.9}$$

or

$$dT_c = \frac{UA}{\dot{m}_c c_c L}\,\Delta T\, dx$$

Similarly,

$$dT_h = \frac{UA}{\dot{m}_h c_h L}\,\Delta T\, dx \tag{9.10}$$

Subtracting Eq. (9.9) from Eq. (9.10) and noting that $d(T_h - T_c) = d(\Delta T)$, gives

$$\frac{d\,\Delta T}{\Delta T} = \frac{UA}{L}\left(\frac{1}{\dot{m}_h c_h} - \frac{1}{\dot{m}_c c_c}\right)dx \tag{9.11}$$

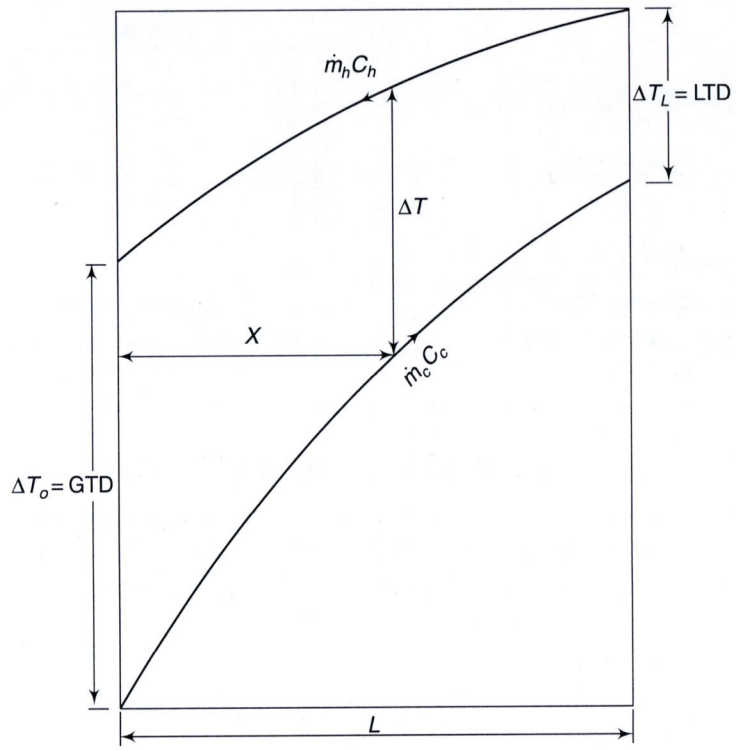

Fig. 9.5(b)   Nomenclature for the axial temperature distribution in a double-pipe counter-flow heat exchanger

Integrating Eq. (9.11) with the boundary condition $\Delta T = \Delta T_0$ at $x = 0$, and assuming that U, c_c, and c_h are independent of x, we obtain

$$\Delta T = \Delta T_0 e^{ax} \tag{9.12}$$

where $$a = \frac{UA}{L}\left(\frac{1}{\dot{m}_h c_h} - \frac{1}{\dot{m}_c c_c}\right) \tag{9.13}$$

For parallel flow,

$$a = -\frac{UA}{L}\left(\frac{1}{\dot{m}_h c_h} + \frac{1}{\dot{m}_c c_c}\right) \tag{9.14}$$

Using the outlet condition $\Delta T = \Delta T_L$ at $x = L$, Eq. (9.12) transforms to

$$\Delta T_L = \Delta T_0 e^{aL} \tag{9.15}$$

Solving for a, we obtain

$$a = \frac{1}{L}\ln\frac{\Delta T_L}{\Delta T_0} \tag{9.16}$$

Substituting Eq. (9.16) into Eq. (9.12) gives

$$\Delta T = \Delta T_0 \exp\left(\frac{x}{L}\ln\frac{\Delta T_L}{\Delta T_0}\right) \tag{9.17}$$

which reduces to [after taking natural logarithm on both sides of Eq. (9.17)]

$$\Delta T = \Delta T_0 \left(\frac{\Delta T_L}{\Delta T_0} \right)^{x/L} \tag{9.18}$$

Note that Eq. (9.18) is applicable to either counter-flow or parallel-flow conditions since it is independent of parameter a.

9.7 Log-mean Temperature Difference

In the design of heat exchangers, the mean effective temperature difference between the two fluid streams is much more useful than the detailed axial distribution of the temperature difference between the hot and cold fluid streams. This mean effective temperature difference involves the natural logarithm of the ratio of the temperature differences at the two ends of the heat exchanger, and thus has come to be known as the log-mean temperature difference, or LMTD, or ΔT_m. This quantity is defined as

$$\text{LMTD} = \frac{1}{L} \int_0^L \Delta T \, dx \tag{9.19}$$

Substituting ΔT from Eq. (9.18) gives

$$\begin{aligned}
\text{LMTD} &= \frac{\Delta T_0}{L} \int_0^L \left(\frac{\Delta T_L}{\Delta T_0} \right)^{x/L} dx \\
&= \frac{\Delta T_0 - \Delta T_L}{\ln \left(\dfrac{\Delta T_0}{\Delta T_L} \right)}
\end{aligned} \tag{9.20}$$

For the temperature distribution of Fig. 9.5(b), ΔT_0 may be referred to as the greatest temperature difference, or GTD, and ΔT_L as the least temperature difference, or LTD. Therefore, Eq. (9.20) can be written as

$$\text{LMTD} = \frac{\text{GTD} - \text{LTD}}{\ln \dfrac{\text{GTD}}{\text{LTD}}} \tag{9.21}$$

Equation (9.21) applies to either parallel-flow or counter-flow heat exchangers.

9.8 Heat Transfer as a Function of LMTD

Once the LMTD is determined, the heat transfer for the heat exchanger as a whole is evaluated from

$$q = UA(\text{LMTD}) \tag{9.22}$$

9.9 Multi-pass and Crossflow Heat Exchangers: Correction Factor Approach

If a heat exchanger other than the double-pipe type is used, the heat transfer is calculated by using a correction factor applied to the LMTD for a counter-flow double-pipe arrangement with the same hot and cold fluid temperatures. Then Eq. (9.22) takes the form

$$q = UAF(\text{LMTD}) \tag{9.23}$$

Values of the correction factor F are plotted in Fig. 9.6 (Bowman et al. 1940) for several different types of heat exchangers. When a phase change is involved, as in boiling or condensation, since the metal surface remains constant at the boiling liquid or condensing vapour temperature [Figs 9.4(c, d)], the correction factors are all 1.0 in this case.

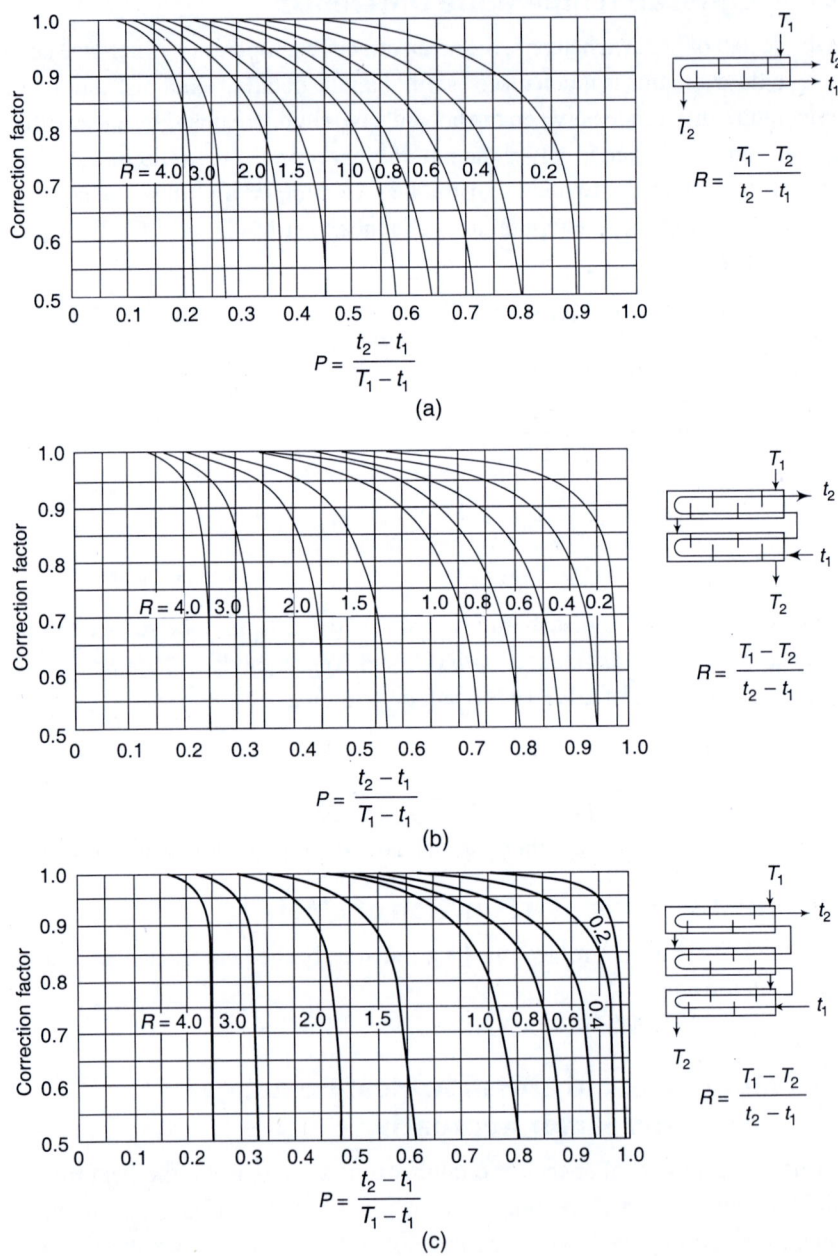

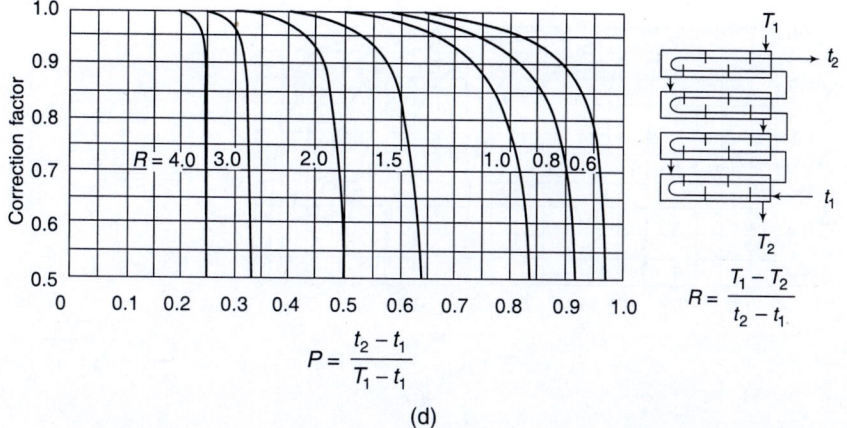

$$P = \frac{t_2 - t_1}{T_1 - t_1}$$

(d)

$$R = \frac{T_1 - T_2}{t_2 - t_1}$$

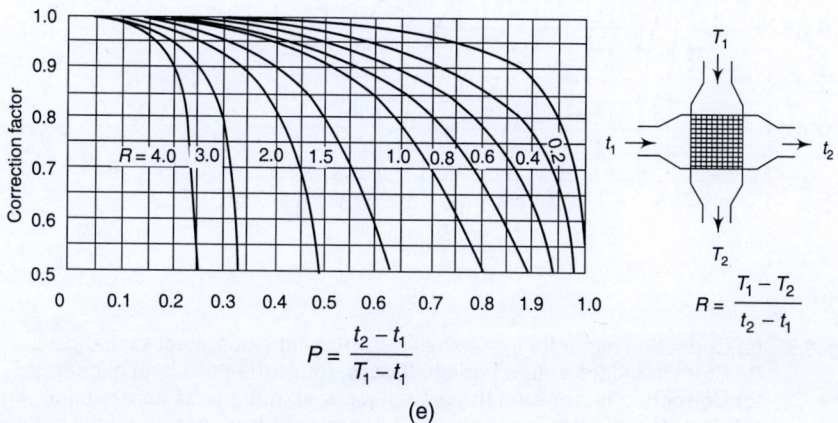

$$P = \frac{t_2 - t_1}{T_1 - t_1}$$

(e)

$$R = \frac{T_1 - T_2}{t_2 - t_1}$$

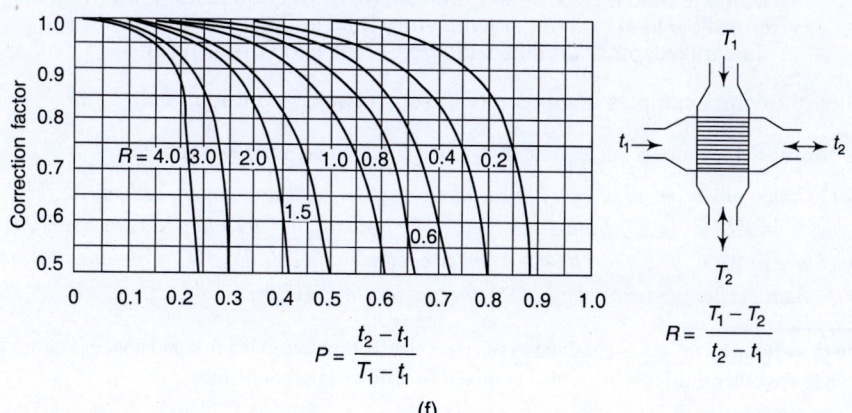

$$P = \frac{t_2 - t_1}{T_1 - t_1}$$

(f)

$$R = \frac{T_1 - T_2}{t_2 - t_1}$$

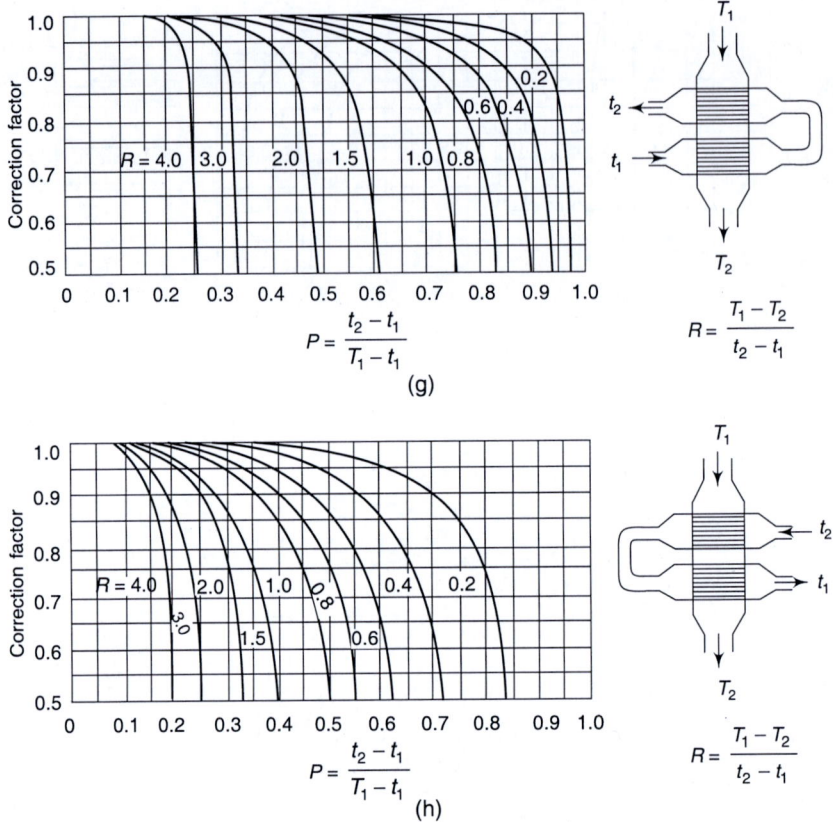

Fig. 9.6 (a) Correction factor for a one-shell-pass, two-tube-pass heat exchanger; (b) Correction factor for a two-shell-pass, four-tube-pass heat exchanger; (c) Correction factor for a three-shell-pass, six-tube-pass heat exchanger; (d) Correction factor for a four-shell-pass, eight-tube-pass heat exchanger; (e) Correction factor for a singe-pass, crossflow heat exchanger with both fluids unmixed; (f) Correction factor for a single-pass, crossflow heat exchanger with one fluid mixed, other unmixed; (g) Correction factor for a two-pass, crossflow heat exchanger (with entry fluid in the lower tube) with one fluid mixed, other unmixed; (h) Correction factor for a two-pass, crossflow heat exchanger (with entry fluid in the upper tube) with one fluid mixed, other unmixed. (*Source:* Fraas and Ozisik, 1965)

The following examples demonstrate the correction factor approach.

Example 9.1 Finned-Tube Cross-flow Heat Exchanger: LMTD Approach

Hot exhaust gases are used in a finned-tube crossflow heat exchanger[1] to heat 2.5 kg/s of water (c = 4.18 kJ/kg °C) from 35 to 85 °C. The gases (c = 1.09 kJ/kg °C) enter at 200 °C and leave at 93 °C. The overall heat transfer coefficient is 180 W/m² °C. Calculate the area of the heat exchanger using the LMTD approach (see Fig. E9.1).

---

[1] In this arrangement gas is confined in separate channels between the fins and hence is unmixed while the other fluid (water) is also unmixed as it flows in separate tubes.

Solution

Assuming pure counter-flow (Fig. E9.2)

$$LMTD = \frac{GTD - LTD}{\ln\left(\dfrac{GTD}{LTD}\right)}$$

$$= \frac{(200-85)-(93-35)}{\ln\left(\dfrac{200-85}{93-35}\right)}$$

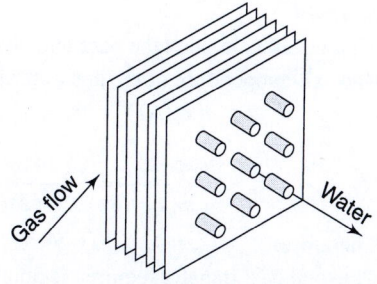

$$= \frac{115-58}{\ln\left(\dfrac{115}{58}\right)}$$

Fig. E9.1 Crossflow heat exchanger in which both fluids are unmixed

$$= 57/0.6844 = 83.27\,°C$$

Now
$$q = \dot{m}_w c_w\,(\Delta T)_{cold}$$
$$= (2.5)(4180)(85-35)$$
$$= 522{,}500 \text{ W}$$

Since this is a crossflow heat exchanger, we have to use a correction factor approach as discussed in Section 9.9. Thus,

$$Q = UA(LMTD)F$$

The correction factor F is obtained from the plot of F versus P with R as a parameter for the case of single-pass crossflow exchanger, both fluids unmixed [Fig. 9.6(e)].

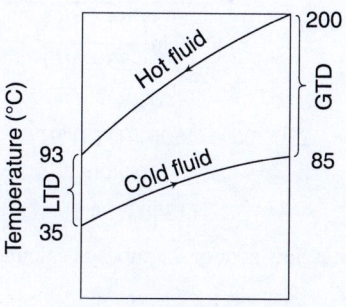

Temperature (°C) / Length of the pure counter-flow heat exchanger

Fig. E9.2 Temperature distribution in the heat exchanger assuming pure counter-flow

$$P = \frac{t_2 - t_1}{T_1 - t_1} = \frac{85-35}{200-35}$$

$$= \frac{50}{165} = 0.303$$

$$R = \frac{T_1 - T_2}{t_2 - t_1} = \frac{200-93}{85-35} = \frac{107}{50} = 2.14$$

$$\Rightarrow \qquad F = 0.92$$

Therefore, from $q = UAF(LMTD)$, we can write

$$522{,}500 = (180)(A)(83.27)(0.92)$$

or
$$A = 522{,}500/13{,}789.512 = 37.89 \text{ m}^2$$

Hence, the area of the heat exchanger is 37.89 m^2.

Example 9.2 Shell-and-Tube Heat Exchanger: LMTD Approach

Water at the rate of 4 kg/s is heated from 40°C to 55°C in a shell-and-tube heat exchanger. On the shell side one pass is used with water as the heating fluid ($\dot{m}_h = 2$ kg/s), entering the exchanger at 95°C. The overall heat transfer coefficient is 1500 W/m²°C and the average water velocity in the 2-cm-diameter tubes is 0.5 m/s. Because of space limitations the tube length must not exceed 3 m. Calculate the number of tube passes, the number of tubes per pass, and the length of the tubes, keeping in mind the design constraint.

Solution

First we assume one tube pass and check whether the design constraint is satisfied or not. The exit temperature of the hot water is calculated from an energy balance

$$q = \dot{m}_c c_c (\Delta T)_c = \dot{m}_h c_h \Delta T_h$$

$$\Delta T_h = \frac{\dot{m}_c c_c \, \Delta T_c}{\dot{m}_h c_h} = \frac{(4)(4180)(55-40)}{(2)(4180)} = 30\,°\text{Ct}$$

Therefore, $T_{h,\,\text{exit}} = 95 - 30 = 65\,°\text{C}$

The total heat transfer required is obtained from

$$q = (4)(4180)(55-40) = 250{,}800 \text{ W} = 250.8 \text{ kW}$$

$$\text{LMTD} = \Delta T_m = \frac{\text{GTD} - \text{LTD}}{\ln\left(\dfrac{\text{GTD}}{\text{LTD}}\right)}$$

$$= \frac{(95-30)-(55-40)}{\ln\left(\dfrac{95-30}{55-40}\right)} = \frac{65-15}{\ln\left(\dfrac{65}{15}\right)} = \frac{15}{0.47} = 31.91\,°\text{C}$$

Now, $\qquad q = UA\Delta T_m$

or $\qquad 250{,}800 = (1500)(A)(31.91)$

$\Rightarrow \qquad A = \dfrac{250{,}800}{(1500)(34.1)} = 5.24 \text{ m}^2$

The total flow area of the tubes is calculated from

$$\dot{m}_c = \rho A_c u$$

or $\qquad A_c = \dfrac{\dot{m}_c}{\rho u}$

$$= 4/(1000)(0.5) = 0.008 \text{ m}^2$$

Now, $\qquad A_c = n\left(\dfrac{\pi d^2}{4}\right)$

where n is the number of tubes.

$$n = \frac{4\,A_c}{\pi d^2}$$

$$= \frac{4\,(0.008)}{\pi\,(0.02)^2} = 25.46 = 25 \text{ tubes}$$

The surface area per tube per metre length is

$$\pi d = \pi(0.02) = 0.06283 \text{ m}^2/\text{tube m}$$

It may now be recalled the total surface area required for one-tube-pass exchanger was calculated as $A = 5.24 \text{ m}^2$. Therefore, the length of the tube for this type of exchanger is computed from

$$n\pi dL = 5.24$$

or $\qquad L = 5.24/(25)(0.06283) = 3.34 \text{ m}$

This length is greater than the allowable 3 m. So we should try more than one tube pass, that is, two-passes. Now, we will have to use a correction factor since the heat exchanger configuration is deviating from the simple double-pipe type. From Fig. 9.6(a), using

$$P = \frac{t_2 - t_1}{T_1 - t_1} = \frac{55 - 40}{95 - 40} = 0.27$$

$$R = \frac{T_1 - T_2}{t_2 - t_1} = \frac{95 - 65}{55 - 40} = 2$$

we get $F = 0.925$

Thus, $A_{\text{total}} = \dfrac{q}{U F \Delta T_m}$

$$= \frac{250,800}{(1500)(0.925)(34.1)} = 5.66 \text{ m}^2$$

For the two-tube-pass exchanger the length is calculated from

$$A_{\text{total}} = 2n\pi dL$$

where n = number of tubes per pass

= 25 (the same as in one-tube-pass case because of velocity requirement)

$$\Rightarrow \qquad L = \frac{5.3}{2\,(25)(0.06283)} = 1.8 \text{ m}$$

This length is within the 3 m limit. Therefore, the final design choice is

Number of tubes per pass = 25

Number of passes = 2

Length of tube per pass = 1.8 m

9.10   Effectiveness–NTU Method

The LMTD method is useful when the inlet and exit temperatures are known or are easily determined. But when the inlet and exit temperatures are to be evaluated for a given heat exchanger, the analysis frequently involves an iterative procedure because of the logarithm function in the LMTD. Thus, if U, A, T_{c_i}, T_{h_i} are known for a counter-flow heat exchanger then T_{h_o}, T_{c_o} can be obtained from the following equations:

$$\dot{m}_c c_c (T_{c_o} - T_{c_i}) = UA \frac{(T_{h_i} - T_{c_o}) - (T_{h_o} - T_{c_i})}{\ln\left(\dfrac{T_{h_i} - T_{c_o}}{T_{h_o} - T_{c_i}}\right)} \tag{9.24}$$

$$\dot{m}_h c_h (T_{h_i} - T_{h_o}) = UA \frac{(T_{h_i} - T_{c_o}) - (T_{h_o} - T_{c_i})}{\ln\left(\dfrac{T_{h_i} - T_{c_o}}{T_{h_o} - T_{c_i}}\right)} \tag{9.25}$$

But, since the denominators of the RHS of Eqs (9.24) and (9.25) contain logarithmic function, a trial-and-error solution procedure is necessary to obtain T_{h_o} and T_{c_o}. In such cases the effectiveness–NTU (number of transfer units) method is more convenient as it is a direct method and no iterations are required to obtain a solution when the overall heat transfer coefficient is known.

The heat transfer effectiveness ε is defined as

$$\varepsilon = \frac{\text{actual heat transfer}}{\text{maximum possible heat transfer}} \tag{9.26}$$

For a parallel-flow heat exchanger,

$$q_{actual} = \dot{m}_h c_h (T_{h_1} - T_{h_2}) = \dot{m}_c c_c (T_{c_2} - T_{c_1}) \tag{9.27}$$

For a counter-flow heat exchanger,

$$q_{actual} = \dot{m}_h c_h (T_{h_1} - T_{h_2}) = \dot{m}_c c_c (T_{c_1} - T_{c_2}) \tag{9.28}$$

Note that the subscripts 1 and 2 represent the inlet and exit of the heat exchanger.

The maximum possible heat transfer is defined as the maximum value of heat transfer that could be attained if one of the fluids were to undergo a temperature change equal to the maximum temperature difference present in the exchanger, which is the difference between the entry temperatures of the hot and cold fluids. The fluid that might undergo this maximum temperature difference is that having $(\dot{m}c)_{min}$ since the energy balance requires that the energy received by one fluid is that given up by the other fluid. If we let the fluid with larger $\dot{m}c$ go through a maximum temperature difference, this would require that the other fluid undergo a temperature difference greater than the maximum and this is impossible. Thus,

$$q_{max} = (\dot{m}c)_{min} \left(T_{h_{inlet}} - T_{c_{inlet}} \right) \tag{9.29}$$

The minimum fluid may be either hot or cold fluid, depending on the product of the mass flow rate and specific heat of the hot and cold fluids.

For the parallel-flow heat exchanger:

$$\varepsilon_h = \frac{\dot{m}_h c_h (T_{h_1} - T_{h_2})}{\dot{m}_h c_h (T_{h_1} - T_{c_1})} = \frac{T_{h_1} - T_{h_2}}{T_{h_1} - T_{c_1}} \tag{9.30}$$

$$\varepsilon_c = \frac{\dot{m}_c c_c (T_{c_2} - T_{c_1})}{\dot{m}_c c_c (T_{h_1} - T_{c_1})} = \frac{T_{c_2} - T_{c_1}}{T_{h_1} - T_{c_1}} \tag{9.31}$$

For the counter-flow heat exchanger:

$$\varepsilon_h = \frac{\dot{m}_h c_h (T_{h_1} - T_{h_2})}{\dot{m}_h c_h (T_{h_1} - T_{c_2})} = \frac{T_{h_1} - T_{h_2}}{T_{h_1} - T_{c_2}} \tag{9.32}$$

$$\varepsilon_c = \frac{\dot{m}_c c_c (T_{c_1} - T_{c_2})}{\dot{m}_c c_c (T_{h_1} - T_{c_2})} = \frac{T_{c_1} - T_{c_2}}{T_{h_1} - T_{c_2}} \tag{9.33}$$

The subscripts on the effectiveness symbols designate the fluid having the minimum value of $\dot{m}c$.

Thus the concept of effectiveness also provides us with a yardstick to compare various types of heat exchangers in order to select the type best suited to accomplish a particular heat transfer objective.

9.10.1 Derivation of an Expression for the Effectiveness in Parallel Flow

For a parallel-flow heat exchanger we know from Eqs (9.12) and (9.14) that

$$\Delta T = \Delta T_0 e^{ax}$$

where

$$a = -\frac{UA}{L} \left(\frac{1}{\dot{m}_h c_h} + \frac{1}{\dot{m}_c c_c} \right)$$

At $x = L$,
$$\Delta T_L = \Delta T_0 e^{aL}$$

Taking natural logarithm on both sides of the above equation, we get

$$\ln\left(\frac{\Delta T_L}{\Delta T_0}\right) = aL$$

or
$$\ln\frac{T_{h_2} - T_{c_2}}{T_{h_1} - T_{c_1}} = -UA\left(\frac{1}{\dot{m}_h c_h} + \frac{1}{\dot{m}_c c_c}\right) \tag{9.34}$$

If the cold fluid is the minimum fluid, then from Eq. (9.31),

$$\varepsilon_c = \frac{T_{c_2} - T_{c_1}}{T_{h_1} - T_{c_1}}$$

Now the temperature ratio of Eq. (9.34) can be rewritten [using Eq. (9.27)] as

$$\frac{(T_{h_2} - T_{c_2})}{(T_{h_1} - T_{c_1})} = \frac{T_{h_1} + \dfrac{\dot{m}_c c_c}{\dot{m}_h c_h}(T_{c_1} - T_{c_2}) - T_{c_2}}{T_{h_1} - T_{c_1}} \tag{9.35}$$

Equation (9.35) can now be rewritten as

$$\frac{T_{h_1} - T_{c_1} + \dfrac{\dot{m}_c c_c}{\dot{m}_h c_h}(T_{c_1} - T_{c_2}) + (T_{c_1} - T_{c_2})}{T_{h_1} - T_{c_1}} = 1 - \left(1 + \frac{\dot{m}_c c_c}{\dot{m}_h c_h}\right)\varepsilon$$

$$\Rightarrow \qquad \frac{T_{h_2} - T_{c_2}}{T_{h_1} - T_{c_1}} = 1 - \left(1 + \frac{\dot{m}_c c_c}{\dot{m}_h c_h}\right)\varepsilon \tag{9.36}$$

Therefore, combining Eqs (9.34) and (9.36) we obtain

$$\ln\left\{1 - \left(1 + \frac{\dot{m}_c c_c}{\dot{m}_h c_h}\right)\varepsilon\right\} = -UA\left(\frac{1}{\dot{m}_h c_h} + \frac{1}{\dot{m}_c c_c}\right)$$

or
$$1 - \left(1 + \frac{\dot{m}_c c_c}{\dot{m}_h c_h}\right)\varepsilon = \exp\left[-UA\left(\frac{1}{\dot{m}_h c_h} + \frac{1}{\dot{m}_c c_c}\right)\right]$$

or
$$\varepsilon = \frac{1 - \exp\left[-UA\left(\dfrac{1}{\dot{m}_h c_h} + \dfrac{1}{\dot{m}_c c_c}\right)\right]}{\left(1 + \dfrac{\dot{m}_c c_c}{\dot{m}_h c_h}\right)}$$

$$= \frac{1 - \exp\left[-\left(\dfrac{UA}{\dot{m}_c c_c}\right)\left(1 + \dfrac{\dot{m}_c c_c}{\dot{m}_h c_h}\right)\right]}{\left(1 + \dfrac{\dot{m}_c c_c}{\dot{m}_h c_h}\right)} \tag{9.37}$$

Since the cold fluid is the minimum fluid in this case,

$$\dot{m}_c c_c = (\dot{m}c)_{\min} = C_{\min}$$

where $C = \dot{m}c$ is defined as the heat capacity. Therefore, effectiveness is expressed in the following form from Eq. (9.37):

$$\varepsilon = \frac{1 - \exp\left[-\left(\dfrac{UA}{C_{min}}\right)\left(1 + \dfrac{C_{min}}{C_{max}}\right)\right]}{\left(1 + \dfrac{C_{min}}{C_{max}}\right)} \tag{9.38}$$

It may be shown that Eq. (9.38) is also valid when the hot fluid is the minimum fluid, that is, when $\dot{m}_h c_h = (\dot{m}c)_{min} = C_{min}$.

A similar analysis may be applied to the counter-flow case, and the expression for effectiveness takes the following form:

$$\varepsilon = \frac{1 - \exp\left[\left(-\dfrac{UA}{C_{min}}\right)\left(1 - \dfrac{C_{min}}{C_{max}}\right)\right]}{1 - \left(\dfrac{C_{min}}{C_{max}}\right)\exp\left[\left(-\dfrac{UA}{C_{min}}\right)\left(1 - \dfrac{C_{min}}{C_{max}}\right)\right]} \tag{9.39}$$

The expression UA/C_{min} is called *number of transfer units* (which is abbreviated as NTU) since it is representative of the size of the heat exchanger.

9.10.2 Physical Significance of NTU

$$NTU = \frac{UA}{C_{min}} = \frac{\text{heat exchanging capacity per degree of mean temperature difference}}{\text{heat transferred per degree of temperature rise, to or from either fluid}}$$

For a fixed U/C_{min}, NTU is a measure of the actual heat transfer area A. Higher the NTU, larger is the physical size. Note that UA has the same unit (W/°C) as C_{min} $(\dot{m}_c c_c$ or $\dot{m}_h c_h)$ and hence the ratio is dimensionless.

9.10.3 Effectiveness–NTU Relations for Some Heat Exchangers

Parallel-flow heat exchanger From Eq. (9.38), we can write
(using $C_R = C_{min}/C_{max}$)

$$\varepsilon = \frac{1 - \exp[-NTU(1 + C_R)]}{1 + C_R} \tag{9.40}$$

For $C_R = 1$,

$$\varepsilon = \frac{1}{2}[1 - \exp(-2NTU)] \tag{9.41}$$

Counter-flow heat exchanger From Eq. (9.39) we can write

$$\varepsilon = \frac{1 - \exp[-NTU(1 - C_R)]}{1 - C_R \exp[-NTU(1 - C_R)]} \tag{9.42}$$

For $C_R = 1$, applying L'Hopital's rule to Eq. (9.42),

$$\varepsilon = \frac{NTU}{1 + NTU} \tag{9.43}$$

Boilers and condensers   In boiling or condensation the fluid temperature stays essentially constant or the fluid behaves as if it had infinite heat capacity (in other words, infinite specific heat). In these cases, $C_{min}/C_{max} \to 0$ and $C_{max} \to \infty$ and all the heat exchanger effectiveness–NTU relations approach a single simple equation,

$$\varepsilon = 1 - \exp(-NTU) \tag{9.44}$$

9.10.4   ε–NTU Charts

Kays and London (1964) have presented effectiveness–NTU (ε–NTU) charts for various heat exchanger arrangements. Some of these charts are shown in Figs 9.7(a)–(f).

While these charts have great practical use in design problems, in the applications demanding more precision in the calculations the design procedures should be computer-based, requiring analytical expressions for these curves. Some effectiveness relations are given in Section 9.10.3. For other types of exchangers, for example, crossflow and shell-and-tube exchangers, readers are referred to Holman (1997).

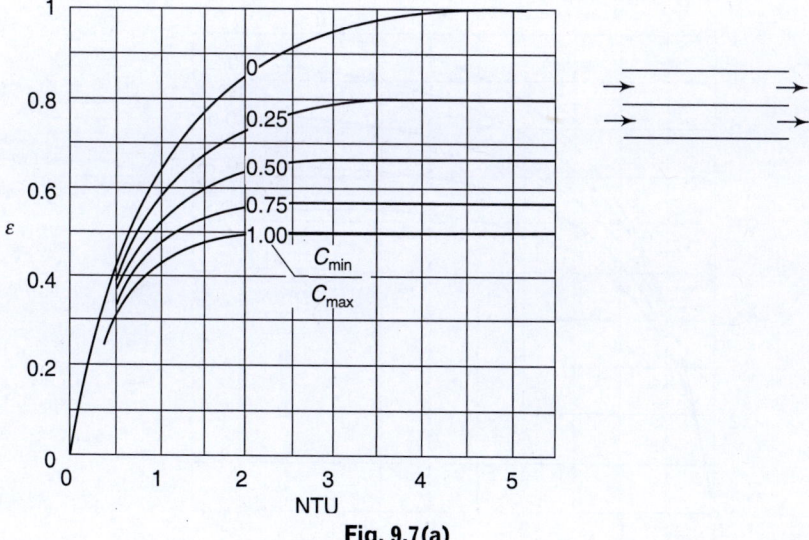

Fig. 9.7(a)

(contd)

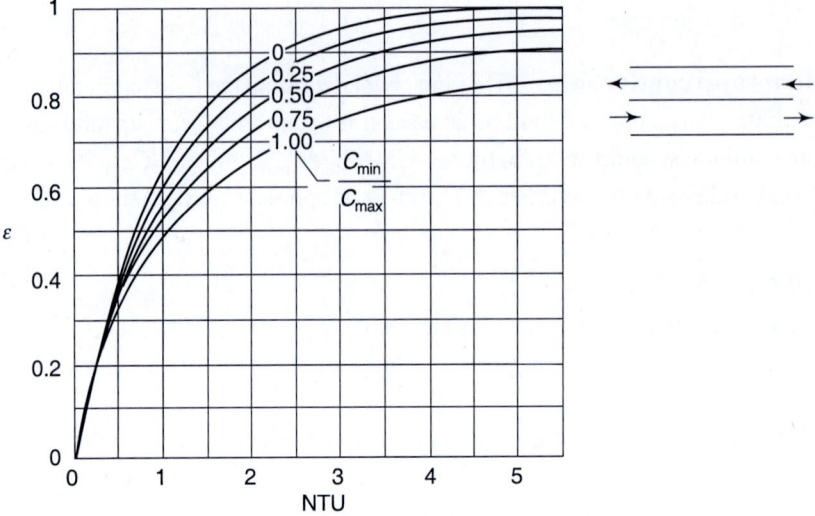

Fig. 9.7(b)

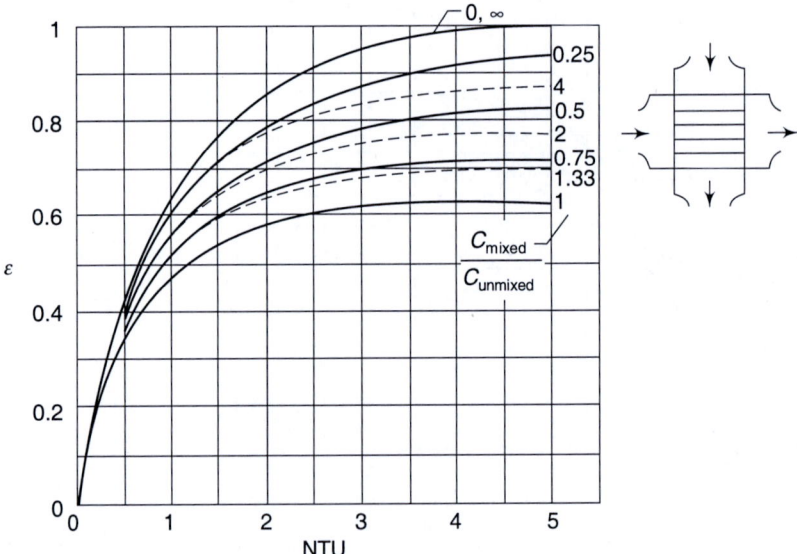

Fig. 9.7(c)

(contd)

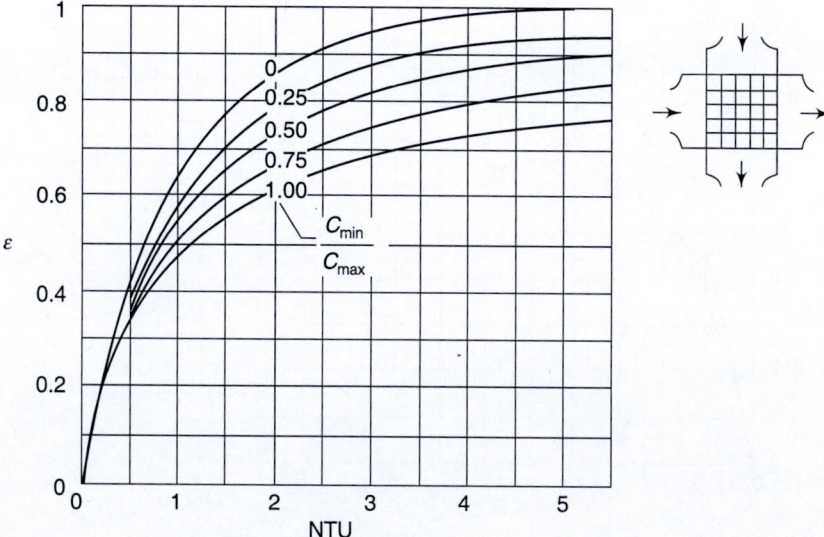

Fig. 9.7(d)

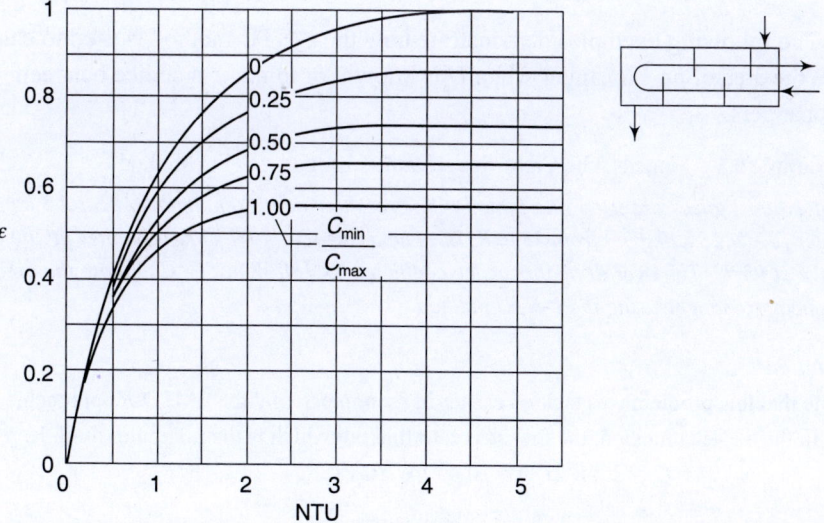

Fig. 9.7(e)

(contd)

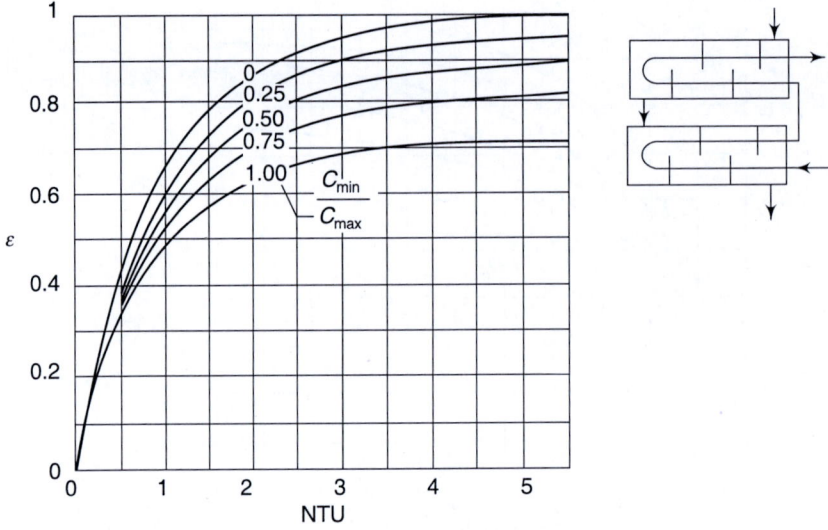

Fig. 9.7(f)

Fig. 9.7 (a) Effectiveness–NTU chart for a parallel-flow heat exchanger; (b) Effectiveness NTU chart for a counter-flow heat exchanger; (c) Effectiveness–NTU chart for a crossflow heat exchanger with one fluid mixed, other fluid unmixed; (d) Effectiveness–NTU chart for a crossflow heat exchanger with both fluids unmixed; (e) Effectiveness–NTU chart for a one-shell-pass, two-tube-pass heat exchanger and any multiple of two-tube passes (2, 4, 6, etc., tube passes); (f) Effectiveness–NTU chart for a two-shell-pass, four tube-pass heat ex-changer and any multiple of four-tube passes (4, 8, 12, etc., tube passes) (*Source:* Bejan, 1993).

The following examples demonstrate how the ε-NTU method is used to calculate the area of an exchanger when U is known, or to make a choice between two exchangers.

Example 9.3 Finned-Tube Cross-flow Heat Exchanger: ε-NTU Method

Hot exhaust gases are used in a finned-tube crossflow heat exchanger to heat 2.5 kg/s of water [c = 4.18 kJ/kg°C] from 35 to 85°C. The gases [c = 1.09 kJ/kg°C] enter at 200 and leave at 93°C. The overall heat transfer coefficient is 180 W/m²°C. Calculate the area of the heat exchanger using the ε-NTU method.

Solution

Note that this problem was tackled earlier in Example 9.1 by the LMTD-F approach.

In the ε - NTU method, the first step is to find out which is the minimum fluid. Here,

$$\dot{m}_c c_c = C_c = 2.5 \times 4180 = 10,450 \text{ W/°C}.$$

$$\dot{m}_h c_h = C_h = \frac{q}{T_{h_1} - T_{h_2}}$$

$$= \frac{\dot{m}_w c_w \Delta T_w}{T_{h_1} - T_{h_2}}$$

$$= \frac{(2.5)(4180)(85-35)}{200 - 93} = 4883 \text{ W/°C}$$

Therefore, $C_{min} = 4883$ W/°C

$C_{max} = 10,450$ W/°C

It is clearly seen that the hot exhaust gas is the minimum fluid. Therefore,

$$\text{NTU} = \frac{UA}{C_{min}} = \frac{180(A)}{4883}$$

$$= 0.03686A$$

$$\frac{C_{min}}{C_{max}} = \frac{4883}{10,450} = 0.4672$$

$$\varepsilon = \varepsilon_h = \frac{T_{h_1} - T_{h_2}}{T_{h_1} - T_{c_1}} = \frac{200 - 93}{200 - 35} = 107/165 = 0.6484$$

From the ε-NTU graph for a crossflow exchanger with both fluids unmixed [Fig. 9.7(d)], we read

$$\text{NTU} = 1.4$$

$$1.4 = 0.03686\, A$$

or $A = 1.4/0.03686 = 37.98\ \text{m}^2$

Example 9.4   Selection of Heat Exchangers: ε-NTU method

It is desired to heat 230 kg/h of water (c = 4.18 kJ/kg °C) from 35 to 93 °C with oil (c = 2.1 kJ/kg °C) having an initial temperature of 175 °C. The mass flow of oil is also 230 kg/h. Two double-pipe heat exchangers are available:

Exchanger 1: U = 570 W/m² °C, A = 0.47 m²
Exchanger 2: U = 370 W/m² °C, A = 0.94 m²

Which exchanger should be used?

Solution

Since we do not know the flow-path configuration of the double-pipe heat exchanger, let us first try parallel flow. On energy balance between the hot and cold fluids,

$$\dot{m}_w c_w (T_{w_2} - T_{w_1}) = \dot{m}_o c_o (T_{o_1} - T_{o_2})$$

or $(230)(4180)(93 - 35) = (230)(2100)(175 - T_{o_2})$

$\Rightarrow$ $T_{o_2} = 59.58\ °C$

But T_{o_2} must be greater than the exit temperature of water, that is, 93 °C. Hence, parallel flow is impossible.

Counter-flow heat exchanger

$$\dot{m}_w c_w = \frac{230 \times 4180}{3600} = 267\ \text{W/ C}$$

$$\dot{m}_o c_o = \frac{230 \times 2100}{3600} = 134.16\ \text{W/ C}$$

Therefore, $C_{min} = 134.16$ W/°C, and hence oil is the minimum fluid.

$$\frac{C_{min}}{C_{max}} = \frac{134.16}{267} = 0.5023$$

Exchanger 1:

$$\text{NTU} = \frac{UA}{C_{min}} = \frac{570 \times 0.47}{134.16} = 1.996$$

From the ε-NTU chart for a counter-flow heat exchanger [Fig. 9.7(b)],

$$\varepsilon = 0.77$$

Exchanger 2:

$$\text{NTU} = \frac{UA}{C_{min}} = \frac{370 \times 0.94}{134.16} = 2.6$$

From the ε-NTU chart for a counter-flow heat exchanger [Fig. 9.7(b)]

$$\varepsilon = 0.82$$

Since $\varepsilon_{exchanger\ 2} > \varepsilon_{exchanger\ 1}$, exchanger 2 should be used.

9.10.5 Advantages of the ε-NTU Method

1. It should now be clear that the ε-NTU method has a decided advantage over the LMTD-F approach only when U is known. In other words, if $T_{c_1}, T_{h_1}, \dot{m}_h, \dot{m}_c, c_c, c_h$, and U are given, specification of total heat transfer area A enables one to find quickly the outlet temperatures T_{c_2} and T_{h_2} without trial and error—or conversely one can find the necessary transfer area to give the desired inlet and exit temperatures of the hot and cold fluids.
2. The curves plotted for ε are not as sensitive to errors as are the F-factor curves.

However, it should be also noted that when U is not known (which is often the case since U is dependent on the inlet and exit temperatures or the average fluid temperatures) then a trial-and-error solution is still necessary. Values of the inlet and outlet temperatures (also called terminal temperatures) must be assumed to calculate U, from which terminal temperatures can be recomputed, using the ε-NTU approach.

9.11 Design Considerations for Heat Exchangers

So far we have discussed the performance characteristics of a given heat exchanger of fixed geometry. That is, if the total heat transfer area and the type of heat exchanger are known, then the exit temperatures of the fluids for various entrance conditions can be determined. Conversely, if a heat transfer rate is specified, then the required surface area can be determined.

In most cases, however, the problem involves designing or selecting a heat exchanger of unspecified dimensions in order to achieve desired transfer of heat between fluids of specified terminal temperatures. One has to understand that the problem of design is particularly complex since there is no unique answer to the problem. Several different heat exchangers may meet the design objectives equally well. The final choice depends on many factors such as cost, space requirements, experience of the designer, etc. Also, it is desirable to follow certain standard practices, such as the use of tubes of standard diameters, lengths, etc.

A detailed discussion on heat exchanger design procedures is beyond the scope of this book. Interested readers are advised to read Fraas and Ozisik (1965) and Janna (1993) in order to get an in-depth knowledge of this topic.

Example 9.5   Design of a Counter-flow, Concentric-tube Heat Exchanger

Design a counter-flow, concentric-tube heat exchanger to use water for cooling hot engine oil from an industrial power station. The mass flow rate of the oil is given as 0.2 kg/s, and its inlet temperature is 90 °C. Water is available at 20 °C, but its temperature rise is restricted to 12.5 °C because of environmental concerns. The outer tube diameter must be less than 5 cm, and the inner tube diameter must be greater than 1.5 cm due to constraints arising from space and piping considerations. The engine oil must be cooled to a temperature below 50 °C. Obtain an acceptable design, if the length of the heat exchanger must not exceed 200 m. c_p (in J/kgK), μ (in kg/sm), and k (W/mK) for oil and water are 2100, 0.03, 0.15 and 4179, 8.55×10^{-4}, 0.613, respectively. Assume the thickness of the inner tube (made of brass) to be small.

Solution

To satisfy the constraints of the problem, the outlet temperature of oil may be taken as 45°C, the inner tube diameter as 2 cm, and the outer tube diameter as 4 cm. These values may have to be adjusted if the length of the heat exchanger turns out to be greater than 200 m. The length L of the heat exchanger is calculated from

$$q = UA(\text{LMTD}) \tag{A}$$

where
$$U = \cfrac{1}{\cfrac{1}{h_i} + \cfrac{1}{h_o}} \tag{B}$$

The above equation for U has been obtained by neglecting the thermal resistance of the inner tube wall as the wall thickness is given as small and also because of high conductivity of the inner tube material.

$$A = \pi D_i L$$

where D_i is the diameter of the inner tube. Thus,

$$L = \frac{q}{\pi D_i U (\text{LMTD})} \tag{C}$$

Total rate of energy loss by oil

$$q = \dot{m}_h c_{p,h} (T_{h,i} - T_{h,o}) = 0.2 \times 2100 \times (90 - 45) = 18,900 \text{ W}$$

Assuming that the outer surface of the outer tube is perfectly insulated, the amount of energy lost by the oil is gained by water. Therefore,

$$\dot{m}_w c_{p,w} (T_{w,o} - T_{w,i}) = 18,900$$

$$T_{w,o} - T_{w,i} \leq 12.5 \degree C$$

$$\Rightarrow \quad \dot{m}_w \geq \frac{18,900}{4179 \times 12.5} = 0.36 \text{ kg/s}$$

Let us choose the mass flow rate of water as 0.4 kg/s. This gives

$$T_{w,o} = 20 + \frac{18,900}{4179 \times 0.4} = 31.3 \degree C$$

Therefore, the log-mean temperature difference is

$$\text{LMTD} = \frac{(T_{h,i} - T_{w,i}) - (T_{h,o} - T_{w,i})}{\ln \left[\dfrac{T_{h,i} - T_{h,o}}{T_{h,o} - T_{w,i}} \right]} = \frac{(90 - 31.3) - (45 - 20)}{\ln \left(\dfrac{90 - 31.3}{45 - 20} \right)} = 39.5 \degree C$$

Our next task is to determine U for which we need h_i, the inside heat transfer coefficient of the inner tube, and h_o, the heat transfer coefficient on the outside surface of the inner tube. To determine h_i we need to know whether the flow inside the tube is laminar or turbulent. For water flow in the tube,

$$\text{Re}_D = \frac{4\,\dot{m}_w}{\pi D_i \mu} = \frac{4 \times 0.4}{\pi \times 0.02 \times 8.55 \times 10^{-4}} = 2.98 \times 10^4 > 2300$$

Therefore, the flow is turbulent and a correlation such as the Dittus–Boelter correlation [see Eq. (5.231) in Chapter 5] may be used to determine the heat transfer coefficient h_i.

$$\text{Nu}_D = \frac{hD}{k_f} = 0.023\,(\text{Re}_D)^{0.8}\,(\text{Pr})^n$$

where Nu_D is the average Nusselt number. $n = 0.4$ if the fluid is being heated, and $n = 0.3$ if it is being cooled

$$\text{Pr} = \frac{\mu c_p}{k} = \frac{8.55 \times 10^{-4} \times 4179}{0.613} = 5.83$$

$$\text{Re}_D = 29{,}800$$

Hence, $\text{Nu}_D = (0.023)(29{,}800)^{0.8}\,(5.83)^{0.4} = 176.8$

$$h_i = \text{Nu}_D\,\frac{k}{D_i} = \frac{(176.8)(0.613)}{0.02} = 5418.9 \text{ W/m}^2 \text{ K}$$

For the flow of oil in the annulus, the hydraulic diameter D_h is

$$D_h = \frac{4A}{P} = \frac{4\,(\pi/4)(D_o^2 - D_i^2)}{\pi D_o + \pi D_i} = D_o - D_i = 0.04 - 0.02 = 0.02 \text{ m}$$

$$\text{Re}_{D_h} = \frac{\rho u_m D_h}{\mu}$$

where $u_m = $ Mean velocity of oil $= \dfrac{\dot{m}_h}{\rho \pi\,(D_o^2 - D_i^2)/4}$

$$\text{Re}_{D_h} = \frac{4\,\dot{m}_h}{\pi\,(D_o + D_i)\,\mu} = \frac{4 \times 0.2}{\pi\,(0.04 + 0.02) \times 0.03} = 141.5$$

Therefore, the flow in the annulus is laminar.

For $D_i/D_o = 0.02/0.04 = 0.5$, $\text{Nu}_{D_h} = 5.74$ which is obtained from Table 8.2, of Incropera and Dewitt (1998) for fully developed annular flow with one surface isothermal (non-adiabatic) and the other insulated. The Nusselt number is based on the heat transfer coefficient on the inner surface of the annulus, that is, on the outer surface of the inner tube. The table is reproduced below.

D_i/D_o	Nu_i
0.05	17.46
0.10	11.56
0.25	7.37
0.50	5.74
1.00	4.86

Therefore, $h_o = \dfrac{5.74\,k_{\text{oil}}}{D_h} = \dfrac{5.74 \times 0.15}{0.02} = 43.1 \text{ W/m}^2\text{K}$

Hence, $\quad U = \dfrac{1}{\dfrac{1}{h_i} + \dfrac{1}{h_o}} = \dfrac{1}{\dfrac{1}{5418.9} + \dfrac{1}{43.1}} = 42.8 \, \text{W/m}^2\text{K}$

Finally, from Eq. (C),

$$L = \dfrac{18{,}900}{42.8 \times \pi \times 0.02 \times 39.5} = 177.9 \, \text{m}$$

This satisfies the given requirement that the length of the heat exchanger be less than 200 m. Therefore, the design is feasible. Clearly, many other acceptable designs could have been obtained since many variables were chosen arbitrarily to satisfy the given constraints and requirements.

9.12   Compact Heat Exchangers

A special category of heat exchangers, not discussed in the earlier sections, is the compact heat exchangers which have a very high surface area per unit volume, typically greater than 700 m^2/m^3. These exchangers are most adaptable to applications where at least one side is gas and low values of h are expected. Four typical configurations of such heat exchangers are shown in Fig. 9.8 (Kays and London, 1964). In Fig. 9.8(a), a finned-tube heat exchanger with flat tubes is shown. Figure 9.8(b) depicts a circular finned-tube array. Figures 9.8(c) and (d) are the examples of heat exchangers with very high surface areas on both sides of the exchanger.

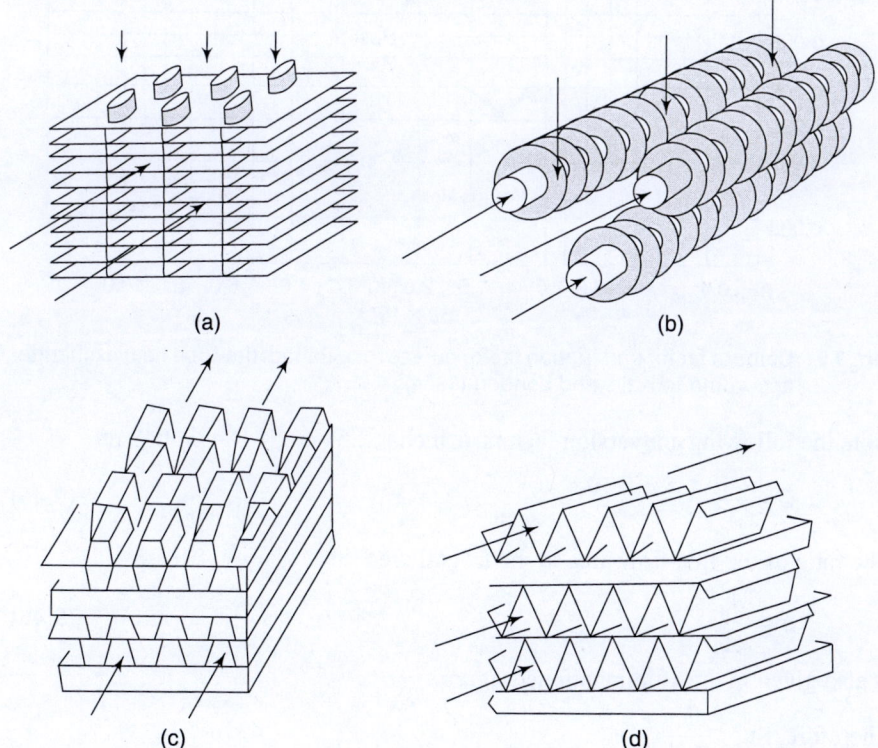

Fig. 9.8   Examples of compact heat exchanger configurations according to Kays and London (1964)

The Colburn and friction factors for a finned, flat-tube exchanger are shown in Fig. 9.9. The lower one is the Colburn factor j_H curve and the upper one is the friction factor f curve. Note that the Colburn factor is $\text{StPr}^{2/3}$. The Stanton and Reynolds numbers are based on the mass velocities in the minimum-flow cross-sectional area and a hydraulic diameter as indicated below.

Fin pitch = 9.68 per inch
Flow passage hydraulic diameter, $D_h = 0.01180$ ft
Fin metal thickness = 0.004 inch, copper
Free-flow area/frontal area, $\sigma = 0.697$
Total heat transfer area/total volume, $\alpha = 229$ ft^2/ft^3
Fin area/total area = 0.795

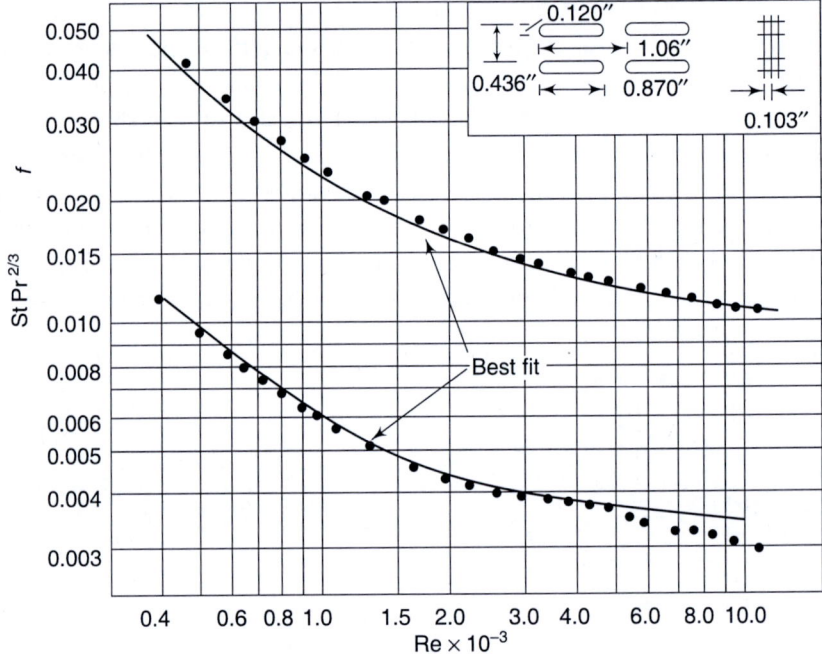

Fig. 9.9 Colburn factor and friction factor curves for a finned, flat-tube heat exchanger according to Kays and London (1964)

Note the following conversion factors: 1 inch = 2.54 cm, 1 ft = 0.3048 m.

$$G = \frac{\dot{m}}{A_c} \tag{9.45}$$

The ratio of the free-flow area to the frontal area

$$\sigma = \frac{A_c}{A} \tag{9.46}$$

is also given in the aforementioned figure.

Therefore, $\text{St} = \dfrac{h}{Gc_p}$

$$\text{Re} = \frac{D_h G}{\mu}$$

Fluid properties are evaluated at the average of the inlet and outlet temperatures. The heat transfer and friction factor *inside* the tubes are evaluated with the hydraulic diameter concept. The pressure drop across the heat exchanger core, that is, the difference between the pressure at the inlet and that at the outlet is calculated from

$$\Delta p = \frac{G^2}{2\rho_{\text{in}}}\left[f\frac{A}{A_c}\frac{\rho_{\text{in}}}{\rho} + (1+\sigma^2)\left(\frac{\rho_{\text{in}}}{\rho_{\text{out}}} - 1\right)\right] \tag{9.47}$$

In Eq. (9.47), ρ is the average density evaluated at the temperature averaged between the inlet and the outlet, $(T_{\text{in}} + T_{\text{out}})/2$. The average density can also be calculated by using the harmonic mean of ρ_{in} and ρ_{out}, that is, $(2\rho_{\text{in}}\rho_{\text{out}})/(\rho_{\text{in}} + \rho_{\text{out}})$. The f factor in the first term of the third bracket of Eq. (9.47) has been obtained experimentally and has been plotted in Fig. 9.9. The f factor accounts for the frictional losses due to fluid friction against the solid walls and for the entrance and exit losses. The second term accounts for the acceleration or deceleration of the flow. This contribution is negligible when density is essentially constant along the passage.

Design calculations of compact exchangers are quite complex and the readers are advised to refer to Kays and London (1964).

Example 9.6 Compact Heat Exchanger

Air enters a finned-tube heat exchanger of the type shown in Fig. 9.8(a) at 1 atm and 300 °C with a velocity of 15 m/s and exits with a mean temperature of 100 °C. The relevant dimensions correspond to the heat exchanger for which Fig. 9.9 was drawn. Calculate the average heat transfer coefficient on the air side.

Solution

The air densities at the inlet and outlet are obtained from Table A1.4 of Appendix A:

$\rho_{\text{in}} = 0.616 \text{ kg/m}^3$, $\rho_{\text{out}} = 0.946 \text{ kg/m}^3$

The other thermophysical properties of air at the average temperature $(300\,°C + 100\,°C)/2 = 200\,°C$ are $\rho = 0.746 \text{ kg/m}^3$, $\mu = 2.58 \times 10^{-5} \text{ kg/sm}$, $c_p = 1.025 \text{ kJ/kgK}$, and $\text{Pr} = 0.68$. The heat transfer coefficient is obtained by using the lower graph of Fig. 9.9, for which we know

$$\frac{A_c}{A_{\text{fr}}} = \sigma = 0.697$$

Note that A_c refers to the minimum free-flow area which corresponds to the smallest cross-section encountered by the fluid. In Fig. 9.8(a) this area occurs between two adjacent tubes aligned on the vertical: A_c is in the same plane of the two centre lines. A_{fr} refers to the frontal area. The dimensionless parameter σ accounts for the contraction and enlargement experienced by the flow stream.

$D_h = 0.0118 \text{ ft} = 0.003597 \text{ m}$

$$G = \frac{\dot{m}}{A_c} = \frac{\rho u_\infty A_{\text{fr}}}{A_c} = \frac{(0.746)(15)}{0.697} = 16.05 \text{ kg/m}^2\text{s}$$

$$\text{Re} = \frac{D_h G}{\mu} = \frac{(0.003597)(16.05)}{2.58 \times 10^{-5}} = 2.237 \times 10^3$$

From Fig. 9.9,

$$j_H = \text{StPr}^{2/3} = 0.0042 = \frac{h}{Gc_p}\text{Pr}^{2/3}$$

Therefore, the heat transfer coefficient is

$$h = (0.0042)(16.05)(1025)(0.68)^{-2/3} = 89.34 \text{ W/m}^2\text{K}$$

Example 9.7 Effectiveness and LMTD of a Heat Exchanger

A double-pipe counter-flow heat exchanger transfers heat between two water streams. If 19 l/s of water is heated in the tubes from 10 to 40°C with 25 l/s of inlet water on the shell side at 46°C, then what are the effectiveness and LMTD of the heat exchanger? The specific heat of water is 4.18 kJ/kg K. 1 l = 1000 cm³. The density of water is 1000 kg/m³.

Solution

T vs. x for the heat exchanger is shown in Fig. E9.7.

$$\dot{m}_c = (19)(1000)(10^{-6})(10^3) = 19 \text{ kg/s}$$

$$\dot{m}_h = (25)(1000)(10^{-6})(10^3) = 25 \text{ kg/s}$$

$$\dot{m}_c\, c = (19)(4180) = 79{,}420 \text{ W/K}$$

$$\dot{m}_h\, c = (25)(4180) = 1{,}04{,}500 \text{ W/K}$$

Thus we see the cold water is the minimum fluid.
Hence, $C_{min} = 79420$ W/K

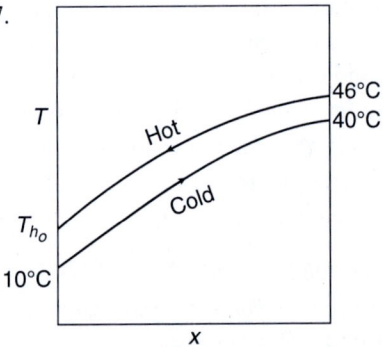

$$\varepsilon_c = \frac{T_{c_o} - T_{c_i}}{T_{h_i} - T_{c_i}} = \frac{40-10}{46-10} = \frac{30}{36} = 0.83$$

From the energy balance,

Fig. E9.7 Axial temperature profiles in the hot and cold water streams in the double-pipe counter-flow heat exchanger

$$\dot{m}_c\, c_c\,(40-10) = \dot{m}_h\, c_h\,(46-T_h)$$

$$\Rightarrow \qquad 79420(30) = (104500)(46-T_{h_o})$$

$$\Rightarrow \qquad\qquad T_{h_o} = 23.2^\circ \text{ C}$$

We also know,

$$LMTD = \frac{GTD - LTD}{\ln\left(\dfrac{GTD}{LTD}\right)}$$

In this problem,

$$GTD = 23.2 - 10 = 13.2°C$$
$$LTD = 46 - 40 = 6°C$$

Therefore, $$LMTD = \frac{13.2-6}{\ln\left(\dfrac{13.2}{6}\right)} = \frac{7.2}{\ln(2.2)} = \frac{7.2}{0.788} = 9.137 \approx 9.14 \text{ °C}$$

Final answers:

$$\varepsilon_c = 0.83$$
$$LMTD = 9.14°C$$

Example 9.8 Exit Temperatures of Hot and Cold Fluids: ε-NTU Method

Water is heated in a building from 20°C at a rate of 84 kg/min by using inlet hot water at 110°C in a single-pass counter-flow heat exchanger.

(a) *Find the heat transfer if the hot water flow is 108 kg/min.*

(b) *Calculate the exit temperatures of both the streams.*

The overall heat transfer coefficient is 320 W/m² K and the heat transfer area is 20 m². The specific heat of water is 4.18 kJ/kg K.

Solution

T vs. x for the heat exchanger is shown in Fig. E9.8.

(a)
$$\dot{m}_c\, c_c = \left(\frac{84}{60}\right)(4180) = 5852 \text{ W/K}$$

$$\dot{m}_h\, c_h = \left(\frac{108}{60}\right)(4180) = 7524 \text{W/K}$$

From the above, we see that cold stream is the minimum fluid.

Hence,
$$\frac{C_{min}}{C_{max}} = \frac{5852}{7524} = 0.778$$

$$NTU = \frac{UA}{C_{min}} = \frac{(320)(20)}{5852} = 1.094$$

From Fig. 9.7(b), we get
$$\varepsilon = 0.52$$

From Eq. (9.42),
$$\varepsilon = 0.55$$

We take the value predicted by Eq.(9.42) as it is more accurate.

Thus,
$$q = \varepsilon C_{min}\,(T_{h_i} - T_{c_i})$$

$$= (0.55)(5852)(110 - 20)$$

$$= 289674 \text{ W} \approx 289.7 \text{ kW}$$

(b) Also, $q = C_{max}\,(T_{h_i} - T_{h_o})$

$$\Rightarrow \qquad T_{h_o} = T_{h_i} - \frac{q}{C_{max}}$$

$$= 110 - \frac{289674}{7524} = 110 - 38.5 = 71.5°C$$

Again,
$$q = C_{min}\,(T_{c_o} - T_{c_i})$$

$$\Rightarrow \qquad T_{c_o} = T_{c_i} + \frac{q}{C_{mi}}$$

$$= 20 + \frac{289674}{5852} = 20 + 49.5 = 69.5°C$$

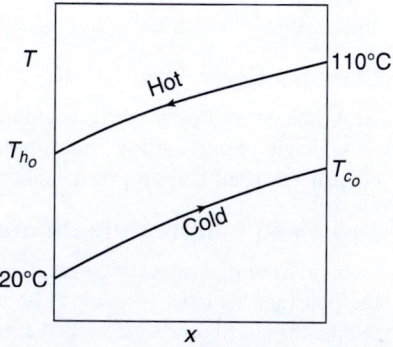

Fig. E9.8 Axial temperature distributions in the hot and cold water streams in the single-pass counter-flow heat exchanger

Important Concepts and Expressions

Heat Exchangers

A heat exchanger is an equipment that is used to transfer heat from one fluid to another, usually through a separating wall. Heat exchangers are employed in a variety of applications, such as, steam power plants, chemical processing plants, building heating, air conditioning, refrigeration systems, and mobile power plants for automotive, marine, and aerospace vehicles.

Classification

Heat exchangers are classified according to fluid flow arrangement (parallel flow, counter-flow, single-pass crossflow, multi-pass crossflow) and types of applications (boilers, condensers, shell-and-tube heat exchangers, radiators, double-pipe heat exchangers).

Log-Mean Temperature Difference

In the design of heat exchangers, the mean effective temperature difference between the two fluid streams is much more useful than the detailed axial distribution of the temperature difference between the hot and cold fluid streams. This mean effective temperature difference involves the natural logarithm of the ratio of the temperature differences at the two ends of the heat exchanger, and thus has come to be known as the log-mean temperature difference, or LMTD.

$$\text{LMTD} = \frac{GTD - LTD}{\ln \dfrac{GTD}{LTD}}$$

GTD is the greatest of the differences between the temperatures of the hot and the cold fluids at the inlet and exit of a heat exchanger.

LTD is the least of the differences between temperatures of the hot and the cold fluids at the inlet and exit of a heat exchanger.

The above expression applies to either parallel-flow or counter-flow heat exchangers.

Heat Transfer as a Function of LMTD

Once the LMTD is determined, the heat transfer for the heat exchanger as a whole is evaluated from

$$q = UA(\text{LMTD})$$

Multi-pass and Crossflow Heat Exchangers: Correction Factor Approach

If a heat exchanger other than the double-pipe type is used, the heat transfer is calculated by using a correction factor applied to the LMTD for a counter-flow double pipe arrangement with the same hot and cold fluid temperatures.

Then $Q = UAF(\text{LMTD})$

Values of the correction factor F are plotted for several different types of heat exchanger.

Effectiveness-NTU (Number of Transfer Units) Method

The LMTD method is useful when the inlet and exit temperatures are known or are easily determined. But when the inlet and exit temperatures are to be evaluated for a given heat exchanger, the analysis frequently involves an iterative procedure because of the logarithm function in the LMTD. In such cases the effectiveness-NTU method is more convenient as it is a direct method and no iterations are required to obtain a solution when the overall heat transfer coefficient is known.

The heat transfer effectiveness ε is defined as

ε = Actual Heat Transfer/Maximum Possible Heat Transfer

The maximum possible heat transfer is defined as the maximum value of heat transfer that could be attained if one of the fluids were to undergo a temperature change equal to the maximum temperature difference present in the heat exchanger, which is the difference between the entry temperatures of the hot and cold fluids. The fluid that might undergo this maximum temperature difference is that having $\left|\, \dot{m}c \,\right|$ since the energy balance requires that the energy received by one fluid is that given up by the other fluid. If we let the fluid with larger $\dot{m}c$ go through a maximum temperature difference, this would require that the other fluid undergo a temperature difference greater than the maximum and this is impossible. Thus,

$$q_{max} = \left(\dot{m}c\right)\ (T_{h_{inlet}} - T_{c_{inlet}})$$

The minimum fluid may be either hot or cold fluid, depending on the product of the mass flow rate and specific heat of the hot and cold fluids.

The concept of effectiveness also provides us with a yardstick to compare various types of heat exchangers in order to select the type best suited to accomplish a particular heat transfer objective.

Expression for the Effectiveness of Parallel Flow Heat Exchangers

$$\varepsilon = \frac{1 - \exp\left[-\left(\dfrac{UA}{C_{min}}\right)\left(1 + \dfrac{C_{min}}{C_{max}}\right)\right]}{\left(1 + \dfrac{C_{min}}{C_{max}}\right)}$$

where $C_{min} = \left|\, \dot{m}c \,\right|$

The expression UA/C_{min} is called number of transfer units (which is abbreviated as NTU) since it is representative of the size of the heat exchanger.

Physical Significance of NTU

$$NTU = \frac{UA}{C_{min}}$$

= (Heat exchanging capacity per degree of mean temperature difference)/
(Heat transferred per degree of temperature rise, to or from either fluid)

For a fixed U/C_{min}, NTU is a measure of the actual heat transfer area A. Higher the NTU, larger is the physical size.

NTU is dimensionless.

Expression for the Effectiveness of Counter-flow Heat Exchangers

$$\varepsilon = \frac{1 - \exp\left[-NTU(1 - C_R)\right]}{1 - C_R \exp\left[-NTU(1 - C_R)\right]}$$

where $C_R = \dfrac{C_{min}}{C_{max}}$

Expression for the Effectiveness of Boilers and Condensers

In boiling or condensation the fluid temperature stays essentially constant or the fluid behaves as if it had infinite heat capacity (in other words, infinite specific heat). In these cases, $C_{min}/C_{max} \to 0$ and $C_{max} \to \infty$ and all the heat exchanger effectiveness-NTU relations approach a single simple equation,

$$\varepsilon = 1 - \exp(-NTU)$$

Effectiveness-NTU Charts

Kays and London (1964) have presented effectiveness-NTU (ε-NTU) charts for various heat exchangers. While these charts have great practical use in design problems, in the applications demanding more precision in the calculations the design procedures should be computer-based, requiring analytical expressions for these curves.

Compact Heat Exchangers

The compact heat exchangers have a very high surface area per unit volume, typically greater than 700 m^2/m^3. These exchangers are most adaptable to applications where at least one side is gas and low values of h are expected.

Review Questions

9.1 Define a heat exchanger.

9.2 How are heat exchangers classified?

9.3 Sketch a shell-tube heat exchanger with baffles.

9.4 What is fouling factor?

9.5 Draw qualitatively axial temperature distributions in parallel-flow, counter-flow heat exchangers, and boilers and condensers.

9.6 Define LMTD.

9.7 How is the correction factor method used?

9.8 Define effectiveness and NTU. Explain the physical significance of NTU.

9.9 What are the advantages of the ε-NTU method?

9.10 What is a compact heat exchanger? Show an example of compact heat exchangers.

Problems

9.1 The overall heat transfer coefficient of a steam condenser (based on the outer surface area of the tube in its early running condition) is 3690 W/m^2K. After the condenser has operated for a long period the fouling factor is evaluated as 0.0002 m^2K/W. What is the present value of the overall heat transfer coefficient?

9.2 In a one-tube-pass, one-shell-pass exchanger, the hot fluid enters at $T_{h_i} = 425°C$ and leaves at $T_{h_o} = 260°C$, while the cold fluid enters at $T_{c_i} = 40°C$ and leaves at $T_{c_o} = 150°C$. Find LMTD if the heat exchanger is arranged for (a) parallel flow, (b) counter-flow. Which arrangement will transfer more heat? How much is the percentage difference?

9.3 A one-shell-pass, two-tube-pass heat exchanger has a total surface area of 5 m^2 and its overall heat transfer coefficient based on that area is known to be 1400 $W/m^2°C$. If 4500 kg/h of water enters the shell side at 315°C while 9000 kg/h of water enters the tube side at 40°C, find the outlet temperature using (a) LMTD, (b) the ε-NTU method. Take specific heat for both fluids to be 4.187 kJ/kg°C. [*Hint:* Use a trial-and-error procedure for part (a).]

9.4 Hot exhaust gases, which enter a finned-tube, crossflow heat exchanger at 300°C and leave at 100°C, are used to heat pressurized water at a flow rate of 1 kg/s from 35°C to 125°C. The exhaust gas specific heat is approximately 1000 J/kgK, and the overall heat transfer coefficient based on the gas side surface area is $U_h = 100$ W/m²K. Determine the required gas side surface area A_h by the ε-NTU method.

9.5 A procedure for open-heart surgery under hypothermic conditions involves cooling patient's blood before the surgery and rewarming it following surgery. It is proposed that a double-pipe, counter-flow heat exchanger of length 0.5 m be used for this purpose with the thin-walled inner pipe having a diameter of 55 mm. If water at a temperature of 60°C and a flow rate of 0.1 kg/s is used to heat blood entering the exchanger with a temperature of 18°C and a flow rate of 0.05 kg/s, what is the temperature of the blood leaving the exchanger? The overall heat transfer coefficient is 500 W/m²K, and the specific heat of blood is 3500 J/kgK and that of water is 4200 J/kgK.

9.6 Suppose the overall heat transfer coefficient U varies greatly from one end of a heat exchanger to the other. The variation of U may be represented by a linear function of the temperature difference, $U = a + b\Delta T$. Show that the total heat transfer is given by

$$q = A \left[\frac{U_2 \Delta T_1 - U_1 \Delta T_2}{\ln\left(\dfrac{U_2 \Delta T_1}{U_1 \Delta T_2}\right)} \right]$$

Subscripts 1 and 2 represent the inlet and exit of the exchanger.

9.7 In terms of parameters R and P show that the correction factor F defined for a parallel-flow heat exchanger is

$$F = \frac{R+1}{R-1} \cdot \frac{\ln\left(\dfrac{1-P}{1-PR}\right)}{\ln\left(\dfrac{1}{1-P(R+1)}\right)}$$

where R = capacity ratio = $\dfrac{\dot{m}_c c_c}{\dot{m}_h c_h} = \dfrac{T_{h_i} - T_{h_o}}{T_{c_o} - T_{c_i}}$

P = effectiveness

$$= \frac{\text{total heat transferred to the cold fluid}}{\begin{array}{c}\text{heat that would be transferred to the cold fluid}\\\text{if it were raised to the inlet temperature of the hot fluid}\end{array}}$$

$$= \frac{\dot{m}_c c_c (T_{c_o} - T_{c_i})}{\dot{m}_c c_c (T_{h_i} - T_{c_i})} = \frac{T_{c_o} - T_{c_i}}{T_{h_i} - T_{c_i}}$$

and $F(P, R) = \dfrac{(\Delta T_m)_{\text{parallel flow}}}{(\Delta T_m)_{\text{counter-flow}}}$

9.8 A heat exchanger is to be designed to heat 6800 kg/h of water from 20°C to 38°C by the use of saturated steam condensing at 240 kN/m². The tubes are to be of 2.5 cm diameter and are not to exceed 2 m in length.

The tube water velocity is 0.45 m/s. The overall heat transfer coefficient is 3700 W/m²K. Determine the number of tubes per pass, the number of tube passes, and the length of the tubes.

9.9   A one-shell-pass, one-tube-pass heat exchanger is made of 60 brass tubes (5/2 inch outer diameter, 0.083 inch thickness). Water enters the shell side at 150°C and leaves at 40°C, flowing at the rate of 9000 kg/h. The tube water enters at 25°C and leaves at 65°C. The inner and outer tube surface heat transfer coefficients are 1700 and 8500 W/m²K, respectively. Find the required length of the tubes.

9.10 An unfinned, crossflow heat exchanger (one fluid mixed, other unmixed) is used to heat water from 35°C to 85°C, flowing in tubes at the rate of 3 kg/s, with atmospheric pressure air being cooled from 225°C to 100°C. If the overall heat transfer coefficient is 210 W/m²°C, find the surface area required. Which fluid is mixed and which fluid is unmixed in this arrangement? Explain.

9.11 A double-pipe counter-flow heat exchanger transfers heat between two water streams. If 19 L/s is heated in the tubes from 10 to 38°C with 25 L/s of inlet water on the shell side at 46°C, then what are the effectiveness and LMTD?

9.12 Solve Question 9.11 if saturated steam at 105°C is used on the shell side. The outlet of the shell side of the heat exchanger is saturated water at 105°C.

9.13 Solve Question 9.11, but assume that the fluid being heated is ethylene glycol.

9.14 A counter-flow heat exchanger has an entering hot water stream at 12 kg/s and 60°C and a cold stream at 15 kg/s and 12°C. What is the effectiveness if the cold water stream leaves at 40°C? What is the overall heat transfer conductance $U_0 A_0$?

9.15 If the heat transfer surface in Question 9.14 becomes fouled by a substance that decreases the overall heat transfer conductance by 15%, what are the effectiveness and heat transfer rate under the same inlet-water conditions? By how much must the hot water inlet temperature be increased to produce the same heat transfer rate as with the clean heat exchanger?

9.16 Suppose the heat exchanger in Question 9.14 is a parallel-flow heat exchanger. What will be the effectiveness if the value of $U_0 A_0$ is unchanged?

9.17 Water is heated in a building from 20°C at a rate of 70 kg/min by using hot water from a boiler at 110°C in a single-pass counter-flow heat exchanger. Find the heat transfer rate if the hot water flow is 90 kg/min. Also find the exit temperatures of both the streams. The overall heat transfer coefficient is 320 W/m²K and the heat transfer area is 20 m².

9.18 A two-shell-pass, four-tube-pass heat exchanger [Fig. 9.6(b)] is used to heat water with hot exhaust gas. Water enters the tubes at 50°C and leaves it at 125°C with a flow rate of 10 kg/s, while the hot exhaust gas enters the shell side at 300°C and leaves at 125°C. The total heat transfer surface area is 800 m². Calculate the overall heat transfer coefficient.

9.19 A single-pass, crossflow heat exchanger with the flow arrangement as shown in Fig. 9.6(e) is used to heat water flowing at a rate of 0.5 kg/s from 25°C to 80°C by using oil ($c_p = 2100$ J/kgK). Oil enters the tubes at 175°C at the same rate as that of water. The overall heat transfer coefficient is 300 W/m²K. Calculate the required heat transfer surface area.

9.20 A crossflow heat exchanger as shown in Fig. 9.7(d), has a heat transfer surface area of 12 m². It is used to heat air entering at 10°C at 3 kg/s with hot water entering at 80°C at 0.4 kg/s. The overall heat transfer coefficient can be taken as 300 W/m²K. Calculate the total heat transfer rate and the exit temperatures of the air and water.

Chapter 10

Finite-difference Methods in Heat Conduction

10.1 Introduction

In the preceding chapters the readers have been initiated into analytical solutions of heat transfer. One clearly sees that there are only a handful of exact solutions available in the classical literature of heat transfer. In actual situations, problems are lot more complex, as in those involving non-linear governing equations and/ or boundary conditions, and irregular geometry. These do not allow analytical solutions to be obtained. Therefore, it is necessary to use numerical techniques for most problems of practical interest. In this chapter only numerical methods in heat conduction by finite-difference techniques are introduced. An interesting application of numerical heat conduction to biomedical engineering is given. The Online Resource Centre of the book contains computer programs and/or solutions to some questions and typical examples in this chapter.

The finite-difference method enables one to integrate a differential equation numerically by evaluating the values of a function at a discrete (finite) number of points. The origin of this method is Taylor series expansion, which assumes that the function is smooth, i.e., continuous and differentiable.

At this point, a reader uninitiated into the numerical methods may ask: What is the role of computer here? Such a query is expected and needs some attention. Most of us seem to forget that some of the numerical schemes (e.g., finite difference) that are extensively used today for the solution of problems on computer were developed when computer was not even invented. Now, to return to the original question, the answer is that with the aid of the algorithm of the solution method translated into a programming language, such as FORTRAN, fed into a computer (which does arithmetic operations at a tremendous speed) one can obtain the solution of mathematical equations in seconds or even in a fraction of a second. A simple example will clarify this point. One can very easily solve a set of three linear simultaneous equations manually through the Gaussian elimination method. Typically, in this method, for a system of n equations, the total number of multiplications and divisions is roughly $1/3\ n^3$. Therefore, for $n = 3$, the number of operations is 9, which is clearly manageable manually. However, for $n = 10$, this number jumps to 333. For $n = 100$, the number skyrockets to 333,000. A mainframe computer with an average megaflops (A *mega* is a million and *flops* is an abbreviation for floating-point operations per second. A floating-point operation is an arithmetic operation on operands which are real numbers with fractional parts. Normally multiplication and division are counted as major arithmetic operations as compared to addition and subtraction on a computer.) rating of 1 (i.e., 10^6 arithmetic operations per second) will solve the aforementioned problem in 0.333 s. In computer simulations, it is possible to handle one hundred or more (even greater than a thousand) number of algebraic equations. This is the reason why computers are used in solving complex problems.

10.2   Introduction to Finite-difference, Numerical Errors, and Accuracy

The basic approach in solving a problem by computer methods using a finite-difference scheme is to discretize the derivatives appearing in the governing differential equation at a finite number of uniformly or non-uniformly spaced grid points that fill the computational domain. The governing differential equation is then transformed into a system of difference equations. This means that if there are 100 grid points (where variables are not known) there will be 100 equations to solve per variable. The necessity of using a computer arises because of the huge number of arithmetic operations that are required to be carried out in a reasonable time for solving a large number of equations. The simplification inherent in the use of algebraic equations rather than differential equations is what makes numerical methods so powerful and widely applicable. In the following subsection three difference schemes, namely, central-, forward-, and backward-difference expressions for first- and second-order derivatives are derived for a uniform grid, that is, when grid points are uniformly spaced.

10.2.1   Central-, Forward-, and Backward-difference Expressions for a Uniform Grid

10.2.1.1   Central difference

Consider a smooth function $y = f(x)$ as shown in Fig. 10.1. The Taylor series for the said function at $x_i + h$ expanded about x_i is

$$y(x_i + h) = y_i + y_i'h + \frac{y_i''h^2}{2!} + \frac{y_i'''h^3}{3!} + \cdots \tag{10.1}$$

where $h = \Delta x$. y_i, y_{i+1}, and y_{i-1} are the ordinates corresponding to $x_i, x_i + h$, and $x_i - h$, respectively. The function at $(x_i - h)$ is similarly given by

$$y(x_i - h) = y_i - y_i'h + \frac{y_i''h^2}{2!} - \frac{y_i'''h^3}{3!} + \cdots \tag{10.2}$$

Subtracting Eq. (10.2) from Eq. (10.1), we obtain

$$y(x_i + h) - y(x_i - h) = 2y_i'h + \frac{2y_i'''h^3}{3!} + \cdots$$

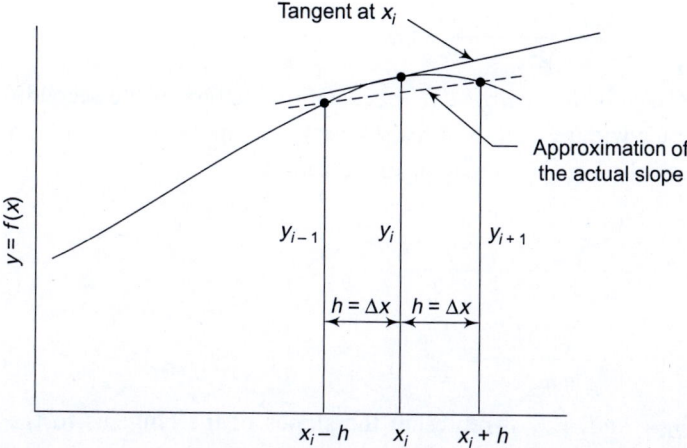

Fig. 10.1 Uniformly spaced grid points on a continuous and differentiable function $y = f(x)$

or
$$2y_i'h = y(x_i + h) - y(x_i - h) - \frac{y_i'''h^3}{3} + \cdots$$

Therefore,
$$y_i' = \frac{y(x_i + h) - y(x_i - h)}{2h} - \frac{1}{6}(y_i'''h^2) + (\text{higher-order terms})$$

or
$$y_i' = \frac{y_{i+1} - y_{i-1}}{2h} + (\text{error of order } h^2)$$

$$y_i' = \frac{y_{i+1} - y_{i-1}}{2h} + 0(h^2) \qquad\qquad (10.3)$$

The notation $0(h^2)$ means that in arriving at Eq. (10.3), terms of the order of h^2 and higher have been neglected. $0(h^2)$ is called the *truncation error*. The truncation error is the difference between the exact mathematical expression and its numerical approximation.

Equation (10.3) is called the *central-difference* approximation of y' (i.e., dy/dx) at x_i with an error of order h^2. In Fig. 10.1, the approximation is depicted by the slope of the dashed line. The actual derivative is shown by the solid line drawn tangent to the curve at x_i. The difference can be viewed as due to the truncation error resulting from using a truncated Taylor series.

Now, adding Eqs (10.1) and (10.2), we get

$$y(x_i + h) + y(x_i - h) = 2y_i + y_i''h^2 + \frac{y_i^{iv}h^4}{12} + \cdots$$

or
$$y_i''h^2 = y(x_i + h) + y(x_i - h) - 2y_i - \frac{y_i^{iv}h^4}{12} + (\text{higher-order terms})$$

or
$$y_i'' = \frac{y(x_i + h) - 2y_i + y(x_i - h)}{h^2} - \frac{y_i^{iv}h^4}{12} + (\text{higher-order terms})$$

$$y_i'' = \frac{y_{i-1} - 2y_i + y_{i-1}}{h^2} + 0(h^2) \tag{10.4}$$

Equation (10.4) is the *central-difference* approximation of the second derivative of the function with respect to x (i.e., d^2y/dx^2) evaluated at x_i with an error of order h^2. Alternatively, Eq. (10.4) may be expressed as

$$y_i'' = \frac{\dfrac{y_{i+1} - y_i}{h} - \dfrac{y_i - y_{i-1}}{h}}{h}$$

or

$$y_i'' = \frac{y_{i+1/2}' - y_{i-1/2}'}{h} \tag{10.5}$$

where $y_{i+1/2}'$ and $y_{i-1/2}'$ represent the slopes of the tangents to the curve at $x_i + h/2$ and $x_i - h/2$, respectively.

The aforederived central-difference expressions reveal that the first and second derivatives of the function involve values of the function on both sides of the x-value at which the derivative of the function is to be evaluated.

10.2.1.2 Forward-difference

From Taylor series expansions, it is also easy to obtain expressions for the derivatives which are entirely in terms of values of the function at x_i and points to the right of x_i. These are called *forward-difference* expressions.

Starting from the Taylor series expansion as given in Eq. (10.1), we get

$$y_i'h = y(x_i + h) - y_i - \frac{y_i''h^2}{2!} - \frac{y_i'''h^3}{3!} - \cdots$$

or

$$y_i'h = \frac{y(x_i + h) - y_i}{h} - \frac{y_i''h^2}{2!} - \frac{y_i'''h^3}{3!} - \cdots$$

Dropping terms of the order of h and higher,

$$y_i' = \frac{y(x_i + h) - y_i}{h} + 0(h) \tag{10.6}$$

Similarly,

$$y(x_i + 2h) = y_i + y_i'(2h) + \frac{y_i''(2h)^2}{2!} + \frac{y_i'''(2h)^3}{3!} + \cdots$$

$$= y_i + y_i'(2h) + 2y_i''h^2 + 0(h^3) \tag{10.7}$$

Also,

$$y(x_i + h) = y_i + y_i'h + \frac{y_i'''h^2}{2!} + 0(h^3) \tag{10.8}$$

Multiplying Eq. (10.8) by 2 and substracting from Eq. (10.7) gives

$$y_i'' = \frac{y(x_i + 2h) - 2y(x_i + h) + y_i}{h^2} + 0(h)$$

$$y_i'' = \frac{y_{i+2} - 2y_{i+1} + y_i}{h^2} + 0(h) \tag{10.9}$$

10.2.1.3 Backward-difference

Following the approach given in Section 10.2.1.2, one can easily obtain derivative expressions which are entirely in terms of the values of the function at x_i and points to the left of x_i. These are known as backward-difference expressions, which are given below for y_i' and y_i''.

$$y_i' = \frac{y_i - y_{i-1}}{h} + 0(h) \tag{10.10}$$

$$y_i'' = \frac{y_i - 2y_{i-1} + y_{i-2}}{h^2} + 0(h) \tag{10.11}$$

The readers are encouraged to derive Eqs (10.10) and (10.11) themselves.

10.2.1.4 Conditions for using forward-, backward-, and central-difference expressions

- Forward-difference expressions are used when data to the left of a point, at which a derivative is desired, are not available.
- Backward-difference expressions are used when data to the right of the desired point are not available.
- Central-difference expressions are used when data on both sides of the desired point are available and are more accurate than either forward- or backward-difference expressions.

10.2.1.5 Difference expressions of higher accuracy

By retaining a greater number of terms in the Taylor series, it is possible to obtain forward-, backward-, and central-difference expressions for a higher accuracy. The following expressions show central-difference expressions for y_i' and y_i'' with an error of $0(h^4)$ and forward- and backward-difference expressions for the same with an error of $0(h^2)$. It is apparent that for a greater accuracy more number of neighbouring points are involved. For example, Eq. (10.10) is a two-point forward-difference scheme for y_i', while Eq. (10.14) is a three-point forward-difference scheme. Central difference with an error of $0(h^4)$:

$$y_i' = \frac{-y_{i+2} + 8y_{i+1} - 8y_{i-1} + y_{i-2}}{12h} \tag{10.12}$$

$$y_i'' = \frac{-y_{i+2} + 16y_{i+1} - 30y_i + 16y_{i-1} - y_{i-2}}{12h^2} \tag{10.13}$$

Forward difference with an error of $0(h^2)$:

$$y_i' = \frac{-y_{i+2} + 4y_{i+1} - 3y_i}{2h} \tag{10.14}$$

$$y_i'' = \frac{-y_{i+3} + 4y_{i+2} - 5y_{i+1} + 2y_i}{h^2} \tag{10.15}$$

Backward difference with an error of $0(h^2)$:

$$y_i' = \frac{3y_i - 4y_{i-1} + y_{i-2}}{2h} \tag{10.16}$$

$$y_i'' = \frac{2y_i - 5y_{i-1} + 4y_{i-2} - y_{i-3}}{h^2} \tag{10.17}$$

Now, the question is: When does one use a higher-order difference scheme? There is no set answer to this. It depends on the accuracy requirement of a problem, and the analyst will have to use his own judgement.

10.2.2 Numerical Errors

Three most important errors that commonly occur in numerical solutions are the (a) round-off error (b) truncation error, and (c) discretization error.

Round-off error The round-off error is introduced because of the inability of the computer to handle a large number of significant digits. Typically, in single-precision, the number of significant digits retained ranges from 7 to 16, although it may vary from one computer system to another. The round-off error arises due to the fact that a finite number of significant digits or decimal places are retained and all real numbers are rounded off by the computer. The last retained digit is rounded off if the first discarded digit is equal to or greater than 5. Otherwise, it is unchanged. For example, if five significant digits are to be kept in place, 5.37527 is rounded off to 5.3753, and 5.37524 to 5.3752.

Truncation error The truncation error is due to the replacement of an exact mathematical expression by a numerical approximation. This error has been discussed earlier in this chapter with respect to finite-difference approximations. Basically, it is the difference between an exact expression and the corresponding truncated form (for example, truncated Taylor series) used in the numerical solution.

Discretization error The discretization error is the error in the overall solution that results from the truncation error assuming the round-off error to be negligible. Therefore,

(Discretization error) = (exact solution) – (numerical solution

with no round-off error)

10.2.3 Accuracy of a Solution: Optimum Step Size

The accuracy of a numerical solution is determined by its total error, which is the sum of the round-off error and truncation error. Hence, (total error) = (round-off error) + (truncation error). However, it is obvious that the round-off error increases as the total number of arithmetic operations increases. Again, the total number of arithmetic operations increases if the step size decreases (that is, when the number of grid points increases). Therefore, the round-off error is inversely proportional to the step size. On the other hand, the truncation error decreases as the step size decreases (or as the number of grid points increases).

Because of the aforementioned opposing effects, an optimum step size is expected, which will produce minimum total error in the overall solution.

10.2.4 Method of Choosing Optimum Step Size: Grid Independence Test

A numerical analyst has to be extremely careful as regards the accuracy of a solution. To get the most accurate solution (i.e., the solution with the least total error), one has to perform a grid independence test. The test is carried out by experimenting with various grid sizes and watching how the solution changes with respect to the changes in the grid size. Finally, a stage will come when changing the grid spacings will not affect the solution. In other words, the solution will become independent of grid spacing. The largest value of grid spacings for which the solution is essentially independent of the step size is chosen so that both the computational time and effort and the round-off error are minimized.

10.3 Numerical Methods for Conduction Heat Transfer

Many difficult problems arise in conduction, for example, variable thermal conductivity, distributed energy sources, radiation boundary conditions, for which analytical solutions are not available. Approximate solutions for these are obtained by numerical methods. The basic approach is to arrive at the relevant governing differential equation based on the physics of the particular problem. They are then converted into the required finite-difference forms. To begin with, the numerical solution procedure for the problem of a simple one-dimensional steady-state heat conduction in a cooling fin is described. It is to be noted that a simple, closed-form straightforward analytical solution for this problem is available. The idea is to show the use of the numerical method and to compare the numerical solution with its analytical counterpart.

10.3.1 Numerical Methods for a One-dimensional Steady-state Problem

Consider one-dimensional, steady-state heat conduction in an isolated rectangular horizontal fin as shown in Fig. 10.2. The base temperature is maintained at $T = T_0$ and the tip of the fin is insulated. The

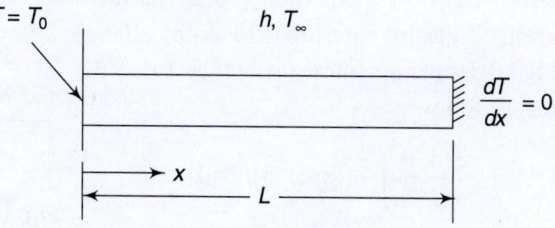

Fig. 10.2 Physical domain of a rectangular fin

fin is exposed to a convective environment (neglecting radiation heat transfer from the fin) which is at T_∞ ($T_\infty < T_0$). The average heat transfer coefficient of the fin to the surroundings is h. The length of the fin is L and the coordinate axis begins at the base of the fin. The one-dimensionality arises from the fact that the thickness of the fin is much small as compared to its length, and the width can be considered either too long or the sides of the fin to be insulated.

Governing differential equation The energy equation for the fin in the steady state (assuming constant k) is

$$\frac{d^2T}{dx^2} - \frac{hP}{kA}(T - T_\infty) = 0 \tag{10.18}$$

where P and A are the perimeter and the cross-sectional area of the fin, respectively.

Boundary conditions Since Eq. (10.18) is a linear, second-order ordinary differential equation, two boundary conditions are needed to completely describe this problem. Boundary conditions are as follows.

$$\text{BC-1: at } x = 0, \ T = T_0 \tag{10.19a}$$

$$\text{BC-2: at } x = L, \ \frac{dT}{dx} = 0 \tag{10.19b}$$

Non-dimensionalization Non-dimensionalizing Eqs (10.18) and (10.19) and using the dimensionless variables

$$\theta = \frac{T - T_\infty}{T_0 - T_\infty}, X = \frac{x}{L},$$

we obtain

$$\frac{d^2\theta}{dx^2} - (mL)^2 \theta = 0 \tag{10.20}$$

where $m^2 = \dfrac{hP}{kA}$

and $\theta(0) = 1$ \hfill (10.21a)

$\theta'(1) = 0$ \hfill (10.21b)

where $\theta' = \dfrac{d\theta}{dX}$

Discretization Equation (10.20) is discretized at any interior grid point i (see Fig.10.3) using central difference for $d^2\theta/dX^2$ as follows:

$$\left(\frac{d^2\theta}{dX^2}\right)_i - [(mL)^2\theta]_i = 0$$

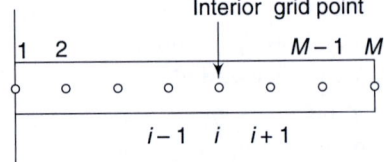

or

$$\frac{\theta_{i-1} - 2\theta_i + \theta_{i+1}}{(\Delta X)^2} - (mL)^2\theta_i = 0$$

Fig. 10.3 Computational domain of the fin with equally spaced grid points

or $\theta_{i+1} - D\theta_i + \theta_{i+1} = 0, i = 2, \ldots, M$ \hfill (10.22)

where $D = 2 + (mL)^2(\Delta X)^2$.

Handling of the boundary condition At $x = L$, i.e., at $i = M$, Eq. (10.22) reduces to

$$\theta_{M-1} - D\theta_M + \theta_{M+1} = 0 \tag{10.23}$$

A careful look at Eq. (10.23) reveals that θ_{M+1} represents a fictitious temperature θ at point $M + 1$, which lies outside the computational domain. There is a remedy to tackle this issue.

Remedy: ***Image point technique***   It is assumed that the θ versus X curve extends beyond $X = 1$ so that at $X = 1$, the condition $d\theta/dX = 0$ is satisfied. In other words, the θ versus X curve can be imagined to look as in Fig. 10.4. The dotted line represents the mirror-image extension of the solid line, indicating that a minimum exists at $X = 1$. Figure 10.4(a) shows a mirror-image extension of the fin. Therefore, the boundary condition at $X = 1$ can be approximately satisfied by taking

$$\theta_{M+1} = \theta_{M-1} \tag{10.24}$$

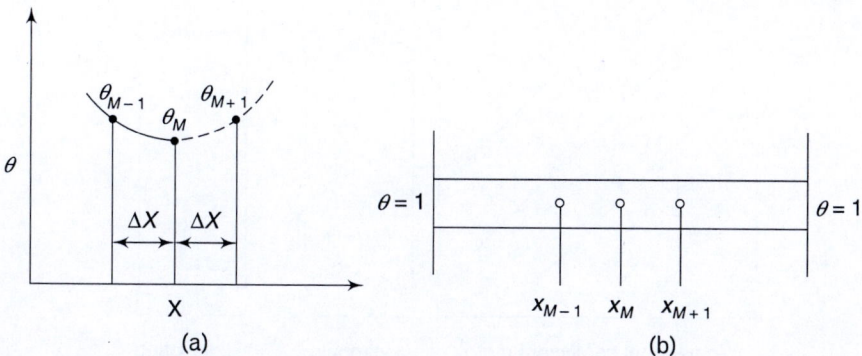

Fig. 10.4   (a) Mirror-image extension of the θ versus X curve near the fin tip, (b) mirror-image extension of the fin

Equation (10.24) also follows from the central-difference approximation of $d\theta/dX$ at $i = M$. Substituting Eq. (10.24) into Eq. (10.23), we get

$$2\theta_{M-1} - D\theta_M = 0 \tag{10.25}$$

Therefore, we can write that

$$\theta_i = 1 \text{ for } i = 1 \text{ (known)}$$
$$\theta_{i+1} - D\theta_i + \theta_{i+1} = 0 \text{ for } i = 2, \ldots, M-1 \tag{10.26}$$
$$2\theta_{M-1} - D\theta_M = 0 \text{ for } i = M$$

Hence, we have a set of $M - 1$ linear simultaneous algebraic equations and $M - 1$ unknowns, which can be easily solved by standard numerical methods.

For the case in which $M = 5$, we have $N = M - 1 = 4$ equations to solve. The four equations can be written in the matrix form as

$$\begin{bmatrix} D & -1 & & \\ -1 & D & -1 & \\ & -1 & D & -1 \\ & & -2 & D \end{bmatrix} \begin{bmatrix} \theta_1 \\ \theta_2 \\ \theta_3 \\ \theta_4 \end{bmatrix} = \begin{bmatrix} 1 \\ 0 \\ 0 \\ 0 \end{bmatrix} \tag{10.27}$$

It should be noted that θ_1 corresponds to the temperature at grid point 2 in Fig. 10.3, and so on for θ_2, θ_3, θ_4.

An alternative to the image-point scheme is to use a second-order backward difference for $d\theta/dX$ at $i = M$.

10.3.1.1   Methods of solution

In Eq. (10.27), the coefficient matrix has three diagonals—the main diagonal, sub-diagonal and super-diagonal, and hence the name tridiagonal matrix (TDM). The set of equations in Eq. (10.27) is called tridiagonal system of equations. See Fig. 10.5 for a pictorial representation of TDM.

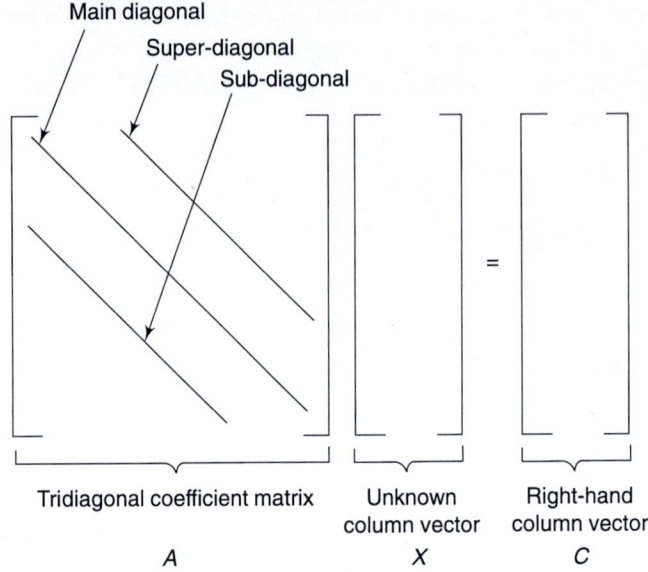

Main diagonal
Super-diagonal
Sub-diagonal

Tridiagonal coefficient matrix   Unknown   Right-hand
column vector   column vector

A   *X*   *C*

Fig. 10.5 Pictorial representation of a tridiagonal coefficient matrix, an unknown vector, and a known right hand vector

The set of equations in Eq. (10.27) can be solved by any of the three methods:
1. Gaussian elimination
2. Thomas algorithm (or tridiagonal matrix algorithm or simply TDMA)
3. Gauss–Seidel iterative method

Gaussian elimination (GE)   This method reduces a given set of N equations to an equivalent triangular set, so that one of the equations has only one unknown. This unknown is determined and the remaining unknowns are obtained by the process of back substitution. The basic approach is shown in a step-by-step form as given. The set of equations to be solved are written in a matrix form in Eq. (10.28):

$$\begin{bmatrix} a_{11} & a_{12} & a_{13} \\ a_{21} & a_{22} & a_{23} \\ a_{31} & a_{32} & a_{33} \end{bmatrix} \begin{bmatrix} x_1 \\ x_2 \\ x_3 \end{bmatrix} = \begin{bmatrix} c_1 \\ c_2 \\ c_3 \end{bmatrix} \tag{10.28}$$

a_{11} is called the *pivot*, below which the terms are to be made zero.

Step I

$$\begin{bmatrix} a_{11} & a_{12} & a_{13} \\ & a_{22}^{(1)} & a_{23}^{(1)} \\ & a_{32}^{(1)} & a_{33}^{(1)} \end{bmatrix} \begin{bmatrix} x_1 \\ x_2 \\ x_3 \end{bmatrix} = \begin{bmatrix} c_1 \\ c_2^{(1)} \\ c_3^{(1)} \end{bmatrix} \tag{10.29}$$

The superscript represents the step number. $a_{22}^{(1)}$ is now the pivot for the next operation.

Step II

$$
\begin{bmatrix}
a_{11} & a_{12} & a_{13} \\
 & a_{22}^{(1)} & a_{23}^{(1)} \\
 & & a_{33}^{(2)}
\end{bmatrix}
\begin{bmatrix}
x_1 \\
x_2 \\
x_3
\end{bmatrix}
=
\begin{bmatrix}
c_1 \\
c_2^{(1)} \\
c_3^{(2)}
\end{bmatrix}
\tag{10.30}
$$

Solution accuracy The round-off error may significantly affect the accuracy if a large number of equations is involved. In addition, the round-off error is cumulative because the errors are carried on from one step to the other during the elimination process. Consequently, GE is generally used if the number of equations is typically less than 20 when the coefficient matrix is dense. For a sparse coefficient matrix, however, a large number of equations can be solved. The TDM system is a good example of a sparse coefficient matrix. If Gaussian elimination is applied to this system, only one of the a's is eliminated from the column containing the pivot element in each step, since the remaining elements below the diagonal are zero. Therefore, only one elimination process is employed at each step. The number of operations needed for solving a tridiagonal system is of order N, that is, $O(N)$ as compared to $O(N^3)$ for a system with a dense coefficient matrix. Therefore, a much smaller number of operations and consequently much lower round-off errors arise in the solution of such systems. Obviously, the computer time is much less for solution by TDMA. Thus, large tridiagonal systems are generally solved by this method.

Thomas algorithm or TDMA The set of equations in Eq. (10.26) can be readily solved by the Gaussian elimination method with a maximum of three variables per equation. The solution can be expressed very concisely.

Equation (10.26) is actually a special form of the system (using $N = M - 1$)

$$
\begin{aligned}
&b_1 T_1 + c_1 T_2 = d_1 \\
&a_2 T_1 + b_2 T_2 + c_2 T_3 = d_2 \\
&a_3 T_2 + b_3 T_3 + c_3 T_4 = d_3 \\
&\qquad\vdots \\
&a_i T_{i-1} + b_i T_i + c_i T_{i+1} = d_i \\
&\qquad\vdots \\
&a_{N-1} T_{N-2} + b_{N-1} T_{N-1} + c_{N-1} T_N = d_{N-1} \\
&a_N T_{N-1} + b_N T_N = d_N
\end{aligned}
\tag{10.31a}
$$

First, let us demonstrate the validity of a recursion solution of the form (Carnahan et al. 1969)

$$
T_i = \gamma_i - \frac{c_i}{\beta_i} T_{i+1}
\tag{10.31b}
$$

in which the constants β_i and γ_i are to be determined. The substitution of Eq. (10.31b) into Eq. (10.31a) gives

$$
a_i \left(\gamma_{i-1} - \frac{c_{i-1}}{\beta_{i-1}} T_i \right) + b_i T_i + c_i T_{i+1} = d_i
\tag{10.32}
$$

Rearranging Eq. (10.32), we obtain

$$
T_i = \frac{d_i - a_i \gamma_{i-1}}{b_i - \dfrac{a_i c_{i-1}}{\beta_{i-1}}} - \frac{c_i T_{i+1}}{b_i - \dfrac{a_i c_{i-1}}{\beta_{i-1}}}
\tag{10.33}
$$

Equation (10.33) verifies the form of Eq. (10.31a), subject to the following recursion relations:

$$\beta_i = b_i - \frac{a_i c_{i-1}}{\beta_{i-1}}$$

$$\gamma_i = \frac{d_i - a_i \gamma_{i-1}}{\beta_i}$$

Also, from the first equation of Eq. (10.31a),

$$T_1 = \frac{d_1}{b_1} - \frac{c_1}{b_1} T_2$$

from which we get

$$\beta_1 = b_1, \gamma_1 = \frac{d_1}{\beta_1}$$

Finally, the substitution of the recursion solution into the last equation of Eq. (10.31a) yields

$$T_N = \frac{d_N - a_N T_{N-1}}{b_N} = \frac{d_N - a_N \left(\gamma_{N-1} - \dfrac{c_{N-1}}{\beta_{N-1}} T_N \right)}{b_N}$$

from which

$$T_N = \frac{d_N - a_N \gamma_{N-1}}{b_N - \dfrac{a_N c_{N-1}}{\beta_{N-1}}} = \gamma_N$$

In a nutshell, the complete algorithm for the solution of the tridiagonal system is

$$T_N = \gamma_N \tag{10.34}$$

$$T_i = \gamma_i - \frac{c_i T_{i+1}}{\beta_i}, i = N-1, N-2, \dots, 1$$

where β's and γ's are determined from the recursion formulae

$$\beta_1 = b_1, \gamma_1 = \frac{d_1}{\beta_1}$$

$$\beta_i = b_i - \frac{a_i c_{i-1}}{\beta_{i-1}}, i = 2, 3, \dots, N \tag{10.35}$$

$$\gamma_i = \frac{d_i - a_i \gamma_{i-1}}{\beta_i}, i = 2, 3, \dots, N$$

The above algorithm is also known as the *Thomas algorithm*.[1]

Finally, it is to be noted that Eq. (10.26) might also be solved by the Gauss-Seidel iteration scheme discussed next.

---

[1] See Listing A8.2 in Appendix A8 for an application of TDMA. The Online Resource Centre also provides the listing (see 1DFIN_Ins.c).

Gauss-Seidel iterative method (GS) For a large number of equations (typically of the order of several hundred) iterative methods such as Jacobi, Gauss-Seidel, which initiate the computations with a guessed solution and iterate to the desired solution of the systems of equations within a specified convergence criterion, using improved guesses in the second and third iterations till the final one, are often more efficient. In the GS method only the values of the latest iteration are stored, and each iterative computation of the unknown employs the most recent values of the other unknowns. In this method, unlike in direct methods, such as Gaussian elimination, the round-off error does not accumulate. The round-off error after each iteration simply produces a less accurate input for the next iteration. Therefore, the resulting round-off error in the numerical solution is only what arises in the computation for the final iteration. However, the solution is not exact but is obtained to an arbitrary, specified, convergence criterion.

Example 10.1 Gauss-Seidel Iterative Method

Solve the set of three simultaneous algebraic equations shown in Eq. (10.37) by the Gauss-Seidel iterative method.

Solution

$$x_2^{(2)} = \frac{c_2 - a_{21} x_1^{(2)} - a_{23} x_3^{(1)} - \cdots - a_{2N} x_N^{(1)}}{a_{22}}$$

Here, the value of x_1 is known after the second iteration, and the others are known only after the first iteration.

Therefore, $x_i^{(p+1)} = \dfrac{c_i - \sum\limits_{j=1}^{i-1} a_{ij} x_j^{(p+1)} - \sum\limits_{j=i+1}^{N} a_{ij} x_j^{(p)}}{a_{ii}}$ for $i = 1, 2, ..., N$ (10.36)

It may be noted that GS iteration consumes $2N$ steps (not to be confused with the number of arithmetic operations) in each iteration.

Convergence criteria for the GS method Typical convergence criteria used are

1. $\left| x_i^{(p+1)} - x_i^{(p)} \right| \le \varepsilon$ for $i = 1, 2, ..., N$

2. $\left| \dfrac{x_i^{(p+1)} - x_i^{(p)}}{x_i^{(p)}} \right| \le \varepsilon$ for $i = 1, 2, ..., N$

where ε is a very small number, e.g., 0.01, 0.001, 0.00001.

Criterion 2 is applicable if an estimate of the magnitude of the unknowns x_i is not available and none of the unknowns is expected to be zero.

Conditions for convergence in the GS method: Scarborough criterion
Convergence is guaranteed for linear systems if

$$|a_{ii}| \ge \sum_{j=1, j \ne i}^{N} |a_{ij}| \text{ for all } i$$

and if $\quad |a_{ii}| > \sum\limits_{j=1,\,j\neq i}^{N} |a_{ij}| \quad$ for at least one i

that is, when the system is diagonally dominant. This is also known as the *Scarborough criterion*. This is a sufficient condition, which means that convergence may still be possible even if the above condition is not satisfied. Fortunately, it turns out that in fluid flow and heat transfer problems, finite-difference formulation indeed leads to a diagonally dominant coefficient matrix, which is the reason why for large systems the Gauss–Seidel method is so widely used.

Application of the GS iterative method   In order to demonstrate the iteration process, the following system of three linear equations is solved by the GS iterative method using a *pocket calculator*:

$$10x_1 + x_2 + 2x_3 = 44$$
$$2x_1 + 10x_2 + x_3 = 51 \qquad\qquad (10.37)$$
$$x_1 + 2x_2 + 10x_3 = 61$$

Clearly, the coefficient matrix in Eq. (10.37) is diagonally dominant because

$$|10| > |1| + |2|$$
$$|10| > |2| + |1|$$
$$|10| > |1| + |2|$$

Therefore, the Scarborough criterion is satisfied and hence, one is certain to get a converged solution using the GS iterative method. As a first guess, let us take

$$[x_1, x_2, x_3]^0 = [0, 0, 0]$$

We take $\varepsilon = 0.02$.
Then,

$$x_1^{(1)} = \frac{1}{10}[44 - x_2^{(0)} - 2x_3^{(0)}]$$

$$= \frac{1}{10}[44 - (0) - 2(0)] = 4.40$$

$$x_2^{(1)} = \frac{1}{10}[51 - 2x_1^{(1)} - x_3^{(0)}]$$

$$= \frac{1}{10}[51 - 2(4.40) - 0] = 4.22$$

$$x_3^{(1)} = \frac{1}{10}[61 - x_1^{(1)} - 2x_2^{(1)}]$$

$$= \frac{1}{10}[61 - 4.40 - 2(4.22)] = 4.81$$

Now, check for convergence after the first iteration:

$$|0 - 4.40| = 4.40 > \varepsilon$$
$$|0 - 4.22| = 4.22 > \varepsilon$$
$$|0 - 4.81| = 4.81 > \varepsilon$$

We find that there is no convergence. One more iteration gives

$$[x_1, x_2, x_3]^2 = [3.01, 4.01, 4.99]$$

Again check for convergence after the second iteration:

$$|3.01 - 4.40| = 1.39 > \varepsilon$$
$$|4.01 - 4.22| = 0.21 > \varepsilon$$
$$|4.81 - 4.99| = 0.18 > \varepsilon$$

Still there is no convergence. One more iteration gives

$$[x_1, x_2, x_3]^3 = [3.00, 4.00, 5.00]$$

We again check for convergence after the third iteration:

$$|3.01 - 3.00| = 0.01 < \varepsilon$$
$$|4.01 - 4.00| = 0.01 < \varepsilon$$
$$|4.99 - 5.00| = 0.01 < \varepsilon$$

We see that convergence is reached. Hence, further computation stops. Therefore, it required three iterations to obtain a converged solution. Incidentally, the converged solution is also the exact solution of this set of equations. The reason is that the number of unknowns is very small in this case.

Relaxation: Over-relaxation and under-relaxation One of the problems with the GS method is that it is relatively slow to converge to the solution. The rate of convergence can often be improved by the relaxation method which is explained next.

Let us consider the example that has been solved by the GS method.

$$x_1^{(1)} = x_1^{(0)} + \left| \frac{\frac{1}{10}\{44 - x_2^{(0)} - 2x_3^{(0)}\} - x_1^{(0)}}{\text{(Change produced by the current iteration)}} \right| \tag{10.38}$$

Now the change produced by the current iteration can be increased if we multiply it by a factor $\alpha\,(\alpha > 1)$. However, α also has an upper limit. For $\alpha > 2$, the change is so great that instead of convergence, divergence occurs, that is, the solution never converges. Therefore, Eq. (10.37) can now be written as

$$x_1^{(1)} = x_1^{(0)} + \alpha\left[\frac{1}{10}\{44 - x_2^{(0)} - 2x_3^{(0)}\} - x_1^{(0)}\right]$$

$$\Rightarrow \qquad x_1^{(1)} = \frac{\alpha}{10}(44 - x_2^{(0)} - 2x_3^{(0)}) + x_1^{(0)}(1 - \alpha) \tag{10.39}$$

From Eq. (10.39), it is readily seen that for

$$\alpha = 0, \ x_1^{(1)} = x_1^{(0)} \qquad \text{(no progress)}$$

$$\alpha = 1, \ x_1^{(1)} = x_{1_{G-S}}^{(1)} \qquad \text{(basic GS iteration)}$$

$0 < \alpha < 1$, under-relaxation $\rightarrow$ interpolation between $x_1^{(0)}$ and $x_{1_{GS}}^{(1)}$

$1 < \alpha < 2 \rightarrow$ over-relaxation $\rightarrow$ extrapolation beyond $x_{1_{GS}}^{1}$

In a compact form, the relaxation method may be written as

$$x_i^{(p+1)} = \frac{\alpha \left[c_i - \sum_{j=1}^{i-1} a_{ij} x_j^{(p+1)} - \sum_{j=j+1}^{N} a_{ij} x_j^{(p)} \right]}{a_{ii}} + (1-\alpha) x_i^{(p)} \qquad (10.40a)$$

for $i = 1, 2, ..., N$

Successive under-relaxation (SUR) or under-relaxation is generally used for non-linear equations and for systems that result in a divergent GS iteration. Successive over-relaxation (SOR) or over-relaxation is widely used for accele rating convergence in linear systems.

Optimum relaxation factor α_{opt}   The question is: What value of the over-relaxation factor should be used? There is no set rule to determine this. One has to do numerical experimentation to find out the relaxation factor which gives the highest rate of convergence. This is called the *optimum relaxation factor* α_{opt}, which lies between 1 and 2 and varies from one problem to another. For the simple case of Laplace's equation ($\nabla^2 T = 0$) in a square with Dirichlet boundary conditions (that is, known temperature on the boundaries), Young (1954) and Frankel (1950) show that α_{opt} equals the smaller root of

$$t^2 \alpha^2 - 16\alpha + 16 = 0 \qquad (10.40b)$$

with $t = 2\cos(\pi/n)$, where n is the total number of increments into which the side of the square is divided. In other words, n is the number of grid spacings. The number of iterations required for a given convergence criterion falls very rapidly when the parameter is in the immediate vicinity of α_{opt}, and it is generally better to overestimate α_{opt} than to underestimate it (Carnahan et al. 1969).

10.3.1.2   Solution of Eq. (10.27) by all three methods

We shall now solve Eq. (10.27) having a tridiagonal coefficient matrix by the Gaussian elimination, TDMA, and Gauss–Seidel iterative method and choose the proper method.

Recall Eq. (10.27) given below:

$$\begin{bmatrix} D & -1 & & \\ -1 & D & -1 & \\ & -1 & D & -1 \\ & & -2 & D \end{bmatrix} \begin{bmatrix} \theta_1 \\ \theta_2 \\ \theta_3 \\ \theta_4 \end{bmatrix} = \begin{bmatrix} 1 \\ 0 \\ 0 \\ 0 \end{bmatrix}$$

where $D = 2 + (mL)^2 (\Delta X)^2$. Let us consider a fin with $mL = 2$. Also, $\Delta X = 1/4 = 0.25$. Therefore, $D = 2 + (2)^2 (0.25)^2 = 2.25$. Hence, Eq. (10.27) now becomes

$$\begin{bmatrix} 2.25 & -1 & & \\ -1 & 2.25 & -1 & \\ & -1 & 2.25 & -1 \\ & & -2 & 2.25 \end{bmatrix} \begin{bmatrix} \theta_1 \\ \theta_2 \\ \theta_3 \\ \theta_4 \end{bmatrix} = \begin{bmatrix} 1 \\ 0 \\ 0 \\ 0 \end{bmatrix} \qquad (10.41)$$

Solution by Gaussian elimination

Step I: The pivot is 2.25. So, first eliminate all terms below the pivot.

$$R1/2.25 + R2 \quad \begin{bmatrix} 2.25 & -1 & & \\ & 1.81 & -1 & \\ & -1 & 2.25 & -1 \\ & & -2 & 2.25 \end{bmatrix} \begin{bmatrix} \theta_1 \\ \theta_2 \\ \theta_3 \\ \theta_4 \end{bmatrix} = \begin{bmatrix} 1 \\ 0.444 \\ 0 \\ 0 \end{bmatrix}$$

Step II: Now, the pivot is 1.81. So eliminate all terms below the pivot.

$$R2/1.81 + R3 \quad \begin{bmatrix} 2.25 & -1 & & \\ & 1.81 & -1 & \\ & & 1.7 & -1 \\ & & -2 & 2.25 \end{bmatrix} \begin{bmatrix} \theta_1 \\ \theta_2 \\ \theta_3 \\ \theta_4 \end{bmatrix} = \begin{bmatrix} 1 \\ 0.444 \\ 0.246 \\ 0 \end{bmatrix}$$

Step III: Now, the pivot is 1.7. So eliminate all terms below the pivot.

$$R3 \times 1.176 + R4 \quad \begin{bmatrix} 2.25 & -1 & & \\ & 1.81 & -1 & \\ & & 1.7 & -1 \\ & & & 1.07 \end{bmatrix} \begin{bmatrix} \theta_1 \\ \theta_2 \\ \theta_3 \\ \theta_4 \end{bmatrix} = \begin{bmatrix} 1 \\ 0.444 \\ 0.246 \\ 0.290 \end{bmatrix}$$

Step III is the last step as now the triangular coefficient matrix is produced. $R1, R2,$ $R3, R4$ are used to denote the first, second, third, and fourth rows, respectively. The unknowns are obtained by back substitution.

Back substitution

$$\theta_4 = \frac{0.290}{1.07} = 0.271$$
$$1.7\theta_3 - \theta_4 = 0.246 \Rightarrow \theta_3 = 0.304$$
$$1.81\theta_2 - \theta_3 = 0.444 \Rightarrow \theta_2 = 0.413$$
$$2.25\theta_1 - \theta_2 = 1 \Rightarrow \theta_1 = 0.628$$

Therefore, the unknown temperatures are

$$\theta_1 = 0.628, \ \theta_2 = 0.413, \ \theta_3 = 0.304, \ \theta_4 = 0.271$$

The total number of arithmetic operations (multiplications and divisions) to obtain the solution is 13 (9 for elimination and 4 for back substitution).

Solution by TDMA Recall the tridiagonal matrix algorithm given by Eqs (10.34) and (10.35). With respect to Eq. (10.41),

$$d_1 = 1, d_2 = 0, d_3 = 0, d_4 = 0$$
$$b_1 = 2.25, b_2 = 2.25, b_3 = 2.25, b_4 = 2.25$$
$$a_2 = -1, a_3 = -1, a_4 = -2$$
$$c_1 = -1, c_2 = -1, c_3 = -1$$

From the TDMA,

$$\theta_4 = \gamma_4,$$
$$\beta_1 = b_1 = 2.25$$

$$\gamma_1 = d_1/\beta_1 = \frac{1}{2.25} = 0.444$$

$$\beta_2 = b_2 - (a_2c_1/\beta_1)$$
$$= 2.25 - [(1)(-1)/2.25] = 1.805$$

$$\beta_3 = b_3 - (a_3c_2/\beta_2)$$
$$= 2.25 - [(-1)(-1)/1.805] = 1.695$$

$$\beta_4 = b_4 - (a_4c_3/\beta_3)$$
$$= 2.25 - [(-2)(-1)/1.695] = 1.07$$

$$\gamma_2 = (d_2 - a_2\gamma_1)/\beta_2$$
$$= [0 - (-1)(0.444)]/1.805 = 0.246$$

$$\gamma_3 = (d_2 - a_3\gamma_2)/\beta_3 = [0 - (-1)(0.246)]/1.695 = 0.145$$

$$\gamma_4 = (d_4 - a_4\gamma_3)/\beta_4$$
$$= [0 - (-2)(0.145)]/1.07 = 0.271$$

$$\theta_4 = \gamma_4 = 0.271$$

$$\theta_3 = \gamma_3 - (c_3\theta_4)/\beta_3 = 0.145 - [(-1)(0.271)]/1.695 = 0.304$$

$$\theta_2 = \gamma_2 - (c_2\theta_3)/\beta_2$$
$$= 0.246 - [(-1)(0.304)]/1.805 = 0.414$$

$$\theta_1 = \gamma_1 - (c_1\theta_2)/\beta_1$$
$$= 0.444 - [(-1)(0.414)]/2.25 = 0.628$$

Therefore, $\theta_1 = 0.628$, $\theta_2 = 0.414$, $\theta_3 = 0.304$, $\theta_4 = 0.271$

The total number of arithmetic operations required to obtain the solution is 10.

Solution by the Gauss–Seidel iteration Let $[\theta_1, \theta_2, \theta_3, \theta_4]^0 = [1, 1, 1, 1]$ and $\varepsilon = 0.001$.

Iteration 1: $[\theta_1, \theta_2, \theta_3, \theta_4]^1 = [0.888, 0.839, 0.817, 0.726]$
Iteration 2: $[\theta_1, \theta_2, \theta_3, \theta_4]^2 = [0.817, 0.726, 0.645, 0.573]$
Iteration 3: $[\theta_1, \theta_2, \theta_3, \theta_4]^3 = [0.767, 0.627, 0.6, 0.533]$
Iteration 4: $[\theta_1, \theta_2, \theta_3, \theta_4]^4 = [0.723, 0.588, 0.498, 0.442]$
Iteration 5: $[\theta_1, \theta_2, \theta_3, \theta_4]^5 = [0.705, 0.535, 0.434, 0.385]$
Iteration 6: $[\theta_1, \theta_2, \theta_3, \theta_4]^6 = [0.682, 0.496, 0.391, 0.347]$
Iteration 7: $[\theta_1, \theta_2, \theta_3, \theta_4]^7 = [0.664, 0.468, 0.362, 0.321]$
Iteration 8: $[\theta_1, \theta_2, \theta_3, \theta_4]^8 = [0.652, 0.45, 0.342, 0.304]$
Iteration 9: $[\theta_1, \theta_2, \theta_3, \theta_4]^9 = [0.664, 0.438, 0.329, 0.292]$
Iteration 10: $[\theta_1, \theta_2, \theta_3, \theta_4]^{10} = [0.639, 0.43, 0.32, 0.284]$
Iteration 11: $[\theta_1, \theta_2, \theta_3, \theta_4]^{11} = [0.635, 0.424, 0.314, 0.279]$
Iteration 12: $[\theta_1, \theta_2, \theta_3, \theta_4]^{12} = [0.632, 0.42, 0.31, 0.275]$
Iteration 13: $[\theta_1, \theta_2, \theta_3, \theta_4]^{13} = [0.631, 0.418, 0.308, 0.273]$
Iteration 14: $[\theta_1, \theta_2, \theta_3, \theta_4]^{14} = [0.63, 0.416, 0.306, 0.272]$
Iteration 15: $[\theta_1, \theta_2, \theta_3, \theta_4]^{15} = [0.629, 0.415, 0.305, 0.271]$

Convergence is reached on the 15th iteration. Hence, further computation stops. Total number of iterations required = 15

Total number of arithmetic operations in each iteration = 5
Therefore, total number of arithmetic operations = 75

10.3.1.3 Number of arithmetic operations for each method: A comparison

As can be seen from Table 10.1, it is obvious that TDMA is the fastest method and the Gauss–Seidel iteration is the worst method for solving a tridiagonal system of equations. Therefore, the choice falls on the TDMA for the solution of Eq. (10.41).

Table 10.1 A comparative study of the number of arithmetic operations needed to solve Eq. (10.41) by three methods

Name of the method	Number of arithmetic operations
Gaussian elimination	13
Tridiagonal matrix algorithm (TDMA)	10
Gauss–Seidel iteration	
(initial guess = [1, 1, 1, 1], $\varepsilon = 0.001$)	75

Hence, *for solving a tridiagonal system of linear equations, Thomas algorithm (TDMA) is preferred.*

10.3.1.4 Checking for accuracy

The accuracy of a numerical solution is usually checked in one of the three ways:

1. *Comparison with the analytical solution:* For most practical problems analytical solutions do not exist. But, this is a good way of checking the accuracy of a new numerical method.

2. *Comparison with the limiting case analytical solution:* This is possible when the analytical solution for some limiting value of a parameter governing the solution is available.

3. *Comparison with experimental results:* This is most desirable for complex problems, such as turbulence, combustion, non-Newtonian fluid flow, and heat transfer, which require many assumptions for the purpose of modelling.

10.3.1.5 Comparison of the present numerical result with the corresponding analytical solution

For the present problem of heat conduction in a fin, an analytical solution is available. Therefore, a comparison of the numerical results with the exact solution (for $mL = 2$) will enable us to obtain an estimate of the numerical error. Table 10.2 gives a comparison of the numerical and analytical solutions.

Table 10.2 clearly reveals that $\Delta X = 0.25$ is not good enough and the grid spacing needs to be finer. In other words, a higher number of grid points is necessary to obtain a more accurate solution. However, one has to be also careful in increasing the number of grid points as this will result in a higher round-off error. Therefore, a grid independence test, which gives an optimum ΔX, is called for. A point to note is that even with a relatively coarse grid the accuracy is quite good. This means that with a slight decrease in the grid spacings, the numerical solution will be even closer to its analytical counterpart. Another interesting feature of Table 10.2 is the gradually increasing error for increasing X. This is possible because of the fact that at the left boundary ($X=0$) the Dirichlet condition is imposed and, therefore, for both numerical and exact solutions the same temperature is used for the calculation of temperature at $X=0.25$. Hence, the temperature of the grid point closest to the left boundary (i.e., at $X = 0.25$) computed by the numerical method is most accurate and the error accumulates as the distance of a grid point with respect to the left boundary increases.

Table 10.2   A comparison of the numerical and analytical solutions of the fin problem

Location X	Temperature θ		Absolute per cent error with respect to the exact solution
	Numerical	Exact $\left(\theta = \dfrac{\cosh mL(1-X)}{\cosh mL} \right)$	
0^{a}	1	1	0
0.25	0.628	0.625	0.48
0.50	0.414	0.410	0.97
0.75	0.304	0.2995	1.50
1	0.271	0.266	1.88

^{a}Dirichlet boundary condition and hence not computed.

10.3.1.6   Convective boundary condition

If the tip of the fin was convective instead of insulated, the discretization equation at $i = M$ would have to be modified. The dimensionless boundary condition at the fin tip in the changed scenario would be written in the mathematical form as

$$\frac{d\theta}{dX} + \frac{h_e L}{k}\theta = 0 \tag{10.42}$$

where h_e is the convection heat transfer coefficient from the tip of the fin to the surroundings.

Using the *image-point technique* as discussed earlier and the central-difference scheme for discretization of $d\theta/dX$ at $i = M$, Eq. (10.42) is expressed as

$$\frac{\theta_{M+1}-\theta_{M-1}}{2\Delta X} + \frac{h_e L}{k}\theta_M = 0$$

or
$$\theta_{M+1} = \theta_{M-1} - \frac{2h_e L \Delta X}{k}\theta_M \tag{10.43}$$

Substituting the expression for θ_{M+1} from Eq. (10.43) into Eq. (10.23), we obtain

$$2\theta_{M-1} - \theta_M\left(D + \frac{2h_e L \Delta X}{k} \right) = 0 \tag{10.44}$$

Therefore, only the last equation is changed. The method of solution remains the same as before.

To check the accuracy of Eq. (10.43), substituting $h_e = 0$ (corresponding to the insulation condition), we obtain

$$2\theta_{M-1} - D\theta_M = 0$$

which is the same as that obtained for the insulated tip.

10.3.2 Numerical Methods for Two-dimensional Steady-state Problems

Consider the case of steady heat conduction in a long square slab ($2L \times 2L$) in which heat is generated at a uniform rate of q''' W/m^3 (Fig. 10.6). The problem can be assumed to be a two-dimensional one as the dimension of the slab is much longer in the direction normal to the cross-sectional plane. Therefore, the end effects can be neglected. All four sides are maintained at $T = T_\infty$, temperature of the surrounding fluid, assuming a large heat transfer coefficient.

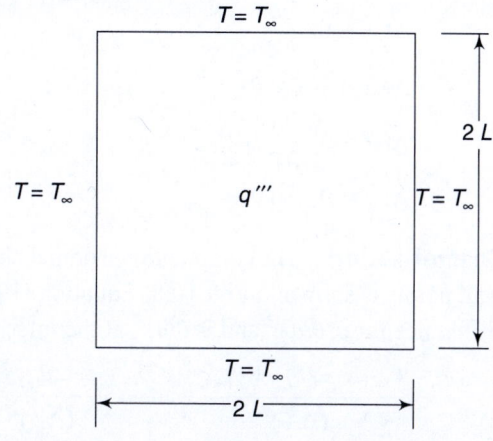

Fig. 10.6 Physical domain of the slab with square cross-section ($2L \times 2L$)

Consideration of symmetry A close look at the physics of the problem reveals that the problem is geometrically and thermally symmetric. Therefore, from the temperature distribution in any quarter of the physical domain, by mirror-imaging one can get the solution for the entire region. Figure 10.7 shows the computational domain (top right-hand quarter). The use of symmetry enables the numerical analyst to obtain the solution much faster as the number of grid points is greatly reduced.

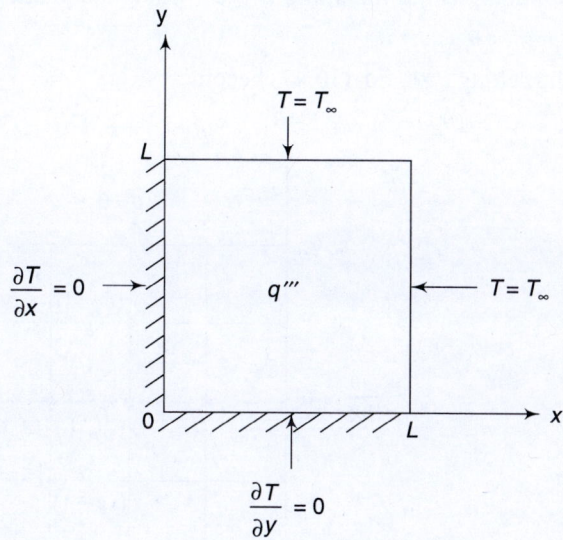

Fig. 10.7 Computational domain (top right-hand quarter) considering symmetry

Governing differential equation The governing non-dimensional energy equation (assuming constant k) is

$$\frac{\partial^2 \theta}{\partial X^2} + \frac{\partial^2 \theta}{\partial Y^2} + 1 = 0 \qquad (10.45)$$

where $\theta = \dfrac{T - T_\infty}{(q''' L^2 / k)}, X = \dfrac{x}{L}, Y = \dfrac{y}{L}$

Boundary conditions The non-dimensional boundary conditions are as follows:

$$\text{At } X = 0, \ \frac{\partial \theta}{\partial X} = 0 \tag{10.46a}$$

$$\text{At } X = 1, \ \theta = 0 \tag{10.46b}$$

$$\text{At } Y = 0, \ \frac{\partial \theta}{\partial Y} = 0 \tag{10.46c}$$

$$\text{At } Y = 0, \ \theta = 0 \tag{10.46d}$$

Discretization The computational domain including the notations for the interior grid points is shown in Fig. 10.8. Equation (10.45) is discretized using the central difference for $\partial^2\theta/\partial X^2$ and $\partial^2\theta/\partial Y^2$ at the interior grid point (i, j) as follows:

$$\frac{\theta_{i-1,j} - 2\theta_{i,j} + \theta_{i+1,j}}{(\Delta X)^2} + \frac{\theta_{i-1,j} - 2\theta_{i,j} + \theta_{i+1,j}}{(\Delta Y)^2} + 1 = 0 \tag{10.47}$$

Taking $\Delta X = \Delta Y$, Eq. (10.47) reduces to

$$-\theta_{i-1,j} - \theta_{i,j-1} + 4\theta_{i,j} - \theta_{i,j+1} - \theta_{i+1,j} = (\Delta X)^2 \tag{10.48}$$

Boundary condition along $X = 0$ Using the image-point technique,

$$\theta_{i-1,j} = \theta_{i+1,j}$$

and setting $i = 1$, Eq. (10.47) becomes

$$-2\theta_{2,j} - \theta_{i,j-1} + 4\theta_{i,j} - \theta_{i,j+1} = (\Delta X)^2 \tag{10.49}$$

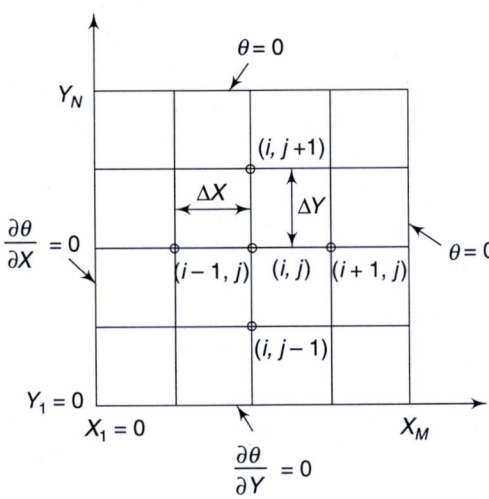

Fig. 10.8 Interior grid points in the computational domain

Boundary condition along $Y = 0$ Using the image-point technique

$$\theta_{i,j-1} = \theta_{i,j+1}$$

and setting $j = 1$, Eq. (10.47) becomes

$$-\theta_{i+1,1} - 2\theta_{i,2} + 4\theta_{i,1} - \theta_{i-1,1} = (\Delta X)^2 \tag{10.50}$$

Handling of corner points Corner points need special attention because they belong to both horizontal and vertical surfaces. Therefore, the boundary conditions at both the surfaces apply there. However, if one or both of the surfaces have the Dirichlet condition (specified temperature), then there is no problem because the corner point can be assumed to have a specified temperature. But, if both surfaces have Neumann (insulation) and/or Robbins (convective) conditions, then the corner point needs to be handled separately since both conditions exist there. For more details about Dirichlet, Neumann, and Robbins conditions see Ghoshdastidar (1998). With respect to the present problem, out of the four corner points, only the bottom-left corner point is exposed to Neumann conditions in X and Y directions. Other three have either one or both surfaces exposed to the Dirichlet condition. Figure 10.9 shows the image points for the bottom-left-hand corner represented by the grid point (1,1).

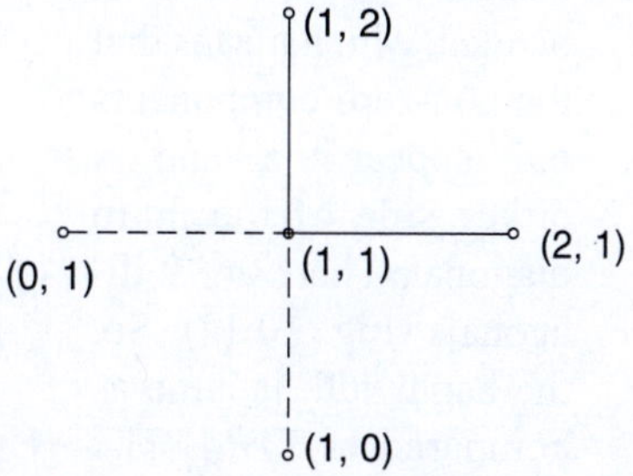

Fig. 10.9 Image points for the bottom left-hand corner point (1, 1)

Using the image-point technique,

$$\theta_{0,1} = \theta_{2,1} \tag{10.51}$$

$$\theta_{1,0} = \theta_{1,2} \tag{10.52}$$

Now,
$$\left(\frac{\partial^2 \theta}{\partial X^2}\right)_{1,1} = \frac{\theta_{0,1} - 2\theta_{1,1} + \theta_{2,1}}{(\Delta X)^2} \tag{10.53}$$

$$\left(\frac{\partial^2 \theta}{\partial Y^2}\right)_{1,1} = \frac{\theta_{1,0} - 2\theta_{1,1} + \theta_{1,2}}{(\Delta Y)^2} \tag{10.54}$$

Substituting Eq. (10.51) into Eq. (10.53), and Eq. (10.52) into Eq. (10.54), we get

$$\left(\frac{\partial^2 \theta}{\partial X^2}\right)_{1,1} = \frac{2\theta_{2,1} - 2\theta_{1,1}}{(\Delta X)^2} \tag{10.55}$$

$$\left(\frac{\partial^2 \theta}{\partial Y^2}\right)_{1,1} = \frac{2\theta_{1,2} - 2\theta_{1,1}}{(\Delta Y)^2} \tag{10.56}$$

Setting $i = 1, j = 1$ and substituting Eqs (10.55) and (10.56) into Eq. (10.47), we obtain for $\Delta X = \Delta Y$,

$$2\theta_{2,1} - 4\theta_{1,1} + 2\theta_{1,2} + (\Delta X)^2 = 0 \tag{10.57}$$

10.3.2.1 Methods of solution

Let us consider an example in which $\Delta X = 1/4$. The grid points that are unlabelled (Fig. 10.10) are all at temperature $\theta = 0$ as imposed by the boundary condition. Thus, there are 16 unknown temperatures to find (Fig. 10.11).

Since there are 16 unknowns, there will be 16 equations to solve. Using the matrix representation for these equations, Eq. (10.58) is obtained from Eqs (10.48)–(10.50) and Eq. (10.57). A close look at Eq. (10.58) reveals the following.

1. The division of X and Y into relatively coarse subdivisions leads to many equations ($4 \times 4 = 16$). In a practical situation, the number of equations may be hundred or more.

2. The coefficient matrix is banded, which means that the non-zero components only appear in a band on either side of the main diagonal. There are 9 diagonals (Fig. 10.12). So, the bandwidth is large as compared to TDM. The advantage of having a banded matrix (this is true also for TDM) is that special sub-routines can be written to solve the problem in less computer time than if the matrix was filled with non-zero components.

3. The zero matrix components outside the band need not be stored in the computer. This is of great significance in large problems where the computer memory size becomes a limiting factor.

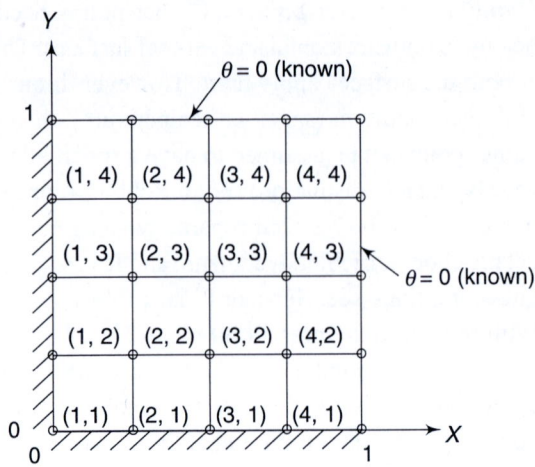

Fig. 10.10 Labelling of grid points using double-subscript notation

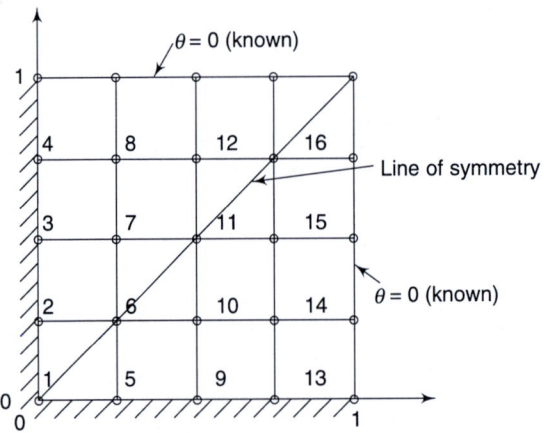

Fig. 10.11 Labelling of the same grid points using single-subscript notation

Choice of the proper method Equation (10.58) can be solved in two ways:

    (i) By Gaussian elimination

    (ii) By Gauss-Seidel iteration[2]

Let us weigh the pros and cons of both methods before we make our final choice.

It is interesting to note that the banded coefficient matrix in Eq. (10.58) has 124 components within the band rather than the 256 spaces that would have been required to store the entire matrix. One could even reduce the bandwidth by recognizing the physical and geometrical symmetry across one of the diagonals of

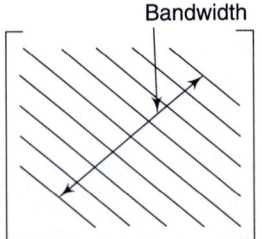

Fig. 10.12 Pictorial representation of the banded coefficient matrix showing nine diagonals

[2] See Listing A8.3 in Appendix A8. The Online Resource Centre also provides the listing (see 2D COND_Sq.c)

the square as shown in Fig. 10.12. This means that $\theta_{i,j} = \theta_{j,i}$ or $\theta_2 = \theta_5$, $\theta_3 = \theta_9$, $\theta_4 = \theta_{13}$, and so on. This would reduce the number of equations from 16 to 10.

Furthermore, it may be noted that many of the components within the band itself are zero. In this case, 60 of 124 band components are zero. These components must still be stored, however, if Gaussian elimination is to be used, because during the elimination process they will, in general, change to non-zero values. If computer storage is critical, one might prefer to use a method that does not require storing these zero components in the band. The Gauss-Seidel iteration method is one way of doing this. In addition to this, the round-off error is minimum in the Gauss–Seidel method. Therefore, in view of the aforesaid two overwhelming merits, in spite of the clear-cut advantage of Gaussian elimination because of its non-iterative nature, the GS method is chosen to solve Eq. (10.58).

$$
\begin{bmatrix}
4 & -2 & 0 & 0 & -2 & & & & & & & & & & & \\
-1 & 4 & -1 & 0 & 0 & -2 & & & & & & & & & & \\
0 & -1 & 4 & -1 & 0 & 0 & -2 & & & & & & & & & \\
0 & 0 & -1 & 4 & 0 & 0 & 0 & -2 & & & & & & & & \\
-1 & 0 & 0 & 0 & 4 & -2 & 0 & 0 & -1 & & & & & & & \\
& -1 & 0 & 0 & 1 & 4 & -1 & 0 & 0 & -1 & & & & & & \\
& & -1 & 0 & 0 & -1 & 4 & -1 & 0 & 0 & -1 & & & & & \\
& & & -1 & 0 & 0 & -1 & 4 & 0 & 0 & 0 & -1 & & & & \\
& & & & -1 & 0 & 0 & 0 & 4 & -2 & 0 & 0 & -1 & & & \\
& & & & & -1 & 0 & 0 & -1 & 4 & -1 & 0 & 0 & -1 & & \\
& & & & & & -1 & 0 & 0 & -1 & 4 & -1 & 0 & 0 & -1 & \\
& & & & & & & -1 & 0 & 0 & -1 & 4 & 0 & 0 & 0 & -1 \\
& & & & & & & & -1 & 0 & 0 & 0 & 4 & -2 & 0 & 0 \\
& & & & & & & & & -1 & 0 & 0 & -1 & 4 & -1 & 0 \\
& & & & & & & & & & -1 & 0 & 0 & -1 & 4 & -1 \\
& & & & & & & & & & & -1 & 0 & 0 & -1 & 4
\end{bmatrix}
\begin{bmatrix}
\theta_1 \\ \theta_2 \\ \theta_3 \\ \theta_4 \\ \theta_5 \\ \theta_6 \\ \theta_7 \\ \theta_8 \\ \theta_9 \\ \theta_{10} \\ \theta_{11} \\ \theta_{12} \\ \theta_{13} \\ \theta_{14} \\ \theta_{15} \\ \theta_{16}
\end{bmatrix}
=
\begin{bmatrix}
1/16 \\ 1/16 \\ 1/16 \\ 1/16 \\ 1/16 \\ 1/16 \\ 1/16 \\ 1/16 \\ 1/16 \\ 1/16 \\ 1/16 \\ 1/16 \\ 1/16 \\ 1/16 \\ 1/16 \\ 1/16
\end{bmatrix}
$$

(10.58)

Check for accuracy For the present problem, the accuracy of the numerical results can be checked by comparing it with the corresponding analytical solution available. Subsequently, a grid independence test must be done to obtain the desired results. The analytical (or exact) solution in the dimensionless form is given below:

$$
\theta(X, Y) = \frac{1}{2}[1 - X^2] - 2\sum_{n=0}^{\infty} \frac{(-1)^n \cosh \lambda_n Y}{\lambda_n^3 \cosh \lambda_n} \cos \lambda_n X
$$

(10.59)

where $\lambda_n = (2n + 1)\,\pi/2$, where $n = 0, 1, 2, \ldots$.

10.4 Transient One-dimensional Problems

Consider a hot infinite plate (Fig. 10.13) of finite thickness $2L$. The plate is suddenly exposed to a cool fluid at T_∞. The initial temperature of the plate is T_i ($T_i > T_\infty$). The heat transfer coefficient is large. We wish to find the temperature of the plate as a function of space and time using a numerical method.

The problem can be modelled as a one-dimensional, unsteady-state problem because $\partial T/\partial y = \partial T/\partial z = 0$ as the plate is infinitely long in y and z directions.

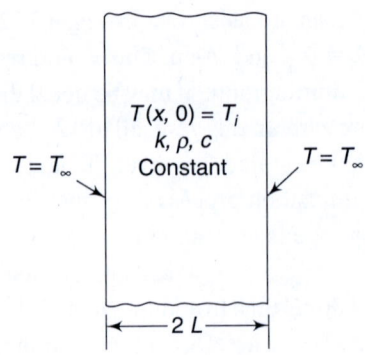

Consideration of symmetry Since the problem is a thermally and geometrically symmetric one, only one-half of the plate can be taken as the computational domain with the insulation boundary condition at $x = L$ (Fig. 10.14).

Governing differential equation For constant thermophysical properties k, ρ, c, the non-dimensional energy equation for the plate is

$$\frac{\partial \theta}{\partial \tau} = \frac{\partial^2 \theta}{\partial X^2} \qquad (10.60a)$$

where $\quad \theta = \dfrac{T - T_\infty}{T_i - T_\infty}, X = \dfrac{x}{L}, \tau = \dfrac{\alpha t}{L^2}$

Fig. 10.13 Physical domain of the one-dimensional transient conduction in an infinite plane slab

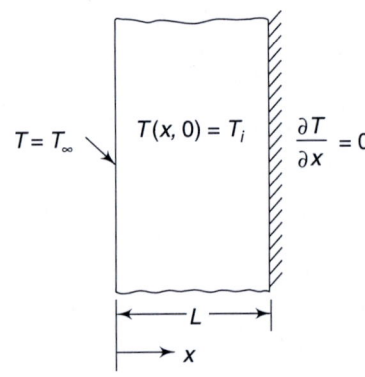

Initial and boundary conditions The initial and boundary conditions are as follows:

IC: at $\tau = 0$, $\theta = 1$ for all X (10.60b)

For $\tau > 0$,

BC-1: at $X = 0$, $\theta = 0$ (10.60c)

BC-2: at $X = 1$, $\dfrac{\partial \theta}{\partial X} = 0$ (10.60d)

Fig. 10.14 Computational domain of the one-dimensional transient conduction problem

Discretization For any interior grid point, the finite-difference formulation gives

$$\frac{\partial \theta_i}{\partial \tau} = \frac{\theta_{i-1} - 2\theta_i + \theta_{i+1}}{(\Delta X)^2} \quad \text{for } i = 1, \dots, M \qquad (10.61)$$

The equation for $X = 1$ is obtained by using the image-point technique, i.e., by substituting $\theta_{M+1} = \theta_{M-1}$ in Eq. (10.61) for $i = M$. Therefore,

$$\frac{\partial \theta_M}{\partial \tau} = \frac{2\theta_{M-1} - 2\theta_M}{(\Delta X)^2} \qquad (10.62)$$

For the sake of demonstration, let us take four equal subdivisions in the X-direction (Fig. 10.15). Therefore, $\Delta X = 1/4$.

At $i = 0$, $\theta = 0$ (known)

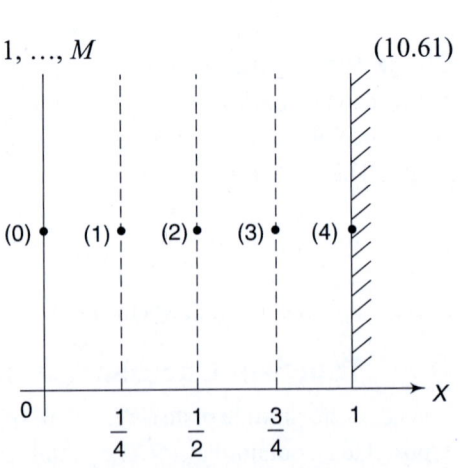

Fig. 10.15 Equally spaced grid points in the x-direction of the computational domain

At $i = 1, 2, 3$, from Eq. (10.61), we obtain

$$\frac{d\theta_1}{d\tau} = \frac{1}{(\Delta X)^2}(\theta_0 - 2\theta_1 - \theta_2) = \frac{1}{(\Delta X)^2}(-2\theta_1 + \theta_2) \tag{10.63}$$

$$\frac{d\theta_2}{d\tau} = \frac{1}{(\Delta X)^2}(\theta_1 - 2\theta_2 + \theta_3) \tag{10.64}$$

$$\frac{d\theta_3}{d\tau} = \frac{1}{(\Delta X)^2}(\theta_2 - 2\theta_3 + \theta_4) \tag{10.65}$$

At $i = 4$, from Eq. (10.62), we obtain

$$\frac{d\theta_4}{d\tau} = \frac{1}{(\Delta X)^2}(2\theta_3 - 2\theta_4) \tag{10.66}$$

Thus, we have four simultaneous ordinary differential equations [Eqs (10.63)–(10.66)] to solve. This system of ordinary differential equations may be classified as initial-value problems. This is because these equations are to be solved for the unknowns as a function of time, beginning with an initial value for each of the unknowns. In this case the initial values are obtained from the initial temperature distribution in the plate, which is given by

$$\theta_1(0) = \theta_2(0) = \theta_3(0) = \theta_4(0) = 1 \tag{10.67}$$

10.4.1 Methods of Solution

There are three methods by which this initial-value problem can be solved. These are the (a) Euler (or explicit), (b) Crank–Nicolson, and (c) pure implicit methods.

Euler method (also known as the explicit method) Since the given problem is an initial-value one, we will know the solution θ^p and will seek θ^{p+1} at some later point in time $\tau^{p+1} = \tau^p + \Delta\tau$. In the Euler method of solution, the solution at a future time τ^{p+1} is obtained by computing the derivative at the present time τ^p and then by moving ahead in time in the following way:

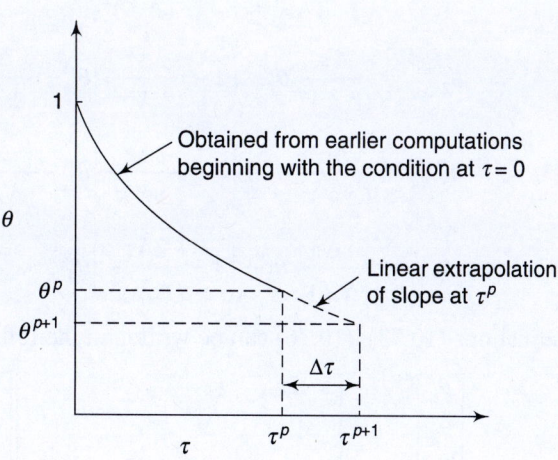

Fig. 10.16 Pictorial representation of the Euler method

$$\theta^{p+1} = \theta^p + \left.\frac{d\theta}{d\tau}\right|^p \Delta\tau \tag{10.68a}$$

The Euler scheme is pictorially represented in Fig. 10.16. For the grid points 1, 2, 3, 4, Eq. (10.68a) can be written as

$$\theta_i^{p+1} = \theta_i^p + \left.\frac{d\theta_i}{d\tau}\right|^p \Delta\tau \tag{10.68b}$$

$$\theta_2^{p+1} = \theta_2^p + \frac{d\theta_2}{d\tau}\bigg|^p \Delta\tau \tag{10.68c}$$

$$\theta_3^{p+1} = \theta_3^p + \frac{d\theta_3}{d\tau}\bigg|^p \Delta\tau \tag{10.68d}$$

$$\theta_4^{p+1} = \theta_4^p + \frac{d\theta_4}{d\tau}\bigg|^p \Delta\tau \tag{10.68e}$$

Substituting Eq. (10.68b) into Eq. (10.63), Eq. (10.68c) into Eq. (10.64), Eq. (10.68d) into Eq. (10.65), Eq. (10.68e) into Eq. (10.66), we obtain Eqs (10.69)–(10.72) as shown below.

$$\frac{\theta_1^{p+1} - \theta_1^p}{\Delta\tau} = \frac{1}{(\Delta X)^2}(-2\theta_1^p + \theta_2^p) \tag{10.69}$$

$$\frac{\theta_2^{p+1} - \theta_2^p}{\Delta\tau} = \frac{1}{(\Delta X)^2}(\theta_1^p - 2\theta_1^p + \theta_3^p) \tag{10.70}$$

$$\frac{\theta_3^{p+1} - \theta_3^p}{\Delta\tau} = \frac{1}{(\Delta X)^2}(\theta_2^p - 2\theta_3^p + \theta_4^p) \tag{10.71}$$

$$\frac{\theta_4^{p+1} - \theta_4^p}{\Delta\tau} = \frac{1}{(\Delta X)^2}(2\theta_3^p - 2\theta_4^p) \tag{10.72}$$

Equations (10.69)–(10.72) are then rearranged to give

$$\theta_1^{p+1} = \left(1 - \frac{2\Delta\tau}{(\Delta X)^2}\right)\theta_1^p + \frac{\Delta\tau}{(\Delta X)^2}\theta_2^p \tag{10.73}$$

$$\theta_2^{p+1} = \frac{\Delta\tau}{(\Delta X)^2}\theta_1^p + \left(1 - \frac{2\Delta\tau}{(\Delta X)^2}\right)\theta_2^p + \frac{\Delta\tau}{(\Delta X)^2}\theta_3^p \tag{10.74}$$

$$\theta_3^{p+1} = \frac{\Delta\tau}{(\Delta X)^2}\theta_2^p + \left(1 - \frac{2\Delta\tau}{(\Delta X)^2}\right)\theta_3^p + \frac{\Delta\tau}{(\Delta X)^2}\theta_4^p \tag{10.75}$$

$$\theta_4^{p+1} = \frac{2\Delta\tau}{(\Delta X)^2}\theta_3^p + \left(1 - \frac{2\Delta\tau}{(\Delta X)^2}\right)\theta_4^p \tag{10.76}$$

Equations (10.73)–(10.76) can be written in the following matrix form:

$$\begin{bmatrix}\theta_1 \\ \theta_2 \\ \theta_3 \\ \theta_4\end{bmatrix}^{p+1} = \begin{bmatrix}1-2r & r & & \\ 4 & 1-2r & r & \\ & r & 1-2r & r \\ & & 2r & 1-2r\end{bmatrix}\begin{bmatrix}\theta_1 \\ \theta_2 \\ \theta_3 \\ \theta_4\end{bmatrix}^p \tag{10.77}$$

where $r = \Delta\tau/(\Delta X)^2$.

The TDM on the right-hand side of Eq. (10.77) is known (and constant) once the size of the time step $\Delta\tau$ is chosen. The known values of θ at τ^p (i.e., θ^p) are multiplied by this TDM to obtain the new values of θ at τ^{p+1}. This matrix multiplication is quite easy to carry out on the computer since only the non-zero terms will contribute to the calculation. Thus, θ^{p+1} values are obtained explicitly in terms of θ^p values and hence the name, *explicit method*. Note that the solution of simultaneous algebraic equations is not necessary in this scheme, which makes it a very attractive method. Once θ^{p+1} are obtained, they are stored in θ^p, and the computation is repeated for the next time step. This procedure continues until the result at the desired time is obtained or till the steady state is reached.

However, a major drawback of this method is that for $r > 0.5$, that is when $1 - 2r$ is negative, the solution becomes unstable. Therefore, a stability limit of $r \leq 0.5$ is imposed, which results in considerable restriction on the time step $\Delta\tau$ for a particular value of ΔX. Thus more computer time is required to obtain the desired solution at a particular point in time. The stability is discussed in detail in Section 10.4.2.

Crank–Nicolson method In the Euler method the value of the derivative at the beginning of the time interval was used to progress in time. A more accurate method would be to use the arithmetic mean value of the derivatives at the beginning and at the end of the time interval, i.e., use the time derivative at $p + 1/2$, a time which is midway between p and $p + 1$. Therefore,

$$\theta^{p+1} = \theta^p + \frac{1}{2}\left[\frac{d\theta}{d\tau}\bigg|^p + \frac{d\theta}{d\tau}\bigg|^{p+1}\right]\Delta\tau \tag{10.78}$$

Substituting Eqs (10.63)–(10.66) in Eq. (10.78), we obtain

$$\frac{\theta_1^{p+1} - \theta_1^p}{\Delta\tau} = \frac{1}{2(\Delta x)^2}[(-2\theta_1^p + \theta_2^p) + (-2\theta_1^{p+1} + \theta_2^{p+1})] \tag{10.79}$$

$$\frac{\theta_2^{p+1} - \theta_2^p}{\Delta\tau} = \frac{1}{2(\Delta x)^2}[(\theta_1^p - 2\theta_2^p + \theta_3^p) + (\theta_1^{p+1} - 2\theta_2^{p+1} + \theta_3^{p+1})] \tag{10.80}$$

$$\frac{\theta_3^{p+1} - \theta_3^p}{\Delta\tau} = \frac{1}{2(\Delta x)^2}[(\theta_2^p - 2\theta_3^p + \theta_4^p) + (\theta_2^{p+1} - 2\theta_3^{p+1} + \theta_4^{p+1})] \tag{10.81}$$

$$\frac{\theta_4^{p+1} - \theta_4^p}{\Delta\tau} = \frac{1}{2(\Delta x)^2}[(2\theta_3^p - 2\theta_4^p) + (2\theta_3^{p+1} - 2\theta_4^{p+1})] \tag{10.82}$$

Equations (10.79)–(10.82) are rearranged, which results in a set of four simultaneous algebraic equations in $\theta_1^{p+1}, \theta_2^{p+1}, \theta_3^{p+1}, \theta_4^{p+1}$, represented in the matrix form as

$$\begin{bmatrix} 1+r & -r/2 & & \\ -r/2 & 1+r & -r/2 & \\ & -r/2 & 1+r & -r/2 \\ & & -r & 1+r \end{bmatrix}\begin{bmatrix} \theta_1 \\ \theta_2 \\ \theta_3 \\ \theta_4 \end{bmatrix}^{p+1}$$

$$= \begin{bmatrix} 1-r & (r/2) & & \\ (r/2) & 1-r & (r/2) & \\ & (r/2) & 1-r & (r/2) \\ & & r & 1-r \end{bmatrix}\begin{bmatrix} \theta_1 \\ \theta_2 \\ \theta_3 \\ \theta_4 \end{bmatrix}^{p} \tag{10.83}$$

Similar to the case in the Euler method, the right-hand side can be computed directly because all the components are known. This results in a column matrix as before. The difference arises in the fact that this does not give an explicit result for the unknowns on the left-hand side; rather an implicit TDM system of algebraic equations results. This system of algebraic equations must then be solved in each time step. The Crank–Nicolson method, although a stable method, gives erroneous results in the early time if the time step is too large. However, the error damps out as time progresses towards the steady state.

Pure implicit method   In contrast with the Euler or the Crank–Nicolson method, in the pure implicit scheme, the time derivative at the new time is used to move ahead in time. Thus,

$$\theta^{p+1} = \theta^{p} + \frac{d\theta}{d\tau}\bigg|^{p+1} \Delta\tau \tag{10.84}$$

From Eqs (10.63)–(10.66) and Eq. (10.84), we obtain

$$\frac{\theta_1^{p+1} - \theta_1^{p}}{\Delta\tau} = \frac{1}{(\Delta X)^2}[(-2\theta_1^{p+1} + \theta_2^{p+1})] \tag{10.85}$$

$$\frac{\theta_2^{p+1} - \theta_2^{p}}{\Delta\tau} = \frac{1}{(\Delta X)^2}[(\theta_1^{p+1} + 2\theta_2^{p+1} + \theta_3^{p+1})] \tag{10.86}$$

$$\frac{\theta_3^{p+1} - \theta_3^{p}}{\Delta\tau} = \frac{1}{(\Delta X)^2}[(\theta_2^{p+1} - 2\theta_3^{p+1} + \theta_4^{p+1})] \tag{10.87}$$

$$\frac{\theta_4^{p+1} - \theta_4^{p}}{\Delta\tau} = \frac{1}{(\Delta X)^2}[(2\theta_3^{p+1} - 2\theta_4^{p+1})] \tag{10.88}$$

$$\begin{bmatrix} 1+2r & -r & & \\ -r & 1+2r & -r & \\ & -r & 1+2r & -r \\ & & -2r & 1+2r \end{bmatrix} \begin{bmatrix} \theta_1 \\ \theta_2 \\ \theta_3 \\ \theta_4 \end{bmatrix}^{p+1} = \begin{bmatrix} \theta_1 \\ \theta_2 \\ \theta_3 \\ \theta_4 \end{bmatrix}^{p} \tag{10.89}$$

Equation (10.89) is an implicit set of equations to solve for the new temperatures at each time step. The pure implicit scheme is an unconditionally stable scheme, that is, there is no restriction on the time step, which is in sharp contrast with the Euler and the Crank–Nicolson method.

However, the Euler and the pure implicit methods have the same order of accuracy, while the Crank–Nicolson method is more accurate than either of the two for the same time step. The accuracy and stability of each of the three methods are detailed in the sections to follow.

10.4.1.1   Accuracy of the Euler, Crank–Nicolson, and pure implicit methods

In the Euler method, at any grid point i, $d\theta/d\tau$ is evaluated at τ^p, that is,

$$\frac{d\theta}{d\tau}\bigg|_i^p = \frac{\theta_i^{p+1} - \theta_i^{p}}{\Delta\tau} \tag{10.90}$$

Equation (10.90) is forward difference in time. Therefore, the order of accuracy in time is $0(\Delta\tau)$.

In the pure implicit method, $d\theta/d\tau$ is evaluated at τ^{p+1}, that is,

$$\frac{d\theta}{d\tau}\Big|_i^{p+1} = \frac{\theta_i^{p+1} - \theta_i^p}{\Delta\tau} \tag{10.91a}$$

Equation (10.91a) actually arises from Eq. (10.91b) shown below.

$$\frac{d\theta}{d\tau}\Big|_i^{p+1} = \frac{\theta_i^{p+1} - \theta_i^{(p+1)-1}}{\Delta\tau} \tag{10.91b}$$

Although the RHS of Eq. (10.91a) looks the same as that of Eq. (10.90), the former is actually backward difference in time, which is obvious from Eq. (10.91b). Therefore, the order of accuracy in time is $0(\Delta\tau)$.

In the Crank–Nicolson method, $d\theta/d\tau$ is evaluated at $p + 1/2$, that is,

$$\frac{d\theta}{d\tau}\Big|_i^{p+1/2} = \frac{\theta_i^{p+1} - p - \theta_i^p}{2\left(\dfrac{\Delta\tau}{2}\right)} \tag{10.92a}$$

$$\Rightarrow \qquad \frac{d\theta}{d\tau}\Big|_i^{p+1/2} = \frac{\theta_i^{p+1} - \theta_i^p}{\Delta\tau} \tag{10.92b}$$

Again, although the RHS of Eq. (10.92b) looks the same as that of Eqs (10.90) and (10.91a), the former is actually central difference in time, which is obvious from Eq. (10.92a). Therefore, the order of accuracy in time is $0(\Delta\tau)^2$. This explains why the Crank–Nicolson scheme is one order more accurate in time as compared to the Euler or pure implicit scheme.

In all the three methods, the space derivatives are discretized using the central-difference scheme. Therefore, the order of accuracy in space in the Euler, pure implicit, and Crank–Nicolson methods is $0(\Delta X)^2$.

The Euler method, the Crank-Nicolson method, and the pure implicit method are also called FTCS (forward-time, central space), CTCS (central-time, central space), and BTCS (backward-time, central space), respectively.

To summarize, the order of accuracy of each method can be written as follows.

Euler or explicit: $0[(\Delta X)^2, (\Delta\tau)]$ FTCS

Crank-Nicolson: $0[(\Delta X)^2, (\Delta\tau)^2]$ CTCS

Pure Implicit: $0[(\Delta X)^2, (\Delta\tau)]$ BTCS

10.4.2 Stability: Numerically Induced Oscillations

From the preceding discussion, it is apparent that all the three schemes will give better results if time steps are made smaller. In practice, however, one would usually like to take as large a time step as one can to reduce the computational effort and time. In addition to decreasing the accuracy of the solution, large time steps can introduce some unwanted, numerically induced oscillations into the solution, making it physically unrealistic. Such solutions are not acceptable and the method that produces such a solution is called *unstable method*. This brings us to the formal definition of a stable numerical scheme, which is the one for which errors from any source (round-off, truncation, mistakes, etc.) are not permitted to grow in the sequence of numerical procedures as the calculation proceeds from one step to the next.

Case of one grid point   Consider the case of only one grid point, that is, the grid point on the insulated boundary of a plate (Fig. 10.17). Therefore, $\Delta X = 1$, $r = \Delta \tau$. Hence, we have only one equation to solve, that is,

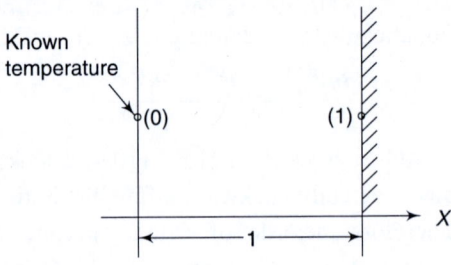

$$\frac{d\theta_1}{d\tau} = -2\theta_1 \qquad (10.93a)$$

subject to the initial condition

Fig. 10.17   The case of only one grid point

$$\theta_1(0) = 1 \qquad (10.93b)$$

The analytical solution of Eq. (10.93a) is

$$\theta_1 = e^{-2\tau} \qquad (10.94)$$

The following three equations are obtained corresponding to the three numerical schemes:

Euler: $\theta_1^{p+1} = (1-2r)\theta_1^{p}$ $\qquad (10.95)$

Crank–Nicolson: $(1+r)\,\theta_1^{p+1} = (1-r)\theta_1^{p}$ $\qquad (10.96)$

Pure implicit: $(1+2r)\,\theta_1^{p+1} = \theta_1^{p}$ $\qquad (10.97)$

Each of these may be put in the following general form:

$$\theta^{p+1} = \lambda\theta^{p} \qquad (10.98)$$

where λ is defined by

Euler: $\lambda = 1 - 2r$ $\qquad (10.99)$

Crank–Nicolson: $\lambda = \dfrac{1-r}{1+r}$ $\qquad (10.100)$

Pure implicit: $\lambda = \dfrac{1}{1+2r}$ $\qquad (10.101)$

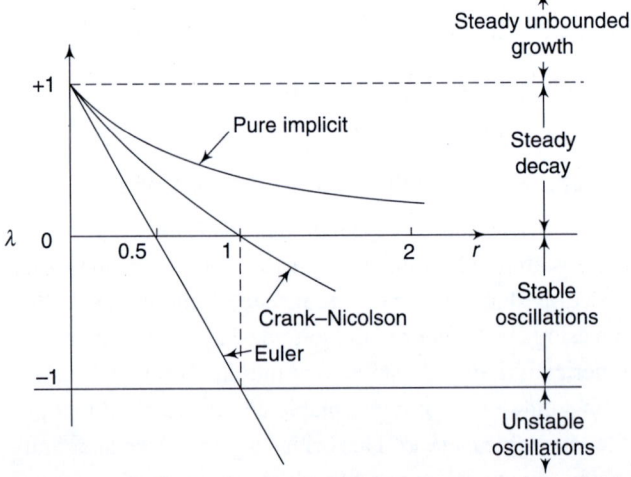

Fig. 10.18   Stability curves for the case of one grid point

The value of λ determines the character of the solution. This is self-explanatory from Fig. 10.18, which shows λ as a function of r ($r = \Delta\tau$ for this special case) for each of the three numerical methods we have considered.

A close inspection of Fig. 10.18 reveals that as $r \to 0$, that is, if the time step is made smaller and smaller, all three schemes become identical. As the time step is increased, in each case, the solutions begin to deviate from one another. The Euler method can have steady decay, stable oscillations, or unstable oscillations. The Crank–Nicolson method can have either steady decay or stable oscillations. The pure implicit method has only a steadily decaying type of solution. From the graph, it is also seen that the stability limit for the Euler method is 0.5 while that for the Crank–Nicolson method is 1.0. The Euler method becomes totally unstable at $r = 1.0$. While the Euler method is called conditionally stable, the Crank–Nicolson method is called unconditionally stable because the oscillations ultimately damp out with time. The pure implicit method is truly unconditionally stable method.

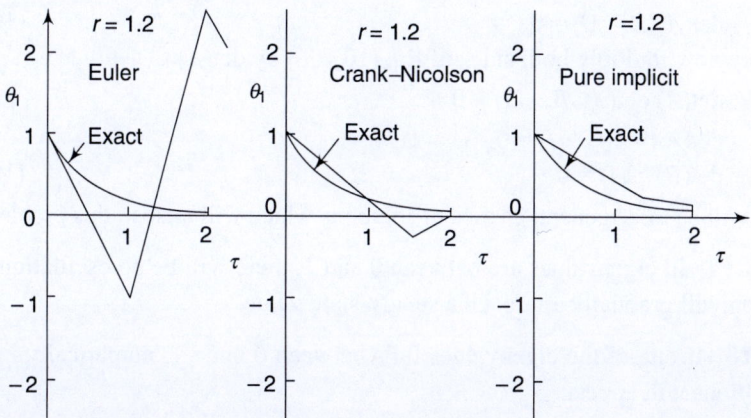

Fig. 10.19 Comparison of the numerical solutions based on the Euler, Crank–Nicolson, and pure implicit methods with the corresponding exact solution for the case of one grid point

Figure 10.19 compares the three numerical solutions to the corresponding exact solution (drawn qualitatively) $\theta_1 = e^{-2\tau}$ for $r = 1.2$ which exceeds the stability limit for both the Euler method and the Crank–Nicolson method.

The figures reveal that oscillations in the Euler solution grow without bound, and oscillations are seen in the Crank–Nicolson solution but gradually damp out for large time. The pure implicit solution does not show any oscillations.

Case of more than one grid point The example of one grid point may be extended to the more general case in which there are more than one grid points, that is, more than one equation. The matrix representation for any of three numerical schemes can be written as

$$A\theta^{p+1} = B\theta^p \tag{10.102}$$

where A and B are matrices that depend on the particular method. Equation (10.102) can be written as

$$\theta^{p+1} = A^{-1}B\theta^p \tag{10.103}$$

Note that in the right-hand side of Eq. (10.103) $A^{-1}B$ is a square matrix. We also know that associated with every square matrix (let us call this matrix S) are a special set of vectors, called *eigenvectors*, and a related set of scalars, called *eigenvalues*. Formally, the vector x is an eigenvector of S if and only if x is a non-zero vector and λ is a scalar (which may be zero), such that

$$Sx = \lambda x \qquad (10.104)$$

The scalar λ is an eigenvalue of S if and only if there exists a non-zero vector x such that Eq. (10.104) holds.

The eigenvalues λ of the matrix $A^{-1}B$ play a similar role to the λ in the case of one grid point [see Eq. (10.98)]. If there are N simultaneous equations being handled, there will be N eigenvalues of $A^{-1}B$. These values will determine the character of the solution. Now,

$$A^{-1}B\theta^p = \lambda\theta^p$$
or $\quad (A^{-1}B - \lambda I)\theta^p = 0 \qquad (10.105)$

where I is an $N \times N$ unit matrix. To get a non-trivial solution of Eq. (10.105),

$$\det(A^{-1}B - \lambda I) = 0 \qquad (10.106)$$

We may now multiply both sides of Eq. (10.106) by $\det(A)$ to get

$$\det(A)\det(A^{-1}B - \lambda I) = 0$$
or $\quad \det(AA^{-1}B - A\lambda I) = 0$
or $\quad \det(B - \lambda A) = 0 \qquad (10.107)$

There will be three general classes of solutions which will arise in this problem.

Case I   If all eigenvalues are between 0 and 1, there will be no oscillations. The solution will gradually approach a steady-state value.

Case II   If one of the eigenvalues falls between 0 and -1, numerically induced oscillations will appear.

Case III   If one of the eigenvalues is less than -1, the oscillations will be unstable.

An example of the two-grid-point case for the Euler method   As an example, let us consider the two-grid-point case for the Euler method. The matrices A and B are then given by

$$A = \begin{bmatrix} 1 & 0 \\ 0 & 1 \end{bmatrix}$$

$$B = \begin{bmatrix} 1 - 2r & r \\ 2r & 1 - 2r \end{bmatrix}$$

Then, $\quad B - \lambda A = \begin{bmatrix} 1 - 2r\lambda & r \\ 2r & 1 - 2r - \lambda \end{bmatrix}$

The determinant of the above matrix is given by

$$\det(B - \lambda A) = (1 - 2r - \lambda)^2 - 2r^2 = 0 \qquad (10.108)$$

Solving Eq. (10.108), we get

$$\lambda_1 = 1 - r(2 + \sqrt{2})$$
$$\lambda_2 = 1 - r(2 - \sqrt{2})$$

The value of λ_1 will determine the character of the solution since it is this value that is most likely to be negative (because of the larger coefficient of r).

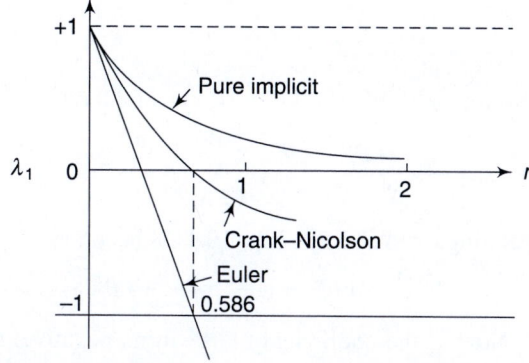

The λ versus r plots (Fig. 10.20) show the same general trend of Fig. 10.18, but the curves have shifted to the left so that the critical values of r are smaller than those for the one-grid-point case. The upper limit for stable oscillations of the Euler method is now $2/(2 + \sqrt{2}) = 0.586$ [since $\lambda_{\text{crit}} = -1 = 1 - r(2 + \sqrt{2})$] as compared to 1.0 in the one-grid-point case.

Fig. 10.20 Stability curves for the case of two grid points

In the limit as the number of nodes becomes infinite, it can be shown that the stability limit for the Euler method approaches 0.5 (Myers 1971).

10.4.2.1 Convection boundary condition

Figure 10.21 shows the same problem but with a finite heat transfer coefficient h. Solving the problem using the Euler method (keeping the grid spacings the same as before), the following matrix equation is obtained:

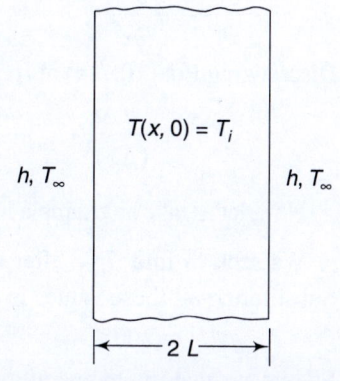

Fig. 10.21 Physical domain of the one-dimensional transient problem with convective boundary conditions

$$\theta^{p+1} = B\theta^p \qquad (10.109)$$

where $B =$

$$\begin{bmatrix} 1 - 2r(1 + \mathrm{Bi}\,\Delta X)2r & & & & \\ r & 1 - 2r & r & & \\ & r & 1 - 2r & r & \\ & & r & 1 - 2r & r \\ & & & 2r & 1 - 2r \end{bmatrix}$$

The critical stability limit can be shown to be (Myers 1971)

$$r_{\text{crit}} = 0.5\,\frac{1}{1 + \mathrm{Bi}\,\Delta X} \qquad (10.110)$$

where Bi is the Biot number and is equal to hL/k. It is obvious from Eq. (10.110) that the stability limit for the convection boundary condition becomes more restrictive.

10.4.3 Stability Limit of the Euler Method from Physical Standpoint

Let us consider the non-dimensional governing equation for a one-dimensional transient condition.

$$\frac{\partial\theta}{\partial t} = \frac{\partial^2\theta}{\partial X^2} \tag{10.111}$$

Discretizing Eq. (10.111) at (p, i) using the Euler method, we obtain

$$\frac{\theta_i^{p+1} - \theta_i^p}{\Delta\tau} = \frac{\theta_{i-1}^p - 2\theta_i^p + \theta_{i+1}^p}{(\Delta X)^2} \tag{10.112}$$

Defining $r = \Delta\tau/(\Delta X)^2$, Eq. (10.112) becomes

$$\theta_i^{p+1} = r\,\theta_{i-1}^p + (1 - 2r)\,\theta_i^p + r\,\theta_{i+1}^p \tag{10.113}$$

Now, if the coefficient of θ_i^p turns negative, that is, if $r > 1/2$ we generate a condition that will violate the second law of thermodynamics. This can be better explained by using the dimensional form of Eq. (10.111), that is,

$$\frac{\partial T}{\partial t} = \alpha\frac{\partial^2 T}{\partial x^2} \tag{10.114}$$

Discretizing Eq. (10.114) at (p, i) using the Euler method, we obtain

$$T_i^{p+1} = \frac{\alpha\,\Delta t}{(\Delta x)^2}[T_{i-1}^p + T_{i+1}^p] + \left[1 - \frac{2\alpha\,\Delta t}{(\Delta x)^2}\right]T_i^p \tag{10.115}$$

Now, let us take an example in which $T_{i+1}^p = 100°C$, $T_i^p = 0°C$, and $T_{i-1}^p = 100°C$.

We seek to find T_{i+1}^p after an interval of Δt. Let us also take $\alpha\Delta t/(\Delta x)^2 = 1$. Substituting all these values in Eq. (10.115), we obtain

$$T_i^{p+1} = 200°C$$

So, we see that the temperature at grid point i at time $p + 1$ will exceed the temperature at the two neighbouring grid points at time level p (Fig. 10.22). This seems unreasonable since the maximum temperature that we expect to find at point i at time $p + 1$ is $100°C$. So, the aforesaid result is in clear violation of the second law of thermodynamics.

The reason for this is that by using $\alpha\Delta t/(\Delta x)^2 = 1$, the coefficient of T_i^p becomes -1, that is, negative. Therefore, a simple way of obtaining the stability limit is to set the coefficient of the temperature at the present time level of the grid point, at which the solution is desired, greater than or equal to zero. Therefore, for stability,

$$1 - \frac{\alpha\,\Delta t}{(\Delta x)^2} \geq 0 \tag{10.116}$$

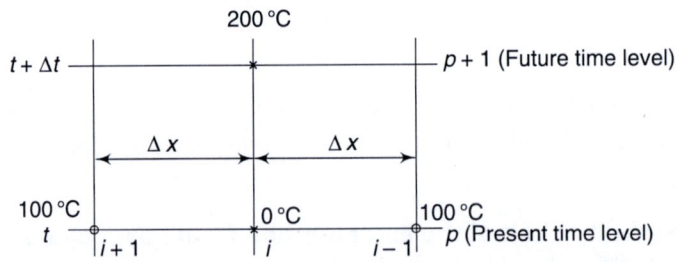

Fig. 10.22 An example showing a physically impossible solution arising out of $\alpha\Delta t/(\Delta x)^2$

> 1/2 in the Euler method

or $\qquad \dfrac{\alpha \, \Delta t}{(\Delta x)^2} \leq \dfrac{1}{2}$ \hfill (10.117)

The above principle is also applicable to two- or three-dimensional transient heat conduction problems.

10.5 Two-dimensional Transient Heat Conduction Problems

The governing differential equation for the two-dimensional transient conduction (assuming constant k, ρ, c) is

$$\frac{\partial T}{\partial t} = \alpha \left(\frac{\partial^2 T}{\partial x^2} + \frac{\partial^2 T}{\partial y^2} \right) \tag{10.118}$$

The Euler or explicit method of solution leads to

$$\frac{T_{i,j}^{p+1} - T_{i,j}^{p}}{\Delta t} = \alpha \left[\frac{T_{i-1,j}^{p} - 2T_{i,j}^{p} + T_{i+1,j}^{p}}{(\Delta x)^2} + \frac{T_{i,j-1}^{p} - 2T_{i,j}^{p} + T_{i,j+1}^{p}}{(\Delta y)^2} \right] \tag{10.119}$$

The solution of Eq. (10.119) presents no difficulties except the following stability criterion has to be satisfied:

$$\Delta t \leq \frac{1}{2\alpha \left[(\Delta x)^{-2} + (\Delta y)^{-2} \right]} \tag{10.120}$$

On the other hand, the pure implicit scheme leads to

$$\frac{T_{i,j}^{p+1} - T_{i,j}^{p}}{\Delta t} = \alpha \left[\frac{T_{i-1,j}^{p+1} - 2T_{i,j}^{p+1} + T_{i+1,j}^{p+1}}{(\Delta x)^2} + \frac{T_{i,j-1}^{p+1} - 2T_{i,j}^{p+1} + T_{i,j+1}^{p+1}}{(\Delta y)^2} \right] \tag{10.121}$$

which, for $\Delta x = \Delta y$, transforms to

$$-rT_{i-1,j}^{p+1} - rT_{i,j-1}^{p+1} + (1 + 4r)T_{i,j}^{p+1} - rT_{i,j+1}^{p+1} - rT_{i+1,j}^{p+1} = T_{i,j}^{p} \tag{10.122}$$

where $r = \alpha \Delta t / (\Delta x)^2$.

As in the one-dimensional case, this scheme is unconditionally stable and independent of r. There are now five unknowns per equation, and Gaussian elimination can still be used but only at the expense of considerable amount of computation and cumulative round-off error. An alternative is to use the Gauss-Seidel iteration, although this method may be slow to converge. The Crank-Nicolson method will involve solving a large number of equations in each time step, and the associated difficulties are more or less similar to that in the pure implicit scheme.

10.5.1 Alternating Direction Implicit Method

The alternating direction implicit method (ADI) circumvents the aforementioned difficulties and yet makes use of the TDMA to obtain a direct solution. The main advantage of this method is its non-iterative character, which enables the user to obtain the solution in much less computer time. In other words, it is a direct and, hence, fast method.

Essentially, the principle is to employ two difference equations, which are used in turn over successive time steps, each of duration $\Delta t/2$. The first equation is implicit only in the x-direction and the second is implicit only in the y-direction. Thus, if $T_{i,j}^*$ is an intermediate value at the end of the first half time step, we have

$$\frac{T_{i,j}^* - T_{i,j}^p}{\Delta t/2} = \alpha \left[\frac{T_{i-1,j}^* - 2T_{i,j}^* + T_{i+1,j}^*}{(\Delta x)^2} + \frac{T_{i,j-1}^p - 2T_{i,j}^p + T_{i,j+1}^p}{(\Delta y)^2} \right] \quad (10.123)$$

followed by

$$\frac{T_{i,j}^{p+1} - T_{i,j}^*}{\Delta t/2} = \alpha \left[\frac{T_{i-1,j}^* - 2T_{i,j}^* + T_{i+1,j}^*}{(\Delta x)^2} + \frac{T_{i,j-1}^{p+1} - 2T_{i,j}^{p+1} + T_{i,j+1}^{p+1}}{(\Delta y)^2} \right] \quad (10.124)$$

in the next half time step.

For simplicity, taking $\Delta x = \Delta y$, Eqs (10.123) and (10.124), respectively, become

$$-T_{i-1,j}^* + 2\left(\frac{1}{r}+1\right)T_{i,j}^* - T_{i+1,j}^* = T_{i,j-1}^p + 2\left(\frac{1}{r}-1\right)T_{i,j}^p + T_{i,j+1}^p \quad (10.125)$$

$$-T_{i,j-1}^{p+1} + 2\left(\frac{1}{r}+1\right)T_{i,j}^{p+1} - T_{i,j+1}^{p+1} = T_{i-1,j}^* + 2\left(\frac{1}{r}-1\right)T_{i,j}^* + T_{i+1,j}^* \quad (10.126)$$

where $r = \alpha \Delta t/(\Delta x)^2$

Equation (10.125) is solved for the intermediate values T^*, which are then used in Eq. (10.126), thus leading to the solution $T_{i,j}^{p+1}$ at the end of the whole time interval Δt. The method is unconditionally stable and the truncation error is $0[(\Delta x)^2, (\Delta y)^2, (\Delta t)^2]$. Thus, it is as accurate as the Crank–Nicolson scheme and more accurate than the Euler and pure implicit methods for the same time step and grid sizes.

10.6 Problems in Cylindrical Geometry: Handling of the Condition at the Centre

In this section numerical methods for the solution of axisymmetric and non-axisymmetric heat conduction problems in cylindrical coordinates are discussed, with special emphasis on the handling of the condition at the centre of the cylindrical domain.

10.6.1 Axisymmetric Problems

Example 10.2 Transient Axisymmetric Heat Conduction in a Long Cylinder: Formulation by Pure Implicit Method[3]

An infinitely long cylinder of radius r_0 is initially at a uniform temperature T_i. It is suddenly immersed in a bath of hot fluid maintained at T_∞. The heat transfer coefficient between the bath and the cylindrical surface is h. Assume constant k, ρ, c of the cylinder. Formulate the problem to find the temperature history T(r, t). Use the pure implicit finite-difference technique.

Solution

Governing differential equation (GDE) Using the non-dimensional parameters

$$\tau = \frac{\alpha t}{r_0^2}, R = \frac{r}{r_0}, \theta = \frac{T - T_i}{T_\infty - T_i} \quad (10.127)$$

the dimensional energy equation of the cylinder transforms to

$$\frac{\partial \theta}{\partial \tau} = \frac{\partial^2 \theta}{\partial R^2} + \frac{1}{R}\frac{\partial \theta}{\partial R} \quad (10.128)$$

---

[3] See the footnote to Problem 10.10 at the end of this chapter.

Note that the dimensional form of the energy equation for the cylinder assuming one-dimensional transient conduction, the problem being axi-symmetric and the z-direction being infinite, is

$$\frac{1}{r}\frac{\partial}{\partial r}\left(r\frac{\partial T}{\partial r}\right) = \frac{1}{\alpha}\frac{\partial T}{\partial t}$$

Initial and boundary conditions

IC: $\tau = 0$, $\theta = 0$ for $0 \le R \le 1$ (10.129a)

For $\tau > 0$,

BC-1: $R = 1$, $H\dfrac{\partial \theta}{\partial R} = 1 - \theta$ (10.129b)

where $H = k/hr_0$.

BC-2: $R = 0$, $\dfrac{\partial \theta}{\partial R} = 0$ (from symmetry about $R = 0$)   (10.129c)

Discretization by the pure implicit scheme   At any interior grid point i (Fig. E10.2), Eq. (10.128) is discretized as

$$\frac{\theta_i^{p+1} - \theta_i^p}{\Delta \tau} = \frac{\theta_{i+1}^{p+1} - 2\theta_i^{p+1} + \theta_{i-1}^{p+1}}{(\Delta R)^2} + \frac{1}{R_i}\frac{\theta_{i+1}^{p+1} - \theta_{i-1}^{p+1}}{2\,\Delta R} \qquad (10.130)$$

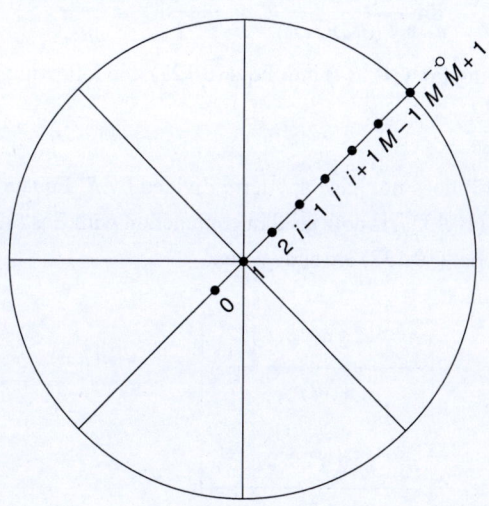

Fig. E10.2   Grid points in the computational domain of an axisymmetric one-dimensional transient heat conduction problem in cylindrical co-ordinates

At the outer boundary, i.e., at R = 1   At $i = M$, using the image-point technique,

$$H\frac{\theta_{M+1}^{p+1} - \theta_{M-1}^{p+1}}{2\,\Delta R} = 1 - \theta_M^{p+1} \qquad (10.131)$$

From Eq. (10.131), we obtain

$$\theta_{M+1}^{p+1} = \frac{2\,\Delta R}{H}\left(1 - \theta_M^{p+1}\right) + \theta_{M-1}^{p+1} \qquad (10.132)$$

Substituting θ_{M+1}^{p+1} from Eq. (10.131) to Eq. (10.130) for $i = M$, we get

$$\Rightarrow \qquad \frac{\theta_M^{p+1} - \theta_M^p}{\Delta \tau} = \frac{\dfrac{\Delta R}{H} - \theta_M^{p+1}\left(1 + \dfrac{\Delta R}{H}\right) + \theta_{M-1}^{p+1}}{\dfrac{(\Delta R)^2}{2}} + (1 - \theta_M^{p+1})/H \qquad (10.133)$$

Treatment of the condition at the centre, i.e., at R = 0   Recall the condition at the centre, i.e., $\partial\theta/\partial R \to 0$ as $R \to 0$. The second term of the GDE [Eq. (10.128)] is $1\,\partial\theta/R\partial R$. This can be written as $(\partial\theta/\partial R)/R$.

At the centre, i.e., at $R = 0$, the second term will give rise to 0/0 condition. However, this difficulty can be alleviated by making use of l'Hôpital's rule, which says that if

$$\lim_{x \to a} f(x) = \lim_{x \to a} g(x) = 0$$

and $$\lim_{x \to a} \frac{g'(x)}{f'(x)} = L$$

then $$\lim_{x \to a} \frac{g(x)}{f(x)} = L$$

Therefore, invoking l'Hôpital's rule to the centre condition, we obtain

$$\lim_{R \to 0} \frac{\partial\theta/\partial R}{R} = \lim_{R \to 0} \frac{(\partial/\partial R)(\partial\theta/\partial R)}{(\partial/\partial R)(R)} = \frac{\partial^2\theta/\partial R^2}{1} = \frac{\partial^2\theta}{\partial R^2} \qquad (10.134)$$

Hence, at $R = 0$, putting Eq. (10.134) into Eq. (10.128), the following equation results:

$$\frac{\partial\theta}{\partial \tau} = 2\frac{\partial^2\theta}{\partial R^2} \qquad (10.135)$$

Note that Eq. (10.135) does not have any term divided by R. Equation (10.136), the discretized form of Eq. (10.135), is now used in conjunction with Eqs (10.130) and (10.133). The steps leading to Eq. (10.138) are shown next.

At any grid point i,

$$\frac{\theta_i^{p+1} - \theta_i^p}{\Delta \tau} = 2\left[\frac{\theta_{i+1}^{p+1} - 2\theta_i^{p+1} + \theta_{i-1}^{p+1}}{(\Delta R)^2}\right]$$

at $i = 1$ (centre)

$$\frac{\theta_i^{p+1} - \theta_i^p}{\Delta \tau} = 2\left[\frac{\theta_2^{p+1} - 2\theta_1^{p+1} + \theta_0^{p+1}}{(\Delta R)^2}\right] \qquad (10.136)$$

Using the image-point technique,

$$\theta_0^{p+1} = \theta_2^{p+1} \qquad (10.137)$$

Substituting Eq. (10.137) into Eq. (10.136),

$$\frac{\theta_i^{p+1} - \theta_i^p}{\Delta \tau} = 2\left[\frac{2\theta_2^{p+1} - 2\theta_1^{p+1}}{(\Delta R)^2}\right]$$

$$= \frac{4}{(\Delta R)^2}[\theta_2^{p+1} - \theta_1^{p+1}] \qquad (10.138)$$

In summary, we have now M sets of linear simultaneous equations to be solved in each time step $\Delta\tau$. These are:

> For $i = 1$, Eq. (10.138)
> For $i = 2, ..., M-1$, Eq. (10.128)
> For $i = M$, Eq. (10.133)

10.6.2 Non-axisymmetric Problems

In Example 10.2, the ϕ-symmetry exists. In other words, there is no circumferential variation of heat flux or temperature because the surrounding fluid is at a uniform temperature T_∞. However, a situation could arise (e.g., due to non-uniform circumferential radiant heating) where there is a non-uniform heat flux or temperature on the periphery of the cylinder (see Figs 10.23 and 10.24). In that case, at $r = 0$,

$$\frac{\partial T}{\partial r} \neq 0$$

In other words, ϕ-symmetry is destroyed. This kind of problem is known as non-axisymmetric problem. However, the difficulty at $r = 0$ still exists. In addition to that, all we know about the condition at $r = 0$ is that at $r = 0$, T = finite and nothing else.

Assuming that $T \neq f(z)$, the difference between the present axisymmetric and the non-axisymmetric problem is that the latter is a two-dimensional problem in space because T now varies in ϕ as well in contrast with the former which is a one-dimensional problem in space (there T is only a function of r).

In the steady state, the governing differential equation for the non-axisymmetric problem is

$$\frac{\partial^2 T}{\partial r^2} + \frac{1}{r}\frac{\partial T}{\partial r} + \frac{1}{r^2}\frac{\partial^2 T}{\partial\phi^2} = 0 \quad (10.139)$$

At $r = 0$, both

$$\frac{1}{r}\frac{\partial T}{\partial r}$$

and

$$\frac{1}{r^2}\frac{\partial^2 T}{\partial\phi^2}$$

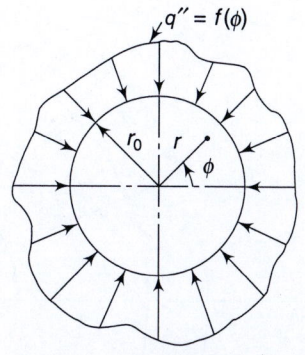

Fig. 10.23 Non-uniform heat flux boundary condition for a $T(r, f)$ problem in cylindrical coordinates

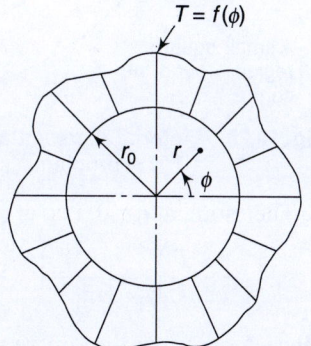

Fig. 10.24 Non-uniform specified temperature boundary condition for a $T(r, \phi)$ problem in cylindrical coordinates

of the LHS of Eq. (10.139) become infinity, leading to an undesirable situation. But, since at $r = 0$, $\partial T/\partial r \neq 0$, l'Hôpital's rule cannot be applied. Furthermore, at $r = 0$, $\partial T/\partial\phi$ does not exist. Therefore, a special treatment at $r = 0$ is obviously needed.

The aforesaid problem can be circumvented by taking a very small, square region around the centre of the cylinder so that the governing differential equation in Cartesian coordinates is valid in that region. The middle points of the top, bottom, right, and left sides of the square will now coincide with the points on the horizontal and vertical lines passing through the centre of the circle (Fig. 10.25).

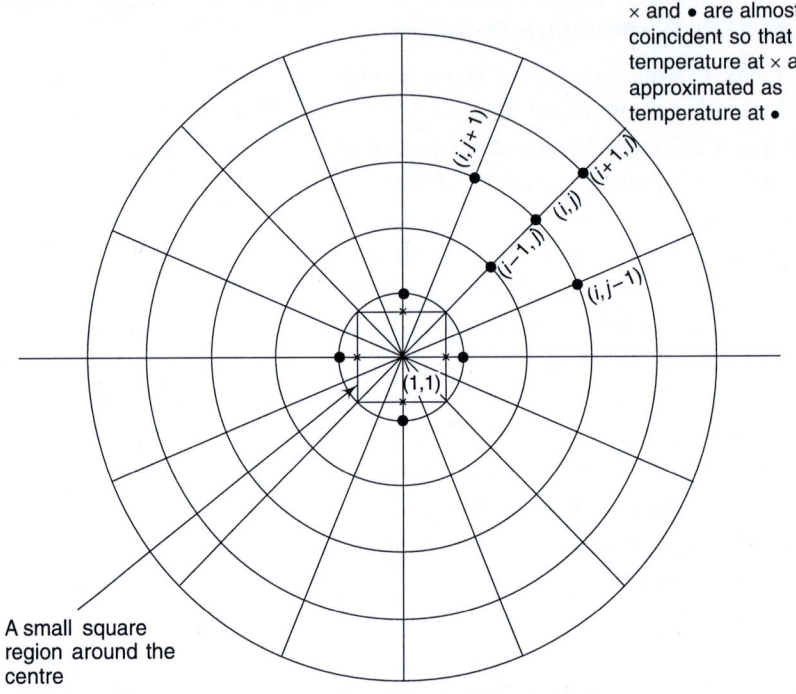

Fig. 10.25   Pictorial representation of the treatment of the condition at the centre for a $T(r, \phi)$ problem

Therefore, at $r = 0$, i.e., at $(1, 1)$,

$$\left(\frac{\partial^2 T}{\partial x^2}\right)_{1,1} + \left(\frac{\partial^2 T}{\partial y^2}\right)_{1,1} = 0 \tag{10.140}$$

Since $\Delta x = \Delta y$ in the square region, $T_{1,1}$ will be the arithmetic mean of the four surrounding grid-point temperatures (denoted by crosses in Fig. 10.25). In the GS iteration, $T_{1,1}$ will be updated until there is no change in $T_{1,1}$, indicating the convergence of the solution.

In the rest of the domain, the r-ϕ system is used. The method is quite accurate, as in a small region around the centre, a circle almost coincides with a square. Hence, the solution in the Cartesian geometry is a good approximation.

Boundary conditions in the ϕ-direction   There are no obvious boundary conditions in the ϕ-direction. Because T is single-valued and continuous,

$$T(r, \phi) = T(r, \phi + 2\pi) \tag{10.141a}$$

$$\frac{1}{r}\frac{\partial T}{\partial \phi}(r, \phi) = \frac{1}{r}\frac{\partial T}{\partial \phi}(r, \phi + 2\pi) \tag{10.141b}$$

Numerically, the above conditions do not pose any problem.

10.7 One-dimensional Transient Heat Conduction in Composite Media

In many engineering applications, parts of equipment are made of two, three, or sometimes more materials having different thermal conductivities. A typical example is one-layer or multilayer insulation over a steampipe. To analyse such problems, a simplifying assumption such as perfect contact between the surfaces of different materials is made. This means that there is assumed to be no air gap between the contacting surfaces. Therefore, at the interface, temperature and heat flux equalities exist. These are known as compatibility conditions.

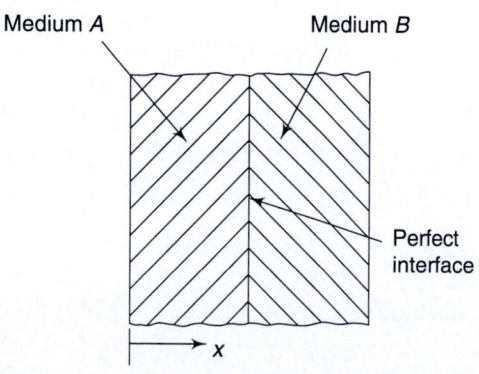

Fig. 10.26 A composite body of two materials having a perfect interface

Consider one-dimensional transient conduction in a plane, composite wall made of medium A and medium B, having thermal properties k_A, ρ_A, c_A and k_B, ρ_B, c_B, respectively (see Fig. 10.26). It is obvious that the main hurdle to overcome is the interface.

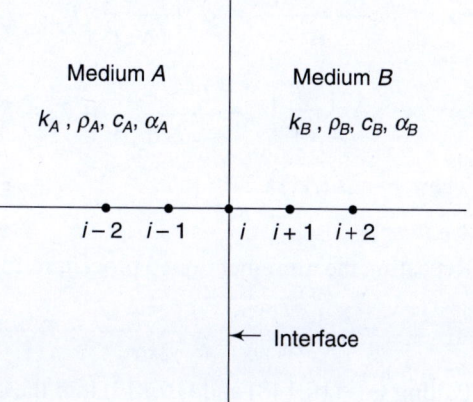

Fig. 10.27 Grid points in the vicinity of the interface

Figure 10.27 shows the grid point i at the interface and the grid points $i-1$, $i-2$ and $i+1$, $i+2$ in its vicinity in medium A and medium B respectively.

The following shows how an expression for the temperature T_i at the grid point i is obtained using the Euler method in conjunction with the Taylor series expansion for temperatures in the immediate neighbourhood of the interface grid point i.

The governing differential equations for material A and material B are

$$\frac{\partial T_A}{\partial t} = \alpha_A \frac{\partial^2 T_A}{\partial x^2} \tag{10.142}$$

$$\frac{\partial T_B}{\partial t} = \alpha_B \frac{\partial^2 T_B}{\partial x^2} \tag{10.143}$$

subject to the compatibility conditions at the interface

$$T_{iA} = T_{iB} \tag{10.144}$$

$$k_A \left(\frac{\partial T}{\partial x} \right)_{iA} = k_B \left(\frac{\partial T}{\partial x} \right)_{iB} \tag{10.145}$$

Medium A

From Taylor series expansion of T_{i-1} about point i, dropping terms beyond second order, we have

$$T_{i-1}^p = T_i^p - \Delta x \left(\frac{\partial T}{\partial x}\right)_{iA} + \frac{(\Delta x)^2}{2} \left(\frac{\partial^2 T}{\partial x^2}\right)_{iA}$$

or $\qquad \left(\frac{\partial^2 T}{\partial x^2}\right)_{iA} = \frac{2}{(\Delta x)^2} \left[T_{i-1}^p - T_i^p + \Delta x \left(\frac{\partial T}{\partial x}\right)_{iA}\right]$ (10.146)

Also, $\qquad \left(\frac{\partial T}{\partial x}\right)_{iA} = \frac{T_i^{p+1} - T_i^p}{\Delta t}$ (10.147)

Substituting Eqs (10.146) and (10.147) into Eq. (10.142), we obtain

$$\left(\frac{\partial T_A}{\partial t}\right)_{iA} = \alpha_A \left(\frac{\partial^2 T_A}{\partial x^2}\right)_{iA}$$

$$\Rightarrow \qquad \frac{T_i^{p+1} - T_i^p}{\Delta t} = \alpha_A \left[\frac{2}{(\Delta x)^2}\left\{T_{i-1}^p - T_i^p + \Delta x \left(\frac{\partial T}{\partial x}\right)_{iA}\right\}\right]$$

$$\Rightarrow \qquad \Delta x \left(\frac{\partial T}{\partial x}\right)_{iA} = \frac{T_i^{p+1} - T_i^p}{2r\alpha_A} + T_i^p - T_{i-1}^p \qquad (10.148)$$

where $r = \Delta t/(\Delta x)^2$.

Medium B

Repeating the aforementioned procedure for medium B, it can be shown that

$$-\Delta x \left(\frac{\partial T}{\partial x}\right)_{iB} = \frac{T_i^{p+1} - T_i^p}{2r\alpha_B} + T_i^p - T_{i+1}^p \qquad (10.149)$$

Putting Eqs (10.148) and (10.149) into the compatibility conditions [Eqs (10.144) and (10.145)] we have

$$T_i^{p+1} = T_i^p + \left[\frac{2r\alpha_A}{\alpha_A/\alpha_B + k_A/k_B}\right]\left\{T_{i+1}^p - \left(1 + \frac{k_A}{k_B}\right)T_i^p + \frac{k_A}{k_B}T_{i-1}^p\right\} \quad (10.150)$$

Equation (10.150) is the required explicit finite-difference representation. The accuracy of Eq. (10.150) can be checked by setting α_A/α_B and k_A/k_B equal to 1. The resulting expression indeed corresponds to the expression for T_i^{p+1} for a single-material body using the explicit scheme.

10.8   Treatment of Non-linearities in Heat Conduction

The non-linearity in the governing differential equation or in the boundary condition does not preclude its solution by one of the basic methods discussed earlier in this chapter. One must not forget that the objective of any simple finite-difference representation is always to approximate the non-linear PDE by an algebraic equation which is linear in its unknowns. Therefore, the non-linearities arising in the difference equation are locally linearized.

10.8.1 Non-linear Governing Differential Equation: Variable Thermal Conductivity

In heat conduction, a non-linear governing differential equation results if the thermal conductivity of the material is not a constant but is a function of temperature. For the definition of a non-linear PDE, see Ghoshdastidar (1998).

Take, for example, the case of one-dimensional transient heat conduction in a plane wall of thickness L having conductivity $k = k(T)$. ρ and c are assumed constants. The problem is pictorially described in Fig. 10.28. The governing differential equation is

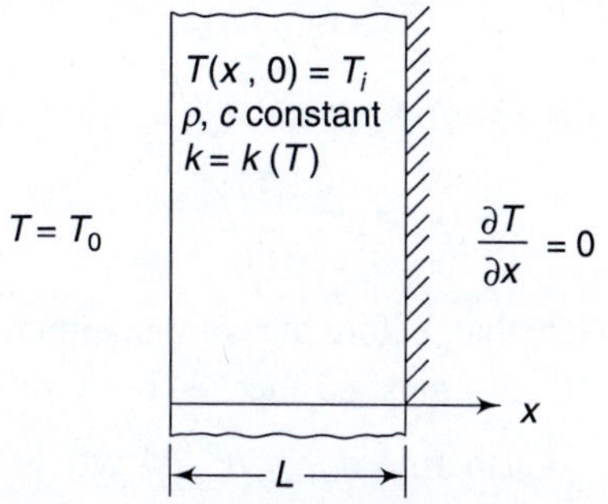

$$\frac{\partial T}{\partial t} = \frac{1}{\rho c}\frac{\partial}{\partial x}\left(k\frac{\partial T}{\partial x}\right) \qquad (10.151)$$

with the initial condition

At $t = 0$, $T = T_i$, for all x \hfill (10.152)

and boundary conditions, for $t > 0$,

At $x = 0$, $T = T_0$ \hfill (10.153)

At $x = L$, $\dfrac{\partial T}{\partial x} = 0$ \hfill (10.154)

Fig. 10.28 Physical domain of the one-dimensional transient heat conduction problem with conductivity as a function of temperature

Discretization

$$\frac{dT_i}{dt} = \frac{1}{(\rho c)_i}\left[k_i\left(\frac{\partial^2 T}{\partial x^2}\right)_i + \left(\frac{\partial T}{\partial x}\right)_i\left(\frac{\partial k}{\partial x}\right)_i\right] \qquad (10.155)$$

or

$$\frac{dT_i}{dt} = \frac{1}{(\rho c)_i}\left[k_i\frac{T_{i+1} - 2T_i + T_{i-1}}{(\Delta x)^2} + \left(\frac{T_{i+1} - T_{i-1}}{2\Delta x}\right)\left(\frac{k_{i+1} - k_{i-1}}{2\Delta x}\right)\right]$$

or

$$\frac{dT_i}{dt} = \frac{1}{(\Delta x)^2(\rho c)_i}[a_i T_{i+1} - 2b_i T_i + c_i T_{i-1}] \qquad (10.156)$$

where

$$a_i = k_i + \frac{k_{i+1}}{4} - \frac{k_{i-1}}{4} \qquad (10.157a)$$

$$b_i = k_i \qquad (10.157b)$$

$$c_i = k_i - \frac{k_{i+1}}{4} + \frac{k_{i-1}}{4} \qquad (10.157c)$$

Euler method

$$\frac{T_i^{p+1} - T_i^p}{\Delta t} = \frac{1}{(\Delta x)^2(\rho c)_i}[a_i^p T_{i+1}^p - 2b_i^p T_i^p + c_i^p T_{i-1}^p]$$

$$\Rightarrow \qquad T_i^{p+1} = \frac{\Delta t}{(\Delta x)^2(\rho c)_i}[a_i^p T_{i+1}^p + c_i^p T_{i-1}^p] + T_i^p\left(1 - \frac{2b_i^p \Delta t}{(\Delta x)^2(\rho c)_i}\right)$$

$$(10.158)$$

For stability,

$$1 - \frac{2b_i^p \Delta t}{(\Delta x)^2 (\rho c)_i} \geq 0$$

Since $(\rho c)_i = \rho c$ and $b_i^p = k_i^p$, from the aforesaid stability criterion, we obtain

$$\Delta t \leq \frac{(\Delta x)^2 (\rho c)}{2 k_i^p} \tag{10.159}$$

Note that, before the computation for the next time step, the stability criterion in Eq. (10.159) must be checked because k_i^p will change from one time step to the next.

Furthermore, a_i^p, b_i^p, c_i^p will have to be evaluated in each time step to calculate T_i^{p+1}.

Pure implicit method   In the pure implicit method, Eq. (10.156) will look like

$$\frac{T_i^{p+1} - T_i^p}{\Delta t} = \frac{1}{(\Delta x)^2 (\rho c)} [a_i^{p+1} T_{i+1}^{p+1} - 2b_i^{p+1} T_i^{p+1} + c_i^{p+1} T_{i-1}^{p+1}] \tag{10.160}$$

But, the problem is that we do not know a_i^{p+1}, b_i^{p+1}, and c_i^{p+1} to start with. To alleviate this difficulty, we can use an iterative method. Let

$$a_i^{p+1} \approx a_i^p \tag{10.161}$$

$$b_i^{p+1} \approx b_i^p \tag{10.162}$$

$$c_i^{p+1} \approx c_i^p \tag{10.163}$$

The substition of Eqs (10.161) – (10.163) into Eq. (10.160) will locally linearize the discretization equation. Once this is done, T_i^{p+1} can be calculated in a usual manner by solving a set of linear equations in each time step. If the time steps are not too large and/or the property variations with temperature are not too strong, the result may be quite satisfactory. However, one can use a better and more accurate solution procedure described next.

Improved solution procedure   To improve the solution, the values of T_i^{p+1} can be used to recalculate a_i^{p+1}, b_i^{p+1}, and c_i^{p+1}. An improved solution can be found by solving again for the unknown temperatures T_i^{p+1} using improved a_i^{p+1}, b_i^{p+1}, and c_i^{p+1}. This process can be continued in the same time step until there is no change in a_i, b_i, and c_i. Then, the computation for the next time step begins, and the aforesaid algorithm is repeated.

10.8.2   Non-linear Boundary Conditions

Case I: Heat transfer coefficient a function of boundary temperatures
Non-linearities in heat conduction problems can also arise if the boundary conditions are non-linear, e.g., $h = f(T)$ even though the governing differential equation is linear. The following example illustrates the method of solution for two-dimensional steady heat conduction in a cooling fin with the heat transfer coefficient being a function of the fin surface temperature. Since the fin temperature is not uniform, h will take different values at different locations on the fin surface.

Illustration 10.1 *Figure 10.29 shows a long cooling fin whose base is at T_0.*

The temperature variation in the z-direction is ignored. Heat is lost from the sides and the tip of the fin by convection to the surrounding air at a local rate $q'' = h(T_s - T_\infty)$ W/m^2, where T_s and T_∞ are the temperatures at a point on the fin surface and of the air, respectively. The heat transfer coefficient h is given by the following correlation $h = 1.2(T_s - T_\infty)^{1/3}$ $W/m^2 K$. Formulate the problem so that you can compute:

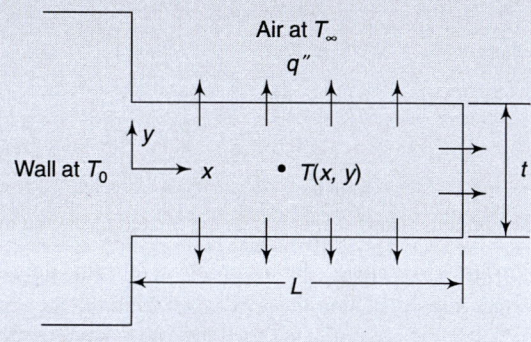

Fig. 10.29 Physical domain of the two-dimensional cooling fin

(a) *the steady-state temperature distribution inside the fin.*

(b) *the total rate of heat loss from the fin to the air.*

Also, indicate the solution algorithm.

Solution

(a) Formulation for obtaining the steady-state temperature distribution

Governing differential equation and boundary conditions The computational domain is shown in Fig. 10.30.

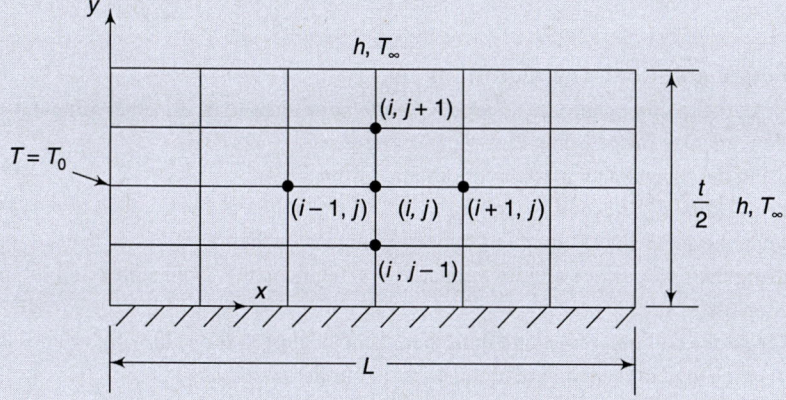

Fig. 10.30 Computational domain (showing interior grid points) of the two-dimensional fin

GDE: $\dfrac{\partial^2 T}{\partial x^2} + \dfrac{\partial^2 T}{\partial y^2} = 0$ (10.164)

BC-1: at $x = 0$, $T = T_0$ (10.165)

BC-2: at $x = L$, $-k\dfrac{\partial T}{\partial x} = h(T - T_\infty)$ (10.166)

BC-3: at $y = 0$, $\dfrac{\partial T}{\partial y} = 0$ (symmetry) (10.167)

$$\text{BC-4: at } y = \frac{t}{2}, \quad -k\frac{\partial T}{\partial y} = h(T - T_\infty) \tag{10.168}$$

where $h = 1.2(T_s - T_\infty)^{1/3}$.

Discretization At any interior point (i, j)

$$\frac{T_{i+1,j} - 2T_{i,j} + T_{i-1,j}}{(\Delta x)^2} + \frac{T_{i,j+1} - 2T_{i,j} + T_{i,j-1}}{(\Delta y)^2} = 0 \tag{10.169}$$

The boundary and corner points are treated in the manner discussed earlier in this chapter.

Hurdle to overcome In this problem, the main hurdle is the boundary condition, which is non-linear. The non-linearity arises from the fact that heat transfer is a function of the fin surface temperature, which varies from point to point. For example, at $x = L$,

$$-k\frac{\partial T}{\partial x} = h(T - T_\infty)$$
$$= 1.2(T - T_\infty)^{1/3}(T - T_\infty)$$
$$= 1.2(T - T_\infty)^{4/3} \tag{10.170}$$

From the definition of a non-linear boundary condition, we know that a boundary condition is non-linear if it contains products of the dependent variable or its derivatives. Since T at $x = L$ (that is, T_s) is not known initially, Eq. (10.170) is a non-linear boundary condition.

Basic approach The basic approach is to assign values for h at the boundaries to locally linearize the boundary condition. Since h is a known function of T_s, a temperature distribution for the boundary points is assumed to start the computation. The overall algorithm is given as:

Method of solution: The algorithm

1. To start the computations, assume a temperature distribution in the computational domain. One may assume $T_{i,j} = T_0$ as a first guess.
2. Using the assumed temperature, evaluate h from
   $h = 1.2(T_s - T_\infty)^{1/3}$
3. Solve the set of linear equations by the GS iterative scheme.
4. Using the new surface and tip temperatures, recalculate h from step 2 and return to Step 3.
5. Repeat the solution procedure until there is no change in the value of h or until there is no change in the temperature distribution in the fin.

(b) Total heat transfer from the fin to the air per unit width of the fin

Method 1

$$\frac{q}{W} = 2\int_0^{t/2} -k\frac{\partial T}{\partial x}\bigg|_{x=0} dy$$

$$= -2k\int_0^{t/2} \frac{\partial T}{\partial x}\bigg|_{x=0} dy \tag{10.171}$$

For a two-dimensional fin, $\partial T/\partial x|_{x=0}$ is a function of y. Evaluate $\partial T/\partial x|_{x=0}$ at each grid point in the y-direction using a suitable forward-difference scheme (2-point or 3-point, etc.) and then evaluate the integral using numerical integration, such as Simpson's rule.

Method 2

$$\frac{q}{W} = 2\left[\int_0^L h(T_s - T_\infty)\,dx + \int_0^{t/2} h(T_s - T_\infty)\,dy\right]$$ (10.172)

where W is the width of the fin in the z-direction.

Case II: Radiation boundary condition

Illustration 10.2 *Consider an infinite slab of thickness L (Fig.10.31).*
Initially the body is at a uniform temperature T_i. Suddenly, one side is exposed to a very hot environment at temperature T_∞, the other side being insulated. We wish to find the temperature distribution in the slab as a function of time.

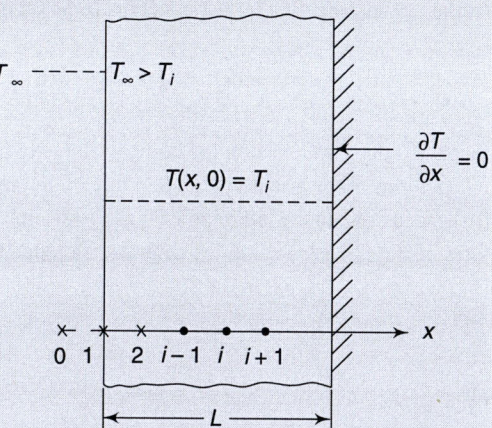

Solution

We make the following assumptions:

1. One-dimensional heat transfer is taking place because the slab is infinite in y and z directions.

2. The surrounding environment is a non-participating medium, that is, it neither emits nor absorbs radiation.

Fig. 10.31 Physical domain of the problem with radiation boundary condition on the left boundary

3. Since the environment is very hot, heat transfer to the solid boundary is occurring by radiation only.

4. The emissivity ε of the boundary receiving radiation is not a function of temperature.

5. The emissivity of the boundary receiving radiation is sufficiently high.

Governing differential equation and initial and boundary conditions The governing differential energy equation for the slab is

$$\frac{\partial T}{\partial t} = \alpha \frac{\partial^2 T}{\partial x^2}$$ (10.173)

subject to the initial condition, at $t = 0$,

$$T = T_i$$ (10.174)

and the boundary conditions, at $t > 0$,

$$x = 0, \quad -k\frac{\partial T}{\partial x} = \sigma\varepsilon(T_\infty^4 - T^4)$$ (10.175)

$$x = L, \quad \frac{\partial T}{\partial x} = 0$$ (10.176)

Note that σ is the Stefan–Boltzmann constant with the value of 5.668×10^{-8} W/m²K⁴.

Formulation The present problem is non-linear in nature because of the non-linearity in the radiation boundary condition [Eq. (10.175)] arising out of the T^4 term. Note that the governing differential equation is *linear*. For succinctness, only the treatment of the radiative boundary condition at $x = 0$ is discussed in this section.

Equation (10.173) is discretized using the explicit scheme as follows:

$$\frac{T_i^{p+1} - T_i^p}{\Delta t} = \alpha \frac{T_{i+1}^p - 2T_i^p + T_{i-1}^p}{(\Delta x)^2}$$

or $\quad T_i^{p+1} = \frac{\alpha \Delta t}{(\Delta x)^2}(T_{i+1}^p + T_{i-1}^p) + T_i^p\left(1 - \frac{2\alpha \Delta t}{(\Delta x)^2}\right)$ (10.177)

At the left surface, that is, at $i = 1$ (Fig. 10.31), Eq. (10.177) becomes

$$T_1^{p+1} = \frac{\alpha \Delta t}{(\Delta x)^2}(T_2^p + T_0^p) + T_1^p\left(1 - \frac{2\alpha \Delta t}{(\Delta x)^2}\right)$$ (10.178)

Since T_0^p does not exist on the body, we have to find a suitable difference expression for T_0^p in terms of the grid point temperature of the slab. This is done by using an image point (0) spaced at Δx to the left of point 1, as shown in Fig. 10.31. Now, we write the discretized form of Eq. (10.175) using the central-difference formulation at point 1, that is,

$$-k\frac{T_2^p - T_0^p}{2\Delta x} = \sigma\varepsilon(T_\infty^4 - T_1^{p^4})$$

or $\quad T_0^p = \frac{2\sigma\varepsilon(\Delta x)}{k}(T_\infty^4 - T_1^{p^4}) + T_2^p$ (10.179)

Equation (10.179) is still non-linear and hence, needs to be *linearized*. Linearization is achieved by writing $T_1^{p^4}$ as a product of $T_1^{p^3}$ and T_1^p as shown next.

$$T_0^p = -T_1^p\left(\frac{2\sigma\varepsilon(\Delta x)}{k}T_1^{p^3}\right) + T_2^p + \frac{2\sigma\varepsilon(\Delta x)}{k}T_\infty^4$$ (10.180)

Now, substituting Eq. (10.180) into Eq. (10.178), we obtain

$$T_1^{p+1} = \frac{\alpha \Delta t}{(\Delta x)^2}\left(2T_2^p + \frac{2\sigma\varepsilon(\Delta x)}{k}T_\infty^4\right) + T_1^p\left(1 - \frac{2\alpha \Delta t}{(\Delta x)^2} - \frac{\alpha \Delta t}{(\Delta x)^2}\frac{2\Delta x\sigma\varepsilon}{k}T_1^{p^3}\right)$$ (10.181)

Replacing $2\sigma\varepsilon(\Delta x)/k$ by R, Eq. (10.181) becomes

$$T_1^{p+1} = \frac{\alpha \Delta t}{(\Delta x)^2}(2T_2^p + RT_\infty^4) + T_1^p\left(1 - \frac{2\alpha \Delta t}{(\Delta x)^2} - \frac{\alpha \Delta t}{(\Delta x)^2}RT_1^{p^3}\right)$$ (10.182)

It may be noted that the coefficient of T_1^p in Eq. (10.182) will depend on time. This does not create much difficulty, because this term will have to be recomputed at each time step before going on to the next time-step.

10.9 Handling of Irregular Geometry in Heat Conduction

So far we have considered surfaces parallel to the x and y axes. For such simple cases, we can usually arrange for certain grid points to lie on the boundaries. However, in practice, there are cases where the surface is curved or irregular and the boundary does not fall on regular grid points (Fig. 10.32). Hence, special treatments are needed for handling the grid points in the neighbourhood of an irregular boundary. Two types of the situation may arise: (1) the temperature of the irregular boundary is specified, (2) the heat flux or normal derivative is specified on the irregular boundary.

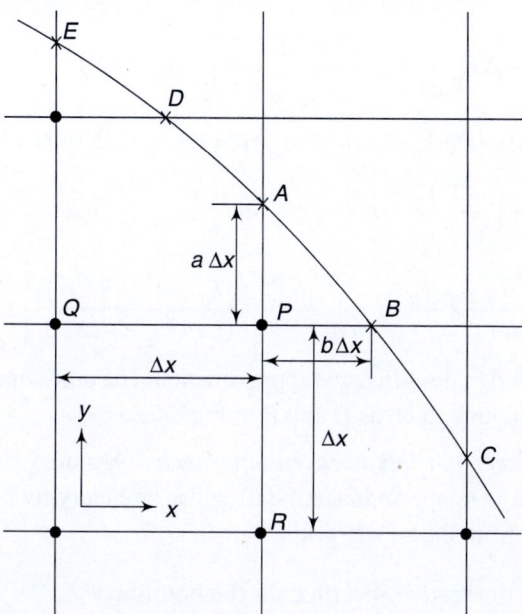

Fig. 10.32 Grid points near an irregular boundary with the Dirichlet condition

Case I: Surface temperature specified

Method 1: The Taylor series approach Consider, for example, a point P in a square mesh as shown in Fig. 10.33. Let the boundary conditions be of the *Dirichlet type*. The temperatures are specified on the irregular boundary C of which we shall use points A and B. The interior grid points are labelled P, Q, and R.

Let us assume that the heat conduction problem we are solving is governed by Laplace's equation, that is, $\nabla^2 T = 0$. Therefore, we have to derive finite difference expressions for $\partial^2 T/\partial x^2$ and $\partial^2 T/\partial y^2$ at point P.

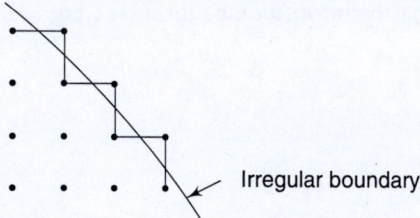

Fig. 10.33 Approximation of an irregular boundary with the Dirichlet condition by steps formed from the square mesh

The procedure is to write the appropriate Taylor series expansions and to eliminate unwanted derivatives. Thus we have, taking $\Delta x = \Delta y$,

$$T_A = T_p + a\Delta x\left(\frac{\partial T}{\partial y}\right)_p + \frac{(a\Delta x)^2}{2!}\left(\frac{\partial^2 T}{\partial y^2}\right)_p + 0[(\Delta x)^3] \qquad (10.183)$$

$$T_R = T_p - \Delta x\left(\frac{\partial T}{\partial y}\right)_p + \frac{(\Delta x)^2}{2!}\left(\frac{\partial^2 T}{\partial y^2}\right)_p + 0[(\Delta x)^3] \qquad (10.184)$$

$$T_B = T_p + b\Delta x \left(\frac{\partial T}{\partial y}\right)_p + \frac{(b\Delta x)^2}{2!}\left(\frac{\partial^2 T}{\partial y^2}\right)_p + 0[(\Delta x)^3] \tag{10.185}$$

$$T_Q = T_p - \Delta x \left(\frac{\partial T}{\partial y}\right)_p + \frac{(\Delta x)^2}{2!}\left(\frac{\partial^2 T}{\partial y^2}\right)_p + 0[(\Delta x)^3] \tag{10.186}$$

The elimination of $(\partial T/\partial x)_p$ and $(\partial T/\partial y)_p$ from Eqs (10.183)–(10.186) leads to

$$\left(\frac{\partial^2 T}{\partial x^2}\right)_p + \left(\frac{\partial^2 T}{\partial y^2}\right)_p$$

$$= \frac{2}{(\Delta x)^2}\left[\frac{T_Q}{b+1} + \frac{T_R}{a+1} + \frac{T_A}{a(a+1)} + \frac{T_B}{b(b+1)} - \frac{(a+b)T_p}{ab}\right] + 0(\Delta x) \tag{10.187}$$

which is the required finite-difference approximation. The procedure is then repeated for other pairs of points, such as D and E in Fig. 10.33.

Method 2: An easier but less accurate approach A more convenient but less accurate approach is to approximate the irregular boundary by a jagged series of steps constructed from the square grid (Fig. 10.33).

Case II: Normal derivative specified on the boundary
The treatment of the Neumann type of the boundary condition on an irregular boundary is more involved, as is demonstrated next.

When $\theta < 45°$: Consider *point P* near an irregular boundary at which $\partial T/\partial n$ is known. To solve the problem, it is necessary to write equations defining variables at all grid points, such as P near the boundary. The value of the normal derivative at G, $\partial T/\partial n|_G$ is known and can be expressed approximately by the following backward-difference form (see Fig. 10.34):

$$\left.\frac{\partial T}{\partial n}\right|_G = \frac{T_p - T_F}{\overline{FP}} \tag{10.188}$$

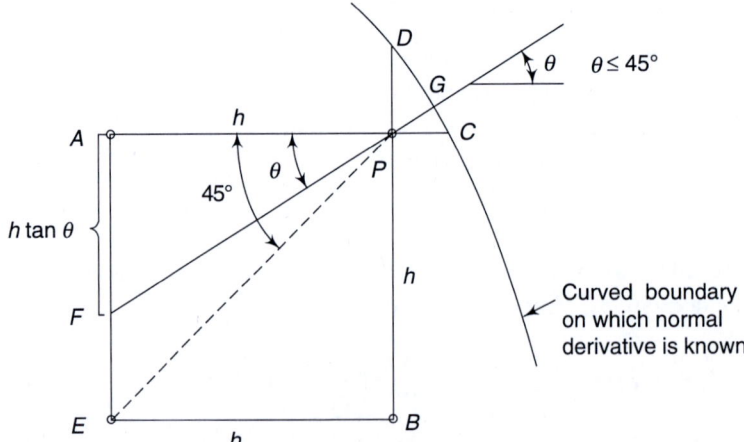

Fig. 10.34 Irregular boundary with the normal derivative specified on it ($\theta \leq 45°$)

When $\theta \leq 45°$: A value of T_F can be obtained by linearly interpolating between the vertically spaced grid points A and E such that

$$T_F = T_A + (T_E - T_A)\frac{h\tan\theta}{h}$$

$$= T_A(1 - \tan\theta) + T_E\tan\theta \qquad (10.189)$$

Combining Eqs (10.188) and (10.189), we obtain

$$T_p = \overline{FP}\left.\frac{\partial T}{\partial n}\right|_G + T_F$$

$$= \left(\frac{h}{\cos\theta}\right)\left.\frac{\partial T}{\partial n}\right|_G + T_A(1 - \tan\theta) + T_E\tan\theta \qquad (10.190)$$

An equation of the above type, that is, Eq. (10.190), must be used for each interior grid point in the immediate neighbourhood of the irregular boundary. The θ for each such grid point will be determined from the geometry of the irregular boundary.

When $\theta > 45°$: In this case, linear interpolation can be used along a horizontal line, such as line EB, to determine an expression for T at F (see Fig. 10.35). Thus,

$$T_F = T_B + \frac{T_E - T_B}{h}h\cot\theta \qquad (10.191)$$

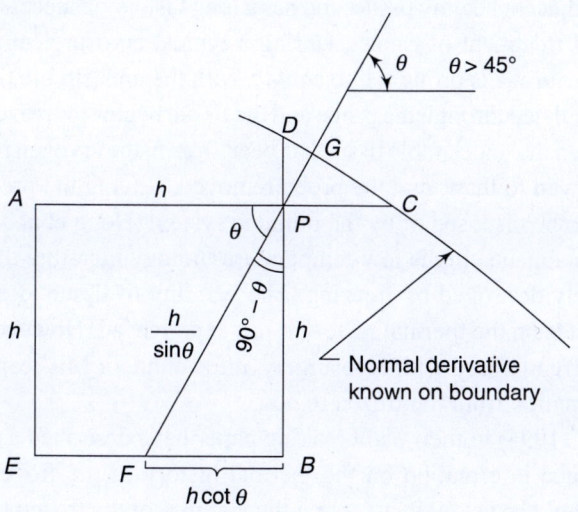

Fig. 10.35 Irregular boundary with the normal derivative specified on it ($\theta > 45°$)

Also, $\left.\dfrac{\partial T}{\partial n}\right|_G = \dfrac{T_p - T_F}{\overline{FP}} = \dfrac{T_p - T_F}{(h/\sin\theta)}$ $\qquad (10.192)$

we get $T_p = \left.\dfrac{\partial T}{\partial n}\right|_G\left(\dfrac{h}{\sin\theta}\right) + T_B(1 - \cot\theta) + T_E\cot\theta$ $\qquad (10.193)$

Points to remember

1. In considering problems with irregular boundaries having the Neumann condition, the boundary equations developed previously for rectangular domains are not applicable.

2. The equations developed for grid points in the vicinity of the irregular boundaries replace the boundary equations used for a rectangular domain.
3. To obtain T at the boundary grid points, they may be computed using linear interpolation between the boundary point desired and the adjacent interior grid points whose values were calculated earlier.

Conclusion

The above discussion on the treatment of irregular geometry reveals that although finite difference is capable of handling irregular or curved boundaries, the procedure is rather tedious as lots of account keeping is necessary for grid points near the boundary. A better alternative is to use a grid-generation technique or finite-element method which is better suited for handling irregular geometry.

10.10 Application of Computational Heat Transfer to Cryosurgery

Cryosurgery is a surgical technique that uses freezing to destroy undesirable tissue. Freezing is accomplished with cryosurgical probes, usually cooled with a cryogen, such as liquid nitrogen, and insulated except at their active tip. The surgical procedure is minimally invasive, selectively destroys undesirable tissue without affecting the adjacent healthy tissue, and has a long history of successful application in the surgical treatment of cancer. During a typical cryosurgical procedure, the active tip of the probe is brought into contact with the undesirable tissue. Then the cryogen is circulated through the probe and the tissue begins to freeze outward from the probe. When the undesirable tissue has been frozen, the cryogen flow is stopped, the tissue allowed to thaw, and the probe removed, leaving the previously frozen tissue *in situ* to be disposed of by the immune system (Hong et al., 1994).

It is well established in the low-temperature-biology literature that tissue is not indiscriminately destroyed by freezing. The viability of tissue after freezing and thawing depends on the thermal history it has experienced. However, the imaging techniques currently used with cryosurgery, ultrasound, or MRI cannot be used to retrieve information from the frozen region.

Hong et al. (1994) in their pathbreaking paper have described a new technique that can produce information on the thermal history of the frozen region. This information could be useful in assessing the viability of the frozen tissue. The new technique combines MR imaging of the frozen region with the numerical solution of the energy equation in the frozen region. In this section the theoretical basis of the new technique is introduced.

10.10.1 Mathematical Model

General problems of cryosurgery are difficult to solve because the location of the change of phase interface (1) is unknown, (2) changes as a function of time, and (3) must be determined as part of the solution to the problem. While the thermal properties of the frozen tissue are known, the heat transfer analysis during cryosurgery is further complicated by the unknown thermal properties of the unfrozen tissue, including blood flow and metabolism. In this work, the authors have developed a novel technique that tracks the freezing interface by MR imaging to simplify the mathematical formulation of the problem. The method is, in general, not restricted to specific geometries, and can be used to determine numerically the temperature history in the frozen region during cryosurgery.

In a phase-change problem, the analyst will have to solve energy equations in both phases. In addition to this, an equation that satisfies the energy balance on the change of phase interface will have to be solved to determine the location of the interface with time.

During the MRI-monitored cryosurgery, the position of the phase interface is given by the MR image. Therefore, there is no need to determine the transient position of the interface. Consequently there is no need to solve for the temperature distribution in the unfrozen region. To determine the temperature distribution in the frozen region, the only equation that remains to be solved is the energy equation in that region. It is important to note that Eq. (10.194) is in the frozen region, where there is no metabolic heat or blood flow. Therefore, the frozen tissue can be modelled accurately as a biologically inert region. The mathematical model can be simplified further by considering the fact that the time scale associated with the motion of the phase interface is much larger than the time scale associated with the transient changes in temperature (Hong et al. 1994). Therefore, the problem can be solved by invoking the quasi-steady approximation, which leads to the quasi-steady energy equation [Eq. (10.194)] in the frozen zone:

$$\frac{\partial}{\partial x}\left(k\frac{\partial T}{\partial x}\right) + \frac{\partial}{\partial y}\left(k\frac{\partial T}{\partial y}\right) + \frac{\partial}{\partial z}\left(k\frac{\partial T}{\partial z}\right) = 0 \tag{10.194}$$

The solution of Eq. (10.194) does not require an initial condition. The boundary conditions are as follows. The temperature of the cryosurgical probe, T_p, serves as the boundary condition on the cryosurgical probe surface $p(x, y, z)$,

$$T(p, t) = T_p(t) \tag{10.195}$$

The temperature on the boundary corresponding to the phase-change interface $f(x, y, z, t)$ is the phase transformation temperature T_{ph},

$$T[f(x, y, z, t), t] = T_{ph} \tag{10.196}$$

The boundary conditions on the free boundaries of the domain are general, and can be of two types:

$$T(x, y, z, t) = T_b(t)$$

or $$-k\frac{\partial T(x, y, z, t)}{\partial n} = h(T - T_\infty) \tag{10.197}$$

where T_b is the surface temperature, h is the heat transfer coefficient, T_∞ is the temperature of the fluid surrounding the tissue, and n is the direction normal to the surface. These boundary conditions can be estimated from the normal physiological temperature of the tissue and the temperature of the environment. The location of the change of phase interface is determined from the MR image.

10.10.2 Finite-difference Formulation

Three-dimensional energy equation—Eq. (10.194)—can be solved for an arbitrary geometry by using the finite-difference technique. The digitized format of the MR images is particularly appropriate for this. The finite-difference formulation of Eq. (10.194) takes the form

$$
k_{i,j,k} \frac{T_{i+1,j,k} - 2T_{i,j,k} + T_{i-1,j,k}}{(\Delta x)^2} + \frac{k_{i+1,j,k} - k_{i-1,j,k}}{2\Delta x} \frac{T_{i+1,j,k} - T_{i-1,j,k}}{2\Delta x}
$$
$$
+ k_{i,j,k} \frac{T_{i,j+1,k} - 2T_{i,j,k} + T_{i,j-1,k}}{(\Delta y)^2} + \frac{k_{i,j+1,k} - k_{i,j-1,k}}{2\Delta y} \frac{T_{i,j+1,k} - T_{i,j-1,k}}{2\Delta y}
$$
$$
+ k_{i,j,k} \frac{T_{i,j,k+1} - 2T_{i,j,k} + T_{i,j,k-1}}{(\Delta z)^2} + \frac{k_{i,j,k+1} - k_{i,j,k-1}}{2\Delta z} \frac{T_{i,j,k+1} - T_{i,j,k-1}}{2\Delta z}
$$
$$
= 0 \tag{10.198}
$$

The temperature distribution in the frozen region is computed by solving the algebraic equation (Eq. 10.198) for all the grid points (also called nodes) in the frozen domain, and implementing the boundary conditions with the standard procedures described earlier in this chapter.

10.10.3 Solution Algorithm

The steps to determine the temperature distribution in the frozen cryolesion with an MRI-assisted analysis are as follows.

1. Place the cryoprobe(s) in contact with the tissue to be frozen.
2. Acquire multiple slice images with the plane of the images oriented in such a way that the longitudinal axis of the probe lies within that plane.

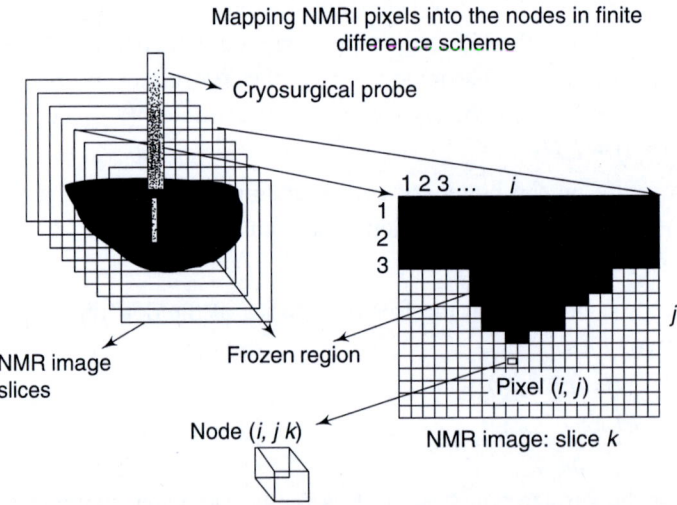

Fig. 10.36 Principles of finite-difference grid (or mesh) generation scheme [Reproduced from Hong et al. (1994) with permission from Professor B. Rubinsky (corresponding author), Biomedical Engineering Laboratory, Department of Mechanical Engineering, University of California, Berkeley.]

3. Delineate a region of interest (ROI) around the cryoprobe(s), and process the image to identify the location of the cryoprobes and the outer boundaries of the ROI.

4. Generate a finite-difference grid (also called mesh) in the ROI. Each MR voxel is mapped into a node in the numerical temperature calculation. See Fig. 10.36.

5. Start freezing.

6. Acquire multiple slice images of the entire volume during freezing.

7. Process the image to identify the location of the change of the phase interface.

8. Draw information from the computer memory on the thermal properties of frozen tissues and input temperature boundary conditions.

9. Calculate the temperature of the frozen region with the finite-difference scheme, Eq. (10.198).

10. Encode the calculated temperature into an 8-bit integer (0–255), and super-impose on the original MR image to provide visual information using a scale of grey (or colour) on the instantaneous temperature distribution in the frozen region during cryosurgery.

10.10.4 Experimental Verification of the Technique

The experimental data were obtained from simulated cryosurgical protocols performed in gelatin phantoms. A schematic diagram of the phantom and the cryosurgical probe is shown in Fig. 10.37. Two thermocouples, TC1 and TC2, are attached to the surface of the cryosurgical probe and are used to input the temperature of the probe in the calculation. The third thermocouple, TC3, is used to verify the numerical temperature calculation. Figure 10.38 shows the comparison between the thermocouple TC3 reading and the calculated temperatures at the same location. The thermocouple readings are plotted as a continuous line and illustrate the complex thermal history that the tissue may experience during cryosurgery. The sharp changes of temperature are caused by random fluctuations in the flow rate of liquid nitrogen. Numerically calculated temperatures are shown on the plot as discrete points (denoted by filled triangles), which demonstrate the very high accuracy of the solution when compared to the thermocouple reading.

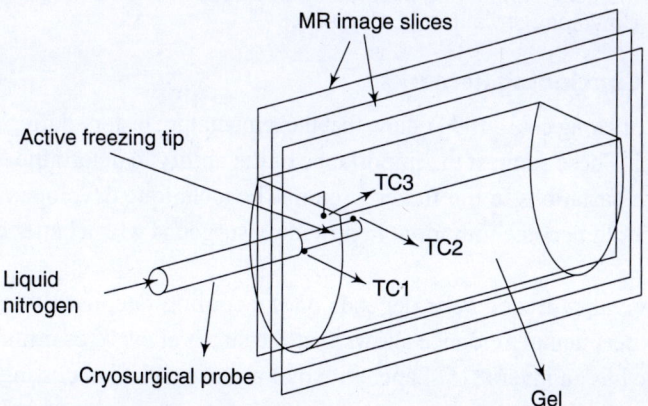

Fig. 10.37 Sketch of the cryosurgical probe and gel experimental set-up [Reproduced from Hong et al. (1994) with permission from Professor B. Rubinsky (corresponding author), Biomedical Engineering Laboratory, Department of Mechanical Engineering, University of California, Berkeley.]

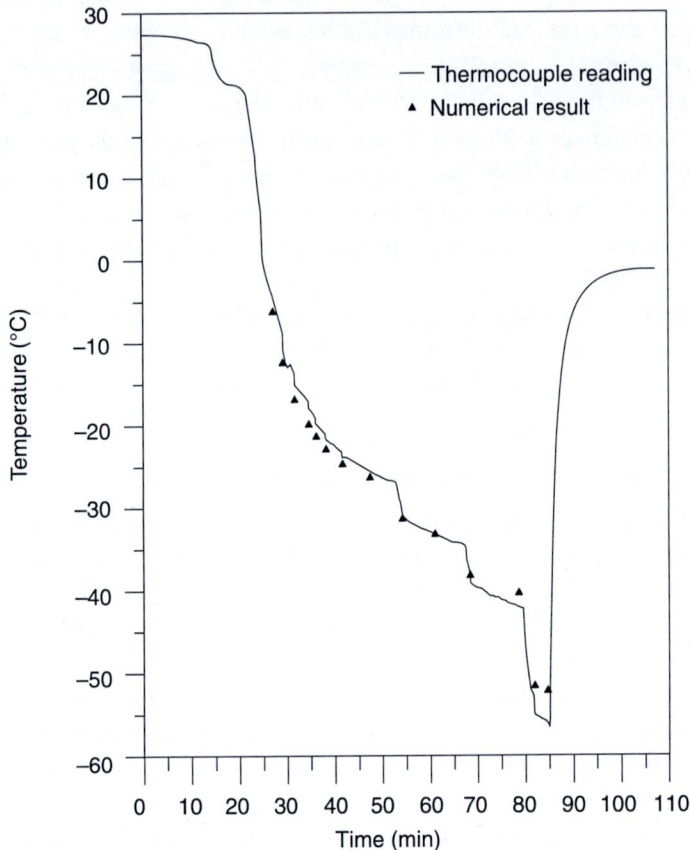

Fig. 10.38   Comparison between the thermocouple TC3 reading and the calculated temperatures at the same location [Reproduced from Hong et al. (1994) with permission from Professor B. Rubinsky, (corresponding author), Biomedical Engineering Laboratory, Department of Mechanical Engineering, University of California, Berkeley.]

10.10.5   Concluding Remarks

The results of Hong et al. (1994) show that the temperature history during cryosurgery is not trivial. These suggest the importance of the ability to determine the complex temperature variations in the frozen region. The technique developed by Hong et al. (1994) could become important in providing surgeons with a better control over cryosurgery.

The survival of frozen cells depends on the cooling rate, usually expressed in °C/min. Experimental evidence shows that the survival curve as a function of the cooling rate has an inverse U-shape, with maximum survival occurring at a certain optimum cooling rate, and with survival decreasing at above and below optimal cooling rates. Another important thermal parameter affecting the survival of frozen cells is the minimum temperature to which the cells are frozen. Physicians practising cryosurgery recommend freezing to at least −50 °C to ensure destruction of the undesirable tissue. The calculated temperatures can be used during cryosurgery in several ways, as suggested by Hong et al. (1994), as follows.

1. The recommended minimum temperature can be highlighted in the MR display to provide the physicians an immediate indication of the extent of the tissue whose destruction by freezing is assured.

2. The temperature history and in particular the cooling rates can be calculated for each voxel and can be used together with physiological data to indicate regions in which the tissue was destroyed by freezing and regions in which it may have survived freezing.

3. The calculated temperature history can be used with a feedback control system for varying the temperature of the cryosurgical probe in a controlled way to ensure that the tissue is destroyed at desired locations.

The ability to display the temperature distribution during cryosurgery visually may become an important diagnostic tool for surgeons performing cryosurgery in internal organs.

Additional Examples

In this section, two additional problems have been solved. The detailed solutions along with computer programs, written in FORTRAN 77 and C, are given.

Example 10.3 1D Steady Heat Conduction in a Rod Fin with Convective Tip: Formulation and Solution[4]

Compare the temperature distribution in a rod fin having a diameter of 2 cm and length of 10 cm and exposed to a convection environment h = 25 W/m² °C, for three fin materials: (a) copper (k = 385 W/m °C), (b) stainless steel (k = 17 W/m °C), and glass (k = 0.8 W/m °C). Assume that the tip is convecting and T_0 = 500°C, T_∞ = 25 °C. Also, calculate the relative heat transfer and fin efficiencies. Check your numerical results with the analytical solution.

Solution

Consider the cylindrical fin with convecting tip as shown in Fig. E10.3(a).

Since, in this case, the diameter of the rod is small as compared with its length and the convection essentially controls the heat flow, there will be no radial temperature distribution in the rod. But there will be a large axial temperature distribution and hence, it will be treated as a one-dimensional heat conduction problem (see Section 2.13).

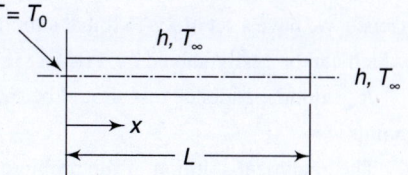

Fig. E10.3(a) Cylindrical fin with convecting tip

The governing differential equation in the non-dimensional form is

$$\frac{d^2\theta}{dX^2} - (mL)^2\theta = 0 \tag{A}$$

and the non-dimensional boundary conditions are

BC-1: At $X = 0$, $\theta = 0$ $\hspace{2cm}$ (A1)

BC-2: At $X = 1$, $\dfrac{d\theta}{dX} = -\dfrac{hL}{k}\theta$ $\hspace{2cm}$ (A2)

---

[4] See the programs convfin.f (Listing 10.1) and 1DFIN_Conv.c (Listing A8.1) in the ORC. Note that the program convfin.f must be compiled along with TDMA.f which is also given in the ORC.

where $\theta = (T - T_\infty)/(T_0 - T_\infty)$, $X = x/L$, and $m^2 = hp/kA$.

Equation (A) is discretized at any interior grid point i [see Fig. E10.3(b)] using central difference for $d^2\theta/dX^2$ as follows:

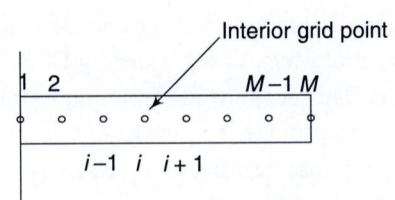

$$\left(\frac{d^2\theta}{dX^2}\right)_i - [(mL^2)\theta]_i = 0$$

Fig. E10.3(b)   Computational domain of the cylindrical fin with equally spaced grid points

or $$\frac{\theta_{i-1} - 2\theta_i + \theta_{i+1}}{(\Delta X)^2} - (mL)^2\,\theta_i = 0$$

or $$\theta_{i-1} - D\theta_i + \theta_{i+1} = 0,\ i = 2, ..., M \qquad (B)$$
where $D = 2 + (mL)^2\,(\Delta X)^2$

and $$\Delta X = \frac{1}{M-1}$$

where M is the total number of grid points.

Using the image point technique at $i = M$. (see Discretization under Section. 10.3.1)

$$\frac{\theta_{M+1} - \theta_{M-1}}{2\,\Delta X} - \frac{hL}{k}\theta_M = 0$$

or $$\theta_{M+1} = \theta_{M-1} - \frac{2\,hL\Delta X}{k}\theta_M \qquad (C)$$

Substituting θ_{M+1} from Eq. (C) into Eq. (B), for $i = M$, we get

$$2\,\theta_{M-1} - \theta_M\left(D + \frac{2\,hL\Delta X}{k}\right) = 0 \qquad (D)$$

Therefore, we can write that

$\theta_1 = 1$ for $i = 1$ (known)

$\theta_{i-1} - D\theta_i + \theta_{i+1} = 0$ for $i = 2, ..., M-1$ \qquad (E)

$$2\theta_{M-1} - \theta_M\left(D + \frac{2\,hL\Delta X}{k}\right) = 0 \quad \text{for } i = M$$

Hence, we have a set of $M-1$ linear simultaneous algebraic equations and $M-1$ unknowns, which can be easily solved by TDMA.

A grid independence test should be carried out to find out the optimum number of grid points.

The analytical solution of the problem is [see Eq. (2.90)]

$$\theta = \frac{T - T_\infty}{T_0 - T_\infty} = \frac{\cosh m\,(L - x) + \left(\dfrac{h}{mk}\right)\sinh m\,(L - x)}{\cosh mL + \left(\dfrac{h}{mk}\right)\sinh mL} \qquad (F)$$

The heat transfer through the base of the rod ($x = 0$) is

$$q = -kA\left.\frac{dT}{dx}\right|_{x=0}$$

$$= mkA\,(T_0 - T_\infty)\ \frac{\dfrac{h}{mk} + \tanh mL}{1 + \left(\dfrac{h}{mk}\right)\tanh mL} \qquad (G)$$

The fin efficiency is

$$\eta_f = \frac{q}{(T_0 - T_\infty)\,hpL + (T_0 - T_\infty)\,hA}$$

$$= \frac{q}{(T_0 - T_\infty)\,h\,(\pi D)\,L + (T_0 - T_\infty)\,h\left(\dfrac{\pi D^2}{4}\right)}$$

$$= \frac{q}{(T_0 - T_\infty)\,\pi Dh\left(L + \dfrac{D}{4}\right)} \tag{H}$$

q is calculated from Eq. (G) and hence η_f can be calculated from Eq. (H). Results are obtained for 10 grid points (excluding the base grid point at which the temperature is already specified) for three different fin materials and fixed fin dimensions, base and ambient temperatures, by executing the computer program convfin·f (Listing 10.1) available in the Online Resource of this book. The input data are the number of grid points where the temperatures are to be computed and the thermal conductivity value of the fin material.

Tables E10.3(a), E10.3(b), and E10.3(c) show the comparison of numerical and analytical solutions of the non-dimensional fin temperature with copper, stainless steel, and glass as the material of the fin, respectively. It may be noted that in these tables the first grid point value is also shown although the same is not computed; it being specified already as a boundary condition.

Table E10.3(a)   Temperature distribution in copper fin

Dimensionless distance	Dimensionless temperature (Numerical)	Dimensionless temperature (Analytical)
0	1.0	1.0
0.1	0.9876	0.9876
0.2	0.9765	0.9765
0.3	0.9667	0.9667
0.4	0.9581	0.9581
0.5	0.9508	0.9508
0.6	0.9447	0.9447
0.7	0.9399	0.9399
0.8	0.9362	0.9362
0.9	0.9338	0.9338
1.0	0.9326	0.9326

Table E10.3(b)   Temperature distribution in stainless steel fin

Dimensionless distance	Dimensionless temperature (Numerical)	Dimensionless temperature (Analytical)
0	1.0	1.0
0.1	0.8518	0.8516
0.2	0.7286	0.7282
0.3	0.6268	0.6263
0.4	0.5435	0.5429
0.5	0.4762	0.4755
0.6	0.4228	0.4222
0.7	0.3819	0.3852
0.8	0.3523	0.3516
0.9	0.3329	0.3322
1.0	0.3234	0.3227

Table E10.3(c)   Temperature distribution in glass fin

Dimensionless distance	Dimensionless temperature (Numerical)	Dimensionless temperature (Analytical)
0	1.0	1.0
0.1	0.4624	0.4536
0.2	0.2138	0.2057
0.3	0.0989	0.0933
0.4	0.0457	0.0423
0.5	0.0211	0.0192
0.6	0.0098	0.0087
0.7	0.0045	0.0040
0.8	0.0021	0.0018
0.9	0.0011	0.0009
1.0	0.0007	0.0005

Figure E10.3(c) shows the dimensionless temperature (θ) versus dimensionless distance (X) curves for copper, stainless steel, and glass fins, respectively. For each fin material, comparison of numerical and analytical solutions is presented graphically.

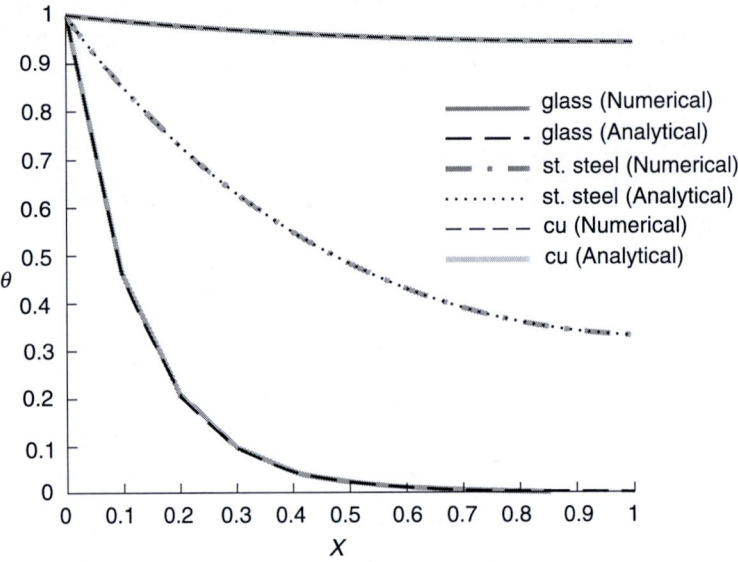

Fig. E10.3(c) θ versus X plots of numerical and analytical solutions for three different fin materials (copper, stainless steel, and glass)

A careful look at the tables as well as the figure reveals that even with 11 grid points, the numerical solution tallies 100% with the analytical solution for the copper fin, whereas the match is nearly 100% for the stainless fin and much worse for the glass fin. This is because in the copper fin, because of very high conductivity of the material, the temperature profile is almost flat. Hence a small number of grid points could capture the gradient very accurately. However, accuracy is slightly less for the stainless steel fin, in which the temperature gradient is steeper. For the glass fin, which has very low conductivity, the temperature gradient is very high and hence the number of grid points used is not sufficient for very accurate prediction of the temperature profile. Hence, the conclusion is that while for the copper fin use of 11 grid points is adequate, slightly more number is required for the stainless steel fin and even larger number of grid points is needed for the glass fin in order to have a better accuracy.

Calculation of heat transfer from the numerical results

$$q = -kA \left. \frac{dT}{dx} \right|_{x=0}$$

$$= -\frac{kA}{L} (T_0 - T_\infty) \left. \frac{d\theta}{dX} \right|_{X=0}$$

$$A = \frac{\pi}{4} (D)^2 = \frac{\pi}{4} (2 \times 10^{-2})^2$$

$$= 0.000314 \text{ m}^2$$

$$\Delta X = \frac{1}{M-1} = \frac{1}{11-1} = \frac{1}{10} = 0.1$$

$$T_0 = 500\,°C$$
$$T_\infty = 25\,°C$$

Using a three-point forward-difference scheme, we obtain [see Eq. (10.145)]

$$\left. \frac{d\theta}{dX} \right|_{X=0} = \frac{-\theta_{i+2} + 4\theta_{i+1} - 3\theta_i}{2\Delta X}$$

In this case, $X = 0$ corresponds to $i = 1$. Therefore,

$$\left. \frac{d\theta}{dX} \right|_{X=0} = \frac{-\theta_3 + 4\theta_2 - 3\theta_i}{2\Delta X}$$

Copper fin (k = 385 W/mK)

$$q = \frac{-385(0.000314)(500-25)}{0.1} \left(\frac{-0.9765 + 4(0.9876) - 3(1)}{2(0.1)} \right)$$

$$= \frac{-57.42}{0.1} \left(\frac{-0.9765 + 3.9504 - 3}{0.2} \right)$$

$$= 74.9 \text{ W}$$

Stainless steel fin (k = 17 W/mK)

$$q = \frac{-17(0.000314)(500-25)}{0.1} \left(\frac{-0.7286 + 4(0.8518) - 3(1)}{2(0.1)} \right)$$

$$= -\left(\frac{2.535}{0.1} \right) \left(\frac{-0.7286 + 3.4072 - 3}{0.2} \right)$$

$$= 40.73 \text{ W}$$

Glass fin (k = 0.8 W/m K)

$$q = \frac{-0.8(0.000314)(500-25)}{0.1} - \left(\frac{-0.2138 + 4(0.4624) - 3(1)}{2(0.1)} \right)$$

$$= -\left(\frac{0.1193}{0.1} \right) \left(\frac{-0.2138 + 1.8496 - 3}{0.2} \right)$$

$$= 8.14 \text{ W}$$

Calculation of fin efficiency [see Eq. (H)] from the numerical results

Copper fin

$$\eta_f = \frac{q}{(T_0 - T_\infty)\pi Dh\left(L + \dfrac{D}{4}\right)}$$

$$= \frac{74.9}{(500 - 25)\pi(0.02)(25)\left(0.1 + \dfrac{0.02}{4}\right)} = \frac{74.9}{78.34} = 95.6\%$$

Stainless steel fin

$$\eta_f = \frac{q}{(T_0 - T_\infty)\pi Dh\left(L + \dfrac{D}{4}\right)}$$

$$= \frac{40.73}{(500 - 25)\pi(0.02)(25)\left(0.1 + \dfrac{0.02}{4}\right)}$$

$$= \frac{40.73}{78.34} = 51.99\%$$

Glass fin

$$\eta_f = \frac{q}{(T_0 - T_\infty)\pi Dh\left(L + \dfrac{D}{4}\right)}$$

$$= \frac{8.14}{(500 - 25)\pi(0.02)(25)\left(0.1 + \dfrac{0.02}{4}\right)} = \frac{8.14}{78.34} = 10.39\%$$

This study clearly shows that high-conductivity fin material gives rise to high fin efficiency.

Note that in convfin.f program the calculation of fin heat transfer and efficiency have not been implemented. The readers may input the numerical and analytical fin heat transfer and efficiency expressions in the program and compare the two results.

Listing 10.1   convfin.f

```
C       THE NAME OF THE PROGRAM IS:convfin.f
C       EXAMPLE 10.3
C       THIS PROGRAM MUST BE COMPILED ALONG WITH tdma.f.
C       COMPILATION COMMAND:f77 convfin.f TDMA.F
C       THE INPUT DATA ARE TO ENTERED IN AN INTERACTIVE MANNER.
        DIMENSION A(50),B(50),C(50),D(50),THETA(50),X(50),ATHETA(50)
        INTEGER P,Q,R,S
        REAL K,M,L
        OPEN(UNIT=1,FILE='N.TXT')
        WRITE(*,*) '    PROGRAM TO CALCULATE TEMPERATURE DISTRIBUTION '
        WRITE(*,*) '         IN A FIN WITH CONVECTING TIP '
        WRITE(*,*) ' '
        WRITE(*,*) ' ENTER THE VALUE OF N (N+1 IS THE NO. OF GRID POINTS)'
        WRITE(*,*) '     '
```

```
      WRITE(*,*) ' ENTER THE VALUE OF K (THERMAL CONDUCTIVITY)'
      WRITE(*,*) '    '
      WRITE(*,*) ' RESULTS ARE STORED IN THE FILE N.TXT          '
      READ *,N,K
      DO 10 P=2,N-1
10    A(P)= -1
      A(N)= -2
      DO 20 Q=1,N-1
      M=SQRT(5000.0/K)
      L=0.1
      O=M*L
      DX=1.0/N
      X(1)=DX
      D1=2.0+(O**2)*(DX**2)
      Z=(5.0*DX)/K
20    B(Q)= D1
      B(N)=D1+Z
      DO 30 R=1,N-1
30    C(R)= -1
      D(1)=1
      DO 40 S=2,N
40    D(S)= 0
      HM=25.0/(M*K)
      DO 5 I=2,N
      X(I)=X(I-1)+DX
      OL=(O*(1-(X(I))))
      ATHETA(I)=(COSH(OL)+HM*SINH(OL))/(COSH(O)+HM*SINH(O))
5     CONTINUE
      OL=(O*(1-(X(1))))
      ATHETA(1)=(COSH(OL)+HM*SINH(OL))/(COSH(O)+HM*SINH(O))
      CALL TDMA(1,N,A,B,C,D,THETA)
      WRITE(1,*)  '  DIMENSIONLESS  DIMENSIONLESS   DIMENSIONLESS '
      WRITE(1,*)  '     LENGTH       TEMPERATURE     TEMPERATURE  '
      WRITE(1,*)  '                  (NUMERICAL)     (ANALYTICAL) '
      WRITE(1,25)(X(I),THETA(I),ATHETA(I),I=1,N)
25    FORMAT(4X,F4.2,4X,F15.4,4X,F15.4)
      CLOSE(UNIT=1)
      STOP
      END
```

Example 10.4   2D Transient Heat Conduction in an Infinitely Long Bar Subjected to Radiative Heating: Formulation and Solution (Listings 10.2 and 10.3).[5]

An infinitely long bar of thermal diffusivity α has a square cross-section of side 2L. It is initially at a uniform temperature T_i when it is placed quickly inside a furnace that behaves as a black-body enclosure at T_b. In terms of the following non-dimensional variables, write the GDE, IC, and BCs, and hence show that the general solution is of the form $\theta = \theta$ (X, Y,τ, β, γ), where the dimensionless parameters β and γ are given by

$$\beta = \frac{k}{\varepsilon \sigma L T_b^3}, \; \gamma = \frac{T_i}{T_b}$$

Also, $X = x/L$, $Y = y/L$, $\tau = \alpha t/L^2$, and $\theta = (T - T_i)/(T_b - T_i)$.

Let τ_{cf} be the dimensionless time taken for the dimensionless temperature θ_c to rise to a specified fractional value f. Write a computer program that will enable plots to be made of τ_{cf} against β, with γ as a parameter.

Suggested test problem: *A steel bar has L = 0.3 m, $\alpha = 1.16 \times 10^{-5}$ m²/s, $T_i = 300$ K, k = 45 W/m°C, $\varepsilon = 0.79$. It is placed inside a black-body enclosure with $T_b = 1273$ K. How*

[5] See the programs 2dtranscond.f and 2dtranscond1.f in the Online Resource Centre.

long will it take for the centre temperature to rise to 786.5 K ($\theta_c = 0.5$)? How long, if L = 0.6 m, with all other quantities unchanged?

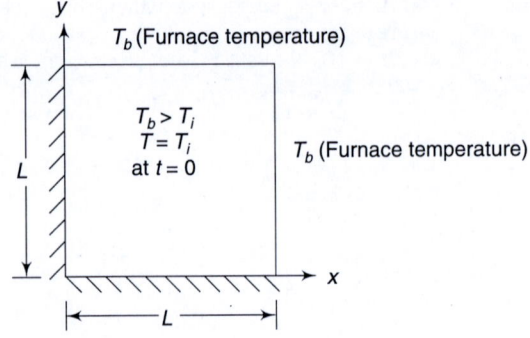

Fig. E10.4(a)   Computational domain

Solution

This is a two-dimensional transient conduction problem. Because of thermal and geometric symmetry, one-fourth of the physical domain is taken as the computational domain [Fig. E10.4(a)].

The governing differential equation for the problem is

$$\text{GDE: } \frac{\partial T}{\partial t} = \alpha \left[\frac{\partial^2 T}{\partial x^2} + \frac{\partial^2 T}{\partial y^2} \right] \tag{A}$$

$$\text{IC: at } t = 0, T = T_i \tag{A1}$$

$$\text{BC-1: at } x = 0, \frac{\partial T}{\partial x} = 0 \tag{A2}$$

$$\text{BC-2: at } y = 0, \frac{\partial T}{\partial y} = 0 \tag{A3}$$

$$\text{BC-3: at } x = L, \frac{\partial T}{\partial x} + k = \sigma \varepsilon (T_b^4 - T^4) \tag{A4}$$

$$\text{BC-4: at } y = L, \frac{\partial T}{\partial y} + k = \sigma \varepsilon (T_b^4 - T^4) \tag{A5}$$

Note the positive sign before the LHSs of Eqs (A4) and (A5) as the temperature will be higher at those boundaries than at the inside. Using the dimensionless parameters given in the problem statement, we obtain the following non-dimensional governing differential equation and initial and boundary conditions.

In the non-dimensional form:

$$\text{GDE: } \frac{\partial \theta}{\partial \tau} = \frac{\partial^2 \theta}{\partial x^2} + \frac{\partial^2 \theta}{\partial y^2} \tag{B}$$

$$\text{IC: at } \tau = 0, \theta = 0 \tag{B1}$$

$$\text{BC-1: at } X = 0, \frac{\partial \theta}{\partial X} = 0 \tag{B2}$$

$$\text{BC-2: at } Y = 0, \frac{\partial \theta}{\partial Y} = 0 \tag{B3}$$

$$\text{BC-3: at } X = 1, \beta (1 - \gamma) \frac{\partial \theta}{\partial X} = [1 - \{\gamma + \theta (1 - \gamma)\}^4] \tag{B4}$$

$$\text{BC-4: at } Y = 1, \beta (1 - \gamma) \frac{\partial \theta}{\partial Y} = [1 - \{\gamma + \theta (1 - \gamma)\}^4] \tag{B5}$$

Therefore, we can clearly see that

$$\theta = \theta (X, Y, \tau, \beta, \gamma)$$

A square grid $\Delta X = \Delta Y$ is used. The explicit finite-difference method has been applied to solve the problem.

The finite-difference representation at any interior grid point (i, j) is given below:

$$\frac{\theta_{i,j}^{p+1} - \theta_{i,j}^{p}}{\Delta \tau} = \frac{\theta_{i+1,j}^{p} - 2\theta_{i,j}^{p} + \theta_{i-1,j}^{p}}{(\Delta X)^2} + \frac{\theta_{i,j+1}^{p} - 2\theta_{i,j}^{p} + \theta_{i,j-1}^{p}}{(\Delta Y)^2}$$

Since $\Delta X = \Delta Y$, the above equation can be written as

$$\theta_{i,j}^{p+1} = \theta_{i,j}^{p} + \frac{\Delta \tau}{(\Delta X)^2} + (\theta_{i+1,j}^{p} - 2\theta_{i,j}^{p} + \theta_{i-1,j}^{p} + \theta_{i,j+1}^{p} - 2\theta_{i,j}^{p} + \theta_{i,j-1}^{p})$$

$$= \theta_{i,j}^{p}\left(1 - \frac{4\,\Delta \tau}{(\Delta X)^2}\right) + \frac{\Delta \tau}{(\Delta X)^2}(\theta_{i+1,j}^{p} + \theta_{i-1,j}^{p} + \theta_{i,j+1}^{p} + \theta_{i,j-1}^{p}) \qquad \text{(C)}$$

For stability,

$$1 - \frac{4\,\Delta \tau}{(\Delta X)^2} \geq 0$$

or

$$\frac{\Delta \tau}{(\Delta X)^2} \leq \frac{1}{4} \qquad \text{(D)}$$

The stability limit should also be checked for boundary and corner points and the minimum $\Delta \tau$ should be taken. Boundary conditions (considering all boundary and corner points) have been implemented using image point techniques (see *Discretization* under Section. 10.3.2). For the implementation of the radiation boundary condition see Case II in Section. 10.8.2. This part is left as an exercise to the readers.

A grid independence test has been performed at three different times ($\tau = 0.25$, $\tau = 1.0$, $\tau = 2.5$). Finally, the optimum grid size chosen is $\Delta X = \Delta Y = 1/10$; that is, 11×11 grid points have been employed to solve this problem.

The computer program 2dtranscond.f has been written to predict the temperature distribution at various times. 2dtranscond1.f, which is a slight modification of 2dtranscond.f, predicts centre temperature, that is, the lower left corner point of the computational domain as a function of the time. The program are written in FORTRAN 77. The input data are to be entered in an interactive manner.

Test problem 1

$$L = 0.3 \text{ m}$$
$$k = 45 \text{ W/m K}$$
$$\alpha = 1.16 \times 10^{-5} \text{ m}^2/\text{s}$$
$$\varepsilon = 0.79$$
$$T_b = 1273 \text{ K}$$
$$T_i = 300 \text{ K}$$
$$\theta_c = \frac{T_c - T_i}{T_b - T_i} = \frac{786.5 - 300}{1273 - 300} = 0.5$$
$$\Delta X = \frac{1}{10}$$
$$\Delta \tau = 0.001 \text{ (within the stability limit)}$$
$$\beta = \frac{k}{\varepsilon \sigma L T_b^3} = 1.6233$$

$$\gamma = \frac{T_i}{T_b} = 0.2357$$

Executing 2dtranscond1.f by using the above input data gives $\tau = 0.537$ for the centre temperature θ_c to reach a value of 0.5. Therefore,

$$t = \frac{L^2 \tau}{\alpha}$$

$$= \frac{(0.3)^2 \ (0.537)}{1.16 \times 10^{-5}}$$

$$= 4166.38 \text{ s}$$

$$= 1.15 \text{ h}$$

Test problem 2

In this case, $L = 0.6$ m. Other conditions are the same as in Test problem 1, except $\beta = k/\varepsilon\sigma L T_b^3 = 0.8116$.

Executing 2dtranscond1.f gives $\tau = 0.379$ for the centre temperature to reach 0.5. Therefore,

$$t = \frac{L^2 \tau}{\alpha}$$

$$= \frac{(0.6)^2 \ (0.379)}{1.16 \times 10^{-5}}$$

$$= 11762.06 \text{ s}$$

$$= 3.27 \text{ h}$$

The readers should try to explain why more time is taken when L is increased to 0.6 m although that results in a larger surface area.

Variation of τ_{cf} with β (γ = constant)   Different values of β can be taken to see the variation in time to reach the dimensionless centre-temperature value of $\theta = 0.5$ for a fixed value of γ. We assume here T_b is not changed to vary β. Table E10.4(a) shows τ_{cf} versus β for $\gamma = 0.2357$ (the same as in test problems 1 and 2).

Table E10.4(a) τ_{cf} [that is, τ at θ (1, 1) = 0.5], corresponding τ versus β for $\gamma = 0.2357$

β	τ at θ (1, 1) = 0.5	Corresponding time (s)
0.2	0.267	2072
0.4	0.3020	2343
0.6	0.339	2630
0.8	0.377	2925
1.0	0.416	3228
1.2	0.454	3523
1.4	0.483	3825
1.6	0.532	4128
1.8	0.571	4430
2.0	0.610	4733

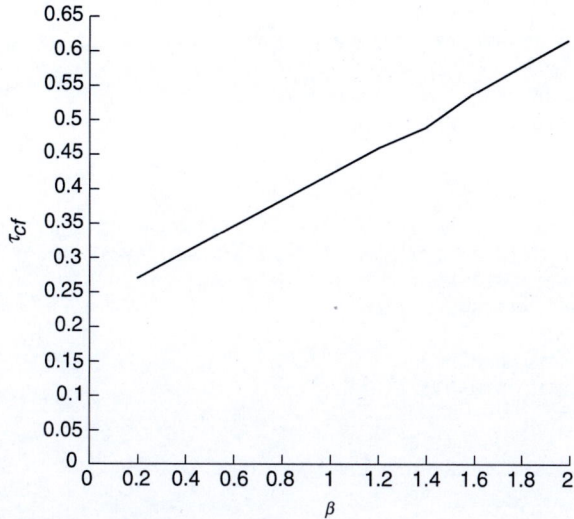

Fig. E10.4(b) τ_{cf} versus β plot for $\gamma = 0.2357$

A plot of τ_{cf} versus β is shown in Fig. E10.4(b). A close look at the table and the plot shows an increase in τ_{cf} with increase in β.

The readers are encouraged to explain the plot physically. Various other values of γ can be taken, for example, $\gamma = 0.125, 0.25, 0.375, 0.5, 0.625$ to see its effect on the τ_{cf} versus β plot.

Listing 10.2 2dtranscond.f

```
C   THE NAME OF THE PROGRAM IS: 2dtranscond.f

C   EXAMPLE 10.4
C   TRANSIENT 2D HEAT CONDUCTION WITH RADIATION BOUNDARY CONDITION
C   THE INPUT DATA ARE TO BE ENTERED IN AN INTERACTIVE MANNER.

    DIMENSION T(150,150), T1(150,150)
    OPEN (UNIT=2,FILE='P.TXT')
    WRITE(*,*)   '
    WRITE(*,*)   'PROGRAM TO CALCULATE TEMPERATURE DISTRIBUTION IN AN'
    WRITE(*,*)   ' INFINITELY LONG BAR HAVING SQUARE CROSS SECTION'
    WRITE(*,*)   '                 '
    WRITE(*,*)   '             ENTER THE VALUE OF N   '
    WRITE(*,*)   ' (N+1, N+1 IS THE NO. OF GRID POINTS)        '
    READ *, N
    WRITE(*,*)   'ENTER THE VALUE OF B (DIMENSIONLESS PARAMETER)        '
    READ *, B
    WRITE(*,*)   'ENTER THE VALUE OF Y (DIMENSIONLESS PARAMETER)        '
    READ *, Y
    WRITE(*,*)   'ENTER THE VALUE OF DT (DIMENSIONLESS TIME STEP)        '
    READ *, DT

    DO 5 I=1,N+1
    DO 5 J=1,N+1
5   T(I,J)=0.0
    DX=1.0/N

    K=0
100 K=K+1
```

```
C  CHECKING OF STABILITY LIMIT FOR INTERIOR AND BOUNDARY POINTS

   DT1=0.25*((DX)**2)
   A=(2.0*DX)/(B*(1.0-Y))
   Y2=Y**2
   Y3=Y**3
   Y4=Y**4
   Z=((T(I,J))**3)*(1.0-4.0*Y+2.0*Y2+4.0*Y3+Y4)+((T(I,J))**2)*(4.0*Y-
   .8.0*Y2+4.0*Y4)+(T(I,J))*(6.0*Y2-12.0*Y3+2.0*Y4)+(4.0*Y3-4.0*Y4)
   DT2=((DX)**2)/(4.0+A*Z)
   DT3=((DX)**2)/(4.0+2.0*A*Z)  .
   DTM=DT1
   IF(DT2.LT.DTM) DTM=DT2
   IF(DT3.LT.DTM) DTM=DT3
   R=DT/(DX)**2

C  CALCULATING THE TEMPERATURE AT INTERIOR GRID POINTS

   DO 10 I=2, N
   DO 20 J=2, N
   T1(I,J)=(1.0-4.0*R)*(T(I,J))+R*(T(I+1,J)+T(I-1,J)+T(I,j+1)+T(I,J-1.)
20 CONTINUE
10 CONTINUE

C  CALCULATING THE TEMPERATURE AT BOTTOM BOUNDARY GRID POINTS

   DO 30 I=1,N+1
   J=1
   IF (I.EQ.1) THEN
   T1(I, J)=(1.0-4.0*R)*(T(I,J))+2.0*R*(T(I+1,J)+T(I,J+1))
   ELSE

   IF(I.EQ.N+1) THEN
   T1(I,J)=(1.0-4.0*R)*(T(I,J))+2.0*R*(T(I,J+1))+2.0*R*(T(I-1,J))
   +A*R.*(1.0-(T(I,J)+Y*(1.0-T(I,J)))**4)
   ELSE
   T1(I,J)=(1.0-4.0*R)*(T(I,J))+2.0*R*(T(I,J+1))+R*(T(I-1,J)+T(I+1,J)
   .)
   ENDIF
   ENDIF
30 CONTINUE

C  CALCULATING THE TEMPERATURE AT LEFT BOUNDARY GRID POINTS

   DO 40 J=2, N+1
   I=1
   IF (J.EQ.N+1) THEN
   T1(I,J)=(1.0-4.0*R)*(T(I,J))+2.0*R*(T(I+1,J))+2.0*R*(T(I,J-1))+A*R
   .*(1.0-(T(I,J)+Y*(1.0-T(I,J)))**4)
   ELSE
   T1(I,J)=(1.0-4.0*R)8(T(I,J))+2.0*R*(T(I+1,J))+R*(T(I,J-1)+T(I,J+1)
   .)
   ENDIF
40 CONTINUE

C  CALCULATING THE TEMPERATURE AT RIGHT BOUNDARY GRID POINTS
```

```
      DO 50 J=2,N+1
      I=N+1
      IF(J.EQ.N+1) THEN
      T1(I,J)=(1.0-4.0*R)*(T(I,J))+2.0*R*(T(T-1,J))+2.0*R*(T(I,J-1))+2.0
     .*A*R*(1.0-(T(I,J)+Y*(1.0-T(I,J))))**4)
      ELSE
      T1(I,J)=(1.0-4.0*R)*(T(I,J))+2.0*R*(T(I-1,J))+R*(T(I,J-1)+T(I,J+1)
     .)+A*R*(1.0-(T(I,J)+Y*(1.0-T(I,J))))**4)
      ENDIF
50    CONTINUE

C   CALCULATING THE TEMPERATURE AT TOP BOUNDARY GRID POINTS

      DO 60 I=2,N
      J=N+1
      T1(I,J)=(1.0-4.0*R)*(T(I,J))+2.0*R*(T(I,J-1))+R*(T(I-1,J)
      +T(I+1,J)
     .)+A*R*(1.0-(T(I,J)+Y*(1.0-T(I,J))))**4)
60    CONTINUE

      E=K*DT
      WRITE(2,65)   E
65    FORMAT (1X, ' TEMPERATURE VALUES AT DIMENSIONLESS TIME =', F10.6)
      DO 70 J=N+1,1,-1
      WRITE(2,75)  (T1(I,J),I=1,N+1)
75    FORMAT(1x,21F8.4)
70    CONTINUE
      DO 80 I=1,N+1
      DO 90 J=1,N+1
      T(I,J)=T1(I,J)
90    CONTINUE
80    CONTINUE

      IF(T(1,1).LE.0.5) GOTO 100
      IF(DT.GT.DTM) THEN
      PRINT *,  ' DT VALUE DOES NOT SATISFY STABILITY CRITERION'
      PRINT *,  '            FOR CHOSEN GRID SIZE '
      ELSE
      PRINT *,  ' THE TEMPERATURE VALUE IS STORED IN THE FILE P.TXT'
      ENDIF

      CLOSE(UNIT=2)
      STOP
      END
```

Listing 10.3 2dtranscond1.f

```
C  THE NAME OF THE PROGRAM IS: 2dtranscond1.f
C  EXAMPLE 10.4
C  TRANSIENT 2D HEAT CONDUCTION WITH RADIATION BOUNDARY CONDITION
C  THE INPUT DATA ARE TO BE ENTERED IN AN INTERACTIVE MANNER.

      DIMENSION T(150,150),T1(150,150)
      OPEN (UNIT=2,FILE='P1.TXT')
      WRITE(*,*)  ' '
      WRITE(*,*)  'PROGRAM TO CALCULATE CENTRE TEMPERATURE IN AN '
      WRITE(*,*)  ' INFINITELY LONG BAR HAVING SQUARE CROSS SECTION '
      WRITE(*,*)  ' '
      WRITE(*,*)  ' ENTER THE VALUE OF N '
```

```
      WRITE(*,*)  ' (N+1,N+1 IS THE NO. OF GRID POINTS) '
      READ *, N
      WRITE(*,*)  'ENTER THE VALUE OF B (DIMENSIONLESS PARAMETER) '
      READ *, B
      WRITE(*,*)  'ENTER THE VALUE OF Y (DIMENSIONLESS PARAMETER) '
      READ *, Y
      WRITE(*,*)  'ENTER THE VALUE OF DT (DIMENSIONLESS TIME STEP) '
      READ *, DT

      DO 5 I=1,N+1
      DO 5 J=1,N+1
    5 T(I,J)=0.0
      DX=1.0/N

      K=0
  100 K=K+1

C  CHECKING OF STABILITY LIMIT FOR INTERIOR AND BOUNDARY POINTS

      DT1=0.25*((DX)**2)
      A=(2.0*DX)/(B*(1.0-Y))
      Y2=Y**2
      Y3=Y**3
      Y4=Y**4
      Z=((T(I,J))**3)*(1.0-4.0*Y+2.0*Y2+4.0*Y3+Y4)+((T(I,J))**2)*(4.0*Y-
     .8.0*Y2+4.0*Y4)+(T(I,J))*(6.0*Y2-12.0*Y3+2.0*Y4)+(4.0*Y3-4.0*Y4)
      DT2=((DX)**2)/(4.0+A*Z)
      DT3=((DX)**2)/(4.0+2.0*A*Z)
      DTM=DT1
      IF(DT2.LT.DTM) DTM=DT2
      IF(DT3.LT.DTM) DTM=DT3
      R=DT/(DX)**2

C  CALCULATING THE TEMPERATURE AT INTERIOR GRID POINTS

      DO 10 I=2,N
      DO 20 J=2,N
      T1(I,J)=(1.0-4.0*R)*(T(I,J))+R*(T(I+1,J)+T(I-1,J)+T(I,J+1)+T(I,J-1
     .))
   20 CONTINUE
   10 CONTINUE

C  CALCULATING THE TEMPERATURE AT BOTTOM BOUNDARY GRID POINTS

      DO 30 I=1,N+1
      J=1
      IF (I.EQ.1) THEN
      T1(I,J)=(1.0-4.0*R)*(T(I,J))+2.0*R*(T(I+1,J)+T(I,J+1))
      ELSE
      IF(I.EQ.N+1) THEN
      T1(I,J)=(1.0-4.0*R)*(T(I,J))+2.0*R*(T(I,J+1))+2.0*R*(T(I-1,J))+A*R
     .*(1.0-(T(I,J)+Y*(1.0-T(I,J)))**4)
      ELSE
      T1(I,J)=(1.0-4.0*R)*(T(I,J))+2.0*R*(T(I,J+1))+R*(T(I-1,J)+T(I+1,J)
     .)
      ENDIF
      ENDIF
   30 CONTINUE
```

```
C   CALCULATING THE TEMPERATURE AT LEFT BOUNDARY GRID POINTS

    DO 40 J=2,N+1
    I=1
    IF (J.EQ.N+1) THEN
    T1(I,J)=(1.0-4.0*R)*(T(I,J))+2.0*R*(T(I+1,J))+2.0*R*(T(I,J-1))+A*R
   .*(1.0-(T(I,J)+Y*(1.0-T(I,J)))**4)
    ELSE
    T1(I,J)=(1.0-4.0*R)*(T(I,J))+2.0*R*(T(I+1,J))+R*(T(I,J-1)+T(I,J+1)
   .)
    ENDIF
40  CONTINUE

C   CALCULATING THE TEMPERATURE AT RIGHT BOUNDARY GRID POINTS

    DO 50 J=2,N+1
    I=N+1
    IF(J.EQ.N+1) THEN
    T1(I,J)=(1.0-4.0*R)*(T(I,J))+2.0*R*(T(I-1,J))+2.0*R*(T(I,J-1))+2.0
   .*A*R*(1.0-(T(I,J)+Y*(1.0-T(I,J)))**4)
    ELSE
    T1(I,J)=(1.0-4.0*R)*(T(I,J))+2.0*R*(T(I-1,J))+R*(T(I,J-1)+T(I,J+1)
   .)+A*R*(1.0-(T(I,J)+Y*(1.0-T(I,J)))**4)
    ENDIF
50  CONTINUE

C   CALCULATING THE TEMPERATURE AT TOP BOUNDARY GRID POINTS

    DO 60 I=2,N
    J=N+1
    T1(I,J)=(1.0-4.0*R)*(T(I,J))+2.0*R*(T(I,J-1))+R*(T(I-1,J)+T(I+1,J)
   .)+A*R*(1.0-(T(I,J)+Y*(1.0-T(I,J)))**4)
60  CONTINUE

    E=K*DT
    WRITE(2,65) E
65  FORMAT(1X,' CENTRE TEMPERATURE AT DIMENSIONLESS TIME =',F10.6)
    WRITE(2,75) (T1(1,1))
75  FORMAT(1X,21F8.4)
70  CONTINUE
    DO 80 I=1,N+1
    DO 90 J=1,N+1
    T(I,J)=T1(I,J)
90  CONTINUE
80  CONTINUE

    IF(T(1,1).LE.0.5) GOTO 100
    IF(DT.GT.DTM) THEN
    PRINT  *, ' DT VALUE DOES NOT SATISFY STABILITY CRITERION '
    PRINT  *, ' FOR CHOSEN GRID SIZE '
    ELSE
    PRINT  *, 'THE TEMPERATURE VALUE IS STORED IN THE FILE P1.TXT '
    ENDIF

    CLOSE(UNIT=2)
    STOP
    END
```

Important Concepts and Expressions

Basic Approach in Solving a Problem by Numerical Method

The basic approach in solving a problem by computer methods using a finite difference scheme is to discretize the derivatives appearing in the governing differential equation at a finite number of uniformly or non-uniformly spaced grid points that fill the computational domain. The governing differential equation is then transformed into a system of difference equations. This means that if there are 100 grid points (where variables are not known) there will be 100 equations to solve per variable. The necessity of using a computer arises because of the huge number of arithmetic operations that are required to be carried out in a reasonable time for solving a large number of equations. The simplification inherent in the use of algebraic equations rather than differential equations is what makes numerical methods so powerful and widely applicable.

Central, Forward, and Backward Difference Expressions for a Uniform Grid

Central Difference

$$y_i' = \frac{y_{i+1} - y_{i-1}}{2h} + O(h^2)$$

$$y_i'' = \frac{y_{i-1} - 2y_i + y_{i+1}}{h^2} + O(h^2)$$

The notation, $O(h^2)$ means that in deriving the above expression from Taylor series expansion, terms of the order of h^2 and higher have been neglected. $O(h^2)$ is called the truncation error. The truncation error is the difference between the exact mathematical expression and its numerical approximation.

Forward Difference

$$y_i' = \frac{y_{i+1} - y_i}{h} + O(h)$$

$$y_i'' = \frac{y_{i+2} - 2y_{i+1} + y_i}{h^2} + O(h)$$

Backward Difference

$$y_i' = \frac{y_i - y_{i-1}}{h} + O(h)$$

$$y_i'' = \frac{y_i - 2y_{i-1} + y_{i-2}}{h^2} + O(h)$$

When can we use Central, Forward, and Backward Difference Expressions?

- Central difference expressions are used when data on both sides of the desired point are available and are more accurate than either forward or backward difference expressions.
- Forward difference expressions are used when data to the left or bottom of a point, at which a derivative is desired, are not available.
- Backward difference expressions are used when data to the right or top of a desired point are not available.

Numerical Errors

Round-off Error: The round-off error is introduced because of the inability of the computer to handle a large number of significant digits. Typically, in single-precision, the number of significant digits retained ranges from 7 to 16, although it may vary from one computer system to another. The round-off error arises due to the fact that a finite number of significant digits or decimal places are retained and all real numbers are rounded off by the computer. The last retained digit is rounded off if the first discarded digit is equal to or greater than 5. Otherwise, it is unchanged. For example, if five significant digits are to be kept in place, 5.37527 is rounded off to 5.3753, and 5.37524 to 5.3752.

Truncation Error: The truncation error is due to the replacement of an exact mathematical expression by a numerical approximation. Basically, it is the difference between an exact expression (e.g., truncated Taylor series) used in the numerical solution.

Discretization Error: The discretization error is the error in the overall solution that results from the truncation error assuming the round-off error to be negligible. Therefore,

Discretzation error = Exact Solution – Numerical Solution with no round-off error

Accuracy of a Solution: Optimum Step Size

Total Error = Round-off Error + Truncation Error

It is obvious that the round-off error increases as the total number of arithmetic operation increases. Again, the total number of arithmetic operations increases if the step size decreases (i.e., when the number of grid points increases). Therefore, the round-off error is inversely proportional to step size. On the other hand, the truncation error decreases as the step size decreases (or as the number of grid points increases).

Because of the aforementioned opposing effects, an optimum step size is expected, which will produce minimum total error in the overall solution.

Method of Choosing Optimum Step Size: Grid Independence Test

A numerical analyst has to be extremely careful as regards the accuracy of a solution. To get the most accurate solution (that is, the solution with the least total error), one has to perform a grid independence test. The test is carried out by experimenting with various grid sizes and watching how the solution changes with respect to the changes in the grid size. Finally, a stage will come when changing the grid spacing will not affect the solution. In other words, the solution will become independent of grid spacing. The largest value of grid spacing for which the solution is essentially independent of the step size is chosen so that both the computational time and effort and the round-off error are minimized.

Finite Difference Formulations of Various Problems

First, the finite difference formulation of the problem of a one-dimensional steady state heat conduction in a cooling fin is introduced. The treatment of insulated and convective boundary conditions by the use of image point technique is discussed. The concept of tri-diagonal matrix is brought forth. The resulting system of equations is solved by Gaussian elimination, Thomas algorithm (tridiagonal matrix algorithm or TDMA), and Gauss-Seidel iterative methods. The superiority of the Thomas algorithm for solving systems having a tridiagonal coefficient matrix is demonstrated. The concepts of over-relaxation, under-relaxation and optimum relaxation factors are introduced with respect to the Gauss-Seidel (GS) iterative method.

Subsequently, the formulation of a two-dimensional steady state problem with heat generation in a square slab is shown. Handling of corner points is demonstrated. It is also shown that for a large system of equations with a banded coefficient matrix, the GS iterative method is superior to the Gaussian elimination because of minimum round-off error in the former.

Then, the finite difference formulation of a one-dimensional transient heat conduction problem in a plane slab is taken up. Three methods of solution, namely, Euler, Crank-Nicolson, and pure implicit methods, are introduced and compared with one another in respect of their stability and accuracy. For two-dimensional transient problems, the superiority of the alternating direction implicit (ADI) method is then established.

The treatments of the condition at the centre for axisymmetric and non-axisymmetric problems in cylindrical geometry are given in detail. Handling of the interface of a composite body is then detailed. Finally, the treatments of non-linearities arising out of non-linear governing differential equations and/or boundary conditions are discussed, followed by handling of irregular geometries with the aid of example problems. The chapter concludes with a very interesting application of computational heat transfer to cryosurgery.

A CD accompanying this book contains computer programs related to the problems in this chapter.

Review Questions

10.1 Define round-off error, truncation error, and total error.

10.2 What is a grid independence test?

10.3 What is an image point technique?

10.4 What is a tri-diagonal matrix?

10.5 Why is cumulative round-off error minimum in Gauss-Seidel iterative method?

10.6 State Scarborough criterion.

10.7 Define relaxation factor.

10.8 Define stability of a numerical scheme.

10.9 Out of Euler, Crank-Nicolson, and pure implicit schemes which one is least stable?

10.10 How do you handle the condition at the centre of cylindrical body having axi-symmetric boundary condition?

10.11 How do you take care of non-linearity in the governing equation in finite-difference method?

Problems

10.1 Derive the following finite-difference approximations at the point (i,j):

(a) $\dfrac{\partial^4 y}{\partial x^4} = \dfrac{y_{i-2,j} - 4\,y_{i-1,j} + 6\,y_{i,j} - 4\,y_{i+1,j} + y_{i+2,j}}{(\Delta x)^4} + 0(\Delta x)^2$

(b) $\dfrac{\partial y}{\partial x} = \dfrac{-3\,y_{i,j} + 4\,y_{i+1,j} - y_{i+2,j}}{2\,\Delta x} + 0(\Delta x)^2$

(c) $\dfrac{\partial^2 y}{\partial x^2} = \dfrac{2\,y_{i,j} - 5\,y_{i+1,j} + 4\,y_{i+2,j} - y_{i+3,j}}{(\Delta x)^2} + 0(\Delta x)^2$

10.2 Suppose you are using a computer which can retain only 7 significant digits in its operations. Indicate how the following numbers will be rounded off by your computer.

(a) 17.453256

(b) 8.2172599999

(c) 4.7123562

(d) 6.8127563

(e) 3.11111111

10.3 Calculate the numerical value of $e^{0.7}$ by using Taylor series for the e^x given below. How many terms are needed if the error from the true value of the quantity $e^{0.7}$ is to be less than 0.1%?

$$e^x = 1 + x + \frac{1}{2!}x^2 + \frac{1}{3!}x^3 + \cdots, \quad -\infty < x < \infty$$

10.4 A steel plate of thickness 5 cm is initially at 500°C. It is suddenly plunged into cold water. As a result of this, the surface temperature drops to 70°C and remains at this value for the rest of the cooling process. Estimate the time elapsed for the mid-plane temperature to cool down to 300°C (a) by dividing the thickness of the plate into six equal divisions and applying a finite-difference method such as the Euler, pure implicit, or Crank–Nicolson method, (b) by using the Heisler chart. For steel, take $\alpha = 1.16 \times 10^{-5}$ m^2/s.[6]

10.5 The cross-section of a hollow square duct is shown in Fig. Q10.5. The sides of the two squares are in the ratio 5:1. The interior and exterior surfaces of the duct are maintained at 800°C and 50°C, respectively. Estimate the steady-state temperature distribution in the duct wall.

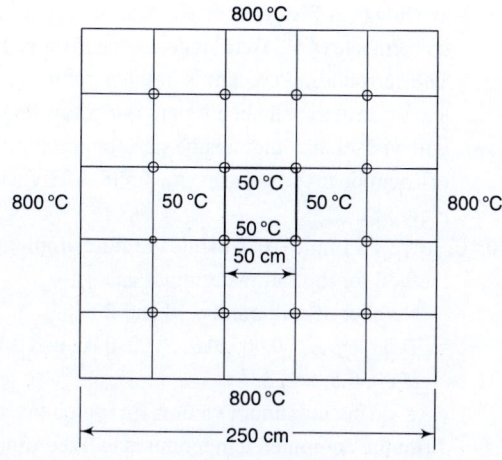

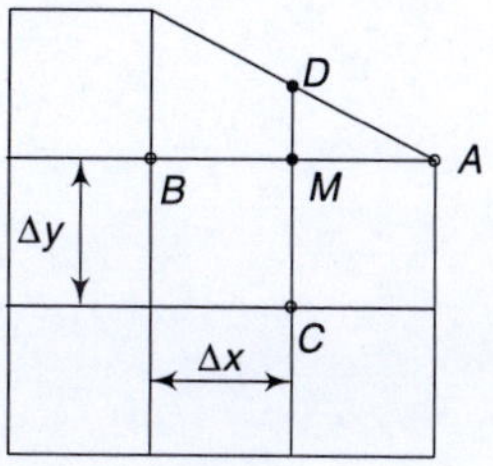

Fig. Q10.5

10.6 Using the finite-difference method, find the steady-state temperature distribution in a square plate, one side of which is maintained at 100°C, with the other three sides maintained at 0°C. Compare your results with the corresponding analytical solution.

10.7 Temperature distribution as a function of time is to be obtained by the Euler method in the computational domain shown in Fig. Q10.7. The boundary conditions are of the Dirichlet type. What formula should be used to compute T_M^{p+1}? Assume $\Delta x = \Delta y$.

Fig. Q10.7

10.8 Consider the cross-section of a circular duct (Fig. Q10.8) that passes through a square block

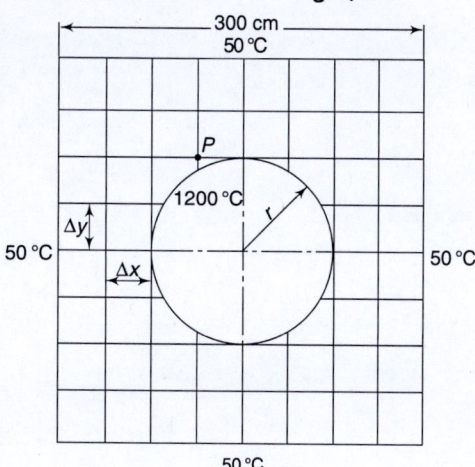

Fig. Q10.8

---

[6] See the C++ programs in the Online Resource Centre— `Ex.10.4_Euler.cpp`, `Ex.10.4_PureIimplicit.cpp`, `Ex.10.4_C-N.cpp`, for solutions of this problem using the Euler, Pure implicit, and Crank-Nicolson methods, respectively.

of refractory material of thermal conductivity 1.04 W/m K. The inner surface is maintained steadily at 1200°C by a hot gas, and the outer surface is constant at 50°C. If $r = 15$ cm, estimate the steady temperatures at the indicated grid points ($\Delta x = \Delta y = 7.5$ cm). What is the finite-difference approximation at point P? Would the answers change for $r = 30$ cm, and $\Delta x = \Delta y = 15$ cm?

10.9 Consider the cross-section of a nuclear fuel rod as shown in Fig. Q10.9. Nuclear energy at a uniform rate of q''' W/m³ is generated in the rod. The surrounding coolant is at temperature T_∞, and the heat transfer coefficient is h. Show how will you obtain the steady-state temperature distribution in the rod using the finite-difference method.

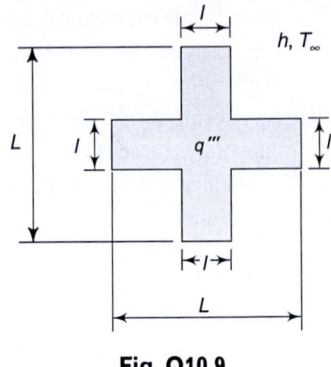

Fig. Q10.9

10.10 Solve Example 10.2 using the pure implicit method for the following input data:[7]

Number of grid points: 10 and 20

Time step $\Delta\tau$: 0.005, 0.01, 0.02, 0.05, and 0.1

H: 0, 0.5, and 2

τ_{max} (the maximum value τ for which the solution is of interest): 1.5

Print the computed temperatures at fixed time intervals.

Chapter 11

Mass Transfer

11.1 Introduction

At the very beginning of this book, in Chapter 1, we have defined heat transfer as the energy interaction due to a temperature difference in a medium or between media. We know that heat is energy in transit and heat transfer occurs whenever there exists a temperature difference in a medium or between media. Similarly, mass transfer takes place whenever there exists an uneven concentration or density of some chemical species in a mixture. Mass transfer is mass in transit as a result of a species concentration difference in a mixture. The concentration gradient is related to mass transfer in the same way as the temperature gradient is to heat transfer. Just like temperature difference is the driving potential for heat transfer, concentration difference is the same for mass transfer.

Let us look at a simple example of mass transfer in a liquid. Drop a small crystal of potassium permanganate ($KMnO_4$) into a jar of water. What you will observe is that very near the crystal there is a dark purple concentrated solution of $KMnO_4$. Because of the concentration gradient that is established, $KMnO_4$ diffuses away from the crystal. The transient diffusion process can then be followed by observing the growth of the purple region—dark purple where the $KMnO_4$ concentration is high and light purple where it is low. An example of mass diffusion in gases in a practical situation is the transport of nitrous oxide from an automobile exhaust through a stagnant atmosphere. There will be transport of nitrous oxide in the direction away from the source, where its concentration is the highest. An example of mass diffusion in solids is the diffusion of helium through Pyrex or carbon dioxide through rubber. Since mass diffusion is strongly affected by molecular spacing, diffusion occurs more readily in gases than in liquids and more effectively in liquids than in solids. The applications of mass transfer are many, such as, psychrometry, drying, evaporative cooling, transpiration cooling, diffusion-controlled combustion.

Mass diffusion is similar to heat conduction. Convection mass transfer occurs when the concentration of some species at a surface differs from its concentration in a gas moving over the surface. However, mass diffusion is more complicated than viscous flow or heat conduction, because here, for the first time, one has to deal with mixtures. In a diffusing mixture, each individual species has a different velocity and, therefore, one has to average the velocities of the species to get a local velocity for the mixture. It is imperative that a local velocity be chosen before the rates of diffusion can be defined. Hence, in the next section the definitions of concentrations, velocities, and fluxes are discussed in some detail.

11.2 Definitions of Concentrations, Velocities, and Mass Fluxes

In a multicomponent system, the concentrations of the various species may be expressed in the following ways.

(a) ρ_i, the mass concentration, is the mass of species i per unit volume of the solution.

(b) c_i, the molar concentration = ρ_i/M_i, is the number of moles of species i per unit volume of the solution.

(c) Y_i, the mass fraction = ρ_i/ρ, is the mass concentration of species i divided by the total mass density of the solution.

(d) x_i, the mole fraction = c_i/c, is the molar concentration of species i divided by the total molar density of the solution.

In a diffusion mixture the various chemical species are moving at different velocities. Note that by 'velocity' we mean the sum of the velocities of the molecules of species i within a small volume element divided by the number of such molecules. Therefore, it should not be confused with the velocity of an individual molecule of species i.

Then, for a mixture of n species, the local mass average velocity v is defined as

$$v = \frac{\sum\limits_{i=1}^{n} \rho_i v_i}{\sum\limits_{i=1}^{n} \rho_i} \tag{11.1}$$

Note that ρv is the local rate at which mass passes through a unit cross-section placed perpendicular to velocity v.

Similarly, local molar average velocity v^* is defined as

$$v^* = \frac{\sum\limits_{i=1}^{n} c_i v_i}{\sum\limits_{i=1}^{n} c_i} \tag{11.2}$$

Note that cv^* is the local rate at which moles pass through a unit cross-section placed perpendicular to velocity v^*.

In flow systems one is more interested in the motion of component i relative to the local motion of the fluid stream rather than with respect to stationary coordinates. This leads to the definition of the *diffusion velocities*:

$$v_i - v = \text{diffusion velocity of } i \text{ with respect to } v \tag{11.3}$$

$$v_i - v^* = \text{diffusion velocity of } i \text{ with respect to } v^* \tag{11.4}$$

Let us now define mass and molar fluxes. The mass and molar fluxes relative to the *stationary* coordinates are

$$n_i = \rho_i v_i \text{ (mass)} \tag{11.5}$$

$$N_i = c_i v_i \text{ (molar)} \tag{11.6}$$

The mass and molar fluxes relative to the mass-averaged velocity v are

$$j_i = \rho_i(v_i - v) \text{ (mass)} \tag{11.7}$$

$$J_i = c_i(v_i - v) \text{ (molar)} \tag{11.8}$$

and the mass and molar fluxes relative to the molar average velocity v^* are

$$j_i^* = \rho_i(v_i - v^*) \text{ (mass)} \tag{11.9}$$

$$J_i^* = c_i(v_i - v^*) \text{ (molar)} \tag{11.10}$$

Example 11.1   Relationship between J_i^* and N_i in an n-component System

(a) Find a relationship between fluxes J_i^* and N_i in an n-component system. (b) Show that the sum of the fluxes J_i^* is zero.

Solution

(a)   Recall from Eq. (11.2) the definition of v^*. Now,

$$J_i^* = c_i(v_i - v^*)$$

$$= c_i v_i - c_i v^*$$

$$= c_i v_i - c_i \frac{\displaystyle\sum_{j=1}^{n} c_j v_j}{\displaystyle\sum_{j=1}^{n} c_j} = c_i v_i - \frac{c_i}{c}\sum_{j=1}^{n} c_j v_j$$

Again, $N_i = c_i v_i$. Therefore,

$$J_i^* = N_i - x_i \sum_{j=1}^{n} N_j \tag{A}$$

(b)   Expanding Eq. (A) from $i = 1$ to $i = n$ and summing up results in

$$J_1^* = N_1 - x_1 \sum_{j=1}^{n} N_j$$

$$J_2^* = N_2 - x_2 \sum_{j=1}^{n} N_j$$

$$\vdots$$

$$J_n^* = N_n - x_n \sum_{j=1}^{n} N_j$$

$$\sum_{i=1}^{n} J_i^* = \sum_{i=1}^{n} N_i - \sum_{j=1}^{n} N_j (x_1 + x_2 + \cdots + x_n)$$

$$= \sum_{i=1}^{n} N_i - \sum_{j=1}^{n} N_j (1) = 0$$

Therefore, $\qquad \sum_{i=1}^{n} J_i^* = 0$ (B)

Equation (B) shows that the sum of the molar diffusion fluxes relative to the molar average velocity is zero in any mixture. Thus, in a binary mixture, $J_A^* = -J_B^*$.

11.3 Fick's Law of Diffusion

The readers may recall that in Chapter 1, the thermal conductivity k is defined as the proportionality factor between the heat flux and temperature gradient. Analogously, the mass diffusivity (also, called the binary diffusivity of A in B) $D_{AB} = D_{BA}$ is defined as

$$J_A^* = -cD_{AB} \nabla x_A$$ (11.11)

Equation (11.11) is Fick's law of diffusion written in terms of the molar diffusion flux J_A^*. This equation states that *species A diffuses (i.e., moves relative to the mixture) in the direction of decreasing mole fraction of A, just as heat flows by conduction in the direction of decreasing temperature.*

Another important form of Fick's law in terms of N_A, the molar flux relative to the stationary coordinates, is

$$N_A = x_A (N_A + N_B) - cD_{AB} \nabla x_A$$ (11.12)

Equation (11.12) shows that the diffusion flux N_A relative to stationary coordinates is the resultant of two vector quantities: the vector $x_A (N_A + N_B)$, which is the molar flux of A resulting from the bulk motion of the fluid, and the vector $J_A^* = -cD_{AB} \nabla x_A$, which is the molar flux of A resulting from the diffusion superimposed on the bulk flow.

The unit of binary diffusivity D_{AB} is m²/s. Note that thermal diffusivity α also has the same unit. Binary diffusivities of various substances at 1 atm are listed in Appendix A6.

It may be noted that $D_{AB} = D_{BA}$. This can be demonstrated by considering 1D mass diffusion of species A into species B and vice versa. From Fick's law of diffusion,

$$J_{A,y}^* = -cD_{AB} \frac{dx_A}{dy}$$ (11.13)

$$J_{B,y}^* = -cD_{BA} \frac{dx_B}{dy}$$ (11.14)

Adding Eqs (11.13) and (11.14),

$$J_{A,y}^* + J_{B,y}^* = -cD_{AB} \frac{dx_A}{dy} + \left(-cD_{BA} \frac{dx_B}{dy} \right)$$ (11.15)

However, since the diffusive medium remains stationary, the net flow rate through a plane at any y must be zero.

Hence, $\quad J_{A,y}^* + J_{B,y}^* = 0$ (11.16)

Also, in a plane at any y,

$$x_A + x_B = 1 \tag{11.17}$$

Substituting Eqs (11.16) and (11.17) into Eq. (11.15), we obtain

$$0 = -cD_{AB} \frac{dx_A}{dy} - cD_{BA} \frac{d}{dy}(1 - x_A)$$

$$= -cD_{AB} \frac{dx_A}{dy} + cD_{BA} \frac{dx_A}{dy}$$

$$= c \frac{dx_A}{dy}(D_{BA} - D_{AB}) \tag{11.18}$$

Hence, from Eq. (11.18)

$$D_{AB} = D_{BA}$$

Since D_{AB} must be equal to D_{BA}, the binary diffusivity values listed in Appendix A6 are also called *mutual diffusion coefficients*.

11.4 Analogy Between Heat Transfer and Mass Transfer

The way in which heat transfer and mass transfer are analogous can be seen from the following equations for the fluxes of mass and energy in one-dimensional systems:

$$j_{A_x} = -D_{AB} \frac{d}{dx}(\rho_A) \text{, Fick's law for constant } \rho \tag{11.19}$$

$$q_x = -\alpha \frac{d}{dx}(\rho c_p T) \text{, Fourier's law for constant } \rho c_p \tag{11.20}$$

Equations (11.19) and (11.20) state, respectively, that (a) mass transport occurs because of a gradient in mass concentration and (b) energy transport occurs because of a gradient in energy concentration.

11.5 Derivation of Various Forms of the Equation of Continuity for a Binary Mixture

The principle of conservation of mass applied to the chemical species is the conservation principle that governs the migration of a chemical species through a gaseous, liquid, or solid medium. In this section the species conservation equation or the equation of continuity for a binary mixture will be derived for the general case where the bulk of the medium moves relative to the boundaries. This general form of the equation of continuity applies directly to the problems of mass transfer by convection.

Referring to Fig. 11.1, we apply the law of conservation of mass of species A to a volume element $\Delta x \Delta y \Delta z$ fixed in space, through which a binary mixture of A and B is flowing. Within this, element A may be produced by chemical reaction $(B \rightarrow A)$ at a rate of r_A (kg/m^{3}s). The various contributions to the mass balance are:

The time rate of change of mass of A in the volume element: $\dfrac{\partial \rho_A}{\partial t}\Delta x\, \Delta y\, \Delta z$

The inflow of A across face at x: $n_{A_x}\big|_x \Delta y\, \Delta z$

The outflow of A across face at $x + \Delta x$: $n_{A_x}\big|_{x+\Delta x} \Delta y\, \Delta z$

The rate of production of A by chemical reaction: $r_A \Delta x \Delta y \Delta z$

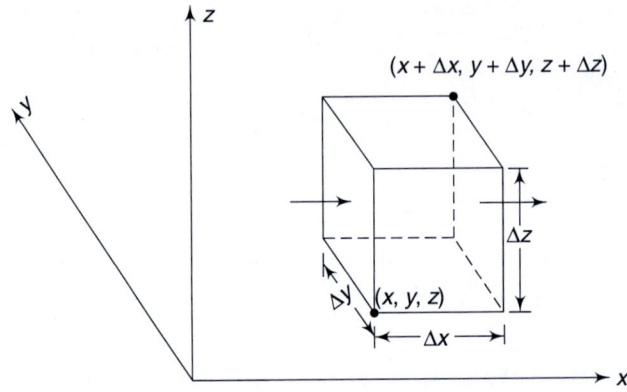

Fig.11.1 Volume element for deriving equation of continuity for a binary mixture

There are also inflow and outflow terms in the y- and z-directions. The conservation of mass for a species states that

$$\begin{matrix} \text{Rate of mass} \\ \text{flow out} \end{matrix} = \begin{matrix} \text{Rate of mass} \\ \text{flow in} \end{matrix} - \begin{matrix} \text{Rate of mass} \\ \text{accumulation} \end{matrix} + \begin{matrix} \text{Rate of mass} \\ \text{generation} \end{matrix}$$

Thus, $\quad n_{A_x}\big|_{x+\Delta x} \Delta y\, \Delta z + n_{A_y}\big|_{y+\Delta y} \Delta x\, \Delta z + n_{A_z}\big|_{z+\Delta z} \Delta x\, \Delta y$

$$= n_{A_x}\big|_x \Delta y\, \Delta z + n_{A_y}\big|_y \Delta x\, \Delta z + n_{A_z}\big|_z \Delta x\, \Delta y - \frac{\partial \rho_A}{\partial t}\Delta x\, \Delta y\, \Delta z + r_A \Delta x\, \Delta y\, \Delta z$$

Dividing by $\Delta x \Delta y \Delta z$, for the volume element shrinking to zero, we obtain

$$\frac{\partial \rho_A}{\partial t} + \frac{\partial n_{A_x}}{\partial x} + \frac{\partial n_{A_y}}{\partial y} + \frac{\partial n_{A_z}}{\partial z} = r_A \qquad (11.21)$$

Equation (11.21) is the equation of continuity for component A in a binary mixture. In the vector notation, Eq. (11.21) takes the form

$$\frac{\partial \rho_A}{\partial t} + (\nabla \cdot n_A) = r_A \qquad (11.22)$$

Similarly, the equation of continuity for species B is

$$\frac{\partial \rho_B}{\partial t} + (\nabla \cdot n_B) = r_B \qquad (11.23)$$

Adding Eqs (11.22) and (11.23), we get

$$\frac{\partial(\rho_A + \rho_B)}{\partial t} + \nabla \cdot (n_A + n_B) = r_A + r_B$$

or $\qquad \dfrac{\partial \rho}{\partial t} + \nabla \cdot (\rho v) = 0 \qquad (11.24)$

[Note: $\rho_A + \rho_B = \rho, r_A + r_B = 0$

$$n_A + n_B = \rho_A v_A + \rho_B v_B = (j_A + \rho_A v) + (j_B + \rho_B v)$$
$$= (j_A + j_B) + v(\rho_A + \rho_B)$$
$$= 0 + v\rho = \rho v]$$

For a fluid of constant density ρ, the equation of continuity [Eq. (11.24)] becomes

$$\nabla \cdot v = 0 \tag{11.25}$$

Now, $n_A = Y_A(n_A + n_B) - \rho D_{AB} \nabla Y_A \tag{11.26}$

Substituting Eq. (11.26) into Eq. (11.22), we obtain

$$\frac{\partial \rho_A}{\partial t} + \nabla \cdot [Y_A(n_A + n_B) - \rho D_{AB} \nabla Y_A] = r_A$$

or $\quad \dfrac{\partial \rho_A}{\partial t} + \nabla \cdot [Y_A(\rho v) - \rho D_{AB} \nabla Y_A] = r_A$

or $\quad \dfrac{\partial \rho_A}{\partial t} + \nabla \cdot Y_A(\rho v) = \nabla \cdot \rho D_{AB} \nabla Y_A + r_A$

or $\quad \dfrac{\partial \rho_A}{\partial t} + (\nabla \cdot \rho_A v) = (\nabla \cdot \rho D_{AB} \nabla Y_A) + r_A \tag{11.27}$

The assumption of constant ρ and D_{AB} yields, from Eq. (11.27),

$$\frac{\partial \rho_A}{\partial t} + \rho_A(\nabla \cdot v) + (v \cdot \nabla \rho_A) = D_{AB} \nabla^2 \rho_A + r_A$$

But $\quad \nabla \cdot v = 0$

Therefore, $\dfrac{\partial \rho_A}{\partial t} + (v \cdot \nabla \rho_A) = D_{AB} \nabla^2 \rho_A + r_A$

Dividing by M_A, we get

$$\frac{\partial c_A}{\partial t} + (v \cdot \nabla c_A) = D_{AB} \nabla^2 c_A + R_A \tag{11.28}$$

where R_A is the molar rate of production of A per unit volume. Equation (11.28) is usually used for diffusion in dilute liquid solutions at constant temperature and pressure. Equation (11.28) can also be written as

$$\frac{Dc_A}{Dt} = D_{AB} \nabla^2 c_A + R_A \tag{11.29}$$

If $R_A = 0$, then Eq. (11.29) takes the form

$$\frac{Dc_A}{Dt} = D_{AB} \nabla^2 c_A \tag{11.30}$$

Equation (11.30) is of the same form as

$$\frac{DT}{Dt} = \alpha \nabla^2 T \tag{11.31}$$

The similarity of Eqs (11.30) and (11.31) is the basis for the analogy of heat transfer and mass transfer in flowing fluids with constant ρ.

Equations (11.29)–(11.31) can also be written in terms of molar units. If R_A is the molar rate of production of A per unit volume, then we can write by analogy the equation of continuity for the mixture,

$$\frac{\partial c_A}{\partial t} + (\nabla \cdot N_A) = R_A \tag{11.32}$$

Similarly, for component B,

$$\frac{\partial c_B}{\partial t} + (\nabla \cdot N_B) = R_B \tag{11.33}$$

Adding Eqs (11.32) and (11.33), we obtain

$$\frac{\partial c}{\partial t} + (\nabla \cdot cv^*) = (R_A + R_B) \tag{11.34}$$

Note that to arrive at Eq. (11.34) from Eq. (11.33) we have used the relation $N_A + N_B = cv^*$. However, $R_A + R_B$ is not equal to zero unless 1 mole of B is produced for every mole of A disappearing (or vice versa).

Finally, for a fluid of constant molar density c, Eq. (11.34) can be written as

$$(\nabla \cdot v^*) = \frac{1}{c}(R_A + R_B) \tag{11.35}$$

Now, substituting Eq. (11.12) into Eq. (11.32), we get

$$\frac{\partial c_A}{\partial t} + (\nabla \cdot c_A v^*) = (\nabla \cdot cD_{AB}\nabla x_A) + R_A \tag{11.36}$$

Assumption of constant c and D_{AB}   Equation (11.36) transforms into

$$\frac{\partial c_A}{\partial t} + c_A(\nabla \cdot v^*) + (v^* \cdot \nabla c_A) = D_{AB}\nabla^2 c_A + R_A \tag{11.37}$$

But, from Eq. (11.35), we know

$$\nabla \cdot v^* = \frac{1}{c}(R_A + R_B)$$

Therefore, Eq. (11.37) becomes

$$\frac{\partial c_A}{\partial t} + (v^* \cdot \nabla c_A) = D_{AB}\nabla^2 c_A + R_A - \frac{c_A}{c}(R_A + R_B) \tag{11.38}$$

Equation (11.38) is generally valid for low-density gases at constant temperature and pressure. Note that the left-hand side of Eq. (11.38) cannot be written as Dc_A/Dt because of the appearance of v^* instead of v.

Assumption of zero velocity   In the special case of no chemical reaction occurring, r_A, r_B, R_A, R_B are all zero. If in addition, v is zero or v^* is zero, then we get, from Eq. (11.28) or (11.38),

$$\frac{\partial c_A}{\partial t} = D_{AB}\nabla^2 c_A \tag{11.39}$$

Equation (11.39) is called Fick's second law of diffusion or simply the diffusion equation. This equation is generally valid for diffusion in solids or stationary liquids or equimolar counter-diffusion in low-density gases. Equation (11.39) is similar to the transient heat conduction equation (i.e., $\partial T/\partial t = \alpha \nabla^2 T$). This similarity forms the basis for the analogous treatment of many heat conduction and mass diffusion problems in solids.

11.6 Analogy Between Special Forms of the Heat Conduction and Mass Diffusion Equations

Table 11.1 shows the analogy between heat conduction and diffusion equations for some special cases.

Table 11.1 Analogy between heat conduction and diffusion equations

Case	*Heat conduction*	*Diffusion*
Unsteady-state non-flow	GDE: $\dfrac{\partial T}{\partial t} = \alpha \nabla^2 T$ *Application*: Heat conduction in solids *Assumptions*: Constant thermal conductivity and stationary medium	GDE: $\dfrac{\partial c_A}{\partial t} = D_{AB} \nabla^2 c_A$ *Application:* Diffusion of traces of A through B *Assumptions:* Constant binary diffusivity, stationary medium, and no chemical reactions or *Application:* Equimolar counter-diffusion in low-density gases *Assumptions:* Constant binary diffusivity and molar mixture density, zero molar velocity of the mixture, no chemical reactions
Steady-state flow	GDE: $v \cdot \nabla T = \alpha \nabla^2 T$ *Application:* Heat conduction in laminar incompressible flow (also called laminar convective heat transfer) *Assumptions:* Constant thermal conductivity and density, no viscous dissipation, steady state	GDE: $v \cdot \nabla c_A = D_{AB} \nabla^2 c_A$ *Application:* Diffusion in laminar flow (dilute solution of A in B) *Assumptions:* Constant binary diffusivity, constant mixture density, steady state, no chemical reactions

(Contd.)

(Contd.)

Case	Heat conduction	Diffusion
Steady-state non-flow	GDE: $\nabla^2 T = 0$	GDE: $\nabla^2 c_A = 0$
	Application: Steady-state conduction in solids	*Application:* Steady diffusion in solids
	Assumptions: Constant thermal conductivity and density, no viscous dissipation, steady state	*Assumptions:* Constant binary diffusivity, constant mixture density, steady state, no chemical reactions, stationary mixture (zero bulk velocity)

11.7   Boundary Conditions in Mass Transfer

The boundary conditions used in mass transfer are very similar to those used in heat transfer. Some frequently applied boundary conditions are given below.

(a) *Specified surface concentration*: The concentration at a surface can be specified; for example, $c_A = c_{A_0}$.

(b) *Specified mass or molar surface flux*: The mass or molar surface flux can be specified; for example, $N_A = N_{A_0}$.

(c) *Convective mass loss at the surface*: At the solid surface, A is lost to the surrounding fluid stream according to the relation

$$N_{A_0} = h_m(c_{A_0} - c_{A_\infty})$$

in which N_{A_0} is the molar flux at the surface, c_{A_0} is the surface concentration, c_{A_∞} is the concentration in the fluid stream, and h_m is the mass transfer coefficient.

(d) *Rate of chemical reaction specified at the surface*: The rate of chemical reaction at the surface can be specified. For example, if component A disappears at a surface by a first-order chemical reaction, then $N_{A_0} = k_1'' c_{A_0}$, that is, the rate of disappearance at a surface is proportional to the surface concentration (the proportionality constant k_1'' being a first-order chemical rate constant). The units: k_1'' in m/s, c_A in moles/m^3, N_{A_0} in moles/m^{2}s, and h_m in moles/m^{2}s (moles/m^3) or m/s. "''' " indicates a rate constant related to the surface source.

Example 11.2   Rate of Burning of a Coal Particle

Calculate the rate of burning of a coal particle (diameter 0.25 cm) in an atmosphere of pure oxygen, at 1000 K and 10^5 N/m^2, assuming that a very large blanketing layer of CO_2 is formed around the particle. Assume that the combustion rate is such that all the oxygen reaching the surface is instantaneously consumed. Hence, the concentration of oxygen at the surface is effectively zero. Also, assume that the concentration of CO_2 far away is zero. The binary diffusivity of oxygen in carbon dioxide may be taken as 10^{-4} m^2/s. Note that the reaction $C + O_2 \rightarrow CO_2$ is an equimolar reaction, that is, the rate of burning of oxygen is equal to the rate of production of carbon dioxide, so that in the *steady state equal number of moles of O_2 and CO_2 are diffusing in either direction. This is equivalent to zero molar average velocity.*

Solution

This problem can be treated as O_2 penetrating through a spherical shell of CO_2 of inner radius $r_1 = 1.25 \times 10^{-3}$m and outer radius $r_2 = \infty$.

Let O_2 be designated as species A and CO_2 be designated as species B. Therefore, the equation of continuity of species A is [see Eq. (11.32)]

$$\frac{1}{r^2} \frac{\partial}{\partial r}(r^2 N_{A_r}) = 0$$

or

$$\frac{\partial}{\partial r}(r^2 N_{A_r}) = 0$$

or

$$D_{AB} \frac{1}{r^2} \left[\frac{\partial}{\partial r}\left(r^2 \frac{\partial c_A}{\partial r} \right) \right] = 0$$

or

$$\frac{\partial}{\partial r}\left(r^2 \frac{\partial c_A}{\partial r} \right) = 0$$

or

$$\frac{\partial}{\partial r}\left(r^2 c \frac{\partial x_A}{\partial r} \right) = 0$$

or

$$\frac{\partial}{\partial r}\left(r^2 \frac{\partial x_A}{\partial r} \right) = 0 \qquad (A)$$

The boundary conditions are:

BC-1: at $r = r_1$, $x_A = 0$

BC-2: at $r = \infty$, $x_A = 1$

Integrating Eq. (A) twice, we get

$$x_A = -\frac{C_1}{r} + C_2$$

Applying BC-1 and BC-2, we obtain

$$C_1 = r_1$$
$$C_2 = 1$$

Therefore, $x_A = 1 - r_1/r$. Hence, $\qquad (B)$

$$N_A = + c D_{AB} \left. \frac{\partial x_A}{\partial r} \right|_{r = r_1}$$

$$= \frac{c D_{AB}}{r_1}$$

Note the positive sign in the RHS of Eq. (B) because O_2 is diffusing in a direction opposite to the increasing r. Now,

$$c = \frac{p}{\Re T} = \frac{10^5}{8.314 \times 10^3 \times 1000} = 0.012 \text{ kmol/m}^3$$

Therefore, $N_A = \dfrac{(0.012)(10^{-4})}{\left(\dfrac{2.5 \times 10^{-3}}{2} \right)} = 0.96 \text{ mol/m}^2\text{s}$

Hence, total gmol/s of O_2 is

$$(0.96)\ (4\pi r_i^2) = (0.96)\ (4\times3.1415)\left(\frac{2.5\times10^{-3}}{2}\right)^2 = 18.85\times10^{-6}$$

Since the reaction is equimolar, 1 kmol of O_2 is consumed by 12 kg of carbon. Therefore,

Rate of burning $= 12\times18.85\times10^{-6}\times10^{-3} = 2.26\times10^{-7}$ kg/s

11.8 One-dimensional Steady Diffusion through a Stationary Medium

In this section simplest type of mass diffusion problems in which the mixture is stationary will be discussed. The problem is steady and one-dimensional. The boundary conditions are of Dirichlet type, that is, surface concentrations are specified. The solutions in three coordinate systems will be given.

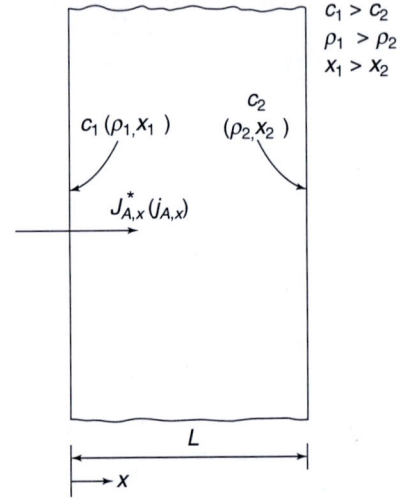

Plane wall

Consider an infinite plane wall of finite thickness L as shown in Fig. 11.2. The continuity of a certain species A across this layer is described by the steady, one-dimensional form of Eq. (11.39) as given below.

$$\frac{d^2 c_A}{dx^2} = 0$$

$\Rightarrow$

$$\frac{d^2 x_A}{dx^2} = 0 \tag{11.40}$$

Fig. 11.2 Steady one-dimensional diffusion of species A through a stationary medium (plane wall)

where x_A is the mole fraction of species A.

The boundary conditions are:

BC-1: at $x = 0$, $x_A = x_1$

BC-2: at $x = L$, $x_A = x_2$

Integrating Eq. (11.40) twice and obtaining the two constants by substituting the boundary conditions in the general solution, we get the particular solution as

$$x_A = (x_2 - x_1)\frac{x}{L} + x_1 \tag{11.41}$$

Hence, the mole fraction distribution is linear. Note the similarity of Eq. (11.41) with the temperature distribution in a plane wall [Eq. (11.38)].

The diffusion flux (which is constant) can be expressed either as a molar flux or mass flux by applying Fick's law of diffusion.

Molar flux

$$J_{Ax}^* = -D_{AB}\frac{dc_A}{dx} = -D_{AB}\frac{\rho}{M}\frac{dx_A}{dx} = D_{AB}\frac{\rho}{M}\frac{x_1 - x_2}{L} = D_{AB}\frac{c_1 - c_2}{L} \tag{11.42}$$

Mass flux:

$$j_{A_x} = -D_{AB} \frac{d\rho_A}{dx} = -D_{AB} \frac{\rho M_A}{M} \frac{dx_A}{dx} = D_{AB} \frac{\rho M_A}{M} \frac{x_1 - x_2}{L} = D_{AB} \frac{\rho_1 - \rho_2}{L}$$

(11.43)

Thus from Eq. (11.42) or (11.43) we see that the flux or flow rate of the species of interest across the slab is proportional to the binary diffusivity D_{AB} and the mole fraction or concentration difference, and is inversely proportional to the wall thickness. The striking similarity of Eq. (11.42) or (11.43) with heat flow across a plane wall [Eq. (2.39)] is worth noting. Using this similarity between heat conduction and mass diffusion the solutions to other steady 1D diffusion problems can be written by an appropriate change of notation in their counterparts in conduction heat transfer in cylindrical and spherical coordinates as well.

Cylindrical shell

$$x_A = x_1 + (x_1 - x_2) \frac{\ln\left(\dfrac{r}{r_1}\right)}{\ln\left(\dfrac{r_1}{r_2}\right)}$$

(11.44)

Equation (11.44) has the same analytical form as the temperature distribution given in Eq. (2.62).

Spherical shell

$$x_A = x_1 + \frac{x_1 - x_2}{\left(\dfrac{1}{r_2} - \dfrac{1}{r_1}\right)} \left(\frac{1}{r_1} - \frac{1}{r}\right)$$

(11.45)

Equation (11.45) has the same analytical form as the temperature distribution given in Eq. (2.69).

11.9 Forced Convection with Mass Transfer over a Flat Plate Laminar Boundary Layer

The topic of this section has applications in mass transfer cooling of a surface from a hot gas stream, for example, the walls of combustion chambers or surfaces of high-speed vehicles. Cooling is accomplished by (a) injecting a foreign gas into the boundary layer fluid, (b) a continuously supplied liquid film which evaporates, or (c) by constructing the walls with a solid substance which sublimes into the hot boundary layer fluid. In all cases, simultaneously with momentum and heat transfer, there is mass transfer normal to the wall.

11.9.1 Exact Solution

Consider steady, two-dimensional flow of an incompressible fluid initially at uniform velocity, concentration, and temperature over a porous flat plate through which gas *A* diffuses upward as shown in Fig. 11.3.

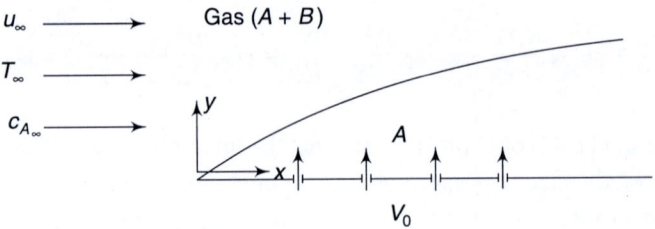

Fig. 11.3   Mass transfer on a flat plate

For constant properties of the fluid mixture (properties of the diffusing gas are identical with the properties of the free stream fluid), the continuity equation of species A and the simplified boundary layer equations of momentum, heat, and mass may be written as follows:

$$\frac{\partial u}{\partial x} + \frac{\partial v}{\partial y} = 0 \tag{11.46}$$

$$u\frac{\partial u}{\partial x} + v\frac{\partial u}{\partial y} = v\frac{\partial^2 u}{\partial y^2} \tag{11.47}$$

$$u\frac{\partial T}{\partial x} + v\frac{\partial T}{\partial y} = \alpha\frac{\partial^2 T}{\partial y^2} \tag{11.48}$$

$$u\frac{\partial c_A}{\partial x} + v\frac{\partial c_A}{\partial y} = D_{AB}\frac{\partial^2 c_A}{\partial y^2} \tag{11.49}$$

Equations (11.47) and (11.48) were earlier obtained in the absence of mass transfer. However, in the present case it has been assumed that any additional momentum and heat fluxes associated with mass transfer are negligible. Equation (11.49) is the boundary-layer-type equation based on the assumption that the diffusion flux in the x-direction is negligible compared with that in the y-direction.

The boundary conditions are:

$$y = 0,\ u = 0,\ v = v_0,\ T = T_0,\ c_A = c_{A_0} \tag{11.50}$$

$$y \rightarrow \infty,\ u = u_\infty,\ T = T_\infty,\ c_A = c_{A_\infty} \tag{11.51}$$

The condition $v = v_0$ accounts for the bulk motion that generally accompanies diffusion from a wall.

Each of the partial differential equations (11.47)–(11.49) can again be transformed into an ordinary differential equation by similarity transformation as shown below.

$$\frac{d^2 f'}{d\eta^2} + \frac{1}{2}f\frac{df'}{d\eta} = 0 \tag{11.52}$$

$$\frac{d^2\theta}{d\eta^2} + \frac{1}{2}\left(\frac{v}{\alpha}\right)f\frac{d\theta}{d\eta} = 0 \tag{11.53}$$

$$\frac{d^2\phi}{d\eta^2} + \frac{1}{2}\left(\frac{v}{D_{AB}}\right)f\frac{d\phi}{d\eta} = 0 \tag{11.54}$$

where $\quad \eta = y\sqrt{\dfrac{u_\infty}{vx}}, \quad \theta = \dfrac{T - T_0}{T_\infty - T_0}, \quad \phi = \dfrac{c_A - c_{A_0}}{c_{A_\infty} - c_{A_0}}$

and $\quad f(\eta) = \dfrac{\psi}{\sqrt{u_\infty vx}}$

Boundary conditions in terms of the new variables are:

$$\eta = 0, f' = 0, \ \theta = 0, \ \phi = 0, \ f = -\frac{2v_0}{u_\infty}\sqrt{\text{Re}_x} = \text{constant} \tag{11.55}$$

$$\eta \to \infty, f' = 1, \ \theta = 1, \ \phi = 1 \tag{11.56}$$

The condition f = constant along the wall requires that v_0 at the wall vary as $1/\sqrt{x}$.

Inspection of Eqs (11.55) and (11.56) reveals that the boundary conditions on f' (dimensionless velocity, u/u_∞), θ (dimensionless temperature), and ϕ (dimensionless concentration) are identical. Therefore, when Pr = Sc = 1 (Sc = v/D_{AB} is the Schmidt number), solutions to the ordinary differential Eqs (11.52)–(11.54) must be the same; in other words, the dimensionless velocity, temperature, and concentration profiles (f', θ, ϕ versus η) within the boundary layer must coincide. These profiles show a strong dependence on the parameter f, or equivalently v_0. With increasing v_0, the profiles become flatter. This implies that mass transfer cooling can be used to protect a surface from hot gas streams. It may be noted that the foregoing analysis applies only at low velocities since compressibility and frictional effects were not taken into account.

Therefore, it may be concluded that at low mass transfer rates, a heat transfer analogue with comparable boundary conditions exists for most problems of mass transfer. This suggests that Nusselt number correlations for heat transfer may be applied to the corresponding mass transfer systems by a simple change of notation such as

$$T \to c_A$$

Pr $\to$ Sc (Schmidt[1] number)

Nu $\to$ Sh (Sherwood[2] number)

where Sc = v/D_{AB} and Sh = $h_m L/D_{AB}$.

For laminar boundary layer flow over a flat plate of length L (dilute solution approximation, that is, for very small bulk velocity v_0) we can write using the heat transfer and mass transfer analogy (for Sc $\geq$ 0.5)

---

[1] Ernst Schmidt (1892–1975) was Professor of Engineering Thermodynamics at the TU Munich (1919–1925 and 1952–1961), the TU Danzig (1925–1937), and U. Braunschweig (1937–1952). His main contributions are in the areas of heat and mass transfer analogy, fin optimization, convective flow visualization, natural convection, the radiative properties of solids, dropwise condensation, and the international steam tables.

[2] Thomas K. Sherwood (1903–1976) was Professor of Chemical Engineering at MIT, the USA (1930–1969). He is best known for his research in convective mass transfer with and without chemical reactions and drying.

$$\overline{Sh}_L = 0.664 \, (Re_L)^{1/2} \, (Sc)^{1/3} \qquad (11.57)$$

In a similar way, for turbulent boundary layer flow over a flat plate, using the heat transfer and mass transfer analogy we can write (for $Sc \geq 0.5$)

$$\overline{Sh}_L = 0.037 \, (Re_L^{4/5} - 23{,}550)(Sc)^{1/3} \qquad (11.58)$$

11.9.2 Concentration Boundary Layer and Mass Transfer Coefficient

We have learnt earlier that the shear stress and heat transfer at the solid–fluid interface are determined by the velocity and the temperature gradient within the momentum and thermal boundary layer, respectively. Similarly, the mass transfer at an interface is determined by the concentration boundary layer profile. Figure 11.4 shows a fluid mixture, with the free stream velocity and concentration designated as u_∞ and c_{A_∞}, respectively, flowing over a flat plate. The plate is maintained at a concentration $c_{A_0} > c_{A_\infty}$ and, therefore, mass A diffuses from the surface into the fluid stream. A concentration boundary layer grows from the leading edge just like velocity or thermal boundary layers. The concentration boundary layer thickness δ_c may be defined as the normal distance from the plate where $(c_{A_0} - c_A)/(c_{A_0} - c_{A_\infty}) = 0.99$. The relative rates of the growth of the velocity and the concentration boundary layers are dictated by the Schmidt number $(Sc = v/D_{AB})$ of the fluid.

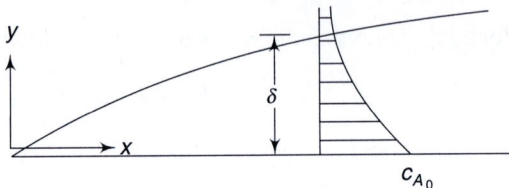

Fig. 11.4 Concentration boundary layer on a flat plate

Similar to the definition of the heat transfer coefficient, the mass transfer coefficient may be defined as

$$h_m = \frac{-cD_{AB} \dfrac{\partial x_A}{\partial y}\bigg|_{y=0}}{c_{A_0} - c_{A_\infty}} \qquad (11.59)$$

Note that in Eq. (11.59) the numerator represents the diffusion mass flux at the surface. Generally, at a solid–liquid interface or a liquid–gas interface, a bulk flow normal to the surface contributes to the additional mass transfer. The mass transfer coefficient is defined in terms of the diffusion flux rather than the total flux, because h_m defined this way depends on a smaller number of parameters and is easier to apply.

11.10 Evaporative Cooling

An important application of the heat transfer and mass transfer analogy is the phenomenon of evaporative cooling of liquid, which occurs when a gas at a higher temperature flows over a liquid surface (Fig. 11.5). Evaporation must occur from the liquid surface, and the energy associated with the phase change is the latent heat of vapourization of the liquid. The energy required for evaporation must come from the internal energy of the liquid, resulting in the lowering of the temperature of the liquid and thus a cooling effect in the liquid is produced. However, in the steady state, the energy lost by the liquid due to evaporation is gained by the convective heat transfer to the liquid from the surroundings, assuming negligible radiation effect.

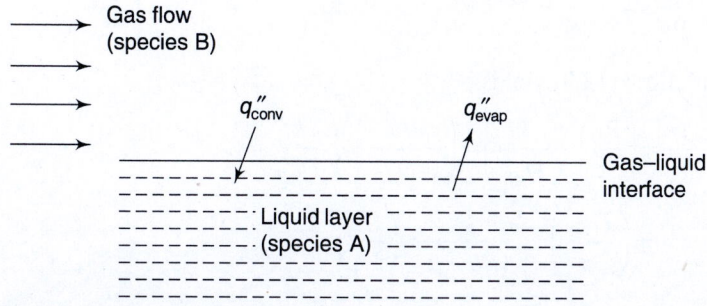

Fig. 11.5 Evaporation at the liquid–gas interface

A common application of evaporative cooling is a device called room cooler (known as desert cooler). Water dripping constantly over porous pads made of straw evaporates, maintaining pads at a lower temperature. The air which is sucked in through the pores of the pads by a fan from the hot surroundings then loses heat to the pads, cooling itself in the process. The cooled air is blown by the fan into the room.

Under the steady-state condition,

$$q''_{conv} - q''_{evap} = 0 \tag{11.60}$$

where

$$q''_{evap} = \frac{\dot{m}}{A} h_{fg} = M_A N_A h_{fg} \tag{11.61}$$

$$q''_{conv} = h(T_\infty - T_0) \tag{11.62}$$

Therefore, from Eq. (11.60), we can write

$$h(T_\infty - T_0) = M_A N_A h_{fg} = M_A h_{fg} N_A = M_A h_{fg} h_m (c_{A_0} - c_{A_\infty})$$
$$= M_A h_{fg} h_m (c_{A,\text{sat}(T_0)} - c_{A,\infty}) \tag{11.63}$$

$c_{A,\text{sat}(T_0)}$ is the vapour concentration at the surface which is that associated with saturation concentration at T_0. Therefore, from Eq. (11.63), we obtain

$$T_\infty - T_0 = h_{fg} M_A \left(\frac{h_m}{h} \right) [c_{A,\text{sat}(T_0)} - c_{A,\infty}] \tag{11.64}$$

Now,

$$c_{A,\text{sat}(T_0)} = \frac{\rho_{A,\text{sat}(T_0)}}{M_A} = \frac{p_{A,\text{sat}(T_0)}}{RT_0 M_A} = \frac{p_{A,\text{sat}(T_0)}}{\Re T_0} \tag{11.65}$$

$$c_{A,\infty} = \frac{p_{A,\infty}}{RT_\infty M_A} = \frac{p_{A,\infty}}{\Re T_\infty} \tag{11.66}$$

where $\Re$ is the universal gas constant and is equal to 8.314 kJ/kmol K. Substituting Eqs (11.65) and (11.66) into Eq. (11.64), we finally get

$$T_\infty - T_0 = \frac{M_A h_{fg}}{\Re} \left(\frac{h_m}{h} \right) \left[\frac{p_{A,\text{sat}(T_0)}}{T_0} - \frac{p_{A,\infty}}{T_\infty} \right] \tag{11.67}$$

Now, from the heat transfer and mass transfer analogy,

$$\frac{\text{Nu}}{\text{Pr}^n} = \frac{\text{Sh}}{\text{Sc}^n}$$

$$\Rightarrow \qquad \frac{hL/k}{\text{Pr}^n} = \frac{h_m L/D_{AB}}{\text{Sc}^n} \tag{11.68}$$

Therefore, Eq. (11.68) yields

$$\frac{h}{h_m} = \left(\frac{\text{Pr}}{\text{Sc}}\right)^n \left(\frac{k}{D_{AB}}\right) = \left(\frac{v/\alpha}{v/D_{AB}}\right)^n \left(\frac{k}{D_{AB}}\right)$$

$$= \left(\frac{D_{AB}}{\alpha}\right)^{n-1} \frac{k}{(k/\rho c_p)} = \left(\frac{1}{\text{Le}}\right)^{n-1} \rho c_p$$

$$= \rho c_p (\text{Le})^{1-n} \tag{11.69}$$

Le is called the Lewis number and is defined as Le $= \alpha/D_{AB}$, which basically means that it is the ratio of the rate of energy transport to the rate of mass transport. For laminar or turbulent flow, usually $n = 1/3$ over any geometry. Therefore, from Eq. (11.69), we can write

$$\frac{h}{h_m} = \rho c_p (\text{Le})^{2/3} \tag{11.70}$$

Finally, putting Eq. (11.70) into Eq. (11.67) the following is obtained:

$$T_\infty - T_0 = \frac{M_A h_{fg}}{\Re} \left(\frac{1}{\rho c_p (\text{Le})^{2/3}}\right) \left[\frac{p_{A,\text{sat}(T_0)}}{T_0} - \frac{p_{A,\infty}}{T_\infty}\right] \tag{11.71}$$

It may be noted that properties of air may be evaluated at $T_m = (T_\infty + T_0)/2$.

11.11 Relative Humidity

Relative humidity (RH) of the atmosphere is an important parameter that controls the mass transfer from a liquid surface to the ambient atmosphere. It is defined as the ratio of the mole fraction of the vapour component to that of the vapour at the saturated state, both corresponding to the mixture temperature and total pressure. Therefore,

$$\text{RH} = \frac{x_v(T, p)}{x_g(T, p)} = \frac{p_v/p}{p_g/p} = \frac{p_v}{p_g} = \frac{p_v(T)}{p_{\text{sat}}(T)} \tag{11.72}$$

11.11.1 Effects of Relative Humidity

Effects of RH can be demonstrated by analysing the following two phenomena.

Phenomenon 1 During winter the relative humidity of the air is usually quite high. We, however, experience dryness of skin in winter. On the other hand, wet clothes do not dry easily in winter. Explain.

Explanation Although RH is very high in the first case, since the atmospheric temperature is below the skin temperature and the concentration of water vapour is lower in the atmosphere (since the mole fraction of vapour is low because of lower vapour pressure) than at the skin, moisture diffuses away from the skin, thus making the skin dry. On the other hand, wet clothes are almost at the same temperature as that of the atmosphere and, therefore, the concentration difference is very small and the mass transfer rate is low. Hence, clothes do not dry easily in winter.

Phenomenon 2 On a winter night the surface of the earth cools faster than the surrounding atmosphere. The relative humidity of the air at every point, including the surface, is 100%. Would water vapour diffuse away to/from the earth's surface under these conditions? Explain.

Explanation Although RH everywhere is 100%, the concentration of water vapour in the atmosphere is higher than in the vicinity of the earth's surface because of higher temperature. Thus there is migration of water vapour from the surroundings to the surface, which condenses as dew since the earth's surface is cooler.

Example 11.3 Rate of Water Evaporation from a Water Surface

Along a horizontal water surface, air stream with velocity $u_\infty = 3$ m/s is flowing. The temperature of the water on the surface is 15°C, the air temperature is 20°C, the total pressure is 1 atm (10^5 N/m²), and the saturation pressure of the water vapour in the air at 20°C is $p_{H_2O,\infty} = 2337$ N/m². The relative humidity of the air is 33%. The water surface along the wind direction has a length of 10 cm. Calculate the amount of water evaporated per hour per metre square from the water surface. The binary diffusivity of water vapour in the air may be taken as 3.3×10^{-5} m²/s. The saturation vapour pressure of water at 15°C is 1705 N/m² and the kinematic viscosity of the air is 1.5×10^{-5} m²/s.

Solution

Let the water vapour be designated as A and the air as B. The Reynolds number for the air flow is

$$Re_L = \frac{u_\infty L}{v} = \frac{(3)(10)(10^{-2})}{1.5 \times 10^{-5}} = 2 \times 10^4$$

Since $Re_L < 5 \times 10^5$, the flow is laminar. Now,

$$Sc = \frac{v}{D_{AB}} = \frac{1.5 \times 10^{-5}}{3.3 \times 10^{-5}} = 0.46$$

Using the heat transfer and mass transfer analogy, from Eq. (11.57) we can write

$$\overline{Sh}_L = 0.664(Re_L)^{1/2}(Sc)^{1/3}$$

$$= 0.664(2 \times 10^4)^{1/2}(0.46)^{1/3} = 72.48$$

Therefore, $\overline{Sh}_L = \dfrac{\overline{h}_m L}{D_{AB}} = 72.48$

Hence, $\overline{h}_m = 72.48 \dfrac{D_{AB}}{L} = \dfrac{72.48 \times 3.3 \times 10^{-5}}{10 \times 10^{-2}} = 239 \times 10^{-4}$ m/s

$$N_A = \bar{h}_m \left(c_{A_0} - c_{A_\infty} \right) = \bar{h}_m \left(\frac{p_{A_0}}{RM_A T_0} - \frac{p_{A_\infty}}{RM_A T_\infty} \right)$$

$$= \bar{h}_m \left(\frac{p_{A_0}}{\Re T_0} - \frac{p_{A_\infty}}{\Re T_\infty} \right) = \bar{h}_m \left(\frac{p_{A,\text{sat}\,(T_0)}}{\Re T_0} - \frac{p_{A,\infty}}{\Re T_\infty} \right)$$

From the definition of relative humidity,

$$\text{RH} = \frac{p_{A,\infty}}{p_{A,\text{sat}\,(T_\infty)}}$$

$$\Rightarrow \qquad 0.33 = \frac{p_{A,\infty}}{2337}$$

Therefore, $P_{A,\infty} = 779$ N/m^2. Also given is $p_{A,\text{sat}\,(15°C)} = 1705$ N/m^2. Hence,

$$N_A = 239 \times 10^{-4} \left(\frac{1705}{8.314 \times 288} - \frac{779}{8.314 \times 293} \right) = 93.69 \times 10^{-7} \text{ kmol/m}^2\text{s}$$

Now, 1 kmol of water weighs 18 kg. Therefore,
$$N_A = (93.69 \times 10^{-7})\,(18)(3600) = 0.607 \text{ kg/m}^2\text{h}$$

Example 11.4   Cooling of Beverages in Hot and Arid Regions

In order to keep beverages cool in hot and arid regions with no refrigeration facility, beverage containers are wrapped in fabrics which are continually moistened with a highly volatile liquid. Such a container of an arbitrary shape, shown in Fig. E11.4, is placed in a dry ambient air at 40 °C, with heat and mass transfer between the wetting agent and the air occurring by forced convection. The wetting agent has a molecular weight of 200 kg/kmol and a latent heat of vapourization of 100 kJ/kg. Its saturation vapour pressure for the prescribed condition is approximately 5000 N/m^2, and the binary diffusivity of vapour in the air is 0.2 $\times 10^{-4}$ m^2/s. What is the steady-state temperature of the beverages?

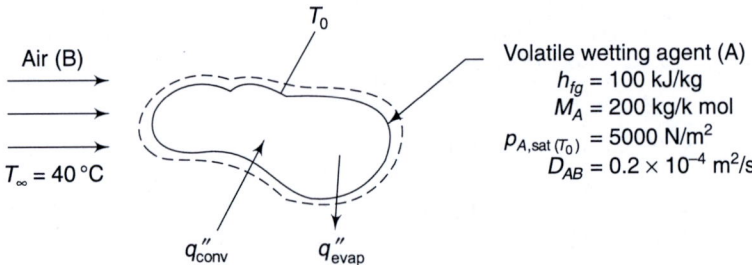

Fig. E11.4   Evaporative cooling of a beverage container

Solution

The properties of the air evaluated at $(T_\infty + T_0)/2 = 300$ K (assumed) are (see Table A1.4):
$$\rho = 1.16 \text{ kg/m}^3, \; c_p = 1.007 \text{ kJ/kg K}, \; \alpha = 22.5 \times 10^{-6} \text{ m}^2/\text{s}$$
Therefore, from Eq. (11.71), we have

$$T_\infty - T_0 = \frac{M_A h_{fg}}{\Re \rho c_p \, (\text{Le})^{2/3}} \left[\frac{p_{A,\text{sat}\,(T_0)}}{T_0} - \frac{p_{A,\infty}}{T_\infty} \right]$$

$P_{A,\infty} = 0$, since the air is dry. Hence,

$$T_\infty - T_0 = \frac{E}{T_0} \tag{A}$$

where $\quad E = \dfrac{M_A h_{fg}\, p_{A,\text{sat}}(T_0)}{\Re \rho c_p\, (\text{Le})^{2/3}}$

From Eq. (A), we can write

$$T_0^2 - T_\infty T_0 + E = 0 \tag{B}$$

The solution of Eq. (B) is

$$T_0 = \frac{T_\infty \pm \sqrt{T_\infty^2 - 4E}}{2}$$

Now, $\quad E = \dfrac{200 \times 100 \times 5000 \times 10^{-3}}{8.314 \times 1.16 \times 1.007 \times \left(\dfrac{22.5 \times 10^{-6}}{20 \times 10^{-6}}\right)^{2/3}} = 9514\ \text{K}^2$

Hence, $\quad T_0 = \dfrac{313 \pm \sqrt{(313)^2 - 4(9514)}}{2} = 278.9\ \text{K} = 5.9\,^\circ\text{C}$

Note that the minus sign has been neglected on physical grounds since T_0 must equal T_∞ if there is no evaporation, in which case $p_{A,\text{sat}} = 0$ and $E = 0$.

It is important to note that the above result is independent of the shape of the container as long as the heat transfer and mass transfer analogy is valid.

Example 11.5 Steady Diffusion of Helium through a Pyrex Plate

Calculate the mass flux for one-dimensional steady state mass diffusion of helium at 400 K through a pyrex plate of thickness 1 mm (Fig.E11.5). The mass concentration of helium at $x = 0$ is 6.085×10^{-7} g/cm³ and zero at the right surface of the plate. The density of pyrex is $\rho_B = 2.6$ g/cm³. The binary diffusivity D_{AB} is 0.2×10^{-7} cm²/s. Show also that the neglect of the mass average velocity of the mixture is valid.

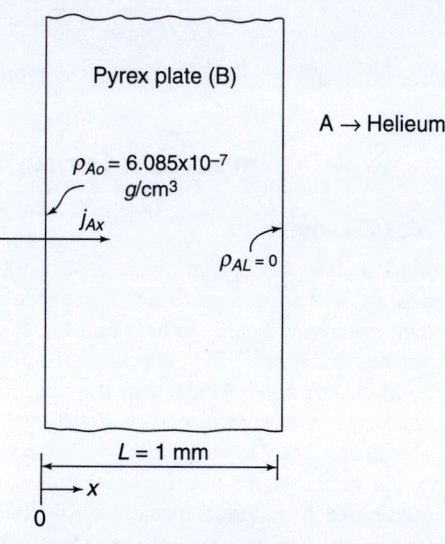

Fig. E11.5 Steady mass diffusion of helium in a pyrex plate

Solution

The mass fraction of helium at the left surface is

$$Y_{A_0} = \frac{\rho_{A_0}}{\rho_{A_0} + \rho_{B_0}} = \frac{6.085 \times 10^{-7}}{6.085 \times 10^{-7} + 2.6} = 2.34 \times 10^{-7}$$

Therefore, the mass flux of helium can be calculated by the application of Fick's law of diffusion.

$$j_{A_x} = -\rho D_{AB} \frac{dY_A}{dx}$$

Since the concentration profile is linear we can write

$$j_{A_x} = \rho D_{AB} \frac{Y_{A_0} - 0}{L} = (2.6)(0.2 \times 10^{-7}) \frac{2.34 \times 10^{-7}}{10^{-1}} = 1.22 \times 10^{-13} \text{ g/cm}^2\text{s}$$

Next the velocity of the helium is calculated from

$$j_{A_x} = \rho_A \left(v_{A_x} - v_x \right)$$

$$\Rightarrow \qquad v_{A_x} = \frac{j_{A_x}}{\rho_A} + v_x$$

At the left surface of the plate ($x = 0$) the magnitude of this velocity is

$$v_{A_x}\Big|_{x=0} = \frac{1.22 \times 10^{-13}}{6.085 \times 10^{-7}} + v_{x_0} = 0.2 \times 10^{-6} + v_{x_0}$$

Also, we know that in general, for a binary mixture

$$v_x = Y_A v_{A_x} + Y_B v_{B_x}$$

Therefore, at $x = 0$

$$v_{x_0} = Y_{A_0} v_A\Big|_{x=0} + Y_{B_0} v_B\Big|_{x=0}$$

$$= Y_{A_0} (0.2 \times 10^{-6} + v_{x_0}) + (1 - Y_{A_0})(0)$$

$$= (2.34 \times 10^{-7})(0.2 \times 10^{-6} + v_{x_0})$$

Thus, $\qquad v_{x_0} = \dfrac{(2.34 \times 10^{-7})(0.2 \times 10^{-6})}{1 - (2.34 \times 10^{-7})} = 0.468 \times 10^{-13}$ cm/s

Hence it is safe to neglect v_x and the assumption of stationary mixture is correct.

Important Concepts and Expressions

Mass Transfer

Mass transfer is mass in transit as a result of a species concentration difference in a mixture. The concentration difference is related to mass transfer in the same way as the temperature gradient is to heat transfer. Just like temperature difference is the driving potential for heat transfer, concentration difference is the same for mass transfer.

Mass diffusion is similar to heat conduction. Convection mass transfer occurs when the concentration of some species at a surface differs from its concentration in a gas moving over the surface. However, mass diffusion is more complicated than viscous flow or heat conduction, because here, for the first time, one has to deal with mixtures. In a diffusing mixture, each individual species has a different velocity and, therefore, one has to average the velocities of the species to get a local velocity for the mixture. It is imperative that a local velocity be chosen before the rates of diffusion can be defined.

Definitions of Concentrations, Velocities, and Mass Fluxes

In a multicomponent system the concentrations of the various species may be expressed in the following ways:

(a) ρ_i, the mass concentration, is the mass of species i per unit volume of the solution.

(b) c_i, the molar concentration $= \rho_i / M_i$, is the number of moles of species i per unit volume of the solution.

(c) Y_i, the mass fraction $= \rho_i / \rho$, is the mass concentration of species i divided by the total mass density of the solution.

(d) x_i, the mole fraction $= c_i / c$, is the molar concentration of species I divided by the total molar density of the solution.

In a diffusion mixture the various chemical species are moving at different velocities. Note that by 'velocity' we mean the sum of the velocities of the molecules of species i within a small volume element divided by the number of such molecules. Therefore, it should not be confused with the velocity of an individual molecule of species i.

Then, for a mixture of n species, the local mass average velocity v is defined as

$$v = \frac{\displaystyle\sum_{i=1}^{N} \rho_i v_i}{\displaystyle\sum_{i=1}^{N} \rho_i}$$

Note that ρv is the local rate at which mass passes through a unit cross-section placed perpendicular to velocity v.

Local molar average velocity v^* is defined as

$$v^* = \frac{\displaystyle\sum_{i=1}^{N} c_i v_i}{\displaystyle\sum_{i=1}^{N} c_i}$$

Note that cv^* is the local rate at which moles pass through a unit cross-section placed perpendicular to velocity v^*.

In flow systems one is more interested in the motion of component i relative to the local motion of the fluid stream rather than with respect to stationary coordinates. This leads to the definition of diffusion velocities:

$$v_i - v = \text{diffusion velocity of } i \text{ with respect to } v$$
$$v_i - v^* = \text{diffusion velocity of } i \text{ with respect to } v^*$$

Let us now define mass and molar fluxes. The mass and molar fluxes relative to the stationary coordinates are

$$n_i = \rho_i v_i \text{ (mass)}$$
$$N_i = c_i v_i \text{ (molar)}$$

The mass and molar fluxes relative to the mass-averaged velocity v are

$$j_i = \rho_i (v_i - v) \text{ (mass)}$$
$$J_i = c_i (v_i - v) \text{ (molar)}$$

And the mass and molar fluxes relative to the molar average velocity v^* are

$$j_i^* = \rho_i (v_i - v) \text{ (mass)}$$
$$J_i^* = c_i (v_i - v^*) \text{ (molar)}$$

Fick's Law of Diffusion

Fick's law of diffusion written in terms of the molar diffusion flux J_A^* is

$$J_A^* = -cD_{AB}\nabla x_A$$

This equation states that species A diffuses (that is, moves relative to the mixture) in the direction of decreasing mole fraction A, just as heat flows by conduction in the direction of decreasing temperature.

The unit of binary diffusivity D_{AB} is m²/s. Also, $D_{AB} = D_{BA}$.

Mass Diffusion Equations:
Various Forms of Equation of Continuity for a Binary Mixture

The principle of conservation of mass applied to the chemical species is the conservation principle that governs the migration of a chemical species through a gaseous, liquid or solid medium.

Case A: Diffusion of traces of A through B

Assumptions: Constant binary diffusivity, stationary medium, and no chemical reactions.

$$\text{GDE: } \frac{\partial c_A}{\partial t} = D_{AB}\nabla^2 c_A$$

Case B: Diffusion in laminar flow (dilute solution of A in B)

Assumptions: Constant binary diffusivity, constant mixture density, steady state, no chemical reactions

$$\text{GDE: } v.\nabla c_A = D_{AB}\nabla^2 c_A$$

Case C: Steady diffusion in solids

Assumptions: Constant binary diffusivity, constant mixture density, steady state, no chemical reactions, stationary mixture (zero bulk velocity)

$$\text{GDE: } \nabla^2 c_A = 0$$

One-dimensional Steady Diffusion through a Stationary Medium

The problem is steady and one-dimensional. The boundary conditions are of Dirichlet type, that is, surface concentrations are specified. The solutions in three coordinate systems will be given.

Plane Wall

$$x_A = (x_2 - x_1)\frac{x}{L} + x_1$$

Hence, the mole fraction distribution is linear.

The diffusion flux (which is constant) can be expressed either as a molar flux or mass flux by applying Fick's law of diffusion.

Molar Flux:

$$J_{Ax}^* = -D_{AB}\frac{dc_A}{dx} = -D_{AB}\frac{\rho}{M}\frac{dx_A}{dx} = D_{AB}\frac{\rho}{M}\frac{x_1 - x_2}{L} = D_{AB}\frac{c_1 - c_2}{L}$$

Mass Flux:

$$j_{Ax} = -D_{AB}\frac{d\rho_A}{dx} = -D_{AB}\frac{\rho M_A}{M}\frac{dx_A}{dx} = D_{AB}\frac{\rho M_A}{M}\frac{x_1 - x_2}{L} = D_{AB}\frac{\rho_1 - \rho_2}{L}$$

Thus we see that the flux or flow rate of the species of interest across the slab is proportional to the binary diffusivity D_{AB} and the mole fraction or concentration difference, and is inversely proportional to the wall thickness.

Cylindrical Shell

$$x_A = x_1 + (x_1 - x_2) \frac{\ln\left(\dfrac{r}{r_1}\right)}{\ln\left(\dfrac{r_1}{r_2}\right)}$$

Spherical Shell

$$x_A = x_1 + \frac{x_1 - x_2}{\left(\dfrac{1}{r_2} - \dfrac{1}{r_1}\right)}\left(\frac{1}{r_1} - \frac{1}{r}\right)$$

Forced Convection with Mass Transfer Over a Flat Plate Boundary Layer

At low mass transfer rates, a heat transfer analogue with comparable boundary conditions exists for most problems of mass transfer. This suggests that Nusselt number correlations for heat transfer may be applied to the corresponding mass transfer systems by a simple change of notation such as

$$T \rightarrow c_A$$

$\text{Pr} \rightarrow \text{Sc}$ (Schmidt number)

$\text{Nu} \rightarrow \text{Sh}$ (Sherwood number)

Using heat and mass transfer analogy we can write for laminar boundary layer flow over a flat plate of length L (dilute solution approximation, that is, for very small bulk velocity v_o) for $\text{Sc} \geq 0.5$,

$$\overline{\text{Sh}}_L = 0.664 \left(\text{Re}_L\right)^{1/2} (\text{Sc})^{1/3}$$

where $\quad \text{Sc} = \dfrac{v}{D_{AB}} \quad$ and $\quad \text{Sh} = \dfrac{h_m L}{D_{AB}}$

In a similar way, for turbulent boundary layer flow over a flat plate, using the heat and mass transfer analogy, we can write (for $\text{Sc} \geq 0.5$)

$$\overline{\text{Sh}}_L = 0.037 \left(\text{Re}_L^{4/5} - 23{,}550\right)(\text{Sc})^{1/3}$$

Note that h_m is the mass transfer coefficient which is defined as

$$h_m = \frac{-cD_{AB} \left.\dfrac{\partial x_A}{\partial y}\right|_{y=0}}{c_{A_o} - c_{A_\infty}}$$

The mass transfer coefficient has a unit of m/s.

Review Questions

11.1 Define mass transfer.

11.2 Define mass concentration, molar concentration, mass fraction, and mole fraction.

11.3 State Fick's law of diffusion.

11.4 Write the governing equation for the steady-state non-flow system in mass transfer. What are the assumptions?

11.5 Define Schmidt number and Sherwood number. What are the corresponding dimensionless numbers in heat transfer?

11.6 Define mass transfer coefficient. What is its unit?

11.7 Define relative humidity.

11.8 Why do we experience dryness of skin in winter?

11.9 Wet clothes do not dry easily in winter. Why?

11.10 Explain the physics of evaporative cooling.

Problems

11.1 The molecular weight of a gas mixture consisting of species A ($M_A = 20$) and B ($M_B = 40$) is 25. The total mixture mass density is 1 kg/m^3. Calculate the following quantities:

(a) Molar fractions of A and B

(b) Mass fractions of A and B

(c) Mass concentrations of A and B

(d) Molar concentrations of A and B

(e) The total pressure if the temperature of the mixture is 300 K.

Assume ideal gas behaviour of species A and B.

11.2 For the gas mixture in Question 11.1 if A moves with a velocity of 1 m/s to the left and B moves with 2 m/s to the right, find

(a) mass and molar average velocities

(b) mass and molar fluxes across a surface which is

(i) stationary,

(ii) moving with mass-average velocity,

(iii) moving with molar-average velocity, and

(iv) moving at 3 m/s to the left.

11.3 Write the governing differential equations for 1D steady-state diffusion in stationary media in Cartesian, cylindrical, and spherical coordinates. Solve the equations for specified surface concentration boundary conditions and obtain the corresponding species diffusion resistances.

11.4 Consider steady, one-dimensional diffusion through a plane membrane. If the diffusivity D varies with concentration, $D = D_0(1 + Ac_a)$, where D_0 and A are constants. Obtain the concentration profile. Plot the profiles for positive and negative values of A.

11.5 A long, porous cylinder of radius r_i is concentric inside another cylinder of radius r_0. A salt solution diffuses through the porous cylinder into the liquid in the annulus. At the outer cylinder the salt is absorbed. If the concentration of salt at r_i is c_i and at r_0 is c_0, and the liquid in the annulus is stagnant, find an expression giving the variation of the salt concentration with the radius in the steady state.

11.6 Write the governing differential equation for the neutron concentration in a long cylindrical element of fissionable material. Assume that the rate of production of the neutrons is proportional to the neutron concentration and the movement of neutrons in the fissionable material obeys Fick's diffusion law. State the conditions for the analogous heat conduction problem.

11.7 Derive an expression for the critical radius for maximum diffusion rate from a cylindrical rod.

11.8 A pulverized coal particle burns in pure oxygen at 1000 K. The process is limited by diffusion of the oxygen counterflow to the CO_2 formed at the particle surface. Assume the coal is pure carbon and has an initial diameter of 0.01 cm. If the binary diffusivity of oxygen in carbon dioxide is 10^{-4} m^{2}s, calculate the time required for 90% of the carbon to burn away.

11.9 Rain left a thin film of water on a tile of the roof of a house in an Indian village. Wind blows over the tile along its 10 cm exposed length. The atmospheric air and the tile surface are both at 25 °C. The relative humidity of the air is 40%. Wind speed = 10 km/h, binary diffusivity = 2.88×10^{-5} m^2/s, Sc = 0.6.

(a) Calculate the average mass transfer coefficient between the tile surface and the air.

(b) What is the mass transfer rate (per unit width of the tile) at which water leaves the tile surface?

11.10 Obtain the concentration profile and molar flux for diffusion through a spherical shell in a non-isothermal film in which temperature changes with distance according to $(T/T_1) = (r/r_1)^n$, where T_1 is the temperature at $r = r_1$. Assume that $(D_{AB}/D_{AB,1}) = (T/T_1)^{3/2}$, where $D_{AB,1}$ is the binary diffusivity at $T = T_1$.

Chapter 12

Solidification and Melting

12.1 Introduction

An important class of heat transfer problems falls in the category of phase change from liquid to solid, that is, *solidification*, or from solid to liquid, that is, *melting*. In solidification, a substance changes its phase by release of heat while in melting, a substance changes its phase by absorption of heat.

The essential feature of such phase change problems is the existence of a moving boundary separating the two phases. The velocity and the shape of this moving front have to be determined. Heat is liberated or absorbed on this boundary, and the thermophysical properties in the solid and liquid phases are different. Hence, the problem is one of the considerable difficulty.

For some materials, there is a distinct line of demarcation between the two phases, called *melt line*. In other words, the solidification or melting takes place at a fixed temperature, and the solid and liquid phases are separated by a definite moving interface. Typical examples are water and pure metals. Other materials, such as alloys, mixtures, and impure materials do not have a definite melt line and melting or solidification occurs over a temperature range. The two phases are separated by a two-phase region, also known as the *mushy region*, which is distinguished by a gradual change in the thermophysical properties of the material from one phase to the other.

So, one of the hurdles in tackling solidification/melting problems is that the interface, either in the form of a melt line or mushy region, is moving. In addition, there may be some circulation in the liquid phase so that heat transfer by natural convection has to be considered. In this chapter, we will concentrate only on the phase-change problems involving a sharp moving front and assume that the heat transfer in the liquid phase is by pure conduction only.

Existing literature on heat transfer offers very few exact solutions. The numerical method is suitable as it can handle the change of thermal properties and the arbitrary shape of the moving front.

We will primarily focus on exact solutions of solidification and melting in this chapter. The following are main applications of solidification:

    (i) Ice formation: It is of great importance in geology and ice manufacture.

    (ii) Solidification of castings: It has important applications in industry. Metal/ alloy casting is an industry which involves high annual investments all over the world. An improvement in the quality of the castings and reduction in the number of rejects would result in a large amount of saving for the casting industry.

(iii) Freezing of food products for preservation, and ice cream making.

Melting has many applications in the steel and metallurgical industries as well as in polymer and food processing. Therefore, understanding the heat transfer mechanisms by which materials solidify or melt and the rates at which solidification or melting takes place are of great interest.

12.2 Exact Solutions of Solidification: One-dimensional Analysis

Two 1D exact solutions of solidification will be discussed in the subsequent sections, namely, the problem of Stefan (1891) and the Neumann problem (1912).

12.2.1 Problem of Stefan

The first published work on solidification was by Stefan (1891). The study was related to the calculation of thickness of polar ice as a function of time. This is the reason why the problem of freezing is frequently referred to as the 'Problem of Stefan'.

12.2.1.1 Analysis

Although Stefan's analysis dealt with freezing of water, the treatment can be extended to the solidification of pure liquid metals as well with the following assumptions.

Assumptions

1. There is no temperature gradient in the liquid phase.
2. The solid phase has negligible heat capacity—that is, steady state heat conduction is occurring.
3. Heat flow is one-dimensional.

Figure 12.1 shows that a layer of solid at some time t has a thickness $x(t)$. The physical domain is treated as semi-infinite and hence 1D heat flow is justified (Assumption 3). The exposed surface (at $x = 0$) is at a temperature T_s, which is lower than the solidification or freezing point, T_∞. The latent heat liberated at the interface must be conducted through the solid, there being no temperature gradient in the liquid (Assumption 1).

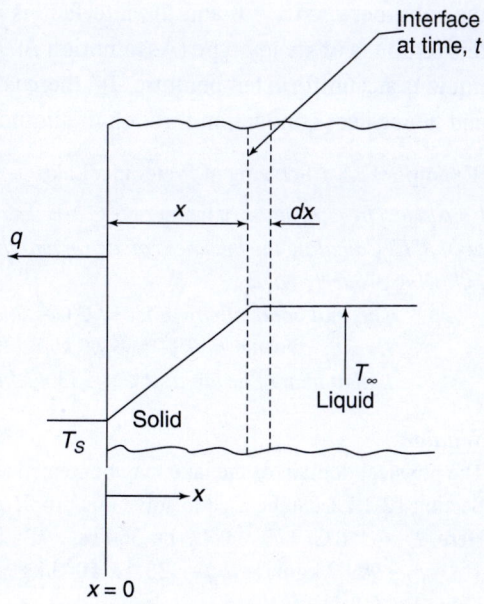

Fig. 12.1 Physical domain of Stefan's problem showing interface and temperature-profile at time, t

Thus the energy balance at the interface results in the following equation, which basically states that when the phase separation surface moves a distance dx, a quantity of heat $\rho_l h_{sf} \dfrac{dx}{dt}$ per unit area is released and must be conducted away through the solid.

$$\rho_l h_{sf} \frac{dx}{dt} = \frac{k_s (T_\infty - T_s)}{x} \tag{12.1}$$

where ρ_l is the density of the liquid (in kg/m^3), k_s is the thermal conductivity of the solid (in W/m K), h_{sf} is the latent heat of fusion (in J/kg), and t denotes the time (in s).

The initial condition is: at $t = 0$, $x = 0$. $\tag{12.2}$

From Eq. (12.1), we get

$$x\,dx = \frac{k_s (T_\infty - T_s)}{\rho_l h_{sf}} dt \tag{12.3}$$

Integrating Eq. (12.3), we get

$$\frac{x^2}{2} = \frac{k_s (T_\infty - T_s)}{\rho_l h_{sf}} t + C \tag{12.4}$$

Applying the initial condition, that is, Eqs (12.2) to (12.4) we obtain $C = 0$. Thus,

$$x = \sqrt{\frac{2 k_s (T_\infty - T_s) t}{\rho_l h_{sf}}} \tag{12.5}$$

Equation (12.5) indicates that $x \sim t^{1/2}$ for a fixed value of $(T_\infty - T_s)$. In other words, the solid layer grows like the square root of time.

It may be recalled that the approximation made here is that the heat capacity of the solid between $x = 0$ and the interface is negligible, that is, heat flow through this region is of steady type (Assumption 2). Another point to note is that since the liquid is at a uniform temperature, T_∞, there is no temperature gradient in the liquid, and hence heat conduction through the liquid is nil.

Example 12.1   Freezing of Water in a Lake

Consider a freezing mass of water at 0°C in a deep lake. The exposed surface is maintained at –1.1°C. Calculate the thickness of ice (in cm) that will be formed after one hour. Use the following property values.

        Thermal conductivity of ice at 0°C = 2.22 W/m K
        Density of water at 0°C = 999.9 kg/m^3
        Latent heat of fusion at 0°C = 333.4 kJ/kg

Solution

The physical domain of the lake may be treated as semi-infinite, and hence the analysis in Section 12.2.1.1 can be used to solve this problem.

Here, $T_s = -1.1°C$, $T_\infty = 0°C$, $t = 3600$ s, $k_s = 2.22$ W/mK

    $\rho_l = 999.9$ kg/m^3, $h_{sf} = 333.4 \times 10^3$ J/kg

Using Eq. (12.5),

$$x = \sqrt{\frac{2 k_s (T_\infty - T_s) t}{\rho_l h_{sf}}} = \sqrt{\frac{2(2.22)(0 - (-1.1))(3600)}{(999.9)(333.4 \times 10^3)}} = 7.26 \times 10^{-3} \text{ m} = 0.726 \text{ cm}$$

Note that $x = 0$ at the exposed surface of the lake and is positive downwards. Therefore, 0.726 cm thick ice will be formed after one hour.

12.2.2 Neumann Problem

A more general result known as Neumann's solution was given by Franz Neumann in his lectures in the 1860s (published in 1912). The details of the solution methodology are given next.

12.2.2.1 Analysis

In this approach the main differences with respect to Stefan's analysis are

 (a) heat capacity in the solid is not neglected and hence the problem is of unsteady type
 (b) the temperature gradient in the liquid is considered, and therefore heat conduction in the liquid is taken into account while considering the energy balance at the interface
 (c) the effect of change in volume during solidification is considered.

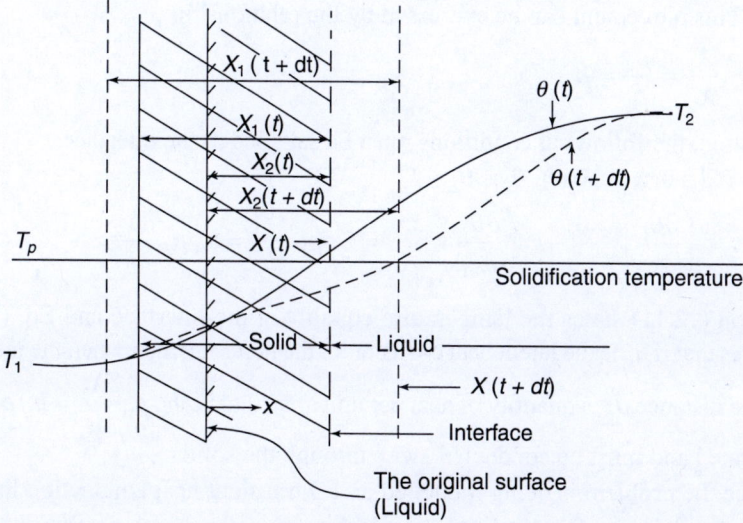

Fig. 12.2 Physical domain and the coordinate system of the Neumann problem

Figure 12.2 shows the physical domain and the coordinate system of the Neumann's problem. The liquid fills the semi-infinite region $x > 0$ initially and is being solidified by the removal of heat at the exposed surface $(x = 0)$, which is maintained at a constant temperature, T_1. At any time t, the surface separating the liquid and solid phases is at $X(t)$. The bulk temperature of the liquid at a large distance away from the interface is T_2 and constant. It may be noted that $T_2 > T_p > T_1$ where T_p is the solidification point of the material (see Fig. 12.2). Hence, heat is conducted from the liquid through the solid phase to the free surface. At the interface, latent heat of fusion is liberated as the liquid is transformed into solid.

At some time t, the region $x < X(t)$ consists of the solid phase with properties, $k_1, \alpha_1, \rho_1, c_1$, and if θ_1 is the temperature within this solid phase, it must satisfy the following 1D transient heat conduction equation:

$$\frac{\partial^2 \theta_1}{\partial x_1^2} = \frac{1}{\alpha_1} \frac{\partial \theta_1}{\partial t} \tag{12.6}$$

The boundary condition is

At $x_1 = 0$, $\theta_1 = T_1$ $\quad$ (12.7)

The region $x > X(t)$ consists of the liquid phase with properties k_2, α_2, ρ_2, c_2, and if θ_2 is the temperature within this liquid phase (neglecting natural convection) it must satisfy the following 1D transient heat conduction equation:

$$\frac{\partial^2 \theta_2}{\partial x_2^2} = \frac{1}{\alpha_2} \frac{\partial \theta_2}{\partial t} \tag{12.8}$$

The boundary condition is

At $x_2 \to \infty$, $\quad$ $\theta_2 = T_2$ $\quad$ (12.9)

During solidification of water to ice, for example, there is an increase in volume (decrease in density) and this effect can be taken care of by noting that the frozen surface will move away from the original surface, dictated by the density of each phase. This movement can be expressed by the relationship

$$\frac{X_1}{X_2} = \frac{\rho_2}{\rho_1} = \beta \tag{12.10}$$

In addition, the following conditions must be satisfied at the interface:

At $x_1 = X_1(t)$ or $x_2 = X_2(t)$, $\theta_1 = \theta_2 = T_p$ $\quad$ (12.11)

$$k_1 \left(\frac{\partial \theta_1}{\partial x_1} \right)_{x_1 = X_1} - k_2 \left(\frac{\partial \theta_2}{\partial x_2} \right)_{x_2 = X_2} = h_{sf} \rho_1 \frac{dX_1}{dt} = h_{sf} \rho_2 \frac{dX_2}{dt} \tag{12.12}$$

Equation (12.11) states the temperature equality at the interface and Eq. (12.12) indicates that if h_{sf} is the latent heat of fusion of the material, then when the interface moves a distance dx, a quantity of heat per unit area, that is, $h_{sf} \rho_1 \dfrac{dX_1}{dt} = h_{sf} \rho_2 \dfrac{dX_2}{dt}$ is liberated and must be conducted away through the solid.

Since the problem is being modelled as 1D transient heat conduction in semi-infinite media, error function and complementary error function solutions of the following form can be assumed:

$$\vartheta_1 = \theta_1 - T_p = (T_1 - T_p) + A \left(erf \frac{x_1}{2\sqrt{\alpha_1 t}} \right) \tag{12.13}$$

$$\vartheta_2 = \theta_2 - T_p = (T_2 - T_p) + B \left(erfc \frac{x_2}{2\sqrt{\alpha_2 t}} \right) \tag{12.14}$$

where A and B are constants, and thus Eqs (12.13) and (12.14) must satisfy Eqs (12.6) and (12.8), respectively. Applying Eq. (12.11) to Eqs (12.13) and (12.14), we get

$$0 = (T_1 - T_p) + A \left(erf \frac{X_1}{2\sqrt{\alpha_1 t}} \right)$$

$$\Rightarrow \qquad T_p - T_1 = A\left(erf\, \frac{X_1}{2\sqrt{\alpha_1 t}} \right) \qquad (12.15)$$

and
$$0 = \left(T_2 - T_p \right) + B\left(erfc\, \frac{X_2}{2\sqrt{\alpha_2 t}} \right) \qquad (12.16)$$

Now, since Eqs (12.15) and (12.16) must be valid for all values of X_1 and X_2, these must be proportional to $\sqrt{t}$. Using Eq. (12.10), therefore, we can write

$$X_1 = K\beta\sqrt{t} \qquad (12.17)$$

$$X_2 = K\sqrt{t} \qquad (12.18)$$

where K is a constant to be determined. When Eqs (12.13), (12.14), (12.17), and (12.18) are used in Eq. (12.12), we obtain

$$\left[Ak_1\, \frac{2}{\sqrt{\pi}}\, \frac{1}{2\sqrt{\alpha_1 t}}\, e^{-\frac{x_1^2}{4\alpha_1 t}} \right]_{x_1 = X_1} + \left[Bk_2\, \frac{2}{\sqrt{\pi}}\, \frac{1}{2\sqrt{\alpha_2 t}}\, e^{-\frac{x_2^2}{4\alpha_2 t}} \right]_{x_2 = X_2} = h_{sf}\rho_1 K\beta \frac{1}{2} t^{-\frac{1}{2}}$$

$$\Rightarrow \qquad \frac{Ak_1}{\sqrt{\pi\alpha_1}}\, e^{-\frac{K^2\beta^2}{4\alpha_1}} + \frac{Bk_2}{\sqrt{\pi\alpha_2}}\, e^{-\frac{K^2}{4\alpha_2}} = \frac{h_{sf}\rho_1 K\beta}{2} \qquad (12.19)$$

Now, from Eqs (12.15) and (12.17),

$$A = \frac{T_p - T_1}{erf\, \dfrac{X_1}{2\sqrt{\alpha_1 t}}} = \frac{T_p - T_1}{erf\, \dfrac{K\beta}{2\sqrt{\alpha_1}}} \qquad (12.20)$$

Again, from Eqs (12.16) and (12.18),

$$B = \frac{T_p - T_2}{erfc\, \dfrac{X_2}{2\sqrt{\alpha_2 t}}} = \frac{T_p - T_2}{erfc\, \dfrac{K}{2\sqrt{\alpha_2}}} \qquad (12.21)$$

Substituting Eqs (12.20) and (12.21) into Eq. (12.19), we get

$$\frac{\left(T_p - T_1 \right)k_1 e^{-\frac{K^2\beta^2}{4\varepsilon_1}}}{\sqrt{\pi\alpha_1}\, erf\left(\dfrac{K\beta}{2\sqrt{\alpha_1}} \right)} - \frac{\left(T_2 - T_p \right)k_2 e^{-\frac{K^2}{4\alpha_2}}}{\sqrt{\pi\alpha_2}\, erfc\left(\dfrac{K}{2\sqrt{\alpha_2}} \right)} = \frac{h_{sf}\rho_1 K\beta}{2} \qquad (12.22)$$

Equation (12.22) is a transcendental equation and can be solved numerically to give K in terms of T_1, T_2, and T_p and the thermal properties of the material. Once K is known, A and B can be found from Eqs (12.20) and (12.21), respectively. Therefore, from Eqs (12.13) and (12.20), the temperature distribution in the solid falls out as

$$\frac{\theta_1 - T_p}{T_1 - T_p} = 1 - \frac{erf\, \dfrac{x_1}{2\sqrt{\alpha_1 t}}}{erf\, \dfrac{K\beta}{2\sqrt{\alpha_1}}} \qquad (12.23)$$

Similarly, from Eqs (12.14) and (12.21),

$$\frac{\theta_2 - T_p}{T_2 - T_p} = 1 - \frac{erfc\dfrac{x_2}{2\sqrt{\alpha_2 t}}}{erfc\dfrac{K}{2\sqrt{\alpha_2}}} \tag{12.24}$$

Recall that $X_1 = K\beta\sqrt{t}$ and $X_2 = K\sqrt{t}$. From Eqs (12.17) and (12.18), it is clear that for $t \to 0$, $X_1 \to 0$ and $X_2 \to 0$, and from Eq. (12.24) it follows that for $x_2 > 0$ and $t \to 0$, $\theta_2 = T_2$; and therefore, the initial condition is that the region $x > 0$ at $t = 0$ is all liquid at T_2. Thus the assumptions of the temperature profiles [Eqs (12.13) and (12.14)] are indeed correct.

12.3   Melting of a Solid: One-dimensional Analysis

Consider a semi-infinite solid, $x \geq 0$, which is initially ($t = 0$) at the melting temperature, T_m (Fig. 12.3). Let the temperature of the exposed surface ($x = 0$) be suddenly raised to T_s ($>T_m$) and maintained at this constant value for time $t > 0$. Thus, the melting begins at the surface, and the liquid–solid interface propagates in the positive x direction. In the solid phase the temperature is uniform (T_m). The change of volume due to melting is neglected in this analysis. Assuming 1D heat conduction in the liquid phase and constant thermophysical properties, the governing differential equation for the temperature distribution $T(x, t)$ in the liquid phase is given by

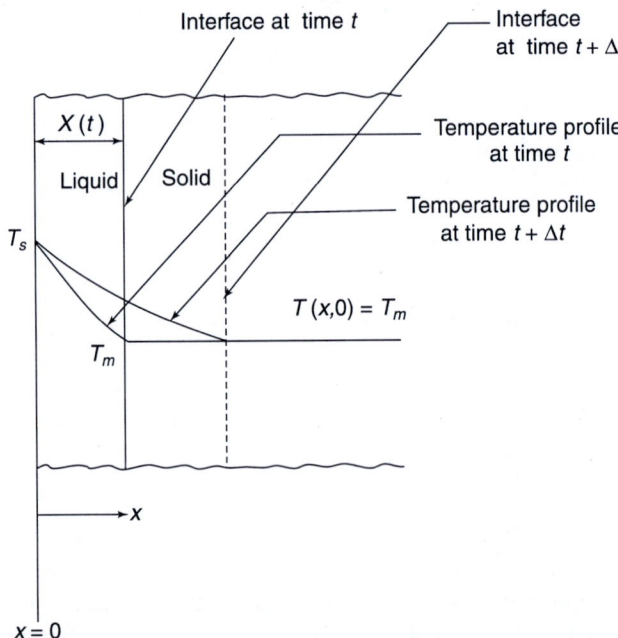

Fig. 12.3   Physical domain of the melting of a semi-infinite solid and the interfaces and temperature profiles at increasing times

$$\frac{\partial^2 T}{\partial x^2} = \frac{1}{\alpha}\frac{\partial T}{\partial t} \tag{12.25}$$

The initial condition is

At $t = 0$, $T = T_m$ (12.26)

The boundary conditions are as follows. At the exposed surface, that is,

At $x = 0$, $T = T_s$ (12.27)

At the interface, that is,

At $x = X(t)$, $T = T_m$ (12.28)

A solution in the following form is now assumed:

$$T(x,t) = A + B\left[erfc\left(\frac{x}{2\sqrt{\alpha t}} \right) \right]$$ (12.29)

where A and B are two arbitrary constants, and $\alpha \left(= \dfrac{k_l}{\rho c_p} \right)$ is the thermal diffusivity of the liquid phase. Equation (12.29) satisfies Eq. (12.25) (see Section 4.6 dealing with heat conduction in a semi-infinite solid). Applying the boundary conditions [Eqs (12.27) and (12.28)], we obtain

$$A = T_m - (T_s - T_m)\frac{erfc(\lambda)}{erf(\lambda)}$$ (12.30)

$$B = \frac{T_s - T_m}{erf(\lambda)}$$ (12.31)

where λ is defined by

$$\lambda = \frac{X(t)}{2\sqrt{\alpha t}} = \text{constant}$$ (12.32)

Putting Eqs (12.30) and (12.31) into Eq. (12.29), we get

$$\frac{T(x,t) - T_m}{T_s - T_m} = 1 - \frac{erf\left(\dfrac{x}{2\sqrt{\alpha t}} \right)}{erf(\lambda)}$$ (12.33)

where the parameter λ is to be estimated.

Determination of λ

It is to be noted that at the interface

$$-k_l \left. \frac{\partial T}{\partial x} \right|_{x = X(t)} = \rho h_{sf} \frac{dX(t)}{dt}$$ (12.34)

Using Eq. (12.34) results in

$$\lambda e^{\lambda^2} erf(\lambda) = \frac{c_p(T_s - T_m)}{h_{sf}\sqrt{\pi}}$$ (12.35)

where c_p is the specific heat of the liquid phase. Equation (12.35) is a transcendental equation and can be solved for λ by a numerical method such as the Newton–Raphson method.

Location of the Interface

Knowing λ, the location of the interface $X(t)$ is obtained from Eq. (12.32) as

$$X(t) = 2\lambda\sqrt{\alpha t}$$ (12.36)

Temperature Distribution in the Liquid Phase

The temperature distribution in the liquid phase $T(x, t)$ is given by Eq. (12.29).

Important Concepts and Expressions

Solidification and Melting

An important class of heat transfer problems falls in the category of phase change from liquid to solid, that is, *solidification*, or from solid to liquid, that is, *melting*. In solidification, a substance changes its phase by release of heat, while in melting, a substance changes its phase by absorption of heat.

The essential feature of such phase change problems is the existence of a moving boundary separating the two phases. The velocity and the shape of this moving front have to be determined. Heat is liberated or absorbed on it, and the thermophysical properties in the solid and liquid phases are different. Hence, the problem is one of considerable difficulty.

Solidification has applications mainly in geology and ice manufacture, the casting industry, and food preservation and ice cream making. Melting has many applications in steel and metallurgical industries as well as in polymer and food processing.

Exact Solutions of Solidification: One-dimensional Analysis

Two 1D exact solutions are presented, namely, Problem of Stefan (1891) and Neumann Problem (1912).

Melting of a Solid: One-dimensional Analysis

Assuming 1D heat conduction in the liquid phase and constant thermophysical properties, the temperature distribution $T(x, t)$ in the liquid phase is given by

$$T(x,t) = A + B\left[erfc\left(\frac{x}{2\sqrt{\alpha t}}\right)\right]$$

where

$$A = T_m - (T_s - T_m)\frac{erfc(\lambda)}{erf(\lambda)}$$

$$B = \frac{T_s - T_m}{erf(\lambda)}$$

where λ is defined by $\lambda = \dfrac{X(t)}{2\sqrt{\alpha t}} = $ constant. Hence,

$$\frac{T(x,t) - T_m}{T_s - T_m} = 1 - \frac{erf\left(\dfrac{x}{2\sqrt{\alpha t}}\right)}{erf(\lambda)}$$

where the parameter λ is to be estimated.

Determination of λ

It is to be noted that at the interface

$$-k_l \left.\frac{\partial T}{\partial x}\right|_{x=X(t)} = \rho h_{sf}\frac{dX(t)}{dt}$$

Using this relation

$$\lambda e^{\lambda^2} erf(\lambda) = \frac{c_p (T_s - T_m)}{h_{sf} \sqrt{\pi}}$$

where c_p is the specific heat of the liquid phase. This is a transcendental equation and can be solved for λ by a numerical method such as the Newton-Raphson method.

Location of the Interface

Knowing λ, the location of the interface $X(t)$ is obtained from

$$X(t) = 2\lambda \sqrt{\alpha t}$$

Review Questions

12.1 Define solidification and melting.

12.2 What is Stefan's problem in solidification? What are the basic assumptions?

12.3 What is Neumann problem in solidification? How does it differ with respect to Stefan's analysis?

12.4 State the 1D formulation of a melting problem.

12.5 How is the location of the interface obtained as a function of time in a melting problem?

Problems

12.1 To produce a bar of Kulfi (Indian ice cream) frozen around a stick, the liquid—which is primarily a mixture of condensed milk, water, and double (heavy) cream with a sprinkle of cardamom, unsalted pistachio nuts, and blanched almonds—is poured into a tapered chilled cavity that gives the Kulfi bar the dimensions shown in Fig. Q12.1. The dimension in the direction normal to the paper is quite large. The slight taper has been provided to allow for the expansion during solidification, and to be pulled out of the mould by the flat stick frozen in the middle.

How long will it take to make the Kulfi? The liquid, which may be

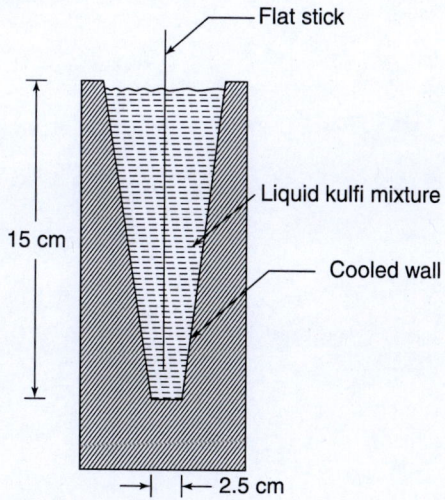

Fig. Q12.1

treated as water, is initially at the freezing point. The properties of the frozen bar are approximately equal to those of ice (see Example Problem 12.1). The cavity wall is maintained at –5°C.

12.2 A 2.5-cm-thick chicken patties are frozen by exposing them to a strong flow of –20°C air. Each patty rests on a 25-m-long perforated conveyor belt. Cold air is blown on both sides of the chicken pieces. The initial temperature of the chicken is equal to the freezing point, that is, –2°C. (a) Calculate the freezing time for each chicken patty. (b) What is the upper limit of the speed of the conveyor belt in order to ensure the complete freezing of each patty? The approximate properties of frozen chicken are $h_{sf} = 225$ kJ/kg, $c = 1.2$ kJ/kg K, $k = 1.6$ W/m K, and $\rho = 1000$ kg/m^3. Note that a chicken contains 60% of water by weight.

12.3 Consider the Neumann problem discussed in Section 12.2.2. For the water–ice system, an approximate expression for K (given below) has been obtained for limited values of T_1 and T_2 around a freezing temperature of $0°$ C. The density ratio of water to ice, that is, β is 1.09.

$$K = \sqrt{\frac{2(T_p - T_1)k_1}{h_{sf}\,\rho_1\,\beta^2}}$$

Will the consideration of the change of density in the ice phase from that of the water phase predict lower or higher freezing time? Justify.

12.4 In the formulation of the Neumann problem (see Section 12.2.2) if the liquid is initially at the phase change temperature T_p then write the modified form of Eq. (12.22).

12.5 A solid, $x \geq 0$, is initially at melting temperature, T_m. At $t > 0$, a constant heat flux q''_s is suddenly applied to the exposed surface ($x = 0$) and is maintained at that value subsequently. Obtain an expression for the location of the liquid–solid interface as a function of time.

12.6 State the Neumann problem discussed in Section 12.2.2 for solidification in terms of melting. Formulate the problem of melting and obtain the solution.

Appendices

Appendix A1 Thermophysical Properties of Matter

Table A1.1 Thermophysical properties of selected metallic solids

Composition	Melting point (K)	Properties at 300 K				Properties at various temperatures (K) k (W/mK)/c_p (J/kgK)									
		ρ (kg/m³)	c_p (J/kgK)	k (W/mK)	$\alpha \times 10^6$ (m²/s)	100	200	400	600	800	1000	1200	1500	2000	2500
Aluminium pure	933	2702	903	237	97.1	302	237	240	231	218					
							482	798	949	1033	1146				
Alloy 2024-T6 (4.5%Cu, 1.5% Mg, 0.6% Mn)	775	2770	875	177	73.0	65	163	186	186						
						473	787	925	1042						
Alloy 195, Cast (4.5% Cu)		2790	883	168	68.2			174	185						
								—							
Beryllium	1550	1850	1825	200	59.2	990	301	161	126	106	90.8	78.7			
						203	1114	2191	2604	2823	3018	3227	3519		
Bismuth	545	9780	122	7.86	6.59	16.5	9.69	7.04							
							112	120	127						
Boron	2573	2500	1107	27.0	9.76	190	55.5	16.8	10.6	9.60	9.85				
							128	600	1463	1892	2160	2338			
Cadmium	594	8650	231	96.8	48.4	203	99.3	94.7							
						198	222	242							
Chromium	2118	7160	449	93.7	29.1	159	111	90.9	80.7	71.3	65.4	61.9	57.2	49.4	
						192	384	484	542	581	616	682	779	937	

(Contd)

Table A1.1 (*Contd*)

Composition	Melting point (K)	Properties at 300 K ρ (kg/m³)	Properties at 300 K c_p (J/kgK)	Properties at 300 K k (W/mK)	Properties at 300 K $\alpha \times 10^6$ (m²/s)	Properties at various temperatures (K) — k (W/mK)/c_p (J/kgK) 100	200	400	600	800	1000	1200	1500	2000	2500
Cobalt	1769	8862	421	99.2	26.6	167 / 236	122 / 379	85.4 / 450	67.4 / 503	58.2 / 550	52.1 / 628	49.3 / 733	42.5 / 674		
Copper pure	1358	8933	385	401	117	482 / 252	413 / 356	393 / 397	379 / 417	366 / 433	352 / 451	339 / 480			
Commercial bronze (90% Cu, 10% Al)	1293	8800	420	52	14		42 / 785	52 / 460	59 / 545						
Phosphor gear bronze (89% Cu, 11% Sn)	1104	8780	355	54	17		41 / —	65 / —	74 / —						
Cartridge brass (70% Cu, 30% Zn)	1188	8530	380	110	33.9	75	95 / 360	137 / 395	149 / 425						
Constantan (55% Cu, 45% Ni)	1493	8920	384	23	6.71	17 / 237	19 / 362								

(*Contd*)

Table A1.1 *(Contd)*

Composition	Melting point (K)	Properties at 300 K				Properties at various temperatures (K) k (W/mK)/c_p (J/kgK)									
		ρ (kg/m³)	c_p (J/kgK)	k (W/mK)	$\alpha \times 10^6$ (m²/s)	100	200	400	600	800	1000	1200	1500	2000	2500
Germanium	1211	5360	322	59.9	34.7	232 / 190	96.8 / 290	43.2 / 337	27.3 / 348	19.8 / 357	17.4 / 375	17.4 / 395			
Gold	1336	19300	129	317	127	327 / 109	323 / 124	311 / 131	298 / 135	284 / 140	270 / 145	255 / 155			
Iridium	2720	22500	130	147	50.3	172 / 90	153 / 122	144 / 133	138 / 138	132 / 144	126 / 153	120 / 161	111 / 172		
Iron pure	1810	7870	447	80.2	23.1	134 / 216	94.0 / 384	69.5 / 490	54.7 / 574	43.3 / 680	32.8 / 975	28.3 / 609	32.1 / 654		
Armco (99.75% pure)		7870	447	72.7	20.7	95.6 / 215	80.6 / 384	65.7 / 490	53.1 / 574	42.2 / 680	32.3 / 975	28.7 / 609	31.4 / 654		
Carbon steels Plain carbon (Mn ≤ 1%, Si ≤ 0.1%)		7854	434	60.5	17.7			56.7 / 487	48.0 / 559	39.2 / 685	30.0 / 1169				
AISI 1010		7832	434	63.9	18.8			58.7 / 487	48.8 / 559	39.2 / 685	31.3 / 1168				

(Contd)

Table A1.1 (*Contd*)

Composition	Melting point (K)	Properties at 300 K				Properties at various temperatures (K) k (W/mK)/c_p (J/kgK)									
		ρ (kg/m³)	c_p (J/kgK)	k (W/mK)	$\alpha \times 10^6$ (m²/s)	100	200	400	600	800	1000	1200	1500	2000	2500
Carbon–silicon (Mn ≤ 1%, 0.1% < Si ≤ 0.6%)		7817	446	51.9	14.9			49.8 501	44.0 582	37.4 699	29.3 971				
Carbon– manganese– silicon (1% < Mn ≤ 0.65%, 0.1% < Si ≤ 0.6%)		8131	434	41.0	11.6			42.2 487	39.7 559	35.0 685	27.6 1090				
Chromium (low) steels 1/2 Cr–1/4Mo–Si (0.18% C, 0.65% Cr, 0.23% Mo, 0.6% Si)		7822	444	37.7	10.9			38.2 492	36.7 575	33.3 688	26.9 969				
1Cr–1/2 Mo (0.16% C, 1% Cr, 0.54% Mo, 0.39% Si)		7858	442	42.3	12.2			42.0 492	39.1 575	34.5 688	27.4 969				

(*Contd*)

Table A1.1 (*Contd*)

Composition	Melting point (K)	Properties at 300 K				Properties at various temperatures (K) k (W/mK)/c_p (J/kgK)									
		ρ (kg/m³)	c_p (J/kgK)	k (W/mK)	$\alpha \times 10^6$ (m²/s)	100	200	400	600	800	1000	1200	1500	2000	2500
1Cr–V (0.2% C, 1.02% Cr, 0.15% V)		7836	443	48.9	14.1			46.8 / 492	42.1 / 575	36.3 / 688	28.2 / 969				
Stainless steels															
AISI 302		8055	480	15.1	3.91			17.3 / 512	20 / 559	22.8 / 585	25.4 / 606				
AISI 304	1670	7900	477	14.9	3.95	9.2 / 272	12.6 / 402	16.6 / 515	19.8 / 557	22.6 / 582	25.4 / 611	28.0 / 640	31.7 / 682		
AISI 316		8238	468	13.4	3.48			15.2 / 504	18.3 / 550	21.3 / 576	24.2 / 602				
AISI 347		7978	480	14.2	3.71			15.8 / 513	18.9 / 559	21.9 / 585	24.7 / 606				
Lead	601	11340	129	35.3	24.1	39.7 / 118	36.7 / 125	34.0 / 132	31.4 / 142						
Magnesium	923	1740	1024	156	87.6	169 / 649	159 / 934	153 / 1074	149 / 1170	146 / 1267					

(*Contd*)

Table A1.1 *(Contd)*

Composition	Melting point (K)	Properties at 300 K				Properties at various temperatures (K) k (W/mK)/c_p (J/kgK)									
		ρ (kg/m³)	c_p (J/kgK)	k (W/mK)	$\alpha \times 10^6$ (m²/s)	100	200	400	600	800	1000	1200	1500	2000	2500
Molybdenum	2894	10240	251	138	53.7	179 / 141	143 / 224	134 / 261	126 / 275	118 / 285	112 / 295	105 / 308	98 / 330	90 / 380	86 / 459
Nickel pure	1728	8900	444	90.7	23.0	164 / 232	107 / 383	80.2 / 485	65.6 / 592	67.6 / 530	71.8 / 562	76.2 / 594	82.6 / 616		
Nichrome (80% Ni, 20% Cr)	1672	8400	420	12	3.4			14 / 480	16 / 525	21 / 545					
Inconel X-750 (73% Ni, 15% Cr, 6.7% Fe)	1665	8510	439	11.7	3.1	8.7 / —	10.3 / 372	13.5 / 473	17.0 / 510	20.5 / 546	24.0 / 626	27.6 / —	33.0 / —		
Niobium	2741	8570	265	53.7	23.6	55.2 / 188	52.6 / 249	55.2 / 274	58.2 / 283	61.3 / 292	64.4 / 301	67.5 / 310	72.1 / 324	79.1 / 347	
Palladium	1827	12020	244	71.8	24.5	76.5 / 168	71.6 / 227	73.6 / 251	79.7 / 261	86.9 / 271	94.2 / 281	102 / 291	110 / 307		
Platinum pure	2045	21450	133	71.6	25.1	77.5 / 100	72.6 / 125	71.8 / 136	73.2 / 141	75.6 / 146	78.7 / 152	82.6 / 157	89.5 / 165	99.4 / 179	

(Contd)

Table A1.1   *(Contd)*

Composition	Melting point (K)	Properties at 300 K				Properties at various temperatures (K) k (W/mK)/c_p (J/kgK)									
		ρ (kg/m³)	c_p (J/kgK)	k (W/mK)	$\alpha \times 10^6$ (m²/s)	100	200	400	600	800	1000	1200	1500	2000	2500
Alloy 60Pt–40Rh (60% Pt, 40% Rh)	1800	16630	162	47	17.4			52	59	65	69	73	76	—	
								—	—	—	—	—	—		
Rhenium	3453	21100	136	47.9	16.7	58.9	51.0	46.1	44.2	44.1	44.6	45.7	47.8	51.9	
						97	127	139	145	151	156	162	171	186	
Rhodium	2236	12450	243	150	49.6	186	154	146	136	121	121	116	110	112	
						147	220	253	274	293	311	327	349	375	
Silicon	1685	2330	712	148	89.2	884	264	98.9	61.9	42.2	31.2	25.7	22.7		
						255	556	790	857	913	946	967	992		
Silver	1235	10500	235	429	174	444	430	425	412	396	379	361			
						187	225	239	250	262	277	292			
Tantalum	3269	16600	140	57.5	24.7	59.2	57.5	57.8	58.6	59.4	60.2	61.0	62.2	64.1	65.6
						110	133	144	146	149	152	155	160	172	189
Thorium	2023	11700	118	54.0	39.1	59.8	54.6	54.5	55.8	56.9	56.9	58.7			
						99	112	124	134	145	156	167			
Tin	505	7310	227	66.6	40.1	85.2	73.3	62.2							
						188	215	243							

(Contd)

Table A1.1 (*Contd*)

Composition	Melting point (K)	Properties at 300 K				Properties at various temperatures (K) k (W/mK)/c_p (J/kgK)									
		ρ (kg/m³)	c_p (J/kgK)	k (W/mK)	$\alpha \times 10^6$ (m²/s)	100	200	400	600	800	1000	1200	1500	2000	2500
Titanium	1953	4500	522	21.9	9.32	30.5	24.5	20.4	19.4	19.7	20.7	22.0	24.5		
						300	465	551	591	633	675	620	686		
Tungsten	3660	19300	132	174	68.3	208	186	159	137	125	118	113	107	100	95
						87	122	137	142	145	148	152	157	167	176
Uranium	1406	19070	116	27.6	12.5	21.7	25.1	29.6	34.0	38.8	43.9	49.0			
						94	108	125	146	176	180	161			
Vanadium	2192	6100	489	30.7	10.3	35.8	31.3	31.3	33.3	35.7	38.2	40.8	44.6	50.9	
						258	430	515	540	563	597	645	714	867	
Zinc	693	7140	389	116	41.8	117	118	111	103						
						297	367	402	436						
Zirconium	2125	6570	278	22.7	12.4	33.2	25.2	21.6	20.7	21.6	23.7	26.0	28.8	33.0	
						205	264	300	322	342	362	344	344	344	

Table A1.2 Thermophysical properties of selected non-metallic solids

Composition	Melting point (K)	Properties at 300 K ρ (kg/m³)	c_p (J/kgK)	k (W/mK)	$\alpha \times 10^6$ (m²/s)	Properties at various temperatures (K) k (W/mK)/c_p (J/kgK) 100	200	400	600	800	1000	1200	1500	2000	2500
Aluminium oxide, sapphire	2323	3970	765	46	15.1	450 —	82 —	32.4 940	18.9 1110	13.0 1180	10.5 1225				
Aluminium oxide, polycrystalline	2323	3970	765	36.0	11.9	133 —	55 —	26.4 940	15.8 1110	10.4 1180	7.85 1225	6.55	5.66	6.00	
Beryllium oxide	2725	3000	1030	272	88.0	—	—	196 1350	111 1690	70 1865	47 1975	33 2055	21.5 2145	15 2750	
Boron	2573	2500	1105	27.6	9.99	190 —	52.5 —	18.7 1490	11.3 1880	8.1 2135	6.3 2350	5.2 2555			
Boron fibre epoxy (30% vol) composite	590	2080													
k, ‖ to fibres				2.29		2.10	2.23	2.28							
k, ⊥ to fibres				0.59		0.37	0.49	0.60							
c_p			1122			364	757	1431							
Carbon Amorphous	1500	1950	—	1.60	—	0.67	1.18	1.89	2.19	2.37	2.53	2.84	3.48	—	

(Contd)

Table A1.2 *(Contd)*

Composition	Melting point (K)	Properties at 300 K				Properties at various temperatures (K) k (W/mK)/c_p (J/kgK)									
		ρ (kg/m³)	c_p (J/kgK)	k (W/mK)	$\alpha \times 10^6$ (m²/s)	100	200	400	600	800	1000	1200	1500	2000	2500
Diamond, type IIa	—	3500	509	2300		10000	4000	1540							
Insulator						21	194	853							
Graphite, pyrolytic	2273	2210													
k, ∥ to layers				1950		4970	3230	1390	892	667	534	448	357	262	
k, ⊥ to layers				5.70		16.8	9.23	4.09	2.68	2.01	1.60	1.34	1.08	0.81	
c_p			709			136	411	992	1406	1650	1793	1890	1974	2043	
Graphite fibre epoxy (25% vol) composite	450	1400													
k, heat flow ∥ to fibres				11.1		5.7	8.7	13.0							
k, heat flow ⊥ to fibres				0.87		0.46	0.68	1.1							
c_p			935			337	642	1216							
Pyroceram, Corning 9606	1623	2600	808	3.98	1.89	5.25	4.78	3.64	3.28	3.08	2.96	2.87	2.79		
							—	909	1038	1122	1197	1264	1498		

(Contd)

Table A1.2   *(Contd)*

Composition	Melting point (K)	ρ (kg/m³)	c_p (J/kgK)	k (W/mK)	$\alpha \times 10^6$ (m²/s)	100	200	400	600	800	1000	1200	1500	2000	2500
						Properties at various temperatures (K) — k (W/mK) / c_p (J/kgK)									
Silicon carbide (k)	3100	3160	675	490	230			—	—	—	87	58	30		
Silicon carbide (c_p)								880	1050	1135	1195	1243	1310		
Silicon dioxide, crystalline (quartz)	1883	2650													
k, ∥ to c axis				10.4		39	16.4	7.6	5.0	4.2					
k, ⊥ to c axis				6.21		20.8	9.5	4.70	3.4	3.1					
c_p			745			—	—	885	1075	1250					
Silicon dioxide, polycrystalline (fused silica) (k)	1883	2220	745	1.38	0.834	0.69	1.14	1.51	1.75	2.17	2.87	4.00			
Silicon dioxide, polycrystalline (fused silica) (c_p)						—	—	905	1040	1105	1155	1195			
Silicon nitride (k)	2173	2400	691	16.0	9.65	—	—	13.9	11.3	9.88	8.76	8.00	7.16	6.20	
Silicon nitride (c_p)							578	778	937	1063	1155	1226	1306	1377	
Sulphur (k)	392	2070	708	0.206	0.141	0.165	0.185								
Sulphur (c_p)						403	606								
Thorium dioxide (k)	3573	9110	235	13	6.1			10.2	6.6	4.7	3.68	3.12	2.73	2.5	
Thorium dioxide (c_p)								255	274	285	295	303	315	330	
Titanium dioxide, polycrystalline (k)	2133	4157	710	8.4	2.8			7.01	5.02	3.94	3.46	3.28			
Titanium dioxide, polycrystalline (c_p)								805	880	910	930	945			

Table A1.3 Thermophysical properties of common materials

Structural building materials			
Description/Composition	Typical properties at 300 K		
	Density ρ (kg/m^3)	Thermal conductivity, k (W/mK)	Specific heat, c_p (J/kgK)
Building boards			
Asbestos–cement board	1920	0.58	—
Gypsum or plaster board	800	0.17	—
Plywood	545	0.12	1215
Sheathing, regular density	290	0.055	1300
Acoustic tile	290	0.058	1340
Hardboard, siding	640	0.094	1170
Hardboard, high density	1010	0.15	1380
Particle board, low density	590	0.078	1300
Particle board, high density	1000	0.170	1300
Woods			
Hardwoods (oak, maple)	720	0.16	1255
Softwoods (fir, pine)	510	0.12	1380
Masonry materials			
Cement mortar	1860	0.72	780
Brick, common	1920	0.72	835
Brick, face	2083	1.3	—
Clay tile, hollow			
1 cell deep, 10 cm thick	—	0.52	—
3 cells deep, 30 cm thick	—	0.69	—
Concrete block, 3 oval cores			
Sand/gravel, 20 cm thick	—	1.0	—
Cinder aggregate, 20 cm thick	—	0.67	—
Concrete block, rectangular core			
2 cores, 20 cm thick, 16 kg	—	1.1	—
Same with filled cores	—	0.60	—
Plastering materials			
Cement plaster, sand aggregate	1860	0.72	—
Gypsum plaster, sand aggregate	1680	0.22	1085
Gypsum plaster, vermiculite aggregate	720	0.25	—
Blanket and batt			
Glass fibre, paper faced	16	0.046	—
	28	0.038	—
	40	0.035	—
Glass fibre, coated; duct liner	32	0.038	835

(*Contd*)

Table A1.3   (*Contd*)

Structural building materials			
Description/Composition	Typical properties at 300 K		
	Density ρ (kg/m^3)	Thermal conductivity, k (W/mK)	Specific heat, c_p (J/kgK)
Board and slab			
Cellular glass	145	0.058	1000
Glass fibre, organic bonded	105	0.036	795
Polystyrene, expanded			
Extruded (R-12)	55	0.027	1210
Moulded beads	16	0.040	1210
Mineral fibre board; roofing material	265	0.049	—
Wood, shredded/cemented	350	0.087	1590
Cork	120	0.039	1800
Loose fill			
Cork, granulated	160	0.045	—
Diatomaceous silica, coarse	350	0.069	—
Powder	400	0.091	—
Diatomaceous silica, fine powder	200	0.052	—
	275	0.061	—
Glass fibre, poured or blown	16	0.043	835
Vermiculite, flakes	80	0.068	835
	160	0.063	1000
Formed/Foamed-in-place			
Mineral wool granules with asbestos/inorganic binders, sprayed	190	0.046	—
Polyvinyl acetate cork mastic; sprayed or troweled	—	0.100	—
Urethane, two-part mixture; rigid foam	70	0.026	1045
Reflective			
Aluminium foil separating fluffy glass mats, 10–12 layers, evacuated; for cryogenic applications (150 K)	40	0.00016	—
Aluminium foil and glass paper laminate, 75–150 layers, evacuated, for cryogenic application (150 K)	120	0.000017	—
Typical silica powder, evacuated	160	0.0017	—

(*Contd*)

Table A1.3 *(Contd)*

Description/ Composition	Maximum service temperature (K)	Typical density (kg/m³)	Industrial insulation													
			Typical thermal conductivity, k (W/mK), at various temperatures (K)													
			200	215	230	240	255	270	285	300	310	365	420	530	645	750
Blankets																
Blanket, mineral fibre, metal reinforced	920	96–192									0.038	0.046	0.056	0.078		
	815	40–96									0.035	0.045	0.058	0.088		
Blanket, mineral fibre, glass; fine fibre, organic bonded	450	10				0.036	0.038	0.040	0.043	0.048	0.052	0.076				
		12				0.035	0.036	0.039	0.042	0.046	0.049	0.069				
		16				0.033	0.035	0.036	0.039	0.042	0.046	0.062				
		24				0.030	0.032	0.033	0.036	0.039	0.040	0.053				
		32				0.029	0.030	0.032	0.033	0.036	0.038	0.048				
		48	0.023	0.025	0.026	0.027	0.029	0.030	0.032	0.033	0.035	0.045				
Blanket, alumina–silica fibre	1530	48												0.071	0.105	0.150
		64												0.059	0.087	0.125
		96												0.052	0.076	0.100
		128												0.049	0.068	0.091
Felt, semirigid; organic bonded	480	50–125						0.035	0.036	0.038	0.039	0.051	0.063			
	730	50						0.030	0.032	0.033	0.035	0.051	0.079			
Felt, laminated; no binder	920	120											0.051	0.065	0.087	

(Contd)

Table A1.3   *(Contd)*

Description/ Composition	Maximum service Temperature (K)	Typical density (kg/m³)	Industrial insulation													
			Typical thermal conductivity, *k* (W/mK), at various temperatures (K)													
			200	215	230	240	255	270	285	300	310	365	420	530	645	750
Blocks, boards, and pipe insulations																
Asbestos paper, laminated and corrugated																
4-ply	420	190								0.078	0.082	0.098				
6-ply	420	255								0.071	0.074	0.085				
8-ply	420	300								0.068	0.071	0.082				
Magnesia, 85%	590	185									0.051	0.55	0.061			
Calcium silicate	920	190									0.055	0.059	0.063	0.075	0.089	0.104
Cellular glass	700	145			0.046	0.048	0.051	0.052	0.055	0.058	0.062	0.069	0.079			
Diatomaceous	1145	345												0.092	0.098	0.104
Silica	1310	385												0.101	0.100	0.115
Polystyrene, rigid																
Extruded (R-12)	350	56	0.023	0.023	0.022	0.023	0.023	0.025	0.026	0.027	0.029					
Extruded (R-12)	350	35	0.023	0.023	0.023	0.025	0.025	0.026	0.027	0.029						
Moulded beads	350	16	0.026	0.029	0.030	0.033	0.035	0.036	0.038	0.040						
Rubber, rigid foamed	340	70						0.029	0.030	0.032	0.033					

(Contd)

Table A1.3 (*Contd*)

Description/ Composition	Maximum service Temperature (K)	Typical density (kg/m³)	Industrial insulation														
			Typical thermal conductivity, k (W/mK), at various temperatures (K)														
			200	215	230	240	255	270	285	300	310	365	420	530	645	750
Insulating cement																
Mineral fibre (rock, slag, or glass)																
With clay binder	1255	430									0.071	0.079	0.088	0.105	0.123	
With hydraulic setting binder	922	560									0.108	0.115	0.123	0.137		
Loose fill																
Cellulose, wood, or paper pulp	—	45							0.038	0.039	0.042					
Perlite, expanded	—	105	0.036	0.039	0.042	0.043	0.046	0.049	0.051	0.053	0.056					
Vermiculite, expanded	—	122			0.056	0.058	0.061	0.063	0.065	0.068	0.071					
		80			0.049	0.051	0.055	0.058	0.061	0.063	0.066					

(*Contd*)

Table A1.3   (*Contd*)

Description/Composition	Temperature (K)	Density ρ (kg/m^3)	Thermal conductivity, k (W/mK)	Specific heat, c_p (J/kgK)
		Other materials		
Asphalt	300	2115	0.062	920
Bakelite	300	1300	1.4	1465
Brick, refractory				
Carborundum	872	—	18.5	—
	1672	—	11.0	—
Chrome brick	473	3010	2.3	835
	823		2.5	
	1173		2.0	
Diatomaceous	478	—	0.25	—
silica, fired	1145	—	0.30	
Fire clay, burnt 1600 K	773	2050	1.0	960
	1073	—	1.1	
	1373	—	1.1	
Fire clay, burnt 1725 K	773	2325	1.3	960
	1073		1.4	
	1373		1.4	
Fire clay brick	478	2645	1.0	960
	922		1.5	
	1478		1.8	
Magnesite	478	—	3.8	1130
	922	—	2.8	
	1478		1.9	
Clay	300	1460	1.3	880
Coal, anthracite	300	1350	0.26	1260
Concrete (stone mix)	300	2300	1.4	880
Cotton	300	80	0.06	1300
Foodstuffs				
Banana (75.7% water content)	300	980	0.481	3350
Apple, red (75% water content)	300	840	0.513	3600
Cake, fully baked	300	280	0.121	
Chicken meat, white	198	—	1.60	
(74.4% water content)	233	—	1.49	
	253		1.35	
	263		1.20	
	273		0.476	
	283		0.480	
	293		0.489	

(*Contd*)

Table A1.3 *(Contd)*

Description/Composition	Temperature (K)	Density ρ (kg/m³)	Thermal conductivity, k (W/mK)	Specific heat, c_p (J/kgK)
Other materials				
Glass				
Plate (soda lime)	300	2500	1.4	750
Pyrex	300	2225	1.4	835
Ice	273	920	1.88	2040
	253	—	2.03	1945
Leather (sole)	300	998	0.159	—
Paper	300	930	0.180	1340
Paraffin	300	900	0.240	2890
Rock				
Granite, Barre	300	2630	2.79	775
Limestone, Salem	300	2320	2.15	810
Marble, Halston	300	2680	2.80	830
Quartzite, Sioux	300	2640	5.38	1105
Sandstone, Berea	300	2150	2.90	745
Rubber, vulcanized				
Soft	300	1100	0.13	2010
Hard	300	1190	0.16	—
Sand	300	1515	0.27	800
Soil	300	2050	0.52	1840
Snow	273	110	0.049	—
		500	0.190	—
Teflon	300	2200	0.35	—
	400		0.45	—
Tissue, human				
Skin	300	—	0.37	—
Fat layer (adipose)	300	—	0.2	—
Muscle	300	—	0.41	—
Wood, cross grain				
Balsa	300	140	0.055	—
Cypress	300	465	0.097	—
Fir	300	415	0.11	2720
Oak	300	545	0.17	2385
Yellow pine	300	640	0.15	2805
White pine	300	435	0.11	—
Wood, radial				
Oak	300	545	0.19	2385
Fir	300	420	0.14	2720

Table A1.4 Thermophysical properties of gases at atmospheric pressure

T (K)	ρ (kg/m³)	c_p (kJ/kgK)	$\mu \times 10^7$ (Ns/m²)	$v \times 10^6$ (m²/s)	$k \times 10^3$ (W/mK)	$\alpha \times 10^6$ (m²/s)	Pr
Air							
100	3.5562	1.032	71.1	2.00	9.34	2.54	0.786
150	2.3364	1.012	103.4	4.426	13.8	5.84	0.758
200	1.7458	1.007	132.5	7.590	18.1	10.3	0.737
250	1.3947	1.006	159.6	11.44	22.3	15.9	0.720
300	1.1614	1.007	184.6	15.89	26.3	22.5	0.707
350	0.9950	1.009	208.2	20.92	30.0	29.9	0.700
400	0.8711	1.014	230.1	26.41	33.8	38.3	0.690
450	0.7740	1.021	250.7	32.39	37.3	47.2	0.686
500	0.6964	1.030	270.1	38.79	40.7	56.7	0.684
550	0.6329	1.040	288.4	45.57	43.9	66.7	0.683
600	0.5804	1.051	305.8	52.69	46.9	76.9	0.685
650	0.5356	1.063	322.5	60.21	49.7	87.3	0.690
700	0.4975	1.075	338.8	68.10	52.4	98.0	0.695
750	0.4643	1.087	354.6	76.37	54.9	109	0.702
800	0.4354	1.099	369.8	84.93	57.3	120	0.709
850	0.4097	1.110	384.3	93.80	59.6	131	0.716
900	0.3868	1.121	398.1	102.9	62.0	143	0.720
950	0.3666	1.131	411.3	112.2	64.3	155	0.723
1000	0.3482	1.141	424.4	121.9	66.7	168	0.726
1100	0.3166	1.159	449.0	141.8	71.5	195	0.728
1200	0.2902	1.175	473.0	162.9	76.3	224	0.728
1300	0.2679	1.189	496.0	185.1	82	238	0.719
1400	0.2488	1.207	530	213	91	303	0.703
1500	0.2322	1.230	557	240	100	350	0.685
1600	0.2177	1.248	584	268	106	390	0.688
1700	0.2049	1.267	611	298	113	435	0.685
1800	0.1935	1.286	637	329	120	482	0.683
1900	0.1833	1.307	663	362	128	534	0.677
2000	0.1741	1.337	689	396	137	589	0.672
2100	0.1658	1.372	715	431	147	646	0.667
2200	0.1582	1.417	740	468	160	714	0.655
2300	0.1513	1.478	766	506	175	783	0.647
2400	0.1448	1.558	792	547	196	869	0.630
2500	0.1389	1.665	818	589	222	960	0.613
3000	0.1135	2.726	955	841	486	1570	0.536

(Contd)

Table A1.4 *(Contd)*

T (K)	ρ (kg/m³)	c_p (kJ/kgK)	$\mu \times 10^7$ (Ns/m²)	$\nu \times 10^6$ (m²/s)	$k \times 10^3$ (W/mK)	$\alpha \times 10^6$ (m²/s)	Pr
Ammonia (NH₃)							
300	0.6894	2.158	101.5	14.7	24.7	16.6	0.887
320	0.6448	2.170	109	16.9	27.2	19.4	0.870
340	0.6059	2.192	116.5	19.2	29.3	22.1	0.872
360	0.5716	2.221	124	21.7	31.6	24.9	0.872
380	0.5410	2.254	131	24.2	34.0	27.9	0.869
400	0.5136	2.287	138	26.9	37.0	31.5	0.853
420	0.4888	2.322	145	29.7	40.4	35.6	0.833
440	0.4664	2.357	152.5	32.7	43.5	39.6	0.826
460	0.4460	2.393	159	35.7	46.3	43.4	0.822
480	0.4273	2.430	166.5	39.0	49.2	47.4	0.822
500	0.4101	2.467	173	42.2	52.5	51.9	0.813
520	0.3942	2.504	180	45.7	54.5	55.2	0.827
540	0.3795	2.540	186.5	49.1	57.5	59.7	0.824
560	0.3708	2.577	193	52.0	60.6	63.4	0.827
580	0.3533	2.613	199.5	56.5	63.8	69.1	0.817
Carbon dioxide (CO₂)							
280	1.9022	0.830	140	7.36	15.20	9.63	0.765
300	1.7730	0.851	149	8.40	16.55	11.0	0.766
320	1.6609	0.872	156	9.39	18.05	12.5	0.754
340	1.5618	0.891	165	10.6	19.70	14.2	0.746
360	1.4743	0.908	173	11.7	21.2	15.8	0.741
380	1.3961	0.926	181	13.0	22.75	17.6	0.737
400	1.3257	0.942	190	14.3	24.3	19.5	0.737
450	1.1782	0.981	210	17.8	28.3	24.5	0.728
500	1.0594	1.02	231	21.8	32.5	30.1	0.725
550	0.9625	1.05	251	26.1	36.6	36.2	0.721
600	0.8826	1.08	270	30.6	40.7	42.7	0.717
650	0.8143	1.10	288	35.4	44.5	49.7	0.717
700	0.7564	1.13	305	40.3	48.1	56.3	0.717
750	0.7057	1.15	321	45.5	51.7	63.7	0.714
800	0.6614	1.17	337	51.0	55.1	71.2	0.716
Carbon monoxide (CO)							
200	1.6888	1.045	127	7.52	17.0	9.63	0.781
220	1.5341	1.044	137	8.93	19.0	11.9	0.753
240	1.4055	1.043	147	10.5	20.6	14.1	0.744
260	1.2967	1.043	157	12.1	22.1	16.3	0.741
280	1.2038	1.042	166	13.8	23.6	18.8	0.733
300	1.1233	1.043	175	15.6	25.0	21.3	0.730
320	1.0529	1.043	184	17.5	26.3	23.9	0.730

(Contd)

Table A1.4 (*Contd*)

T (K)	ρ (kg/m^3)	c_p (kJ/kgK)	$\mu \times 10^7$ (Ns/m^2)	$v \times 10^6$ (m^2/s)	$k \times 10^3$ (W/mK)	$\alpha \times 10^6$ (m^2/s)	Pr
340	0.9909	1.044	193	19.5	27.8	26.9	0.725
360	0.9357	1.045	202	21.6	29.1	29.8	0.725
380	0.8864	1.047	210	23.7	30.5	32.9	0.729
400	0.8421	1.049	218	25.9	31.8	36.0	0.719
450	0.7483	1.055	237	31.7	35.0	44.3	0.714
500	0.67352	1.065	254	37.7	38.1	53.1	0.710
550	0.61226	1.076	271	44.3	41.1	62.4	0.710
600	0.56126	1.088	286	51.0	44.0	72.1	0.707
650	0.51806	1.101	301	58.1	47.0	82.4	0.705
700	0.48102	1.114	315	65.5	50.0	93.3	0.702
750	0.44899	1.127	329	73.3	52.8	104	0.702
800	0.42095	1.140	343	81.5	55.5	116	0.705
Helium (He)							
100	0.4871	5.193	96.3	19.8	73.0	28.9	0.686
120	0.4060	5.193	107	26.4	81.9	38.8	0.679
140	0.3481	5.193	118	33.9	90.7	50.2	0.676
160	—	5.193	129	—	99.2	—	—
180	0.2708	5.193	139	51.3	107.2	76.2	0.673
200	—	5.193	150	—	115.1	—	—
220	0.2216	5.193	160	72.2	123.1	107	0.675
240	—	5.193	170	—	130	—	—
260	0.1875	5.193	180	96.0	137	141	0.682
280	—	5.193	190	—	145	—	—
300	0.1625	5.193	199	122	152	180	0.680
350	—	5.193	221	—	170	—	—
400	0.1219	5.193	243	199	187	295	0.675
450	—	5.193	263	—	204	—	—
500	0.09754	5.193	283	290	220	434	0.668
550	—	5.193	—	—	—	—	—
600	—	5.193	320	—	252	—	—
650	—	5.193	332	—	264	—	—
700	0.06969	5.193	350	502	278	768	0.654
750	—	5.193	364	—	291	—	—
800	—	5.193	382	—	304	—	—
900	—	5.193	414	—	330	—	—
1000	0.04879	5.193	446	914	354	1400	0.654

(*Contd*)

Table A1.4 *(Contd)*

T (K)	ρ (kg/m³)	c_p (kJ/kgK)	$\mu \times 10^7$ (Ns/m²)	$v \times 10^6$ (m²/s)	$k \times 10^3$ (W/mK)	$\alpha \times 10^6$ (m²/s)	Pr
Hydrogen (H_2)							
100	0.24255	11.23	42.1	17.4	67.0	24.6	0.707
150	0.16156	12.60	56.0	34.7	101	49.6	0.699
200	0.12115	13.54	68.1	56.2	131	79.9	0.704
250	0.09693	14.06	78.9	81.4	157	115	0.707
300	0.08078	14.31	89.6	111	183	158	0.701
350	0.06924	14.43	98.8	143	204	204	0.700
400	0.06059	14.48	108.2	179	226	258	0.695
450	0.05386	14.50	117.2	218	247	316	0.689
500	0.04848	14.52	126.4	261	266	378	0.691
550	0.04407	14.53	134.3	305	285	445	0.685
600	0.04040	14.55	142.4	352	305	519	0.678
700	0.03463	14.61	157.8	456	342	676	0.675
800	0.03030	14.70	172.4	569	378	849	0.670
900	0.02694	14.83	186.5	692	412	1030	0.671
1000	0.02424	14.99	201.3	830	448	1230	0.673
1100	0.02204	15.17	213.0	966	488	1460	0.662
1200	0.02020	15.37	226.2	1120	528	1700	0.659
1300	0.01865	15.59	238.5	1279	568	1955	0.655
1400	0.01732	15.81	250.7	1447	610	2230	0.650
1500	0.01616	16.02	262.7	1626	655	2530	0.643
1600	0.0152	16.28	273.7	1801	697	2815	0.639
1700	0.0143	16.58	284.9	1992	742	3130	0.637
1800	0.0135	16.96	296.1	2193	786	3435	0.639
1900	0.0128	17.49	307.2	2400	835	3730	0.643
2000	0.0121	18.25	318.2	2630	878	3975	0.661
Nitrogen (N_2)							
100	3.4388	1.070	68.8	2.00	9.58	2.60	0.768
150	2.2594	1.050	100.6	4.45	13.9	5.86	0.759
200	1.6883	1.043	129.2	7.65	18.3	10.4	0.736
250	1.3488	1.042	154.9	11.48	22.2	15.8	0.727
300	1.1233	1.041	178.2	15.86	25.9	22.1	0.716
350	0.9625	1.042	200.0	20.78	29.3	29.2	0.711
400	0.8425	1.045	220.4	26.16	32.7	37.1	0.704
450	0.7485	1.050	239.6	32.01	35.8	45.6	0.703
500	0.6739	1.056	257.7	38.24	38.9	54.7	0.700
550	0.6124	1.065	274.7	44.86	41.7	63.9	0.702
600	0.5615	1.075	290.8	51.79	44.6	73.9	0.701
700	0.4812	1.098	321.0	66.71	49.9	94.4	0.706
800	0.4211	1.22	349.1	82.90	54.8	116	0.715

(Contd)

Table A1.4 (*Contd*)

T (K)	ρ (kg/m³)	c_p (kJ/kgK)	$\mu \times 10^7$ (Ns/m²)	$\nu \times 10^6$ (m²/s)	$k \times 10^3$ (W/mK)	$\alpha \times 10^6$ (m²/s)	Pr
900	0.3743	1.146	375.3	100.3	59.7	139	0.721
1000	0.3368	1.167	399.9	118.7	64.7	165	0.721
1100	0.3062	1.187	423.2	138.2	70.0	193	0.718
1200	0.2807	1.204	445.3	158.6	75.8	224	0.707
1300	0.2591	1.219	466.2	179.9	81.0	256	0.701
Oxygen (O₂)							
100	3.945	0.962	76.4	1.94	9.25	2.44	0.796
150	2.585	0.921	114.8	4.44	13.8	5.80	0.766
200	1.930	0.915	147.5	7.64	18.3	10.4	0.737
250	1.542	0.915	178.6	11.58	22.6	16.0	0.723
300	1.284	0.920	207.2	16.14	26.8	22.7	0.711
350	1.100	0.929	233.5	21.23	29.6	29.0	0.733
400	0.9620	0.942	258.2	26.84	33.0	36.4	0.737
450	0.8554	0.956	281.4	32.90	36.3	44.4	0.741
500	0.7698	0.972	303.3	39.40	41.2	55.1	0.716
550	0.6998	0.988	324.0	46.30	44.1	63.8	0.726
600	0.6414	1.003	343.7	53.59	47.3	73.5	0.729
700	0.5498	1.031	380.8	69.26	52.8	93.1	0.744
800	0.4810	1.054	415.2	86.32	58.9	116	0.743
900	0.4275	1.074	447.2	104.6	64.9	141	0.740
1000	0.3848	1.090	477.0	124.0	71.0	169	0.733
1100	0.3498	1.103	505.5	144.5	75.8	196	0.736
1200	0.3206	1.115	532.5	166.1	81.9	229	0.725
1300	0.2960	1.125	588.4	188.6	87.1	262	0.721
Water vapour (Steam)							
380	0.5863	2.060	127.1	21.68	24.6	20.4	1.06
400	0.5542	2.014	134.4	24.25	26.1	23.4	1.04
450	0.4902	1.980	152.5	31.11	29.9	30.8	1.01
500	0.4405	1.985	170.4	38.68	33.9	38.8	0.998
550	0.4005	1.997	188.4	47.04	37.9	47.4	0.993
600	0.3652	2.026	206.7	56.60	42.2	57.0	0.993
650	0.3380	2.056	224.7	66.48	46.4	66.8	0.996
700	0.3140	2.085	242.6	77.26	50.5	77.1	1.00
750	0.2931	2.119	260.4	88.84	54.9	88.4	1.00
800	0.2739	2.152	278.6	101.7	59.2	100	1.01
850	0.2579	2.186	296.9	115.1	63.7	113	1.02

Table A1.5 Thermophysical properties of saturated fluids

				Saturated liquids				
T (K)	ρ (kg/m³)	c_p (kJ/kgK)	$\mu \times 10^2$ (Ns/m²)	$\nu \times 10^6$ (m²/s)	$k \times 10^3$ (W/mK)	$\alpha \times 10^6$ (m²/s)	Pr	$\beta \times 10^3$ (K⁻¹)
Engine oil (unused)								
273	899.1	1.796	385	4.280	147	0.910	47,000	0.70
280	895.3	1.827	217	2,430	144	0.880	27,500	0.70
290	890.0	1.868	99.9	1,120	145	0.872	12,900	0.70
300	884.1	1.909	48.6	550	145	0.859	6,400	0.70
310	877.9	1.951	25.3	288	145	0.847	3,400	0.70
320	871.8	1.993	14.1	161	143	0.823	1,965	0.70
330	865.8	2.035	8.36	96.6	141	0.800	1,205	0.70
340	859.9	2.076	5.31	61.7	139	0.779	793	0.70
350	853.9	2,118	3.56	41.7	138	0.763	546	0.70
360	847.8	2.161	2.52	29.7	138	0.753	395	0.70
370	841.8	2.206	1.86	22.0	137	0.738	300	0.70
380	836.0	2.250	1.41	16.9	136	0.723	233	0.70
390	830.6	2.294	1.10	13.3	135	0.709	187	0.70
400	825.1	2.337	0.874	10.6	134	0.695	152	0.70
410	818.9	2.381	0.698	8.52	133	0.682	125	0.70
420	812.1	2.427	0.564	6.94	133	0.675	103	0.70
430	806.5	2.471	0.470	5.83	132	0.662	88	0.70
Ethylene glycol [C₂H₄(OH)₂]								
273	1,130.8	2.294	6.51	57.6	242	0.933	617	0.65
280	1,125.8	2.323	4.20	37.3	244	0.933	400	0.65
290	1,118.8	2.368	2.47	22.1	248	0.936	236	0.65
300	1,114.4	2.415	1.57	14.1	252	0.939	151	0.65
310	1,103.7	2.460	1.07	9.65	255	0.939	103	0.65
320	1,096.2	2.505	0.757	6.91	258	0.940	73.5	0.65
330	1,089.5	2.549	0.561	5.15	260	0.936	55.0	0.65
340	1,083.8	2.592	0.431	3.98	261	0.929	42.8	0.65
350	1,079.0	2.637	0.342	3.17	261	0.917	34.6	0.65
360	1,074.0	2.682	0.278	2.59	261	0.906	28.6	0.65
370	1,066.7	2.728	0.228	2.14	262	0.900	23.7	0.65
373	1,058.5	2.742	0.215	2.03	263	0.906	22.4	0.65
Glycerine [C₃H₅(OH)₃]								
273	1,276.0	2.261	1,060	8,310	282	0.977	85,000	0.47
280	1,271.9	2.298	534	4,200	284	0.972	43,200	0.47
290	1,265.8	2.367	185	1,460	286	0.955	15,300	0.48
300	1,259.9	2.427	79.9	634	286	0.935	6,780	0.48

(Contd)

Table A1.5 *(Contd)*

				Saturated liquids				
T (K)	ρ (kg/m^3)	c_p (kJ/kgK)	$\mu \times 10^2$ (Ns/m^2)	$v \times 10^6$ (m^2/s)	$k \times 10^3$ (W/mK)	$\alpha \times 10^6$ (m^2/s)	Pr	$\beta \times 10^3$ (K^{-1})
310	1,253.9	2.490	35.2	281	286	0.916	3,060	0.49
320	1,247.2	2.564	21.0	168	287	0.897	1,870	0.50
Freon (Refrigerant-12) (CCl$_2$F$_2$)								
230	1,528.4	0.8816	0.0457	0.299	68	0.505	5.9	1.85
240	1,498.0	0.8923	0.0385	0.257	69	0.516	5.0	1.90
250	1,469.5	0.9037	0.0354	0.241	70	0.527	4.6	2.00
260	1,439.0	0.9163	0.0322	0.224	73	0.554	4.0	2.10
270	1,407.2	0.9301	0.0304	0.216	73	0.558	3.7	2.35
280	1,374.4	0.9450	0.0283	0.206	73	0.562	3.7	2.35
290	1,340.5	0.9609	0.0265	0.198	73	0.567	3.5	2.55
300	1,305.8	0.9781	0.0254	0.195	72	0.564	3.5	2.75
310	1,268.9	0.9963	0.0244	0.192	69	0.546	3.4	3.05
320	1,228.6	1.0155	0.0233	0.190	68	0.545	3.5	3.5
Mercury (Hg)								
273	13,595	0.1404	0.1688	0.1240	8,180	42.85	0.0290	0.181
300	13,529	0.1393	0.1523	0.1125	8,540	45.30	0.0248	0.181
350	13,407	0.1377	0.1309	0.0976	9,180	49.75	0.0196	0.181
400	13,287	0.1365	0.1171	0.0882	9,800	54.05	0.0163	0.181
450	13,167	0.1357	0.1075	0.0816	10,400	58.10	0.0140	0.181
500	13,048	0.1353	0.1007	0.0771	10,950	61.90	0.0125	0.182
550	12,929	0.1352	0.0953	0.0737	11,450	65.55	0.0112	0.184
600	12,809	0.1355	0.0911	0.0711	11,950	68.80	0.0103	0.187

		Saturated liquid–vapour, 1 atm			
Fluid	T_{sat} (K)	h_{fg} (kJ/kg)	ρ_f (kg/m^3)	ρ_g (kg/m^3)	$\sigma \times 10^3$ (Nm)
Ethanol	351	846	757	1.44	17.7
Ethylene glycol	470	812	1,111	—	32.7
Glycerine	563	974	1,260	—	63.0
Mercury	630	301	12,740	3.90	417
Refrigerant R-12	243	165	1,488	6.32	15.8
Refrigerant R-113	321	147	1,411	7.38	15.9

Table A1.6 Thermophysical properties of saturated water

Temperature T(K)	Pressure P (bars)[a]	Specific volume (m^3/kg) $v_f \times 10^3$	v_g	Heat of vapourization h_{fg} (kJ/kg)	Specific heat (kJ/kgK) $c_{p,f}$	$c_{p,g}$	Viscosity N s/m^2 $\mu_f \times 10^6$	$\mu_g \times 10^6$	Thermal conductivity (W/mK) $k_f \times 10^3$	$k_g \times 10^3$	Prandtl number Pr_f	Pr_g	Surface tension, $\sigma_f \times 10^3$ (N/m)	Expansion coefficient, $\beta_f \times 10^6$ (K^{-1})	Temperature T(K)
273.15	0.00611	1.000	206.3	2502	4.217	1.854	1750	8.02	569	18.2	12.99	0.815	75.5	−68.05	273.15
275	0.00697	1.000	181.7	2497	4.211	1.855	1652	8.09	574	18.3	12.22	0.817	75.3	−32.74	275
280	0.00990	1.000	130.4	2485	4.198	1.858	1422	8.29	582	18.6	10.26	0.825	74.8	46.04	280
285	0.01387	1.000	99.4	2473	4.189	1.861	1225	8.49	590	18.9	8.81	0.833	74.3	114.1	285
290	0.01917	1.001	69.7	2461	4.184	1.864	1080	8.69	598	19.3	7.56	0.841	73.7	174.0	290
295	0.02617	1.002	51.94	2449	4.181	1.868	959	8.89	606	19.5	6.62	0.849	72.7	227.5	295
300	0.03531	1.003	39.13	2438	4.179	1.872	855	9.09	613	19.6	5.83	0.857	71.7	276.1	300
305	0.04712	1.005	29.74	2426	4.178	1.877	769	9.29	620	20.1	5.20	0.865	70.9	320.6	305
310	0.06221	1.007	22.93	2414	4.178	1.882	695	9.49	628	20.4	4.62	0.873	70.0	361.9	310
315	0.08132	1.009	17.82	2402	4.179	1.888	631	9.69	634	20.7	4.16	0.883	69.2	400.4	315
320	0.1053	1.011	13.98	2390	4.180	1.895	577	9.89	640	21.0	3.77	0.894	68.3	436.7	320
325	0.1351	1.013	11.06	2378	4.182	1.903	528	10.09	645	21.3	3.42	0.901	67.5	471.2	325
330	0.1719	1.016	8.82	2366	4.184	1.911	489	10.29	650	21.7	3.15	0.908	66.6	504.0	330
335	0.2167	1.018	7.09	2354	4.186	1.920	453	10.49	656	22.0	2.88	0.916	65.8	535.5	335
340	0.2713	1.021	5.74	2342	4.188	1.930	420	10.69	660	22.3	2.66	0.925	64.9	566.0	340
345	0.3372	1.024	4.683	2329	4.191	1.941	389	10.89	668	22.6	2.45	0.933	64.1	595.4	345
350	0.4163	1.027	3.846	2317	4.195	1.954	365	11.09	668	23.0	2.29	0.942	63.2	624.2	350
355	0.5100	1.030	3.180	2304	4.199	1.968	343	11.29	671	23.3	2.14	0.951	62.3	652.3	355

(Contd)

Table A1.6   (*Contd*)

Temperature T(K)	Pressure P(bars)[a]	Specific volume (m³/kg)		Heat of vapourizationt h_{fg} (kJ/kg)	Specific heat (kJ/kgK)		Viscosity Ns/m²		Thermal conductivity (W/mK)		Prandtl number		Surface tension, $\sigma_f \times 10^3$ (N/m)	Expansion coefficient $\beta_f \times 10^6$ (K⁻¹)	Temperature T(K)
		$v_f \times 10^3$	v_g		$c_{p,f}$	$c_{p,g}$	$\mu_f \times 10^6$	$\mu_g \times 10^6$	$k_f \times 10^3$	$k_g \times 10^3$	Pr_f	Pr_g			
360	0.6209	1.034	2.645	2291	4.203	1.983	324	11.49	674	23.7	2.02	0.960	61.4	697.9	360
365	0.7514	1.038	2.212	2278	4.209	1.999	306	11.69	677	24.1	1.91	0.969	60.5	707.1	365
370	0.9040	1.041	1.861	2265	4.214	2.017	289	11.89	679	24.5	1.80	0.978	59.5	728.7	370
373.15	1.0133	1.044	1.679	2257	4.217	2.029	279	12.02	680	24.8	1.76	0.984	58.9	750.1	373.15
375	1.0815	1.045	1.574	2252	4.220	2.036	274	12.09	681	24.9	1.70	0.987	58.6	761	375
380	1.2869	1.049	1.337	2239	4.226	2.057	260	12.29	683	25.4	1.61	0.999	57.6	788	380
385	1.5233	1.053	1.142	2225	4.232	2.080	248	12.49	685	25.8	1.53	1.004	56.6	814	385
390	1.794	1.058	0.980	2212	4.239	2.104	237	12.69	686	26.3	1.47	1.013	55.6	841	390
400	2.455	1.067	0.731	2183	4.256	2.158	217	13.05	688	27.2	1.34	1.033	53.6	896	400
410	3.302	1.077	0.553	2153	4.278	2.221	200	13.42	688	28.2	1.24	1.054	51.5	952	410
420	4.370	1.088	0.425	2123	4.302	2.291	185	13.79	688	29.8	1.16	1.075	49.4	1010	420
430	5.699	1.099	0.331	2091	4.331	2.369	173	14.14	685	30.4	1.09	1.10	47.2		430
440	7.333	1.110	0.261	2059	4.36	2.46	162	14.50	682	31.7	1.04	1.12	45.1		440
450	9.319	1.123	0.208	2024	4.40	2.56	152	14.85	678	33.1	0.99	1.14	42.9		450
460	11.71	1.137	0.167	1989	4.44	2.68	143	15.19	673	34.6	0.95	1.17	40.7		460
470	14.55	1.152	0.136	1951	4.48	2.79	136	15.54	667	36.3	0.92	1.20	38.5		470
480	17.90	1.167	0.111	1912	4.53	2.94	129	15.88	660	38.1	0.89	1.23	36.2		480

(*Contd*)

Table A1.6 *(contd)*

Temperature T(K)	Pressure P (bars)[a]	Specific volume (m³/kg)		Heat of vapourization, h_{fg} (kJ/kg)	Specific heat (kJ/kgK)		Viscosity Ns/m²		Thermal conductivity (W/mK)		Prandtl number		Surface tension, $\sigma_f \times 10^3$ (N/m)	Expansion coefficient $\beta_f \times 10^6$ (K⁻¹)	Temperature T(K)
		$v_f \times 10^3$	v_g		c_{pf}	c_{pg}	$\mu_f \times 10^6$	$\mu_g \times 10^6$	$k_f \times 10^3$	$k_g \times 10^3$	Pr_f	Pr_g			
490	21.83	1.184	0.0922	1870	4.59	3.10	124	16.23	651	40.1	0.87	1.25	33.9	—	490
500	26.40	1.203	0.0766	1825	4.66	3.27	118	16.59	642	42.3	0.86	1.28	31.6	—	500
510	31.66	1.222	0.0631	1779	4.74	3.47	113	16.95	631	44.7	0.85	1.31	29.3	—	510
520	37.70	1.244	0.0525	1730	4.84	3.70	108	17.33	621	47.5	0.84	1.35	26.9	—	520
530	44.58	1.268	0.0445	1679	4.95	3.96	104	17.72	608	50.6	0.85	1.39	24.5	—	530
540	52.38	1.294	0.0375	1622	5.08	4.27	101	18.1	594	54.0	0.86	1.43	22.1	—	540
550	61.19	1.323	0.0317	1564	5.24	4.64	97	18.6	580	58.3	0.87	1.47	19.7	—	550
560	71.08	1.355	0.0269	1499	5.43	5.09	94	19.1	563	63.7	0.90	1.52	17.3	—	560
570	82.16	1.392	0.0228	1429	5.68	5.67	91	19.7	548	76.7	0.94	1.59	15.0	—	570
580	94.51	1.433	0.0193	1353	6.00	6.40	88	20.4	528	76.7	0.99	1.68	12.8	—	580
590	108.3	1.482	0.0163	1274	6.41	7.35	84	21.5	513	84.1	1.05	1.84	10.5	—	590
600	123.5	1.541	0.0137	1176	7.00	8.75	81	22.7	497	92.9	1.14	2.15	8.4	—	600
610	137.3	1.612	0.0115	1068	7.85	11.1	77	24.1	467	103	1.30	2.60	6.3	—	610
620	159.1	1.705	0.0094	941	9.35	15.4	72	25.9	444	114	1.52	3.46	4.5	—	620
625	169.1	1.778	0.0085	858	10.6	18.3	70	27.0	430	121	1.65	4.20	3.5	—	625
630	179.7	1.856	0.0075	781	12.6	22.1	67	28.0	412	130	2.0	4.8	2.6	—	630
635	190.9	1.935	0.0066	683	16.4	27.6	64	30.0	392	141	2.7	6.0	1.5	—	635
640	202.7	2.075	0.0057	560	26	42	59	32.0	367	155	4.2	9.6	0.8	—	640
645	215.2	2.351	0.0045	361	90	—	54	37.0	331	178	12	26	0.1	—	645
647.3[b]	221.2	3.170	0.0032	0	00	00	45	45.0	238	238	00	00	0.0	—	647.3[b]

[a] 1 bar = 10^5 N/m².
[b] Critical temperature.

Table A1.7 Thermophysical properties of liquid metals

Composition	Melting point (K)	T (K)	ρ (kg/m³)	c_p (kJ/kgK)	$v \times 10^7$ (m²/s)	k (W/mK)	$\alpha \times 10^5$ (m²/s)	Pr
Bismuth	544	589	10,011	0.1444	1.617	16.4	0.138	0.0142
		811	9,739	0.1545	1.133	15.6	1.035	0.0110
		1033	9,467	0.1645	0.8343	15.6	1.001	0.0083
Lead	600	644	10,540	0.159	2.276	16.1	1.084	0.024
		755	10,412	0.155	1.849	15.6	1.223	0.017
		977	10,140	—	1.347	14.9	—	—
Potassium	337	422	807.3	0.80	4.608	45.0	6.99	0.0066
		700	741.7	0.75	2.397	39.5	7.07	0.0034
		977	674.4	0.75	1.905	33.1	6.55	0.0029
Sodium	371	366	929.1	1.38	7.516	86.2	6.71	0.011
		644	860.2	1.30	3.270	72.3	6.48	0.0051
		977	778.5	1.26	2.285	59.7	6.12	0.0037
NaK, (45%/55%)	292	366	887.4	1.130	6.522	25.6	2.552	0.026
		644	821.7	1.055	2.871	27.5	3.17	0.0091
		977	740.1	1.043	2.174	28.9	3.74	0.0058
NaK, (22%/78%)	262	366	849.0	0.946	5.797	24.4	3.05	0.019
		672	775.3	0.879	2.666	26.7	3.92	0.0068
PbBi, (44.5%/55.5%)		1033	690.4	0.883	2.118	—	—	—
	398	422	10,524	0.147	—	9.05	0.586	—
		644	10,236	0.147	1.496	11.86	0.790	0.189
		922	9,835	—	1.171	—	—	—
Mercury	234				See Table A1.5			

Table A1.8   Total, hemispherical emissivity of selected surfaces

Metallic solids and their oxides

Description/Composition	Emissivity ε at various temperatures (K)										
	100	200	300	400	600	800	1000	1200	1500	2000	2500
Aluminium											
Highly polished, film	0.02	0.03	0.04	0.05	0.06						
Foil, bright	0.06	0.06	0.07								
Anodized			0.82	0.76							
Chromium (polished or plated)	0.05	0.07	0.10	0.12	0.14						
Copper											
Highly polished			0.03	0.03	0.04	0.04	0.04				
Stably oxidized					0.50	0.58	0.80				
Gold											
Highly polished or film	0.01	0.02	0.03	0.03	0.04	0.05	0.06				
Foil, bright	0.06	0.07	0.07								
Molybdenum											
Polished					0.06	0.08	0.10	0.12	0.15	0.21	0.26
Shot-blasted, rough					0.25	0.28	0.31	0.35	0.42		
Stably oxidized					0.80	0.82					
Nickel											
Polished					0.09	0.11	0.14	0.17			
Stably oxidized					0.40	0.49	0.57				
Platinum (polished)						0.10	0.13	0.15	0.18		

(Contd)

Table A1.8 (*Contd*)

Metallic solids and their oxides

Description/Composition	Emissivity ε at various temperatures (K)										
	100	200	300	400	600	800	1000	1200	1500	2000	2500
Silver (polished)			0.02	0.02	0.03	0.05	0.08				
Stainless steels											
Typical, polished			0.17	0.17	0.19	0.23	0.30				
Typical, cleaned			0.22	0.22	0.24	0.28	0.35				
Typical, lightly oxidized						0.33	0.40				
Typical, highly oxidized						0.67	0.70	0.76			
AISI 347, stably oxidized					0.87	0.88	0.89	0.90			
Tantalum (polished)								0.11	0.17	0.23	0.28
Tungsten (polished)							0.10	0.13	0.18	0.25	0.29

(*Contd*)

Table A1.8   *(Contd)*

Non-metallic substances[a]		
Description/Composition	Temperature (K)	Emissivity ε
Aluminium oxide	600	0.69
	1000	0.55
	1500	0.41
Asphalt pavement	300	0.85–0.93
Building materials		
Asbestos sheet	300	0.93–0.96
Brick, red	300	0.93–0.96
Gypsum or plaster board	300	0.90–0.92
Wood	300	0.82–0.92
Cloth	300	0.75–0.90
Concrete	300	0.88–0.93
Glass, window	300	0.90–0.95
Ice	273	0.95–0.98
Paints		
Black (Parsons)	300	0.98
White, acrylic	300	0.90
White, zinc oxide	300	0.92
Paper, white	300	0.92–0.97
Pyrex	300	0.82
	600	0.80
	1000	0.71
	1200	0.62
Pyroceram	300	0.85
	600	0.78
	1000	0.69
	1500	0.57
Refractories (furnace liners)		
Alumina brick	800	0.40
	1000	0.33
	1400	0.28
	1600	0.33
Magnesia brick	800	0.45
	1000	0.36
	1400	0.31
	1600	0.40
Kaolin insulating brick	800	0.70
	1200	0.57

(Contd)

Table A1.8 *(Contd)*

Description/Composition	Temperature (K)	Emissivity ε
Non-metallic Substances[a]		
	1400	0.47
	1600	0.53
Sand	300	0.90
Silicon carbide	600	0.87
	1000	0.87
	1500	0.85
Skin	300	0.95
Snow	273	0.82–0.90
Soil	300	0.93–0.96
Rocks	300	0.88-0.95
Teflon	300	0.85
	400	0.87
	500	0.92
Vegetation	300	0.92–0.96
Water	300	0.96

[a]The emissivity values in this table correspond to a surface temperature of approximately 300 K.
(*Source:* Incropera and Dewitt, 1998)

Appendix A2   Numerical Values of Bessel Functions

Table A2.1   Numerical values of $J_n(x)$, $Y_n(x)$, $I_n(x)$, and $K_n(x)$

x	$J_0(x)$	$J_1(x)$	$Y_0(x)$	$Y_1(x)$	$I_0(x)$	$I_1(x)$	$K_0(x)$	$K_1(x)$
0.0	1.0000	0.000	$-\infty$	$-\infty$	1.000	0.000	∞	∞
0.1	0.9975	0.0499	−1.5342	−6.4590	1.0025	0.0501	2.4271	9.8538
0.2	0.9900	0.0995	−1.0811	−3.3238	1.0100	0.1005	1.7527	4.7760
0.3	0.9776	0.1483	−0.8073	−2.2931	1.0226	0.1517	1.3725	3.0560
0.4	9.9604	0.1960	−0.6060	−1.7809	1.0404	0.2040	1.1145	2.1844
0.5	0.9385	0.2423	−0.4445	−1.4715	1.0635	0.2579	0.9244	1.6564
0.6	0.9120	0.2867	−0.3085	−1.2604	1.0920	0.3137	0.7775	1.3028
0.7	0.8812	0.3290	−0.1907	−1.1032	1.1263	0.3719	0.6605	1.0503
0.8	0.8463	0.3688	−0.0868	−0.9781	1.1665	0.4329	0.5653	0.8618
0.9	0.8075	0.4059	0.0056	−0.8731	1.2130	0.4971	0.4867	0.7165
1.0	0.7652	0.4401	0.0883	−0.7812	1.2661	0.5652	0.4210	0.6019
1.1	0.7196	0.4709	0.1622	−0.6981	1.3262	0.6375	0.3656	0.5098
1.2	0.6711	0.4983	0.2281	−0.6211	1.3937	0.7147	0.3185	0.4346
1.3	0.6201	0.5520	0.2865	−0.5485	1.4693	0.7973	0.2782	0.3725
1.4	0.5669	0.5419	0.3379	−0.4791	1.5534	0.8861	0.2437	0.3208
1.5	0.5118	0.5579	0.3824	−0.4123	1.6467	0.9817	0.2138	0.2774
1.6	0.4554	0.5699	0.4204	−0.3476	1.7500	1.0848	0.1880	0.2406
1.7	0.3980	0.5778	0.4520	−0.2847	1.8640	1.1963	0.1655	0.2094
1.8	0.3400	0.5815	0.4774	−0.2237	1.9896	1.3172	0.1459	0.1826
1.9	0.2818	0.5812	0.4968	−0.1644	2.1277	1.4482	0.1288	0.1597
2.0	0.2239	0.5767	0.5104	−0.1070	2.2796	1.5906	0.1139	0.1399
2.1	0.1666	0.5683	0.5183	−0.0517	2.4463	1.7455	0.1008	0.1228
2.2	0.1104	0.5560	0.5208	0.0015	2.6291	1.9141	0.0893	0.1079
2.3	0.0555	0.5399	0.5181	0.0523	2.8296	2.0978	0.0791	0.0950
2.4	0.0025	0.5202	0.5104	0.1005	3.0493	2.2981	0.0702	0.0837
2.5	−0.0484	0.4971	0.4981	0.1459	3.2898	2.5167	0.0624	0.0739
2.6	−0.0968	0.4708	0.4813	0.1884	3.5533	2.7554	0.0554	0.0653
2.7	−0.1424	0.4416	0.4605	0.2276	3.8417	3.0161	0.0493	0.0577
2.8	−0.1850	0.4097	0.4359	0.2635	4.1573	3.3011	0.0438	0.0511
2.9	−0.2243	0.3754	0.4079	0.2959	4.5027	3.6126	0.0390	0.0453
3.0	−0.2601	0.3391	0.3769	0.3247	4.8808	3.9534	0.0347	0.0402
3.1	−0.2921	0.3009	0.3431	0.3496	5.2945	4.3262	0.0310	0.0356
3.2	−0.3202	0.2613	0.3071	0.3707	5.7472	4.7343	0.0276	0.0316
3.3	−0.3443	0.2207	0.2691	0.3879	6.2426	5.1810	0.0246	0.0281
3.4	−0.3643	0.1792	0.2296	0.4010	6.7848	5.6701	0.0220	0.0250
3.5	−0.3801	0.1374	0.1896	0.4102	7.3782	6.2058	0.0196	0.0222
3.6	−0.3918	0.0955	0.1477	0.4154	8.0277	6.7927	0.0175	0.0198
3.7	−0.3992	0.0538	0.1061	0.4167	8.7386	7.4357	0.0156	0.0176
3.8	−0.4226	0.0128	0.0645	0.4141	9.5169	8.1404	0.0140	0.0157
3.9	−0.4018	−0.0272	0.0234	0.4078	10.369	8.9128	0.0125	0.0140
4.0	−0.3971	−0.0660	−0.0169	0.3979	11.302	9.7595	1.1160	1.2484
4.2	−0.3766	−0.1386	−0.0938	0.3680	13.442	11.706	0.8927	0.9938
4.4	−0.3423	−0.2028	−0.1633	0.3260	16.010	14.046	0.7149	0.7923

(Contd)

Table A2.1   (*Contd*)

x	$J_0(x)$	$J_1(x)$	$Y_0(x)$	$Y_1(x)$	$I_0(x)$	$I_1(x)$	$10^{-2} \times K_0(x)$	$10^{-2} \times K_1(x)$
4.6	−0.2961	−0.2566	−0.2235	0.2737	19.093	16.863	0.5730	0.6325
4.8	−0.2404	−0.2985	−0.2723	0.2136	22.794	20.253	0.4597	0.5055
5.0	−0.1776	−0.3276	−0.3085	0.1479	27.240	24.336	0.3691	0.4045
5.2	−0.1103	−0.3432	−0.3313	0.0792	32.584	29.254	0.2966	0.3239
5.4	−0.0412	−0.3453	−0.3402	0.0101	39.009	35.182	0.2385	0.2597
5.6	0.0270	−0.3343	−0.3354	−0.0568	46.738	42.328	0.1918	0.2083
5.8	0.0917	−0.3110	−0.3177	−0.1192	56.038	50.946	0.1544	0.1673
6.0	0.1506	−0.2767	−0.2882	−0.1750	67.234	61.342	0.1244	0.1344
6.2	0.2017	−0.2329	−0.2483	−0.2223	80.718	73.886	0.1003	0.1081
6.4	0.2433	−0.1816	−0.1999	−0.2596	96.962	89.026	0.0808	0.0869
6.6	0.2740	−0.1250	−0.1452	−0.2857	116.54	107.30	0.0652	0.0700
6.8	0.2931	−0.0652	−0.0864	−0.3002	140.14	129.38	0.0526	0.0564
7.0	0.3001	−0.0047	−0.0259	−0.3027	168.59	156.04	0.0425	0.0454
7.2	0.2951	0.0543	0.0339	−0.2934	202.92	188.25	0.0343	0.0366
7.4	0.2786	0.1096	0.0907	−0.2731	244.34	227.17	0.0277	0.0295
7.6	0.2516	0.1592	0.1424	−0.2428	294.33	274.22	0.0224	0.0238
7.8	0.2154	0.2014	0.1872	−0.2039	354.68	331.10	0.0181	0.0192
8.0	0.1717	0.2346	0.2235	−0.1581	427.56	399.87	0.0146	0.0155
8.2	0.1222	0.2580	0.2501	−0.1072	515.59	483.05	0.0118	0.0126
8.4	0.0692	0.2708	0.2662	−0.0535	621.94	583.66	0.0096	0.0101
8.6	0.0146	0.2728	0.2715	0.0011	750.46	705.38	0.0078	0.0082
8.8	−0.0392	0.2641	0.2659	0.0544	905.80	852.66	0.0063	0.0066
9.0	−0.0903	0.2453	0.2499	0.1043	1093.6	1030.9	0.0051	0.0054
9.2	−0.1367	0.2174	0.2245	0.1491	1320.7	1246.7	0.0041	0.0043
9.4	−0.1768	0.1816	0.1907	0.1871	1595.3	1507.9	0.0033	0.0035
9.6	−0.2090	0.1395	0.1502	0.2171	1927.5	1824.1	0.0027	0.0028
9.8	−0.2323	0.0928	0.1045	0.2379	2329.4	2207.1	0.0022	0.0023
10.0	−0.2459	0.0435	0.0557	0.2490				
10.5	−0.2366	−0.0789	0.0675	0.2337				
11.0	−0.1712	−0.1768	−0.1688	0.1637				
11.5	−0.0677	−0.2284	−0.2252	0.0579				
12.0	0.0477	−0.2234	−0.2252	−0.0571				
12.5	0.1469	−0.1655	−0.1712	−0.1538				
13.0	0.2069	−0.0703	−0.0782	−0.2101				
13.5	0.2150	0.0380	−0.0301	−0.2140				
14.0	0.1711	0.1334	0.1272	−0.1666				
14.5	0.0875	0.1934	0.1903	−0.0810				
15.0	−0.0142	0.2051	0.2055	0.0211				
15.5	−0.1092	0.1672	0.1706	0.1148				

(*Source:* Kakac and Yener, 1993)

Appendix A3   Laplace Transforms

Table A3.1   Laplace transforms

Transform No.	$\bar{f}(p)$	$f(t)\ (t > 0)$
1	$\dfrac{1}{p}$	1
2	$\dfrac{1}{p^2}$	t
3	$\dfrac{1}{p^n},\ n = 1, 2, 3, \ldots$	$\dfrac{t^{n-1}}{(n-1)!},\ 0! = 1$
4	$\dfrac{1}{\sqrt{p}}$	$\dfrac{1}{\sqrt{\pi t}}$
5	$\dfrac{1}{p - a}$	e^{at}
6	$\dfrac{1}{(p-a)^n},\ n = 1, 2, 3, \ldots$	$\dfrac{1}{(n-1)!}t^{n-1}e^{at},\ 0! = 1$
7	$\dfrac{a}{p^2 + a^2}$	$\sin at$
8	$\dfrac{p}{p^2 + a^2}$	$\cos at$
9	$\dfrac{a}{p^2 - a^2}$	$\sinh at$
10	$\dfrac{p}{p^2 - a^2}$	$\cosh at$
11	$\dfrac{2ap}{(p^2 + a^2)^2}$	$t \sin at$
12	$\dfrac{p^2 - a^2}{(p^2 + a^2)^2}$	$t \cos at$
13	$\dfrac{2ap}{(p^2 + a^2)^2}$	$t \sinh at$
14	$\dfrac{p^2 + a^2}{p^2 - a^2}$	$t \cosh at$
15	$\dfrac{p^2 + a^2}{p^2 - a^2}$	$\sin at - at \cos at$
16	$\dfrac{4a^3}{p^4 + 4a^4}$	$at \cosh at - \sinh at$
17	$\dfrac{2a^3}{(p^2 + a^2)^2}$	$\sin at \cosh at - \cos at \sinh at$

(Contd)

Table A3.1   *(Contd)*

Transform No.	$\bar{f}(p)$	$f(t)$ $(t > 0)$
18	$\dfrac{2a^2 p}{p^4 + 4a^4}$	$\sin at \sinh at$
19	$\dfrac{1}{\sqrt{p^2 + a^2}}$	$J_0(at)$
20	$\dfrac{1}{\sqrt{p^2 - a^2}}$	$I_0(at)$
21	$\dfrac{a}{\left(p^2 + a^2\right)^{3/2}}$	$tJ_1(at)$
22	$\dfrac{a}{(p^2 + a^2)^{3/2}}$	$tJ_0(at)$
23	$\dfrac{a}{(p^2 - a^2)^{3/2}}$	$tJ_1(at)$
24	$\dfrac{a}{(p^2 - a^2)^{3/2}}$	$tI_0(at)$
25	$e^{-x\sqrt{p/a}}$	$\dfrac{x}{2(\pi a t^3)^{1/2}} e^{-x^2/4at}$
26	$\dfrac{e^{-x\sqrt{p/a}}}{\sqrt{p/a}}$	$\left(\dfrac{a}{\pi t}\right)^{1/2} e^{-x^2/4at}$
27	$\dfrac{e^{-x\sqrt{p/a}}}{p}$	$\mathrm{erfc}\left[\dfrac{x}{2(at)^{1/2}}\right]$
28	$\dfrac{e^{-x\sqrt{p/a}}}{p\sqrt{p/a}}$	$2\left(\dfrac{at}{\pi}\right)^{1/2} e^{-x^2/4at} - x\,\mathrm{erfc}\left[\dfrac{x}{2(at)^{1/2}}\right]$
29	$\dfrac{e^{-x\sqrt{p/a}}}{p^2}$	$t + \dfrac{x^2}{2a}\mathrm{erfc}\left[\dfrac{x}{2(at)^{1/2}}\right] - x\left(\dfrac{t}{\pi a}\right)^{1/2} e^{-x^2/4at}$
30	$\dfrac{e^{-a/p}}{p^{3/2}}$	$\dfrac{\sin 2\sqrt{at}}{\sqrt{\pi a}}$
31	$\dfrac{e^{-a/p}}{\sqrt{p}}$	$\dfrac{\cos 2\sqrt{at}}{\sqrt{\pi t}}$
32[a]	$\dfrac{(\gamma + \ln p)}{p}$	$\ln t$
33	$\dfrac{\sinh px}{p \sinh pa}$	$\dfrac{x}{a} + \dfrac{2}{\pi}\sum_{n=1}^{\infty}\dfrac{(-1)^n}{n}\sin\dfrac{n\pi x}{a}\cos\dfrac{n\pi t}{a}$

[a] γ = Euler's constant = 0.5772156...

(Contd)

Table A3.1   *(Contd)*

Transform No.	$\overline{f}(p)$	$f(t)$ $(t>0)$
34	$\dfrac{\sinh px}{p\cosh pa}$	$\dfrac{4}{\pi}\displaystyle\sum_{n=1}^{\infty}\dfrac{(-1)^n}{2n-1}\sin\dfrac{(2n-1)\pi x}{2a}\sin\dfrac{(2n-1)\pi t}{2a}$
35	$\dfrac{\cosh px}{p\sinh pa}$	$\dfrac{t}{a}+\dfrac{2}{\pi}\displaystyle\sum_{n=1}^{\infty}\dfrac{(-1)^n}{2n-1}\cos\dfrac{n\pi x}{a}\sin\dfrac{n\pi t}{a}$
36	$\dfrac{\cosh px}{p\cosh pa}$	$1+\dfrac{4}{\pi}\displaystyle\sum_{n=1}^{\infty}\dfrac{(-1)^n}{2n-1}\cos\dfrac{(2n-1)\pi x}{2a}\cos\dfrac{(2n-1)\pi t}{2a}$
37	$\dfrac{\sinh x\sqrt{p}}{\sinh a\sqrt{p}}$	$\dfrac{2\pi}{a^2}\displaystyle\sum_{n=1}^{\infty}(-1)^n\,n\,e^{-n^2\pi^2 t/a^2}\sin\dfrac{n\pi x}{a}$
38	$\dfrac{\cosh x\sqrt{p}}{\cosh a\sqrt{p}}$	$\dfrac{\pi}{a^2}-\displaystyle\sum_{n=1}^{\infty}(-1)^{n-1}(2n-1)e^{-(2n-1)^2\pi^2 t/4a^2}\cos\dfrac{(2n-1)\pi x}{2a}$
39	$\dfrac{\sinh x\sqrt{p}}{p\sinh a\sqrt{p}}$	$\dfrac{x}{a}+\dfrac{2}{\pi}\displaystyle\sum_{n=1}^{\infty}\dfrac{(-1)^n}{n}e^{-n^2\pi^2 t/a^2}\sin\dfrac{n\pi x}{a}$
40	$\dfrac{\cosh x\sqrt{p}}{p\cosh a\sqrt{p}}$	$1+\dfrac{4}{\pi}\displaystyle\sum_{n=1}^{\infty}\dfrac{(-1)^n}{2n-1}e^{-(2n-1)^2\pi^2 t/4a^2}\cos\dfrac{(2n-1)\pi x}{2a}$

(*Source:* Kakac and Yener, 1993)

Appendix A4 Numerical Values of Error Function

Table A4.1 Numerical values of error function

z	erf z	z	erf z	z	erf z	z	erf z
0.00	0.00000	0.35	0.37938	0.70	0.67780	1.05	0.86243
0.01	0.01128	0.36	0.38932	0.71	0.68466	1.06	0.86614
0.02	0.02256	0.37	0.39920	0.72	0.69143	1.07	0.86977
0.03	0.03384	0.38	0.40900	0.73	0.69810	1.08	0.87332
0.04	0.04511	0.39	0.41873	0.74	0.70467	1.09	0.87680
0.05	0.05637	0.40	0.42839	0.75	0.71115	1.10	0.87020
0.06	0.06762	0.41	0.43796	0.76	0.71753	1.11	0.88353
0.07	0.07885	0.42	0.44746	0.77	0.72382	1.12	0.88378
0.08	0.09007	0.43	0.45688	0.78	0.73001	1.13	0.88997
0.09	0.10128	0.44	0.46622	0.79	0.73610	1.14	0.89308
0.10	0.11246	0.45	0.47548	0.80	0.74210	1.15	0.89612
0.11	0.12362	0.46	0.48465	0.81	0.74800	1.16	0.89909
0.12	0.13475	0.47	0.49374	0.82	0.75381	1.17	0.90200
0.13	0.14586	0.48	0.50274	0.83	0.75952	1.18	0.90483
0.14	0.15694	0.49	0.51166	0.84	0.76514	1.19	0.90760
0.15	0.16799	0.50	0.52049	0.85	0.77066	1.20	0.91031
0.16	0.17901	0.51	0.52924	0.86	0.77610	1.21	0.91295
0.17	0.18999	0.52	0.53789	0.87	0.78143	1.22	0.91553
0.18	0.20093	0.53	0.54646	0.88	0.78668	1.23	0.91805
0.19	0.21183	0.54	0.55493	0.89	0.79184	1.24	0.92050
0.20	0.22270	0.55	0.56332	0.90	0.79690	1.25	0.92290
0.21	0.23352	0.56	0.57161	0.91	0.80188	1.26	0.92523
0.22	0.24429	0.57	0.57981	0.92	0.80676	1.27	0.92751
0.23	0.25502	0.58	0.58792	0.93	0.81156	1.28	0.92973
0.24	0.26570	0.59	0.59593	0.94	0.81627	1.29	0.93189
0.25	0.27632	0.60	0.60385	0.95	0.82089	1.30	0.93400
0.26	0.28689	0.61	0.61168	0.96	0.82542	1.31	0.93606
0.27	0.29741	0.62	0.61941	0.97	0.82987	1.32	0.93806
0.28	0.30788	0.63	0.62704	0.98	0.83423	1.33	0.94001
0.29	0.31828	0.64	0.63458	0.99	0.83850	1.34	0.94191
0.30	0.32862	0.65	0.64202	1.00	0.84270	1.35	0.94376
0.31	0.33890	0.66	0.64937	1.01	0.84681	1.36	0.94556
0.32	0.34912	0.67	0.65662	1.02	0.85083	1.37	0.94731
0.33	0.35927	0.68	0.66378	1.03	0.85478	1.38	0.94901
0.34	0.36936	0.69	0.67084	1.04	0.85864	1.39	0.95067
1.40	0.95228	1.60	0.97634	1.80	0.98909	2.00	0.99432
1.41	0.95385	1.61	0.97720	1.81	0.98952	2.20	0.99814
1.42	0.95537	1.62	0.97803	1.82	0.98994	2.40	0.99931
1.43	0.95685	1.63	0.97884	1.83	0.99034	2.60	0.99976
1.44	0.95829	1.64	0.97962	1.84	0.99073	2.80	0.99992

(Contd)

Table A4.1 *(Contd)*

z	erf z	z	erf z	z	erf z	z	erf z
1.45	0.95829	1.65	0.98037	1.85	0.99111	3.00	0.999978
1.46	0.96105	1.66	0.98110	1.86	0.99147	3.20	0.999994
1.47	0.96237	1.67	0.98181	1.87	0.99182	3.40	0.999998
1.48	0.96365	1.68	0.98249	1.88	0.99215	3.60	1.000000
1.49	0.96489	1.69	0.98315	1.89	0.99247		
1.50	0.96610	1.70	0.98379	1.90	0.99279		
1.51	0.96772	1.71	0.98440	1.91	0.99308		
1.52	0.96841	1.72	0.98500	1.92	0.99337		
1.53	0.96951	1.73	0.98557	1.93	0.99365		
1.54	0.97058	1.74	0.98613	1.94	0.99392		
1.55	0.97162	1.75	0.98667	1.95	0.99417		
1.56	0.97262	1.76	0.98719	1.96	0.99442		
1.57	0.97360	1.77	0.98769	1.97	0.99466		
1.58	0.97454	1.78	0.98817	1.98	0.99489		
1.59	0.97546	1.79	0.98864	1.99	0.99511		

(*Source:* Kakac and Yener, 1993)

Appendix A5　Radiation View Factor Charts

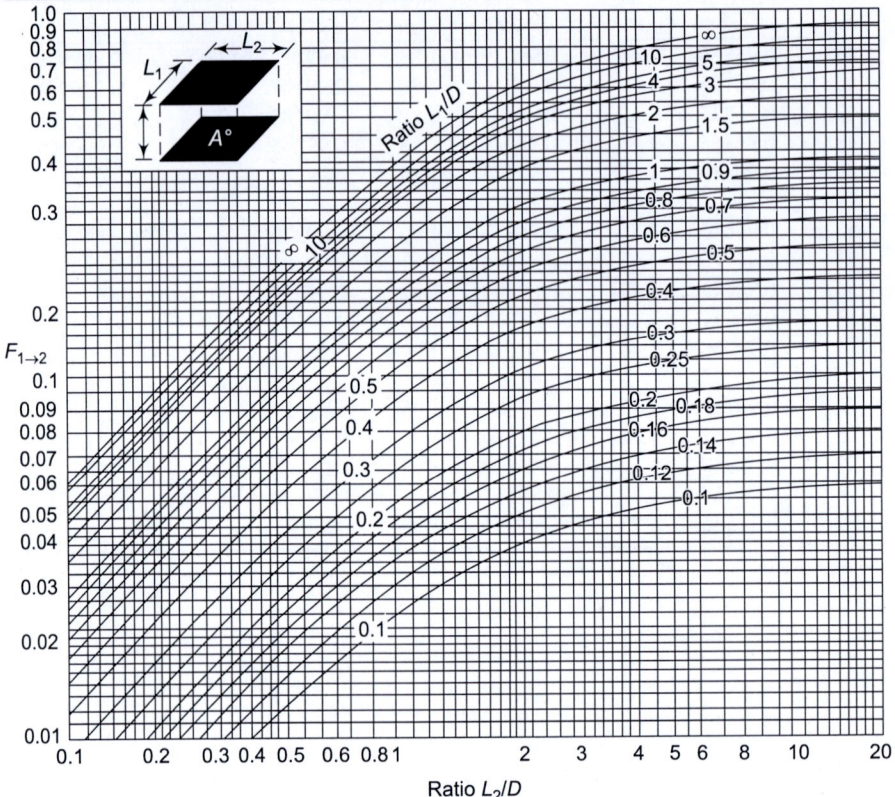

Fig. A5.1　View factor between two aligned parallel rectangles of equal size
(*Source*: Cengel, 2003)

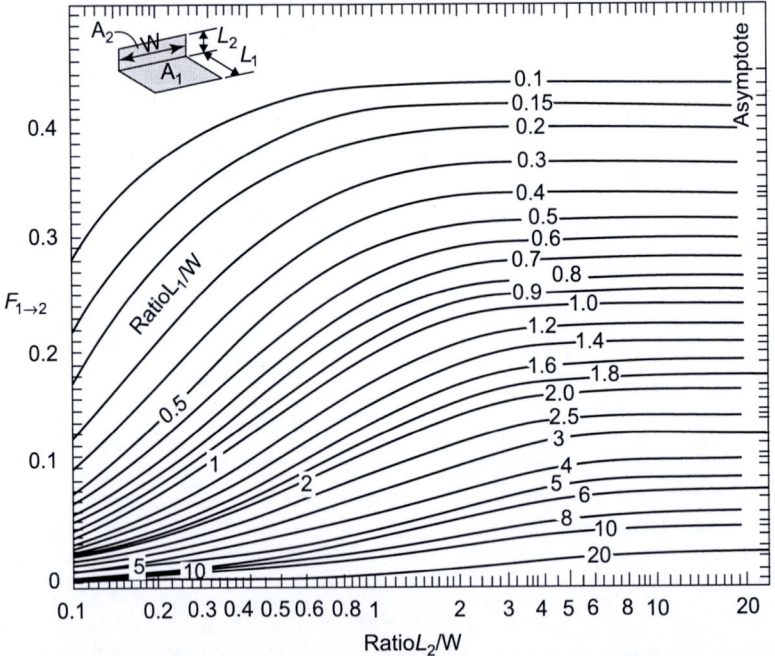

Fig. A5.2　View factor between two perpendicular rectangles with a common edge
(*Source*: Cengel, 2003)

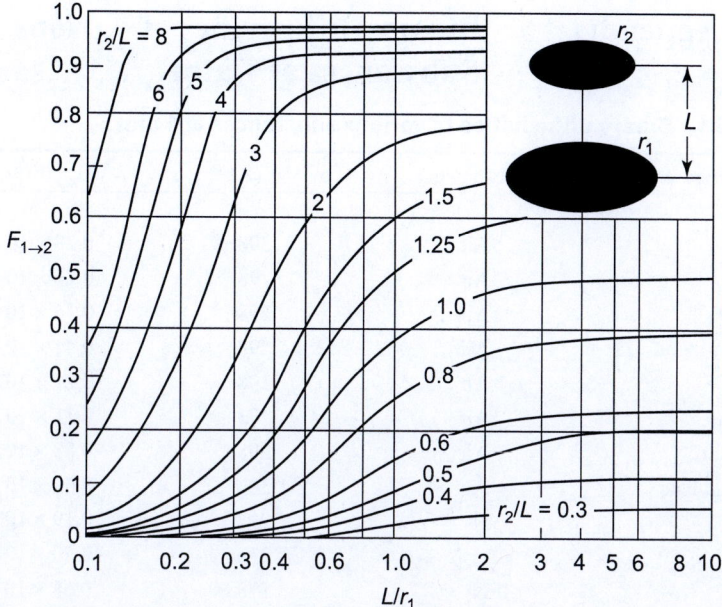

Fig. A5.3 View factor between two coaxial parallel discs (*Source*: Cengel, 2003)

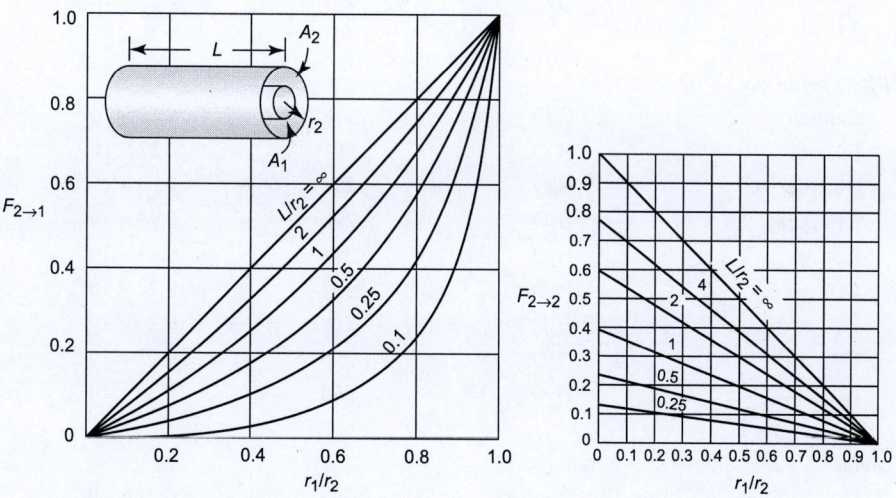

Fig. A5.4 View factors for two concentric cylinders of finite length : (a) outer cylinder to inner cylinder, (b) outer cylinder to itself (*Source*: Cengel, 2003)

Appendix A6   Binary Diffusivities of Various Substances at 1 atm

Table A6.1   Binary diffusivities of various substances at 1 atm

Substance A	Substance B	T (K)	D_{AB} (m²/s)
Gases			
NH_3	Air	298	0.28×10^{-4}
H_2O	Air	298	0.26×10^{-4}
CO_2	Air	298	0.16×10^{-4}
H_2	Air	298	0.41×10^{-4}
O_2	Air	298	0.21×10^{-4}
Acetone	Air	273	0.11×10^{-4}
Benzene	Air	298	0.88×10^{-5}
Naphthalene	Air	300	0.62×10^{-5}
Ar	N_2	293	0.19×10^{-4}
H_2	O_2	273	0.70×10^{-4}
H_2	N_2	273	0.68×10^{-4}
H_2	CO_2	273	0.55×10^{-4}
CO_2	N_2	293	0.16×10^{-4}
CO_2	O_2	273	0.14×10^{-4}
O_2	N_2	273	0.18×10^{-4}
Dilute solutions			
Caffeine	H_2O	298	0.63×10^{-9}
Ethanol	H_2O	298	0.12×10^{-8}
Glucose	H_2O	298	0.69×10^{-9}
Glycerol	H_2O	298	0.94×10^{-9}
Acetone	H_2O	298	0.13×10^{-8}
CO_2	H_2O	298	0.20×10^{-8}
O_2	H_2O	298	0.24×10^{-8}
H_2	H_2O	298	0.63×10^{-8}
N_2	H_2O	298	0.26×10^{-8}
Solids			
O_2	Rubber	298	0.21×10^{-9}
N_2	Rubber	298	0.15×10^{-9}
CO_2	Rubber	298	0.11×10^{-9}
He	SiO_2	298	0.4×10^{-13}
H_2	Fe	293	0.26×10^{-12}
Cd	Cu	293	0.27×10^{-18}
Al	Cu	293	0.13×10^{-33}

(*Source:* Incropera and Dewitt, 1998)

Appendix A7 Thermophysical Properties of Water at Atmospheric Pressure

T (°C)	ρ (g/cm³)	c_p (kJ/kg K)	c_v (kJ/kg K)	h_{fg} (kJ/kg)	β (K⁻¹)
0	0.9999	4.217	4.215	2501	-0.6×10^{-4}
5	1	4.202	4.202	2489	$+0.1 \times 10^{-4}$
10	0.9997	4.192	4.187	2477	0.9×10^{-4}
15	0.9991	4.186	4.173	2465	1.5×10^{-4}
20	0.9982	4.182	4.158	2454	2.1×10^{-4}
25	0.9971	4.179	4.138	2442	2.6×10^{-4}
30	0.9957	4.178	4.118	2430	3.0×10^{-4}
35	0.9941	4.178	4.108	2418	3.4×10^{-4}
40	0.9923	4.178	4.088	2406	3.8×10^{-4}
50	0.9881	4.180	4.050	2382	4.5×10^{-4}
60	0.9832	4.184	4.004	2357	5.1×10^{-4}
70	0.9778	4.189	3.959	2333	5.7×10^{-4}
80	0.9718	4.196	3.906	2308	6.2×10^{-4}
90	0.9653	4.205	3.865	2283	6.7×10^{-4}
100	0.9584	4.216	3.816	2257	7.1×10^{-4}

T (°C)	μ (g/cm s)	ν (cm²/s)	k (W/m K)	α (cm²/s)	Pr	$\dfrac{g\beta}{\alpha\nu} = \dfrac{\mathrm{Ra_H}}{H^3 \Delta T}$ (K⁻¹ cm⁻³)
0	0.01787	0.01787	0.56	0.00133	13.44	-2.48×10^3
5	0.01514	0.01514	0.57	0.00136	11.13	$+0.47 \times 10^3$
10	0.01304	0.01304	0.58	0.00138	9.45	4.91×10^3
15	0.01137	0.01138	0.59	0.00140	8.13	9.24×10^3
20	0.01002	0.01004	0.59	0.00142	7.07	14.45×10^3
25	0.00891	0.00894	0.60	0.00144	6.21	19.81×10^3
30	0.00798	0.00802	0.61	0.00146	5.49	25.13×10^3
35	0.00720	0.00725	0.62	0.00149	4.87	30.88×10^3
40	0.00654	0.00659	0.63	0.00152	4.34	37.21×10^3
50	0.00548	0.00554	0.64	0.00155	3.57	51.41×10^3
60	0.00467	0.00475	0.65	0.00158	3.01	66.66×10^3
70	0.00405	0.00414	0.66	0.00161	2.57	83.89×10^3
80	0.00355	0.00366	0.67	0.00164	2.23	101.3×10^3
90	0.00316	0.00327	0.67	0.00165	1.98	121.8×10^3
100	0.00283	0.00295	0.68	0.00166	1.78	142.2×10^3

(*Source:* Bejan, 1993)

Appendix A8 Solutions Finite-difference Problems in Heat Conduction Using C

Listing A8.1 `1DFIN_Conv.c`

```c
#include<stdio.h>
#include<conio.h>
#include<math.h>
void main()
{
  float D,dx,m,x,y,z,h,o,w,t,b[40],beta[40],gamma[40],temp[40];
  int i,j,k,l,n,s,q,p,a[40],c[40],d[40];
  printf("\n                    **  SEE EXAMPLE 10.3 OF HEAT TRANSFER ** ");
  printf("\n                        FIN WITH CONVECTING TIP ");
  printf("\n                        THIS IS A C++ PROGRAM.");
  printf("\n                        INPUT DATA TO BE ENTERED IN AN INTERAC-
                                    TIVE MANNER");
  printf("\n\n\n\n\n       **TEMPERATURE DISTRIBUTION IN AN 1D PIN FIN WITH
CONVECTING TIP**");
  printf("\n\nenter the value of cross-sectional area A(sq.m)  of the fin");
  scanf("%f",&x);
  printf("\n\nenter the value of thermal conductivity K(W/mK) for the fin");
  scanf("%f",&y);
  printf("\n\nenter the value of perimeter P(m)  of the fin");
  scanf("%f",&z);
  printf("\n\nenter the value of the heat transfer coefficient H(W/sq.mK)");
  scanf("%f",&h);
  printf("\n\nenter the length of the pin fin L(m)");
  scanf("%f",&t);
  printf("\nenter the heat transfer co-efficient(W/sq.mK) at the tip");
  scanf("%f",&o);
  m=sqrt((h*z)/(y*x));
  printf("\n\nenter the value of n where n+1 is the number of grid points
including the base grid point");
  scanf("%d",&n);
  dx=1.0/n;
  D=2+(m*m*t*t*dx*dx);
  w=(2*o*t*dx)/y;
  printf("\n\n\n\n\n\n\nThe value of D for the TDM is %f",D);
  /*STORING THE SUBDIAGONAL(A),DIAGONAL(B) AND SUPERDIAGONAL(C) ELEMENTS IN
1D ARRAY
  FOR SOLUTION BY THE THOMAS ALGORITHM*/
  for(i=2;i<=n-1;i++)
  a[i]=-1;
  a[n]=-2;
  for(j=1;j<=n-1;j++)
  c[j]=-1;
  for(k=2;k<=n;k++)
  d[k]=0;
  d[1]=1;
  for(l=1;l<=n-1;l++)
  b[l]=D;
  b[n]=D+w;
```

```
/*THOMAS ALGORITHM STARTS FROM HERE*/
  beta[1]=b[1];
  gamma[1]=(d[1])/(beta[1]);
  for(s=2;s<=n;s++)
  {
    beta[s]=(b[s])-((a[s]*c[s-1])/(beta[s-1]));
    gamma[s]=(d[s]-(a[s]*gamma[s-1]))/(beta[s]);
  }


/*CALCULATION OF GRID POINT TEMPERATURES STARTING FROM LAST GRID POINT TO
LAST BUT ONE*/
  temp[n]=gamma[n];
  printf("\n\n\n\n\n**THE TEMPERATURE DISTRIBUTION STARTING FROM FIN TIP
EXCLUDING BASE TEMPERATURE**");
  printf("\n\n\n\n\n The dimensionless temperature THETA=%f",temp[n]);
  for(q=n-1;q>=1;q--)
  {
    temp[q]=gamma[q]-(c[q]*temp[q+1])/(beta[q]);
    printf("\n The dimensionless temperature THETA=%f",temp[q]);
  }
  getch();
}
```

Listing A8.2 `1DFIN_Ins.c`

```c
#include<stdio.h>
#include<conio.h>
#include<math.h>
void main()
{
  float D,dx,m,x,y,z,h,t,b[40],beta[40],gamma[40],temp[40];
  int i,j,k,l,n,s,q,p,a[40],c[40],d[40];
  printf("\n                   ** See Section 10.3.1 of HEAT TRANSFER ** ");
  printf("\n                   APPLICATION OF TDMA(THOMAS ALGORITHM) ");
  printf("\n                   THIS IS A C++ Program.");
  printf("\n                   INPUT DATA TO BE ENTERED IN AN INTERACTIVE
MANNER");

  printf("\n\n\n\n\n      **TEMPERATURE DISTRIBUTION IN AN 1D RECTANGULAR
FIN WITH INSULATED TIP**");
  printf("\n\nenter the value of cross-sectional area A(sq.m)  of the fin");
  scanf("%f",&x);
  printf("\n\nenter the value of thermal conductivity K(W/mK) for the fin");
  scanf("%f",&y);
  printf("\n\nenter the value of perimeter P(m)  of the fin");
  scanf("%f",&z);
  printf("\n\nenter the value of the heat transfer coefficient H(W/sq.mK)");
  scanf("%f",&h);
  printf("\n\nenter the length of the pin fin L(m)");
  scanf("%f",&t);
  m=sqrt((h*z)/(y*x));
  printf("\n\nenter the value of n where n+1 is the number of grid points
including the base grid point");
  scanf("%d",&n);
  dx=1.0/n;
  D=2+(m*m*t*t*dx*dx);
  printf("\n\n\n\n\n\nThe value of D for the TDM is %f",D);
  /*STORING THE SUBDIAGONAL(A),DIAGONAL(B) AND SUPERDIAGONAL(C) ELEMENTS IN
1D ARRAY
```

```
FOR SOLUTION BY THE THOMAS ALGORITHM*/
for(i=2;i<=n-1;i++)
a[i]=-1;
a[n]=-2;
for(j=1;j<=n-1;j++)
c[j]=-1;
for(k=2;k<=n;k++)
d[k]=0;
d[1]=1;
for(l=1;l<=n;l++)
b[l]=D;
/*THOMAS ALGORITHM STARTS FROM HERE*/
beta[1]=b[1];
gamma[1]=(d[1])/(beta[1]);
for(s=2;s<=n;s++)
{
  beta[s]=(b[s])-((a[s]*c[s-1])/(beta[s-1]));
  gamma[s]=(d[s]-(a[s]*gamma[s-1]))/(beta[s]);
}


/*CALCULATION OF GRID POINT TEMPERATURES STARTING FROM LAST GRID POINT TO
LAST BUT ONE*/
temp[n]=gamma[n];
printf("\n\n\n\n\n**THE TEMPERATURE DISTRIBUTION STARTING FROM TIP EX-
CLUDING BASE TEMPERTURE**");
printf("\n\n\n\n\n The dimensionless temperature THETA=%f",temp[n]);
for(q=n-1;q>=1;q--)
{
  temp[q]=gamma[q]-(c[q]*temp[q+1])/(beta[q]);
  printf("\n The dimensionless temperature THETA=%f",temp[q]);
}
getch();
}
```

Listing A8.3 2DCOND_Sq.c

```
#include<stdio.h>
#include<conio.h>
#include<math.h>
int main()
{
  float t[50][50],ta[50][50],dx,error;
  int i,j,k,n,l;
  printf("\n           ** See Section 10.3.2 of HEAT TRANSFER **");
  ;printf("\n    2D Steady State Heat Conduction with Heat Generation in a
Square Rod");
  ;printf("\n           SOLUTION BY POINT-BY-POINT GAUSS-SEIDEL ITERATIVE
METHOD ");
  ;printf("\n           THIS IS A C++ PROGRAM.");
  ;printf("\n           INPUT DATA TO BE ENTERED IN AN INTERACTIVE MAN-
NER");
  ;printf("\n           The results are stored in the file named 'p.txt'.")
  ;FILE *fp;
  printf("\nEnter the number of grid points including the boundary points
in x-direction");
  scanf("%d",&n);
  dx=1.0/(n-1);
  for(i=1;i<=n;i++)
```

```
        {
          for(j=1;j<=n;j++)
            {
               t[i][j]=0;
               ta[i][j]=0;
            }
        }
     for(k=1;k<=1000;k++)
     {
       for(j=1;j<=n;j++)
     {
     ta[n][j]=0;
     ta[j][n]=0;
     }
     for(i=n-1;i>=2;i--)
     {
        for(j=n-1;j>=2;j--)
          {
            t[i][j]=((dx*dx)+ta[i+1][j]+ta[i][j+1]+ta[i][j-1]+ta[i-1][j])/4;
            ta[i][j]=t[i][j];
          }
     }
     for(j=n-1;j>=2;j--)
       {
         t[1][j]=((dx*dx)+ta[1][j+1]+ta[1][j-1]+2*ta[2][j])/4;
         ta[1][j]=t[1][j];
       }
       for(i=n-1;i>=2;i--)
       {
         t[i][1]=((dx*dx)+ta[i+1][1]+ta[i-1][1]+2*ta[i][2])/4;
         ta[i][1]=t[i][1];
       }
       t[1][1]=((dx*dx)+2*ta[1][2]+2*ta[2][1])/4.0;
       ta[1][1]=t[1][1];
     }
     printf("\n\n\nTHE TEMPERATURE DISTRIBUTION IN THE RIGHT HAND QUARTER\n\
n\n");
     printf("\n");
     fp=fopen("p.txt","w");
     for(j=n;j>=1;j--)
     {
       for(i=1;i<=n;i++)
         {
         printf("%f",ta[i][j]);
         fprintf(fp,"%f",ta[i][j]);
         printf("   ");
         }
       printf("\n");
     }
   fclose(fp);
   getch();
}
```

Answers and Hints to Problems

Chapter 2

2.1 $k = 0.1032$ W/m K

2.2 The basic approach is similar to that given for Cartesian coordinates. Apply law of conservation of energy on an elemental volume element (see Fig. 2.8). Write q_r, q_ϕ, and q_z from Fourier's law. Write $q_{r+\Delta r}$, $q_{\phi+\Delta\phi}$, and $q_{z+\Delta z}$ using Taylor series expansion dropping terms beyond second.

2.3 The general energy equation for a moving solid is

$$\frac{\partial T}{\partial t} + u\frac{\partial T}{\partial x} + v\frac{\partial T}{\partial y} + w\frac{\partial T}{\partial z} = \alpha\left[\frac{\partial^2 T}{\partial x^2} + \frac{\partial^2 T}{\partial y^2} + \frac{\partial^2 T}{\partial z^2} + \frac{q'''}{k}\right]$$

Note that u, v, w are constants since the body is a solid.

Given below is a derivation of the energy equation for steady 1D heat flow without heat generation in a moving solid. The same can be easily extended to 3D heat flow.

Take an elemental control volume of size $\Delta x \times 1 \times 1$.

Now, Rate of energy in = Rate of energy out

The difference between the steady heat flows in stationary and moving solids is that in the latter the enthalpy (also called thermal energy) will have to be considered in addition to the heat conduction. The kinetic energy will cancel out because the velocity is same at the inlet and exit of the control volume.

Thus, on energy balance

$$\frac{\partial}{\partial x}\left(-k\frac{\partial T}{\partial x}\right) + \frac{\partial}{\partial x}(\rho u i) = 0$$

Now, $\partial i = c\partial T$

Therefore, $\quad -k\dfrac{\partial^2 T}{\partial x^2} + \rho u c\dfrac{\partial T}{\partial x} = 0$

or $\qquad\qquad u\dfrac{\partial T}{\partial x} = \dfrac{k}{\rho c}\dfrac{\partial^2 T}{\partial x^2}$

Since $\qquad\qquad \alpha = \dfrac{k}{\rho c}$

Therefore, $\qquad u\dfrac{\partial T}{\partial x} = \alpha\dfrac{\partial^2 T}{\partial x^2}$

Note that i is the enthalpy (or thermal energy) per unit mass and hence, $i = cT$. u is constant.

2.4 3500 W/m^2

2.5 The derivation for cylindrical coordinates is shown below.

$$q = \frac{2\pi k L (T_1 - T_2)}{\ln\left(\dfrac{r_2}{r_1}\right)} = \frac{k(T_1 - T_2) 2\pi L}{\ln\left(\dfrac{2\pi r_2 L}{2\pi r_1 L}\right)}$$

$$= \frac{k(T_1 - T_2)(2\pi r_2 - 2\pi r_1)L}{\ln\left(\dfrac{A_2}{A_1}\right)(r_2 - r_1)} = \frac{A_2 - A_1}{\ln\left(\dfrac{A_2}{A_1}\right)} k \left(\frac{T_1 - T_2}{r_2 - r_1}\right)$$

$$= \bar{A} k \left(\frac{T_1 - T_2}{r_2 - r_1}\right)$$

Therefore, $\bar{A} = \dfrac{A_2 - A_1}{\ln\left(\dfrac{A_2}{A_1}\right)}$

2.6 Series-parallel circuit

2.7 GDE: $\dfrac{d^2 T}{dr^2} + \dfrac{1}{r}\dfrac{dT}{dr} + \dfrac{q'''}{k} = 0$

BC-1: at $r = r_i$, $k\dfrac{dT}{dr} = h_1 \left(T - T_{\infty_i}\right)$

BC-2: at $r = r_o$, $-k\dfrac{dT}{dr} = h_2 \left(T - T_{\infty_o}\right)$

General solution: $T = -\dfrac{q''' r^2}{4k} + C_1 \ln r + C_2$

2.8 GDE: $\dfrac{1}{r^2}\dfrac{d}{dr}\left[r^2 \dfrac{dT}{dr}\right] + \dfrac{q'''}{k} = 0$

BC-1: at $r = 0$, $\dfrac{dT}{dr} = 0$ (axi-symmetric)

BC-2: at $r = r_o$, $-k\dfrac{dT}{dr} = h(T - T_{\infty})$

The temperature distribution in the sphere is

$$T = \frac{q''' r_o^2}{6k}\left[1 - \left(\frac{r}{r_o}\right)^2\right] + T_{\infty} + \frac{q''' r_o}{3h}$$

The temperature at $r = 0$ can then be found from the above expression.

2.9 $T_{F,\max} - T_{\infty} = \dfrac{q_0''' r_i^2}{4k_F}\left(1 + \dfrac{b}{4}\right) + \dfrac{q_0''' r_i^2}{2k_C}\left(1 + \dfrac{b}{2}\right)\left(\dfrac{k_C}{r_C h} + \ln\dfrac{r_C}{r_i}\right)$

2.10 126°C

2.11 $r = 6.12$ cm (obtained iteratively)

2.12 97 W/mK

2.13 $L = 13$ cm. As this length is greater than the tube diameter (which is 9 cm), it is necessary to locate the well obliquely in the tube.

2.14 $\eta_f = \dfrac{\sqrt{hpkA}\left(\sinh mL + N\cosh mL\right)}{h\left(pL + A\right)\left(\cosh mL + N\sinh mL\right)}$

where $N = \dfrac{h}{mk}$

2.15   The rate of heat transfer from the fins on a wall is

$$q_f = \eta_f \, hA_f \left(T_0 - T_\infty\right)$$

The rate of heat removed from the wall between the fins, and the exposed areas above the first and below the last fin in the array is

$$q_w = hA_w \left(T_0 - T_\infty\right)$$

Therefore, the total rate of heat transfer is

$$q_T = q_f + q_w$$

$$\phi_{fin-array} = \dfrac{q_T}{q_{no-fin-array}}$$

2.16 $\theta^* = \dfrac{\left(\theta_1^* e^{2\,mL} - \theta_2^* e^{mL}\right)e^{-mx} + \left(\theta_2^* e^{mL} - \theta_1^*\right)e^{mx}}{e^{2\,mL} - 1}$

where $\theta^* = \theta - \dfrac{q'''}{km^2}$

and $\theta = T - T_\infty$

$$m^2 = \dfrac{hp}{kA}$$

2.17   Thickness of the wall = 0.345 m

2.18 $q = \dfrac{2\pi L\left(T_1 - T_2\right)\left[k_0 + \dfrac{b}{2}\left(T_1 + T_2\right)\right]}{\ln\dfrac{r_2}{r_1}}$

2.19 $q = \dfrac{k_0 A\left(T_1 - T_2\right)}{L}\left[1 + \dfrac{b}{2}\left(T_2 + T_1\right) + \dfrac{c}{3}\left(T_1^2 + T_1 T_2 + T_2^2\right)\right]$

2.20   GDE: $\dfrac{d^2 T}{dx^2} + \dfrac{q'''}{k} = 0$

Take half of the physical domain as the computational domain, since the problem is thermally and geometrically symmetric.

BC-1: at $x = 0$, $\dfrac{dT}{dx} = 0$

BC-2: at $x = L$, $T = 25$

General solution: $T = -\dfrac{q'''}{2k}x^2 + C_1 x + C_2$

2.21 $I_{max} = 216$ A

2.22 $I_{max} = 8555$ A

2.23   (a) GDE: $\dfrac{d}{dx}\left(A\dfrac{d\theta}{dx}\right) - \dfrac{hp}{k}\theta = 0$

where $\quad \theta = T - T_\infty$

$$A = \pi \left(\frac{R}{L} x \right)^2$$

$$p = 2\pi r \sqrt{1 + \left(\frac{R}{L} \right)^2}$$

Substitute A and p into the GDE to obtain

$$\frac{d^2\theta}{dx^2} + \frac{2}{x}\frac{d\theta}{dx} - l^2 \frac{\theta}{x} = 0$$

where $\quad l^2 = \frac{2hL}{kR}\sqrt{1 + \left(\frac{R}{L}\right)^2}$

(b) Use a suitable change of variable.

2.24 BC-1: at $x = 0$, $T =$ finite or $\theta =$ finite

BC-2: at $x = L$, $T = T_0$ or $\theta = \theta_0$

General solution: $\theta = \dfrac{BI_1\left(2lx^{1/2}\right) + DK_1\left(2lx^{1/2}\right)}{x^{1/2}}$

Applying BC-1: $D = 0$

Applying BC-2: $B = \dfrac{\theta_0 L^{1/2}}{I_1\left(2lL^{1/2}\right)}$

2.25 $q = -\left[-kA\left(\dfrac{d\theta}{dx}\right)_{x=L} \right] = kA\left(\dfrac{d\theta}{dx}\right)_{x=L}$

2.26 $q = \sqrt{hpkA}\,\theta_0\,\tanh mL$

where $\quad \theta_0 = T_0 - T_\infty$

$\qquad A = (t.1) = t$

$\qquad p = 2$ since $t \ll 1$

$$mL = \sqrt{\frac{hp}{kA}}L = \sqrt{\frac{2h}{k}\frac{(Lt)}{t^{3/2}}} = L\sqrt{\frac{2h}{kt}} = \xi$$

$\qquad (Lt) =$ Volume per unit width

Therefore, $\quad q = \sqrt{2hkt}\,\theta_0\,\tanh \xi$

or $\quad \dfrac{q}{\theta_0} = \sqrt{2hkt}\,\tanh\xi$

For maximum heat transfer from the fin (taking h, k, and Lt as constants)

$$\frac{d}{dt}\left(\frac{q}{\theta_0}\right) = 0$$

or $\quad \dfrac{d}{dt}\left(\sqrt{2hkt}\,\tanh\xi\right) = 0$

$\Rightarrow \qquad\qquad\qquad \xi = \dfrac{1}{6}\sinh 2\xi$

The above is a transcendental equation and can be solved graphical method or numerical method such as Newton-Raphson.

2.27 GDE: Material '*a*'

$$\frac{d^2\theta_a}{dx^2} - m_a^2\theta_a = 0$$

GDE: Material '*b*'

$$\frac{d^2\theta_b}{dx^2} - m_b^2\theta_b = 0$$

where $\theta = T_0 - T_\infty$

BCs:

At $x = 0$, $\theta_a = \theta_0$

At $x = 2L$, $\theta_b = \theta_0$

At $x = L$, $\theta_a = \theta_b = \theta_j$

At $x = L$, $-k_a \dfrac{d\theta_a}{dx} = k_b \dfrac{d\theta_b}{dx}$ (Note that $\dfrac{d\theta_b}{dx}$ is positive.)

2.28 $U_o = \dfrac{1}{\dfrac{r_o^2}{h_i r_i^2} + \dfrac{r_o^2}{k_1}\left(\dfrac{1}{r_i} - \dfrac{1}{r_1}\right) + \dfrac{r_o^2}{k_2}\left(\dfrac{1}{r_1} - \dfrac{1}{r_o}\right) + \dfrac{1}{h_o}}$

2.29 $\phi_{\text{finned-wall}} = \dfrac{A_w + \eta_f A_f}{A_b + A_w}$

where $\eta_f = \dfrac{\tanh mL}{mL}$

2.30 Thickness of magnesia = 3.15 in., thickness of insulation = 2.28 in.

Hint: Assume the insulation–magnesia contact temperature as 300 °C as the starting guess and solve the problem iteratively.

2.31 (a) Energy balance:

$$\rho A U c T - kA\frac{dT}{dx} = \rho A U c T + \frac{d}{dx}(\rho A U c T)\Delta x - kA\frac{dT}{dx} + \frac{d}{dx}\left(-kA\frac{dT}{dx}\right)\Delta x$$
$$+ hp\Delta x(T - T_\infty)$$

$$\Rightarrow \quad \frac{d^2\theta}{dx^2} + \frac{U}{\alpha}\frac{d\theta}{dx} - m^2\theta = 0$$

where $m^2 = \sqrt{\dfrac{hp}{kA}}$

$$\theta = T - T_\infty$$

(b) GDE: $\dfrac{d^2\theta}{dx^2} + \dfrac{U}{\alpha}\dfrac{d\theta}{dx} - m^2\theta = 0$

General solution:

$$\theta = e^{-\frac{U}{2\alpha}x}\left[C_1 e^{\sqrt{\frac{U^2}{4\alpha^2} + m^2}\,x} + C_2 e^{-\sqrt{\frac{U^2}{4\alpha^2} + m^2}\,x} \right]$$

2.33 (a) No. It is not a long fin. $mL = 1.55$, (b) Top edge temperature = 32.1 °C

2.34 $q''' = \dfrac{I^2 R_1}{V} = \dfrac{I^2 R_1}{\left(\dfrac{\pi D^2}{4}\right) L}$ where $R_1 = R_L$

$$T_c = T_\infty + \dfrac{q''' r_o^2}{4k}\left[1 + \dfrac{2k}{h r_o}\right]$$

$$T_s = T_c - \dfrac{q''' r_o^2}{4k}$$

2.35 (a) $T = -\dfrac{A}{12k}\left(x^4 - L^3 x\right)$

    (b) $T_{max} = 0.0818°\ C$

2.36 GDE: $\dfrac{1}{r^2}\dfrac{d}{dr}\left[r^2 \dfrac{dT}{dr}\right] + \dfrac{q'''}{k} = 0$

    BC-1: at $r = 0$, $\dfrac{dT}{dr} = 0$

    BC-2: at $r = b$, $T = T_w$

2.38 $T_F - T_0 = \dfrac{q_0''' r_1^2}{6 k_F}\left\{\left[1 - \left(\dfrac{r}{r_1}\right)^2\right] + \dfrac{3}{10}b\left[1 - \left(\dfrac{r}{r_1}\right)^4\right]\right\} + \dfrac{q_0''' r_1^2}{3 k_C}\left(1 + \dfrac{3}{5}b\right)\left(1 - \dfrac{r_1}{r_2}\right)$

$T_C - T_0 = \dfrac{q_0''' r_1^2}{3 k_C}\left(1 + \dfrac{3}{5}b\right)\left(\dfrac{r_1}{r} - \dfrac{r_1}{r_2}\right)$

The subscripts F and C stand for fuel and cladding, respectively.

2.39 Voltage drop = 40 V

2.40 (a) $\dfrac{T - T_\infty}{T_R - T_\infty} = \dfrac{R}{r}$,   (b) $q = \dfrac{k}{R}(T_R - T_\infty)$

2.41 13.7 A

2.42 (a) $q = 4\pi r_0 r_1\left(\dfrac{k_0 + k_1}{2}\right)\left(\dfrac{T_1 - T_0}{r_1 - r_0}\right)$,   (b) 19.5 kg/h

2.43 594.4 W

Chapter 3

3.1 (a) $T = T_A \sin\dfrac{\pi x}{L}\exp\left(-\dfrac{\pi y}{L}\right)$,   (b) $q'' = \dfrac{\pi k T_A}{L}\sin\dfrac{\pi x}{L}$

3.2 GDE: $\dfrac{\partial^2 T}{\partial x^2} + \dfrac{\partial^2 T}{\partial y^2} = 0$

    Let $\theta = T - T_\infty$.

    BC-1: at $x = 0$, $-k\dfrac{\partial \theta}{\partial x} = q_2'' - h\theta$

BC-2: at $x = L$, $-k\dfrac{\partial \theta}{\partial x} = h\theta$

BC-3: at $y = 0$, $-k\dfrac{\partial \theta}{\partial x} = q_1''$

BC-4: at $y = 1$, $\theta = T_0 - T_\infty = \theta_i$

Although the governing equation is linear and homogeneous, the problem in its original form cannot be solved by the method of separation of variables because both x and y directions are non-homogeneous. However, it can be solved by the method of superposition as follows:

$$\theta = \theta_1 + \theta_2 + \theta_3$$

Subproblem 1

GDE: $\nabla^2 \theta_1 = 0$

BC-1: at $x = 0$, $-k\dfrac{\partial \theta_1}{\partial x} = -h\theta_1$

BC-2: at $x = L$, $-k\dfrac{\partial \theta_1}{\partial x} = h\theta_1$

BC-3: at $y = 0$, $\dfrac{\partial \theta_1}{\partial y} = 0$

BC-4: at $y = l$, $\theta_1 = \theta_0$

In this subproblem, x is the homogeneous direction while y is the non-homogeneous direction and hence this subproblem can be solved by the method of separation of variables.

Subproblem 2

GDE: $\nabla^2 \theta_2 = 0$

BC-1: at $x = 0$, $-k\dfrac{\partial \theta_2}{\partial x} = -h\theta_2$

BC-2: at $x = L$, $-k\dfrac{\partial \theta_2}{\partial x} = h\theta_2$

BC-3: at $y = 0$, $-k\dfrac{\partial \theta_2}{\partial x} = q_1''$

BC-4: at $y = l$, $\theta_2 = 0$

In this subproblem, x is the homogeneous direction while y is the non-homogeneous direction and hence this subproblem can be solved by the method of separation of variables.

Subproblem 3

GDE: $\nabla^2 \theta_3 = 0$

BC-1: at $x = 0$, $-k\dfrac{\partial \theta_3}{\partial x} = q_2'' - h\theta_3$

BC-2: at $x = L$, $-k\dfrac{\partial \theta_3}{\partial x} = h\theta_3$

BC-3: at $y = 0$, $-k\dfrac{\partial \theta_3}{\partial x} = 0$

BC-4: at $y = l$, $\theta_3 = 0$

In this subproblem, x is the non-homogeneous direction while y is the homogeneous direction and hence this subproblem can be solved by the method of separation of variables.

3.3 GDE: $\nabla^2\theta = 0$, where $\theta = T - T_\infty$.

BC-1: at $x = 0$, $\theta = \theta_0$

BC-2: at $x = L$, $-k\dfrac{\partial\theta}{\partial x} = h\theta$

BC-3: at $y = 0$, $\theta = \theta_0$

BC-4: at $y = L$, $-k\dfrac{\partial\theta}{\partial x} = h\theta$

Clearly, the problem cannot be solved by the separation of variables because both x and y directions are non-homogeneous. However, the problem of non-homogeneity in both directions can be alleviated by breaking the problem into two subproblems so that $\theta = \theta_1 + \theta_2$ and each subproblem can be solved by the separation of variables method.

3.4 Use $\theta = T - T_\infty$

Split the problem into three subproblems. Then, $\theta = \theta_1 + \theta_2 + \theta_3$. Each subproblem can be solved by the method of separation of variables.

3.6 Split the problem into two subproblems.

3.8 GDE: $\dfrac{1}{r}\dfrac{\partial}{\partial r}\left(r\dfrac{\partial\theta}{\partial r}\right) + \dfrac{1}{r^2}\dfrac{\partial^2\theta}{\partial\phi^2} + \dfrac{\partial^2\theta}{\partial z^2} =$

BC-1: $\theta(0, \phi, z) = $ finite

BC-2: $\theta(R, \phi, z) = \theta_0$ in $0 < \phi < \pi$

$= 0$ in $\pi < \phi < 2\pi$

BC-3: $\theta(r, \phi, z) = \theta(r, \phi + 2\pi, z)$

BC-4: $\dfrac{\partial\theta(r, \phi, z)}{r\partial\phi} = \dfrac{\partial\theta(r, \phi + 2\pi, z)}{r\partial\phi}$

BC-5: $\theta(r, \phi, 0) = 0$

BC-6: $\theta(r, \phi, L) = 0$

Here, z and ϕ are the homogeneous directions while r is the non-homogeneous direction and hence separation of variables can be used.

Let us use the product solution of the form

$$\theta(r, \phi, z) = \mathfrak{R}(r)\,\Phi(\phi)\,Z(z)$$

Finally, $\dfrac{\theta(r, \phi, z)}{\theta_0} = \displaystyle\sum_{n=1}^{\infty}\left(\dfrac{2}{n\pi}\right)\dfrac{I_0(\lambda_n r)}{I_0(\lambda_n R)}\sin\lambda_n z$

$$+2\sum_{n=1}^{\infty}\sum_{\mu=1}^{\infty}\left(\dfrac{2}{n\pi}\right)\left(\dfrac{2}{\mu\pi}\right)\dfrac{I_\mu(\lambda_n r)}{I_\mu(\lambda_n R)}\sin\lambda_n z\,\sin\mu\varsigma$$

where $\lambda_n = n\pi/L$, $n = 1, 2, 3, \dots$.

3.9 This problem can be solved by the method of superposition by breaking it up into four subproblems.

3.10 GDE: $\nabla^2 T = 0$

BC-1: at $x = 0$, $T=0$

BC-2: at $x = a$, $\dfrac{\partial T}{\partial x} = 0$

BC-3: at $y = 0$, $T = T_0$

BC-4: at $y = \infty$, $T = T_\infty = 0$

x is the homogeneous direction.

y is the non-homogeneous direction.

The problem can be solved directly by the method of separation of variables.

3.11 Split the problem into three sub-problems.

3.12 $T = \dfrac{T_2 - T_1}{H} z + T_1$

3.13 (a) Use $\theta = T - T_0$. Divide the problem into two sub-problems.

(b) The method of separation of variables can be directly used to obtain the solution for θ.

3.14 (a) Use $\theta = T - T_\infty$. The method of separation of variables can be directly applied to obtain $\theta(x,y)$.

$$q_y \Big|_{y=0} = (Lx1) \int_0^L -k \frac{\partial T}{\partial y} \Big|_{y=0} dx$$

$$q_x \Big|_{x=L} = (Lx1) \int_0^L -k \frac{\partial T}{\partial x} \Big|_{x=L} dy$$

(b) Once $T(x,y)$ is obtained, $T(0,0)$ can be easily calculated.

3.15 Use $\theta = T - T_\infty$. Split the problem into two sub-problems.

3.16 Except for the symbols of the plate dimensions the problem is same as Prob.3.4.

3.17 Use $\theta = T - T_0$

Write GDE in cylindrical coordinates (r, ϕ).

BC-1: at $r = r_i$, $\theta = 0$

BC-2: at $r = r_0$, $+k \dfrac{\partial \theta}{\partial r} = q''(\phi)$

BC-3: at $\phi = 0$, $\dfrac{\partial \theta}{\partial \phi} = 0$

BC-4: at $\phi = \phi_0$, $\dfrac{\partial \theta}{\partial \phi} = 0$

We see that ϕ is the homogeneous direction and r is the non-homogeneous direction. The problem can be solved by the method of separation of variables.

3.18 Use $\theta = T - T_\infty$

Write GDE in cylindrical coordinates (r, z).

BC-1: at $r = 0$, $\dfrac{\partial \theta}{\partial r} = 0$ (Axi-symmetry)

BC-2: at $r = R$, $-k \dfrac{\partial \theta}{\partial r} = h\theta$

BC-3: at $z = 0$, $\theta = T_0 - T_\infty$

BC-4: at $z = \infty$, $\theta = 0$ (as $T = T_\infty$)

Clearly, r is the homogeneous direction and z is the non-homogeneous direction. Therefore, the problem can be solved by the method of separation of variables.

Chapter 4

4.1 (a) Lumped system approximation can be made in both cases, since $Bi < 0.1$.

(b) For copper: $T = 83.21°C$

For aluminium: $T = 62.64°C$

Hint: Check whether the Biot number $(= hr_0/k$, where r_0 is the radius of the cylinder) is less than 0.1. If it is, then the lumped system approximation can be made and Eq. (4.5) can be used.

4.2 $Bi = \dfrac{hL}{k} = \dfrac{(12)(1.25 \times 10^{-2})}{19} = 0.0079$

Since $Bi < 0.1$, therefore, lumped system approximation can be made. Equation (4.5) can be used to obtain T_i.

4.3 $\theta = T - T_\infty = \dfrac{2}{R^2} \sum_{n=1}^{\infty} e^{-\lambda_n^2 \alpha t} \dfrac{J_0(\lambda_n r)}{J_0^2(\lambda_n R) + J_1^2(\lambda_n R)} \int_0^R r \left[f(r) - T_\infty \right] J_0(\lambda_n r)\, dr$

where the defining relation for the λ_n's is

$$\lambda_n R \dfrac{J_1(\lambda_n R)}{J_0(\lambda_n R)} = \lambda_n R \dfrac{J_1(\lambda_n R)}{J_0(\lambda_n R)} \qquad\qquad (A)$$

and one finds that the λ_n's are the roots of Eq. (A).

4.4 *Hint:*

(a) Use the Heisler chart for the centre-line temperature of an infinitely long cylinder (Fig. 4.10).

(b) In this case the cylinder can be modelled as a short cylinder since $L/D < 3$. Formulate the problem as a product solution of the centre-line temperatures of an infinitely long cylinder and infinitely long plate. Use Figs 4.8 and 4.10.

4.5 10.42 min

4.6 $t = 100$ s. The solution procedure is iterative.

4.7 1.92 s

4.8 $T(x, t) - T_i = \dfrac{2q''(\alpha t/\pi)^{1/2}}{k} \exp\left(-\dfrac{x^2}{4\alpha t}\right) - \dfrac{q''x}{k} \operatorname{erfc}\left(\dfrac{x}{2\sqrt{\alpha}}\right)$

4.9 At $t = 0$, $T = T_0 + T_m \sin\dfrac{\pi x}{L}$

The problem can be solved by the method of separation of variables using $\theta = T - T_0$.

4.10 At $t = 0$, $T = T_1 + \dfrac{T_2 - T_1}{L} x$

The problem can be solved by the method of separation of variables method by using $\theta = T - T_0$.

4.11 Since $\dfrac{2L}{D} = \dfrac{12}{8} = 1.5$ is less than 3, it is a short cylinder. The solution procedure is similar to that of Prob. 4.4(b).

(a) Maximum temperature will occur at the centre of the short cylinder, that is, at $(0, 0)$.

(b) Minimum temperature will occur at (r_0, L) and $(r_0, -L)$.

(c) Plot $T\big|_{r=0}$ vs. z at the time when the centre temperature $T(0, 0)$ reaches 500°C.

Note that z varies from $-L$ to $+L$.

4.12 This is a 3D transient conduction problem in Cartesian coordinates, $T(x, y, z, t)$.

$$\left(\frac{T_c - T_\infty}{T_i - T_\infty}\right)_{2L \times 2L \times 2L} = \left(\frac{T_c - T_\infty}{T_i - T_\infty}\right)_{2L} \left(\frac{T_c - T_\infty}{T_i - T_\infty}\right)_{2L} \left(\frac{T_c - T_\infty}{T_i - T_\infty}\right)_{2L} \quad \text{(A)}$$

The maximum temperature will occur at the centre of the cube. Note that $2L = 20$ cm.

LHS of Eq. (A) is $\dfrac{55 - 50}{100 - 50} = 0.1$

Use Fig. 4.8 corresponding to $\dfrac{k}{hL} \to \infty$ since $h = \infty$ in this case.

The time required for the centre temperature to reach 55°C will have to be obtained iteratively so that LHS of Eq. (A) is satisfied.

4.13 $\theta = \dfrac{mV}{n\beta - 1}\left[e^{-\beta t} - e^{-\frac{t}{n}}\right]$

$$\theta_{max} = \frac{mV}{n\beta - 1}\left[e^{-\frac{\beta n \ln(n\beta)}{n\beta - 1}} - e^{-\frac{\ln(n\beta)}{n\beta - 1}}\right]$$

where $\theta = T - T_\infty$

θ_{max} is reached at

$$t = \frac{n \ln(n\beta)}{n\beta - 1}$$

4.14 This is a 1D transient conduction problem, $T(x, t)$.

$2L = 1.5$ cm

Use Fig. 4.8 and Fig. 4.9 to obtain the required solution.

4.15 This is a 2D transient conduction problem, $T(x, y, t)$.

$2L = 10$ cm

$2l = 20$ cm

$$\left(\frac{T_c - T_\infty}{T_i - T_\infty}\right)_{2L \times 2l} = \left(\frac{T_c - T_\infty}{T_i - T_\infty}\right)_{2L} \left(\frac{T_c - T_\infty}{T_i - T_\infty}\right)_{2l}$$

Use Fig. 4.8 to obtain the solution.

4.16 This is a problem of transient conduction in a semi-infinite solid.

Use $\dfrac{T - T_\infty}{T_i - T_\infty} = \text{erf}\left[\dfrac{x}{2(\alpha t)^{1/2}}\right]$ \quad (A)

where $\alpha = \dfrac{k}{\rho c}$

(a) $T_i = 0°C$

$T_\infty = 1500°C$

$\alpha = \dfrac{0.814}{(1906)(879)} = 4.86 \times 10^{-7}\ \text{m}^2/\text{s}$

$x = 0.2$ m

$t = 8 \times 3600 = 2.88 \times 10^4\ \text{s}$

Thus, $T(0.2\ \text{m}, 2.88 \times 10^4\ \text{s})$ can be calculated from Eq. (A).

(b) $T(0.2\ \text{m}, 2.88 \times 10^4\ \text{s}) = \text{erf}\left[\dfrac{0.4}{2(4.86 \times 10^{-7}\ t)^{1/2}}\right]$ \quad (B)

Find t from Eq. (B) by interpolating from the values in the error function table in Appendix A4.

4.17 GDEs: $\dfrac{\partial^2 T_A}{\partial x^2} = \dfrac{1}{\alpha}\dfrac{\partial T_A}{\partial t}$

$$\dfrac{\partial^2 T_B}{\partial x^2} = \dfrac{1}{\alpha}\dfrac{\partial T_B}{\partial t}$$

ICs: at $t = 0$, $T_A = T_1$
   at $t = 0$, $T_B = T_2$

BCs: at $x = 0$, $\dfrac{\partial T_A}{\partial x} = 0$

at $x = a$, $-k\dfrac{\partial T_A}{\partial x} = -k\dfrac{\partial T_B}{\partial x}$ $\Rightarrow$ $\dfrac{\partial T_A}{\partial x} = \dfrac{\partial T_B}{\partial x}$

at $x = a$, $T_A = T_B$

at $x = a + b$, $\dfrac{\partial T_B}{\partial x} = 0$

Use Laplace transform to obtain $T_A\,(x, t)$ and $T_B\,(x, t)$. The steady temperature is obtained at $t = \infty$.

4.18 The solution procedure is same as in Prob. 4.17. The only difference is that materials are different. Thus we will have α_A, α_B and k_A, k_B. Hence, in the first GDE, $\alpha = \alpha_A$ and in the second GDE, $\alpha = \alpha_B$. At $x = a$, the continuity of heat flux will give

$$-k_A\dfrac{\partial T_A}{\partial x} = -k_B\dfrac{\partial T_B}{\partial x}.$$

4.19 $\dfrac{T - T_\infty}{T_0 - T_\infty} = \mathrm{erf}\left[\dfrac{x}{2\,(\alpha t)^{1/2}}\right]$

$$\dfrac{\partial T}{\partial x} = (T_0 - T_\infty)\dfrac{2}{\sqrt{\pi}}\left[e^{-\frac{x^2}{4\alpha t}}\right]\dfrac{1}{2\,(\alpha t)^{1/2}}$$

$$\left.\dfrac{\partial T}{\partial x}\right|_{x=0} = G = (T_0 - T_\infty)\dfrac{1}{\sqrt{\pi\alpha t}}$$

Since, $T_\infty = 0$ K

$$G = \dfrac{T_0}{\sqrt{\pi\alpha t}} \tag{1}$$

$$G = \dfrac{1}{2700}\ \text{°C/cm}$$

$$\alpha = 0.0118\ \text{cm}^2/\text{s}$$

$$T_0 = 3870\text{°C}$$

From Eq. (1), t, that is, the age of the earth can be obtained.
$t = 29.46 \times 10^{14}$ s $= 93.4 \times 10^6$ years

4.20 *Hint:* See Example 4.3.

Contact temperature, T_c
Copper wall: 149.64 °C
Brick wall: 72.78 °C
Asbestos wall: 39.45 °C

Chapter 5

5.1 Use $h_{x,avg} = \dfrac{1}{x}\displaystyle\int_0^x h_x\, dx$ and $Nu_x = 0.332\, Re_x^{1/2}\, Pr^{1/3}$

5.2 For very low Pr the momentum boundary layer thickness is very small as compared to the thermal boundary thickness. Therefore, little error will be introduced if the velocity everywhere in the thermal boundary layer is assumed to be the free stream velocity, u_∞. The similarity form of the energy equation is

$$\frac{\theta''}{\theta'} + \frac{Pr}{2} f = 0 \qquad\qquad (A)$$

Differentiating Eq. (A) with respect to η,

$$\frac{d\left(\dfrac{\theta''}{\theta'}\right)}{d\eta} + \frac{Pr}{2}\frac{df}{d\eta} = 0 \qquad\qquad (B)$$

But, $f' = \dfrac{df}{d\eta} = \dfrac{u}{u_\infty} = 1$ (since $u = u_\infty$ everywhere in the thermal boundary layer)

Thus, Eq. (B) now becomes

$$\frac{d\left(\dfrac{\theta''}{\theta'}\right)}{d\eta} + \frac{Pr}{2} = 0 \qquad\qquad (C)$$

Equation (C) can now be integrated three times with respect to η.
An additional boundary condition is necessary, that is, $\theta''(0) = 0$ which is obtained by the application of $u = v = 0$ at $y = 0$ in the original form of the energy equation.

5.3 $\bar{h}_L = 2476.64\ \text{W/m}^2\ \text{K}$

5.4 (a) Check Re_L. If $Re_L > 5 \times 10^5$, then a mixed boundary layer type flow exists on the flat plate.
If $EcPr \ll 1$ then the viscous dissipation can be neglected.

(b) For a mixed boundary layer type flow over a flat plate $(5 \times 10^5 < Re_L < 10^7)$ the following expressions for $\overline{Nu}_L$ and $\bar{c}_f$.

$$\overline{Nu}_L = \frac{\bar{h}_L L}{k_f} = 0.037\, Pr^{1/3}\left(Re_L^{4/5} - 23550\right)$$

$$\bar{c}_f = \frac{0.074}{Re_L^{1/5}} - \frac{1740}{Re_L}$$

$$q = \bar{h}_L A(T_s - T_\infty)$$

$$F_D = \frac{1}{2}\rho u_\infty^2\, \bar{c}_f\, A$$

(c) If the boundary layer is assumed to be turbulent from the leading edge, then

$$Nu_x = 0.0296\, Re_x^{4/5}\, Pr^{1/3}$$

$$\overline{Nu}_L = 1.25\, Nu\big|_{x=L}$$

$$c_f = \frac{0.0594}{Re_L^{1/5}}$$

$$\bar{c}_f = \frac{0.074}{Re_L^{1/5}}$$

5.5 The properties of liquid sodium evaluated at $T_f = \dfrac{370 + 200}{2} = 285°C$ or 558 K are obtained from Table A1.7. At 558 K, $v = 4.584 \times 10^{-7} \, m^2/s$

$$Re_L = \frac{u_\infty L}{v} = \frac{(0.6)(7.5 \times 10^{-2})}{4.584 \times 10^{-7}}$$

$$= 0.9816 \times 10^5$$

Since $Re_L < 5 \times 10^5$, the flow is laminar.

For laminar liquid, metal flow over a flat plate maintained at constant temperature, the local Nusselt number can be expressed as

$$Nu_x = 0.564 \, Re_x^{\frac{1}{2}} \, Pr^{\frac{1}{2}}$$

Also, $\overline{Nu}_L = 2 \, Nu\big|_{x=L}$

Since $\overline{Nu}_L = \dfrac{\bar{h}_L L}{k_f}$, $\bar{h}_L$ can be easily obtained.

Thus, $q''_{avg} = \bar{h}_L (T_\infty - T_s)$

Note that $T_\infty > T_s$ in this case

5.6 Use the expression $T_e = T_s - (T_s - T_i) e^{-\frac{h\pi DL}{\dot{m}c_p}}$ to evaluate L.

h is to be calculated from

$$Nu_D = \frac{hD}{k_f} = 3.66$$

Properties are to be evaluated at $T_{mean} = \dfrac{T_i + T_e}{2}$.

5.7 $T_e = 53.27°C$

5.8 Assume fully developed flow and heat transfer. Since the tube length is not specified, outlet temperature cannot be calculated. Hence, use fluid properties at the inlet temperature.

5.9 From Appendix A7 evaluate water properties at 35°C (308 K) by linear interpolation.

(a) $Re_D = \dfrac{\rho v_m D}{\mu}$

$\Rightarrow v_m = \dfrac{Re_D \, \mu}{\rho D}$

$v_{max} = 2 v_m$

(b) $Nu_D = 4.364$ (Laminar, constant heat flux case)

$\Rightarrow \dfrac{hD}{k_f} = 4.364 \quad \Rightarrow h = \dfrac{4.364 \, k_f}{D}$

5.10 $F_D = 0.0686$ N

$q \approx 2.62$ kW

5.11 $Nu_x = 72.078$

5.12 *Hint:*

$$\bar{h}_L = \frac{1}{L}\left(\int_0^{x_c} 0.332 \frac{k}{x} Re_x^{1/2} Pr^{1/3}\, dx + \int_{x_c}^{L} 0.0296 \frac{k}{x} Re_x^{4/5} Pr^{1/3}\, d\right.$$

x_c stands for the length of the plate after which transition from laminar to turbulent flow takes place. It is based on $Re_{crit} = 5 \times 10^5$.

5.13 $u_\infty = 2.54$ m/s

5.14 *Hint:*

The following assumptions are made for solving the problem.
1. The bottom surface of the iceberg is plane.
2. The drift is steady.

Solution procedure:
- First, calculate the Reynolds number to determine whether the flow is laminar or turbulent.
- Calculate the average Nusselt number based on an appropriate correlation.
- Calculate the average heat transfer coefficient and average heat flux q''_a.
- Now, the average heat flux is absorbed by the melting process. Therefore,

$$q''_s = \rho h_{sf} \frac{dH}{dt} \quad \text{where } \rho_{ice} \text{ is } 917 \text{ kg/m}^3 \text{ (density of ice at } 0°C)$$

where h_{sf} is the latent heat of melting for ice, $h_{sf} = 333.4$ kJ/kg

The average melting rate (expressed in m/s or cm/h) is therefore

$$\frac{dH}{dt} = \frac{q''_s}{\rho h_{sf}}$$

5.15 Use the expression $T_e = T_s - (T_s - T_i)e^{-\frac{h\pi DL}{\dot{m}c_p}}$.

5.16 First check whether the flow is laminar or turbulent.

Based on that calculate $z_{fd,h}$.

If $z_{fd,h} < L$, then the flow is hydrodynamically fully developed.

$$q = \dot{m}c_p (T_e - T_i)$$

$$q'' = \frac{q}{\pi DL}$$

Also, $q'' = h(T_s - T_m)$

$$\Rightarrow T_s = T_m + \frac{q''}{h}$$

where $T_m = 20°C$

h is to obtained for a suitable Nusselt number correlation for this case.

5.17 Check first whether the flow is laminar or turbulent.

(a) If the flow is laminar then see whether $\dfrac{z_{fd,h}}{D} < \dfrac{L}{D}$ where $\dfrac{z_{fd,h}}{D} = 0.05 Re_D$.

If yes, then the flow is hydrodynamically fully developed.

If the flow is turbulent then see whether $\dfrac{z_{fd,h}}{D} < \dfrac{L}{D}$ where $\dfrac{z_{fd,h}}{D} = 60$.

If yes, then the flow is hydrodynamically fully developed.

(b) Check whether $\dfrac{z_{fd,t}}{D} < \dfrac{L}{D}$.

where $\dfrac{z_{fd,t}}{D} = 0.05\,\mathrm{Re}_D\,\mathrm{Pr}$ for laminar flow

$\dfrac{z_{fd,t}}{D} = 60$ for turbulent flow

(c) Use the expression $T_e = T_s - (T_s - T_i)\,e^{-\frac{h\pi DL}{\dot{m}c_p}}$.

5.18 (a) $T = T_s + \dfrac{C}{16}\left(r_0^4 - r^4\right)$

where $C = \dfrac{1}{4\,\mu k}\left(\dfrac{dp}{dz}\right)^2$

(b) $q = \dfrac{\pi r_0^4\, LkC}{2}$

where C is as given in (a).

5.19 (a) Calculate v_m from

$$v_m = \dfrac{\dot{m}}{\rho\left(\dfrac{\pi D^2}{4}\right)}$$

Now, $\mathrm{Re}_D = \dfrac{\rho v_m D}{\mu}$

Based on the value of Re_D check whether the flow is laminar or turbulent. Obtain Nu_D accordingly and hence, h.

(b) $T_s - T_m = \dfrac{q''}{h}$.

(c) $T_e - T_i = \dfrac{q''\pi DL}{\dot{m}c_p}$

Chapter 6

6.1 $\mathrm{Gr}_x = 8.5 \times 10^8$

6.2 $u_0 = 1.1$ m/s

6.3 For the case of constant wall heat flux, Grashof number is defined as

$$\mathrm{Gr}_x = \dfrac{g\beta\left(\dfrac{q'' x}{k_f}\right)x^3}{\nu^2} \tag{A}$$

In Eq. (A), $\dfrac{q'' x}{k_f}$ plays the same role as the temperature difference. Since in this problem the average wall temperature is not known, the air properties are evaluated at 20°C in the first iteration.

6.4 The procedure is similar to that given in Section 5.7.2. Starting equations are Eqs (6.32) and (6.33).

6.5 The air properties are to be evaluated at $T_f = \dfrac{45+25}{2} = 35°C$ (308 K)

Vertical Face

$$Gr_L = \frac{g\beta(T_s - T_\infty)L^3}{v^2} \tag{A}$$

Note that: $L = 10$ cm $= 0.1$ m

Check whether $Gr_L <$ or $> 10^9$, that is, whether the flow is laminar or turbulent. Based on that appropriate Nusselt number correlations can be used. Thus, $\bar{h}_L$ can be obtained.

Face inclined 45° to the vertical (downward-facing heated surface)

Replace 'g' by 'gcos(-θ)', that is, gcos(-45°) in the expression for Gr_L and calculate the modified Gr_L. The rest of the procedure is same as before.

Horizontal face (heated surface facing upward)

First, calculate Ra_L where

$$L = \frac{A_s}{p} = \frac{(10\times10)}{40} = \frac{100}{40} = 2.5 \text{ cm}$$

Then, depending on the value of Ra_L apply either Eq. (6.67) or Eq. (6.68) to obtain $\overline{Nu}_L$ and hence, $\bar{h}_2$.

6.6 This is a thin cylinder case.

Horizontal

Apply Eq. (6.73).

Vertical

Apply Eq. (6.70).

Air properties are evaluated at $T_f = \dfrac{250+25}{2} = 137.5°C$.

6.7 The weight of paper is expected to decrease due to the buoyancy force arising out of the density difference of air due to the temperature difference in a body force field (in this case, gravity).

Neglecting drag, the force balance on the sheet of paper gives

$$W_{\text{apparent}} = W_{\text{actual}} - F_{\text{buoyancy}}$$

Therefore,

$$\text{Change in weight} = W_{\text{actual}} - W_{\text{apparent}} = F_{\text{buoyancy}}$$

Hence, the percent change in weight

$$\frac{F_{\text{buoyancy}}/\text{Mass}}{W_{\text{actual}}/\text{Mass}} \times 100 = \frac{2g\beta(T_s-T_\infty)}{g} \times 100 = 2\beta(T_s - T_\infty) \times 1$$

Note the factor 2 as both sides of the paper are considered. But,

$$\beta = \frac{1}{\dfrac{T_s+T_\infty}{2}+273} = \frac{1}{\dfrac{60+20}{2}+273} = \frac{1}{313} \text{ K}^{-1}$$

Therefore, the percent change in weight

$$= 2\left(\frac{1}{313}\right)(60-20)(100) = 25.5$$

Hence, the weight of the paper will show a decrease of 25.56% when the sun shines on the paper.

6.8 This is Case B of Section 6.5.5.

Evaluate properties of the fluid at $\dfrac{T_h + T_c}{2} = \dfrac{50 + 20}{2} = 35°$ C

Now, based on the value of $Gr_\delta \, Pr$, use the appropriate correlation. $\delta = 50$ cm $= 0.5$ m in the present problem.

6.9 $Gr_L = \dfrac{g\beta(T_\infty - T_s)L^3}{v^2}$

Here, $L = 0.9$ m, $T_\infty = 30°$ C, $T_s = 20°$C.

Evaluate air properties at $\dfrac{30 + 20}{2} = 25°C$

If $Gr_L < 10^9$ then the flow is laminar. Then use either Eq. (6.61) or correlations obtained from similarity or integral analysis.

$$q = \bar{h}_L \, A(T_\infty - T_s)$$

where $A = 0.9 \times 0.6 = 0.54$ m^2

6.10 Since the vertical plate is hotter than water, the buoyancy will push the fluid upward.

Water properties are evaluated at $\dfrac{60 + 20}{2} = 40°C$ (see Appendix A7).

Calculate Gr_L. If $Gr_L < 10^9$, the flow is laminar. Based on the flow regime use appropriate correlation for $\overline{Nu}_L$. Calculate average Nusselt number and hence, average heat transfer coefficient and total heat transfer.

6.11 $\left(Ra_H\right)_{air} = \left(Ra_H\right)_{water}$

$$Pr_{air} \, \dfrac{g\beta_{air} \, (T_\infty - T_s) H_{air}^3}{v_{air}^2} = Pr_{water} \, \dfrac{g\beta_{water} \, (T_\infty - T_s) H_{water}^3}{v_{water}^2}$$

$$H_{water} = \left[\dfrac{Pr_{air}}{Pr_{water}} \left(\dfrac{v_{water}}{v_{air}} \right)^2 \left(\dfrac{\beta_{air}}{\beta_{water}} \right) \right]^{1/3} H_{air}$$

For property values of air see Table A1.4 in Appendix A1.
For property values of water see Appendix A7.
Note: $H_{air} = 4$ m

v_{water} and v_{air} are evaluated at $\dfrac{25 + 15}{2} = 20°C$.

$$\beta_{air} = \dfrac{1}{T_f} = \dfrac{1}{20 + 273} = \dfrac{1}{293} \text{ K}^{-1}$$

ρ_{water} is to be obtained from Appendix A7.

6.12 This is Case A of Section 6.4.2.

$$L = \dfrac{D}{4}$$

From $\overline{Nu}_L$, $\bar{h}_L$ is calculated.

$$q = \bar{h}_L \, A(T_s - T_\infty)$$

The water properties are evaluated at the film temperature. See Appendix A7.

6.13 Air properties are evaluated at $T_f = \dfrac{150+25}{2} = 87.5°$ C .

Calculate Gr_D.

If Gr_D Pr $< 10^9$, apply Eq. (6.71).

If Gr_D Pr $> 10^9$, apply Eq. (6.72).

Thus, $\overline{Nu}_D$ and hence, $\bar{h}_D$ is obtained.

$$q = \bar{h}_D\left(\pi DL\right)\left(T_s - T_\infty\right)$$

6.14 If $\dfrac{D}{L} \geq \dfrac{35}{Gr_L^{1/4}}$, the vertical plate Nusselt number correlations are valid. Otherwise,

Use Eq. (6.70).

$$q_{conv} = \bar{h}_L\left(\pi DL\right)\left(T_s - T_\infty\right) \tag{A}$$

$$q_{rad} = \varepsilon\sigma\left(\pi DL\right)\left(T_s^4 - T_\infty^4\right) \tag{B}$$

Note that in Eq. (B)

$\varepsilon = 1$

$\sigma = 5.668 \times 10^{-8}$ W/m^{2}K^4

$T_s = 100 + 273 = 373$ K

$T_\infty = 27 + 273 = 300$ k

Therefore, $q_{total} = q_{conv} + q_{rad}$

Air properties are evaluated at the film temperature.

6.15 See Section 6.5.6 for the appropriate Nusselt number correlation to be used.

6.16 *Hint:* This is a Benard convection problem. Use Eq. (6.74).

6.17 No, heat transfer will not be the same.

6.18 (a) This is Case A of Section 6.5.2.

$$L = \dfrac{A_s}{p} = \dfrac{0.2 \times 0.2}{(0.2)(4)} = 0.05 \text{ m}$$

(b) This is Case B of Section 6.5.2.

Air properties are evaluated at the film temperature.

6.19 Air properties are evaluated at $T_f = \dfrac{400+25}{2} = 212.5°\text{C}$.

Since the bottom surface is insulated, there is no heat loss from that side. Thus, there are four vertical surfaces and one horizontal surface (top side) from where heat transfer is taking place.

For the vertical side, $L = 10$ cm $= 0.1$ m. Calculate Gr_L. Use appropriate correlation to compute $\overline{Nu}_L$ and hence, $\bar{h}_{vertical}$. For the horizontal side which represents hot surface facing upward (Case A of Section 6.5.2) use appropriate correlation to find $\overline{Nu}_L$ and hence, $\bar{h}_{horizontal}$. Note that, here, $L = \dfrac{A_s}{p} = \dfrac{15 \times 20}{2(15+20)} = \dfrac{300}{70} = 4.28$ cm $= 0.0428$ m.

Therefore, $q_{total} = q_{vertical} + q_{horizontal}$

6.20 Check the value of Gr_D Pr.

If Gr_D Pr $< 10^9$, use Eq. (6.71).

If Gr_D Pr $< 10^9$, use Eq. (6.72).

$$\dfrac{q}{L} = \bar{h}_L\left(\pi D\right)\left(T_\infty - T_s\right)$$

Air properties are evaluated at $\dfrac{35+20}{2} = 27.5°\text{C}$.

Chapter 7

7.1 $\Delta T_w = T_w - T_{sat} = 220-100=120°C$

At $\Delta T_w = 120°$ C, minimum heat flux occurs. Use Eq. (7.32) to obtain q''_{min}. Evaluate the properties at $p = 1$ atm $= 1.0133$ bar from Table A1.6.

7.2 $\Delta T_w = 10°$ C

This corresponds to the nucleate boiling regime since $\Delta T_w < 30°$ C.

Use the correlation of Rohsenow (1952), that is, Eq. (7.20) to calculate q''_w.

$$q_w = q''_w (\pi DL)$$

The properties are to be evaluated at 5 atm (1 atm $= 1.0133$ bar).

7.3 From the energy balance at the steady state,

$$q_w = q''_w A = \dot{m}h_{fg} \tag{A}$$

Assume nucleate boiling.

Now, from the correlation of Fritz [Eq. (7.23)], we can write

$$h = 1.95(q''_w)^{0.72} (p)^{0.24}$$

$$\Rightarrow \quad q''_w = \left[\frac{h}{1.95(p)^{0.24}}\right]^{\frac{1}{0.72}} \tag{B}$$

From Eqs (A) and (B), we get

$$\left[\frac{h}{1.95(p)^{0.24}}\right]^{\frac{1}{0.72}} \left(\frac{\pi D^2}{4}\right) = \dot{m}h_{fg} \tag{C}$$

From Eq. (C), h can be obtained.

Now, $q = hA(T_w - T_{sat}) = \dot{m}h_{fg}$

$$\Rightarrow \quad T_w = \frac{\dot{m}h_{fg}}{hA} + T_{sat}$$

where $A = \dfrac{\pi D^2}{4}$

7.4 Use Eq. (7.22) to calculate $q''_{w.max}$.

7.5 Use Rohsenow's nucleate boiling correlation, that is, Eq. (7.20) to calculate q''_w.

Now, $h_{boil} = \dfrac{q''_w}{\Delta T_w}$ $\tag{A}$

Since h_{boil} is very large only turbulent forced convection can give h of that order.

$$h_{boil} = h_{turbulent} \tag{B}$$

$h_{turbulent}$ can be expressed from the Dittus-Boelter correlation [Eq. (5.211)] as given below.

$$\mathrm{Nu}_D = 0.023\,\mathrm{Re}_D^{0.8}\,\mathrm{Pr}^n \quad (n = 0.4 \text{ since the fluid is heated})$$

Therefore, $\dfrac{h_{turbulent}\,D}{k_f} = 0.023\,\mathrm{Re}_D^{0.8}\,\mathrm{Pr}^{0.4}$ $\tag{C}$

Using Eqs (A), (B), and (C), Re_D can be obtained.

Now, $\quad Re_D = \dfrac{\rho v_m D}{\mu}$ (D)

Thus, from Eq. (D) v_m can be calculated.

Note that in Eq. (C) water property values are to be taken from Appendix A7.

7.6 Each case corresponds to stable film boiling condition since $\Delta T_w > 120°C$. In addition, heat transfer is not only convection but also by radiation as $T_w > 300°C$.

$$q_w'' = \bar{h} \Delta T_w$$ (A)

where $\bar{h}^{-4/3} = \bar{h}_{conv}^{-4/3} + \bar{h}_{rad} \, \bar{h}^{-1/3}$ (B)

Eq. (B) is a transcendental equation which has to be solved iteratively for $\bar{h}$.

Use Eq. (7.33) for $\bar{h}_{conv}$ and Eq. (7.35) for $\bar{h}_{rad}$.

7.7 Assumption: Laminar film condensation on tubes.

The average condensation rate for a single tube of the array may be obtained, where for a unit length of the tube,

$$\dot{m}_1' = \dfrac{\bar{h}_{D,n} \, (\pi D)(T_{sat} - T_w)}{h_{fg}'}$$ (A)

Use Eq. (7.67) to obtain $\bar{h}_{D,n}$. While using Eq. (7.67) use h_{fg}' instead of h_{fg} to take into account the effect of sub-cooling. Since the array is 20×20, in each vertical tier there are 20 tubes. So, in Eq. (7.67), $n = 20$.

Knowing, $\bar{h}_{D,n}$ $\dot{m}_1$ can be evaluated from Eq.(A).

For the complete array,

$$\dot{m}' = n^2 \, \dot{m}_1'$$

The same problem can also be solved by applying Chen (1961) correlation for calculation of $\bar{h}_{D,n}$.

7.8 Assume laminar flow.

Obtain $\bar{h}_L$ from Eq. (7.43). Use h_{fg}' instead of h_{fg} in Eq. (7.43).

$$q = \bar{h}_L \, A (T_{sat} - T_w)$$

$$\dot{m} = \dfrac{q}{h_{fg}'}$$

To check the validity of the laminar flow assumption, calculate Re_δ from

$$Re_\delta = \dfrac{4 \dot{m}}{\mu_l b}$$

If $Re_\delta > 30$ but < 1800, then some portion of the condensate is in the wavy-laminar region and the solution using Kutateladze's correlation [Eq. (7.48)] will give a more accurate result. If $Re_\delta > 1800$, then the solution using Labunstov correlation [Eq. (7.49)] will be appropriate.

7.9 (a) Use Eq. (7.61) to calculate $\bar{h}$, and hence, $\dot{m}$.

(b) Use Eq. (7.43) assuming $R \gg \delta$ to calculate $\bar{h}$, and hence, $\dot{m}$.

In both cases assume laminar flow.

For the case (b), check the validity of laminar flow and the use of vertical plate correlation by calculating Re_δ and $\delta(L)$, respectively.

7.10 In a zero-gravity condition, condensate forms at the cold outer surface of the tube and does not run off. The transient rate of condensation on the outer surface of the tube may be approximated by assuming that a quasi-steady-state conduction heat transfer takes place in the condensed liquid. Making an energy balance

$$q = \frac{T_{sat} - T_w}{\dfrac{1}{2\pi kL}\ln\left(\dfrac{R_\delta}{R}\right)} = \dot{m}h_f.$$

where L is the length of the tube.

$\dot{m}$ = rate of condensation

$$= \frac{d}{dt}\left[\pi(R_\delta^2 - R^2)L\rho\right] = 2\pi R_\delta L\rho\frac{dR_\delta}{dt}$$

or

$$\frac{q}{L} = \frac{2\pi k(T_{sat} - T_w)}{\ln\left(\dfrac{R_\delta}{R}\right)} = 2\pi R_\delta \rho h_{fg}\frac{dR_\delta}{dt}$$

Integrating

$$\int R_\delta \ln\left(\frac{R_\delta}{R}\right)dR_\delta = \frac{(T_{sat} - T_w)k}{\rho h_{fg}}\int dt$$

or

$$\frac{R_\delta^2}{2}\ln\left(\frac{R_\delta}{R}\right) - \frac{R_\delta^2}{4} = \frac{(T_{sat} - T_w)k}{\rho h_{fg}}t + C$$

Now, $R_\delta = R$ at $t = 0$. Therefore,

$$C = -\frac{R^2}{4}$$

Hence,

$$\frac{R_\delta^2}{2}\ln\left(\frac{R_\delta}{R}\right) - \frac{R_\delta^2}{4} + \frac{R^2}{4} = \frac{(T_{sat} - T_w)k}{\rho h_{fg}}t$$

Dividing both sides of the above equation by R^2 and rearranging, we obtain

$$\left(\frac{R_\delta}{R}\right)^2\left[\ln\left(\frac{R_\delta}{R}\right) - \frac{1}{2}\right] + \frac{1}{2} = \left(\frac{R_\delta}{R}\right)^2\left[\ln\left(\frac{R_\delta}{R}\right) - \frac{1}{2}\right] + \frac{1}{2}$$

7.11 $p = 500$ kN/m$^2 = \dfrac{500 \times 10^3}{10^5} = 5$ bar

$\Delta T_w = 15°$ C which corresponds to nucleate boiling regime.

q_w'' is calculated from Eq. (7.10).

7.12 $\Delta T_w = 240 - 100 = 140°$ C

Since $\Delta T_w > 120°$ C, stable film boiling is occurring.

$\bar{h}$ is obtained from Eq. (7.33).

If the effect of radiation is to be considered then $\bar{h}$ will have to be estimated from Eq. (7.34).

7.13 Use Eq. (7.22).

7.14 For both case nucleate boiling is occurring since $\Delta T_w < 30°$ C.
Use Eq. (7.20).

7.15 The approach is same as in Q.7.15 except that $C_{s,f}$ is different (see Table 7.1).

7.16 Burning of foods occurs when the rate of heat transfer to the cooking medium (oil, water, etc.) is very high, which results in rapid depletion of the cooking fluid due to high rate of evaporation and hence the food gets stuck to the surface of the pan and is burnt eventually.

In the case of ordinary uncoated utensils there are a large number of small cavities on the surface of the pan. These cavities act as nucleation sites for the bubbles to form. As heat is added, more and more bubbles form, grow, and collapse. A few escape to the surface of the fluid and burst. Due to this rapid growth and collapse of the bubbles, the liquid undergoes an intense agitation and hence more circulation of liquid near the pan surface takes place. This results in a very high heat transfer rate and hence high rate of evaporation of the cooking medium.

On the other hand, in the case of non-stick (Teflon-coated) pans these cavities are covered by the coating. Also because of its smoothness, the coated surface by itself has very less number of cavities. So, effectively, there is less number of nucleation sites on the pan surface. So, bubble formation is much less and hence, less heat transfer to the cooking medium which results in a lower rate of evaporation. So the risk of burning of foods is lower.

7.17 Assume laminar flow. Use Eq. (7.43) to obtain $\bar{h}_L$ and hence, q and $\dot{m}$. Note that in this case areas of both sides of the plate are to be considered. Check the validity of the laminar flow assumption by calculating Re_δ.

7.18 Assuming laminar flow use Eq. (7.42) to calculate δ and Eq. (7.38) to compute u_{max} by evaluating $u|_{y=\delta}$. Three x-locations are:

$$x = 1m$$
$$x = 0.5 \text{ m}$$
$$x = 0.75 \text{ m}$$

7.19 Obtain saturated Freon-12 properties from Table A1.5.

$$T_{sat} = 310 \text{ K}$$
$$T_w = 273 + 12 = 285 \text{ K}$$

Use Eq. (7.61) to obtain $\bar{h}_D$. Replace h_{fg} by h'_{fg} in Eq. (7.61). Neglect ρ_v as compared to ρ_l.

$$q = \bar{h}_D \left(\pi DL\right)\left(T_{sat} - T_w\right)$$

$$\dot{m} = \frac{q}{h'_{fg}}$$

At 310 K, for saturated Freon-12, $h_{fg} = 130.64$ kJ/kg (Source: Arora, 2001)

$$h'_{fg} = h_{fg} + 0.68 c_{p,l} \left(T_{sat} - T_w\right)$$

7.20 For the case of Freon-12 condensing on the vertical tier of 5 horizontal tubes use Eq. (7.68), that is,

$$\bar{h}_{D,n} = \bar{h}_D \, n^{-1/4}$$

where $\bar{h}_D$ is obtained from the solution of Q.7.19.

Note that $n = 5$ here.

$$q = \bar{h}_{D,n} \left(n\pi DL \right) \left(T_{sat} - T_w \right)$$

$$\dot{m} = \frac{q}{h'_{fg}}$$

Chapter 8

8.1 6.8×10^{-5} st

8.2 1353 W/m^2

8.3 0.75

8.4 $\dfrac{r \tan^{-1}(b/c)}{b}$

8.5 (a) $F_{1-2} = \sqrt{1 + \left(\dfrac{D}{L}\right)^2} - \dfrac{D}{L}$

    (b) $F_{1-2} = \sqrt{1 + \left(\dfrac{D}{2L}\right)^2} + \dfrac{D}{2L} - \sqrt{1 + \left(\dfrac{D}{L}\right)^2}$

8.6 $F_{11} = 0$, $F_{12} = 1$, $F_{21} = \dfrac{A_1}{A_2}$, $F_{22} = 1 - \dfrac{A_1}{A_2}$

8.7 Apply the concept of view factor algebra (see Section 8.10.3).

8.8 780 W

8.9 $q_1 = -321{,}350$ W; $q_2 = 321{,}350$ W

    $T_3 = 244.9\,°C$; $T_4 = 240.4\,°C$

8.10 (a) $q_1 = 16473$ W, (b) No change in the result.

8.11 (a) $\dfrac{q}{A} = \dfrac{\sigma\left(T_1^4 - T_2^4\right)}{\dfrac{1}{\varepsilon_1} + \dfrac{1}{\varepsilon_2} - 1} = \dfrac{q_1}{A_1} = -\dfrac{q_2}{A_2}$ (A)

where $\varepsilon_1 = 0.6$ and $\varepsilon_2 = 0.7$, $T_1 = 573$ K and $T_2 = 673$ K.

In this problem even if emissivity values are interchanged the result will be unchanged since the denominator of Eq. (A) will remain unaltered. The approximation that has been made in deriving Eq. (A) is that emissivity is independent of temperature.

(b) In Eq. (A), put $\varepsilon_1 = \varepsilon_2 = 1$. Thus we get

$$\frac{q}{A} = \sigma\left(T_1^4 - T_2^4\right)$$

(c) In Eq. (A) put $T_2 = 500 + 273 = 773$ K and obtain $\dfrac{q}{A}$.

8.12 $\left(\dfrac{q}{A}\right)_{withshield} = \dfrac{\sigma\left(\dfrac{AT_1^4 - T_2^4}{1 + A}\right)}{\dfrac{1}{\varepsilon_1} + \dfrac{1}{\varepsilon_3} - 1}$

where $A = \dfrac{\dfrac{1}{\varepsilon_1} + \dfrac{1}{\varepsilon_3} - 1}{\dfrac{1}{\varepsilon_3} + \dfrac{1}{\varepsilon_2} - 1}$

Here, $T_1 = 673$ K, $T_2 = 573$ K

$\varepsilon_1 = 0.8$, $\varepsilon_3 = 0.05$

$\varepsilon_2 = 0.4$,

$\left(\dfrac{q}{A}\right)_{\text{without shield}} = \dfrac{\sigma\left(T_1^4 - T_2^4\right)}{\dfrac{1}{\varepsilon_1} + \dfrac{1}{\varepsilon_2} - 1}$

Here, $T_1 = 673$ K $T_2 = 573$ K

$\varepsilon_1 = 0.8$ $\varepsilon_2 = 0.4$

8.13 69.5 cal/s cm^2

8.14 This is due to the night-time radiation loss occurring between the water-covered surface and the very cold sink of the outer space.

8.15 Ist part: $q_1' = -q_2' = 37$ kW/m

2nd part: $T_R = 1102$ K

8.16 24.94 kW

8.17 −744 W

8.18 It is a three-surface enclosure problem. Use electrical network analogy and apply Kirchhoff's current law at each node.

8.19 The opening is treated as a hypothetical surface of area, A_N. Because the surroundings are large as compared to the opening, radiation exchange between the enclosure and the surroundings may be treated as approximating the surface as a blackbody at $T_N = T_\infty$. The heat inputs to the surface i's are dictated by Eq. (8.77).

8.20 This is a three-surface enclosure problem. Apply Kirchhoff's current law at each node. Solve three equations simultaneously to get J_1, J_2, and J_3.

$$q_2'' = \dfrac{E_{b2} - J_2}{\dfrac{1 - \varepsilon_2}{\varepsilon_2}}$$

$$T_2 = \left(\dfrac{E_{b2}}{\sigma}\right)^{1/4}$$

The subscript '2' indicates the bottom surface.

8.21 The basic approach is same as in the plane radiation shield problem.

8.22 17157 W/m^2

Chapter 9

9.1 $U_{\text{dirty}} = 2127.66$ W/m^2 K

9.2 (a) 219.5°C, (b) 246.5°C

For equal values of U and A, the counterflow heat exchanger would transfer 12% more heat than the parallel flow one.

9.3 (a) $T_{h_o} = 146.5°C$, $T_{c_o} = 124.3°C$
 (b) $T_{h_o} = 146.4°C$, $T_{c_o} = 124.2°C$

9.4 $A_h = 20.93$ m^2

9.5 T_{blood} leaving the exchanger is 27.24°C.

9.6
$$\frac{d(\Delta T)}{\Delta T} = \frac{U(\Delta T_2 - \Delta T_1)}{q_{total}} dA \qquad \text{(A)}$$

where $U = a + b\Delta T$
Integrate Eq. (A).

9.7 Write the expressions of LMTD for parallel flow and counterflow heat exchangers. Substitute them into the expression of F and get the final form using R and P.

9.8 Follow the procedure of Example 9.2.

9.9 $L = 21.33$ m

9.10 Use ε-*NTU* method. The relevant chart is Fig. 9.7(c). In this case, air is mixed while water is unmixed. This is because water is flowing in separate tubes and has no chance to mix whereas no such restriction is there for air flow.

9.11 Cold water is the minimum fluid.
 $\varepsilon_c = 0.77$
 LMTD = 11.02°C

9.12 Cold water is the minimum fluid. $C_{max} \rightarrow \infty$.
 $\varepsilon_c = 0.295$
 LMTD = 80.22°C

9.13 Use the properties of ethylene glycol at $\frac{10+38}{2} = 24°C$ (297 K). See Table A1.5.

The rest of the procedure is same as in Q.9.12.

9.14 Hot water is the minimum fluid.
 $\varepsilon_h = 0.73$
 $U_0 A_0 = 107992.63$ W/K

9.15 $\varepsilon_{dirty} = 0.62$

 $q_{dirty} = 1494401.4$ W

9.16 $\varepsilon_{dirty} = 0.53$

9.17 $q = 263.4$ kW, $T_{h_o} = 68°C$, $T_{c_o} = 74°C$

9.18 35.6 W/m^{2}K

9.19 Use ε-NTU method

9.20 Use ε-NTU method

Chapter 10

10.1 (a) Let $A = \dfrac{\partial^2 y}{\partial x^2}$

Then $\dfrac{\partial^4 y}{\partial x^4} = \dfrac{\partial^2 A}{\partial x^2}$

$$\left(\frac{\partial^2 A}{\partial x^2}\right)_{i,j} = \frac{A_{i+1,j} - 2A_{i,j} + A_{i-1,j}}{(\Delta x)^2} + 0(\Delta x)^2 \qquad \text{(A)}$$

Substitute $A_{i+1, j}$, $A_{i,j}$, and $A_{i-1,j}$ into Eq. (A).

(b) Express $y_{i+1,j}$ and $y_{i+2,j}$ using Taylor series expansion with the truncation error $0(\Delta x)^4$.

(c) Express $y_{i+1,j}$, $y_{i+2,j}$, and $y_{i+3,j}$ using Taylor series expansion with the truncation error $0\left(\Delta x\right)^4$.

10.2 (a) 17.45236

(b) 8.217260

(c) 4.712356

(d) 6.812756

(e) 3.111111

10.3 Using Taylor series

2 terms $\Rightarrow e^{0.7} = 1.7$

3 terms $\Rightarrow e^{0.7} = 1.945$

4 terms $\Rightarrow e^{0.7} = 2.002$

5 terms $\Rightarrow e^{0.7} = 2.012$

Actual value of $e^{0.7} = 2.013$

5 terms $\Rightarrow$ Error $= 0.049\% < 0.1\%$

4 terms $\Rightarrow$ Error $= 0.546\% > 0.1\%$

Therefore, five terms are needed in the series to get an accuracy of less than 0.1%.

10.4 Because of thermal and geometric symmetry, the computational domain of the problem is half of its physical domain. Thus the half-thickness of the plate has three equal subdivisions. Hence, there are four grid points including the one on the outer surface.

Hence, $\Delta x = \dfrac{2.5 \times 10^{-2}}{3} = 0.833 \times 10^{-2}$ m

GDE: $\dfrac{\partial T}{\partial t} = \alpha \dfrac{\partial^2 T}{\partial x^2}$

IC: at $t = 0$, $T = 500°C$

BC-1: at $x = 0$, $\dfrac{\partial T}{\partial x} = 0$

BC-2: at $x = 2.5$ cm $= 0.0025$ m, $T = 70°C$

The problem is to be solved by three methods: (i) Euler, (ii) pure implicit, (iii) Crank-Nicolson. See Section 10.4 for details.

For Euler method, numerical stability has to be considered. For this problem, to ensure stability in Euler solution, $\Delta t \leq 2.993$s. Take $\Delta t = 2$ s. Use the same time step for other two methods as well.

Euler method predicts that it will take approximately 20 s for the mid-plane temperature to reach $300°C$.

Using Heisler Chart (Fig. 4.10), $t = 19.4$ s. For this case, $h \to \infty$ and hence, $\dfrac{1}{Bi} = 0$.

Therefore, the numerical solution is quite close to the Heisler chart solution.

The computer programs written in C++ for this problem showing application of each of the three methods are available in the Online Resource Centre of this book.

10.5 The problem is thermally and geometrically symmetric about x- and y-axes. Hence, only one-quarter of the duct wall is sufficient for the purpose of computing steady-state temperature distribution.

GDE: $\dfrac{\partial^2 T}{\partial x^2} + \dfrac{\partial^2 T}{\partial y^2} = 0$

Use second-order central difference scheme for discretization at the interior points and image-point technique for symmetry (i.e., insulation) boundaries. A total of 21 simultaneous linear algebraic equations will have to be solved. Gauss-Seidel iteration is the suggested method.

10.6 GDE: $\dfrac{\partial^2 T}{\partial x^2} + \dfrac{\partial^2 T}{\partial y^2} = 0$

Use central difference scheme to discretize the GDE at the interior grid points. Corner points do not come in the finite-difference equations. The resulting set of simultaneous linear algebraic equations can be solved by Gauss-Seidel iteration scheme. A grid independence test has to be performed to obtain optimum number of grid points. The analytical solution of this problem is given in Eq. (3.25).

10.7 $T_M^{p+1} = T_M^p (1-6r) + r(T_A^p + T_B^p) - \dfrac{8r}{3}\left(T_D^p + \dfrac{T_c^p}{2} \right)$

where $r = \dfrac{\alpha \Delta t}{(\Delta x)^2}$

For stability, $1 - 6r \geq 0$.

10.8 GDE: $\dfrac{\partial^2 T}{\partial x^2} + \dfrac{\partial^2 T}{\partial y^2} = 0$

Finite-difference equation at point p:

$$\frac{1}{(\Delta x)^2}\left[T_B + T_Q - 2T_p \right] + \frac{2}{b(b+1)(\Delta y)^2}\left[bT_A + T_R - (b+1)T_p \right] = 0$$

where $b = 0.268$
Since $\Delta x = \Delta y$, the final equation at point p is

$$T_p = 873.24 + 0.1667 T_A + 0.1057 T_Q$$

2nd part:
The equation at the point p will be same as the value of b will remain unchanged. However, the interior grid point temperatures will be different because T_A and T_Q will not be evaluated at the earlier locations.

10.9 Because of thermal and geometrical symmetry about x- and y-axes only one-quarter of the physical domain is sufficient for computation. The computational domain is L-shaped. There will be five corner points. There will be also another corner point but this will be treated as an interior point because the neighbouring grid points will fall in the computational domain itself.

GDE: $\dfrac{\partial^2 T}{\partial x^2} + \dfrac{\partial^2 T}{\partial y^2} + \dfrac{q'''}{k} = 0$

Use central difference scheme for discretization at the interior points and image-point technique at boundaries and corner points.

Carry out a grid independence test. Finally, solve the resulting simultaneous linear algebraic equations by Gauss-Seidel iteration scheme.

10.10 See Section 10.6.1 for the solution methodology. The computer program written in C++ for this problem is included in the Online Resource Centre.

Chapter 11

11.1 (a) $x_A = 0.75$ and $x_B = 0.25$

(b) $Y_A = 0.6$ and $Y_B = 0.4$

(c) $\rho_A = 0.6$ kg/m^3 and $\rho_B = 0.4$ kg/m^3

(d) $c_A = 0.03$ kgmol/m^3 and $c_B = 0.01$ kgmol/m^3

(e) Total pressure $= 9.98 \times 10^4$ N/m^2

11.2 (a) $v = 0.2$ m/s (to the right)

$v^* = -0.25$ m/s (to the left)

(b) (i) $n_A = -0.6$ kg/m^{2}s

$n_B = 0.8$ kg/m^{2}s

$N_A = -0.03$ kgmol/m^{2}s

$N_B = 0.02$ kgmol/m^{2}s

(ii) $j_A = -0.72$ kg/m^{2}s

$j_B = 0.72$ kg/m^{2}s

$J_A = -0.036$ kgmol/m^{2}s

$J_B = 0.018$ kg mol/m^{2}s

(iii) $j_A^* = -0.45$ kg/m^{2}s

$j_A^* = 0.9$ kg/m^{2}s

$j_B^* = -0.0225$ kgmol/m^{2}s

$J_A^* = 0.0225$ kgmol/m^{2}s

(iv) Mass flux of A: 1.2 kg/m^{2}s

Mass flux of B: 2.0 kg/m^{2}s

Molar flux of A: 0.06 kgmol/m^{2}s

Molar flux of B: 0.05 kgmol/m^{2}s

11.3 Cartesian: $\dfrac{d^2 c_A}{dx^2} = 0$

Species diffusion resistance $= \dfrac{d^2 c_A}{dx^2} =$

Cylindrical: $\dfrac{1}{r}\dfrac{d}{dr}\left(c D_{AB} r \dfrac{dx_A}{dr}\right) = 0$

Species diffusion resistance $= \dfrac{\ln(r_2/r_1)}{2\pi L D_{AB}}$

Spherical: $\dfrac{1}{r^2}\dfrac{d}{dr}\left(c D_{AB} r^2 \dfrac{dx_A}{dr}\right) = 0$

Species diffusion resistance $= \dfrac{1}{r^2}\dfrac{d}{dr}\left(c D_{AB} r^2 \dfrac{d.}{c}\right)$

11.4 GDE: $\dfrac{d}{dx}\left(D \dfrac{dc_a}{dx}\right) = 0$ (A)

where $D = D_0 \left(1 + A c_a\right)$

Equation (A) is a non-linear equation as $D = f(c_a)$.

Integrating Eq. (A) from $x = 0$ $(c_a = c_{a1})$ to $x = L$ $(c_a = c_{a2})$ gives

$$(c_a - c_{a1}) + \frac{A}{2}(c_a^2 - c_{a1}^2) + \frac{D_m}{D_0}\left(\frac{c_{a1} - c_{a2}}{L}\right)x = 0 \tag{B}$$

where $D_m = \dfrac{1}{c_{a1} - c_{a2}} \displaystyle\int_{c_{a1}}^{c_{a2}} D_0(1 + Ac_a)dc_a = D_0\left[1 + \dfrac{A(c_{a1} + c_{a2})}{2}\right]$

As seen from Eq.(B), the concentration distribution is not linear.

11.5 $c_A(r) = \dfrac{c_i - c_o}{\ln\left(\dfrac{r_i}{r_o}\right)} \ln\left(\dfrac{r}{r_i}\right) + c_i$

11.6 $\dfrac{\partial c_A}{\partial t} = D_{AB} \dfrac{1}{r}\dfrac{\partial}{\partial r}\left(r\dfrac{\partial c_A}{\partial r}\right) + k_0 c_A$

where k_0 is a constant. D_{AB} is assumed as constant.
The analogous heat conduction problem:

$$\frac{\partial T}{\partial t} = k\frac{1}{r}\frac{\partial}{\partial r}\left(r\frac{\partial T}{\partial r}\right) + AT$$

where A is a constant. k is the thermal conductivity of the rod, assumed constant.

11.7 This problem is very similar to that of critical thickness of insulation in heat conduction (see Section 2.10).

$$\text{Molar diffusion rate} = \frac{c_i - c_o}{\left(\dfrac{\ln\left(\dfrac{r}{r_i}\right)}{2\pi LD_{AB}} + \dfrac{1}{h_{m_o}(2\pi rL)}\right)} \tag{A}$$

The molar diffusion rate is maximum if the denominator is minimum.

Thus, $\dfrac{d}{dr}\left[\dfrac{\ln\dfrac{r}{r_i}}{D_{AB}} + \dfrac{1}{h_{m_o}r}\right] = 0$

$\Rightarrow$ $\qquad\qquad r_c = \dfrac{D_{AB}}{h_{m_o}}$

Note that the unit of h_m is m/s (see Section 11.7) and that of D_{AB} is m²/s.

11.8 Time required for 90% carbon to burn away is 0.0703 s.

11.9 (a) 0.0216 m/s, (b) 3×10^{-5} kg/sm

11.10 $N_{A_r}\Big|_{r=r_1} = \dfrac{\left(pD_{AB,1} / R\bar{T_1}\right)\left[1 + \left(\dfrac{n}{2}\right)\right]}{r_1^2\left[r_1^{-1-(n/2)} - r_2^{-1-(n/2)}\right]r_1^{n/2}} \ln\dfrac{x_{B_2}}{x_{B_1}}$

Note that: $x_A = 1 - x_B$ and $n \neq -2$.

Chapter 12

12.1 The freezing time for Kulfi $= 0.65$ h

12.2 (a) The freezing time for each chicken patty $= 0.169$ h

(b) Upper limit on the speed of the conveyor belt $= 4.1$ cm/s

12.3 A lower freezing time is calculated if the change of density in the ice phase from that of water phase is considered.

12.4
$$\frac{\left(T_p - T_1\right)k_1 e^{-K^2 \beta^2 / 4\alpha_1}}{\sqrt{\pi\alpha_1}\, erf\left(K\beta / 2\sqrt{\alpha_1}\right)} = \frac{h_{sf}\, \rho_1 K\beta}{2}$$

12.5 The problem is similar to that discussed in Section 12.3 except that at $x = 0$,

$$q'' = -k_1 \frac{\partial T}{\partial x}.$$

12.6 The problem is now that in which region $x > 0$ is initially solid at temperature T_1 and for $t > 0$, the plane $x = 0$ is maintained at constant temperature, $T_2 > T_p$. The solution for solidification as discussed in Section 12.2.2 can be used with thermophysical properties of the solid and liquid interchanged and $(T_p - T_1)$ and $(T_2 - T_p)$ also interchanged in the LHS of Eq. (12.22).

References

Arora, C.P. 2001, h_{fg} : *Refrigeration and Air Conditioning,* Tata McGraw-Hill, New Delhi, p. 948.

Akers, W. W., H. A. Deans, and O. K. Crosser 1958, 'Condensing heat transfer within horizontal tubes', *Chemical Engineering Progress Symposium Series,* vol. 55, no. 29, p.71.

Anderson, D. A., J. C. Tannehill, and R. H. Pletcher 1984, *Computational Fluid Mechanics and Heat Transfer*, Hemisphere, Washington, DC.

Arpaci, V. S. 1966, *Conduction Heat Transfer*, Addison-Wesley, Reading, MA.

Asako, Y., H. Nakamura, and M. Faghri 1988, 'Developing laminar flow and heat transfer in the entrance region of regular polygonal ducts', *Int. J. Heat Mass Transfer*, vol. 31, pp. 2590–2593.

Bayazitoglu, Y. and O. M. Necati 1988, *Elements of Heat Transfer*, McGraw-Hill Book Company, New York.

Bejan, A. 1993, *Heat Transfer*, John Wiley & Sons, Inc., New York.

Bejan, A. and J. L. Lage 1990, 'The Prandtl number effect on the transition in natural convection along a vertical surface', *ASME Journal of Heat Transfer*, vol. 112, pp. 787–790.

Berenson, P. J. 1961, 'Film boiling heat transfer for a horizontal surface', *ASME Journal of Heat Transfer*, vol. 83, pp. 351–358.

Bird, R. Byron, W. E. Stewart, and E. N. Lightfoot 1960, *Transport Phenomena*, Wiley International Edition, John Wiley & Sons, New York.

Bird, B. R., W. E. Stewart, and E. N. Lightfoot 2002, *Transport Phenomena*, 2nd edn, John Wiley & Sons (Asia), Singapore 2009, Wiley India Reprint.

Bowman, R. A., A. E. Mueller, and W. M. Nagle 1940, 'Mean temperature difference in design', *Transactions of ASME*, vol. 62, p. 283.

Bromley, A. L. 1950, 'Heat transfer in stable film boiling', *Chemical Engineering Progress*, vol. 46, pp. 221–227.

Burmeister, L. C. 1993, *Convective Heat Transfer*, 2nd edn, John Wiley & Sons, New York.

Carey, V. P. 2008, *Liquid-Vapor Phase-Change Phenomena*, 2nd edn, Taylor & Francis, New York.

Carnahan, B., H. A. Luther, and J. O. Wilkes 1969, *Applied Numerical Methods*, John Wiley & Sons, New York.

Carslaw, H. S. and J. C. Jaeger 1959, *Conduction of Heat in Solids*, 2nd edn, Clarendon Press, Oxford.

Çengel, Y. A. 2003, *Heat Transfer*, 2nd edn, Tata McGraw-Hill Publishing Co. Ltd, New Delhi.

Cengel, Y. A. and M. A. Boles 2003, *Thermodynamics*, 4th edn, Tata McGraw-Hill Publishing Company, New Delhi.

Chapman, A. J. 1989, *Heat Transfer*, 4th edn, Macmillan Publishing Company, New York.

Chato, J. C. 1962, 'Laminar condensation inside horizontal and inclined tubes', *Journal of ASHRAE*, vol. 4, no. 52.

Chen, J. C. 1966, 'A correlation for boiling heat transfer to saturated fluids in convective flow', *Industrial and Engineering Chemistry Process Design and Development*, vol. 5, no. 3, pp. 322–333.

Chen, M. M. 1961, *Journal of Heat Transfer, Trans. ASME*, C83:55.

Churchill, S. W. and H. H. S. Chu 1975, 'Correlating equations for laminar and turbulent free convection from a vertical plate', *International Journal of Heat Mass Transfer*, vol.18, pp.1323–1329.

Churchill, S. W. and H. Ozoe 1973, 'Correlations for forced convection with uniform heating in flow over a plate and in developed flow in a tube', *ASME Journal of Heat Transfer*, vol. 95, pp. 78–84.

Churchill, S. W. and M. Bernstein 1977, 'A correlating equation for forced convection from gases and liquids to a circular cylinder in cross-flow', *ASME Journal of Heat Transfer*, vol. 99, pp. 300–306.

Cole, R. 1967, 'Bubble frequencies and departure volumes at subatmospheric pressures', *AIChE Journal*, vol. 13, pp. 779–783.

Collier, J. G. 1972, *Convective Boiling and Condensation*, McGraw-Hill, New York.

Constantinides, A. 1987, *Applied Numerical Methods with Personal Computers*, McGraw-Hill Inc., New York.

Diller, K. R. and T. P. Ryan 1998, 'Heat transfer in living systems: Current opportunities', *ASME Journal of Heat Transfer*, vol. 120, pp. 810–829.

Dittus, P. W. and L. M. K. Boelter 1930, *University of California Publications in Engineering*, vol. 2, no. 13, pp. 443–461; reprinted in *International Communications in Heat and Mass Transfer*, vol. 12, pp. 3–22, 1985.

Drew, T. B. and C. Mueller 1937, 'Boiling', *Transactions of AIChE*, vol. 33, p. 449.

Eckert, E. R. G. and R. M. Drake Jr. 1959, *Heat and Mass Transfer*, 2nd edn, McGraw-Hill Book Company, Inc., New York.

Eckert, E. R. G. and R. M. Drake 1972, *Analysis of Heat and Mass Transfer*, McGraw-Hill Book Company, New York.

Eckert, E. R. G. and T. W. Jackson 1950, *NACA (now NASA) TN 2207*, Washington.

Fox, R. W. and A. T. McDonald 1995, *Introduction to Fluid Mechanics*, 4th edn, John Wiley & Sons, New York.

Fraas, A. P. and M. Necati Ozisik 1965, *Heat Exchanger Design*, John Wiley & Sons, Inc., New York.

Frankel, S. P. 1950, 'Convergence rates of iterative treatments of partial differential equations', *Maths Tables Aids Comput.*, vol. 4, pp. 65–75.

Fritz, W. 1935, 'Berechnung des Maximalvolume von Dampfblasen', *Phys. Z.*, vol. 36, pp. 379–388.

Fritz, W. 1963, *VDI-Warmeatlas*, Chap. Hb2, VDI Verlag, Dusseldorf.

Fujii, T. and M. Fujii 1976, 'The dependence of local Nusselt number on Prandtl number in the case of free convection along a vertical surface with constant heat flux', *Int. J. Heat Mass Transfer*, vol. 19, pp. 121–122.

Gebhart, B. 1971, *Heat Transfer*, 2nd edn, McGraw-Hill Book Co., New York.

Gebhart, B. 1993, *Heat Conduction and Mass Diffusion*, McGraw-Hill, Inc., New York.

Ghoshdastidar, P. S. 1998, *Computer Simulation of Flow and Heat Transfer*, Tata McGraw-Hill Publishing Co. Ltd, New Delhi.

Gupta, V. 1995, *Elements of Heat and Mass Transfer*, New Age International Publisher, New Delhi.

Heisler, M. P. 1947, 'Temperature charts for induction and constant temperature heating', *Transactions of ASME*, vol. 69, pp. 227–236.

Hewitt, G. F. 1998, 'Name, number and unit', Plenary Lecture, *11th International Heat Transfer Conference,* Kyongju, Korea.

Holman, J. P. 1997, *Heat Transfer*, 8th edn, McGraw-Hill, Inc., New York.

Holman, J. P. 1981, *Heat Transfer*, International Student Edition, McGraw-Hill International Book Company, New York.

Hong, J. S., S. W. Grant Pease, and B. Rubinsky 1994, 'MR imaging assisted temperature calculations during cryosurgery', *Magnetic Resonance Imaging*, vol. 12, no. 7, pp. 1021–1031.

Hottel, H. C. and R. B. Egbert 1942, 'Radiant heat transmission from water vapor', *Transactions of AIChE,* vol. 38, p. 531.

Hottel, H. C. 1954, 'Radiation heat transfer' in: W. H. McAdams (ed.), *Heat Transmission*, 3rd edn, Chap. 4, McGraw-Hill, New York.

Incropera, F. P. and D. P. Dewitt 1998, *Fundamentals of Heat and Mass Transfer*, 4th edn, John Wiley & Sons, New York.

Jakob, M. and W. Fritz 1931, 'Versuche über den Verdampfungsvorgang', *Forsch. Ingenieurwes*, vol. 2, pp. 435–447.

Jaluria, Y. and K. E. Torrance 1986, *Computational Heat Transfer*, Hemisphere, Washington, DC.

Jaluria, Y. 1988, *Computer Methods in Engineering*, Allyn and Bacon, Inc., Boston.

Jaluria, Y. 1998, *Design and Optimization of Thermal Systems*, McGraw-Hill, New York.

James, M. L., G. M. Smith, and J. C. Wolford 1967, *Applied Numerical Methods for Digital Computers with FORTRAN*, International Text Book Co., Scranton, PA.

Janna, W. S. 1993, *Design of Fluid Thermal Systems*, PWS-Kent Publishing Co., Boston.

Kakac, S. and Y. Yener 1993, *Heat Conduction*, 3rd edn, Taylor & Francis, London.

Karel, M. and D. B. Lund 2003, *Physical Principles of Food Preservation*, 2nd edn, CRC Press.

Karlekar, B. V. and R. M. Desmond 1989, *Heat Transfer*, 2nd edn, Prentice-Hall of India Pvt. Ltd, New Delhi.

Kays, W. M. and M. E. Crawford 1993, *Convective Heat and Mass Transfer*, 3rd edn, McGraw-Hill, Inc., New York.

Kays, W. M. and A. L. London 1964, *Compact Heat Exchangers*, 2nd edn, McGraw-Hill Book Company, New York.

Kreider, J. F. and A. Rabl 1994, *Heating and Cooling of Buildings*, McGraw-Hill Inc., New York.

Kutateladze, S. S. 1948, 'On the transition of film boiling under natural convection', *Kotloturbostroenie*, vol. 3, p.10.

Kutateladze, S. S. 1963, *Fundamentals of Heat Transfer*, Academic Press, New York.

Labunstov, D. A. 1957, 'Heat transfer in film condensation of pure steam on vertical surfaces and horizontal tubes', *Teploenergetica*, vol. 4, p. 72.

Lefevre, E. J. and A. J. Ede 1956, 'Laminar free convection from the outer surface of a vertical circular cylinder', *Proceedings of the Ninth International Congress on Applied Mechanics*, vol. 4, pp.175–183.

Lienhard, J. H. 1987, *A Heat Transfer Textbook*, 2nd edn, Prentice Hall, Englewood Cliffs, NJ.

Lockhart, R. W. and R. C. Martinelli 1949, 'Proposed correlation of data for isothermal two-phase two-component flow in pipes', *Chemical Engineering Progress*, vol. 45, p. 39.

Michelsen, M. L. and J. Viladsen 1974, 'The Graetz problem with axial heat conduction', *International Journal of Heat and Mass Transfer*, vol. 17, no. 11, pp. 1391–1402.

Modest, M. F. 1993, *Radiation Heat Transfer*, McGraw-Hill, Inc., New York.

Myers, G. E. 1971, *Analytical Methods in Conduction Heat Transfer*, McGraw-Hill, New York.

Notter, R. H. and C. A. Sleicher 1972, 'A solution to the turbulent Graetz problem III. Fully developed and entry region heat transfer rates, *Chem. Eng. Sci.*, vol. 27, pp. 2073–2093.

Nukiyama, S. 1934, 'The maximum and minimum values of heat transmitted from metal to boiling water under atmospheric pressure', *Journal of Japanese Society of Mechnical Engineering*, vol. 37, p. 367 [Translation: *International Journal of Heat and Mass Transfer*, vol. 9, p. 1419, 1966].

Nusselt, W. 1916, 'Die oberflaechenkondensation des wasserdampfer', *Zeitschrift des Vereines Deutscher Ingeneieure*, vol. 60, pp. 541–569.

Ostrach, S. 1952, *National Advisory Commission on Aeronautics Technical Note, 2635.*

Patankar, S. V. 1980, *Numerical Heat Transfer and Fluid Flow*, Hemisphere, Washington, DC.

Pohlhausen, E. 1921, *Zeitschrift fuer Angewandte Mathematic und Mechanik*, vol. 1, p. 115.

Ralston, A. and P. Rabinowitz 1978, *A First Course in Numerical Analysis*, McGraw-Hill, New York.

Rohsenow, W. M. 1952, 'A method of correlating heat transfer data for surface boiling liquids', *Transactions of ASME*, vol. 74, pp. 969–976.

Rohsenow, W. M. and H. Y. Choi 1961, *Heat, Mass and Momentum Transfer*, Prentice-Hall, Inc., New Jersey.

Rohsenow, W. M. 1973, 'Film condensation' in: W.M. Rohsenow and J.P. Hartnett (Eds.), *Handbook of Heat Transfer*, Section 12, Part A, McGraw-Hill Book Company, New York.

Rohsenow, W. M. and J. P. Hartnett 1973, *Handbook of Heat Transfer*, McGraw-Hill Book Company, New York.

Schlichting, H. 1968, *Boundary-Layer Theory,* 6th edn, McGraw-Hill Book Company, New York.

Siegel, R. and J. R. Howell 1981, *Thermal Radiation Heat Transfer*, 2nd edn, Hemisphere, New York.

Sleicher, C. A. and M. W. Rouse 1975, *International Journal of Heat and Mass Transfer*, vol.18, pp. 677–683.

Sparrow, E. M. and J. L. Gregg 1956, *Transactions of ASME*, vol. 78, pp. 435–440.

Sparrow, E. M. and J. L. Gregg 1958, *Trans. ASME*, vol. 80, pp. 379–386.

Sparrow, E. M. and R. D. Cess 1978, *Radiation Heat Transfer*, Hemisphere, New York.

Stephan, K. 1992, *Heat Transfer in Condensation and Boiling,* Springer-Verlag, Berlin.

Thomas, G. B. and R. L. Finney 1984, *Calculus and Analytic Geometry*, 6th edn, Narosa Publishing House, New Delhi.

Touloukian, Y. S., G. A. Hawkins, and M. Jacob 1948, *Transactions of ASME*, vol. 70, p. 13.

Vliet, G. C. and C. K. Liu 1969, 'An experimental study of turbulent natural convection boundary layers', *ASME J. Heat Transfer*, vol. 91, pp. 517–531.

Whitaker, S. 1972, 'Forced convection heat transfer correlations for flow in pipes, past flat plates, single cylinders, single spheres, and flow in packed beds and tube bundles', *AIChE Journal,* vol. 18, pp. 361–371.

Yang, K. T. 1960, *J. Appl. Mech.*, vol. 27, p. 230.

Young, D. 1954, 'Iterative methods for solving partial differential equations of elliptic type', *Transactions of American Mathematical Society,* vol. 76, pp. 92–111.

Zuber, N. 1958, 'On the stability of boiling heat transfer', *Transactions of ASME*, vol. 80, p. 711.

Zuber, N. 1959, 'Hydrodynamic aspects of boiling heat transfer', *U.S. AEC Report AECU 4439*, June.

Zuber, N. 1963, 'Nucleate boiling—The region of isolated bubbles—Similarity with natural convection', *International Journal of Heat and Mass Transfer*, vol. 6, pp. 53–65.

Zukauskas, A. A. 1987, 'Convective heat transfer in crossflow', in: S. Kakac, R.K. Shah, and W. Aung (eds), *Handbook of Single-Phase Convective Heat Transfer*, Chap. 6 , Wiley, New York.

Index

About the Author

P.S. Ghoshdastidar is currently Professor, Department of Mechanical Engineering, Indian Institute of Technology Kanpur. A Ph D from the University of South Carolina, he has over 27 years of teaching and research experience.

Dr Ghoshdastidar has published numerous research papers in reputed international journals and conference proceedings. He is a reviewer for many prestigious international journals and is an elected fellow of World Innovation Foundation, UK. He is also an Associate Editor of *Heat Transfer Research,* an international journal published by Begell House, Inc., USA.

Related Titles

Fluid Mechanics And Machinery | 9780195699630
C.S.P. Ojha, IIT *Roorkee*
R. Berndtsson, *Lund University, Sweden*
P.N. Chandramouli, *NIE, Mysore*

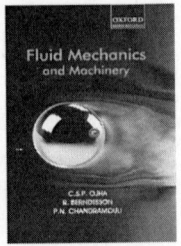

Fluid Mechanics and Machinery is designed to serve as a textbook for undergraduate and postgraduate students of mechanical and civil engineering. It will also be useful for students of electrical, aerospace, chemical, and production engineering. The book provides an exhaustive coverage of the essential concepts of the mechanics of fluids, both static as well as dynamic. In addition, it provides an overview of the design and operation of various hydraulic machines such as pumps and turbines.

Key Features
♦ Lays emphasis on the practical applicability of fluid mechanics to real-life situations
♦ Discusses dimensional analysis and flow of fluids through orifices, mouthpieces, and pipes, and over notches and weirs

Engineering Thermodynamics (Revised First Edition) | 9780198078876
Parthasarathi Chattopadhyay, *Techno India College of Technology, Kolkata*

This revised edition of *Engineering Thermodynamics* is designed as a textbook for undergraduate students of mechanical engineering. It provides an in-depth coverage of the fundamental principles of thermodynamics. While providing the mathematical representation, it lays emphasis on the physical aspects of the subject. Starting with the basic concepts, the book gradually discusses important topics such as entropy, thermodynamic availability, properties of steam, real and ideal gas, power cycles and chemical equilibrium in increasing order of complexity.

Key Features
♦ Includes numerous refrigeration tables and conversion tables along with basic data on fuels and combustion
♦ Provides refrigeration charts for some commonly used refrigerants

Engineering Mechanics | 9780195696554
Basudeb Bhattacharyya, *Bengal Engineering and Science University, Shibpur (previously known as Bengal Engineering College)*

Engineering Mechanics is specially designed for undergraduate students of engineering for an introductory course on engineering mechanics. It covers in detail the basic theories and principles of both statics and dynamics.

Key Features
♦ Includes appendices containing common formulae of trigonometry, differentiation, integration, and properties of geometrical figures and homogeneous solids
♦ Covers the various analytical aspects of rigid body dynamics

Other Related Titles

1. 9780195687811   Appukuttan: *Introduction to Mechatronics*
2. 9780198003155   Behera and Kar: *Intelligent Systems and Control*
3. 9780195689051   Russel and Adebiyi: *Engineering Thermodynamics*
4. 9780198061106   Subramanian: *Strength of Materials, 2e*
5. 9780198062240   Vela Murali: *Engineering Mechanics*
6. 9780198062325   Uicker et al.: *Theory of Machines and Mechanisms, 3e*
7. 9780195673913   Ashitava Ghosal: *Robotics—Fundamental Concepts and Analysis*